MODEL PREDICTIVE CONTROL OF HIGH POWER CONVERTERS AND INDUSTRIAL DRIVES

MODEL PREDICTIVE CONTROL OF HIGH POWER CONVERTERS AND INDUSTRIAL DRIVES

Tobias Geyer

ABB Corporate Research Center, Baden-Dättwil, Switzerland

This edition first published 2017

Registered office
John Wiley & Sons Ltd, The Atrium, Southern Gate, Chichester, West Sussex, PO19 8SQ, United Kingdom

For details of our global editorial offices, for customer services and for information about how to apply for permission to reuse the copyright material in this book please see our website at www.wiley.com.

Library of Congress Cataloging-in-Publication Data

Names: Geyer, Tobias, 1975- author.
Title: Model predictive control of high power converters and industrial drives / Tobias Geyer.
Description: First edition. | Chichester, West Sussex, United Kingdom : John Wiley & Sons, Inc., 2016. | Includes bibliographical references and index.
Identifiers: LCCN 2016014286 (print) | LCCN 2016015090 (ebook) | ISBN 9781119010906 (cloth) | ISBN 9781119010869 (pdf) | ISBN 9781119010890 (epub)
Subjects: LCSH: Electric drives–Automatic control. | Power converter–Automatic control. | Predictive control.
Classification: LCC TK4058 .G49 2016 (print) | LCC TK4058 (ebook) | DDC 621.46–dc23
LC record available at http://lccn.loc.gov/2016014286

A catalogue record for this book is available from the British Library.

Set in 10/12pt, TimesLTStd by SPi Global, Chennai, India.

10 9 8 7 6 5 4 3 2 1

To Luci, David, and Jan

Contents

Preface

This book focuses on model predictive control (MPC) schemes for industrial power electronics. The emphasis is on three-phase ac–dc and dc–ac power conversion systems for high-power applications of 1 MVA and above. These systems are predominantly based on multilevel voltage source converters that operate at switching frequencies well below 1 kHz. The book mostly considers medium-voltage (MV), variable-speed drive systems and, to a lesser extent, MV grid-connected converters. The proposed control techniques can also be applied to low-voltage power converters when operated at low pulse number, that is, at small ratios between the switching frequency and the fundamental frequency.

For high-power converters, the pulse number typically ranges between 5 and 15. As a result, the concept of averaging, which is commonly applied to power electronic systems to conceal the switching aspect from the control problem, leads to performance deterioration. In general, to achieve the highest possible performance for a high-power converter, averaging is to be avoided, and the traditionally used current control loop and modulator should be replaced by one single control entity.

This book proposes and reviews control methods that fully exploit the performance potential of high-power converters, by ensuring fast control at very low switching frequencies and low harmonic distortions. To achieve this, the control and modulation problem is addressed in one computational stage. Long prediction horizons are required for the MPC controllers to achieve excellent steady-state performance. The resulting optimization problem is computationally challenging, but can be solved in real time by branch-and-bound methods. Alternatively, the optimal switching sequence to be applied during steady-state operation—the so-called optimized pulse pattern (OPP)—can be precomputed offline and refined online to achieve fast closed-loop control.

To this end, the research vision is to combine the benefits of deadbeat control methods (such as direct torque control) with the optimal steady-state performance of OPPs, by resolving the antagonism between the two. Three such MPC methods are presented in detail.

Acknowledgments

First and foremost, I would like to thank Georgios Papafotiou, who introduced me to the exciting field of power electronics, Stefan Schröder, who taught me the fundamentals of high power electronics, and Manfred Morari, who showed me how to achieve meaningful research results.

I am deeply grateful to my wife Jan for her love and support, which allowed me to work during long nights, weekends, and holidays to finalize this book. I apologize to my children Luci and David for having spared little time for them. I thank my parents for being role models of diligence and scientific curiosity.

Over the years, I had the privilege of collaborating with many outstanding control and power electronics researchers. I would like to particularly mention Daniel Quevedo, Toit Mouton, Stefan Schröder, Georgios Papafotiou, and Petros Karamanakos. I would also like to thank Johann Kolar, José Rodríguez, and Ralph Kennel for their support and encouragement.

I am very grateful to Andrew Paice, Jan-Henning Fabian, John Boys, Grant Covic, Udaya Madawala, and Keith Jones for facilitating and supporting my affiliation with the University of Auckland. The majority of the concepts and results in this book are due to this very productive 3.5 years, during which I had few obligations and the unique opportunity to exclusively focus on research.

At ABB, I have had the pleasure of working with many able colleagues on control-related topics that are part of this book. I am particularly grateful to Nikolaos Oikonomou, Wim van der Merwe, Peter Al Hokayem, Vedrana Spudić, Christian Stulz, Eduardo Rohr, Andrea Rüetschi, Thomas Burtscher, Christof Gutscher, Andrey Kalygin, Rick Kieferndorf, Silvia Mastellone, Helfried Peyrl, Jan Poland, Tobias Thurnherr, and Michail Vasiladiotis. A special word of gratitude is reserved for Gerald Scheuer for his steadfast support of and conviction in modern control methodologies.

Many MSc and PhD students contributed through their thesis work to this book. Among them, I would like to specifically thank James Scoltock, Baljit Riar, Thomas Burtscher, Aleksandar Paunović, Georgios Darivianakis, Yashar Zeinaly, and Joël Vallone. It was a pleasure to work with them, and I am grateful for their hard work, curiosity, and ideas.

Petros Karamanakos and my wife Jan meticulously proofread the manuscript. Toit Mouton, Thomas Besselmann, Mario Schweizer, Vedrana Spudić, Baljit Riar, and James Scoltock provided helpful suggestions and pointed out mistakes in early versions of the manuscript. Thank you very much.

I am grateful to Peter Mitchell and John Wiley & Sons, Ltd for publishing this book. I would also like to thank Richard Davies, Prachi Sahay, Liz Wingett, Shruthe Mothi, and Ella Mitchell for their guidance and support.

List of Abbreviations

Abbreviations

ac	alternating current
A/D	analog-to-digital
AFE	active front end
ANPC	active neutral-point-clamped
CB-PWM	carrier-based pulse width modulation
CPU	central processing unit
DB	deadbeat
dc	direct current
DFE	diode front end
DFT	discrete Fourier transform
DPC	direct power control
DSC	direct self-control
DSP	digital signal processor
DTC	direct torque control
EMF	electromotive force
FACTS	flexible ac transmission system
FC	flying capacitor
FCS	finite control set
FOC	field-oriented control
FPGA	field-programmable gate array
GCT	gate-commutated thyristor
IGBT	insulated-gate bipolar transistor
IGCT	integrated-gate-commutated thyristor
IM	induction machine
LQR	linear quadratic regulator
MIMO	multiple-input multiple-output
MLD	mixed logical dynamical
MMC	modular multilevel converter
MPC	model predictive control
MPDBC	model predictive direct balancing control
MPDCC	model predictive direct current control

MPDPC	model predictive direct power control
MPDTC	model predictive direct torque control
MP^3C	model predictive pulse pattern control
MV	medium-voltage
NPC	neutral-point-clamped
OPP	optimized pulse pattern
PCC	point of common coupling
PI	proportional–integral
PMSM	permanent magnet synchronous machine
pu	per unit
PWM	pulse width modulation
QP	quadratic program
rms	root-mean-square
SHE	selective harmonic elimination
SISO	single-input single-output
SVM	space vector modulation
TDD	total demand distortion
THD	total harmonic distortion
VC	vector control
V/f	volts per frequency
VOC	voltage-oriented control
VSD	variable-speed drive
VSI	voltage source inverter

Variables

i, v	instantaneous value of variables that are functions of time
$\vec{i}$, $\vec{v}$	space vectors
I, V	rms values
$\boldsymbol{x}$	column vector
$\boldsymbol{x}^T$	row vector
$\boldsymbol{X}$	matrix
$\mathcal{S}$	set

Symbols

$\mathbf{0}_{n \times m}$	zero matrix of dimensions $n \times m$
$\boldsymbol{A}$	system matrix (discrete time)
$\boldsymbol{B}$	input matrix (discrete time)
c	coefficient
C	capacitance (F)
$\boldsymbol{C}$	input matrix (continuous or discrete time)
d	pulse number
D	determinant
e, E	energy (J or pu)
f	frequency (Hz or pu)
$\boldsymbol{F}$	system matrix (continuous time)
$\boldsymbol{G}$	input matrix (continuous time)

$\boldsymbol{H}$	Hessian matrix
$i, \boldsymbol{i}, I$	current (A or pu)
$\boldsymbol{I}_n$	identity matrix of dimensions $n \times n$, $\boldsymbol{I}_n = \text{diag}(1, 1, \ldots, 1)$
j	imaginary unit, $\sqrt{-1}$
J	cost function
k	discrete time step
$\boldsymbol{K}$	transformation matrix
ℓ	discrete time step (relative to k)
L	inductance (H)
m	modulation index
M	moment of inertia (kg m^2 or pu)
n	order of harmonic, number of modules
N	length of switching sequence
p	number of pole pairs
pf	power factor
P	(instantaneous) real power (W or pu)
Q	(instantaneous) reactive power (Var or pu)
$\boldsymbol{q}, \boldsymbol{Q}$	penalty vector or matrix
R	resistance (Ω or pu)
sl	slip
S	apparent power (V A or pu)
t	time (s or pu)
T	torque (N m or pu)
$u, \boldsymbol{u}$	switch position, input (or manipulated) variable
$\Delta u, \Delta \boldsymbol{u}$	change in switch position
$U, \boldsymbol{U}$	sequence of switch positions (switching sequence)
$v, \boldsymbol{v}, V$	voltage (V or pu)
$\boldsymbol{V}$	generator matrix
$x, \boldsymbol{x}$	state variable
X	reactance (pu)
$y, \boldsymbol{y}$	output variable
Z	impedance (Ω or pu)
α	switching angle in pulse pattern (rad)
γ	load angle, that is, angle between the stator and rotor flux vectors (rad)
δ	(half of the) bound width
$\varepsilon, \boldsymbol{\varepsilon}$	degree of bound violation (at a time step)
$\epsilon, \boldsymbol{\epsilon}$	rms bound violation (over the prediction horizon)
λ	scalar penalty weight
$\boldsymbol{\lambda}$	flux linkage vector (Wb)
ϕ	phase angle (rad)
ρ	radius of sphere
σ	total leakage factor
θ	angle (argument) in pulse pattern (rad)
$\nu, \boldsymbol{\nu}$	insertion index
φ	angular position of a reference frame (rad)
$\psi, \boldsymbol{\psi}$	flux (linkage) (pu)

Ψ	flux (linkage) magnitude (pu)
τ	time constant (s or pu)
ω	rotational speed or angular frequency (rad/s or pu)
ξ, $\boldsymbol{\xi}$, ζ, $\boldsymbol{\zeta}$	slack or auxiliary variable

Subscripts

c_{on}, c_{off}, c_{rr}	turn–on, turn–off and reverse recovery energy loss coefficients (J/(VA))
C_m	module capacitance (F)
f_c	carrier frequency
f_{DL}	frequency of deadlocks
f_{sw}	switching frequency
i_1	fundamental current
i_a, i_b, i_c	phase a, b, and c currents
i_α, i_β	real and imaginary parts of the current (in the stationary reference frame)
i_B	base current
$\boldsymbol{i}_c$	converter current vector
$\boldsymbol{i}_{\text{circ}}$	circulating current vector
i_d, i_q	real and imaginary parts of the current (in the rotating reference frame)
$\boldsymbol{i}_{\text{err}}$	current error vector
$\boldsymbol{i}_g$	grid current vector
i_n	neutral point current
$\boldsymbol{i}_r$	rotor current vector
i_R	rated current
$\boldsymbol{i}_{\text{rip}}$	ripple current vector
$\boldsymbol{i}_s$	stator current vector
i_T	anode current
I_{TDD}	total demand distortion (TDD) of the current
L_{br}	branch inductor
L_{ls}	stator leakage inductance
L_{lr}	rotor leakage inductance
L_m	main (or magnetizing) inductance
L_σ	total leakage inductance
N_p	prediction horizon (number of time steps)
N_s	switching horizon (number of switching events)
T_e	electromechanical torque
$T_{e,\text{min}}$	lower bound on the electromagnetic torque
$T_{e,\text{max}}$	upper bound on the electromagnetic torque
T_ℓ	load torque
T_p	prediction horizon (length in time)
θ_p	prediction horizon (angular interval)
T_s	sampling interval
u_{opt}	optimal control input (or manipulated variable)
v_{dc}	instantaneous dc-link voltage
V_{dc}	nominal dc-link voltage
$v_{\text{dc,lo}}$, $v_{\text{dc,up}}$	instantaneous voltage of the lower and upper dc-link half, respectively
v_n	neutral point potential

v_{ph}	phase voltage
ω_1	fundamental frequency
ω_{fr}	angular speed of the reference frame
ω_g	electrical grid frequency
ω_m	mechanical shaft speed
ω_r	electrical angular speed of the rotor
ω_s	stator frequency
ω_{sl}	slip frequency

Superscripts

i^*	current reference
$\vec{i}$	current space vector
$\hat{i}_n$	amplitude of harmonic current of order n
i'	scaled version of i, e.g., when turned into the per unit system
$\bar{\boldsymbol{u}}$	switch position multiplied with the generator matrix $\boldsymbol{V}$

Operations

$\mathrm{d}x/\mathrm{d}t$	time derivative of the variable x
$\exp(x)$, e^x	exponential of the variable x
$\Re\{x\}$	real part of the complex variable x
$\Im\{x\}$	imaginary part of the complex variable x
$\mathrm{conj}\{x\}$	complex conjugate of the complex variable x
$\boldsymbol{x} \times \boldsymbol{y}$	cross product of the vectors **x** and **y**
$x \in \mathcal{S}$	variable x belongs to the set $\mathcal{S}$
$\boldsymbol{x}^T$	transpose of the vector $\boldsymbol{x}$
$\boldsymbol{X}^{-1}$	inverse of the matrix $\boldsymbol{X}$
$\lvert x \rvert$	absolute value of the scalar x
$\lVert \boldsymbol{x} \rVert_1$	1-norm of the vector **x** (sum of the absolute values)
$\lVert \boldsymbol{x} \rVert_2$	2-norm or length of the vector **x** (square root of the sum of the squared values, Euclidian norm). To simplify the notation, we will often simply write $\lVert \boldsymbol{x} \rVert$
$\lVert \boldsymbol{x} \rVert_\infty$	infinity-norm of the vector **x** (largest absolute value)

About the Companion Website

Don't forget to visit the companion website for this book:

www.wiley.com/go/geyermodelpredictivecontrol

There you will find valuable material designed to enhance your learning, including animations of the control concepts.

Scan this QR code to visit the companion website

Part One

Introduction

1

Introduction

Power electronics applications with power levels in excess of 1 MVA, such as medium-voltage (MV) drives, are introduced in this chapter along with their market and technology trends. The commonly used control and modulation schemes are summarized. An introduction to model predictive control (MPC) is provided, which focuses on the control principle of MPC, and its advantages and challenges. This chapter concludes with a summary of the main results of the book and an outline.

1.1 Industrial Power Electronics

1.1.1 Medium-Voltage, Variable-Speed Drives

A typical representative of an industrial power electronic system is a variable-speed drive (VSD). The block diagram of such a system is shown in Fig. 1.1. It consists of an optional step-down transformer connected to the grid, an (active) rectifier, a dc-link, an inverter, and an electrical machine that drives the mechanical load. Additional components such as the controller, cooling, protection, and switchgear are part of the VSD system, but are not shown in the figure.

A VSD allows the operation of an electrical machine at an adjustable speed and at an adjustable electromagnetic torque. This is achieved by decoupling the grid electrically from the machine. The grid's *fixed* frequency ac quantities, which are either 50 or 60 Hz, are rectified to dc quantities, using either a diode rectifier or an active front end. An inverter transforms these dc quantities back to ac at a *variable* frequency, which is proportional to the rotational speed of the mechanical load. The dc-link acts as an energy storage element and decouples the rectifier from the inverter.

By adjusting the phase and amplitude of the rectifier voltages, the power flow between the grid and the dc-link can be manipulated. Similarly, on the machine side, by adjusting the phase and amplitude of the inverter voltages, the machine currents and thus the electromagnetic torque and magnetization of the machine are controlled.

MV VSDs use line-to-line rms voltages between 690 V and 20 kV, with typical voltages in the range 2.4–6.9 kV. Power ratings are usually in excess of 1 MVA. Because of the

Model Predictive Control of High Power Converters and Industrial Drives, First Edition. Tobias Geyer.

Companion Website: www.wiley.com/go/geyermodelpredictivecontrol

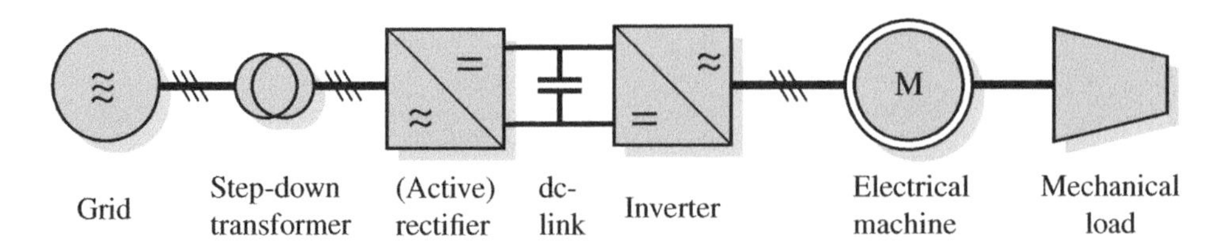

Figure 1.1 Variable-speed drive system

high currents and voltages, high-power semiconductor switches are used in the rectifier and inverter to commutate and control the currents. The semiconductor switches are operated such that the resulting currents approximate, albeit in a coarse manner, sinusoidal waveforms at steady-state operation.

As a well-known example for an MV VSD system, Fig. 1.2 depicts the ACS6000 and a typical MV induction machine. The ACS6000 is based on the three-level neutral-point-clamped (NPC) topology with water-cooled integrated-gate-commutated thyristors (IGCTs).

(a) ACS6000 with an active front end, the terminal and control unit, the inverter unit, the dc-link capacitor bank, a voltage limiter and the water cooling unit

(b) MV induction machine

Figure 1.2 Medium-voltage VSD system. Source: ABB Image Bank. Reproduced with permission of ABB Ltd

It is rated at an output voltage between 2.3 and 3.3 kV. In the single-drive configuration shown in Fig. 1.2(a), the ACS6000 provides 5–12 MVA. Up to 36 MVA is available in the multidrive configuration.

1.1.2 Market Trends

Sale of industrial high-power electronics is experiencing high annual growth rates. For MV drives, for example, the growth rate is consistently above 10% per year, with worldwide revenues of 3.7 billion USD in 2014 [1]. The high growth is driven by four major trends:

1.1.2.1 Electrification

Combustion engines are increasingly being augmented or replaced by electrical drives with the aims of increasing efficiency, reducing emissions, reducing fuel consumption, and removing the clutch and gear box to simplify the mechanical drive train. Examples of this include diesel-electric propulsion systems for trains, large mining trucks, tug boats, and large ships. In the oil and gas industry, gas turbines in compressor trains have traditionally required a starter motor, which—if designed accordingly—may also act as a helper motor, thus augmenting the gas turbine [2]. Furthermore, drives are about to fully replace gas turbines in large liquefied natural gas (LNG) compressor trains. In the low-voltage range, (hybrid) electric automotive vehicles constitute a major and rapidly growing trend.

1.1.2.2 Renewable Power Generation and Energy Storage

Wind turbines have traditionally relied on low-voltage, doubly fed induction machines. Modern wind turbines for the offshore market often exceed 3 MW and have adopted full back-to-back power conversion stages. For higher power ratings, MV generators are used [3]. Pumped hydro storage systems are typically based on MV doubly fed induction machines [4]. Utility-scale photovoltaic plants and large battery energy storage systems are based on MV power converters.

1.1.2.3 Industrial Drive Applications

For industrial drive applications, a distinction between general-purpose and special-purpose drives is generally made [5]. The latter term refers to highly demanding variable-speed and variable-torque applications, such as rolling steel mills, for which a back-to-back power conversion system is mandatory. General-purpose loads, however, such as large pumps, fans, blowers, and compressors, are predominantly connected directly to the grid using standard direct online—rather than inverter-duty—electrical machines. In order to increase their efficiency during partial load operation, many grid-coupled electrical machines are upgraded to VSDs in an retrofit effort [6].

1.1.2.4 Utility-Scale Power Electronics for the Grid

The power system is being overhauled by adding flexible ac transmission systems (FACTS) to it to achieve smart grid capabilities and to enhance the power flow [7, 8]. High-voltage dc (HVDC) systems are installed for bulk power transmission and to connect large offshore wind farms to the grid [9, 10]. Other notable examples for utility-scale, grid-connected power electronics include active voltage conditioners (AVCs) and uninterruptible power supplies (UPSs).

1.1.3 Technology Trends

Industrial power electronic systems in the MV range are influenced by four major technology trends:

1.1.3.1 Multilevel Converter

Over the past 50 years, there has been a continuous shift toward converters with a higher number of output voltage levels [11]. Starting from the two-level converter, the three-level NPC converter was introduced in the early 1980s [12]. Five-level topologies followed around the year 2000. Topologies with a higher number of levels have been available for a few years, which are either based on cascaded H-bridges or the modular multilevel converter (MMC) [13]. The main motivation to adopt these topologies is to achieve higher output power ratings. To keep the currents at bay, this implies increased voltage ratings. Another incentive is to avoid the step-down transformer on the grid side.

1.1.3.2 Product Business

The MV converter business is turning from a project business into a product business, in which MV converters can be bought off the shelf, and installed and commissioned quickly. A relatively large number of competitors coexist, competing with similar products and technologies.

1.1.3.3 Efficiency

Close to 100% efficiency and low losses are paramount for some applications, such as FACTS and photovoltaic systems.

1.1.3.4 Computational Power

The computational power of control hardware is growing exponentially. The observation underlying Moore's law, that the transistor count of integrated circuits doubles every 2 years, still holds true [14]. In industrial power electronics, a transition from relatively small digital signal processors (DSPs) to high-performance DSPs, often augmented by a large field-programmable gate array (FPGA), can be observed. In some cases, multicore processors are adopted as a corner piece of the control hardware.

1.2 Control and Modulation Schemes

1.2.1 Requirements

For industrial power electronic systems, the three pivotal requirements for control and modulation schemes are the following:

1.2.1.1 Low Harmonic Distortions per Switching Losses (or Frequency)

The trade-off between harmonic distortions on one hand and switching frequency or switching losses on the other hand is well known and fundamental to power electronics. The objective is to move this trade-off curve toward the origin, rather than to optimize along the curve; see Fig. 1.3. Lower harmonic distortions allow the reduction or removal of harmonic filters, or the use of standard direct online machines without derating them. Lower switching losses enable either boosting the inverter efficiency or increasing the rating of the inverter. On the grid side, low grid current distortions and compliance with grid codes are required.

1.2.1.2 High Controller Bandwidth

Fast closed-loop control is required to quickly control electrical machines in applications with rapidly changing loads or speed setpoints. This translates into the requirement of fast torque responses of a few milliseconds. Grid-connected power converters often require similarly fast current responses, particularly during power reference steps and faults.

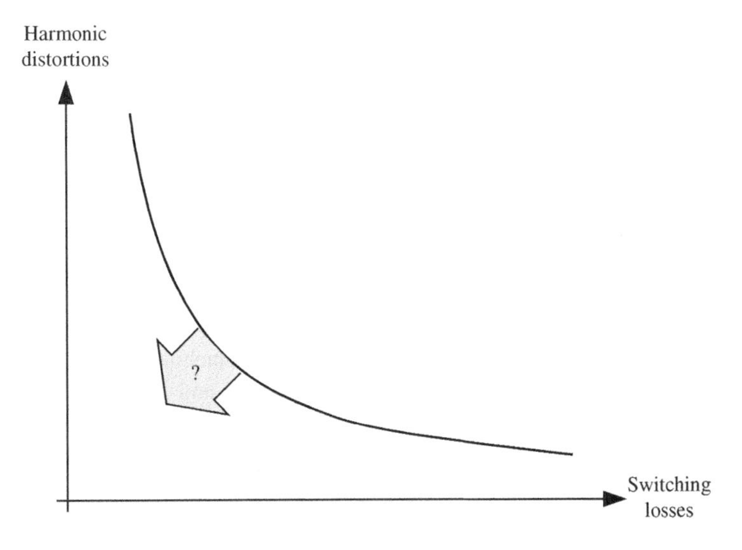

Figure 1.3 Fundamental trade-off between harmonic distortions and switching losses (frequency)

1.2.1.3 Accurate Load Power Control

The load power must be controlled. On the machine side, this implies control of the speed and/or electromagnetic torque of the ac machine. On the grid side, the real and reactive power must be controlled. Typically, the real power is manipulated such that the dc-link voltage is maintained at its nominal level, while the reactive power is set to zero.

Additional requirements include robustness to parameter variations, insensitivity to measurement and observer noise, as well as a high degree of fault tolerance. The computational burden of the control and modulation scheme must be sufficiently low to enable a successful implementation on the available control hardware.

1.2.2 State-of-the-Art Schemes

Almost universally, the controller of an industrial power electronic system is split into a load-side and a grid-side controller. Each controller is subdivided into two cascaded control loops. On the grid side, an outer loop controls the dc-link voltage and manipulates the real power, which is a setpoint for the inner loop. The latter controls the real and reactive power of the converter by manipulating the three-phase converter voltage.

On the load side of a VSD system, the outer loop controls the machine's speed by manipulating the torque reference. The inner loop controls the machine's electromagnetic torque and degree of magnetization by manipulating the voltage applied to the stator windings of the machine. For a grid-connected power electronic system, the load-side controller needs to be designed according to the attached load.

The voltage command of the inner control loop is typically translated into gating signals for the semiconductor switches using a carrier-based pulse width modulator (CB-PWM) [15] or a space vector modulator (SVM) [16]. In both cases, a fast inner control loop is often used, which is typically formulated in a rotating orthogonal reference frame. On the machine side, the reference frame is aligned with a flux linkage vector, leading to the concept of the so-called field-oriented control (FOC) [17, 18]. On the grid side, the reference frame can be aligned with the grid voltage, resulting in voltage-oriented control (VOC), or with a virtual flux vector, giving rise to virtual-flux-oriented control [19].

Lower harmonic distortions per switching frequency can be achieved by using OPPs. Since the related control problem is difficult to solve with a high-bandwidth controller, the commonly used approach is to resort to a slow inner control method, such as scalar or volts per frequency (V/f) control.

A third alternative is to replace the inner control loop by a hysteresis controller. Instead of a modulator, a look-up table is used, which decides the inverter switch positions. Noteworthy examples include direct torque control (DTC) [20] on the machine side, which controls the electromagnetic torque and the magnetization of the machine, and direct power control (DPC) [21] on the grid side, which controls the real and reactive power components. DTC and DPC lead to very fast responses of the controlled variables, but tend to give rise to pronounced harmonic distortions.

Figure 1.4 qualitatively characterizes these three standard control methodologies according to the two control requirements outlined in the previous section. A more comprehensive introduction to the requirements of control and modulation schemes and the state-of-the-art control methods is provided in Chap. 3.

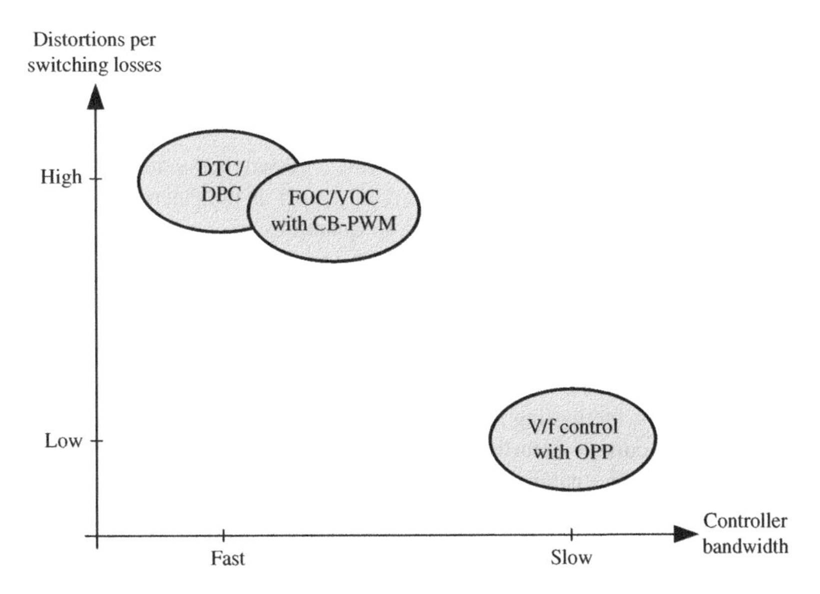

Figure 1.4 State-of-the-art control and modulation schemes for high power converters and industrial drives. These include hysteresis-based control schemes such as direct torque control (DTC) and direct power control (DPC), field-oriented control (FOC) or voltage-oriented control (VOC) with carrier-based pulse width modulation (CB-PWM) or space vector modulation (SVM), and volts per frequency (V/f) control with optimized pulse patterns (OPPs)

In summary, the majority of control and modulation methods used in industry today have the following three attributes: First, the overall multiple-input multiple-output (MIMO) control problem is divided into multiple control loops with single-input single-output (SISO) controllers. These control loops are arranged in a cascaded manner according to the dominant time constant of their loop. Second, the switching behavior of the power converter is neglected through the use of averaging. This allows the use of linear controllers, such as proportional-integral (PI) controllers. These controllers are typically augmented by an additional anti-windup mechanism and a rate limiter. Third, a pulse width modulation (PWM) stage is used to translate the averaged reference quantities into switching signals.

1.2.3 Challenges

Three major challenges can be identified for the design and real-time computation of high-performance control and modulation schemes:

1.2.3.1 Challenge 1: Switched Nonlinear Systems

The main building blocks of power electronic systems are linear circuit elements, such as inductors, capacitors, and resistors, which are complemented by semiconductor switches, which are either active (or controlled) switches or (passive) diodes. For different combinations of switch positions, different system dynamics arise, which can be described by linear functions of time for each combination. As a result, when controlling currents, fluxes, and voltages and manipulating the switch positions, power electronic systems constitute switched *linear* systems, provided that saturation effects of magnetic material, delays, and safety constraints can be neglected [22, 23].

In general, however, power electronic systems represent switched *nonlinear* systems. Nonlinearities arise, for example, when machine variables such as the electromagnetic torque or stator flux magnitude are directly controlled; both quantities are nonlinear functions of currents or flux linkages. For grid-connected converters, the real and reactive power is nonlinear in terms of the currents and voltages. Saturation effects in inductors and current constraints lead to additional nonlinearities.

Averaging [24, 25] is a viable way to conceal the switching behavior, provided that the pulse number is high. A paramount property of CB-PWM is that the ripple current is zero at regular sampling instants, facilitating the use of averaging. For pulse numbers well above 15, CB-PWM results in low current distortions. For low pulse numbers, however, averaging should be avoided and the switching nature of the power electronic system should be addressed by the control and modulation scheme to achieve low current distortions despite the low pulse number. For this, OPPs are the preferred choice. Since sampling instants at which the ripple current is zero in all three phases generally do not exist for OPPs, the concept of averaging is not suitable in the context of OPPs.

1.2.3.2 Challenge 2: MIMO Systems

The decomposition of the MIMO control problem into multiple SISO loops and the use of cascaded control loops greatly simplifies the controller design. This approach works well when the time constants of the cascaded control loops differ by at least an order of magnitude and while operating at (quasi) steady-state operating conditions. During transients and faults, however, the different loops often start interacting with each other in an adverse manner, limiting the achievable performance in terms of controller bandwidth and robustness, and complicating the tuning of the control loops.

For converters with LC filters, for example, the current controller is typically augmented by an active damping loop, with the purpose of dampening the system resonance introduced by the LC filter [26, 27]. To avoid large overshoots during transients, the current response has to be slowed down, for example, by rate-limiting the current reference. For an MMC, a plethora of quantities have to be either regulated along their references or kept at their nominal values. Because of the physical coupling of these quantities, the commonly used approach to control the MMC using multiple SISO loops leads to satisfactory performance only during steady-state operation. Interestingly, few results are available in the literature that showcase a fast dynamic operation of the MMC.

Therefore, for demanding applications, the MIMO characteristic of the power electronic system needs to be addressed by a MIMO controller. The benefit of doing so is a faster dynamic response during transients with less overshoot, as well as a simpler tuning and commissioning procedure.

1.2.3.3 Challenge 3: Short Computation Times

The third challenge results from the short sampling intervals of 1 ms and less that are typically used in power electronic systems. These short sampling intervals limit the time available to compute the control actions. To reduce the cost of power electronic converters sold in high volumes, cheap computational hardware is usually deployed as the control platform. Replacing existing control loops with low computational requirements by new and computationally more demanding methods exasperates the challenge of short sampling intervals. This is particularly the case for direct control methods that avoid the use of a modulator. These methods benefit from very short sampling such as 25 μs.

1.3 Model Predictive Control

Modern control theory formulated in the time domain emerged in the 1960s with the Kalman filter and the linear quadratic regulator [28, 29]. The state-feedback control law of the latter is obtained by minimizing a quadratic cost function over an infinite horizon, subject to the dynamic evolution of a linear system model. The first variants of MPC emerged in the process industry in the 1970s, focusing on nonlinear systems with physical constraints and on a finite horizon formulation.

Traditionally, since its inception 40 years ago, MPC has received little attention from the power electronics community and has been underutilized in this field. Other communities, such as the process industry, had already adopted this concept in the 1980s with great success [30]. Qin and Badgwell report in the late 1990s more than 4500 applications of linear MPCs in various industries, predominantly in refining, petrochemicals, and chemicals. Some applications can also be found in the areas of food processing, aerospace and defence, mining and metallurgy, and the automotive industry [30].

The reasons for the late adoption of MPC by the power electronics community include the limited processing power that was available in the last century to solve the control problem in real time and the very short time constants of power electronic systems necessitating the use of short sampling intervals. The switched (non)linear characteristic of power electronic systems complicates the controller design, analysis, and verification. Nevertheless, some initial investigations in MPC-related concepts for power converters were accomplished in the 1980s. Most importantly, these methods have been successfully implemented and experimentally verified [31, 32].

Over the past decade, however, MPC has rapidly emerged in power electronics. This progress has been facilitated not only by the tremendous increase of the computational power available in the controller hardware but also by the equally significant speed-up of the solvers that compute the solution to the underlying optimization problem. At the same time, complicated, new, multilevel topologies have emerged that require sophisticated control algorithms, the requirements imposed on power electronic systems have become more stringent, and, in the globalized world, companies are facing considerable pressure to retain or regain a competitive edge over their competitors.

1.3.1 Control Problem

Consider a general (power electronic) system with the input vector $\boldsymbol{u} \in \mathbb{R}^{n_u}$ and the output vector $\boldsymbol{y} \in \mathbb{R}^{n_y}$, as shown in Fig. 1.5. Both vectors may contain real-valued and integer components. Physical constraints in the form of actuator limits usually exist on the input. We refer to the system input $\boldsymbol{u}$ as the *manipulated* variable and the system output $\boldsymbol{y}$ as the *controlled* variable.

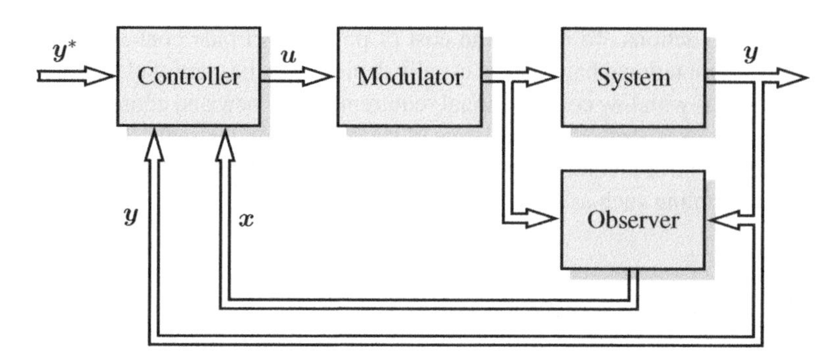

Figure 1.5 Controller regulating the system output $\boldsymbol{y}$ along its reference $\boldsymbol{y}^*$ by manipulating the system input $\boldsymbol{u}$. An optional modulator translates $\boldsymbol{u}$ into the converter switch position. The observer reconstructs the system state $\boldsymbol{x}$

We distinguish between two varieties of the control problem. When a modulation stage is added to the system, the manipulated variable is real-valued and typically a voltage reference. We refer to this as the *indirect* control problem. Averaging can be used to mask the switching phenomenon, and the use of integer variables in the system model can be avoided. On the other hand, when the modulator is removed, the *direct* control problem arises, with the manipulated variable corresponding to the converter switch positions. As a result, averaging cannot be employed, and the system model contains integer variables.

MPC requires the state vector $\boldsymbol{x} \in \mathbb{R}^{n_x}$ of the system. Components of $\boldsymbol{x}$ that cannot be measured, such as the rotor flux linkage, need to be reconstructed by an observer. Using a model of the system that is fed with the system input, the state and the output of the system can be estimated. By feeding back the difference between the measured and the estimated system outputs, observers can be designed such that the estimated states converge to the real states, provided that the observer is asymptotically stable and the system is observable.

The general control problem is to design a controller that achieves the following control objectives: The system output $\boldsymbol{y}$ must be regulated along its reference $\boldsymbol{y}^*$. This can be achieved by feeding back the measured output $\boldsymbol{y}$, comparing it with its reference $\boldsymbol{y}^*$, and manipulating the input $\boldsymbol{u}$ accordingly. This feeding back of the output to the input closes the loop and provides the feedback. The controller also has to guarantee stability and ensure that the constraints are met at all times (constraint satisfaction). These three objectives must be achieved despite disturbances and model uncertainties, necessitating a certain degree of controller robustness.

1.3.2 Control Principle

Over the past decades, MPC has evolved from a collection of control methods into a coherent control paradigm, perhaps even a control philosophy. Several thousand articles have been published on MPC. Despite the different MPC formulations and variations, five key attributes can be identified that are common to the MPC framework. These features are summarized in the following.

1.3.2.1 Internal Dynamic Model

MPC incorporates a dynamic model of the system to be controlled. Let $\boldsymbol{x} \in \mathbb{R}^{n_x}$ denote the state vector of the system, which—in general—includes real-valued and integer components. Starting from the current state, the internal dynamic model enables MPC to predict the sequence of future system states and outputs for a given sequence of manipulated variables.

The dynamic evolution of the system can be described in the continuous-time domain by the state-space representation

$$\frac{\mathrm{d}\boldsymbol{x}(t)}{\mathrm{d}t} = \boldsymbol{f}(\boldsymbol{x}(t), \boldsymbol{u}(t)) \tag{1.1a}$$

$$\boldsymbol{y}(t) = \boldsymbol{h}(\boldsymbol{x}(t), \boldsymbol{u}(t)), \tag{1.1b}$$

where (1.1a) is a nonlinear first-order differential equation that captures the evolution of the state vector over the time $t \in \mathbb{R}$. The outputs $\boldsymbol{y}$ are a nonlinear function $\boldsymbol{h}(\cdot, \cdot)$ of the state and input vectors.

In power electronics, when choosing voltages, currents, or flux linkages as state and output variables, the state-space representation (1.1) is usually linear and we can write it in the following well-known matrix form

$$\frac{\mathrm{d}\boldsymbol{x}(t)}{\mathrm{d}t} = \boldsymbol{F}\boldsymbol{x}(t) + \boldsymbol{G}\boldsymbol{u}(t) \tag{1.2a}$$

$$\boldsymbol{y}(t) = \boldsymbol{C}\boldsymbol{x}(t), \tag{1.2b}$$

with the system matrix $\boldsymbol{F}$, input matrix $\boldsymbol{G}$, and output matrix $\boldsymbol{C}$.

Most linear MPC strategies are formulated in the discrete-time domain, using a constant sampling interval T_s. The manipulated variable is restricted to changing its value only at the discrete sampling instants, that is at the time instants $t = kT_s$, where $k \in \mathbb{N} = \{0, 1, 2, \ldots\}$ denotes the time steps. For the continuous-time state-space model (1.2), the discrete-time representation can easily be computed. Specifically, by integrating (1.2a) from $t = kT_s$ to $t = (k+1)T_s$ and observing that $\boldsymbol{u}(t)$ is constant during this time interval and equal to $\boldsymbol{u}(k)$, we obtain the discrete-time state-space equation

$$\boldsymbol{x}(k+1) = \boldsymbol{A}\boldsymbol{x}(k) + \boldsymbol{B}\boldsymbol{u}(k) \tag{1.3a}$$

$$\boldsymbol{y}(k) = \boldsymbol{C}\boldsymbol{x}(k). \tag{1.3b}$$

The matrices $\boldsymbol{A}$ and $\boldsymbol{B}$ can be computed from their continuous-time counterparts according to

$$\boldsymbol{A} = \mathrm{e}^{\boldsymbol{F}T_s} \quad \text{and} \quad \boldsymbol{F}\boldsymbol{B} = -(\boldsymbol{I} - \boldsymbol{A})\boldsymbol{G}, \tag{1.4}$$

where e denotes the matrix exponential, and $\boldsymbol{I}$ is the identity matrix of appropriate dimensions. We refer to this as *exact* discretization.

If the matrix exponentials were to pose computational difficulties, the forward Euler approximation is often sufficiently accurate for short sampling intervals of up to several tens of microseconds in combination with short prediction horizons. In this case, the discrete-time system matrices are given by

$$\boldsymbol{A} = \boldsymbol{I} + \boldsymbol{F}T_s \quad \text{and} \quad \boldsymbol{B} = \boldsymbol{G}T_s. \tag{1.5}$$

The output matrix $\boldsymbol{C}$ remains the same when deriving the discrete-time system representation.

1.3.2.2 Constraints

Even in cases when the state-space equations are linear as in (1.3), constraints on inputs, states, and outputs

$$u(k) \in \mathcal{U} \subseteq \mathbb{R}^{n_u} \tag{1.6a}$$

$$x(k) \in \mathcal{X} \subseteq \mathbb{R}^{n_x} \tag{1.6b}$$

$$y(k) \in \mathcal{Y} \subseteq \mathbb{R}^{n_y} \tag{1.6c}$$

are usually present, which make the system nonlinear.

For the indirect control problem, when a modulator is added to the system, the real-valued manipulated variable is typically the voltage reference for the PWM. In this case, it is restricted to a bounded continuous set, such as

$$\mathcal{U} = [-1, 1]^{n_u}. \tag{1.7}$$

In contrast to this, for the direct control problem, the switch position of the converter constitutes the manipulated variable, which is constrained to a finite set of integers. A three-level converter, for example, is capable of synthesizing three voltage levels per phase. This characteristic can be captured by the input constraint

$$\mathcal{U} = \{-1, 0, 1\}^{n_u}. \tag{1.8}$$

For a five-level converter, one would have $\mathcal{U} = \{-2, -1, 0, 1, 2\}^{n_u}$. In a three-phase system, the dimension of the input vector is usually $n_u = 3$. The constraints on u are of a physical nature and thus *hard*, implying that they cannot be relaxed.

Constraints on states are sometimes added to prevent the system from operating outside of its safe operating limits. On the converter currents, for example, upper constraints on the absolute value of the currents can be imposed slightly below the trip level to avoid trips and damages due to overcurrents. These constraints are typically imposed in the form of *soft constraints*, which can be slightly violated, albeit at a high cost. Imposing soft rather than hard constraints on state variables is preferable to avoid numerical issues such as the control problem becoming infeasible.

Rather than regulating the controlled variables along their references, controlled variables can be kept within upper and lower bounds by imposing soft constraints on them. In the context of an ac machine, for example, upper and lower bounds can be imposed on the electromagnetic torque and the stator flux magnitude, similar to the hysteresis bounds in DTC.

1.3.2.3 Cost Function

The control objectives are translated into the cost function, which maps the sequences of future states, outputs, and manipulated variables into a scalar cost value. The cost function facilitates the assessment and comparison of the predicted impact the different sequences of manipulated variables (or scenarios) have on the system. This enables MPC to choose the most suitable scenario, which is the one that minimizes the value of the cost function.

A general definition of the cost function is

$$J(\boldsymbol{x}(k), \boldsymbol{U}(k)) = \sum_{\ell=k}^{k+N_p-1} \Lambda(\boldsymbol{x}(\ell), \boldsymbol{u}(\ell)), \tag{1.9}$$

which is the sum of the *stage costs* (or weighting functions) $\Lambda(\cdot, \cdot)$ over the *finite horizon* of N_p time steps. The stage cost penalizes the predicted system behavior, such as the deviation of controlled variables from their references and the control effort, such as the switching frequency. The stage cost is required to be nonnegative. The cost function uses the current state vector $\boldsymbol{x}(k)$ and the *sequence* of manipulated variables

$$\boldsymbol{U}(k) = [\boldsymbol{u}^T(k)\, \boldsymbol{u}^T(k+1)\, \dots\, \boldsymbol{u}^T(k+N_p-1)]^T \tag{1.10}$$

as arguments. Based on these two arguments, and by using the internal dynamic system model, the future states and controlled variables can be predicted over the prediction horizon and penalized accordingly.

1.3.2.4 Optimization Stage

Minimizing the cost function subject to both the evolution of the discrete-time internal system model over the prediction horizon and the constraints gives rise to a constrained finite-time optimal control problem. The argument of the result is the optimal sequence of manipulated variables, $\boldsymbol{U}_{\text{opt}}(k)$. The control problem predominantly used in this book is based on a linear state-update equation, a nonlinear output equation, and constraints on the manipulated variable, which can be stated as

$$\boldsymbol{U}_{\text{opt}}(k) = \arg \underset{\boldsymbol{U}(k)}{\text{minimize}}\; J(\boldsymbol{x}(k), \boldsymbol{U}(k)) \tag{1.11a}$$

$$\text{subject to} \quad \boldsymbol{x}(\ell+1) = \boldsymbol{A}\boldsymbol{x}(\ell) + \boldsymbol{B}\boldsymbol{u}(\ell) \tag{1.11b}$$

$$\boldsymbol{y}(\ell+1) = \boldsymbol{h}(\boldsymbol{x}(\ell+1)) \tag{1.11c}$$

$$\boldsymbol{u}(\ell) \in \boldsymbol{\mathcal{U}} \quad \forall \ell = k, \dots, k+N_p-1. \tag{1.11d}$$

In its most general form, with the system model being nonlinear and the system variables containing integers, the optimization problem underlying MPC is a mixed-integer nonlinear program (MINLP). Traditionally, the optimization problem has exclusively been solved online, requiring the solution to be available in real time.

Rather than solving the mathematical optimization problem for the given state vector at the current time step, the optimization problem can be solved offline for *all possible* states. Specifically, the so-called *state-feedback control law* can be computed for all states $\boldsymbol{x}(k) \in \boldsymbol{\mathcal{X}}$ [33–35], by treating the state vector as a parameter and using *multi-parametric programming*, which is akin to a generalization of sensitivity analysis. Time-varying references $\boldsymbol{y}^*$ and additional time-varying parameters can be treated similarly. The explicit control law can be stored in a look-up table, and the optimal manipulated variable can be read from the look-up table in a computationally efficient manner. We refer to this methodology as *explicit MPC*, in contrast to standard MPC.

Explicit MPC might appear to be an attractive choice for systems with very short sampling intervals, such as power electronic systems. It is computationally viable, however, only for systems with a low-dimensional state vector and with few time-varying references and parameters. This makes explicit MPC an inflexible approach that is ill suited to address problems of higher dimensions. The use of integer manipulated variables further complicates the solution. This approach is therefore not pursued in this book. For a summary on the literature on explicit MPC for power electronic systems, the reader is referred to [36]. For an in-depth review of explicit MPC, see [37].

1.3.2.5 Receding Horizon Policy

The solution to the optimization problem (1.11) yields at time step k an *open-loop* optimal sequence of manipulated variables $\boldsymbol{U}_{\text{opt}}(k)$ from time step k to $k + N_p - 1$. To provide feedback, only the first element of this sequence, namely $\boldsymbol{u}_{\text{opt}}(k)$, is applied to the system. At the next time step $k + 1$, a new state estimate is obtained and the optimization problem is solved again over the shifted horizon from $k + 1$ to $k + N_p$. This policy is referred to as *receding horizon control*. It is illustrated in Fig. 1.6.

In summary, the principle of MPC is that at each sampling instant the manipulated variable is obtained by solving a constrained optimal control problem over a finite prediction horizon. An internal dynamic model of the system is used to predict future states and controlled variables, using the current state of the system as the initial state. The control objectives are captured by a cost function, which is minimized subject to the evolution of the internal model and system constraints. The solution to the underlying optimization problem yields an optimal sequence of manipulated variables. A receding horizon policy is employed, that is, only the first element of this sequence is applied to the system, and the sequence of manipulated variables is recomputed at the next sampling instant over a shifted horizon. Hence, MPC combines (open-loop) constrained optimal control with the receding horizon policy that provides feedback and closes the control loop.

In this section, some fundamental principles of MPC have been introduced. For more details on MPC and its mathematical underpinnings, the reader is referred to the vast literature on MPC that has been accrued in the control community. Prominent survey papers include [30, 38–41], and the classic MPC textbooks are [37, 42–45].

1.3.3 Advantages and Challenges

In Sect. 1.2.3, we have identified three major challenges for the design and implementation of high-performance control and modulation schemes when applied to industrial power electronics. In light of the MPC principle outlined in the previous section, we discuss in this section the aforementioned challenges and the ability of MPC to address them. Of the three challenges, the characteristics of power electronic systems being switched nonlinear systems as well as MIMO systems can easily be addressed by MPC, while the third challenge, namely the short computation times available in power electronics, persists as a profound challenge for MPC.

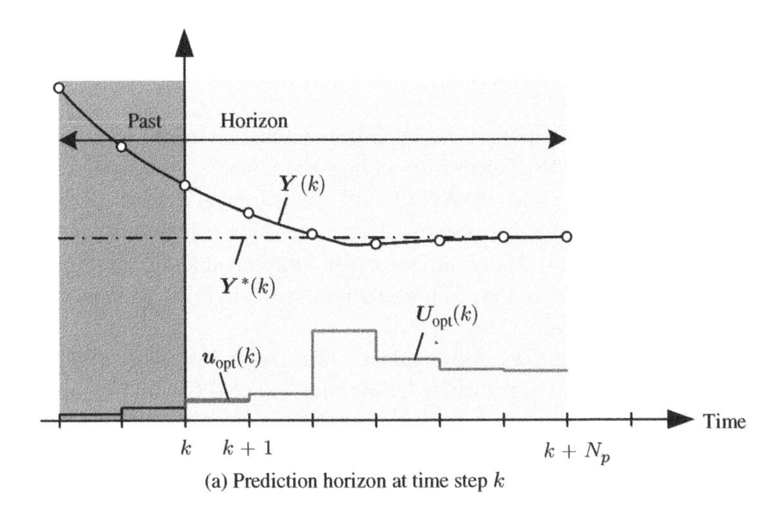

(a) Prediction horizon at time step k

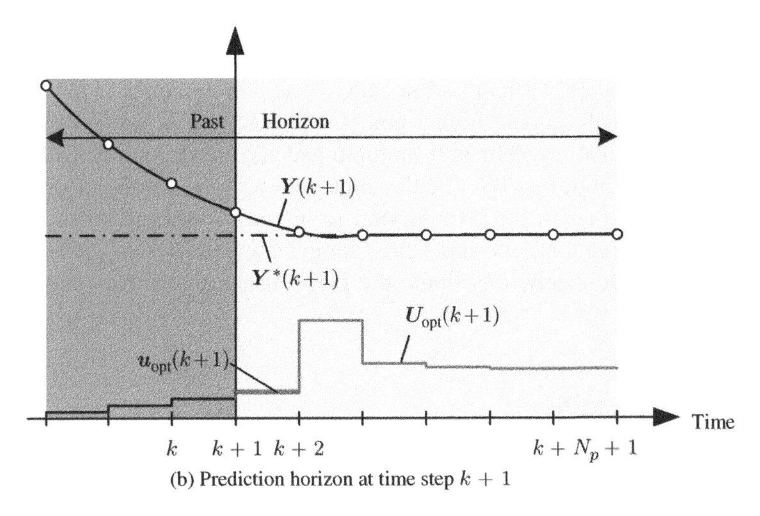

(b) Prediction horizon at time step $k + 1$

Figure 1.6 Receding horizon policy exemplified for the prediction horizon $N_p = 6$. The optimal sequence of manipulated variables $\boldsymbol{U}_{\text{opt}}$ is chosen such that the predicted output sequence $\boldsymbol{Y}$ tracks the output reference $\boldsymbol{Y}^*$. Out of the sequence $\boldsymbol{U}_{\text{opt}}$, only the first element $\boldsymbol{u}_{\text{opt}}$ is applied to the system

1.3.3.1 Advantages of MPC

First, MPC is formulated in the time domain rather than in the frequency domain. This enables MPC to address nonlinear systems in general—and switched nonlinear systems in particular—in a systematic way. This is achieved by incorporating the nonlinear system behavior into the MPC formulation in the form of an internal dynamic model. Averaging is not required, and the modulation stage can be included in the controller. Moreover, MPC is unique in its ability to systematically cope with hard constraints on manipulated variables, states, and controlled variables.

A vast body of literature has emerged on MPC for switched systems, which are sometimes referred to as *hybrid systems* [46]. Various modeling frameworks exist to describe such systems. This includes piecewise affine (PWA) [47] and mixed logical dynamical (MLD) systems [48] for the modeling of linear hybrid systems. These and other frameworks are reviewed and compared with each other in [49]. MPC can be readily formulated and solved for linear hybrid systems, as shown, for example, in [34, 37]. Many nonlinear (hybrid) systems can be approximated by linear hybrid systems.

The use of a cost function allows one to address diverse and possibly conflicting control objectives. These objectives can be prioritized, thus endowing MPC with the capability of—in effect—incorporating multiple control modes in one MPC controller. Furthermore, soft as well as rate constraints can be added to the control problem formulation.

Second, unlike PI-type controllers, MPC is a multivariable control method that is ideally suited for MIMO systems, particularly for complicated systems such as the various MMC topologies or converter systems with additional passive elements such as *LC* filters. Contrary to traditional frequency-domain control methods, additional active damping loops or anti-windup mechanisms are not required in MPC—one current control loop suffices. This simplifies the design, analysis, and tuning process. This benefit is sometimes overlooked. Breaking down the control problem into multiple and ideally decoupled SISO loops and designing individual PI loops for each of them might appear to be a straightforward and easy endeavor. In practise, however, these loops tend to interact with each other in an adverse manner, particularly during transients and faults, complicating the design and commissioning of the control loops. Ultimately, this limits the performance that can be achieved by the closed-loop system.

1.3.3.2 Challenges for MPC

Third, however, the effort required to solve the optimization problem underlying MPC is often considerable. Solving the optimization problem in the given time (usually within a part of the sampling interval) constitutes a major challenge. To extend the applicability of MPC from its traditional application domain of systems with long sampling intervals (e.g., in the process industry) to systems with short sampling intervals (e.g., in the automotive industry or power electronics) has spurred significant research effort along three avenues:

- The computation of the state-feedback control law, the explicit solution, for all possible states, references, and parameters. In many cases, however, the parameter space has proven to be of too high a dimension, leading to computationally intractable problems [50].
- The inception of optimization procedures and solvers with fast convergence rates and a low computational burden. Solvers are investigated that are well suited for implementation on embedded systems. For quadratic programs, for example, the fast gradient method appears to be particularly promising when executed on an FPGA, see, for example, [51–53].
- The investigation of new MPC problem formulations and solution methods that are tailored to the specific control problems that arise from power electronic systems. This is the research direction that is predominantly pursued in this book.

We conclude that the effort to formulate MPC control problems is often quite small, while the effort to solve the underlying optimization problem can be daunting. Unfortunately, the computational burden associated with solving the optimization problem underlying MPC increases exponentially with the length of the prediction horizon. Long prediction horizons yield, in general, a better closed-loop performance than short horizons. In particular, the infinite horizon case often ensures closed-loop stability, provided that a solution with a finite cost exists [42, 43]. However, long horizons exasperate the computational issue.

1.4 Research Vision and Motivation

The research vision behind this book is to devise control algorithms so as to maximize the effectiveness of power electronic systems—or equivalently—to design software to fully utilize the capability of power electronic hardware. For a three-phase, three-level inverter topology for example, the proposed control schemes are capable of reducing the switching losses in the semiconductor switching devices by up to 50% when compared to state-of-the-art schemes. In the MV arena, the switching losses are typically of a magnitude similar to the conduction losses; in some cases, they dominate over the latter. When the thermal cooling capability is the limiting factor, lower switching losses enable one to increase the current accordingly. As a result, the power rating of the hardware can be increased; for example, a 5 MVA inverter can be uprated to 6 MVA or more, and sold at an accordingly higher price tag, thus boosting the sales margin.

Alternatively, such control algorithms allow one to reduce the hardware requirements, for example, to reduce or remove harmonic filters, reduce dc-link filter capacitors, and allow standard direct online machines to be used instead of more expensive machines designed specifically for inverters. Moreover, the safe operating limits of the power electronic system can be translated into safety constraints, which can be added to the MPC problem formulation. Such constraints include, for example, upper constraints on the absolute value of the phase current. MPC ensures that these constraints are always met, thus ensuring a safe and reliable operation of the power converter.

Even more importantly, for MPC, a major part of the control effort is shifted from the design stage to the computational stage. As a result, on one hand, the design effort, the time to market, and the commissioning time are significantly reduced. On the other hand, however, a more powerful control hardware is sometimes required that consists not only of a DSP but often also of an additional FPGA that runs computationally intensive calculations. Nevertheless, the cost of an additional FPGA is in most cases negligible when compared to the cost savings that can be achieved when adopting MPC in the context of MV power electronic systems.

1.5 Main Results

The research objective underlying this book is to combine the advantages of DTC or DPC during transients with the benefits of offline computed OPPs during steady-state operation. As shown in Fig. 1.7, the aim is to devise fast current controllers that generate very low switching losses and distortions. To achieve an OPP-like performance at steady-state operation, very long prediction horizons are required. Smart algorithms are needed to solve the underlying

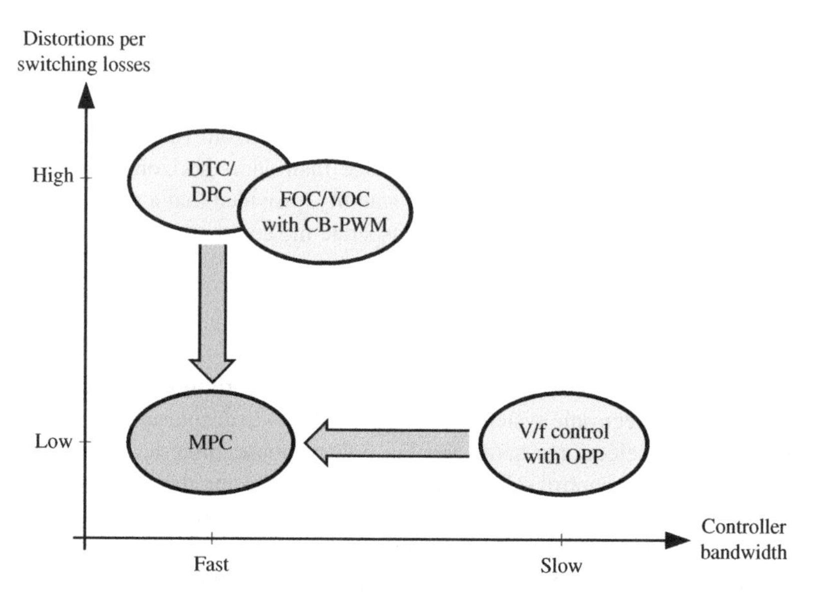

Figure 1.7 Model predictive control combining the merits of DTC or DPC during transients with those of OPPs during steady-state operation

optimization problem in real time, despite the combinatorial explosion of the number of possible solutions in the search space.

Over the past few years, three such schemes have been developed that combine the modulator and inner (current) control loop in one computational stage. These control schemes are based on the key notions of MPC—namely an internal model of the power electronic system to predict the system's response over a prediction horizon, a cost function to assess the predictions, an optimization stage to compute the optimal control action, and a receding horizon policy to provide feedback and robustness [43]. Despite these common characteristics, the three control schemes constitute complementary approaches.

The most commonly used MPC approach in power electronics is to directly manipulate the switch positions of the semiconductors and to formulate the control problem as a reference tracking problem [54]. This approach is often referred to as the *finite control set* (FCS) MPC. Any quantity of a power electronic system, such as a current, electromagnetic torque, angular speed, flux linkage, neutral point potential, real and reactive power, and so on, can be regulated along a given reference, as summarized in Chap. 4. The trade-off between tracking accuracy and switching effort can be adjusted by a tuning parameter. Favorable distortions per switching losses can be achieved when using long prediction horizons. The underlying optimization problem can be solved efficiently by adopting a branch-and-bound method from communication theory called *sphere decoding*, as shown in Chap. 5.

Model predictive direct torque control (MPDTC) is an advance on DTC, where the look-up table is replaced by an online MPC-type optimization stage. MPDTC was developed in early 2004, see [55, 56], experimentally verified on a 2 MVA drive in 2007 [57], and generalized in 2009 to further boost the performance by using even longer prediction horizons, see Chap. 7. Branch-and-bound methodologies can be used to reduce the computational burden by an order of magnitude, as will be shown in Chap. 10. Model predictive direct current control (MPDCC) is a derivative of MPDTC, see Sect. 11.1. Another derivative called model predictive direct balancing control (MPDBC) can be used to balance the internal inverter voltages in multilevel topologies [58].

Model predictive pulse pattern control (MP^3C) is based on OPPs [59] that are controlled in an MPC manner. Specifically, offline computed OPPs are modified online to account for transients and model uncertainties, as well as to provide feedback and robustness. MP^3C was originally devised for the machine-side inverter in electrical drives [60]. For those, MP^3C yields very fast control responses while drastically lowering the switching losses in the converter and/or the current distortions with respect to schemes based on CB-PWM, see Chap. 12.

Although these schemes are based on complementary approaches, they yield very similar closed-loop performances in terms of distortions per switching losses and controller bandwidth. In particular, for three-level MV inverters, the distortions per switching losses are reduced by up to 50% with respect to DTC and CB-PWM or SVM, while for five-level topologies they are reduced by 60% and more, see [61–63]. The current and torque response times are in the range of 1–2 ms. Therefore, at steady-state operating conditions, the resulting distortions per switching losses are similar to those obtained with OPPs. During transients, however, very fast current and torque response times are achieved, similar to deadbeat control.

In all three cases, the key to success was to devise control algorithms that are computationally highly tailored to the specific control problem at hand while utilizing the theoretical foundations of MPC. The standard optimal control approach provides only relatively small performance improvements and is computationally prohibitively demanding, as evidenced by some of the early publications, see, for example, [50, 64, 65]. Very long prediction horizons are required to achieve low distortions per switching losses. A particular effort was required to achieve the solution of these computationally very demanding MPC problems in real time on the commonly available drive control hardware.

1.6 Summary of this Book

The 15 chapters of the book are arranged in five parts.

Part I: Introduction

The first part of this book serves as an introduction, recalling basic power electronic terminology, concepts, and methods. This includes electrical machines, semiconductors, topologies, control, and modulation.

More specifically, following this introductory chapter, industrial power electronic systems are described in detail in Chap. 2. The chapter starts by reviewing some fundamental concepts that will be used throughout the book, such as the per unit system, orthogonal reference frames, and space vectors. State-space models of induction machines are derived, which describe the machines both during steady-state operation and transients. Power semiconductors, such as IGCTs and power diodes, are introduced and their loss models are stated. Three- and five-level voltage source inverters are described and modeled, and four industrial power electronics case studies are defined. The latter refer to MV VSDs and grid-connected converter systems.

After summarizing the requirements that electrical machines, the grid, and converters impose on control and modulation schemes, Chap. 3 reviews the major industrial control and modulation schemes that are used in high-power applications. CB-PWM is explained, its harmonic spectrum is analyzed, and the equivalence with SVM is recalled. A detailed account of OPPs is provided, which includes the derivation of the optimization problem and techniques to solve it in view of multilevel converters. The trade-off between harmonic distortions and the switching effort is shown analytically. In a last step, scalar control, FOC, and DTC are reviewed, which constitute the prevailing control methods used for machine-side converters. An appendix provides an introduction to mathematical optimization.

Part II: Direct Model Predictive Control with Reference Tracking

The second part of this book focuses on direct MPC methods with reference tracking of the output variables.

Chapter 4 introduces the concept of direct MPC by means of a predictive current controller with reference tracking and a prediction horizon of one step. This method is also commonly referred to as *FCS* MPC. Starting with a single-phase inverter with an RL load, the notions of the prediction model, cost function, optimization problem, and enumeration are reviewed. The MPC algorithm is subsequently generalized to the current control problem in three-phase inverter systems. A derivative of this method can be used to solve the torque and flux control problem of VSDs. The similarity between the current and torque controllers is shown by analyzing their cost functions. Moreover, the impact of the tracking error norm on stability is highlighted, and a method is reviewed to compensate for system delays.

Chapter 5 revisits the control problem of regulating the three-phase currents along their references by generalizing the control problem to long prediction horizons. For linear systems with integer inputs, an integer quadratic program results. It is shown that the optimal integer solution lies in a sphere centered on the unconstrained solution—the latter is obtained by relaxing the integer variables to real-valued variables. A branch-and-bound algorithm called *sphere decoding* is adopted that exploits this fact and allows one to quickly solve the underlying optimization problem even for relatively long prediction horizons, such as 10. The sphere decoding principle is illustrated with the help of two examples.

In the next chapter, the performance of long-horizon, direct MPC with reference tracking is evaluated. For an NPC inverter drive system with an induction machine, a horizon of 10 steps reduces the current distortions by 20%, when compared to the horizon 1 case. As a result, long-horizon direct MPC can outperform SVM and CB-PWM during steady-state operation. When an LC filter is added between the inverter and the electrical machine, the performance benefits of long prediction horizons become even more pronounced. Increasing the horizon from 1 to 20, for example, reduces the stator current distortions by up to a factor of 7.

Part III: Direct Model Predictive Control with Bounds

Direct MPC methods that maintain their output variables within upper and lower bounds are described in the third part of the book.

Chapter 7 is devoted to MPDTC. Similar to DTC, MPDTC manipulates the three-phase switch position to keep the controlled quantities, such as the electromagnetic torque, stator flux magnitude, and neutral point potential, within upper and lower bounds. A cost function that captures either the switching frequency or the switching losses is minimized. To render the underlying optimization problem computationally tractable for long prediction horizons, switching is only considered close to these bounds. In between the switching events, the switch position is frozen and the trajectories of the controlled variables are extended in an approximate manner, for example, by using linear or quadratic extrapolation. The cost function, the basic MPDTC algorithm based on enumeration, the corresponding search tree, and different methods of performing extrapolation are described and analyzed.

The closed-loop performance of MPDTC is investigated in Chap. 8. The benefit of adopting long prediction horizons is first shown for an MV NPC inverter drive system operating at steady state. Compared to DTC, the switching losses can be reduced by up to 60% for the same harmonic distortions. In the second part of this chapter, MPDTC is adapted to a five-level inverter drive system, for which the reduction of the harmonic distortions is the main focus. Compared to those of DTC, the harmonic current and torque distortions can be halved for the same—or a slightly lower—switching frequency. During torque transients, both DTC and MPDTC provide excellent results in both case studies.

The next chapter focuses on advanced topics regarding MPDTC. The bounds on the torque and stator flux magnitude form a target set, within which DTC and MPDTC maintain the stator flux vector. The offline computation of the control law facilitates the analysis and illustration of the decision-making process underlying MPDTC as well as the impact of different cost function formulations. The phenomenon of infeasible states or deadlocks is analyzed, and an effective deadlock resolution scheme is proposed. With the aim of inhibiting MPDTC from running into deadlocks, several such methods are proposed and their effectiveness is analyzed in the last part of the chapter.

The focus of Chap. 10 is on reducing the computational burden of MPDTC by an order of magnitude to enable the use of very long prediction horizons in real-time implementations. To this end, a branch-and-bound method is proposed that extends the MPDTC algorithm and computes the optimal switching sequence while exploring only a small part of the search tree. Upper and lower bounds on the cost function are introduced that allow the algorithm to identify and prune suboptimal parts of the search tree without explicitly exploring them. To limit the maximum number of computations, the optimization procedure can be stopped if the number of computational steps exceeds a certain threshold. Despite the possibility of suboptimal solutions, the performance impact is shown to be small, provided that the threshold is chosen carefully.

Chapter 11 generalizes the MPDTC concept and presents two derivatives. MPDCC controls the currents rather than the torque and stator flux magnitude. It is suitable for machine-side and grid-side converters. Thanks to the shape of its current bounds, it tends to achieve lower current distortions per switching losses than MPDTC. Model predictive direct power control (MPDPC) controls the real and reactive power components in a grid-connected converter setup. Both MPDCC and MPDPC are introduced in this chapter along with detailed performance evaluations. The chapter concludes with a comparison of the shape of the bounds of MPDTC, MPDCC, and MPDPC.

Part IV: Model Predictive Control based on Pulse Width Modulation

The fourth part of this book focuses on MPC methods that are based on PWM. These methods are complementary in their approach to the direct MPC techniques discussed in the previous two parts.

Chapter 12 proposes the concept of MP^3C. By definition, offline computed OPPs provide the minimal current distortion for a given switching frequency. Integrating the three-phase voltage waveform of the OPP over time leads to the optimal stator flux trajectory. By manipulating the switching instants of the OPP, the stator flux vector of the electrical machine can be regulated along the optimal flux trajectory, thus achieving fast closed-loop control of the machine. Adopting the notion of MPC, in particular the receding horizon policy, two computational variations of MP^3C are proposed. The first one is based on a quadratic program and uses a long prediction horizon. The second variation is a deadbeat controller that is computationally simple and achieves almost as fast a torque response as DTC. To improve the performance during transients, additional switching transitions can be inserted when the stator flux error exceeds a certain threshold.

The performance of MP^3C is evaluated in Chap. 13 through simulations and experiments on MV drives. Simulation results are provided for an NPC inverter drive system during steady-state operation and transients. When compared to SVM operating at the same switching frequency, MP^3C reduces the current distortions by up to 50%. The benefit of inserting pulses during transients is illustrated. Experimental results for a five-level active NPC inverter drive system are shown in the second part of the chapter, with the MV induction machine operating at up to 1 MVA. A summary and discussion of the main benefits and characteristics of MP^3C is provided at the end of the chapter.

Chapter 14 focuses on an MMC that is controlled by an indirect MPC scheme with CB-PWM. The nonlinear MMC model is derived and linearized, based on which a linear MPC scheme is formulated. By manipulating the reference voltages of the modulator, the controller regulates the phase currents along their references, controls the branch energies, and imposes soft constraints on the branch currents, dc-link current, and capacitor voltages. A subsequent balancing controller maintains the capacitor voltages of the modules around their nominal values. The main benefit of this two-tiered controller is its ability to provide very fast responses during transients while operating the converter within its safe operating limits.

Part V: Summary

The last part of this book provides a performance comparison, summary of the results, conclusions, and an outlook for MPC of high power converters and industrial drives.

An extensive performance comparison is provided in Chap. 15, in which the principal direct MPC schemes discussed in this book are benchmarked with SVM. These direct control schemes include one-step predictive current control, MPDTC, MPDCC, and MP^3C. When minimizing the switching losses in the cost function and adopting long prediction horizons, MPDCC tends to slightly outperform MP^3C, albeit only in terms of harmonic current distortions per switching losses. Correspondingly, long-horizon MPDTC achieves lower torque distortions per switching losses than MP^3C. An in-depth assessment of the proposed control and modulation follows, which discusses their benefits and challenges and highlights promising application areas for each method. The outlook proposes a number of possible future research directions.

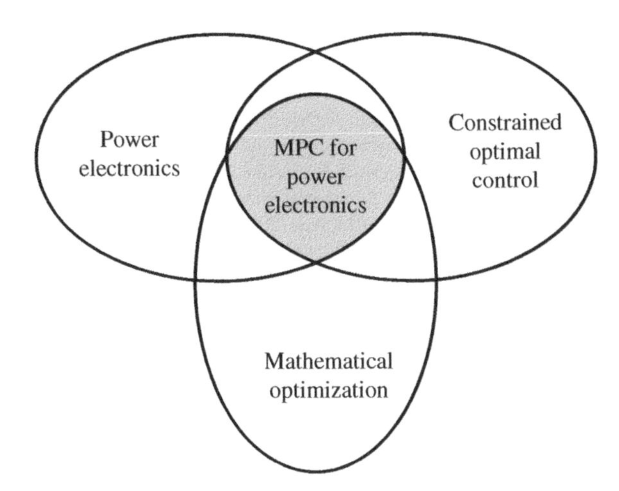

Figure 1.8 The focus of this book is on MPC for power converters and industrial drives, which is a field at the intersection of power electronics, constrained optimal control theory, and mathematical optimization

1.7 Prerequisites

This book is intended for researchers in academia and industry who are interested in an introduction to and a summary of the MPC methods available today for industrial power electronic systems. This includes university students at or above the MSc level, academics, and engineers in industry focusing on research and development. As shown in Fig. 1.8, the field of MPC for power electronics is at the intersection of power electronics, constrained optimal control theory, and mathematical optimization. Specifically, strong domain knowledge in power electronics is required to understand the system and the control problem at hand, MPC theory is required to formulate the control problem, and mathematical optimization is needed to solve it.

The reader is expected to be familiar with power electronics, modern control methods, and the basic notions of mathematical optimization. This includes three-phase machines, multilevel voltage source inverters, PWM, linear systems, linear algebra, state-space representation, discrete-time systems, optimal control, MPC, and quadratic programming.

Some of these prerequisites are covered by the following textbooks. For an introduction to high-power electronics and ac drives, the reader is referred to [66]. PWM is described and analyzed in depth in [15]. Detailed dynamic models of three-phase machines are derived in [67]. For a survey on multilevel converters, the reader is referred to [5, 11].

Linear systems and the state-space representation are described in detail in [68]. Discrete-time systems are explained in [69], while [70] is an excellent textbook on linear algebra. Regarding MPC, [43, 37] are recommended. An introduction to convex optimization is provided in [71]. For an encyclopedia on optimization, the reader is referred to [72].

References

[1] A. Chausovsky, "Industrial motors and drives global market update," tech. rep., IHS Technology, 2012.

[2] H. Kuemmlee, P. Wearon, and F. Kleiner, "Large electric drives—setting trends for oil & gas applications," in *Proceedings of IEEE Petroleum and Chemical Industry Technical Conference*, Sep. 2008.

[3] J. M. Carrasco, L. G. Franquelo, J. T. Bialasiewicz, E. Galván, R. C. P. Guisado, A. M. Prats, J. I. León, and N. Moreno-Alfonso, "Power-electronic systems for the grid integration of renewable energy sources: A survey," *IEEE Trans. Ind. Electron.*, vol. 53, pp. 1002–1016, Aug. 2006.

[4] A. Bocquel and J. Janming, "Analysis of a 300 MW variable speed drive for pump-storage plant applications," in *Proceedings of the European on Power Electronics Conference* (Dresden, Germany), Sep. 2005.

[5] J. Rodríguez, S. Bernet, B. Wu, J. Pontt, and S. Kouro, "Multilevel voltage-source-converter topologies for industrial medium-voltage drives," *IEEE Trans. Ind. Electron.*, vol. 54, pp. 2930–2945, Dec. 2007.

[6] B. P. Schmitt and R. Sommer, "Retrofit of fixed speed induction motors with medium voltage drive converters using NPC three-level inverter high-voltage IGBT based topology," in *Proceedings of the IEEE International Symposium on Industry Electronics* (Pusan, Korea), 2001.

[7] Y. H. Song and A. T. Johns, eds., *Flexible ac transmission systems (FACTS)*. London, UK: Institution of Engineering and Technology, 1999.

[8] X.-P. Zhang, C. Rehtanz, and B. Pal, *Flexible AC transmission systems: Modelling and control*. Springer, 2006.

[9] J. Arrillaga, Y. H. Liu, and N. R. Watson, *Flexible power transmission: The HVDC options*. John Wiley & Sons, Inc., 2007.

[10] N. Flourentzou, V. G. Agelidis, and G. D. Demetriades, "VSC-based HVDC power transmission systems: An overview," *IEEE Trans. Power Electron.*, vol. 24, pp. 592–602, Mar. 2009.

[11] J. Rodríguez, J.-S. Lai, and F. Peng, "Multilevel inverters: A survey of topologies, controls, and applications," *IEEE Trans. Ind. Electron.*, vol. 49, pp. 727–738, Aug. 2002.

[12] A. Nabae, I. Takahashi, and H. Akagi, "A new neutral-point-clamped PWM inverter," *IEEE Trans. Ind. Appl.*, vol. IA-17, pp. 518–523, Sep./Oct. 1981.

[13] A. Lesnicar and R. Marquardt, "An innovative modular multilevel converter topology suitable for a wide power range," in *Proceedings of IEEE Power Tech Conference* (Bologna, Italy), Jun. 2003.

[14] G. E. Moore, "Moore's law at 40," *Understanding Moore's law: Four decades of innovation*, pp. 67–84. Chemical Heritage Press, 2006.

[15] D. G. Holmes and T. A. Lipo, *Pulse width modulation for power converters: Principles and practice*. IEEE Press, 2003.

[16] N. Celanovic and D. Boroyevich, "A fast space-vector modulation algorithm for multilevel three-phase converters," *IEEE Trans. Ind. Appl.*, vol. 37, pp. 637–641, Mar./Apr. 2001.

[17] W. Leonhard, *Control of electrical drives*. Springer, 3rd ed., 2001.

[18] D. W. Novotny and T. A. Lipo, *Vector control and dynamics of AC drives*. Oxford Univ. Press, 1996.

[19] M. Malinowski, M. P. Kazmierkowski, and A. M. Trzynadlowski, "A comparative study of control techniques for PWM rectifiers in AC adjustable speed drives," *IEEE Trans. Power Electron.*, vol. 18, pp. 1390–1396, Nov. 2003.

[20] I. Takahashi and T. Noguchi, "A new quick response and high efficiency control strategy for the induction motor," *IEEE Trans. Ind. Appl.*, vol. 22, pp. 820–827, Sep./Oct. 1986.

[21] T. Noguchi, H. Tomiki, S. Kondo, and I. Takahashi, "Direct power control of PWM converter without power-source voltage sensors," *IEEE Trans. Ind. Appl.*, vol. 34, pp. 473–479, May/Jun. 1998.

[22] M. Senesky, G. Eirea, and T. J. Koo, "Hybrid modelling and control of power electronics," in *Hybrid systems: Computation and control* (A. Pnueli and O. Maler, eds.), vol. 2623 of LNCS, pp. 450–465, Springer, 2003.

[23] T. Geyer, G. Papafotiou, and M. Morari, "Model predictive control in power electronics: A hybrid systems approach," in *Proceedings of the IEEE Conference on Decision and Control* (Sevilla, Spain), Dec. 2005.

[24] R. D. Middlebrook and S. Čuk, "A general unified approach to modeling switching power converter stages," in *Proceedings of IEEE Power Electronics Specialists Conference*, pp. 18–34, 1976.

[25] R. W. Erickson, S. Čuk, and R. D. Middlebrook, "Large signal modeling and analysis of switching regulators," in *Proceedings of IEEE Power Electronics Specialists Conference*, pp. 240–250, 1982.

[26] P. A. Dahono, "A control method to damp oscillation in the input LC filter of AC-DC PWM converters," in *Proceedings of IEEE Power Electronics Specialists Conference*, pp. 1630–1635, Jun. 2002.

[27] J. Dannehl, F. Fuchs, S. Hansen, and P. Thøgersen, "Investigation of active damping approaches for PI-based current control of grid-connected pulse width modulation converters with *LCL* filters," *IEEE Trans. Ind. Appl.*, vol. 46, pp. 1509–1517, Jul./Aug. 2010.

[28] R. E. Kalman, "Contributions to the theory of optimal control," *Bulletin da la Societe Mathematique de Mexicana*, vol. 5, pp. 102–119, 1960.

[29] R. E. Kalman, "A new approach to linear filtering and prediction problems," *Trans. ASME (J. Basic Engineering)*, vol. 87, pp. 35–45, 1960.

[30] S. J. Qin and T. A. Badgwell, "A survey of industrial model predictive control technology," *Control Eng. Pract.*, vol. 11, pp. 733–764, Jul. 2003.

[31] J. Holtz and S. Stadtfeld, "A predictive controller for the stator current vector of AC machines fed from a switched voltage source," in *Proceedings of IEEE International Power Electronics Conference* (Tokyo, Japan), pp. 1665–1675, Apr. 1983.

[32] J. Holtz and S. Stadtfeld, "Field-oriented control by forced motor currents in a voltage fed inverter drive," in *Proceedings of IFAC Symposium* (Lausanne, Switzerland), pp. 103–110, Sep. 1983.

[33] A. Bemporad, M. Morari, V. Dua, and E. N. Pistikopoulos, "The explicit linear quadratic regulator for constrained systems," *Automatica*, vol. 38, pp. 3–20, Jan. 2002.

[34] F. Borrelli, *Constrained optimal control of linear and hybrid systems*, vol. 290 of LNCIS. Springer, 2003.

[35] F. Borrelli, M. Baotić, A. Bemporad, and M. Morari, "Dynamic programming for constrained optimal control of discrete-time linear hybrid systems," *Automatica*, vol. 41, pp. 1709–1721, Oct. 2005.

[36] D. E. Quevedo, R. P. Aguilera, and T. Geyer, "Predictive control in power electronics and drives: Basic concepts, theory and methods," *Advanced and intelligent control in power electronics and drives*, vol. Studies in Computational Intelligence, pp. 181–226. Springer, 2014.

[37] F. Borrelli, A. Bemporad, and M. Morari, *Predictive control for linear and hybrid systems*. Springer, www.mpc.berkeley.edu/mpc-course-material.

[38] C. E. Garcia, D. M. Prett, and M. Morari, "Model predictive control: Theory and practice—A survey," *Automatica*, vol. 25, pp. 335–348, Mar. 1989.

[39] M. Morari and J. H. Lee, "Model predictive control: Past, present and future," *Comput. and Chemical Eng.*, vol. 23, pp. 667–682, 1999.

[40] D. Q. Mayne, J. B. Rawlings, C. V. Rao, and P. O. M. Scokaert, "Constrained model predictive control: Stability and optimality," *Automatica*, vol. 36, pp. 789–814, Jun. 2000.

[41] A. Alessio and A. Bemporad, "A survey on explicit model predictive control," in *Nonlinear model predictive control*, vol. 384 of LNCIS, pp. 345–369. Springer, 2009.

[42] J. M. Maciejowski, *Predictive control with constraints*. Prentice Hall, 2002.

[43] J. B. Rawlings and D. Q. Mayne, *Model predictive control: Theory and design*. Madison, WI, USA: Nob Hill Publ., 2009.

[44] L. Grüne and J. Pannek, *Nonlinear model predictive control: Theory and algorithms*. Springer, 2011.

[45] E. F. Camacho and C. Bordons, *Model predictive control*. Springer, 2nd ed., 2013.

[46] A. J. van der Schaft and J. M. Schumacher, *An introduction to hybrid dynamical systems*, vol. 251 of LNCIS. Springer, 2000.

[47] E. D. Sontag, "Nonlinear regulation: The piecewise linear approach," *IEEE Trans. Automat. Contr*, vol. 26, pp. 346–358, Apr. 1981.

[48] A. Bemporad and M. Morari, "Control of systems integrating logic, dynamics and constraints," *Automatica*, vol. 35, pp. 407–427, March 1999.

[49] W. P. M. H. Heemels, B. De Schutter, and A. Bemporad, "Equivalence of hybrid dynamical models," *Automatica*, vol. 37, pp. 1085–1091, Jul. 2001.

[50] G. Papafotiou, T. Geyer, and M. Morari, "A hybrid model predictive control approach to the direct torque control problem of induction motors," *Int. J. of Robust Nonlinear Control*, vol. 17, pp. 1572–1589, Nov. 2007.

[51] S. Richter, C. Jones, and M. Morari, "Real-time input-constrained MPC using fast gradient methods," in *Proceedings of the IEEE Conference on Decision and Control* (Shanghai, China), pp. 7387–7393, Dec. 2009.

[52] S. Richter, S. Mariéthoz, and M. Morari, "High-speed online MPC based on fast gradient method applied to power converter control," in *Proceedings of the American Control Conference* (Baltimore, MD, USA), 2010.

[53] H. Peyrl, J. Liu, and T. Geyer, "An FPGA implementation of the fast gradient method for solving the model predictive pulse pattern control problem," in *Workshop on Predictive Control of Electrical Drives and Power Electronics* (Munich, Germany), Oct. 2013.

[54] P. Cortés, M. P. Kazmierkowski, R. M. Kennel, D. E. Quevedo, and J. Rodríguez, "Predictive control in power electronics and drives," *IEEE Trans. Ind. Electron.*, vol. 55, pp. 4312–4324, Dec. 2008.

[55] T. Geyer, Low complexity model predictive control in power electronics and power systems. PhD thesis, Autom. Control Lab. ETH Zurich, 2005.

[56] T. Geyer, G. Papafotiou, and M. Morari, "Model predictive direct torque control—Part I: Concept, algorithm and analysis," *IEEE Trans. Ind. Electron.*, vol. 56, pp. 1894–1905, Jun. 2009.

[57] G. Papafotiou, J. Kley, K. G. Papadopoulos, P. Bohren, and M. Morari, "Model predictive direct torque control—Part II: Implementation and experimental evaluation," *IEEE Trans. Ind. Electron.*, vol. 56, pp. 1906–1915, Jun. 2009.

[58] F. Kieferndorf, P. Karamanakos, P. Bader, N. Oikonomou, and T. Geyer, "Model predictive control of the internal voltages of a five-level active neutral point clamped converter," in *Proceedings of IEEE Energy Conversion Congress and Exposition* (Raleigh, NC, USA), Sep. 2012.

[59] G. S. Buja, "Optimum output waveforms in PWM inverters," *IEEE Trans. Ind. Appl.*, vol. 16, pp. 830–836, Nov./Dec. 1980.

[60] T. Geyer, N. Oikonomou, G. Papafotiou, and F. Kieferndorf, "Model predictive pulse pattern control," *IEEE Trans. Ind. Appl.*, vol. 48, pp. 663–676, Mar./Apr. 2012.

[61] T. Geyer, "Generalized model predictive direct torque control: Long prediction horizons and minimization of switching losses," in *Proceedings of the IEEE Conference on Decision and Control* (Shanghai, China), pp. 6799–6804, Dec. 2009.

[62] T. Geyer, "A comparison of control and modulation schemes for medium-voltage drives: Emerging predictive control concepts versus field oriented control," in *Proceedings of the IEEE Energy Conversion Congress and Exposition* (Atlanta, GA, USA), pp. 2836–2843, Sep. 2010.

[63] T. Geyer and G. Papafotiou, "Model predictive direct torque control of a variable speed drive with a five-level inverter," in *Proceedings of the IEEE Industrial Electronics Society Annual Conference* (Porto, Portugal), pp. 1203–1208, Nov. 2009.

[64] G. Papafotiou, T. Geyer, and M. Morari, "Optimal direct torque control of three-phase symmetric induction motors," in *Proceedings of the IEEE Conference on Decision and Control* (Atlantis, Bahamas), Dec. 2004.

[65] T. Geyer and G. Papafotiou, "Direct torque control for induction motor drives: A model predictive control approach based on feasibility," in *Hybrid systems: Computation and control* (M. Morari and L. Thiele, eds.), vol. 3414 of LNCS, pp. 274–290, Springer, Mar. 2005.

[66] B. Wu, *High-power converters and AC drives*. Hoboken, NJ: John Wiley & Sons, Inc., 2006.

[67] P. C. Krause, O. Wasynczuk, and S. D. Sudhoff, *Analysis of electric machinery and drive systems*. Hoboken, NJ: John Wiley & Sons, Inc., 2nd ed., 2002.

[68] P. Antsaklis and A. Michel, *A linear systems primer*. Birkhäuser Verlag, 2007.

[69] G. F. Franklin, J. D. Powell, and M. L. Workman, *Digital control of dynamic systems*. Addison-Wesley, 3rd ed., 1998.

[70] G. Strang, *Introduction to linear algebra*. Wellesley-Cambridge Press, 4th ed., 2009.

[71] S. Boyd and L. Vandenberghe, *Convex optimization*. Cambridge Univ. Press, 2004.

[72] C. A. Floudas and P. M. Pardalos, *Encyclopedia of optimization*. Springer, 2nd ed., 2009.

2

Industrial Power Electronics

Industrial power electronic systems in the high power range are described in detail in this chapter. The chapter starts by reviewing some fundamental concepts that will be used throughout the book. State-space models of induction machines are derived, which describe the machine both during steady-state operation and transients. Power semiconductors are introduced and their loss models are stated. Three- and five-level voltage source inverters are described and modeled, and four industrial power electronics case studies are defined. The latter refer to medium-voltage (MV) variable-speed drives (VSDs) and grid-connected converter systems.

2.1 Preliminaries

In this section, four concepts fundamental to power electronics are reviewed: three-phase systems, the per unit (pu) system, orthogonal reference frames, and space vectors. To assist readers who are not familiar with these concepts, more space than might be considered absolutely necessary has been devoted to summarize these notions and to provide illustrating examples. The aim is to provide an intuitively accessible yet concise introduction and summary.

2.1.1 Three-Phase Systems

We start by introducing three-phase systems with alternating currents (ac). Consider a balanced three-phase voltage source. In each phase, the voltage is sinusoidal with the peak value $\sqrt{2}V_{\text{ph}}$, where V_{ph} denotes the root-mean-square (rms) value of the phase voltage. The voltage waveforms have the same frequency f in the three phases a, b, and c, but their phases are shifted by $2\pi/3$ with respect to each other. The three instantaneous phase voltages at time t are given by

$$v_a(t) = \sqrt{2}V_{\text{ph}} \sin(\omega t) \tag{2.1a}$$

$$v_b(t) = \sqrt{2}V_{\text{ph}} \sin\left(\omega t - \frac{2}{3}\pi\right) \tag{2.1b}$$

Model Predictive Control of High Power Converters and Industrial Drives, First Edition. Tobias Geyer.

Companion Website: www.wiley.com/go/geyermodelpredictivecontrol

$$v_c(t) = \sqrt{2}V_{\rm ph}\sin\left(\omega t - \frac{4}{3}\pi\right), \quad (2.1c)$$

where $\omega = 2\pi f$ denotes the angular frequency.

As shown in Fig. 2.1, the phase voltages are the voltages between the phase terminals A, B, and C of the voltage source and its star point N. In contrast to this, the line-to-line voltages

$$v_{ab}(t) = v_a(t) - v_b(t) = \sqrt{2}V\sin\left(\omega t + \frac{1}{6}\pi\right) \quad (2.2a)$$

$$v_{bc}(t) = v_b(t) - v_c(t) = \sqrt{2}V\sin\left(\omega t - \frac{1}{2}\pi\right) \quad (2.2b)$$

$$v_{ca}(t) = v_c(t) - v_a(t) = \sqrt{2}V\sin\left(\omega t - \frac{7}{6}\pi\right) \quad (2.2c)$$

refer to the voltages between the phase terminals, which have the rms value $V = \sqrt{3}V_{\rm ph}$. Throughout this book, we will use upper case letters to denote rms quantities, and lower case letters to refer to instantaneous quantities. We will often drop the time t from the instantaneous quantities to simplify the notation.

A three-phase star-connected resistive–inductive load (see Fig. 2.1) is connected to the three-phase voltage source via the phase terminals. The star point S of the load is usually not connected to N. The load's phase resistances R and inductors L have the same value in each phase. We say that the load is balanced. The resulting three-phase sinusoidal current waveforms are of the same magnitude and thus also balanced.

Example 2.1 *Consider a voltage source with the line-to-line rms voltage $V = 3.3$ kV and the frequency $f = 50$ Hz. Adding a load with the impedance $Z = R + {\rm j}\omega L$ with $R = 2\ \Omega$ and $L = 2$ mH leads to the rms phase currents $I_{\rm ph} = V/\sqrt{3}/|Z| = 0.91$ kA. The phase currents are shifted with respect to the phase voltages by the angle $\angle Z = 17.4°$. The three-phase voltage and current waveforms are shown in Fig. 2.2.*

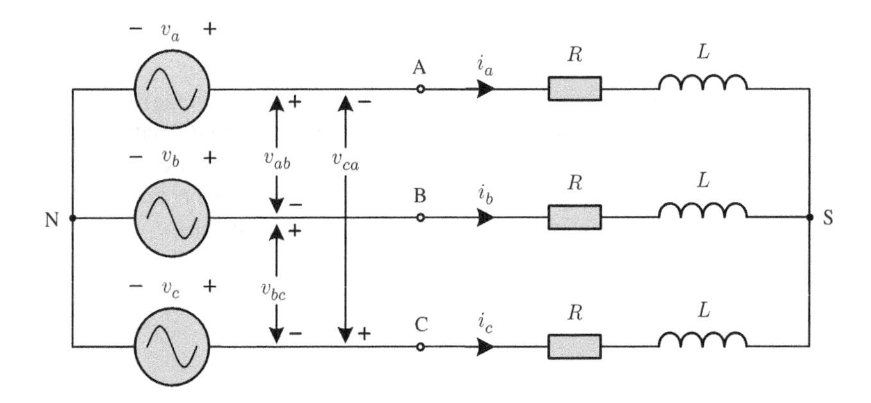

Figure 2.1 Balanced three-phase system with a voltage source and a resistive–inductive load

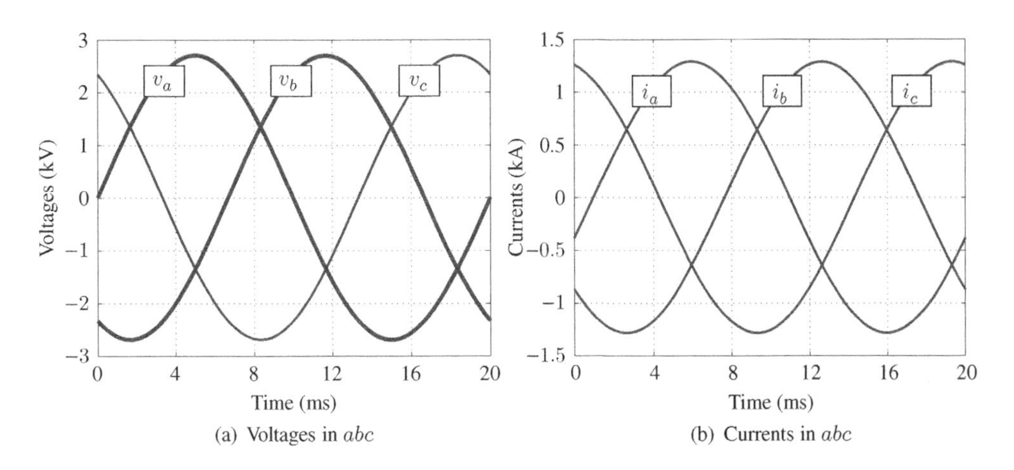

(a) Voltages in abc

(b) Currents in abc

Figure 2.2 Balanced three-phase voltages and currents in SI units

For a balanced three-phase system, the real power P, the reactive power Q, and the apparent power S are given by

$$P = 3V_{\text{ph}} I_{\text{ph}} \cos(\phi) \tag{2.3a}$$

$$Q = 3V_{\text{ph}} I_{\text{ph}} \sin(\phi) \tag{2.3b}$$

$$S = 3V_{\text{ph}} I_{\text{ph}}, \tag{2.3c}$$

where ϕ denotes the phase angle between the voltage and current waveforms. Note that P has the unit watt (W), Q has the unit reactive volt-ampere (Var), and S is measured in volt-ampere (VA). Moreover, $S^2 = P^2 + Q^2$ holds.

The power factor is defined as

$$\text{pf} = |\cos(\phi)| = \frac{P}{S}. \tag{2.4}$$

Example 2.2 *In Example 2.1, the apparent power is equal to* $S = V^2/|Z| = 5.19\,MVA$. *The phase angle between the voltage and current waveforms is* $\phi = \angle Z = 17.4°$, *leading to a power factor of* 0.954.

2.1.2 *Per Unit System*

In the fields of power electronics and power systems, it is common practise to normalize all variables and parameters that are used. Normalization is typically performed such that the normalized variables are equal to 1 when operating at nominal voltage, full power, and nominal frequency. To this end, the so-called pu system is established with three *primary* base quantities.

Table 2.1 Definition of the base values of the per unit system in terms of the rated line-to-line voltage V_R, the rated current I_R, and the rated angular stator or grid frequency ω_{sR} or ω_{gR}

Base quantity	Base value
Voltage	$V_B = \sqrt{2/3}V_R$
Current	$I_B = \sqrt{2}I_R$
Angular frequency	$\omega_B = \omega_{sR}$ or $\omega_B = \omega_{gR}$
Resistance, reactance, impedance	$Z_B = V_B/I_B$
Inductance	$L_B = Z_B/\omega_B$
Capacitance	$C_B = 1/(\omega_B Z_B)$
Apparent power	$S_B = 3/2 V_B I_B$
Flux linkage	$\lambda_B = V_B/\omega_B$
Torque	$T_B = \text{pf}\, p S_B/\omega_B$

We first consider a drive system with a single electrical machine. Assume that the machine is configured in star connection and let V_R denote its rated line-to-line voltage. The base voltage

$$V_B = \sqrt{\frac{2}{3}}V_R \tag{2.5}$$

is defined as the peak value of the machine's rated phase voltage. Accordingly, the peak value of the rated current I_R is selected as the base current

$$I_B = \sqrt{2}I_R. \tag{2.6}$$

The third base quantity is the base angular frequency, which is set equal to the rated angular stator frequency ω_{sR}, that is,

$$\omega_B = \omega_{sR}. \tag{2.7}$$

From the three primary base quantities defined here, additional base quantities can be derived easily. The commonly used base quantities are summarized in Table 2.1. The torque equation is derived in Sect. 2.2. Note that pf denotes the power factor and p is the number of pole pairs in the electrical machine.

When normalizing the drive system with these base quantities and when operating at nominal speed and rated torque, the normalized stator currents have an amplitude of 1. Furthermore, the stator flux magnitude, the angular stator frequency, the electrical power, and the electromagnetic torque are 1.

Example 2.3 *For Example 2.1, a pu system can be established by assuming $V_R = V$, $I_R = I_{\text{ph}}$ and $\omega_R = \omega$. This yields the base quantities $V_B = 2694\,V$, $I_B = 1285\,A$ and $\omega_B = 2\pi 50$ rad/s. As a result, the amplitudes of the three-phase voltage and current waveforms are scaled such that their peak values coincide with 1 (see Fig. 2.3).*

For grid-connected power converters, the pu system is typically established using quantities at the secondary side of the transformer. The base voltage is chosen as the peak value of the

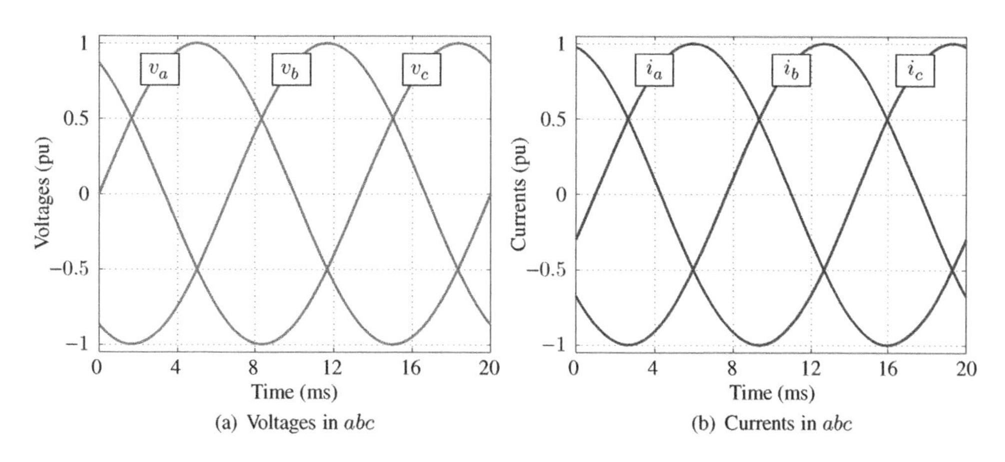

Figure 2.3 Balanced three-phase voltages and currents in the per unit system

rated transformer phase voltage. The definition of the base voltage remains the same as in (2.5), but V_R refers now to the line-to-line voltage at the transformer's secondary winding.

The rated apparent power S_R is used as the second base quantity. It is defined as

$$S_B = S_R. \tag{2.8}$$

The base angular frequency is again chosen as the third base quantity. It is equal to the rated angular grid frequency ω_{gR}, that is,

$$\omega_B = \omega_{gR}. \tag{2.9}$$

This definition of the pu system is based on the peak values of the voltage and current waveforms. Another possibility is to choose the rms values as base values. In this case, (2.5) and (2.6) are replaced by $V_B = V_R/\sqrt{3}$ and $I_B = I_R$, respectively. The apparent base power in Table 2.1 is then given by $S_B = 3V_B I_B$.

In addition to this, the time axis will also be normalized by multiplying the time by ω_B. The notion of the pu system will be illustrated in more detail when applying it to electrical machines in Sect. 2.2.3, to multilevel inverters in Sect. 2.4, and to grid-connected converters in Sect. 2.5.4.

2.1.3 *Stationary Reference Frame*

To simplify the modeling and analysis of balanced three-phase circuits, it is common practice to transform all variables from the three-phase abc system to an orthogonal reference frame, which is either stationary or rotating. We will use the terms *reference frame* and *coordinate system* interchangeably.

The stationary, orthogonal coordinate system is established by the three axes α, β, and 0 (or γ), which are perpendicular to each other, as shown in Fig. 2.4. The so-called Clarke transformation [1] maps the vector $\boldsymbol{\xi}_{abc} = [\xi_a \; \xi_b \; \xi_c]^T$ from the balanced three-phase abc system to the vector $\boldsymbol{\xi}_{\alpha\beta 0} = [\xi_\alpha \; \xi_\beta \; \xi_0]^T$ in the $\alpha\beta 0$ reference frame, and vice versa, via the following transformations:

$$\boldsymbol{\xi}_{\alpha\beta 0} = \boldsymbol{K}\boldsymbol{\xi}_{abc} \text{ and } \boldsymbol{\xi}_{abc} = \boldsymbol{K}^{-1}\boldsymbol{\xi}_{\alpha\beta 0}. \tag{2.10}$$

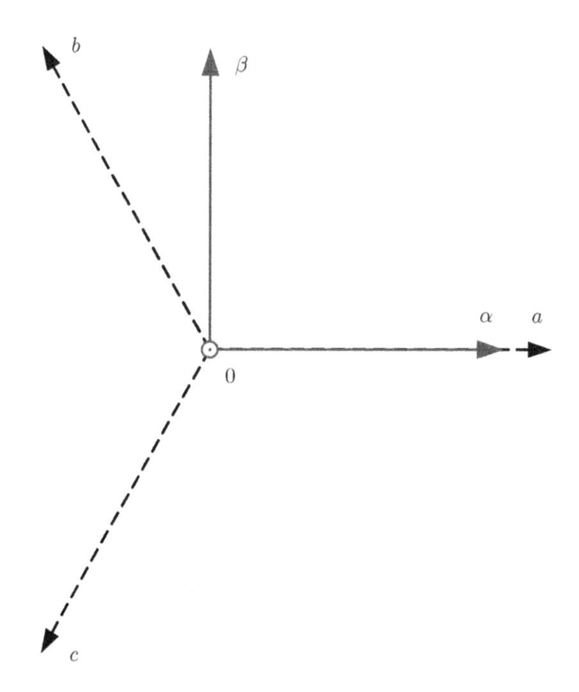

Figure 2.4 Definition of the stationary $\alpha\beta 0$ reference frame

The transformation matrices are given by

$$\boldsymbol{K} = \frac{2}{3}\begin{bmatrix} 1 & -\frac{1}{2} & -\frac{1}{2} \\ 0 & \frac{\sqrt{3}}{2} & -\frac{\sqrt{3}}{2} \\ \frac{1}{2} & \frac{1}{2} & \frac{1}{2} \end{bmatrix} \quad \text{and} \quad \boldsymbol{K}^{-1} = \begin{bmatrix} 1 & 0 & 1 \\ -\frac{1}{2} & \frac{\sqrt{3}}{2} & 1 \\ -\frac{1}{2} & -\frac{\sqrt{3}}{2} & 1 \end{bmatrix}. \tag{2.11}$$

The factor $2/3$ in $\boldsymbol{K}$ in (2.11) ensures that the amplitudes of the (balanced) three-phase signals are preserved. As a result, the Clarke transformation as defined previously is peak or amplitude invariant.

Example 2.4 *Consider again Example 2.3 with its normalized three-phase voltages and currents, which are shown in Fig. 2.3. Using* (2.10) *yields the* $\alpha\beta 0$ *voltages and currents shown in Fig. 2.5, which have the same amplitudes and the same fundamental frequency as the corresponding waveforms in the abc coordinate system.*

The 0*-component is zero in both cases, because the three-phase quantities in this example are sinusoidal waveforms with the same amplitude in each phase and a phase shift of* $2\pi/3$ *between them. More generally, in a three-phase system with its star point not connected to ground, the* 0*-component of the current is always zero, that is,* $i_a + i_b + i_c = 0$ *holds at all time instants.*

When transforming three-phase quantities into the stationary orthogonal reference frame, we often require only the α- and β-components, but not the 0-component. To address this case, we introduce $\boldsymbol{\xi}_{\alpha\beta} = [\xi_\alpha \; \xi_\beta]^T$ and the reduced Clarke transformations

$$\boldsymbol{\xi}_{\alpha\beta} = \tilde{\boldsymbol{K}}\boldsymbol{\xi}_{abc} \text{ and } \boldsymbol{\xi}_{abc} = \tilde{\boldsymbol{K}}^{-1}\boldsymbol{\xi}_{\alpha\beta}. \tag{2.12}$$

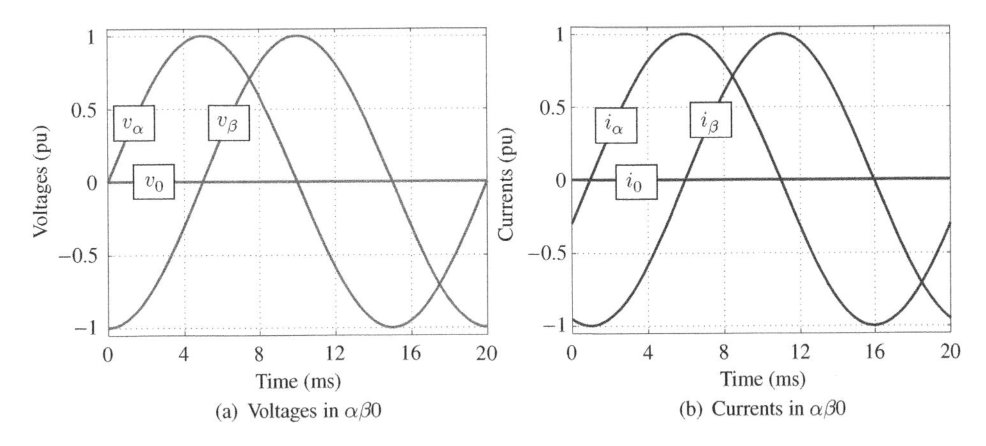

Figure 2.5 Balanced three-phase voltages and currents represented in the stationary $\alpha\beta 0$ reference frame

The corresponding matrices are defined as

$$\tilde{\boldsymbol{K}} = \frac{2}{3}\begin{bmatrix} 1 & -\frac{1}{2} & -\frac{1}{2} \\ 0 & \frac{\sqrt{3}}{2} & -\frac{\sqrt{3}}{2} \end{bmatrix} \text{ and } \tilde{\boldsymbol{K}}^{-1} = \begin{bmatrix} 1 & 0 \\ -\frac{1}{2} & \frac{\sqrt{3}}{2} \\ -\frac{1}{2} & -\frac{\sqrt{3}}{2} \end{bmatrix}. \tag{2.13}$$

Note that $\tilde{\boldsymbol{K}}$ holds the first two rows of $\boldsymbol{K}$, and $\tilde{\boldsymbol{K}}^{-1}$ holds the first two columns of $\boldsymbol{K}^{-1}$. $\tilde{\boldsymbol{K}}^{-1}$ is the pseudo-inverse of $\tilde{\boldsymbol{K}}$, and the 0-component is implicitly assumed to be zero.

2.1.3.1 Remarks

The three-phase *abc* system can be interpreted as a coordinate system with its three axes displaced by $2\pi/3$. The *abc* coordinate system can be described by the unit vectors

$$\boldsymbol{e}_a = \begin{bmatrix} 1 \\ 0 \end{bmatrix}, \boldsymbol{e}_b = \frac{1}{2}\begin{bmatrix} -1 \\ \sqrt{3} \end{bmatrix} \text{ and } \boldsymbol{e}_c = \frac{1}{2}\begin{bmatrix} -1 \\ -\sqrt{3} \end{bmatrix} \tag{2.14}$$

in the orthogonal $\alpha\beta$ coordinate system. By definition, the unit vectors are of the magnitude 1. Note that the α-axis is aligned with the a-axis, as shown in Fig. 2.4.

The Clarke transformation can be interpreted geometrically as a projection of the three phase quantities onto orthogonal and stationary axes. Specifically, the three unit vectors in (2.14) are multiplied with the to-be-transformed *abc* quantities and projected onto the $\alpha\beta$-axes according to

$$\boldsymbol{\xi}_{\alpha\beta} = \frac{2}{3}(\xi_a \boldsymbol{e}_a + \xi_b \boldsymbol{e}_b + \xi_c \boldsymbol{e}_c). \tag{2.15}$$

It is clear that the transformation (2.15) is identical to the reduced Clarke transformation in (2.12). The factor $2/3$ is required in both variations of the transformation to ensure that it is amplitude invariant.

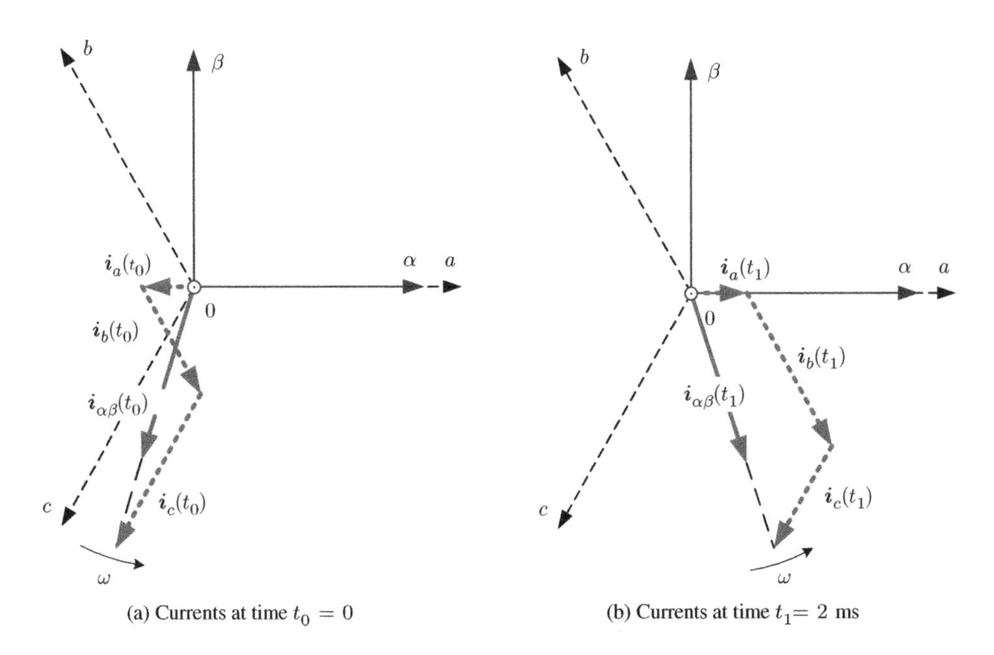

(a) Currents at time $t_0 = 0$

(b) Currents at time $t_1 = 2$ ms

Figure 2.6 Current vectors $\boldsymbol{i}_{\alpha\beta}$ in stationary orthogonal coordinates (solid lines) and their three per-phase contributions (dotted lines)

The third axis, the 0-axis, is orthogonal to the $\alpha\beta$-plane and points out of it, as indicated in Fig. 2.4. A common choice for the 0-component is

$$\xi_0 = \frac{1}{3}(\xi_a + \xi_b + \xi_c). \tag{2.16}$$

Combining (2.15) with (2.16) directly results in the Clarke transformation (2.10).

Example 2.5 *To visualize the projection underlying the Clarke transformation, we focus again on Example 2.3. From Fig. 2.3(b), we read out the instantaneous three-phase currents at time $t_0 = 0$ as $\boldsymbol{i}_{abc}(t_0) = [-0.30\ \ -0.68\ 0.98]^T$. For phase a, we define the per-phase vector $\boldsymbol{i}_a(t_0) = i_a(t_0)\boldsymbol{e}_a$. The vectors $\boldsymbol{i}_b(t_0)$ and $\boldsymbol{i}_c(t_0)$ are defined accordingly. Summing up these three vectors and scaling their vectorial sum by $2/3$ results in the equivalent representation of $\boldsymbol{i}_{\alpha\beta}(t_0) = [-0.30\ \ -0.96]^T$ in the stationary orthogonal coordinate system. This process is visualized in Fig. 2.6(a).*

Accordingly, Fig. 2.6(b) depicts the three-phase currents at time $t_1 = 2\,ms$, $\boldsymbol{i}_{abc}(t_1) = [0.32\ \ -0.98\ 0.66]^T$, which are equivalent to $\boldsymbol{i}_{\alpha\beta}(t_1) = [0.32\ \ -0.95]^T$. We also observe that the current vector rotates counterclockwise at the angular velocity ω. As before, the 0-component of the current is zero.

2.1.4 Rotating Reference Frame

The stationary orthogonal reference frame can be generalized to the rotating orthogonal reference frame with the direct (d), quadrature (q), and zero (0) axis. As shown in Fig. 2.7, the q-axis precedes the d-axis by 90°, that is, the q-axis is in quadrature to the d-axis. As for the

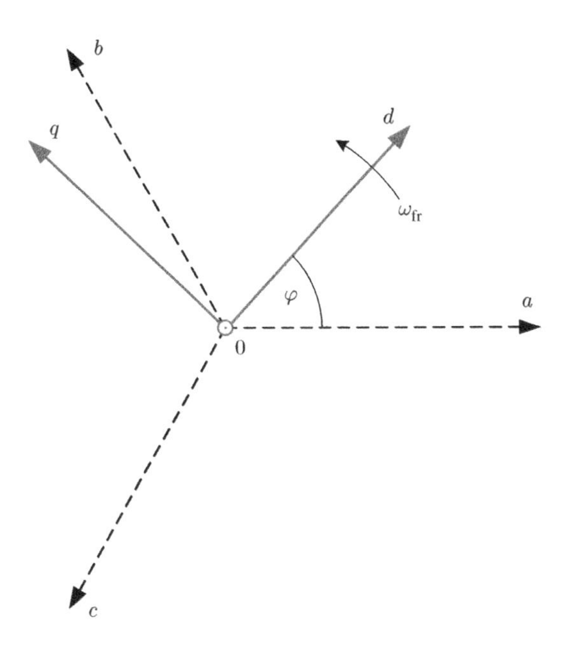

Figure 2.7 Definition of the rotating $dq0$ reference frame

stationary frame, the 0-axis points out of the dq-plane. The angular position of the $dq0$ reference frame is defined by φ, which is the angle between the d-axis of the rotating reference frame and the a-axis of the three-phase system, with

$$\varphi(t) = \int_0^t \omega_{\text{fr}}(\tau)\text{d}\tau + \varphi(0). \tag{2.17}$$

Here, ω_{fr} denotes the angular speed of the reference frame.

The so-called Park transformation [2] transforms the vector $\boldsymbol{\xi}_{abc}$ from the three-phase system to $\boldsymbol{\xi}_{dq0} = [\xi_d\ \xi_q\ \xi_0]^T$ in the rotating reference frame, and vice versa, using

$$\boldsymbol{\xi}_{dq0} = \boldsymbol{K}(\varphi)\boldsymbol{\xi}_{abc} \text{ and } \boldsymbol{\xi}_{abc} = \boldsymbol{K}^{-1}(\varphi)\boldsymbol{\xi}_{dq0} \tag{2.18}$$

with the matrices

$$\boldsymbol{K}(\varphi) = \frac{2}{3}\begin{bmatrix} \cos\varphi & \cos(\varphi - \frac{2\pi}{3}) & \cos(\varphi + \frac{2\pi}{3}) \\ -\sin\varphi & -\sin(\varphi - \frac{2\pi}{3}) & -\sin(\varphi + \frac{2\pi}{3}) \\ \frac{1}{2} & \frac{1}{2} & \frac{1}{2} \end{bmatrix} \tag{2.19}$$

and

$$\boldsymbol{K}^{-1}(\varphi) = \begin{bmatrix} \cos\varphi & -\sin\varphi & 1 \\ \cos(\varphi - \frac{2\pi}{3}) & -\sin(\varphi - \frac{2\pi}{3}) & 1 \\ \cos(\varphi + \frac{2\pi}{3}) & -\sin(\varphi + \frac{2\pi}{3}) & 1 \end{bmatrix}. \tag{2.20}$$

We distinguish between the Clarke and the Park transformation through the dependency on the angle φ; $\boldsymbol{K}$ and $\boldsymbol{K}^{-1}$ refer to the Clarke transformation and its inverse, while the pair $\boldsymbol{K}(\varphi)$ and $\boldsymbol{K}^{-1}(\varphi)$ refers to the Park transformation and its inverse, respectively.

Equations (2.17)–(2.20) summarize the concept of the orthogonal reference frame. The angular speed ω_{fr}, at which this reference frame rotates, can be arbitrary. For electrical drives, two special cases are commonly used. On one hand, the *synchronous* (or synchronously rotating) reference frame is obtained by aligning the d-axis with the machine's stator or rotor flux vector and by setting ω_{fr} to the flux's angular speed ω_s. As a result, during steady-state operation, the machine's interdependent ac quantities are transformed into independent (orthogonal) dc quantities, as will be shown in Sect. 3.6.2.

On the other hand, the *stationary* (i.e., nonrotating) $\alpha\beta0$ reference frame is the result of setting both φ and ω_{fr} to zero. The d- and q-axes are then referred to as α- and β-axes, respectively, with the 0-axis remaining unchanged. The Clarke transformation can thus be interpreted as a special case of the Park transformation.

For grid-connected converters, the d-axis of the rotating $dq0$ reference frame is usually aligned with the grid voltage, or—more precisely—with the voltage at the point of common coupling (PCC). A phase-locked loop is used to ensure that the reference frame rotates in synchronism with the PCC voltage. Alternatively, the stationary $\alpha\beta0$ reference frame can be used also on the grid side.

Example 2.6 *To highlight the notion of the synchronous reference frame, consider the three-phase voltages and currents shown in Fig. 2.3. These quantities were defined in Example 2.3. We set* $\omega_{\text{fr}} = \omega = 2\pi 50$ *rad/s and align the d-axis with the voltage vector by setting* $\varphi(0) = 3\pi/2$. *The instantaneous voltages and currents at time instant* $t = 2$ *ms are illustrated as the vectors* $\boldsymbol{v}$ *and* $\boldsymbol{i}$ *in Fig. 2.8.*

The $dq0$ *reference frame rotates in synchronism with the voltage and current vectors. When represented in the* $dq0$ *frame, their ac quantities turn into dc quantities while their vectorial*

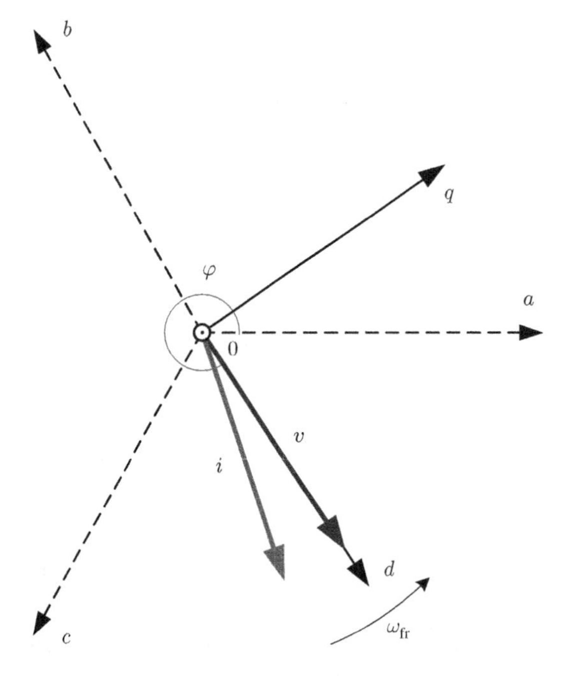

Figure 2.8 Voltage and current vectors of Example 2.3 at time instant $t = 2$ ms represented in the rotating $dq0$ reference frame

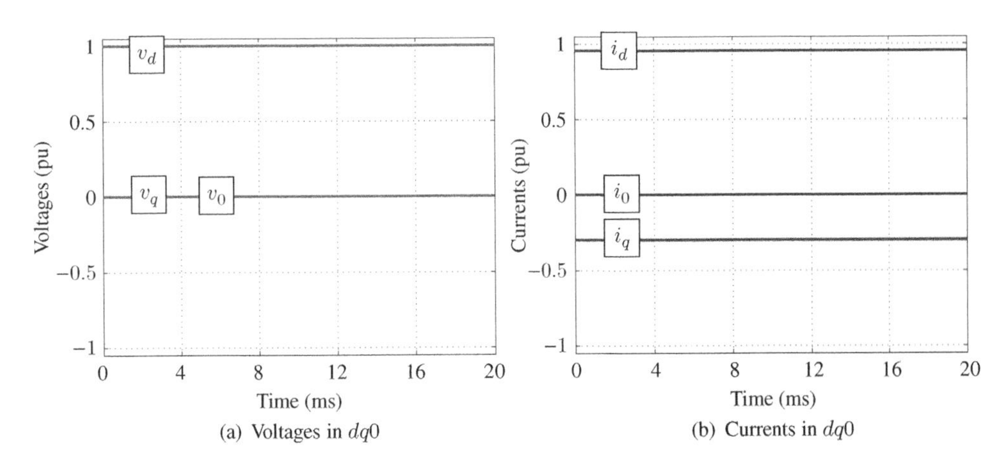

(a) Voltages in $dq0$ (b) Currents in $dq0$

Figure 2.9 Balanced three-phase voltages and currents represented in the rotating $dq0$ reference frame

representation remains stationary. This can be confirmed algebraically by inserting the instantaneous three-phase quantities, as defined in (2.1) *for the voltage, into the Park transformation* (2.18). *The resulting dq0 components are shown in Fig. 2.9.*

The amplitudes of the voltage and current vectors, $\sqrt{v_d^2 + v_q^2 + v_0^2}$ *and* $\sqrt{i_d^2 + i_q^2 + i_0^2}$*, are* 1 *and are thus invariant under the transformation. The* 0*-components of the three-phase voltages and currents remain zero.*

Similar to (2.12), the 0-component can be ignored, which yields the reduced Park transformations

$$\boldsymbol{\xi}_{dq} = \tilde{\boldsymbol{K}}(\varphi)\boldsymbol{\xi}_{abc} \text{ and } \boldsymbol{\xi}_{abc} = \tilde{\boldsymbol{K}}^{-1}(\varphi)\boldsymbol{\xi}_{dq} \tag{2.21}$$

with the transformation matrix

$$\tilde{\boldsymbol{K}}(\varphi) = \frac{2}{3}\begin{bmatrix} \cos\varphi & \cos(\varphi - \frac{2\pi}{3}) & \cos(\varphi + \frac{2\pi}{3}) \\ -\sin\varphi & -\sin(\varphi - \frac{2\pi}{3}) & -\sin(\varphi + \frac{2\pi}{3}) \end{bmatrix} \tag{2.22}$$

and its pseudo-inverse

$$\tilde{\boldsymbol{K}}^{-1}(\varphi) = \begin{bmatrix} \cos\varphi & -\sin\varphi \\ \cos(\varphi - \frac{2\pi}{3}) & -\sin(\varphi - \frac{2\pi}{3}) \\ \cos(\varphi + \frac{2\pi}{3}) & -\sin(\varphi + \frac{2\pi}{3}) \end{bmatrix}. \tag{2.23}$$

For the inverse transformation, the 0-component is implicitly assumed to be zero.

2.1.4.1 Remarks

The transformation of the vector $\boldsymbol{\xi}_{\alpha\beta}$ from the stationary reference frame to the vector $\boldsymbol{\xi}_{dq}$ in the rotating reference frame, and vice versa,

$$\boldsymbol{\xi}_{dq} = \boldsymbol{R}(\varphi)\boldsymbol{\xi}_{\alpha\beta} \text{ and } \boldsymbol{\xi}_{\alpha\beta} = \boldsymbol{R}^{-1}(\varphi)\boldsymbol{\xi}_{dq} \tag{2.24}$$

is carried out through the rotation matrices

$$\boldsymbol{R}(\varphi) = \begin{bmatrix} \cos(\varphi) & \sin(\varphi) \\ -\sin(\varphi) & \cos(\varphi) \end{bmatrix} \text{ and } \boldsymbol{R}^{-1}(\varphi) = \begin{bmatrix} \cos(\varphi) & -\sin(\varphi) \\ \sin(\varphi) & \cos(\varphi) \end{bmatrix}. \quad (2.25)$$

As the orthogonal reference frame is rotated by φ from $\alpha\beta$ to dq in a counterclockwise direction (compare Figs. 2.4 and 2.7), the vector $\boldsymbol{\xi}_{\alpha\beta}$ must be rotated in a clockwise direction to transform it into $\boldsymbol{\xi}_{dq}$. Therefore, the matrix $\boldsymbol{R}$ performs a clockwise rotation. It is also easy to show that $\tilde{\boldsymbol{K}}(\varphi) = \boldsymbol{R}(\varphi)\tilde{\boldsymbol{K}}$ and $\tilde{\boldsymbol{K}}^{-1}(\varphi) = \tilde{\boldsymbol{K}}^{-1}\boldsymbol{R}^{-1}(\varphi)$ hold.

As Examples 2.4 and 2.6 confirm, the Clarke and Park transformations as defined previously are peak or amplitude invariant, implying that the amplitude of the fundamental waveforms remains unchanged. However, this entails that the transformations are not power invariant. Specifically, the instantaneous power expressed in abc variables is given by

$$S_{abc} = v_a i_a + v_b i_b + v_c i_c. \quad (2.26)$$

Substituting (2.18) in (2.26) and imposing the instantaneous power in $dq0$ quantities to be equal to the power expressed in abc variables leads to

$$S_{dq0} = S_{abc} = \frac{3}{2}(v_d i_d + v_q i_q + 2v_0 i_0). \quad (2.27)$$

The same holds for the abc to $\alpha\beta0$ transformation. In both cases, the factor 1.5 is mandatory when translating the power from an orthogonal reference frame to the three-phase system.

For an in-depth review of reference frame theory, the reader is referred to Krause's excellent book [3, Chap. 3].

2.1.5 Space Vectors

Directly related to the concept of orthogonal reference frames is the notion of space vectors, which are a widely used and convenient approach to represent three-phase quantities. Specifically, a space vector is a representation of a three-phase instantaneous quantity in the complex plane with real and imaginary components. We define the complex number $\vec{a} = \exp(\mathrm{j}2\pi/3) = -\frac{1}{2} + \mathrm{j}\sqrt{3}/2$, where j denotes the imaginary unit. We will use complex numbers only in the context of space vectors and we use arrows to indicate them.

Aligning the a-axis with the real axis, the position of the b-axis in the complex plane can be described by $\vec{a}$. Accordingly, $\vec{a}^2$ describes the position of the c-axis. The space vector of the three-phase quantity $\boldsymbol{\xi}_{abc}$ is then defined as

$$\vec{\xi} = \frac{2}{3}(\xi_a + \xi_b\vec{a} + \xi_c\vec{a}^2). \quad (2.28)$$

The factor 2/3 in (2.28) is required to ensure that the space vector has the same amplitude as the three-phase quantity it represents. As a result, the space vector representation is peak or amplitude invariant.

Example 2.7 *Consider once again Example 2.3 with the three-phase current waveforms shown in Fig. 2.3(b). By measuring the currents at time instant* $t_0 = 0$ *and applying* (2.28),

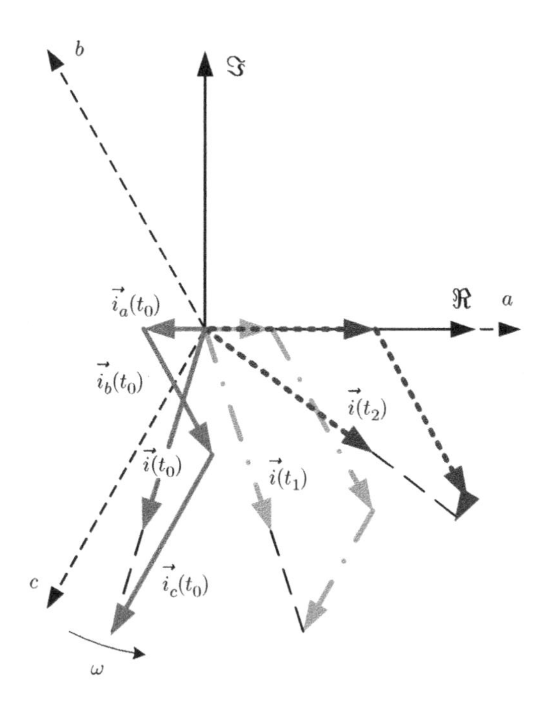

Figure 2.10 Current space vector $\vec{i}$, its phase components, and evolution over time. Straight lines refer to the time instant $t_0 = 0$, dash-dotted lines to $t_1 = 2$ ms and dashed lines to $t_2 = 4$ ms

the space vector of the current $\vec{i}(t_0)$ is obtained. This space vector is shown in Fig. 2.10 along with the three per-phase components of which it is composed. For completeness, these per-phase vectors are given by $\vec{i}_a(t_0) = i_a(t_0)$, $\vec{i}_b(t_0) = i_b(t_0)\vec{a}$, and $\vec{i}_c(t_0) = i_c(t_0)\vec{a}^2$. The space vectors at times $t_1 = 2$ ms and $t_2 = 4$ ms are derived accordingly and are indicated by dashed-dotted and dashed lines, respectively. Note that the space vector's absolute value is the same as the amplitude of the abc current waveforms. Moreover, the space vector rotates with the angular velocity ω.

The space vector representation (2.28) is equivalent to the projection (2.15) onto the α- and β-axes. In particular, the complex numbers $\vec{a}^0 = 1$, $\vec{a}$, and $\vec{a}^2$ can be interpreted as three unit vectors, which are displaced by $2\pi/3$ and are associated with the three phases. Therefore, the space vector representation $\vec{\xi}$ is equivalent to a two-dimensional vector in the $\alpha\beta$-plane. The space vector's real and imaginary parts correspond to the α- and β-components, respectively, according to

$$\xi_\alpha = \Re\{\vec{\xi}\} \text{ and } \xi_\beta = \Im\{\vec{\xi}\}. \tag{2.29}$$

The α- and β-components of the space vector can be explicitly derived as a function of the corresponding a-, b-, and c-components. Substituting (2.28) in (2.29) leads to

$$\boldsymbol{\xi}_{\alpha\beta} = \frac{2}{3}\begin{bmatrix} 1 & -\frac{1}{2} & -\frac{1}{2} \\ 0 & \frac{\sqrt{3}}{2} & -\frac{\sqrt{3}}{2} \end{bmatrix} \boldsymbol{\xi}_{abc}, \tag{2.30}$$

which is the same as the abc to $\alpha\beta$ transformation (2.12), that is, the reduced Clarke transformation.

In addition to that, multiplying the space vector in (2.28) by $\exp(-\mathrm{j}\varphi)$ transforms it from the stationary complex plane to a rotating complex plane. The latter is equivalent to the dq reference frame. In matrix form, this clockwise rotation is given by $\boldsymbol{R}(\varphi)$, which was defined in (2.25).

Based on this, one might be tempted to conclude that orthogonal reference frames and space vectors provide the same representation. Strictly speaking, despite their strong similarities, these two concepts are not the same. Space vectors were originally proposed to describe sinusoidal distributions in space. Such distributions arise, for example, for voltages, currents, and flux linkages in electrical machines because of the allotted windings (see [4] and the references therein). In contrast, orthogonal reference frames assume lumped parameters. For an extensive introduction to the notion of space vectors, the reader is referred to [5, Chap. 4].

2.2 Induction Machines

The modeling process of induction machines is typically performed in three stages. Starting from the three-phase abc quantities and using fundamental physical laws such as Faraday's law of induction and the Lorentz force, the machine's differential equations and its torque equation can be derived. In a second stage—to simplify the representation—the model is then transformed into an orthogonal reference frame as introduced in Sect. 2.1.4. Matrix notation is the natural choice for such a model. Alternatively, electrical machines can be mathematically represented using space vector notation. Space vectors, as introduced in Sect. 2.1.5, lead to a very compact model description. In a last step, the machine model is often translated from SI quantities to normalized quantities, using the pu system (see Sect. 2.1.2).

This section summarizes the dynamic induction machine model using both notations—space vectors with complex numbers and matrix representation with real numbers. The derivation of the differential equations is beyond the scope of this book. The interested reader is referred to the first three chapters of Krause's excellent text book on electrical machines [3].

It is important to note that the dynamic model presented hereafter is based on a number of assumptions and simplifications. These include the following:

- The machine's magnetic material is linear and thus the saturation of the main inductance is neglected.
- Magnetic losses and changes to the rotor resistance because of the skin effect are neglected.
- All machine parameters are time invariant. In particular, changes of the stator resistance because of the temperature variations are neglected.
- The machine is assumed to be symmetrical in its three phases and in the rotor. Specifically, saliency in the rotor geometry is not considered.
- The machine windings are sinusoidally distributed.

2.2.1 *Machine Model in Space Vector Notation*

Hereafter, the dynamic model of a three-phase induction machine in SI units is provided adopting the space vector notation (see [4] and the references therein). This standard dynamic model of an induction machine can be used to describe steady-state as well as transient phenomena. The machine model is formulated in an orthogonal reference frame rotating at an arbitrary angular speed. All rotor quantities are referred to the stator side.

Following the line of thought in [6], the machine model is grouped into three sets of equations. The first set is constituted by the voltage equations

$$\vec{v}_s = R_s\vec{i}_s + \frac{\mathrm{d}\vec{\lambda}_s}{\mathrm{d}t} + \mathrm{j}\omega_{\mathrm{fr}}\vec{\lambda}_s \tag{2.31a}$$

$$\vec{v}_r = R_r\vec{i}_r + \frac{\mathrm{d}\vec{\lambda}_r}{\mathrm{d}t} + \mathrm{j}(\omega_{\mathrm{fr}} - \omega_r)\vec{\lambda}_r\ , \tag{2.31b}$$

where $\vec{v}_s$ ($\vec{v}_r$) denotes the stator (rotor) voltage space vector, $\vec{i}_s$ ($\vec{i}_r$) is the stator (rotor) current space vector, $\vec{\lambda}_s$ ($\vec{\lambda}_r$) is the stator (rotor) flux linkage space vector, and the stator (rotor) winding resistance is given by R_s (R_r). The arbitrary angular speed of the reference frame is denoted by ω_{fr}, while the (electrical) angular speed of the rotor is given by ω_r.

The terms $R_s\vec{i}_s$ and $R_r\vec{i}_r$ represent the resistive voltage drop in the stator and rotor winding, respectively. The terms $\omega_{\mathrm{fr}}\vec{\lambda}_s$ and $(\omega_{\mathrm{fr}} - \omega_r)\vec{\lambda}_r$ are commonly referred to as speed voltages. In this book, if not otherwise stated, induction machines with squirrel-cage rotors are assumed. For such a machine, the left-hand side of (2.31b) is set to zero, that is, $\vec{v}_r = 0$. The equivalent circuit representation of the squirrel-cage induction machine model is depicted in Fig. 2.11.

The flux linkage equations

$$\vec{\lambda}_s = L_s\vec{i}_s + L_m\vec{i}_r \tag{2.32a}$$

$$\vec{\lambda}_r = L_r\vec{i}_r + L_m\vec{i}_s \tag{2.32b}$$

represent the second set of equations, where

$$L_s = L_{ls} + L_m \tag{2.33a}$$

$$L_r = L_{lr} + L_m \tag{2.33b}$$

denote the stator and rotor self-inductance, respectively. Moreover, L_{ls} (L_{lr}) is the stator (rotor) leakage inductance and L_m is the main inductance. The latter is often referred to as the magnetizing inductance.

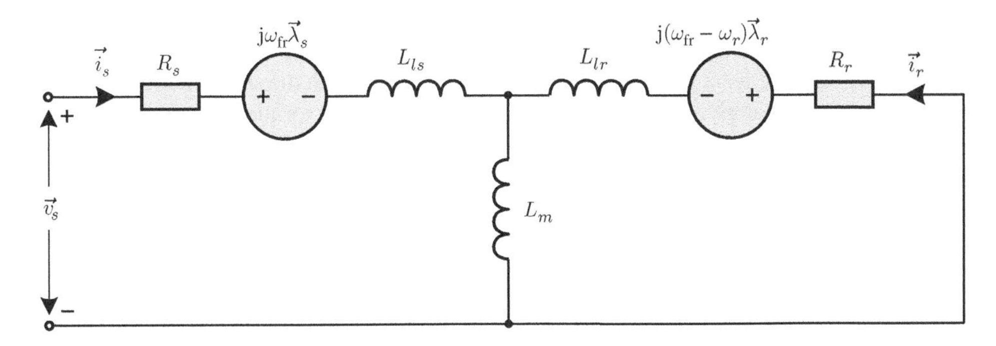

Figure 2.11 Equivalent circuit of a squirrel-cage induction machine in the arbitrary orthogonal reference frame based on space vector notation

The third set of equations includes the torque and the rotational motion equation given by

$$T_e = \frac{3}{2} p\, \Re\{\mathrm{j}\vec{\lambda}_s \operatorname{conj}\{\vec{i}_s\}\} \tag{2.34a}$$

$$M \frac{\mathrm{d}\omega_m}{\mathrm{d}t} = T_e - T_\ell\,, \tag{2.34b}$$

where T_e (T_ℓ) is the electromagnetic (load) torque, p is the number of pole pairs, $\operatorname{conj}\{\vec{i}_s\}$ denotes the complex conjugate of $\vec{i}_s$, M is the moment of inertia of the mechanical load including the machine's rotor, and ω_m is the mechanical angular speed of the shaft. The units in (2.34b) are as follows: the inertia M is given in kg m^2, the angular speed ω_m in rad/s, and the torque T is given in Nm, with N = kg m/s^2.

The electrical rotor speed is

$$\omega_r = p\, \omega_m, \tag{2.35}$$

and the mechanical power is

$$P_m = \omega_m T_e. \tag{2.36}$$

The torque T_e and the power P_m are both positive during the motoring operation, while they are both negative in the generation mode.

The different angular velocities and frequencies are summarized below:

- ω_s is the angular frequency of the stator.
- ω_r is the electrical angular speed of the rotor with $\omega_r = p\, \omega_m$.
- ω_m is the mechanical angular speed of the rotor and shaft.
- ω_{sR} is the rated (or nominal) angular frequency of the stator.
- ω_B is the base angular frequency of the machine voltages and currents. Typically, $\omega_B = \omega_{sR}$ is chosen for the pu system.
- ω_{fr} is the angular speed of the arbitrary reference frame.

A fundamental characteristic of induction machines is the slip, which is the normalized difference between the electrical frequencies of the stator and rotor:

$$\mathrm{sl} = \frac{\omega_s - \omega_r}{\omega_s}. \tag{2.37}$$

In motoring operation—in order to produce an electromagnetic torque—the rotor windings must revolve slightly more slowly than the stator field. This motion of the stator field relative to the rotor induces ac voltages in the rotor—hence the name induction machine. These induced voltages drive ac currents in the rotor, which, in turn, together with the stator field, produce an electromotive force that gives rise to the electromagnetic machine torque. In generation mode, the inverse holds, that is, the electrical angular speed of the rotor is slightly higher than that of the stator field. When zero torque is produced or absorbed, the slip is zero.

2.2.2 Machine Model in Matrix Notation

In modern control theory, including model predictive control, the models are predominantly given in the state-space representation. For this, a matrix representation of the induction machine model appears to be better suited [3]. This representation can be directly obtained by rewriting the dynamic model in space vector notation derived previously. As shown in Sect. 2.1.5, in the arbitrary reference frame, the d-axis is associated with the real part of a space vector, while the q-axis is associated with the imaginary part of a space vector. Specifically, using the space vector of the stator voltage as an example, the real-valued d- and q-components of the stator voltage vector $\boldsymbol{v}_s$ are obtained as

$$\boldsymbol{v}_s = \begin{bmatrix} v_{sd} \\ v_{sq} \end{bmatrix} = \begin{bmatrix} \Re\{\vec{v}_s\} \\ \Im\{\vec{v}_s\} \end{bmatrix}. \tag{2.38}$$

Accordingly, the other space vectors are translated into (real-valued) vectors with d- and q-components. This gives rise to the rotor voltage vector $\boldsymbol{v}_r$, the stator (rotor) current vector $\boldsymbol{i}_s$ ($\boldsymbol{i}_r$), and the stator (rotor) flux linkage $\boldsymbol{\lambda}_s$ ($\boldsymbol{\lambda}_r$).

With these definitions, the voltage equations (2.31) can be rewritten in matrix notation, yielding

$$\boldsymbol{v}_s = R_s \boldsymbol{i}_s + \frac{\mathrm{d}\boldsymbol{\lambda}_s}{\mathrm{d}t} + \omega_{\mathrm{fr}} \begin{bmatrix} 0 & -1 \\ 1 & 0 \end{bmatrix} \boldsymbol{\lambda}_s \tag{2.39a}$$

$$\boldsymbol{v}_r = R_r \boldsymbol{i}_r + \frac{\mathrm{d}\boldsymbol{\lambda}_r}{\mathrm{d}t} + (\omega_{\mathrm{fr}} - \omega_r) \begin{bmatrix} 0 & -1 \\ 1 & 0 \end{bmatrix} \boldsymbol{\lambda}_r . \tag{2.39b}$$

Recall that for the squirrel-cage induction machine $\boldsymbol{v}_r = 0$ holds.

Accordingly, the flux linkage equations (2.32) in matrix form are

$$\boldsymbol{\lambda}_s = L_s \boldsymbol{i}_s + L_m \boldsymbol{i}_r \tag{2.40a}$$

$$\boldsymbol{\lambda}_r = L_r \boldsymbol{i}_r + L_m \boldsymbol{i}_s \tag{2.40b}$$

and the electromagnetic torque equation is

$$T_e = \frac{3}{2} p \, (\boldsymbol{\lambda}_s \times \boldsymbol{i}_s). \tag{2.41}$$

Note that the expanded form of the cross product $\boldsymbol{\lambda}_s \times \boldsymbol{i}_s$ is $\lambda_{sd} i_{sq} - \lambda_{sq} i_{sd}$.

This leads to the equivalent circuit representation of the squirrel-cage induction machine model in the dq reference frame, which is shown in Fig. 2.12.

2.2.3 *Machine Model in the Per Unit System*

It is common practice to normalize the electrical variables and machine parameters by expressing them in the pu system. As the process of normalization often causes a certain degree of confusion, we have devoted more space to this subject matter and provide the reader with a step-by-step derivation of the normalized machine equations.

Recapitulating Sect. 2.1.2, the SI variables and parameters are normalized by dividing them by their respective *base* quantities. We chose the voltage and the current as primary base quantities and set them to the peak machine phase voltage and current, respectively. As the rated machine voltage V_R is the line-to-line rms voltage, this leads to

$$V_B = \sqrt{\frac{2}{3}} V_R \text{ and } I_B = \sqrt{2} I_R . \tag{2.42}$$

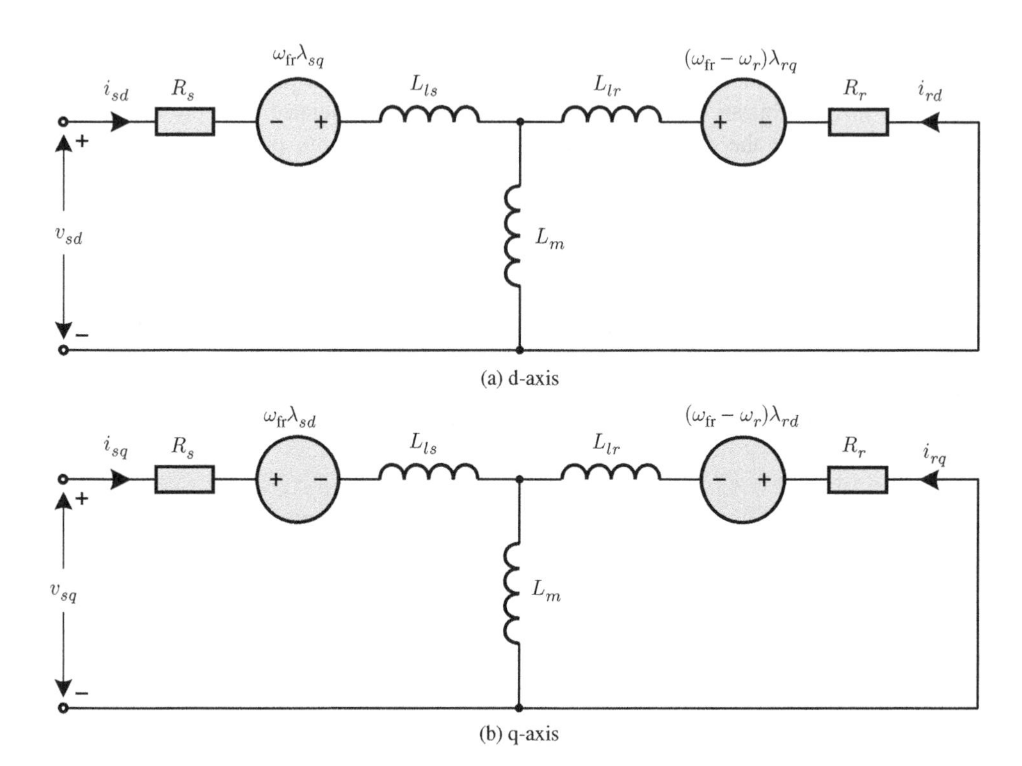

Figure 2.12 Equivalent circuit representation of a squirrel-cage induction machine in the arbitrary reference frame based on matrix notation

The third primary base quantity is the base angular frequency of the machine, which we chose as the rated stator frequency:

$$\omega_B = \omega_{sR} . \tag{2.43}$$

Table 2.1 provides a summary of the base quantities.

The normalization process consists of four steps. First, the flux linkages and inductances are scaled with the base frequency. This leads to the flux linkages per second ψ (with the unit volt) and to the reactances X(with the unit ohm):

$$\psi = \omega_B \lambda \text{ and } X = \omega_B L . \tag{2.44}$$

With these definitions, (2.39) can be rewritten as

$$\boldsymbol{v}_s = R_s \boldsymbol{i}_s + \frac{1}{\omega_B}\frac{\mathrm{d}\boldsymbol{\psi}_s}{\mathrm{d}t} + \frac{\omega_{fr}}{\omega_B}\begin{bmatrix} 0 & -1 \\ 1 & 0 \end{bmatrix} \boldsymbol{\psi}_s \tag{2.45a}$$

$$\boldsymbol{v}_r = R_r \boldsymbol{i}_r + \frac{1}{\omega_B}\frac{\mathrm{d}\boldsymbol{\psi}_r}{\mathrm{d}t} + \frac{\omega_{fr} - \omega_r}{\omega_B}\begin{bmatrix} 0 & -1 \\ 1 & 0 \end{bmatrix} \boldsymbol{\psi}_r . \tag{2.45b}$$

Multiplying (2.40) with ω_B leads to

$$\boldsymbol{\psi}_s = X_s \boldsymbol{i}_s + X_m \boldsymbol{i}_r \tag{2.46a}$$

$$\boldsymbol{\psi}_r = X_r \boldsymbol{i}_r + X_m \boldsymbol{i}_s . \tag{2.46b}$$

Second, the voltage equation (2.45) is divided by the base voltage V_B:

$$\boldsymbol{v}'_s = R'_s \boldsymbol{i}'_s + \frac{1}{\omega_B}\frac{\mathrm{d}\boldsymbol{\psi}'_s}{\mathrm{d}t} + \omega'_{\mathrm{fr}} \begin{bmatrix} 0 & -1 \\ 1 & 0 \end{bmatrix} \boldsymbol{\psi}'_s \tag{2.47a}$$

$$\boldsymbol{v}'_r = R'_r \boldsymbol{i}'_r + \frac{1}{\omega_B}\frac{\mathrm{d}\boldsymbol{\psi}'_r}{\mathrm{d}t} + (\omega'_{\mathrm{fr}} - \omega'_r) \begin{bmatrix} 0 & -1 \\ 1 & 0 \end{bmatrix} \boldsymbol{\psi}'_r \,. \tag{2.47b}$$

The superscript $'$ denotes pu quantities. Specifically, the normalized quantities are defined as

$$\boldsymbol{v}' = \frac{\boldsymbol{v}}{V_B}\,, \quad \boldsymbol{\psi}' = \frac{\boldsymbol{\psi}}{V_B}\,, \quad \boldsymbol{i}' = \frac{\boldsymbol{i}}{I_B}\,, \tag{2.48a}$$

$$R' = \frac{R}{Z_B}\,, \quad X' = \frac{X}{Z_B}\,, \quad \omega' = \frac{\omega}{\omega_B}\,, \tag{2.48b}$$

where we have used the base impedance $Z_B = V_B/I_B$. Similarly, by dividing (2.46) by the base voltage yields

$$\boldsymbol{\psi}'_s = X'_s \boldsymbol{i}'_s + X'_m \boldsymbol{i}'_r \tag{2.49a}$$

$$\boldsymbol{\psi}'_r = X'_r \boldsymbol{i}'_r + X'_m \boldsymbol{i}'_s \,. \tag{2.49b}$$

Third, the time axis can be normalized by defining $t' = \omega_B t$. With this, we can rewrite (2.47) as

$$\boldsymbol{v}'_s = R'_s \boldsymbol{i}'_s + \frac{\mathrm{d}\boldsymbol{\psi}'_s}{\mathrm{d}t'} + \omega'_{\mathrm{fr}} \begin{bmatrix} 0 & -1 \\ 1 & 0 \end{bmatrix} \boldsymbol{\psi}'_s \tag{2.50a}$$

$$\boldsymbol{v}'_r = R'_r \boldsymbol{i}'_r + \frac{\mathrm{d}\boldsymbol{\psi}'_r}{\mathrm{d}t'} + (\omega'_{\mathrm{fr}} - \omega'_r) \begin{bmatrix} 0 & -1 \\ 1 & 0 \end{bmatrix} \boldsymbol{\psi}'_r \,. \tag{2.50b}$$

Equation (2.49) remains unaffected by this.

Fourth, the electromagnetic torque and the rotational motion equation (2.34) are normalized. For the torque, we use the base torque $T_B = \mathrm{pf}\; pS_B/\omega_B$. For the moment of inertia, the corresponding base value is $M_B = T_B/\omega_B^2$. The normalization turns (2.41) and (2.34b) into

$$T'_e = \frac{T_e}{T_B} = \frac{1}{\mathrm{pf}} \boldsymbol{\psi}'_s \times \boldsymbol{i}'_s \tag{2.51a}$$

$$M' \frac{\mathrm{d}\omega_m}{\mathrm{d}t'} = T'_e - T'_\ell \,. \tag{2.51b}$$

Note that pf denotes the power factor. By scaling the cross product $\boldsymbol{\psi}'_s \times \boldsymbol{i}'_s$ by the inverse of the power factor, the normalized torque $T'_e = 1$ pu corresponds to operation at rated torque.

It is important to point out that the normalized machine equations (2.49), (2.50), and (2.51) are structurally the same as the equations in SI units (2.39), (2.40), (2.41), and (2.34b). The only difference is that all variables and parameters have been replaced by their normalized counterparts. The same holds true for the equivalent circuit representation of the squirrel-cage induction machine model in the *dq* reference frame, which was given in Fig. 2.12. However, unlike in the voltage and flux linkage equations, the structure of the torque equation is changed during the normalization operation.

We will drop the superscript $'$ in the remainder of the book to simplify the notation. It will be obvious from the text whether a variable or parameter refers to an SI or pu quantity. In general, pu quantities will be adopted, and all variables and parameters will be normalized, including the time axis. Nevertheless, waveforms of the pu variables will be plotted versus the time axis in seconds rather than versus the normalized time in pu to simplify the exposition.

2.2.4 Machine Model in State-Space Representation

It is convenient to rewrite the model of the squirrel-cage induction machine in state-space representation. This will facilitate formulating and solving the model predictive control problems stated later in this book. In the state-space representation

$$\frac{\mathrm{d}\boldsymbol{x}(t)}{\mathrm{d}t} = \boldsymbol{F}\boldsymbol{x}(t) + \boldsymbol{G}\boldsymbol{u}(t) \tag{2.52a}$$

$$\boldsymbol{y}(t) = \boldsymbol{C}\boldsymbol{x}(t), \tag{2.52b}$$

$\boldsymbol{x}(t)$ denotes the state vector, $\boldsymbol{u}(t)$ the input vector, and $\boldsymbol{y}(t)$ the output vector. $\boldsymbol{F}$, $\boldsymbol{G}$, and $\boldsymbol{C}$ are matrices of appropriate dimensions. In the case of nonlinear output equations, (2.52b) is replaced by $\boldsymbol{y}(t) = \boldsymbol{h}(\boldsymbol{x}(t))$. To simplify the notation, we will often drop the time-dependency from $\boldsymbol{x}$, $\boldsymbol{u}$, and $\boldsymbol{y}$.

We assume that the mechanical speed is constant: that is, the left-hand side of (2.51b) is zero and thus not required. ω_r can be then considered to be a parameter rather than a state variable. This avoids bilinear terms in the differential equations and ensures that the state-space equation is linear.

The machine is modeled in the *dq* reference frame rotating at the arbitrary angular velocity ω_{fr}. In order to represent the dynamic state of the machine's stator and rotor circuits in this reference frame, four state variables are required. Common choices include the stator current, stator flux linkage, rotor current, or the rotor flux linkage, each with *d*- and *q*-components. Any pair of these four *dq* state vectors can be adopted to model the dynamic state of the stator and rotor circuits.

In the following, two state-space representations are derived. In the first one, the stator circuit is represented by the *stator flux linkages*, while for the second representation, the *stator currents* are used as state variables. The rotor circuit is characterized in both cases by the *rotor flux linkages*. Normalized quantities are used.

2.2.4.1 Stator and Rotor Flux Linkages

In a first step, (2.49) is inverted, and the stator and rotor current vectors are represented as a function of the stator and rotor flux vectors:

$$\begin{bmatrix} \boldsymbol{i}_s \\ \boldsymbol{i}_r \end{bmatrix} = \frac{1}{D} \begin{bmatrix} \boldsymbol{I}_2 X_r & -\boldsymbol{I}_2 X_m \\ -\boldsymbol{I}_2 X_m & \boldsymbol{I}_2 X_s \end{bmatrix} \begin{bmatrix} \boldsymbol{\psi}_s \\ \boldsymbol{\psi}_r \end{bmatrix} \tag{2.53}$$

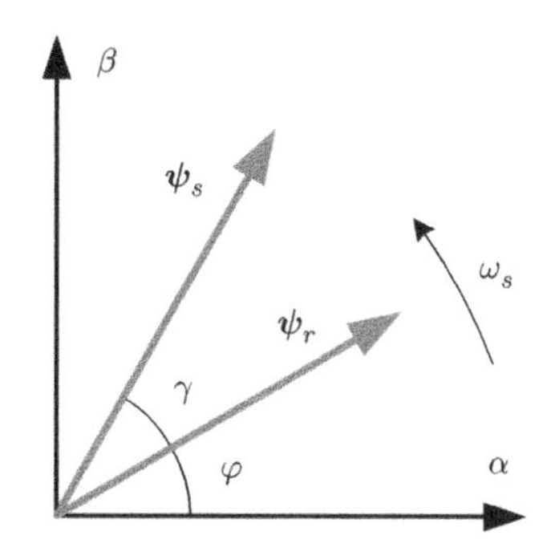

Figure 2.13 Stator and rotor flux vectors in the stationary reference frame

with the determinant

$$D = X_s X_r - X_m^2 \tag{2.54}$$

and $\boldsymbol{I}_2$ denoting the 2×2 identity matrix.

Inserting (2.53) in (2.50) and (2.51a), respectively, leads to the dynamic representation in state-space form

$$\frac{\mathrm{d}\boldsymbol{\psi}_s}{\mathrm{d}t} = -R_s \frac{X_r}{D} \boldsymbol{\psi}_s - \omega_{\mathrm{fr}} \begin{bmatrix} 0 & -1 \\ 1 & 0 \end{bmatrix} \boldsymbol{\psi}_s + R_s \frac{X_m}{D} \boldsymbol{\psi}_r + \boldsymbol{v}_s \tag{2.55a}$$

$$\frac{\mathrm{d}\boldsymbol{\psi}_r}{\mathrm{d}t} = R_r \frac{X_m}{D} \boldsymbol{\psi}_s - R_r \frac{X_s}{D} \boldsymbol{\psi}_r - (\omega_{\mathrm{fr}} - \omega_r) \begin{bmatrix} 0 & -1 \\ 1 & 0 \end{bmatrix} \boldsymbol{\psi}_r + \boldsymbol{v}_r \tag{2.55b}$$

and to the electromagnetic torque

$$T_e = \frac{1}{\mathrm{pf}} \frac{X_m}{D} \boldsymbol{\psi}_r \times \boldsymbol{\psi}_s. \tag{2.56}$$

Expanding the cross product results in $T_e = \frac{1}{\mathrm{pf}} \frac{X_m}{D} (\psi_{rd} \psi_{sq} - \psi_{rq} \psi_{sd})$.

This model is formulated in the arbitrary reference frame with d- and q-components. By setting ω_{fr} to zero, a corresponding model in the stationary reference frame results with α- and β-components.

The stator and rotor flux vectors are shown in Fig. 2.13. During steady-state operation, both flux vectors rotate at the constant angular velocity ω_s. In motoring operation, the stator flux vector lies ahead of the rotor flux vector. The angle γ between the two vectors defines the torque

$$T_e = \frac{1}{\mathrm{pf}} \frac{X_m}{D} \, \|\boldsymbol{\psi}_s\| \, \|\boldsymbol{\psi}_r\| \, \sin(\gamma). \tag{2.57}$$

2.2.4.2 Stator Currents and Rotor Flux Linkages

An alternative, and sometimes more convenient, machine model uses the stator current and the rotor flux vector as state variables. Such a model can be derived by reformulating (2.53) to

$$\boldsymbol{\psi}_s = \frac{D}{X_r} \boldsymbol{i}_s + \frac{X_m}{X_r} \boldsymbol{\psi}_r \tag{2.58}$$

and replacing in (2.55a) the stator flux vector by the stator current and the rotor flux vector using this statement. After some lengthy algebraic manipulations, this leads to

$$\frac{\mathrm{d}\boldsymbol{i}_s}{\mathrm{d}t} = -\frac{1}{\tau_s}\boldsymbol{i}_s - \omega_{\mathrm{fr}}\begin{bmatrix}0 & -1\\ 1 & 0\end{bmatrix}\boldsymbol{i}_s + \left(\frac{1}{\tau_r}\boldsymbol{I}_2 - \omega_r\begin{bmatrix}0 & -1\\ 1 & 0\end{bmatrix}\right)\frac{X_m}{D}\boldsymbol{\psi}_r + \frac{X_r}{D}\boldsymbol{v}_s - \frac{X_m}{D}\boldsymbol{v}_r \tag{2.59a}$$

$$\frac{\mathrm{d}\boldsymbol{\psi}_r}{\mathrm{d}t} = \frac{X_m}{\tau_r}\boldsymbol{i}_s - \frac{1}{\tau_r}\boldsymbol{\psi}_r - (\omega_{\mathrm{fr}} - \omega_r)\begin{bmatrix}0 & -1\\ 1 & 0\end{bmatrix}\boldsymbol{\psi}_r + \boldsymbol{v}_r, \tag{2.59b}$$

where we have introduced the transient stator time constant and the rotor time constant

$$\tau_s = \frac{X_r D}{R_s X_r^2 + R_r X_m^2} \quad \text{and} \quad \tau_r = \frac{X_r}{R_r} \tag{2.60}$$

to allow for a more compact representation. Note that $\boldsymbol{I}_2$ denotes the two-dimensional identity matrix. The electromagnetic torque (2.56) can be expressed in terms of the stator current and the rotor flux vector

$$T_e = \frac{1}{\mathrm{pf}}\frac{X_m}{X_r}\boldsymbol{\psi}_r \times \boldsymbol{i}_s = \frac{1}{\mathrm{pf}}\frac{X_m}{X_r}(\psi_{rd} i_{sq} - \psi_{rq} i_{sd}). \tag{2.61}$$

2.2.5 *Harmonic Model of the Machine*

The machine models derived in Sects. 2.2.1–2.2.4 are dynamic models that describe induction machines during dynamic and steady-state operation. In particular, these models describe voltages, currents, and flux linkages that are—in general—the superposition of fundamental as well as harmonic components.

When assessing the impact of voltage harmonics on the machine, however, a separate model can be derived. This compact model is applicable to harmonic stator quantities with frequencies that are significantly higher than the rated frequency. Adopting the pu system, the voltage equation of such a harmonic model is given by

$$\boldsymbol{v}_s = R_s \boldsymbol{i}_s + X_\sigma \frac{\mathrm{d}\boldsymbol{i}_s}{\mathrm{d}t}, \tag{2.62}$$

where we have also normalized the time and omitted again the superscripts $'$. The equivalent circuit representation of the harmonic model is provided in Fig. 2.14.

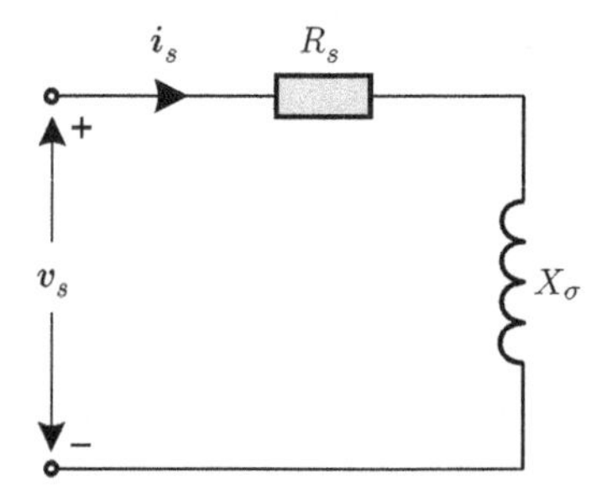

Figure 2.14 Harmonic model of an induction machine in the per unit system

Table 2.2 Machine parameters in the per unit system

Parameter	Symbol
Stator winding resistance	R_s
Rotor winding resistance	R_r
Stator leakage reactance	X_{ls}
Rotor leakage reactance	X_{lr}
Main (or magnetizing) reactance	X_m
Number of pole pairs	p
Stator self-reactance	$X_s = X_{ls} + X_m$
Rotor self-reactance	$X_r = X_{lr} + X_m$
Determinant	$D = X_s X_r - X_m^2$
Transient stator time constant	$\tau_s = \dfrac{X_r D}{R_s X_r^2 + R_r X_m^2}$
Rotor time constant	$\tau_r = \dfrac{X_r}{R_r}$
Total leakage reactance	$X_\sigma = \sigma X_s = \dfrac{D}{X_r}$
Total leakage factor	$\sigma = 1 - \dfrac{X_m^2}{X_s X_r}$

The reactance representing the machine's harmonic characteristic is the total leakage reactance X_σ, which is given by

$$X_\sigma = \sigma X_s, \tag{2.63}$$

with

$$\sigma = 1 - \frac{X_m^2}{X_s X_r} \tag{2.64}$$

denoting the total leakage factor. An alternative representation of (2.63) is $X_\sigma = D/X_r$. As a summary, the machine parameters in the pu system and some of the major deduced quantities are listed in Table 2.2.

2.3 Power Semiconductor Devices

Power semiconductor devices constitute the key building blocks of the power electronic topologies. Integrated-gate-commutated thyristors (IGCTs) and insulated-gate bipolar transistors (IGBTs) are used as active switches in MV voltage source inverters, while power diodes constitute the passive switches. This section provides a brief introduction to IGCTs and power diodes, focusing mostly on their switching and conduction losses. More details on IGCTs are provided in the application note [7].

2.3.1 Integrated-Gate-Commutated Thyristors

The gate driver of gate-commutated thyristors (GCTs) is integrated with the semiconductor switch in one module to provide a very low inductive path. The module, including the

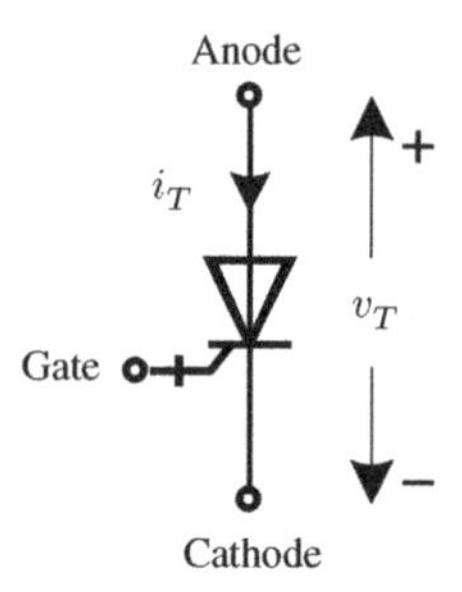

Figure 2.15 Schematic of a gate-commutated thyristor (GCT)

gate driver with the thyristor, is called an integrated-gate-commutated thyristor (IGCT) [8, 9]. GCTs are an advancement on gate turn-off (GTO) thyristors. Unlike GTOs, GCTs do not require turn-off dv/dt snubbers and are therefore commonly referred to as snubberless devices, even though they still require a turn-on di/dt snubber to limit the slope of the rising current.

The schematic of a GCT is shown in Fig. 2.15. GCTs are available with high blocking voltages and current ratings. In the *on* state, when the GCT is conducting, the anode–cathode voltage v_T is equivalent to the on-state voltage, which is typically below 2.5 V. The anode current i_T is limited by the maximum on-state current, which is typically in the range of several kiloamperes. In the *off* state, when the GCT is blocking, v_T is equal to the blocking voltage, which amounts to a few kilovolts. The anode current of a few milliamperes is effectively zero.

The GCT losses can be divided into switching losses and conduction losses. Switching losses arise when the GCT is turned on or off, while conduction losses are due to the on-state resistance. Both types of losses depend on the blocking voltage, the commutated current, and the semiconductor characteristics. The conduction (or on-state) losses of GCTs are—similar to thyristors—very low, while the switching losses are moderate to high. Thus, for high-power applications, the switching frequency of GCTs is typically restricted to a few hundred hertz [7]. Even when operated at such low switching frequencies, their switching losses dominate over the conduction losses.

2.3.1.1 Switching Losses

For GCTs, the turn-on and turn-off losses can be well approximated as being linear in the anode–cathode voltage v_T and in the anode current i_T flowing through the device. This leads to the GCT turn-off (energy) loss

$$e_{\text{off}} = c_{\text{off}} v_T i_T , \tag{2.65}$$

where c_{off} is a coefficient. For the GCT turn-on losses, the corresponding equation

$$e_{\text{on}} = c_{\text{on}} v_T i_T , \tag{2.66}$$

results with the coefficient c_{on}. When using pu quantities, the unit of c_{on} and c_{off} is the joule. Typically, c_{on} is an order of magnitude smaller than c_{off} and hence often neglected.

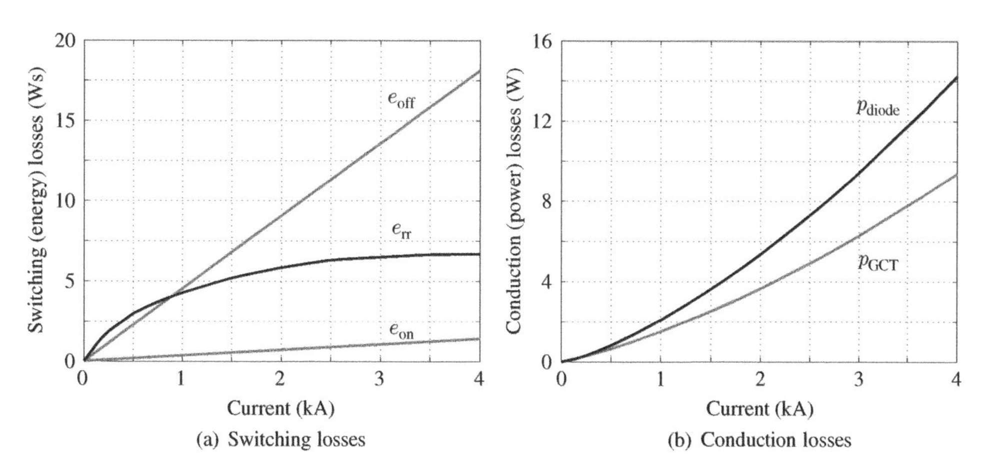

Figure 2.16 Switching and conduction losses for a GCT and a power diode as a function of the current. e_{on} and e_{off} denote the GCT turn-on and turn-off losses, respectively, while e_{rr} is the diode's reverse recovery losses. The conduction losses of the GCT and the diode are given by p_{GCT} and p_{diode}, respectively

2.3.1.2 Conduction Losses

The conduction losses are due to the on-state resistance, causing a voltage drop over the device, which is a function of the on-state current and can be written as

$$v_T = a_{\text{GCT}} + b_{\text{GCT}} i_T + c_{\text{GCT}} \log(i_T + 1) + d_{\text{GCT}} \sqrt{i_T}, \tag{2.67}$$

where the coefficients a_{GCT}, b_{GCT}, c_{GCT}, and d_{GCT} are device-specific parameters. The conduction losses are given by

$$p_{\text{GCT}} = v_T(i_T)\, i_T. \tag{2.68}$$

Using ABB's 35L4510 4.5 kV 4 kA IGCT as an example, the switching losses as a function of the commutated current are depicted in Fig. 2.16, assuming a blocking voltage of 2600 V and a nominal junction temperature of 125°C.

2.3.2 *Power Diodes*

2.3.2.1 Switching Losses

For a diode, the turn-on losses are effectively zero. The turn-off losses, however, which are the so-called *reverse recovery* losses, are often considered to be linear in the voltage but nonlinear in the commutated current i_T. They are given by

$$e_{\text{rr}} = c_{\text{rr}} v_T f_{\text{rr}}(i_T), \tag{2.69}$$

where c_{rr} is the coefficient for the reverse recovery losses. Usually, the value of c_{rr} lies in the interval c_{on} and c_{off}. In (2.69), $f_{\text{rr}}(\cdot)$ is a nonlinear function between 0 and 1, which is typically concave and saturates at 1.

2.3.2.2 Conduction Losses

The modeling of the conduction losses of the diode follows the same principles as for the GCT. Accordingly, the on-state voltage drop is given by

$$v_T = a_{\text{diode}} + b_{\text{diode}} i_T + c_{\text{diode}} \log(i_T + 1) + d_{\text{diode}} \sqrt{i_T} \tag{2.70}$$

with the diode-specific parameters a_{diode}, b_{diode}, c_{diode}, and d_{diode}. The conduction losses are

$$p_{\text{diode}} = v_T(i_T)\, i_T\,. \tag{2.71}$$

The switching and conduction losses of a power diode, ABB's 10H4520 fast recovery diode, are shown in Fig. 2.16. As for the GCT, a blocking voltage of 2600 V and a nominal operating temperature of 125°C are assumed.

2.4 Multilevel Voltage Source Inverters

Two voltage source inverters are presented in this section, starting with the three-level neutral-point-clamped (NPC) inverter. The second topology is an active NPC inverter that features in each phase an additional flying capacitor to increase the per-phase voltage levels from three to five.

2.4.1 NPC Inverter

The NPC inverter topology was originally proposed by Nabae et al. in 1981 [10]. This diode clamped inverter provides three voltage levels per phase. Today, it constitutes the most widely used voltage source inverter in MV drive applications. It is offered as a commercial product by all major drive companies (see also [6, Chap. 1]). A phase leg of ABB's ACS6000 inverter is shown in Fig. 2.17. This NPC phase leg is arranged in three stacks and it is based on water-cooled GCTs.

2.4.1.1 Topology

The equivalent representation of an NPC inverter including the dc-link stage is shown in Fig. 2.18. The dc-link is comprised of two identical dc-link capacitors C_{dc}, which form the neutral point N in between them. The total (instantaneous) dc-link voltage is

$$v_{\text{dc}} = v_{\text{dc,up}} + v_{\text{dc,lo}}\,, \tag{2.72}$$

where $v_{\text{dc,up}}$ and $v_{\text{dc,lo}}$ denote the voltages over the upper and lower dc-link capacitors, respectively. The potential

$$v_n = \frac{1}{2}(v_{\text{dc,lo}} - v_{\text{dc,up}}) \tag{2.73}$$

of the neutral point N floats.

Each phase leg consists of four pairs of active semiconductor switches with freewheeling diodes. GCTs constitute the active switches in Fig. 2.18. The upper and lower pairs are clamped

Figure 2.17 Phase module of an NPC inverter. Source: ABB Image Bank. Reproduced with permission of ABB Ltd

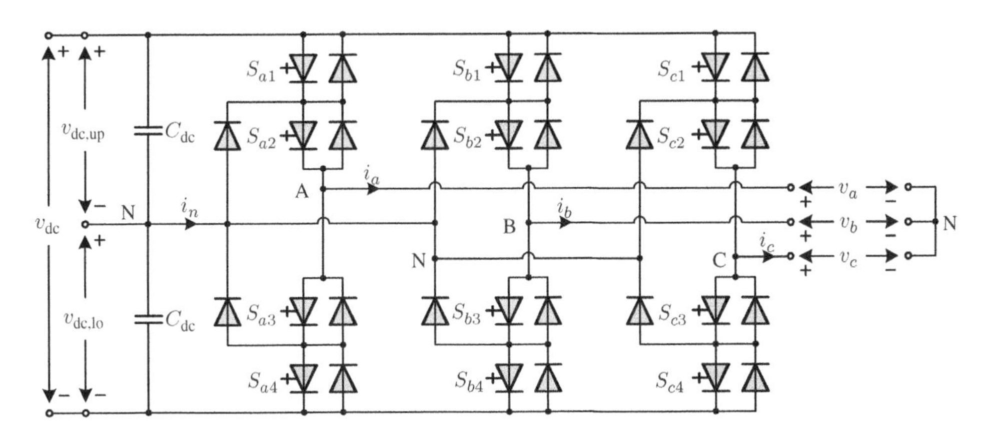

Figure 2.18 Neutral-point-clamped voltage source inverter

to the neutral point with the so-called clamping diodes. The phase terminals A, B, and C are connected to the center points of the respective phase legs.

2.4.1.2 Switch Positions and Voltage Vectors

Let the integer variable $u_x \in \{-1, 0, 1\}$ denote the switch position in one phase leg, with $x \in \{a, b, c\}$. At each phase leg, the inverter can produce three voltage levels. The phase voltages,

Table 2.3 Correspondence between the phase switch positions u_x, the phase voltages v_x, and the switching states S_{x1}–S_{x4}, for phase x, $x \in \{a, b, c\}$

Switch position u_x	Phase voltage v_x	Switching state $S_{x1}\ S_{x2}\ S_{x3}\ S_{x4}$	Effect on neutral point potential v_n
1	$v_{\mathrm{dc,up}}$	1 1 0 0	0
0	0	0 1 1 0	$-i_x$
−1	$-v_{\mathrm{dc,lo}}$	0 0 1 1	0

The effect on the neutral point potential v_n is shown on the right-hand side as a function of the phase current i_x.

which are defined with respect to the dc-link midpoint N, are given by

$$v_x = \begin{cases} v_{\mathrm{dc,up}}, & \text{if } u_x = 1 \\ 0, & \text{if } u_x = 0 \\ -v_{\mathrm{dc,lo}}, & \text{if } u_x = -1\,, \end{cases} \tag{2.74}$$

as summarized in Table 2.3. When neglecting the fluctuations of the neutral point potential, (2.74) can be approximated by

$$v_x \approx \frac{v_{\mathrm{dc}}}{2} u_x\,. \tag{2.75}$$

The three-phase voltage is equal to $\boldsymbol{v}_{abc} = [v_a\ v_b\ v_c]^T$.

Consider the phase leg x and let S_{x1}–S_{x4} denote the four active switches, with S_{x1} referring to the top switch and S_{x4} to the bottom switch. As shown in Table 2.3, the four active switches are operated dually in each phase. The switch position $u_x = 1$, for example, corresponds to the top switches S_{x1} and S_{x2} being on and the lower switches S_{x3} and S_{x4} being off.

There exist $3^3 = 27$ different vectors of the form $\boldsymbol{u}_{abc} = [u_a\ u_b\ u_c]^T$. Using (2.10), these vectors can be transformed into the stationary orthogonal reference frame

$$\boldsymbol{u}_{\alpha\beta 0} = \boldsymbol{K}\boldsymbol{u}_{abc}\,. \tag{2.76}$$

The vectors $\boldsymbol{u}_{\alpha\beta 0} = [u_\alpha\ u_\beta\ u_0]^T$ are commonly referred to as *voltage vectors*, whereas $\boldsymbol{u}_{abc}$ denotes the three-phase *switch position*. The voltage vectors are shown in Fig. 2.19 with their 0-component neglected.

The voltage vectors of the three-level inverter can be divided into four groups: 6 long vectors form the outer hexagon, 6 vectors of intermediate length are located between the long vectors, 12 short vectors span the inner hexagon, and 3 zero vectors are located at the origin of the $\alpha\beta$ plane. The 12 short vectors form six pairs on the $\alpha\beta$ plane, where each pair comprises vectors with the same α- and β-components, but whose 0-components have opposite signs. The zero vectors short-circuit the load connected to the inverter.

The actual voltage at the inverter terminals is calculated from

$$\boldsymbol{v}_{\alpha\beta 0} = \boldsymbol{K}\boldsymbol{v}_{abc} \approx \frac{v_{\mathrm{dc}}}{2}\boldsymbol{K}\boldsymbol{u}_{abc}, \tag{2.77}$$

where we neglected the fluctuations of the neutral point potential in the second part of the equation.

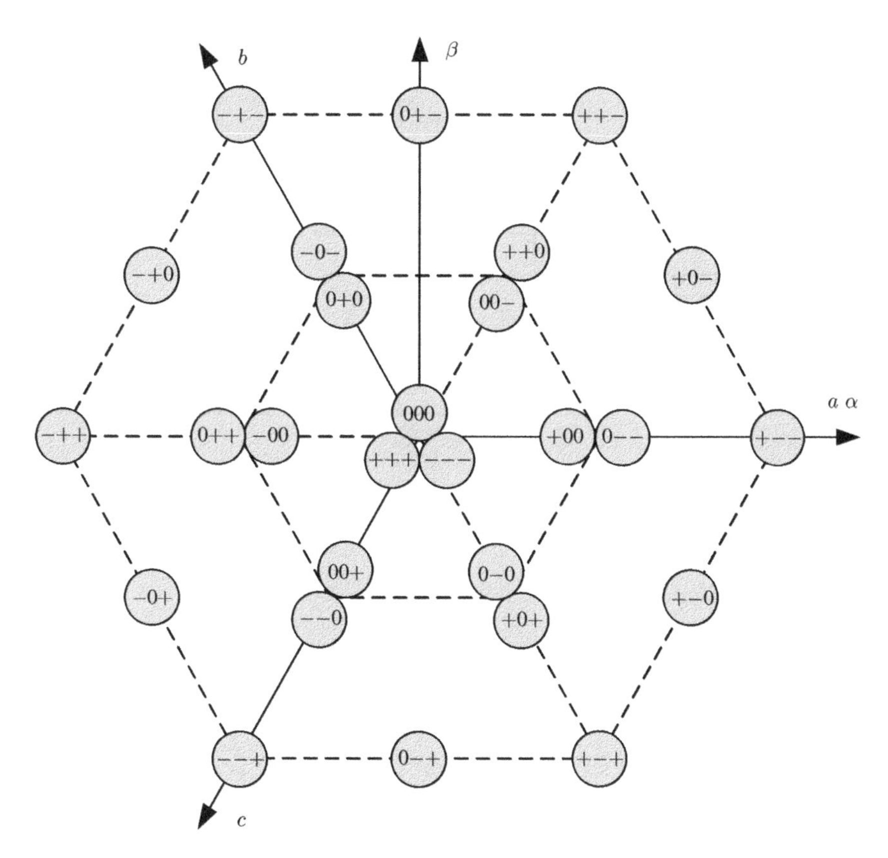

Figure 2.19 Voltage vectors produced by a three-level inverter. The voltage vectors are shown in the $\alpha\beta$ plane along with the corresponding values of the switch positions $\boldsymbol{u}_{abc}$ (where "+" refers to "1" and "−" to "−1")

2.4.1.3 Current Paths

Consider one of the three phase legs with the single-phase switch position u_x. Assume the phase current i_x to be positive, that is, i_x is directed out of the inverter into the load.

- For $u_x = 1$, the upper two active switches are on while the two lower switches are off. The positive phase current flows from the upper dc-link rail through the upper two active switches to the phase terminal, as shown in Fig. 2.20(a).
- For $u_x = 0$, the two middle switches are on while the top and bottom switches are off. The positive current flows from the neutral point through the upper clamping diode and the center top switch to the phase terminal, as can be seen in Fig. 2.20(b).
- For $u_x = -1$, the two lower switches are on while the upper ones are off. The positive current flows from the lower dc-link rail through the lower freewheeling diodes to the phase terminal (see Fig. 2.20(c)).

The current paths for negative phase currents can also be easily derived.

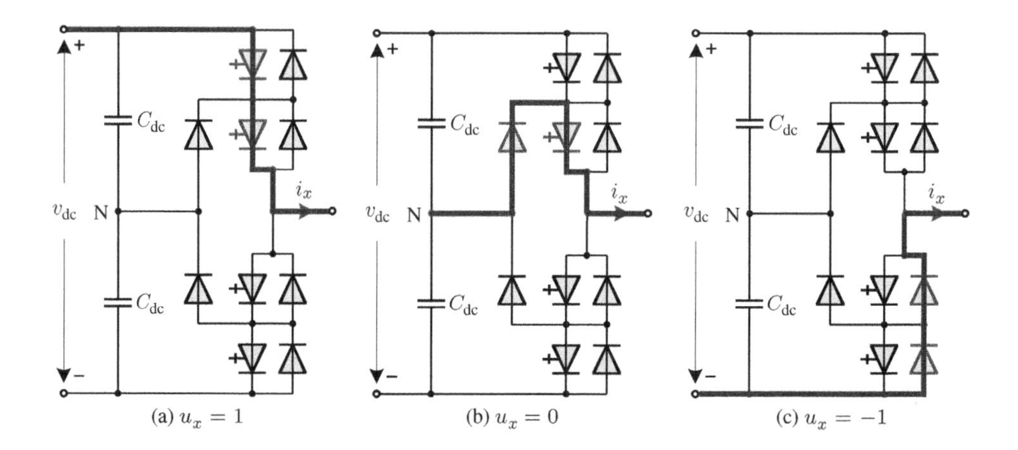

Figure 2.20 Current paths in an NPC phase leg as a function of the switch position u_x, with $x \in \{a, b, c\}$, for a positive phase current i_x

2.4.1.4 Neutral Point Potential

The potential of the neutral point (2.73) evolves as a function of the neutral point current i_n according to

$$\frac{\mathrm{d}v_n}{\mathrm{d}t} = -\frac{1}{2C_{\mathrm{dc}}} i_n. \tag{2.78}$$

Specifically, i_n is the weighted sum of the phase currents i_a, i_b, and i_c, for the phases whose corresponding switch position is zero:

$$i_n = (1 - |u_a|)i_a + (1 - |u_b|)i_b + (1 - |u_c|)i_c. \tag{2.79}$$

As a result, neutral point current is drawn by a phase when its switch position is zero. For a three-phase load, whose star point is not connected, $i_a + i_b + i_c = 0$ holds, and

$$\frac{\mathrm{d}v_n}{\mathrm{d}t} = \frac{1}{2C_{\mathrm{dc}}} |\boldsymbol{u}_{abc}|^T \, \boldsymbol{i}_{abc} \tag{2.80}$$

follows directly, where $\boldsymbol{i}_{abc} = [i_a \; i_b \; i_c]^T$ is the three-phase current and $|\boldsymbol{u}_{abc}| = [|u_a| \; |u_b| \; |u_c|]^T$ is the componentwise absolute value of the inverter switch positions. For more details about the nature of the neutral point potential and methods employed to tackle the related balancing problem, the reader is referred to [11] and [12].

Next, we translate the differential equation (2.78) into the pu system. For this, we use the base voltage V_B, the base current I_B, and the base frequency ω_B as defined in (2.42) and (2.43), respectively. As before, we also define the base impedance $Z_B = V_B/I_B$ and the base capacitance $C_B = 1/(\omega_B Z_B)$ (see also Table 2.1). Dividing the left-hand side of (2.78) by $\omega_B V_B$ and the right-hand side by $\omega_B Z_B I_B$ yields

$$\frac{\mathrm{d}v_n'}{\mathrm{d}t'} = -\frac{1}{2X_{\mathrm{dc}}'} i_n', \tag{2.81}$$

where the superscript $'$ denotes pu quantities, as stated previously in Sect. 2.2.3. In (2.81), we have also normalized the time axis by using $t' = \omega_B t$ and introduced the pu equivalence of the capacitor[1]

$$X'_{\mathrm{dc}} = \frac{C_{\mathrm{dc}}}{C_B}. \tag{2.82}$$

Accordingly, (2.80) turns into

$$\frac{\mathrm{d}v'_n}{\mathrm{d}t'} = \frac{1}{2X'_{\mathrm{dc}}} |\boldsymbol{u}_{abc}|^T \, \boldsymbol{i}'_{abc}. \tag{2.83}$$

As for the induction machine in Sect. 2.2.3, we will drop the superscript $'$ in the remainder of the book to simplify the notation.

2.4.1.5 Switching Transitions

It is a characteristic of the NPC topology that for each switching transition one active switch is turned off and another one is turned on. To avoid a potential short-circuit over one of the dc-link capacitors, a time delay is introduced between the turn-off and the turn-on transition. This time delay is commonly referred to as the *interlocking time*.

By inspecting Table 2.3, it is straightforward to identify the switches that are turned on and off. Table 2.4 summarizes the switching transitions for one phase leg. The active switch that is turned on and the one that is turned off are stated as a function of the single-phase switching transition from the switch position $u_x(k-1)$ to $u_x(k)$.

2.4.1.6 Switching Constraints

Switching a phase leg from 1 to -1 runs the risk of turning on all four active switches in the phase leg—albeit for a short time (see also Table 2.3). This would lead to a short-circuit (or shoot-through) between the upper and the lower dc-link rails. Even more important is that the dynamic voltage sharing across the two blocking switches would not be guaranteed, potentially creating an overvoltage over one of the inner active switches. For these reasons, switching between 1 and -1 is prohibited.

Table 2.4 Switching transitions in the NPC phase leg x, with $x \in \{a, b, c\}$

Switching transition	Switch with the *on* transition	Switch with the *off* transition
$0 \to 1$	S_{x1}	S_{x3}
$1 \to 0$	S_{x3}	S_{x1}
$0 \to -1$	S_{x4}	S_{x2}
$-1 \to 0$	S_{x2}	S_{x4}

[1] Note that the notation of X'_{dc} is slightly misleading, because X'_{dc} is not a reactance, but rather its inverse. Specifically, we have $\frac{1}{X'_{\mathrm{dc}}} = \frac{X_{\mathrm{dc}}}{Z_B}$, with the SI reactance $X_{\mathrm{dc}} = \frac{1}{\omega_B C_{\mathrm{dc}}}$.

This restriction can be described by the constraint

$$\max_{x} |u_x(t) - u_x(t - \mathrm{d}t)| \leq 1, \tag{2.84}$$

where t is the switching instant and $\mathrm{d}t$ is an infinitesimally small time step.

Therefore, switching is allowed only by one step up or down. Switching in a phase leg between 1 and -1, and vice versa, is only possible via an intermediate zero switch position. When doing so, an inverter-specific minimum on-time needs to be adhered to. For MV inverters, these minimum on-times are in the range of several tens of microseconds.

Typically, six di/dt snubbers are used in an NPC inverter with one snubber per upper and lower half of each phase leg. Some manufacturers, however, use only two such snubbers for such an inverter—one in the upper half and another one in the lower half of the inverter. This further limits the set of admissible switching transitions. Specifically, simultaneous switching is allowed in at most two phase legs and only if the switching occurs in opposite inverter halves.

The admissible switching transitions are illustrated in Fig. 2.21. As can be seen, from $[1\ 1\ 1]^T$, for example, switching is possible only to $[0\ 1\ 1]^T$, $[1\ 0\ 1]^T$, or $[1\ 1\ 0]^T$ and not to any of the other 23 switch positions.

2.4.1.7 Switching Losses

Consider one phase leg with the phase current i_x and the single-phase switch position $u_x \in \{-1, 0, 1\}$. During switching transitions, the current is commutated by turning semiconductor devices on and off. As was shown in Sect. 2.3, this gives rise to switching energy losses. The resulting switching losses per switching transition can be derived by inspecting the current paths in Fig. 2.20. As the commutation depends on the polarity of the phase current, the cases with positive and negative phase currents need to be treated separately.

Table 2.5 summarizes the switching energy losses, where the indices 1–4 refer to the pairs of active semiconductor switches and freewheeling diodes (from top to bottom), while the indices 5 and 6 refer to the clamping diodes (from top to bottom). Recall that e_{on} (e_{off}) refers to the turn-on (turn-off) energy losses of active semiconductor switches. Assuming GCTs, the switching losses are given in (2.65) and (2.66), respectively. Accordingly, e_{rr} denotes the reverse recovery losses of diodes (see also (2.69)).

Note that, when commutating a positive phase current from $u_x = 0$ to -1, the voltage over the upper clamping diode remains zero. As a result, this clamping diode incurs no reverse recovery losses. Similarly, when switching from $u_x = -1$ to 0 for a positive phase current, the third freewheeling diode experiences no reverse recovery losses. The same applies to the lower clamping diode and the second freewheeling diode, respectively, when switching from 0 to 1, and vice versa for negative phase currents. As a result, two types of switching transitions exist. In case the current is commutated from a diode to a GCT, reverse recovery and turn-on losses arise, while when the current is commutated from a GCT to a diode, only turn-off losses are generated.

To simplify the computation of the switching losses, one typically assumes the total dc-link voltage to be constant and the fluctuations of the neutral point potential to be small. Both assumptions are usually well justified. As a result, for an NPC inverter, the blocking voltage of each semiconductor is one-half the total dc-link voltage, and the switching losses depend

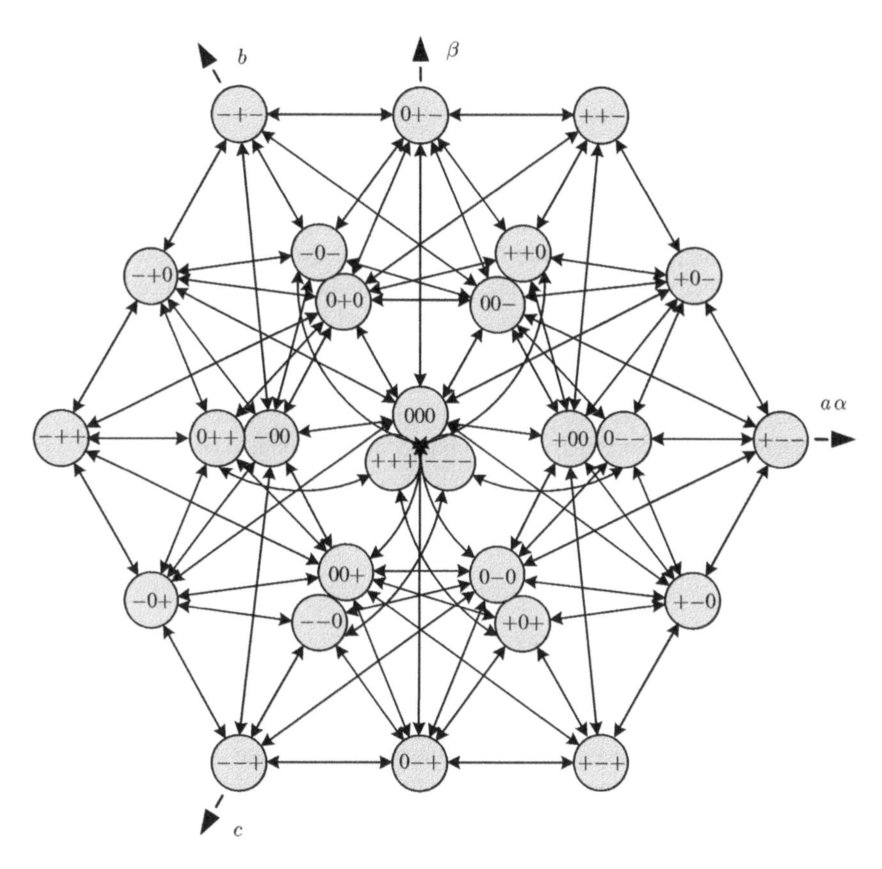

Figure 2.21 Voltage vectors produced by a three-level inverter. The voltage vectors are shown in the $\alpha\beta$ plane along with the corresponding values of the switch positions $\boldsymbol{u}_{abc}$ (where "+" refers to "1" and "−" to "−1"). The switching transitions that are allowed when using two di/dt snubbers are indicated by bidirectional arrows

Table 2.5 Switching energy losses in an NPC phase leg

Polarity of the phase current i_x	Switching transition	Switching energy losses
> 0	$0 \rightarrow 1$	$e_{1,\text{on}} + e_{5,\text{rr}}$
	$1 \rightarrow 0$	$e_{1,\text{off}}$
	$0 \rightarrow -1$	$e_{2,\text{off}}$
	$-1 \rightarrow 0$	$e_{2,\text{on}} + e_{4,\text{rr}}$
< 0	$0 \rightarrow 1$	$e_{3,\text{off}}$
	$1 \rightarrow 0$	$e_{1,\text{rr}} + e_{3,\text{on}}$
	$0 \rightarrow -1$	$e_{4,\text{on}} + e_{6,\text{rr}}$
	$-1 \rightarrow 0$	$e_{4,\text{off}}$

only on the commutated current. As a last step, the switching power losses are obtained by averaging the switching energy losses over time.

2.4.1.8 Conduction Losses

Similar to the switching losses, the conduction losses also depend on the phase current and the switch position, as summarized in Table 2.6 for one phase leg. As previously, the indices 1–4 refer to the semiconductor switches and their freewheeling diodes. The indices 5 and 6 denote the clamping diodes. The conduction losses of GCTs and power diodes, p_{GCT} and p_{diode}, are given in (2.68) and (2.71), respectively.

With the dc-link voltage being effectively constant, the conduction losses depend only on the phase current, which is the sum of the ripple current and its fundamental component. As the ripple current is small compared to the fundamental component of the current (typically in the range of 10% for a three-level inverter), the conduction losses can be considered to be independent of the switching pattern. Hence, when formulating model predictive control problems, the conduction losses are usually not included in the cost function.

2.4.2 *Five-Level ANPC Inverter*

The active neutral-point-clamped (ANPC) topology is a five-level inverter that has been proposed in 2005 [13] and introduced as a commercial product in 2010 [14]. With power ratings of 1 and 2 MVA, this inverter addresses the low power range of the MV drives market. Using high-voltage IGBTs, output voltages of up to 6.9 kV can be achieved. At the same time, very low harmonic distortions in the stator currents result, along with acceptable dv/dt and common-mode voltages. This makes the ANPC inverter particularly suitable for the retrofit market, in which direct online machines are upgraded to VSDs. Four-quadrant operation is achieved by using an active front end (AFE), which is connected via an optional transformer to the grid.

The five-level ANPC topology extends the classic three-level NPC inverter [10] in two ways. The NPC diodes are replaced by active switches as in [15], and floating phase capacitors are added to each phase, similar to a flying capacitor (FC) inverter [16]. This innovative topology combines the advantages of the reliable and conceptually simple NPC inverter with the

Table 2.6 Conduction power losses in an NPC phase leg

Polarity of the phase current i_x	Switch position	Conduction power losses
> 0	1	$p_{1,\text{GCT}} + p_{2,\text{GCT}}$
	0	$p_{2,\text{GCT}} + p_{5,\text{diode}}$
	-1	$p_{3,\text{diode}} + p_{4,\text{diode}}$
< 0	1	$p_{1,\text{diode}} + p_{2,\text{diode}}$
	0	$p_{3,\text{GCT}} + p_{6,\text{diode}}$
	-1	$p_{3,\text{GCT}} + p_{4,\text{GCT}}$

versatility of the flying capacitor inverter. The control and modulation problem is, however, significantly more complex than for the NPC inverter. Balancing the four internal inverter voltages, specifically the neutral point potential and the three phase capacitors, around their references while maintaining a low switching frequency is challenging, particularly when the phase capacitors are small [13].

In the following, we summarize the five-level ANPC topology, its switching restrictions, commutation paths, and the mathematical model of the internal voltages.

2.4.2.1 Topology and Phase Levels

Consider the five-level ANPC inverter depicted in Fig. 2.22. In phase leg x, with $x \in \{a, b, c\}$, the switches S_{x1}–S_{x4} consist of two series-connected IGBTs, while the switches S_{x5}–S_{x8} are single IGBTs. Thus each phase consists of 12 IGBTs. We refer to the switches S_{x1}–S_{x4} as the ANPC switches, and to the switches S_{x5}–S_{x8} as the FC switches.

The dc-link is divided into an upper and a lower half with the two dc-link capacitors C_{dc}. The potential

$$v_n = \frac{1}{2}(v_{\mathrm{dc,lo}} - v_{\mathrm{dc,up}}) \tag{2.85}$$

of the neutral point N floats, with $v_{\mathrm{dc,lo}}$ and $v_{\mathrm{dc,up}}$ denoting the voltages over the lower and the upper dc-link half, respectively. The inverter's total (instantaneous) dc-link voltage is $v_{\mathrm{dc}} = v_{\mathrm{dc,lo}} + v_{\mathrm{dc,up}}$. Neglecting the phase capacitors, this inverter effectively resembles a three-level ANPC inverter with series-connected IGBTs, producing at each phase the three voltage levels $\{-\frac{v_{\mathrm{dc}}}{2}, 0, \frac{v_{\mathrm{dc}}}{2}\}$.

The available number of phase voltage levels is increased to 5 by adding to each phase an FC C_{ph}, which is placed between the outer pairs of the existing series-connected switches

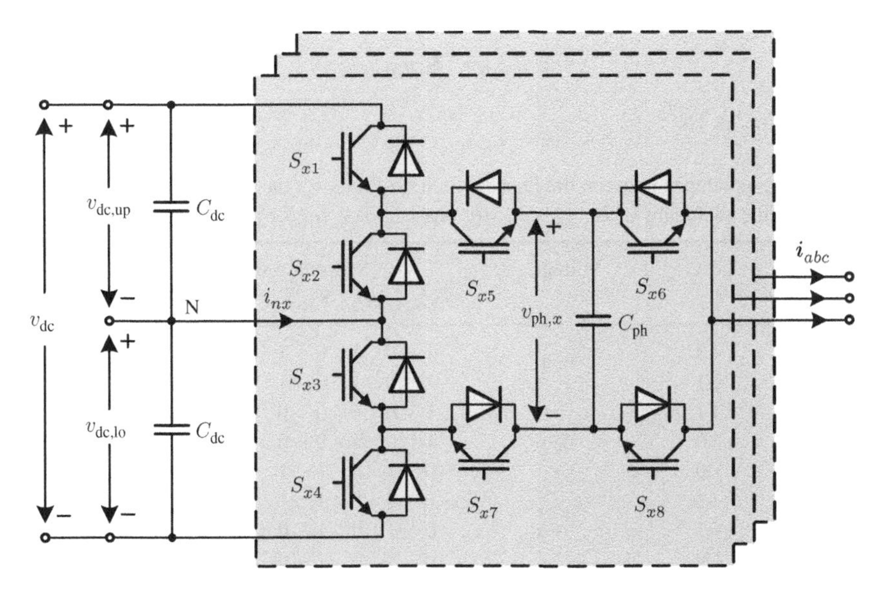

Figure 2.22 Equivalent representation of the five-level active neutral-point-clamped (ANPC) voltage source inverter

S_{x5}–S_{x8}. Let the voltages across the phase capacitors be denoted by $v_{\text{ph},x}$, with $x \in \{a, b, c\}$. The phase capacitor voltages are maintained at half the voltage levels of the individual dc-link capacitors, that is, at $v_{\text{ph},x} = 0.25 v_{\text{dc}}$. This adds the two additional voltage levels $\{-\frac{v_{\text{dc}}}{4}, \frac{v_{\text{dc}}}{4}\}$ and ensures that each IGBT can be rated for the same voltage blocking capability. As a result, at each phase, the inverter produces the five voltage levels $\{-\frac{v_{\text{dc}}}{2}, -\frac{v_{\text{dc}}}{4}, 0, \frac{v_{\text{dc}}}{4}, \frac{v_{\text{dc}}}{2}\}$. These voltages can be described by the integer variables u_a, u_b, $u_c \in \{-2, -1, 0, 1, 2\}$, which we refer to as *phase levels*.

2.4.2.2 Switch Positions and Voltage Vectors

The phase levels −1, 0, and 1 can each be synthesized by two different *switch positions*, described by the integer variables s_a, s_b, $s_c \in \{0, 1, \dots, 7\}$. The phase level $u_x = 1$, for example, with $x \in \{a, b, c\}$, can be generated either with the FC switch configuration $S_{x5} = 1$, $S_{x6} = 0$, $S_{x7} = 0$, and $S_{x8} = 1$, or with $S_{x5} = 0$, $S_{x6} = 1$, $S_{x7} = 1$, and $S_{x8} = 0$. The ANPC switches are in both cases set to $S_{x1} = 1$, $S_{x2} = 0$, $S_{x3} = 1$, and $S_{x4} = 0$. Similarly, two switch positions are available to synthesize the phase level $u_x = -1$, as summarized in Table 2.7. Each pair of switch positions produces effectively the same voltage at the phase terminal. This redundancy can be used to regulate the phase capacitor voltages when the phase level is $u_x = \pm 1$. However, these pairs affect the neutral point potential differently, adding significant complexity to the system to be handled by the control scheme.

The phase voltage is defined with respect to the dc-link midpoint N. It is approximately

$$v_x \approx \frac{v_{\text{dc}}}{4} u_x, \tag{2.86}$$

with $x \in \{a, b, c\}$, provided that the fluctuations on the neutral point potential and the phase capacitor voltages are small. The precise phase voltage depends on the switch position s_x, as detailed in Table 2.7. The three-phase voltage at the inverter terminals is given by

$$\boldsymbol{v}_{\alpha\beta 0} = \boldsymbol{K} \boldsymbol{v}_{abc}, \tag{2.87}$$

with $\boldsymbol{v}_{\alpha\beta 0} = [v_\alpha \ v_\beta \ v_0]^T$.

Table 2.7 Correspondence between the phase switch positions s_x, the phase levels u_x, the phase voltages v_x, and the switching states S_{x1}–S_{x8}, for phase x, $x \in \{a, b, c\}$

Switch position	Level	Voltage	Switching state								Effect on	
s_x	u_x	v_x	S_{x1}	S_{x2}	S_{x3}	S_{x4}	S_{x5}	S_{x6}	S_{x7}	S_{x8}	$v_{\text{ph},x}$	v_n
7	+2	$v_{\text{dc,up}}$	1	0	1	0	1	1	0	0	0	0
6	+1	$v_{\text{dc,up}} - v_{\text{ph},x}$	1	0	1	0	1	0	0	1	i_x	0
5	+1	$v_{\text{ph},x}$	1	0	1	0	0	1	1	0	$-i_x$	$-i_x$
4	0	0	1	0	1	0	0	0	1	1	0	$-i_x$
3	0	0	0	1	0	1	1	1	0	0	0	$-i_x$
2	−1	$-v_{\text{ph},x}$	0	1	0	1	1	0	0	1	i_x	$-i_x$
1	−1	$-v_{\text{dc,lo}} + v_{\text{ph},x}$	0	1	0	1	0	1	1	0	$-i_x$	0
0	−2	$-v_{\text{dc,lo}}$	0	1	0	1	0	0	1	1	0	0

The effect on the phase capacitor voltage $v_{\text{ph},x}$ and on the neutral point potential v_n is shown on the right hand side, as a function of the phase current i_x.

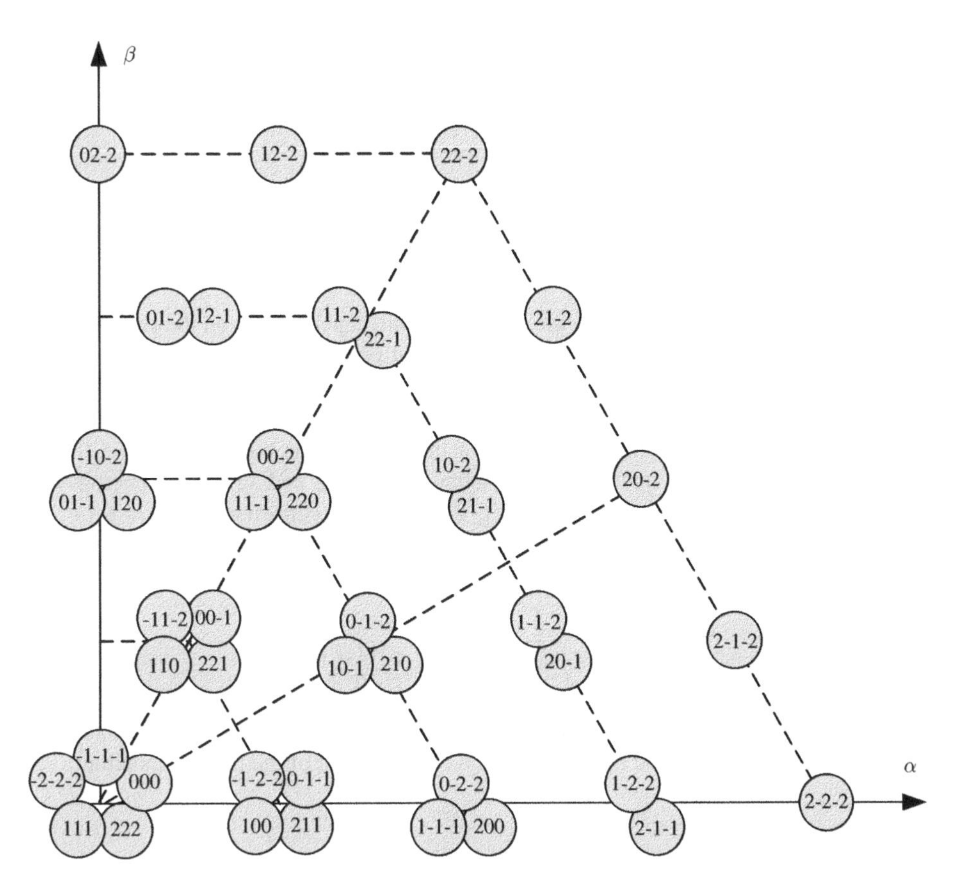

Figure 2.23 Voltage vectors produced by a five-level inverter. The voltage vectors are shown in the $\alpha\beta$-plane along with the corresponding values of the phase levels $\boldsymbol{u}_{abc}$

Neglecting the voltage fluctuations in the dc-link and the phase capacitors, the inverter produces 61 different voltage vectors. These can be synthesized by $5^3 = 125$ different phase levels $\boldsymbol{u}_{abc} = [u_a \; u_b \; u_c]^T$, which in turn are established based on $8^3 = 512$ distinct switch positions $\boldsymbol{s}_{abc} = [s_a \; s_b \; s_c]^T$. The 0-vector $\boldsymbol{v}_{\alpha\beta} = [0 \; 0]^T$, for example, can be synthesized by 26 different switch positions $\boldsymbol{s}$. The voltage vectors and their corresponding phase levels are shown in Fig. 2.23. Only the first quadrant is shown.

2.4.2.3 Dynamics of the Internal Inverter Voltages

The evolution of the capacitor voltage in phase x, with $x \in \{a, b, c\}$, is described by the differential equation

$$\frac{\mathrm{d}v_{\mathrm{ph},x}}{\mathrm{d}t} = \frac{1}{C_{\mathrm{ph}}} \begin{cases} i_x, & \text{if } s_x \in \{2, 6\} \\ -i_x, & \text{if } s_x \in \{1, 5\} \\ 0, & \text{if } s_x \in \{0, 3, 4, 7\}\,, \end{cases} \tag{2.88}$$

which involves the product of the phase capacitance and phase current. The dynamics of the neutral point potential is given by

$$\frac{\mathrm{d}v_n}{\mathrm{d}t} = -\frac{1}{2C_{\mathrm{dc}}}\,(i_{na} + i_{nb} + i_{nc}), \tag{2.89}$$

with i_{nx} denoting the current drawn from the neutral point:

$$i_{nx} = \begin{cases} i_x, & \text{if } s_x \in \{2,3,4,5\} \\ 0, & \text{if } s_x \in \{0,1,6,7\} \,. \end{cases} \tag{2.90}$$

Note that the capacitor voltage of phase a, for example, depends only on the switch position and phase current of phase a, whereas the neutral point potential depends on all three switch positions and all three phase currents.

As previously done for the NPC inverter, we translate in the next step the differential equations of the inverter into the pu system. For this, we use the base voltage V_B, the base current I_B, and the base frequency ω_B as defined in (2.42) and (2.43), respectively. We also define the base impedance $Z_B = V_B/I_B$ and the base capacitance $C_B = 1/(\omega_B Z_B)$ (see also Table 2.1). Following the procedure in Sect. 2.4.1, the differential equation of the capacitor voltages (2.88) in the pu system is given by

$$\frac{\mathrm{d}v'_{\mathrm{ph},x}}{\mathrm{d}t'} = \frac{1}{X'_{\mathrm{ph}}} \begin{cases} i'_x, & \text{if } s_x \in \{2,6\} \\ -i'_x, & \text{if } s_x \in \{1,5\} \\ 0, & \text{if } s_x \in \{0,3,4,7\} \,, \end{cases} \tag{2.91}$$

where the superscript $'$ denotes pu quantities. In this equation, we have also normalized the time axis by using $t' = \omega_B t$ and introduced the pu equivalence of the phase capacitor[2]

$$X'_{\mathrm{ph}} = \frac{C_{\mathrm{ph}}}{C_B}. \tag{2.92}$$

Accordingly, (2.89) turns into

$$\frac{\mathrm{d}v'_n}{\mathrm{d}t'} = -\frac{1}{2X_{\mathrm{dc}}}\,(i'_{na} + i'_{nb} + i'_{nc}), \tag{2.93}$$

with

$$X'_{\mathrm{dc}} = \frac{C_{\mathrm{dc}}}{C_B}. \tag{2.94}$$

As mentioned before, we will drop the superscript $'$ in the remainder of the book to simplify the notation.

2.4.2.4 Switching Constraints

A number of switching restrictions are present in the five-level ANPC topology, both at the single-phase and three-phase levels. The allowed single-phase switching transitions are shown in Fig. 2.24. Only switching by one voltage level up or down is possible. To rule out the possibility of voltage glitches, switching from $s_x = 2$ to $s_x = 4$ and from $s_x = 5$ to $s_x = 3$ are not allowed. The minimum on-time of an IGBT is 30 μs. We will typically adopt the sampling interval $T_s = 25\,\mu s$. In case the switching is restricted to the sampling instants, which is the

[2] Note that, as before for the NPC inverter, the notation of X'_{ph} is slightly misleading, because X'_{ph} is not a reactance, but rather its inverse. Specifically, we have $\frac{1}{X'_{\mathrm{ph}}} = \frac{X_{\mathrm{ph}}}{Z_B}$, with the SI reactance $X_{\mathrm{ph}} = \frac{1}{\omega_B C_{\mathrm{ph}}}$.

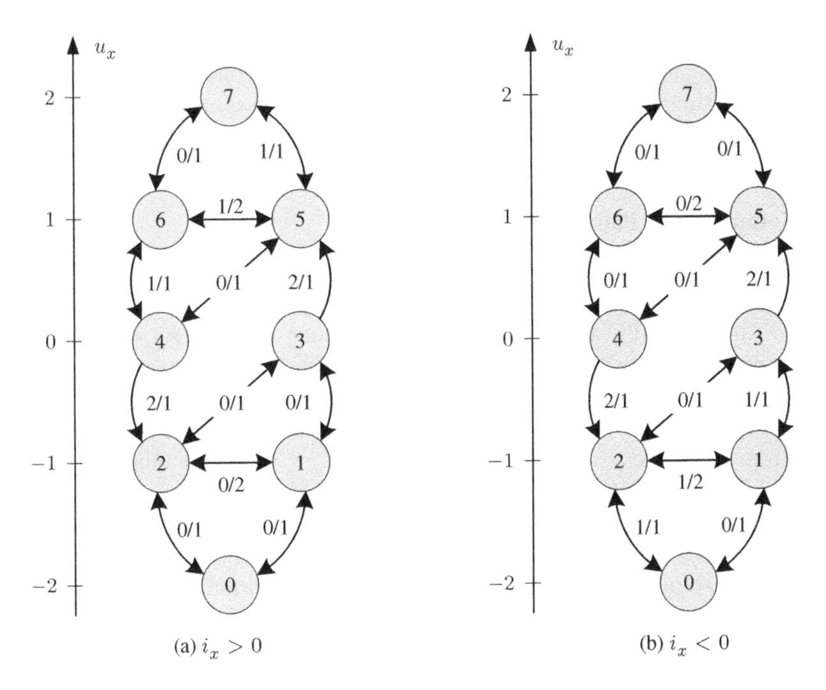

Figure 2.24 Allowed per-phase switching transitions between the single-phase switch positions $s_x \in \{0, 1, \ldots 7\}$, $x \in \{a, b, c\}$, along with the number of *on* transitions of the IGBTs in the ANPC and FC part, respectively. The allowed switching transitions depend on the sign of the phase current i_x. The corresponding phase levels u_x are shown on the left-hand side

case for some of the control and modulation schemes discussed in this book, this effectively leads to a minimum on–time of 50 μs.

Owing to the fact that the inverter uses only two di/dt clamps (or snubbers)—one in the upper dc-link half and another one in the lower half, restrictions on the allowed three-phase switching transitions arise. Switching transitions that lead to transients in the clamp diode can be divided into two categories: transitions that turn the clamp diode on, which we refer to as *on* transitions, and switching transitions that turn the clamp off, called *off* transitions, which result in the reverse recovery effect in the clamp diode. Table 2.8 summarizes the switching transitions that turn the clamps on and off.

Table 2.8 Transitions between single-phase switch positions s_x that turn a di/dt clamp on or off, depending on the sign of the phase current i_x

	Phase current	Transitions $s_x \to s_x$ that turn the clamp on	Transitions $s_x \to s_x$ that turn the clamp off
Upper clamp	$i_x > 0$	$6 \to 4, 6 \to 5, 7 \to 5$	$4 \to 6, 5 \to 6, 5 \to 7$
Upper clamp	$i_x < 0$	$4 \to 6, 5 \to 6, 5 \to 7$	$6 \to 4, 6 \to 5, 7 \to 5$
Lower clamp	$i_x > 0$	$2 \to 0, 2 \to 1, 3 \to 1$	$0 \to 2, 1 \to 2, 1 \to 3$
Lower clamp	$i_x < 0$	$0 \to 2, 1 \to 2, 1 \to 3$	$2 \to 0, 2 \to 1, 3 \to 1$

The transitions in the upper (lower) half of the table affect the upper (lower) clamp.

After a transition that turns the upper (lower) clamp on, at least 50 μs has to pass before the upper (lower) clamp may be turned off. From this requirement, one can deduce the following. Simultaneous *on* transitions are allowed, while simultaneous *on* and *off* transitions are not allowed. An *on* transition is allowed a few microseconds after an *on* or an *off* transition.

2.4.2.5 Commutation Paths

The commutation paths for this topology are rather complex. Figure 2.24 summarizes the number of *on* transitions per switching transition, distinguishing between the *on* transitions of the IGBTs in the ANPC and in the FC part. Switching between $s_x = 6$ and $s_x = 7$, for example, incurs no *on* transition in the ANPC part, but one in the FC part.

It is clear that the number of *on* transitions always equals the number of *off* transitions. It is apparent from Table 2.7 that two *on* and two *off* transitions occur in the ANPC part when switching from $s_x = 4$ to $s_x = 2$ and from $s_x = 3$ to $s_x = 5$. On the other hand, one would expect that no IGBT is turned on or off in the ANPC part when the transitions occur within the group $s_x \in \{0, 1, 2, 3\}$ or $s_x \in \{4, 5, 6, 7\}$. However, in order to balance the switching load and to shift some switching losses from the FC to the ANPC part, switching in the ANPC part does occur also in these cases, depending on the phase current. These additional ANPC switchings shift the commutation of the current from the FC to the ANPC part.

From Table 2.7, it is also clear that in the FC part, for each transition, one IGBT is turned on (and another one is turned off), except for transitions occurring between $s_x = 1$ and $s_x = 2$, as well as between $s_x = 5$ and $s_x = 6$, when two devices are turned on and off.

2.5 Case Studies

Throughout this book, four case studies for industrial power electronic systems are considered. These case studies are introduced and summarized in this section. They include the NPC inverter and a five-level active NPC inverter. In three cases, an MV VSD system is considered with an induction machine. The fourth case study relates to a grid-connected NPC converter.

2.5.1 NPC Inverter Drive System

As a case study, consider a three-level NPC voltage source inverter driving an induction machine, as shown in Fig. 2.25. A 3.3 kV, 50 Hz squirrel-cage induction machine rated at 2 MVA is used as an example for a commonly used MV induction machine. The rated values of the machine are summarized in Table 2.9.

The pu system is established using the base quantities $V_B = \sqrt{2/3}V_R = 2694$ V, $I_B = \sqrt{2}I_R = 503.5$ A, and $\omega_B = \omega_{sR} = 2\pi 50$ rad/s. The machine and inverter parameters are provided in Table 2.10 as SI quantities and pu values, along with their respective symbols. V_{dc} denotes the nominal dc-link voltage, in contrast to the instantaneous and fluctuating voltage v_{dc}. Note that the value of the dc-link capacitance refers to one half of the dc-link (upper or lower half). The model of the induction machine is derived and summarized in Sect. 2.2. The NPC inverter is described in detail in Sect. 2.4.1.

The semiconductors used are the 35L4510 4.5 kV 4 kA IGCT and the 10H4520 fast recovery diode, which are both manufactured by ABB. Recall that the turn-off and turn-on losses of the GCTs are proportional to the product of the anode–cathode voltage v_T with the anode current

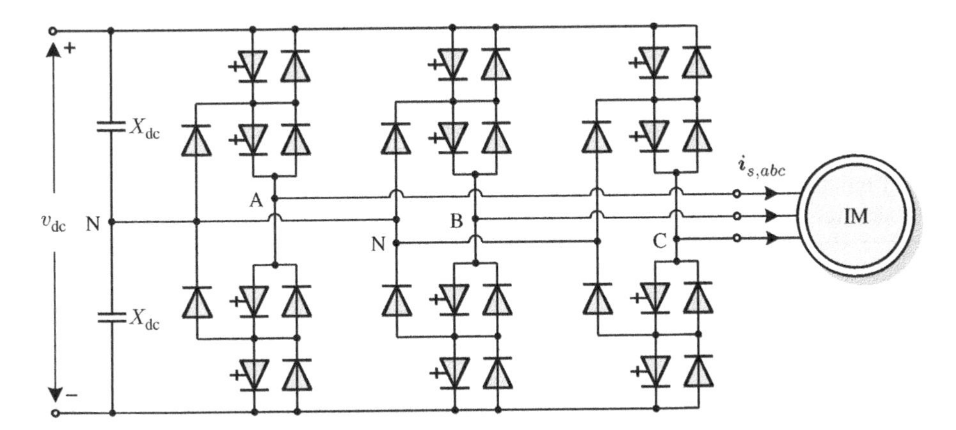

Figure 2.25 Three-level neutral-point-clamped (NPC) voltage source inverter driving an induction machine (IM)

Table 2.9 Rated values of the induction machine

Parameter	Symbol	SI value
Voltage	V_R	3300 V
Current	I_R	356 A
Real power	P_R	1.587 MW
Apparent power	S_R	2.035 MVA
Angular stator frequency	ω_{sR}	$2\pi 50$ rad/s
Rotational speed	ω_{mR}	596 rpm
Air-gap torque	T_R	26.2 kNm

Table 2.10 Drive parameters in the SI (left) and per unit system (right) of the three-level NPC inverter drive system

Parameter	SI symbol	SI value	pu symbol	pu value
Stator resistance	R_s	57.61 mΩ	R_s	0.0108
Rotor resistance	R_r	48.89 mΩ	R_r	0.0091
Stator leakage inductance	L_{ls}	2.544 mH	X_{ls}	0.1493
Rotor leakage inductance	L_{lr}	1.881 mH	X_{lr}	0.1104
Main inductance	L_m	40.01 mH	X_m	2.349
Number of pole pairs	p	5		
dc-link voltage	V_{dc}	5.2 kV	V_{dc}	1.930
dc-link capacitance	C_{dc}	7 mF	X_{dc}	11.77

Table 2.11 Switching losses at the maximum rated values and the turn-off, turn-on, and reverse recovery coefficients

	Losses at maximum rated values	Loss coefficient
GCT turn-off	$e_{\text{off}} = 19.5\,\text{J}$ at $v_T = 2.8\,\text{kV}$ and $i_T = 4\,\text{kA}$	$c_{\text{off}} = 2.362\,\text{s}$
GCT turn-on	$e_{\text{on}} = 1.5\,\text{J}$ at $v_T = 2.8\,\text{kV}$ and $i_T = 4\,\text{kA}$	$c_{\text{on}} = 0.182\,\text{s}$
Diode reverse recovery	$e_{\text{rr}} = 7.2\,\text{J}$ at $v_T = 2.8\,\text{kV}$ and $i_T = 4\,\text{kA}$	$c_{\text{rr}} = 3.058\,\text{s}$

i_T (see Sect. 2.3.1). In an NPC converter, v_T is nearly constant and equal to half the total dc-link voltage. For the turn-off losses, for example, using (2.65), we obtain

$$e_{\text{off}} = c_{\text{off}} \frac{v_{\text{dc}}}{2} i_x, \tag{2.95}$$

with i_x being the commutated phase current, with $x \in \{a, b, c\}$. Note that i_x is always non-negative when being commutated. According to the GCT's data sheet, at the maximum rated values of $v_T = 2.8\,\text{kV}$ and $i_T = 4\,\text{kA}$, the typical turn-off and turn-on losses are $e_{\text{off}} = 19.5\,\text{J}$ and $e_{\text{on}} = 1.5\,\text{J}$, respectively, at the nominal operating temperature of 125°C. Assuming that the voltages and currents in (2.95) are given in the pu system and that the losses are in joules, the coefficients c_{off} and c_{on} can easily be derived, as summarized in Table 2.11. To simplify the computations, the nominal dc-link voltage V_{dc} is usually assumed. The term $c_{\text{off}} \frac{V_{\text{dc}}}{2}$ can then be replaced by one coefficient.

The diode's reverse recovery losses are nonlinear in the commutated current. Following the procedure for the calculation of the GCT switching losses and rewriting (2.69), the reverse recovery losses in pu are given by

$$e_{\text{rr}} = c_{\text{rr}} \frac{v_{\text{dc}}}{2} f_{\text{rr}}(i_x). \tag{2.96}$$

The data sheet of the fast recovery diode states that the reverse recovery losses are $e_{\text{rr}} = 7.2\,\text{J}$ at the maximum rated values of $v_T = 2.8\,\text{kV}$ and $i_T = 4\,\text{kA}$, assuming a di/dt of -400 A/μs. Recall that $f_{\text{rr}}(.)$ is a nonlinear function of the phase current in pu, which can be reconstructed from the data sheet. As shown in Fig. 2.26, we define $f_{\text{rr}}(.)$ such that it is 1 at the base current I_B. With this definition at hand, the coefficient c_{rr} can easily be computed (see Table 2.11).

Similar to the switching losses, the conduction losses also depend on the dc-link voltage and the phase current. The dc-link voltage is constant despite the neutral point fluctuations. The phase current is the sum of the ripple current and the fundamental component, which in turn depends only on the operating point given by the torque and the speed, but not on the switching pattern. As the ripple is small compared to the fundamental current (typically in the range of 10% for an NPC inverter), the conduction losses can be considered to be invariant under the control and modulation scheme used. Therefore, they are not addressed in the model predictive control problems formulated in this book and are not further considered here.

2.5.2 NPC Inverter Drive System with Snubber Restrictions

This second case study is the same as the previous one, except for the following differences: Only two di/dt snubbers are used in the inverter (one per converter half). As a result, only the switching transitions shown in Fig. 2.21 are allowed. The converter is fed either by a 12-pulse diode front end or by an AFE. In the case of the diode front end, the nominal dc-link voltage

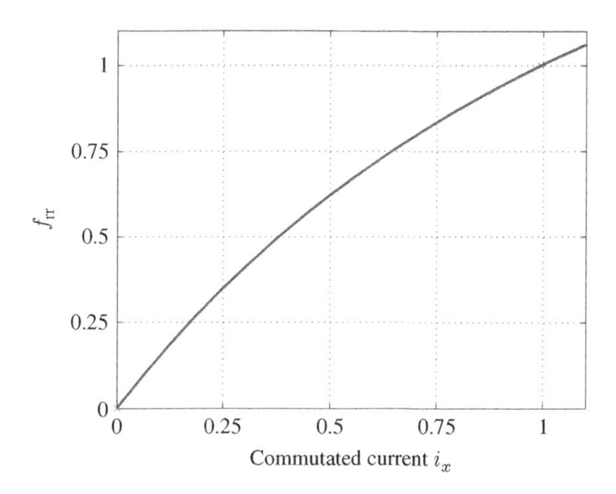

Figure 2.26 Nonlinear function $f_{rr}(i_x)$ of the commutated diode current in per unit

is $V_{dc} = 4294$ V, while for the AFE it is $V_{dc} = 4840$ V. All the other drive parameters are the same as in the previous section. For a summary of the parameters, the reader is referred to Tables 2.9–2.11.

2.5.3 Five-Level ANPC Inverter Drive System

The third case study relates to a five-level MV drive system, as shown in Fig. 2.27. The drive encompasses a 6 kV, 50 Hz squirrel-cage induction machine rated at 1 MVA with a total leakage reactance of $X_\sigma = 0.18$ pu. The rated values of the machine are summarized in Table 2.12.

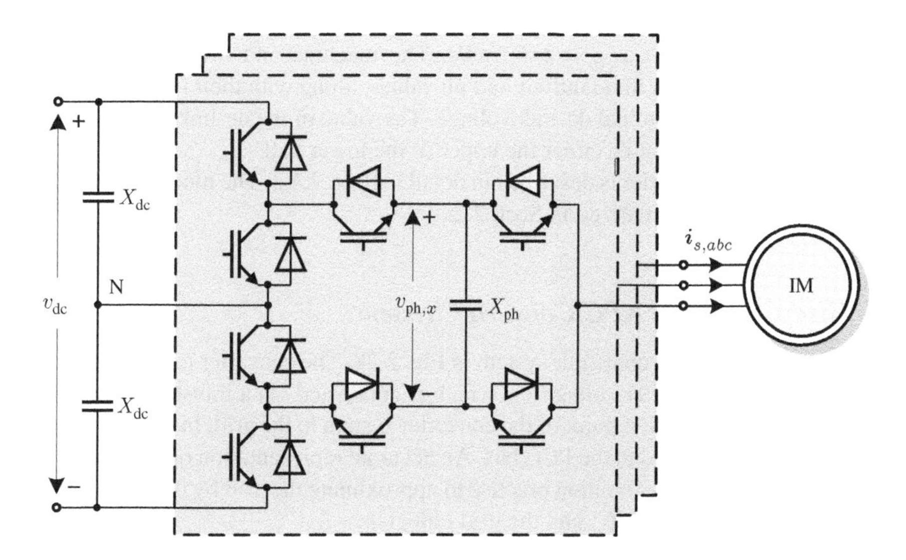

Figure 2.27 Equivalent representation of the five-level ANPC voltage source inverter driving an induction machine (IM)

Table 2.12 Rated values of the induction machine

Parameter	Symbol	SI value
Voltage	V_R	6000 V
Current	I_R	98.9 A
Real power	P_R	850 kW
Apparent power	S_R	1.028 MVA
Angular stator frequency	ω_{sR}	$2\pi 50$ rad/s
Rotational speed	ω_{mR}	1494 rpm
Air-gap torque	T_R	5.568 kNm

Table 2.13 Drive parameters in the SI (left) and per unit system (right) of the five-level ANPC inverter drive system

Parameter	SI symbol	SI value	pu symbol	pu value
Stator resistance	R_s	203 mΩ	R_s	0.0057
Rotor resistance	R_r	158 mΩ	R_r	0.0045
Stator leakage inductance	L_{ls}	9.968 mH	X_{ls}	0.0894
Rotor leakage inductance	L_{lr}	10.37 mH	X_{lr}	0.0930
Main inductance	L_m	277.8 mH	X_m	2.492
Number of pole pairs	p	2		
dc-link voltage	V_{dc}	9.8 kV	V_{dc}	2.000
dc-link capacitor	C_{dc}	200 µF	X_{dc}	2.201
Phase capacitor	C_{ph}	140 µF	X_{ph}	1.541

The pu system is established using the base quantities $V_B = \sqrt{2/3}V_R = 4899\,\text{V}$, $I_B = \sqrt{2}I_R = 139.9\,\text{A}$, and $\omega_B = \omega_{sR} = 2\pi 50\,\text{rad/s}$. The machine and inverter parameters are summarized in Table 2.13 as SI quantities and pu values, along with their respective symbols. Note that V_{dc} denotes the nominal dc-link voltage. The value of the dc-link capacitance refers to one half of the dc-link, that is, either the upper or the lower half.

The five-level ANPC inverter is described in detail in Sect. 2.4.2. The model of the induction machine is derived and summarized in Sect. 2.2.

2.5.4 *Grid-Connected NPC Converter System*

Consider the grid-connected converter system in Fig. 2.28. The converter is represented by the switched three-phase converter voltage $\boldsymbol{v}_c$, which is connected via a transformer to the PCC. The PCC acts as the connection point of the converter system to the grid. In general, additional industrial loads are connected to the PCC bus. An accurate representation of the grid is usually not available. Therefore, it is common practise to approximate the grid by the three-phase grid voltage $\boldsymbol{v}_g$, the grid resistance R_g, and the grid inductance L_g.

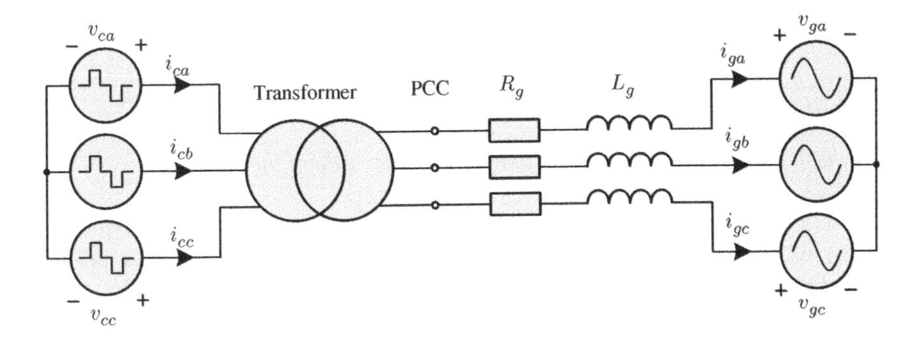

Figure 2.28 Grid-connected converter system in SI units

The short-circuit power

$$S_{sc} = 3\left(\frac{V_g}{\sqrt{3}}\right)^2 / |Z_g| = V_g^2/|Z_g| \tag{2.97}$$

is defined as the power supplied to the PCC in case of a three-phase fault at the PCC, where V_g denotes the rms grid (line-to-line) voltage and $Z_g = \mathrm{j}\omega_g L_g + R_g$ is the grid impedance. The absolute value of the latter is given by $|Z_g| = \sqrt{(\omega_g L_g)^2 + R_g^2}$. The grid impedance is typically dominated by the distribution transformer that connects the PCC to the transmission grid. The short-circuit power can be interpreted as the maximum power that the grid can provide to the PCC.

The short-circuit ratio

$$k_{sc} = S_{sc}/S_c \tag{2.98}$$

relates the short-circuit power of the grid to the rated power S_c of the power converter. Ratios above 20 indicate a strong grid, with the power of the converter being small compared to the maximum power available from the grid. Short-circuit ratios of less than 8 relate to a weak grid, in which the grid impedance dominates over the impedance of the converter system. In general, this reduces the stability margin of the converter system and requires tighter limits on the harmonics the converter may inject into the PCC.

Another characteristic quantity of the grid is the grid impedance ratio

$$k_{XR} = X_g/R_g \tag{2.99}$$

between the grid reactance $X_g = \omega_g L_g$ and the grid resistance R_g. A value of 10 is often assumed.

Based on the grid voltage, converter power, short-circuit ratio, and grid impedance ratio, the grid inductance and resistance can easily be computed as shown in the following example.

Example 2.8 *Consider the rms grid voltage $V_g = 3.3\,kV$, the converter power $S_c = 9\,MVA$, and the short-circuit ratio $k_{sc} = 20$. It follows from* (2.97) *and* (2.98) *that the absolute value of the grid impedance is $|Z_g| = 60.5\,m\Omega$. Assuming the grid impedance ratio $k_{XR} = 10$, the* *following grid inductance and resistance result:*

$$L_g = \frac{|Z_g|}{\omega_g\sqrt{1+1/k_{XR}^2}} = 0.192\ mH\ and\ R_g = \frac{|Z_g|}{\sqrt{1+k_{XR}^2}} = 6.019\ m\Omega. \tag{2.100}$$

Note that these parameters are referred to the secondary side of the transformer.

The rated power and voltage of the transformer at its secondary winding are typically used as the basis for the pu system. Assuming the rated values of the transformer provided in Table 2.14, the pu system is established using the base quantities $V_B = \sqrt{2/3}V_R = 2694$ V, $S_B = S_R = 9$ MVA, and $\omega_B = \omega_{gR} = 2\pi 50$ rad/s. The grid, transformer, and converter parameters are summarized in Table 2.15 as SI quantities and pu values, along with their respective symbols. All quantities are referred to the secondary side of the transformer. The transformer can be represented by its series leakage reactance X_t and the series resistance R_t. The grid is described by the grid reactance X_g and the grid resistance R_g.

We combine the reactance and the resistance of the transformer and the grid to

$$X = X_g + X_t \text{ and } R = R_g + R_t. \tag{2.101}$$

Table 2.14 Rated values of the step-down transformer of the grid-connected converter system

Parameter	Symbol	SI value
Voltage (secondary side)	V_R	3300 V
Current (secondary side)	I_R	1575 A
Apparent power	S_R	9 MVA
Angular grid frequency	ω_{gR}	$2\pi 50$ rad/s

Table 2.15 System parameters in the SI (left) and per unit system (right) of the NPC grid-connected converter system

Parameter	SI symbol	SI value	pu symbol	pu value
Short-circuit ratio	k_{sc}	20		
Grid impedance ratio	k_{XR}	10		
Grid inductance	L_g	0.192 mH	X_g	0.050
Grid resistance	R_g	6.019 mΩ	R_g	0.005
Transformer leakage inductance	L_t	0.385 mH	X_t	0.100
Transformer resistance	R_t	12.10 mΩ	R_t	0.010
dc-link voltage	V_{dc}	5.2 kV	V_{dc}	1.930
dc-link capacitance	C_{dc}	15 mF	X_{dc}	5.702

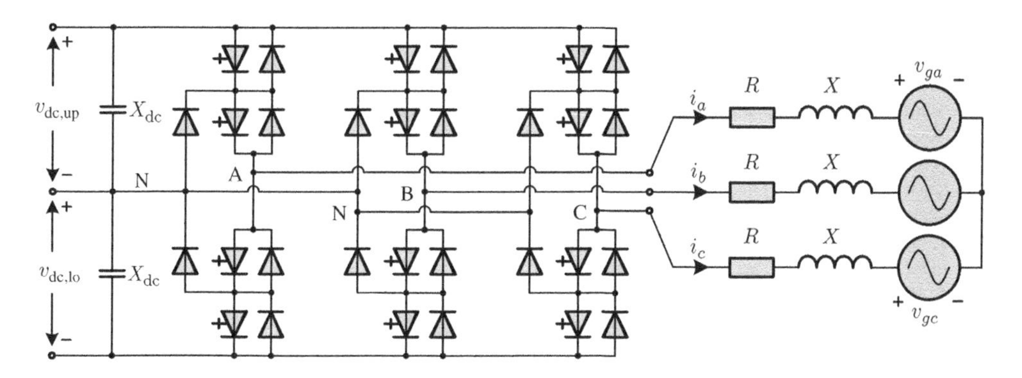

Figure 2.29 Grid-connected NPC converter system in the per unit system

This enables the very compact representation of the grid-connected converter system, which is shown in Fig. 2.29. The NPC converter is shown on the left-hand side with the floating neutral point potential. As before, the value of the dc-link capacitance refers to one half of the dc-link, that is, either to the upper or the lower half. Currents flowing from the converter to the grid are assumed to be positive.

References

[1] W. Duesterhoeft, M. Schulz, and E. Clarke, "Determination of instantaneous currents and voltages by means of alpha, beta, and zero components," *AIEE Trans.*, vol. 70, pp. 1248–1255, Jul. 1951.

[2] R. Park, "Two-reaction theory of synchronous machines—generalized method of analysis—part I," *AIEE Trans.*, vol. 48, pp. 716–727, Jul. 1929.

[3] P. C. Krause, O. Wasynczuk, and S. D. Sudhoff, *Analysis of electric machinery and drive systems*. Hoboken, NJ: John Wiley & Sons, Inc., 2nd ed., 2002.

[4] J. Holtz, "The representation of AC machine dynamics by complex signal graphs," *IEEE Trans. Ind. Electron.*, vol. 42, pp. 263–271, Jun. 1995.

[5] A. Veltman, D. W. J. Pulle, and R. W. De Doncker, *Fundamentals of electrical drives*. Heidelberg: Springer, 2007.

[6] B. Wu, *High-power converters and AC drives*. New York Hoboken, NJ: John Wiley & Sons, Inc., 2006.

[7] ABB Asea Brown Boveri Ltd, "Applying IGCTs, application note 5SYA 2032-03." Online document. www.abb .com/semiconductors.

[8] P. Steimer, H. Grüning, J. Werninger, E. Carroll, S. Klaka, and S. Linder, "IGCT—a new emerging technology for high power, low cost inverters," in *Proceedings of IEEE Industry Applications Society Annual Meeting*, pp. 1592–1599, Oct. 1997.

[9] P. Steimer, O. Apeldoorn, E. Carroll, and A. Nagel, "IGCT technology baseline and future opportunities," in *Proceedings of IEEE Transmission and Distribution Conference and Exposition* (Atlanta, GA, USA), Oct./Nov. 2001.

[10] A. Nabae, I. Takahashi, and H. Akagi, "A new neutral-point-clamped PWM inverter," *IEEE Trans. Ind. Appl.*, vol. IA-17, pp. 518–523, Sep./Oct. 1981.

[11] N. Celanovic and D. Boroyevich, "A comprehensive study of neutral-point voltage balancing problem in three-level neutral-point-clamped voltage source PWM inverters," *IEEE Trans. Power Electron.*, vol. 15, pp. 242–249, Mar. 2000.

[12] H. du Toit Mouton, "Natural balancing of three-level neutral-point-clamped PWM inverters," *IEEE Trans. Ind. Electron.*, vol. 49, pp. 1017–1025, Oct. 2002.
[13] P. Barbosa, P. Steimer, J. Steinke, L. Meysenc, M. Winkelnkemper, and N. Celanovic, "Active neutral-point-clamped multilevel converters," in *Proceedings of IEEE Power Electronics Specialists Conference* (Recife, Brasil), pp. 2296–2301, Jun. 2005.
[14] F. Kieferndorf, M. Basler, L. Serpa, J.-H. Fabian, A. Coccia, and G. Scheuer, "A new medium voltage drive system based on ANPC-5L technology," in *Proceedings of IEEE International Conference on Industrial Technology* (Viña del Mar, Chile), pp. 605–611, Mar. 2010.
[15] T. Brückner, S. Bernet, and H. Guldner, "The active NPC converter and its loss-balancing control," *IEEE Trans. Ind. Electron.*, vol. 52, pp. 855–868, Jun. 2005.
[16] T. Meynard and H. Foch, "Multilevel conversion: High voltage choppers and voltage source inverters," in *Proceedings of IEEE Power Electronics Specialists Conference*, pp. 397–403, Jun. 1992.

3

Classic Control and Modulation Schemes

This chapter provides an introduction to state-of-the-art control and modulation schemes for industrial power electronic systems. Specifically, the requirements of control and modulation schemes are stated, and the almost universally used cascaded control structure is described. The concept of pulse width modulation (PWM) is introduced, and two notable schemes—carrier-based pulse width modulation (CB-PWM) and optimized pulse patterns (OPPs)—are explained in detail. The equivalence of space vector modulation (SVM) and CB-PWM with an appropriate common-mode voltage injection is highlighted. The performance trade-off between current distortions and switching losses is analyzed. The three control schemes that are commonly used for high-power inverters driving electrical machines are reviewed in detail; these control schemes are scalar control, field-oriented control (FOC), and direct torque control (DTC). This chapter concludes with an appendix that introduces mathematical optimization.

3.1 Requirements of Control and Modulation Schemes

The control problem of a medium-voltage (MV) power electronic system presents a high degree of complexity with multiple and conflicting objectives. The requirements for control and modulation schemes can be grouped into requirements relating to the converter and requirements relating to the three-phase component connected to the converter, which is either an electrical machine or the grid. Note that the term "converter" refers either to the active front end on the grid side or to the inverter on the machine side.

3.1.1 Requirements Relating to the Electrical Machine

3.1.1.1 Torque and Flux

Regarding the machine, the electromagnetic torque must be kept close to its reference. During torque transients, a high dynamic performance should be achieved, that is, the torque should be

Model Predictive Control of High Power Converters and Industrial Drives, First Edition. Tobias Geyer.

Companion Website: www.wiley.com/go/geyermodelpredictivecontrol

quickly adjusted within a short transient response time of a few milliseconds. Such transients include torque changes of the load; if the machine's rotational speed is to be kept constant, the machine torque must match the changing load torque to avoid speed fluctuations. On the other hand, to achieve fast speed changes, the machine torque must be quickly adjusted. Fast torque changes are also crucial during grid faults to facilitate low-voltage ride-through of the drive system. As an example, consider a grid fault lasting for several hundred milliseconds. To avoid the drive system from tripping because of too high a dc-link voltage, the machine torque must be almost instantaneously reduced to zero.

The mechanical load usually requires a smooth torque. A low harmonic torque distortion corresponds to a small torque ripple that limits the mechanical stress and wear of the shaft, the bearings, and the load. Moreover, the risk of exciting torsional eigenmodes of the drive train is minimized (see, e.g., [1] and the references therein). For very high-power applications with particularly stiff shafts such as large compressor drive trains used in the oil and gas industry, certain low-frequency torque harmonics should be avoided altogether.

A suitable measure of the harmonic distortion of electric waveforms is the total demand distortion (TDD). For the electromagnetic torque T_e, the TDD is defined as

$$T_{\text{TDD}} = \frac{1}{T_{e,\text{nom}}} \sqrt{\sum_{n \neq 0} (\hat{T}_{e,n})^2}, \tag{3.1}$$

where $T_{e,\text{nom}}$ denotes the nominal torque.[1] The quantities $\hat{T}_{e,n}$, $n > 0$, refer to the amplitudes of the harmonic components of the torque spectrum at the frequencies nf_1, where f_1 is the fundamental frequency. Note that harmonics at any positive n are considered here, not just harmonics at integer multiples of f_1.

The total harmonic distortion (THD) is defined similar to (3.1), but it is related to the dc component of the actual torque rather than the nominal one. As a result, for a torque close to zero, the THD tends to infinity, while the TDD remains more or less constant. Apart from this, the practical impact of torque harmonics is largely independent of the actual torque. For these reasons, we adopt the TDD rather than the THD to assess harmonic distortions. For more details on the TDD and THD, the reader is referred to [2].

Second, to ensure that the machine is appropriately magnetized, the machine's air-gap flux should be controlled such that the desired magnitude of the rotor flux vector is achieved. In particular, saturation of the machine's rotor or demagnetization is to be avoided. When operating below the rated speed, the rotor flux is usually maintained at its nominal value, while beyond the rated speed, field weakening is required to avoid saturation.

3.1.1.2 Stator Currents

The switched voltage waveform of the inverter causes harmonic current distortions, which give rise to iron and copper losses and thus to thermal losses. As the capability of cooling the rotor is limited, thermal losses in the rotor are of particular concern, particularly at low-speed operation. The effect of harmonics on electrical machines is explained in [2, Sect. 6.2]. The losses and temperature rise of squirrel-cage induction machines that are caused by harmonics

[1] When working with SI quantities, the nominal torque is equal to the rated torque T_R (see Sect. 2.1.2). When adopting the per unit (pu) system, the nominal torque is 1.

is estimated in [3]. For a comprehensive review of the literature on and theory of harmonic machine losses, the interested reader is referred to [4, Chap. 2].

To avoid thermal problems, the harmonic distortions of the stator current have to be kept small. For the stator current, the TDD is defined as

$$I_{\text{TDD}} = \frac{1}{\sqrt{2}I_{s,\text{nom}}} \sqrt{\sum_{n \neq 1} (\hat{i}_{s,n})^2}, \tag{3.2}$$

where $I_{s,\text{nom}}$ refers to the nominal rms stator current.[2] The harmonic components $\hat{i}_{s,n}$, $n \geq 0$, are the amplitudes of the current harmonics at frequencies nf_1. The fundamental component $\hat{i}_{s,1}$ is excluded from the sum. Note that, as in (3.1), n is a real number, not a natural number.

The amplitudes $\hat{i}_{s,n}$ of the harmonics are peak (rather than rms) values. The factor $\sqrt{2}$ is required to translate the nominal rms current $I_{s,\text{nom}}$ into its amplitude, ensuring that amplitudes are related to each other in (3.2). Moreover, this definition holds for a single-phase current only. To compute the TDD of a three-phase current, the TDD is computed for each of the a, b, and c phase currents separately. The overall TDD is then determined by taking the mean value of the three phases.

An alternative measure of the current distortions is the THD. For the same reasons as discussed previously for the torque, we prefer to adopt the TDD rather than the THD. For more details on the current TDD and THD, the reader is referred to [2] and [5]. The second reference includes a discussion explaining why the TDD is the preferred choice.

For electrical machines designed and built specifically for use with inverters, current TDDs of up to 10% are often deemed acceptable. In recent years, however, so-called direct online machines, which were previously connected directly to the grid, have been increasingly augmented with back-to-back converter systems. This retrofitting enables the machine to operate at variable speeds, often yielding significant efficiency gains, particularly at partial load operation. As direct online machines are not designed to withstand significant harmonics in the stator currents, inverters retrofitted to such machines must meet stringent requirements on the current TDD. Typically, an upper bound of 5% on the current TDD is required, even though current TDDs below 3% are desirable and often requested by the customer.

3.1.1.3 Common-Mode Voltage and dv/dt

By definition, the common-mode voltage applied to an electrical machine with a floating star point neither affects the (differential-mode) stator current nor the electromagnetic torque produced by the machine. Nevertheless, common-mode voltages establish a common-mode current path through parasitic capacitances from the stator windings via the motor bearings to ground. To avoid damaging the bearings, the rms common-mode voltage is typically limited. More details on the source of bearing currents and their modeling are provided in [6, 7].

In addition, the insulation of the stator windings has to be rated for the resulting dv/dt. The latter mainly depends on the voltage per semiconductor and its switching characteristic

[2] When working with SI quantities, the nominal rms stator current is equal to the rated current I_R of the machine (see Sect. 2.1.2). When adopting the pu system, the nominal peak current is 1, because the pu system is usually based on the machine. As the electrical machine is often slightly overrated in a variable-speed drive (VSD), a stator current below $I_{s,\text{nom}}$ is achieved when operating the inverter at its full power capability.

(the slope thereof). Depending on the machine design, dv/dt is limited to values between 500 V/μs for direct online machines and 3 kV/μs for inverter-rated machines.

3.1.2 Requirements Relating to the Grid

The requirements imposed on grid-connected converters can be broadly classified into requirements that apply during nominal grid conditions and requirements that apply during grid disturbances. More specifically, during nominal grid conditions stringent limits on the emitted voltage and current harmonics are imposed, while during grid disturbances and faults the continued operation of the converter must be ensured, thus requiring immunity to a wide range of grid disturbances.

3.1.2.1 Harmonic Emissions

The harmonics injected by power electronic converters into the grid must meet harmonic standards. These standards are imposed at the point of common coupling (PCC), at which the harmonics are measured and assessed. According to the IEEE 519 working group, the "PCC with the consumer/utility interface is the closest point on the utility side of the customer's service where another utility customer is or could be supplied" (see [5, 8]). Harmonic standards are not intended to be applied within a customer's subsystem, but they are meant to prevent one grid customer harming another.

Several harmonic standards exist nowadays for industrial power electronics. These standards specify the limits on the current and voltage distortions. In the following, we consider two widely adopted standards, namely the IEEE 519 [2] and the IEC 61000-2-4 standard [9].

- *Current distortion limits.* In Table 10.3 of the IEEE 519 standard, current distortion limits are defined at the PCC for general distribution systems with voltages of up to 69 kV [2]. The maximum harmonic current distortion levels are given as a percentage of the nominal fundamental frequency component.[3] These limits depend on the harmonic order and on the short-circuit ratio. The latter was defined in (2.98) in Sect. 2.5.4 as $k_{sc} = S_{sc}/S_c$, which is the short-circuit grid power S_{sc} divided by the rated power S_c of the converter. For $k_{sc} = 20$, for example, the limits on the harmonic current distortions are shown in Fig. 3.1. Stricter limits are imposed on current harmonics that are of a higher or an even order. Dc offsets are not permitted. Larger loads and/or weaker grids correspond to smaller k_{sc} and result in stricter limits on the harmonic current distortions (see Table 10.3 in [2]).
- *Voltage distortion limits.* Table 11.1 of the IEEE 519 standard specifies voltage distortion limits for PCC bus voltages below 69 kV. Individual voltage harmonics are limited to 3% of the nominal fundamental frequency component and the voltage TDD is limited to 5%.

  The IEC 61000-2-4 standard [9] focuses on voltage distortions. Limits up to the 50th harmonic are specified in Tables 2–4 of the IEC standard. Assuming a Class 2 electromagnetic

[3] Note that the nominal fundamental frequency current relates to the maximum demand load current of the power converter. The rated current of the pu system, however, is usually determined by the transformer and is larger than the maximum demand load current. This discrepancy is to be taken into account when imposing limits on current harmonics given in the pu system.

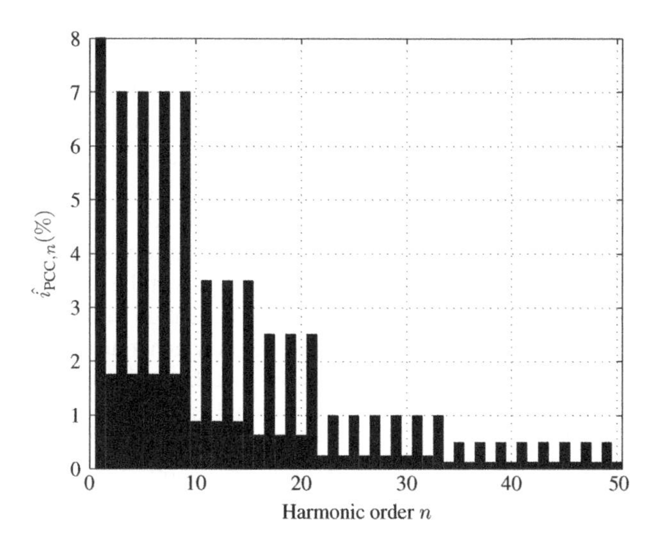

Figure 3.1 Current harmonic limits at the PCC according to the IEEE 519 standard, assuming the short-circuit ratio $k_{sc} = 20$

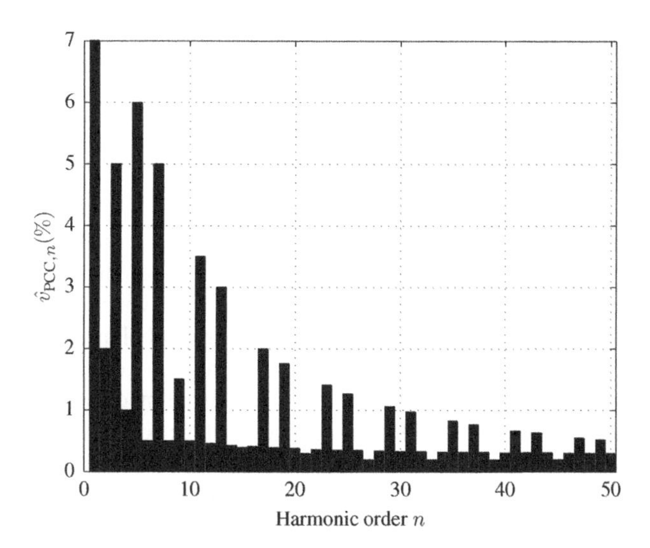

Figure 3.2 Voltage harmonic limits at the PCC according to the IEC 61000-2-4 standard, assuming a Class 2 electromagnetic environment

environment, which applies to general PCCs, these limits are shown in Fig. 3.2. The limits on the non-triplen odd harmonics (the differential-mode harmonics) are relatively loose. The limits on the even-order harmonics are significantly tighter, and the limits on the triplen odd harmonics (the common-mode harmonics) of higher order are particularly stringent. Limits between the 50th harmonic and 9 kHz are provided in Annex C.3 of [9].

For Class 2 electromagnetic environments, the voltage TDD at the PCC is limited to 8%[4].

When interpreting the converter as a harmonic voltage source, the voltage harmonics at the PCC depend on the ratio $Z_g/(Z_g + Z_c)$, where Z_g denotes the grid impedance and Z_c the impedance between the converter and the PCC. The latter relates to the step-down transformer and an optional LC filter. These two impedances act as a voltage divider. For strong grids, when Z_g is small compared to Z_c, the voltage harmonics injected by the power converter are significantly attenuated at the PCC. As a result, relatively large voltage harmonics may be injected by the converter without violating the harmonic limits at the PCC. On the other hand, when the grid is weak, Z_g dominates over Z_c, and the attenuation of the harmonics is minor, severely limiting the harmonic distortions the converter may generate.

Note that the grid standards on harmonic emissions and the limits they impose on the voltage and current distortions are applicable only to nominal grid conditions. In the presence of major disturbances, such as significant voltage imbalances, grid under- or overvoltages, grid frequency deviations, or faults, these grid standards are not applicable.

3.1.2.2 Immunity to Grid Disturbances and Faults

Several IEC standards specify grid phenomena under which the continuous operation of the converter must be ensured. These phenomena can be grouped into variations of the voltage amplitude, voltage imbalances, and frequency variations. We distinguish between continuous and transient phenomena, the latter ranging from several fundamental cycles up to 10 min. During continuous disturbances, the converter must be able to provide full power, while during transient phenomena the power may be reduced but the converter must continue to control the converter currents by modulating the converter voltage. Typical values for the requirements are summarized in the following.

- *Voltage amplitude variations.* Most standards require operation during continuous voltage amplitude variations of up to ±10% and operation during transient voltage amplitude variations of up to ±20%. Ride-through operation is required for voltage dips of up to 100%.
- *Voltage imbalances.* Operation under continuous voltage imbalances of ±2% to ±5% is required, while the converter must be able to tolerate transient voltage imbalances of up to 8% without tripping. The notion of voltage imbalance is defined as the negative sequence voltage component divided by the positive sequence voltage component (see, e.g., [10] and the references therein).
- *Frequency variations.* The requirements for continuous frequency variations vary from ±2% to ±5%, with transient variations of up to ±10%.

The relevant grid standards that specify immunity requirements include the IEC standards 60146-1-1, 61000-2-4, 61800-3, 61800-4, and 61892-1.

[4] Note that [9] specifies a limit on the voltage THD, not on the TDD. Nevertheless, because the voltage at the PCC is effectively constant during nominal operation, the distinction between voltage THD and voltage TDD can be neglected.

3.1.3 Requirements Relating to the Converter

As a consequence of the high voltages and currents typically encountered in MV converters, the switching and conduction losses in the semiconductors are substantial. The limited cooling capability in the converter imposes an upper bound on the tolerable losses of the semiconductors. Despite the fact that the semiconductors are often water-cooled, the switching frequency achievable with the semiconductor devices available today is typically limited to a few hundred hertz.

Significant losses also reduce the efficiency of the converter. Very high converter efficiencies are of particular importance for customers acquiring renewable energy converters and power quality products, such as flexible ac transmission systems (FACTS). During the lifetime of a converter, lowering the losses by several kilowatts will save a substantial amount of money to the customer.

Assuming an almost constant blocking voltage across the semiconductors, the conduction losses are upper bounded by the rated current. In MV applications, the switching losses typically dominate over the conduction losses. An indirect way of limiting the switching losses is to limit the switching frequency by imposing a maximum device switching frequency. Even though the switching frequency is a convenient metric, the switching losses constitute a more direct and thus more meaningful measure. In practise, however, measuring the switching losses during operation is usually too complicated, unreliable, and costly. Nevertheless, the switching losses can be well modeled and reconstructed using the commutated current, voltage, and semiconductor characteristics. Moreover, the minimization of the switching losses provides a degree of freedom that some control and modulation schemes can exploit by shifting switching transitions to switching instants with low current magnitudes, thus reducing the switching losses.

Besides the switching frequency and losses, additional requirements often arise for multilevel inverters, such as the balancing of a neutral point potential around zero. Moreover, in active neutral-point-clamped (NPC) five-level converters (see Sect. 2.5.3), the voltages of the three phase capacitors must be maintained close to their references. For modular multilevel converters, the phase capacitor voltages must be maintained close to their nominal value and the circulating current must be limited. This will be discussed in Chap. 14.

3.1.4 Summary

It is apparent that the requirements relating to the converter and the load (electrical machine or grid) are—to a significant extent—conflicting. The minimization of the switching frequency or losses on the one side and the harmonic distortions on the other side are opposing objectives. This gives rise to a fundamental trade-off, which will be exemplified for carrier-based PWM in Sect. 3.5.

In the following chapters, the requirements for control and modulation schemes will be translated into control objectives, based on which model predictive control problems will be formulated. In general, the control problem is complicated by the fact that the control objectives comprise phenomena of very different time scales. Specifically, the control objectives relating to the electrical machine depend on the very fast stator current dynamic, which is driven by the applied stator voltage and which can be manipulated within several tens of μs. This requires very short sampling intervals, which are often as short as $T_s = 25$ μs. The same applies when controlling grid currents in grid-connected converters.

On the other hand, when minimizing the switching frequency or losses, these quantities need to be evaluated over a time frame that exceeds the sampling interval by several orders of magnitude. In high-power applications, where the device switching frequency is in the range 200–400 Hz, each semiconductor switch is turned on roughly every 2.5–5 ms. This implies time frames of at least 10 ms, over which the switching frequency is to be evaluated.

3.2 Structure of Control and Modulation Schemes

Consider the variable-speed drive (VSD) system shown in Fig. 1.1, which is replicated on the right-hand side of Fig. 3.3. We distinguish between the (grid-side) converter and the (machine-side) inverter. The dc-link capacitor acts as a decoupling element. The overall control task of the VSD system is accordingly decomposed into the grid-side controller and the machine-side controller.

The grid-side controller maintains the dc-link voltage v_{dc} at its reference value v_{dc}^* using a cascaded control loop. The outer loop—the voltage controller—regulates the dc-link voltage by adjusting the reference of the real power P^*. The reference for the reactive power Q^* is usually set to zero. The real and reactive power references are translated into the grid current reference $\boldsymbol{i}_g^*$, which is a two-dimensional vector either in the stationary or in the rotating orthogonal coordinate system. The inner loop, which is the current controller, regulates the grid currents by manipulating the voltage applied by the converter to the grid. In most cases, the current controller consists of two proportional-integral (PI) controllers in a rotating

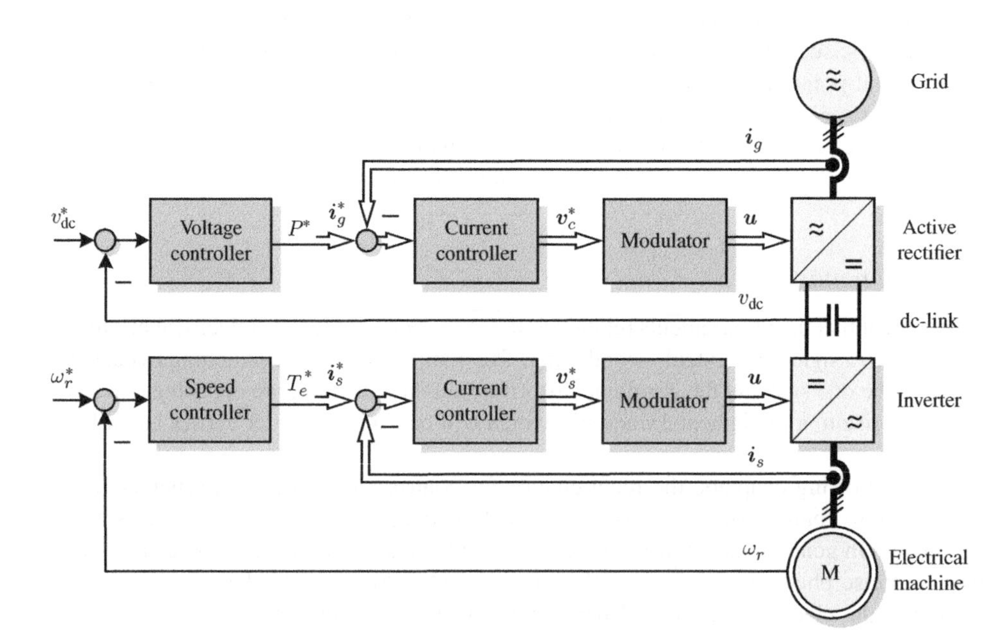

Figure 3.3 VSD system with cascaded control loops for the grid and the machine side

orthogonal reference frame. The PI controllers adjust the real-valued voltage reference $\boldsymbol{v}_c^*$, which the PWM translates into the switching signal $\boldsymbol{u}$.

Similarly, a cascaded control structure is used on the machine side. The outer speed control loop regulates the angular speed of the rotor ω_r along its reference ω_r^*, by manipulating the reference of the electromagnetic torque T_e^*. Another outer control loop, which is omitted in Fig. 3.3, controls the rotor flux magnitude. The outputs of the outer control loops are translated into the stator current reference $\boldsymbol{i}_s^*$. The inner control loop regulates the stator current $\boldsymbol{i}_s$ along its reference, by manipulating the inverter and thus the stator voltage $\boldsymbol{v}_s$. As on the grid side, the (inner) current controller is typically based on two PI control loops with a subsequent PWM stage.

In summary, the grid and the machine side are treated separately. Coupling between the two might be considered through a feedforward term, for example, a power feedforward term from the inverter to the grid-side converter. One or two outer control loops are used, which are single-input single-output (SISO) loops. The inner current control loop is a multiple-input multiple-output (MIMO) control problem, which is often split into two orthogonal SISO loops. To mask the switching characteristic of the power converter, a PWM is usually added to the inner control loop. PWM is explained in detail in Sect. 3.3. The two control schemes predominantly used for the inner current loop on the machine side—FOC and DTC—are summarized in Sect. 3.6.

3.3 Carrier-Based Pulse Width Modulation

PWM is used pervasively in power electronics. For an early reference on PWM, see [11]. As shown in Fig. 3.4, PWM translates a real-valued input signal $\boldsymbol{u}^*$ into a discrete-valued output signal $\boldsymbol{u}$, using pulses of fixed amplitude but variable width. The output waveform $\boldsymbol{u}$ approximates $\boldsymbol{u}^*$ with regard to the magnitude and phase of its fundamental component. However, the switching nature of the PWM implies that undesired harmonic content is added to $\boldsymbol{u}$.

A converter with the dc-link voltage v_{dc} is used as actuator to translate the switching signal $\boldsymbol{u}$ into the switched voltage waveform $\boldsymbol{v}$ at the converter terminals. By appropriately scaling the reference voltage $\boldsymbol{v}^*$, the converter voltage $\boldsymbol{v}$ approximates its reference $\boldsymbol{v}^*$ (see Fig. 3.4). As this principle applies to both (machine-side) inverters and (grid-side) active rectifiers, we used the term *converter* and dropped the subindex from the converter voltage $\boldsymbol{v}$.

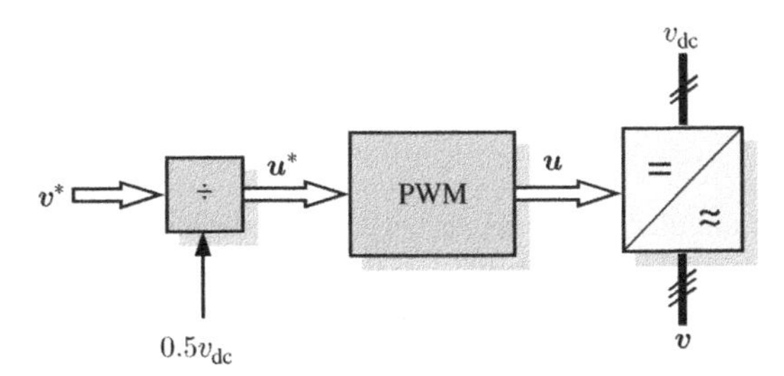

Figure 3.4 The reference voltage $\boldsymbol{v}^*$ is scaled to the modulating signal $\boldsymbol{u}^*$ and translated via PWM to the switching signal $\boldsymbol{u}$, which drives the converter to synthesize the voltage $\boldsymbol{v}$ at the converter terminals

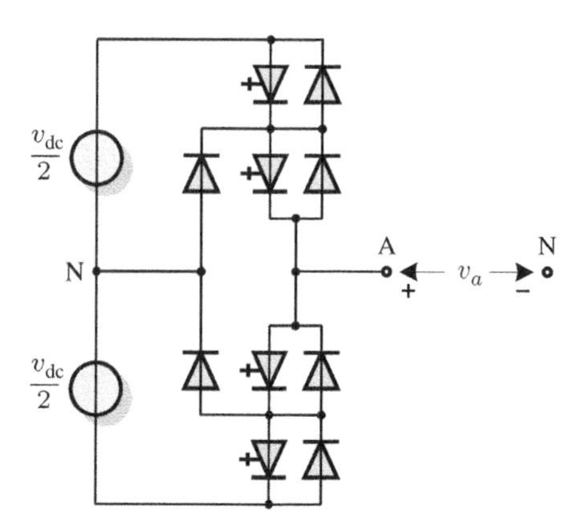

Figure 3.5 Single-phase, three-level converter with the output voltage v_a between the phase leg terminal A and the dc-link midpoint N

Commonly used PWM schemes include CB-PWM, SVM, and OPPs. CB-PWM and OPPs are introduced in the remainder of this section, and the relationship between CB-PWM and SVM is shown.

3.3.1 Single-Phase Carrier-Based Pulse Width Modulation

The simplest and most commonly used type of modulation is CB-PWM. Here we introduce CB-PWM for single-phase converters and generalize it in a subsequent section to the three-phase setup. To this end, consider the single-phase, three-level NPC converter shown in Fig. 3.5 with the dc-link voltage v_{dc}. At its phase terminal, this converter produces the discrete voltage levels $-0.5v_{dc}$, 0, and $0.5v_{dc}$. When assuming the neutral point potential to be zero, the phase voltage with respect to the dc-link midpoint N is given by

$$v_a = \frac{v_{dc}}{2} u_a. \tag{3.3}$$

The phase voltage is a function of the switch position of the phase leg $u_a \in \{-1, 0, 1\}$ (see also (2.75)).

The PWM problem at hand is to translate the voltage reference v_a^* into an appropriate switching signal u_a, such that the phase voltage v_a approximates its reference v_a^*. Assuming steady-state operation, we define the sinusoidal voltage reference

$$v_a^*(t) = \hat{v}_1 \sin(\omega_1 t + \phi_1) \tag{3.4}$$

with the amplitude $\hat{v}_1$, angular frequency $\omega_1 = 2\pi f_1$, and phase ϕ_1. We often refer to f_1 as the fundamental frequency. It is clear from (3.3) that the achievable phase voltage is limited to $|v_a| \leq 0.5v_{dc}$. Therefore, for the time being, we restrict the amplitude of the reference voltage to $\hat{v}_1 \leq 0.5v_{dc}$, but will later lift this restriction when discussing common-mode injection and overmodulation.

3.3.1.1 Modulation

In a first step, as shown in Fig. 3.4, the voltage reference is scaled by half the dc-link voltage, which yields

$$u_a^*(t) = \frac{2}{v_{\mathrm{dc}}} v_a^*(t) = \hat{u}_1 \sin(\omega_1 t + \phi_1). \tag{3.5}$$

We often refer to u_a^* as the *modulating* signal. The magnitude of the modulating signal is the so-called *modulation index*

$$m = \frac{2}{v_{\mathrm{dc}}} \hat{v}_1 = \hat{u}_1. \tag{3.6}$$

In the single-phase case and when operating in the linear modulation range, the modulation index is limited to $m \in [0, 1]$. In the next section, when extending CB-PWM to three-phase systems, we will see that m can be increased beyond 1.

For three-level CB-PWM, two triangular carrier signals are defined with the carrier frequency f_c. The carrier frequency is (significantly) higher than the fundamental frequency, that is, $f_c \gg f_1$. Both carrier signals have the peak-to-peak magnitude 1. The carriers are arranged such that they cover the range from -1 to 1 without overlapping. The phase shift between the two carrier signals is a design parameter. When choosing *phase disposition*, the two carrier signals are in phase, while in the *phase-opposite disposition*, their phases are shifted by 180° with respect to each other. The former is commonly used, because it results in lower harmonic distortions (see also [12]). Lastly, we define the carrier interval

$$T_c = \frac{1}{f_c} \tag{3.7}$$

as the time interval between the upper (or lower) peaks of the carrier signal.

CB-PWM is achieved by comparing the modulating signal u_a^* with the two carrier signals. The switch position u_a is selected based on the following three rules:

- When u_a^* is less than both carrier signals, choose $u_a = -1$.
- When u_a^* is less than the upper carrier signal, but exceeds the lower one, select $u_a = 0$.
- When u_a^* is greater than both carrier signals, choose $u_a = 1$.

In the case of an analog implementation, the *instantaneous* value of the modulating signal u_a^* is compared with the carrier signals. We refer to this as *natural sampling*. Analog CB-PWM can easily be implemented using two comparators.

Example 3.1 *CB-PWM is exemplified in Fig. 3.6 for a single-phase, three-level converter. The upper and lower triangular carrier signals are shown in Fig. 3.6(a). The carrier signals with the frequency $f_c = 450$ Hz are in phase (phase disposition). The sinusoidal modulating signal has the amplitude (modulation index) $m = 0.8$ and the fundamental frequency $f_1 = 50$ Hz. This results in a ratio between the carrier frequency f_c and the fundamental frequency f_1 of 9. Using natural sampling, the intersections of the modulating signal with the carrier signals define the switching instants, which are shown as vertical dashed lines. The resulting sequence of switching commands is shown in Fig. 3.6(b).*

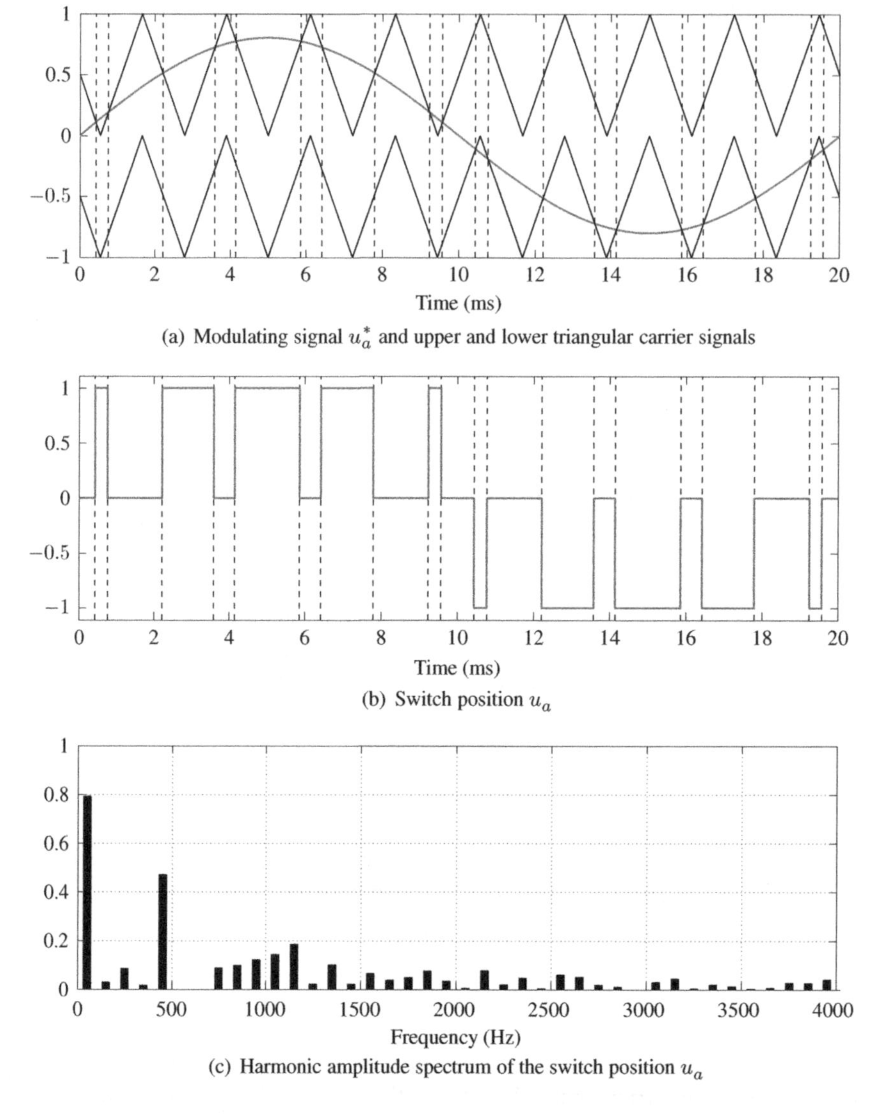

(a) Modulating signal u_a^* and upper and lower triangular carrier signals

(b) Switch position u_a

(c) Harmonic amplitude spectrum of the switch position u_a

Figure 3.6 Natural sampled single-phase CB-PWM with $f_c/f_1 = 9$

In a digital implementation, u_a^* is a sampled signal, resulting in the so-called PWM with *regular sampling*. Two regular sampling techniques are commonly used:

- *Symmetric sampling*. The voltage reference is sampled once per carrier interval T_c, for example, at the upper triangular peaks. Throughout the remainder of the carrier interval, the voltage reference is held constant.
- *Asymmetric sampling*. The voltage reference is sampled twice per carrier interval, that is, at the upper and at the lower peaks of the carrier. The voltage reference is held constant for half the carrier interval.

Table 3.1 Switching instant and switching transition for asymmetric regularly sampled CB-PWM

Polarity of the modulating signal u_a^*	Carrier slope	Switching instant Δt	Switching transition for u_a
≥ 0	Falling	$(1 - u_a^*)\frac{T_c}{2}$	$0 \rightarrow 1$
≥ 0	Rising	$u_a^* \frac{T_c}{2}$	$1 \rightarrow 0$
< 0	Falling	$-u_a^* \frac{T_c}{2}$	$-1 \rightarrow 0$
< 0	Rising	$(1 + u_a^*)\frac{T_c}{2}$	$0 \rightarrow -1$

For regularly sampled CB-PWM, the switching rules can be refined and the explicit switching instants can easily be provided. To this end, consider again phase disposition and asymmetric sampling. Switching is performed when the sampled modulating signal intersects with a carrier slope. We refer to this time instant (relative to the sampling instant) as the switching instant Δt. By definition, the switching instant Δt is bounded by zero and $0.5T_c$. The switching instants and the new switch positions can be derived as a function of the polarity of the modulating signal and the carrier slope, as summarized in Table 3.1.

Example 3.2 *The previous example is repeated here with asymmetric regular sampling instead of natural sampling. The modulating signal, which is shown as the dash-dotted sinusoid, is sampled twice per carrier interval (see Fig. 3.7(a)). The intersections of the sampled modulating signal with the carrier signals define the switching instants.*

We draw four conclusions from these two examples. First, regardless of the sampling method, in general, one switching transition occurs within each carrier half-interval. This follows from the previously stated switching rules. In overmodulation and in degenerate cases, which will be discussed later, switching transitions can be skipped.

Second, CB-PWM can achieve a high degree of symmetry in the resulting switching pattern, which has the period $T_1 = 1/f_1$. In the two previous examples, when shifting the switching pattern by half of its period, it is the negative of the original switching pattern. This is known as *half-wave* symmetry, and it is formally stated as $u_a(t - 0.5T_1) = -u_a(t)$. The switching patterns also exhibit symmetry about the quarter points of the period. We refer to this as *quarter-wave* symmetry. Note that a quarter-wave symmetric signal is also half-wave symmetric. Quarter-wave symmetry is the result of the non-even integer ratio between the carrier and the fundamental frequency. Moreover, the phase shift of the modulating signal with respect to the carrier signal was chosen carefully.

In the first example with natural sampling, the modulating signal is aligned with the negative carrier half-wave ($\phi_1 = \pi f_1/f_c$ phase shift), resulting in five switching transitions within a quarter of the fundamental period. Note that, when aligning the modulating signal with the positive carrier half-wave ($\phi_1 = 0$ phase shift), symmetry in the switching pattern is preserved, but the number of switching transitions is reduced to 4. This is a degenerate case. In the carrier half-intervals around the zero crossings of the modulating signal, the modulating signal does not intersect with a carrier signal, thus preventing switching.

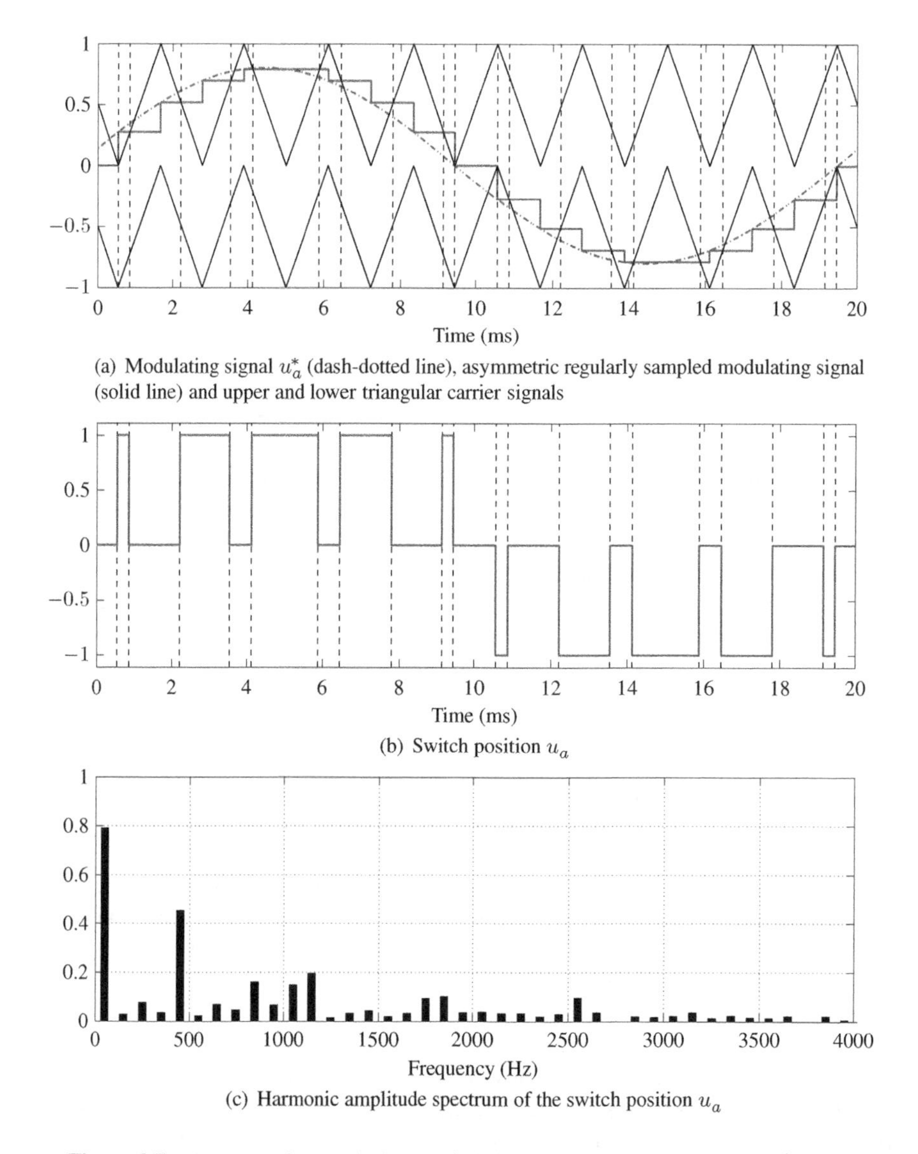

(a) Modulating signal u_a^* (dash-dotted line), asymmetric regularly sampled modulating signal (solid line) and upper and lower triangular carrier signals

(b) Switch position u_a

(c) Harmonic amplitude spectrum of the switch position u_a

Figure 3.7 Asymmetric, regularly sampled single-phase CB-PWM with $f_c/f_1 = 9$

As a general comment, the impact of the phase difference between the modulating signal and the carrier signals is sometimes ignored. Nevertheless, at f_c/f_1 ratios below 10, the influence on the switching pattern, harmonic performance, and power capability of the converter can be significant, as shown in [13].

Third, the sampling of the modulating signal introduces a time delay that manifests itself as a phase shift between the voltage reference and the synthesized converter voltage. For an asymmetric, regularly sampled CB-PWM, this time delay amounts to a quarter of the carrier interval, which is equivalent to a phase shift of $\phi_1 = 0.5\pi f_1/f_c$. In Example 3.2, to compensate for this delay, the phase of the modulating signal was shifted accordingly (see Fig. 3.7(a)) and set to $\phi_1 = 1.5\pi f_1/f_c$. For symmetric, regularly sampled CB-PWM, the time delay and phase shift are twice as large. It is obvious that for low carrier-to-fundamental frequency ratios this issue becomes more prominent.

Finally, the PWM switching pattern resulting from asymmetric regular sampling tightly resembles that of natural sampling, albeit the small phase shift that is introduced by the sampling process. This strong similarity can be observed when comparing Figs. 3.6(b) and 3.7(b) with each other.

3.3.1.2 Harmonic Analysis

Any periodic signal can be represented as the weighted infinite sum (or series) of sinusoidal signals, the so-called *Fourier series*. The computation and analysis of Fourier series is commonly known as *harmonic analysis*.

Consider CB-PWM with the carrier frequency f_c. The modulating signal has the fundamental frequency f_1. The switching decisions made by the PWM depend on these two signals—the modulating signal and the carrier signal(s). These signals have two different fundamental frequencies and periods. To account for this, the Fourier series is formulated as a function of both, and the Fourier coefficients employ two sets of integrals rather than one. The notion of double Fourier series integrals was introduced in [14] for power electronic converters, enabling the derivation of analytical expressions for the harmonic spectrum of PWM.

More recently, by formulating and solving the double Fourier series integrals, this analytical approach has been made popular by the book [15]. The two-level converter case is analyzed and described in detail in [15] and the references therein.[5] For multilevel converters, the analytical expressions turn out to be quite intricate. As a result, closed-form expressions for the magnitude of the harmonic components are available only for natural sampling [17]. For regular sampling, the outer integrals of the Fourier coefficients need to be evaluated numerically [17].

The mathematical details of the harmonic analysis are beyond the scope of this book. The interested reader is referred to [15], particularly to its Appendix 1, for an excellent introduction to this type of analysis. In the following, a brief summary of some of the main results is provided.

The switching nature of PWM leads to harmonics in the output waveform. It is well known that these harmonics are located at the frequencies

$$f_{\mu\nu} = \mu f_c + \nu f_1, \quad \mu \in \mathbb{N}, \nu \in \mathbb{Z}, \tag{3.8}$$

which are integer multiples of the carrier frequency f_c and the fundamental frequency f_1, where μ refers to the integer multiples of the carrier frequency and ν refers to the sidebands. The harmonics at multiples of the carrier frequency result from the triangular carrier signals. In particular, when the modulating signal is a dc signal ($f_1 = 0$), the harmonic spectrum is limited to harmonics at integer multiples of the carrier frequency. In the general case of a sinusoidally

[5] Recently, another promising harmonic analysis approach has been proposed in [16], which avoids the notion of the double Fourier series and relies on superposition and convolution instead. This method tends to be numerically more robust and allows one to also address in a comprehensive manner the case where common-mode voltage harmonics are added to the modulating signal.

varying modulating signal with the fundamental frequency $f_1 > 0$, sideband harmonics are added around the carrier multiple harmonics. These sideband harmonics are the result of the carrier multiple harmonics convoluted with the fundamental component.

More specifically, for natural and asymmetric regular sampling, harmonics exist only at frequencies $f_{\mu\nu}$ for which $\mu + \nu$ is an odd number. Taking this fact into account, the following list summarizes the frequencies of these harmonics and introduces the terminology commonly used to describe them.

- Fundamental component: $f_{01} = f_1$
- Baseband harmonics: $f_{0\nu} = \nu f_1$ with $\nu \in \{3, 5, 7, \dots\}$
- Carrier multiple harmonics: $f_{\mu 0} = \mu f_c$ with $\mu \in \{1, 3, 5, \dots\}$
- Sideband harmonics: $f_{\mu\nu} = \mu f_c + \nu f_1$ with

$$\begin{cases} \mu \in \{1,3,5,\dots\} \text{ and } \nu \in \{\pm 2, \pm 4, \pm 6, \dots\} \\ \mu \in \{2,4,6,\dots\} \text{ and } \nu \in \{\pm 1, \pm 3, \pm 5, \dots\} \end{cases}.$$

Note that for odd multiples of the carrier frequency, the sidebands are located at even multiples of the fundamental frequency around the carrier, and vice versa.

For three-level converters and naturally sampled CB-PWM with phase disposition, the pairs of symmetrical sideband harmonics around the carriers have the same magnitude. Baseband harmonics do not exist when operating in the linear modulation range. In overmodulation, the frequencies of the harmonics remain unchanged, except for the addition of baseband harmonics of the order $\nu = 3, 5, 7$, and so on. For more details on this, the reader is referred to [15, Chap. 11].

As shown in [17], subtle differences arise when considering regular sampling. As a result of the sampling process, baseband harmonics do appear, but they are of low magnitude. Moreover, the magnitudes of the pairs of sideband harmonics are no longer identical. Nevertheless, these differences tend to be small.

Example 3.3 *The harmonic spectrum of single-phase CB-PWM is shown in Fig. 3.8(c). As before, the modulation index is* $m = 0.8$, *and asymmetric regular sampling and phase disposition are used. The ratio* $f_c/f_1 = 21$ *is chosen. Using the discrete Fourier transform, the magnitude of the harmonic spectrum of the switch position* u_a *is computed. To obtain the harmonic spectrum of the phase voltage* v_a, *as defined in (3.3) and Fig. 3.5, the spectrum of the switch position needs to be multiplied with* $0.5v_{dc}$. *Note that the amplitudes of the harmonic spectrum are peak values rather than rms values.*

In Fig. 3.8(c), the fundamental component has the magnitude 0.8, which corresponds to the desired modulation index $m = 0.8$. *As the integer* f_c/f_1 *is odd, harmonics exist only at odd multiples of the fundamental frequency. The carrier frequency is clearly visible at 1050 Hz. As its magnitude is 0.461, it is the dominant non-fundamental component in the spectrum. The even-numbered lower and upper sideband harmonics of the carrier are located at 950 and 1150 Hz, 850 and 1250 Hz, and so on. Note that the amplitudes of the upper sideband harmonics are slightly larger than those of the lower ones. This is a characteristic of regular sampling. For natural sampling, the corresponding upper and lower sideband harmonics have the same amplitudes.*

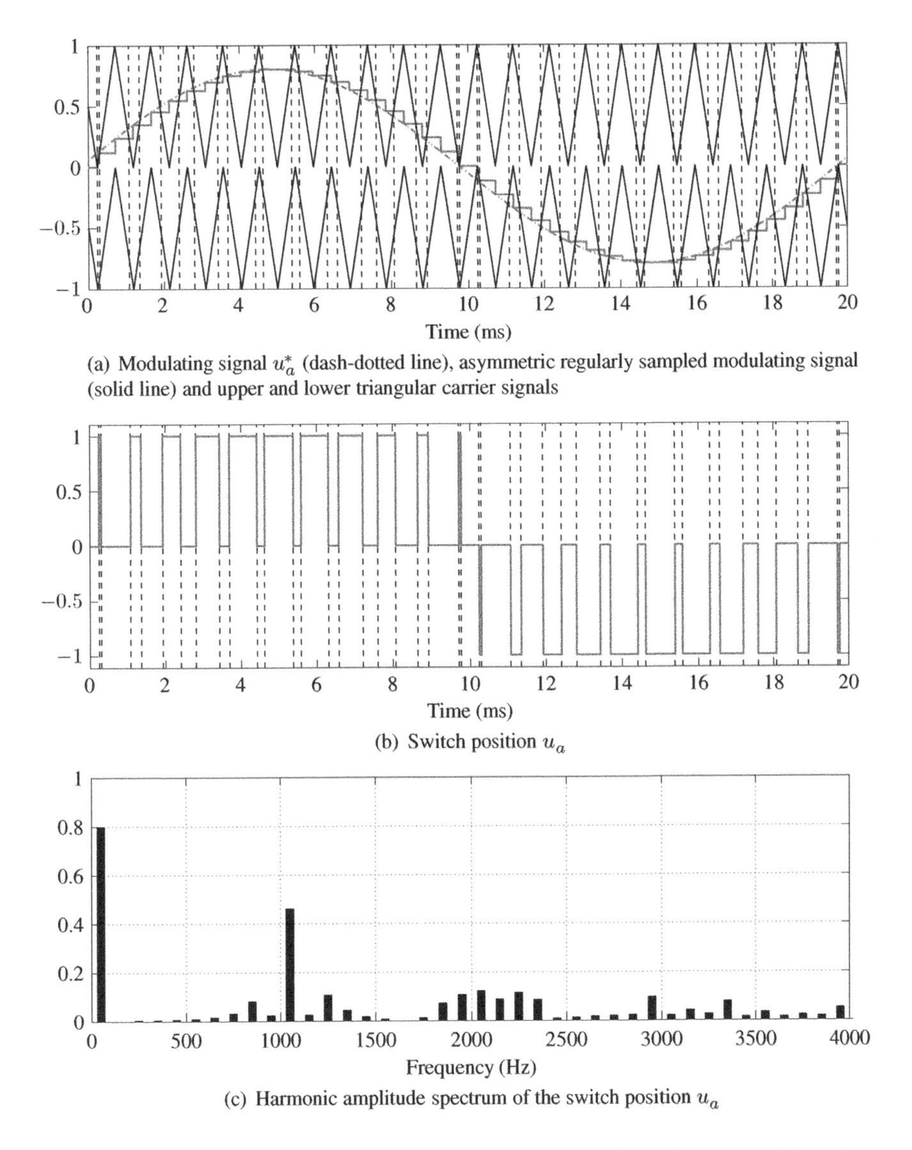

(a) Modulating signal u_a^* (dash-dotted line), asymmetric regularly sampled modulating signal (solid line) and upper and lower triangular carrier signals

(b) Switch position u_a

(c) Harmonic amplitude spectrum of the switch position u_a

Figure 3.8 Asymmetric, regularly sampled, single-phase CB-PWM with $f_c/f_1 = 21$

A harmonic component at 2 times the carrier frequency (at $2f_c = 2100$ Hz) does not exist. Its odd-numbered sidebands tend to be of higher magnitudes than the even-numbered sidebands around the carrier. The harmonics of the third carrier and its sidebands are of low magnitudes and widely spread out.

As shown in [15], the phase of the harmonic component at the frequency $f_{\mu\nu}$ is given by

$$\phi_{\mu\nu} = \mu\phi_c + \nu\phi_1 + \text{const}, \quad \mu \in \mathbb{N}, \nu \in \mathbb{Z}, \tag{3.9}$$

where ϕ_c refers to the phase of the triangular carrier signal, while ϕ_1 is the phase of the modulating signal (see also (3.5)). The constant reflects shifts by $180°$, which are due to Fourier coefficients with negative signs. The expression for the phase of the harmonics (3.9) corresponds to the one for its frequency (3.8).

The sideband harmonics of different carrier multiples tend to overlap when the ratio f_c/f_1 is small. This can be seen in Fig. 3.7(c), where $f_c/f_1 = 9$. Depending on the relationship between the phases of the overlapping harmonics, such overlaps increase or decrease their amplitudes. In particular, the eighth lower sideband harmonic of the carrier is located at $f_c - 8f_1 = f_1$, overlapping with the fundamental component, slightly reducing its magnitude from the desired 0.8 to 0.791.

3.3.2 Three-Phase Carrier-Based Pulse Width Modulation

In this section, single-phase CB-PWM is extended to the three-phase case. As will be shown, because of the harmonic cancellation between the phases, the harmonic content in the output voltages is significantly reduced when compared to the single-phase case. The common-mode voltage represents an additional degree of freedom in the three-phase system, which allows one to extend the linear modulation range from 1 to 1.155.

Consider the three-phase, three-level NPC converter shown in Fig. 3.9 with a fixed neutral point potential. The voltage of phase a with respect to the dc-link midpoint was defined in (3.3). Similarly, the three-phase voltage $\boldsymbol{v}_{abc} = [v_a \;\; v_b \;\; v_c]^T$ with respect to the dc-link midpoint is defined as

$$\boldsymbol{v}_{abc} = \frac{v_{\text{dc}}}{2}\boldsymbol{u}_{abc}, \tag{3.10}$$

where $\boldsymbol{u}_{abc} = [u_a \;\; u_b \;\; u_c]^T$ denotes the three-phase switch position with $\boldsymbol{u}_{abc} \in \{-1, 0, 1\}^3$. With the help of the Clarke transformation (2.11), the voltage $\boldsymbol{v}_{abc}$ is transformed into the stationary orthogonal coordinate system according to $\boldsymbol{v}_{\alpha\beta 0} = \boldsymbol{K}\boldsymbol{v}_{abc}$.

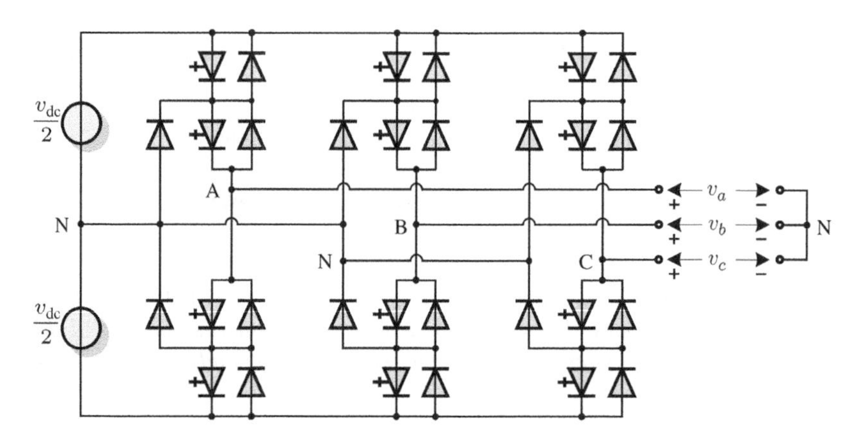

Figure 3.9 Three-phase, three-level converter with the output voltage v_a between the phase leg terminal A and the dc-link midpoint N. The output voltages v_b and v_c of the phase legs B and C are defined accordingly

In a three-phase system with a floating (i.e., not grounded) star point, the third component of the vector $\boldsymbol{v}_{\alpha\beta 0}$, the so-called *common-mode* voltage

$$v_0 = \frac{1}{3}(v_a + v_b + v_c) \tag{3.11}$$

does not drive a phase current. This voltage represents a degree of freedom that can be exploited by the PWM to increase the modulation range and thus the achievable output voltage. The first two orthogonal voltage components v_α and v_β form the *differential-mode* voltage, which does drive a phase current. For a three-phase system, the task of the PWM is to translate the voltage reference $\boldsymbol{v}^*_{\alpha\beta}$ into the three-phase switch position $\boldsymbol{u}_{abc}$, such that the resulting differential-mode voltage $\boldsymbol{v}_{\alpha\beta}$ approximates its reference $\boldsymbol{v}^*_{\alpha\beta}$.

3.3.2.1 Modulation

As in the single-phase case, we scale the voltage reference by half the dc-link voltage and define the modulating three-phase vector

$$\boldsymbol{u}^*_{abc}(t) = \frac{2}{v_{\mathrm{dc}}}\,\boldsymbol{v}^*_{abc}(t) = m \begin{bmatrix} \sin(\omega_1 t + \phi_1) \\ \sin(\omega_1 t - \frac{2\pi}{3} + \phi_1) \\ \sin(\omega_1 t - \frac{4\pi}{3} + \phi_1) \end{bmatrix}. \tag{3.12}$$

Note that phases b and c are phase shifted by 120° and 240°, respectively, with respect to phase a.

The two carrier signals used for the single-phase case are replicated for phases b and c. This results in three upper and three lower carrier signals. In the case of phase disposition, all six carrier signals are in phase. For phase-opposite disposition, the three lower carrier signals are phase shifted by 180° with respect to the upper ones. Three-phase modulation is achieved by three single-phase PWM units that operate in parallel and modulate according to the switching rules stated in the previous section. For asymmetric, regularly sampled CB-PWM, for example, the switching rules summarized in Table 3.1 are used for all three phases.

Example 3.4 *CB-PWM for a three-level three-phase converter is illustrated in Fig. 3.10. The same parameters are used as in Example 3.2, namely the carrier frequency is $f_c = 450$ Hz, the carriers are in phase, asymmetric, regular sampling is used, the modulation index is $m = 0.8$, and the fundamental frequency is $f_1 = 50$ Hz. The phase shift between the modulating signal and the carriers is set again to $\phi_1 = 1.5\pi f_1/f_c$ to compensate for the sampling delay and to align the sampled modulating signal with the negative half-wave of the carrier signals—see also the discussion after Example 3.2. The resulting sequence of switching commands is shown in Fig. 3.10(b).*

For the single-phase CB-PWM shown in Figs. 3.6–3.8, we had observed quarter-wave symmetry for phase a. The modulating signal of phase b is delayed by one-third of the fundamental period with respect to phase a. To preserve the phase relationship between the modulating signal and the carrier signals in phases b and c, integer multiples of the carrier interval must be equal to one-third of the fundamental period $T_1 = 1/f_1$, that is, $T_1/3 = nT_c$, with $n \in \mathbb{N}$. This is equivalent to $f_c/f_1 = 3n$. In the previous example, we had chosen the non-even triplen ratio $f_c/f_1 = 9$. This choice permits quarter-wave symmetry in all three phases.

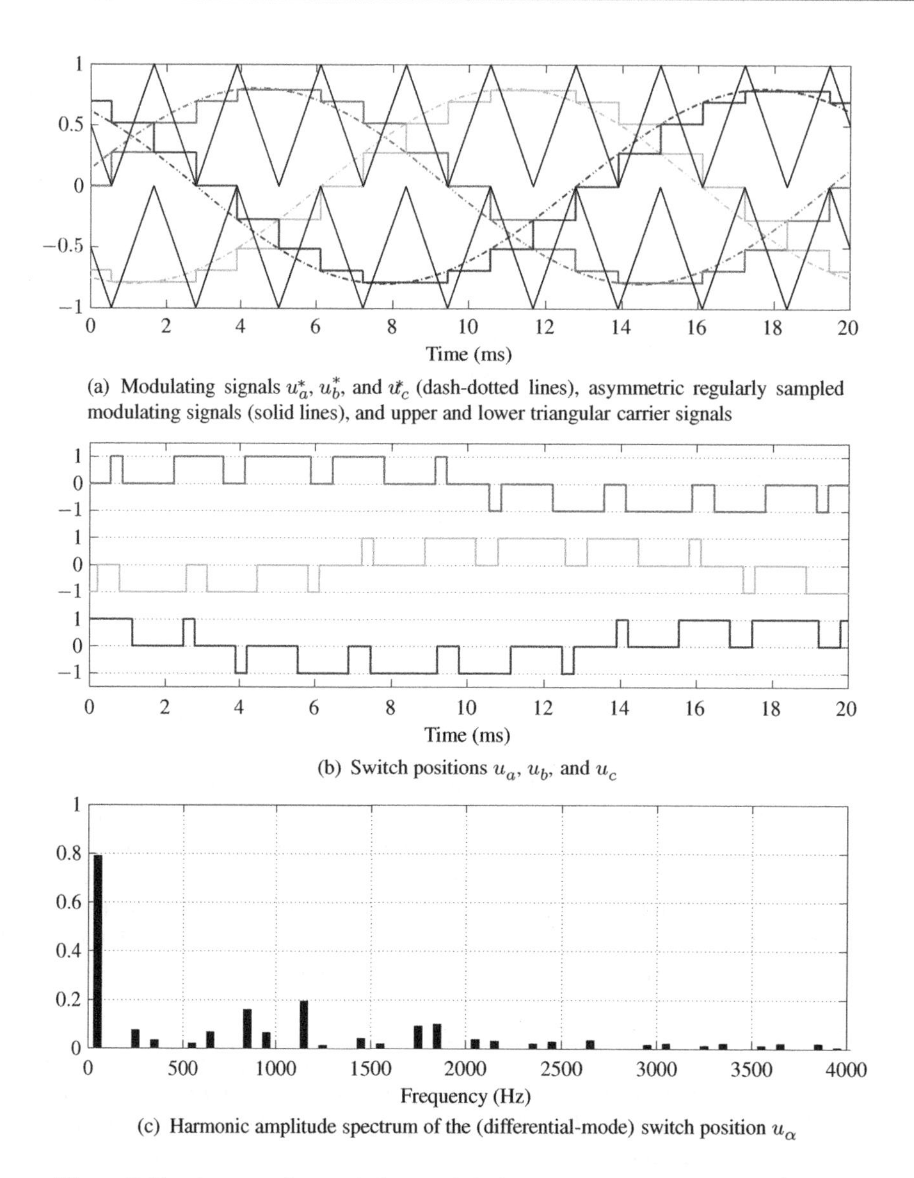

(a) Modulating signals u_a^*, u_b^*, and u_c^* (dash-dotted lines), asymmetric regularly sampled modulating signals (solid lines), and upper and lower triangular carrier signals

(b) Switch positions u_a, u_b, and u_c

(c) Harmonic amplitude spectrum of the (differential-mode) switch position u_α

Figure 3.10 Asymmetric, regularly sampled, three-phase CB-PWM with $f_c/f_1 = 9$

3.3.2.2 Harmonic Analysis

In (3.11), we had defined the notion of the (fundamental) common-mode voltage. Common-mode voltage harmonics are defined accordingly as the harmonic spectrum of the average value of the three phase voltages. Specifically, common-mode voltage harmonics relate to phase voltage harmonics that feature the same amplitude, frequency, and phase in all three phases.

Consider the three voltage harmonics in phases a, b, and c that originate from the same carrier multiple μ and fundamental frequency multiple ν. Clearly, these three harmonics have the same amplitude and the same frequency. According to the phase relation (3.9) and the definition of the modulating vector (3.12), the phases of the harmonics in the three phases are given by

$$\phi_{a,\mu\nu} = \mu\phi_c + \nu\phi_1 + \text{const}, \tag{3.13a}$$

$$\phi_{b,\mu\nu} = \mu\phi_c + \nu(\phi_1 - \frac{2\pi}{3}) + \text{const}, \tag{3.13b}$$

$$\phi_{c,\mu\nu} = \mu\phi_c + \nu(\phi_1 - \frac{4\pi}{3}) + \text{const}. \tag{3.13c}$$

For triplen ν, that is, $\nu \in \{0, 3, 6, \ldots\}$, the three harmonics are in phase. We conclude that voltage harmonics at the frequencies $f_{\mu\nu}$ with $\nu \in \{0, 3, 6, \ldots\}$ are common-mode harmonics, regardless of μ. In particular, all carrier multiple harmonics, triplen sideband harmonics, and triplen baseband harmonics are common-mode harmonics.

Common-mode harmonics cancel out in the line-to-line voltage and do not cause any harmonic current, provided that the star point floats (i.e., is not grounded). We conclude that a three-phase CB-PWM features only the following (differential-mode) voltage harmonics:

- Fundamental component: $f_{01} = f_1$
- Baseband harmonics: $f_{0\nu} = \nu f_1$ with $\nu \in \{5, 7, 11, 13, \ldots\}$
- Sideband harmonics: $f_{\mu\nu} = \mu f_c + \nu f_1$ with

$$\begin{cases} \mu \in \{1, 3, 5, \ldots\} \text{ and } \nu \in \{\pm 2, \pm 4, \pm 8, \pm 10, \ldots\} \\ \mu \in \{2, 4, 6, \ldots\} \text{ and } \nu \in \{\pm 1, \pm 5, \pm 7, \pm 11, \ldots\} \end{cases}.$$

This list can easily be compiled by identifying and removing the common-mode harmonics from the list of single-phase harmonics in the previous section. For naturally sampled CB-PWM in the linear modulation range, the baseband harmonics are all zero. As a result, except for the fundamental component, the only remaining harmonic content relates to the sideband harmonics.

Example 3.5 *Asymmetric, regularly sampled CB-PWM for a three-level, three-phase converter was considered in Example 3.4. Figure 3.10(b) shows the corresponding three-phase switching pattern* $\boldsymbol{u}_{abc} = [u_a \; u_b \; u_c]^T$. *Transforming* $\boldsymbol{u}$ *with the help of the Clarke transformation* $\boldsymbol{K}$ *into stationary orthogonal coordinates results in the differential-mode switching patterns* u_α *and* u_β. *The harmonic amplitude spectra of these two components are identical and are shown in Fig. 3.10(c). As the Clarke transformation is amplitude invariant, the amplitude of the modulating waveform (the modulation index* m*) is preserved. When comparing the three-phase spectrum with the single-phase one in Fig. 3.7(c), one can appreciate the significant content of common-mode voltage harmonics that is present in the single-phase system but cancelled out in the three-phase system.*

3.3.2.3 Common-Mode Voltage Injection

The fact that common-mode voltage harmonics cease to exist in the line-to-line voltages can be exploited to increase the linear modulation region from 1 to $2/\sqrt{3} = 1.155$ by adding an appropriate common-mode term to the three modulating signals.

Consider the three-phase modulating signal $\boldsymbol{u}^*_{abc}$ in Fig. 3.10(a). The upper and lower peaks of the signal are at ± 0.8, in line with the modulation index of $m = 0.8$. The difference between a pair of modulating signals relates to the line-to-line voltage. The difference between $\max(\boldsymbol{u}^*_{abc})$ and $\min(\boldsymbol{u}^*_{abc})$ corresponds to the *peak* line-to-line voltage. When inspecting the figure, one can observe that at any time instant t, this difference is always less than $\sqrt{3}m$. Within a fundamental period, however, the difference between $\max(\boldsymbol{u}^*_{abc})$ and $\min(\boldsymbol{u}^*_{abc})$ is $2m$ by definition (see (3.12)). This implies that the available dc-link voltage is not fully utilized.

To boost the line-to-line voltage, an appropriate offset can be applied to all three modulating signals. As the offset is identical for all three phases, it constitutes a common-mode term. By definition, the addition of a common-mode signal to the modulating signal $\boldsymbol{u}^*_{abc}$ does not affect the line-to-line voltage.

It is standard practise to add one of the two common-mode terms

$$u_0^* = \frac{m}{6}\sin(3\omega_1 t + \phi_1) \tag{3.14}$$

$$u_0^* = -\frac{1}{2}(\min(\boldsymbol{u}^*_{abc}) + \max(\boldsymbol{u}^*_{abc})) \tag{3.15}$$

to the three-phase modulating signal in the form of $\boldsymbol{u}^*_{abc} + u_0^*$. The first term (3.14) is a sinusoidal signal with three times the fundamental frequency and the same phase as the modulating signal. One can show that one-sixth of the modulating signal's amplitude results in the maximum voltage boost. An example for this is shown in Fig. 3.11(a), assuming the modulation index $m = 0.8$. The addition of the third harmonic flattens the peaks of the modulating signal.

The addition of the second term (3.15) centers the three-phase modulating signal around zero. As a result, at any given time instant, $-\min(\boldsymbol{u}^*_{abc} + u_0^*) = \max(\boldsymbol{u}^*_{abc} + u_0^*)$ (see Fig. 3.11(b)). Both common-mode terms increase the linear modulation range by 15.5% from $m = 1$ to $m = 2/\sqrt{3}$, thus fully exploiting the available dc-link voltage.

3.3.2.4 Equivalence with SVM

A popular alternative to CB-PWM is SVM [18]. As shown in [19], CB-PWM with phase disposition can be modified such that it becomes equivalent to SVM, in the sense that both

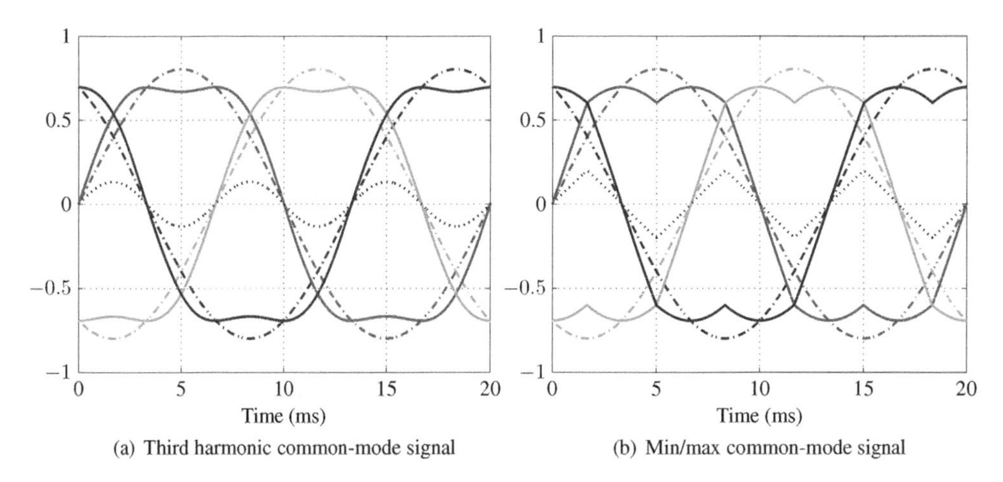

(a) Third harmonic common-mode signal (b) Min/max common-mode signal

Figure 3.11 Injection of the common-mode signal u_0^* (dotted line) to the three-phase modulating signal $\boldsymbol{u}^*_{abc}$ (dash-dotted lines). The three-phase sum $\boldsymbol{u}^*_{abc} + u_0^*$ of the two signals is shown as solid lines

methods yield the same switching pattern when supplied with the same modulating signal. This equivalence can be achieved by the addition of an appropriate common-mode term u_0^* to the modulating signal $\boldsymbol{u}_{abc}^*$. As explained in [19], the required common-mode term is

$$u_0^* = \bar{u}_0^* + \frac{1}{2} - \frac{1}{2}(\min(\bar{\boldsymbol{u}}_{abc}^*) + \max(\bar{\boldsymbol{u}}_{abc}^*)), \tag{3.16}$$

with the scalar and three-phase terms

$$\bar{u}_0^* = -\frac{1}{2}(\min(\boldsymbol{u}_{abc}^*) + \max(\boldsymbol{u}_{abc}^*)) \tag{3.17}$$

$$\bar{\boldsymbol{u}}_{abc}^* = (\boldsymbol{u}_{abc}^* + \bar{u}_0^* + 1) \mod 1. \tag{3.18}$$

Note that the term (3.17) is the same as in (3.15). Also note that for two-level converters, only this min/max common-mode term is required to achieve equivalence between CB-PWM and SVM (see [20]). The expression ξ mod 1 in (3.18) is defined as the remainder of the Euclidean division of ξ by 1. The result is bounded between 0 and 1.

The common-mode signal u_0^* that turns CB-PWM into SVM is illustrated in Fig. 3.12. For modulation indices in the linear operating range (i.e., below 1.155), this common-mode signal differs distinctively from the ones commonly used to boost the line-to-line voltage, that is, (3.14) and (3.15). At $m = 1.155$, however, the SVM common-mode signal is equal to the min/max term (3.15).

Example 3.6 *The equivalence of CB-PWM with SVM is illustrated in Fig. 3.13. The common-mode signal (3.16) is added to the three-phase sinusoidal modulating signal. Asymmetric, regular sampling is applied to this sum. The intersections with the upper and lower triangular carriers determine the switching transitions. The modulation index* $m = 1.155$ *corresponds to the upper end of the linear modulation regime. This is confirmed in Fig. 3.13(a), in which one can observe that the maxima of the dash-dotted lines are 1. Very narrow pulses are formed in Fig. 3.13(b) when the modulating signals are close to 1. The amplitude spectrum of a differential-mode component is shown in Fig. 3.13(c).*

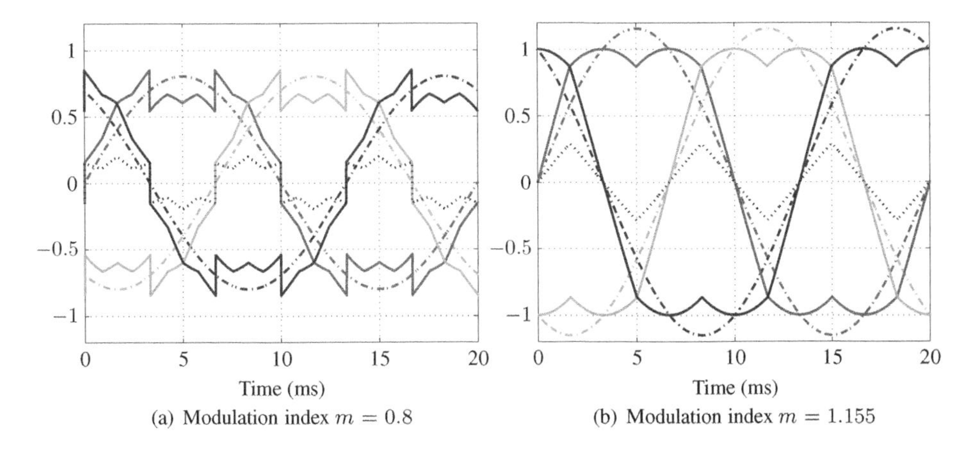

Figure 3.12 Equivalence of CB-PWM with SVM. Addition of the common-mode signal u_0^* (dotted line) to the three-phase modulating signal $\boldsymbol{u}_{abc}^*$ (dash-dotted lines). The new modulating signal $\boldsymbol{u}_{abc}^* + u_0^*$ (solid lines) when applied to CB-PWM results in the SVM switching pattern

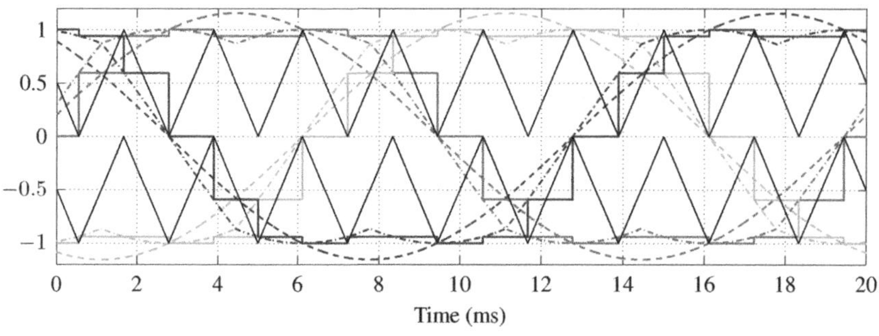

(a) Modulating signals u_a^*, u_b^*, and u_c^* (dashed lines), modulating signals including the common-mode signal (dash-dotted lines), asymmetric regularly sampled modulating signals (solid lines), and upper and lower triangular carrier signals

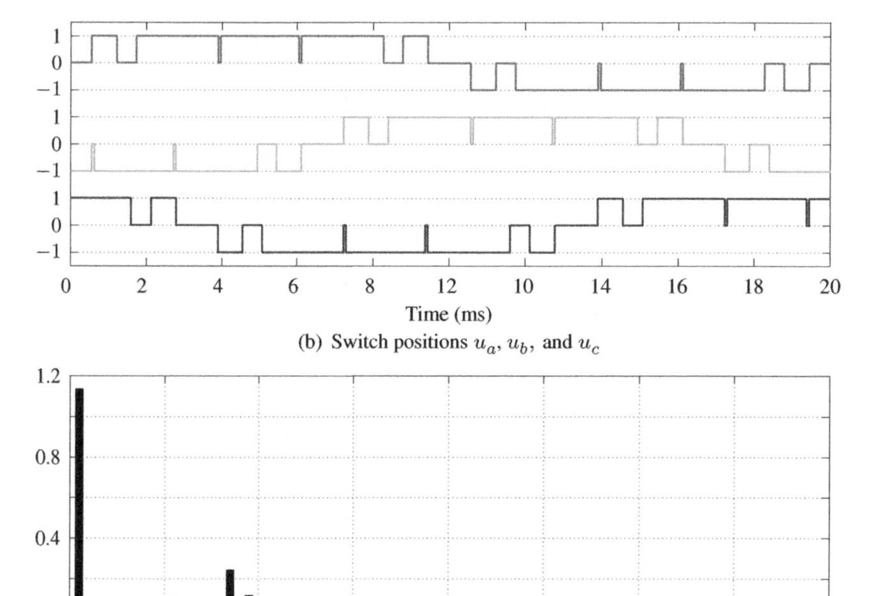

(b) Switch positions u_a, u_b, and u_c

(c) Harmonic amplitude spectrum of the switch position u_α

Figure 3.13 Asymmetric, regularly sampled CB-PWM with $f_c/f_1 = 9$ and the injection of a common-mode voltage such that the resulting switching pattern is the same as for SVM. The modulation index is $m = 1.155$

3.3.2.5 Overmodulation and Six-Step Operation

When increasing the modulation index above $m = 1.155$, CB-PWM enters the so-called *non-linear* modulation regime or *overmodulation*. The linear relationship between the demanded m and the resulting magnitude of the fundamental component $\hat{u}_1$ ceases to exist. To increase $\hat{u}_1$ further, the modulation index m needs to be increased disproportionately. As m is increased beyond 1.155, pulses are removed from the switching pattern. This process starts in the vicinity

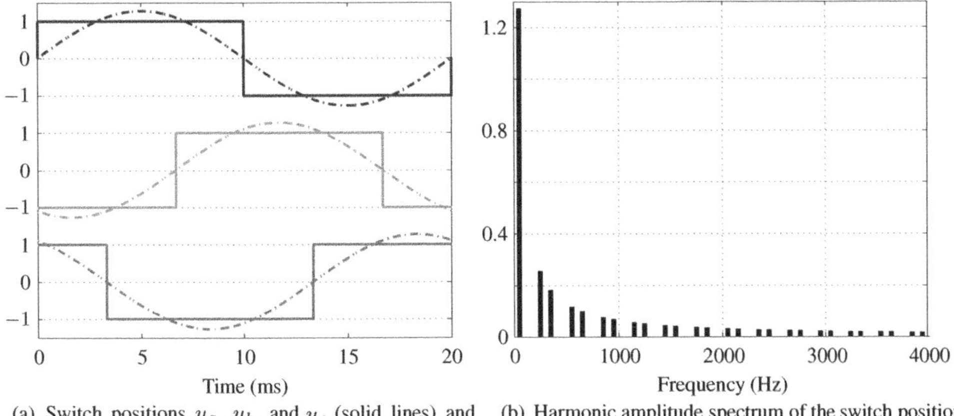

(a) Switch positions u_a, u_b, and u_c (solid lines) and three-phase fundamental waveform (dash-dotted lines) with magnitude $\hat{u}_1 = 4/\pi$

(b) Harmonic amplitude spectrum of the switch position u_α

Figure 3.14 Six-step operation (at the maximum modulation index $m = 4/\pi$)

of the peaks of the modulating signal, that is, around 90° and 270°. The removal of pulses reduces the switching frequency, but also increases the harmonic distortions, namely the current TDD.

For very large m, only one pulse per fundamental half-wave remains, as shown in Fig. 3.14(a). This switching regime, which is called *square-wave* or *six-step* operation, fully utilizes the available dc-link voltage of the converter and maximizes the line-to-line voltage. The magnitude of the fundamental waveform is $\hat{u}_1 = 4/\pi = 1.273$. This statement will be easy to prove when performing Fourier analysis of OPPs in Sect. 3.4.1. In that section, we will also derive the amplitudes of the harmonic spectrum of the six-step operation.

The harmonic spectrum during the six-step operation is shown in Fig. 3.14(b). As the carrier signals do not influence the switching decisions, sidebands around carrier multiple harmonics are nonexistent. Instead, only baseband harmonics are present in the spectrum. In line with the harmonic analysis performed earlier in this section, the baseband harmonics are restricted to non-triplen odd multiples of the fundamental frequency. More specifically, during the six-step operation, harmonic components are located at the frequencies νf_1 with $\nu \in \{5, 7, 11, 13, \ldots\}$.

3.3.3 *Summary and Properties*

The different modulation regimes for three-phase CB-PWM are summarized in Fig. 3.15. Four distinct regions exist for the modulation index m.

- $0 \leq m \leq 1$: linear modulation region
- $1 < m \leq 1.155$: extended linear modulation region provided that an appropriate common-mode signal is added, such as (3.14) or (3.15); overmodulation without common-mode injection
- $1.155 < m < 1.273$: overmodulation
- $m = 1.273$: six-step operation

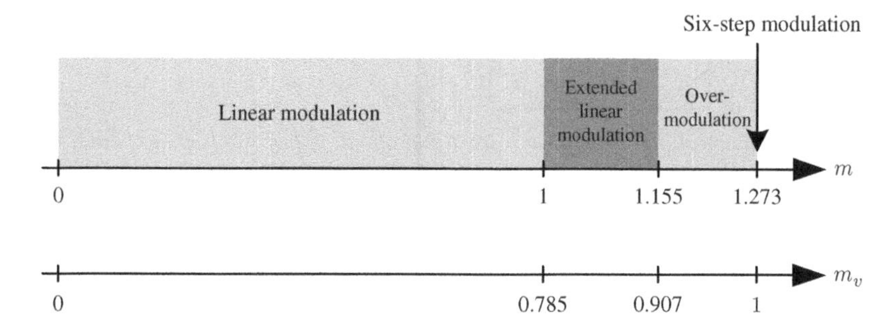

Figure 3.15 Modulation regimes for CB-PWM with the modulation index m and voltage index m_v

Note that $2/\sqrt{3} = 1.155$ and $4/\pi = 1.273$. An alternative definition of the modulation index is the so-called *voltage index*

$$m_v = \frac{\pi}{4} m, \tag{3.19}$$

which is the normalized modulation index. By definition, the voltage index is bounded by 0 and 1, that is, $m_v \in [0, 1]$ (see also Fig. 3.15).

So far, we have only considered PWM schemes in which the ratio f_c/f_1 between the carrier frequency and the fundamental frequency is an integer value, the so-called *synchronous* PWM. Owing to this synchronism, the sideband harmonics are located at integer multiples of the fundamental frequency. As a result, the so-called *subharmonic* spectral components, that is, harmonics below the fundamental frequency, do not exist. For *asynchronous* PWM, the ratio f_c/f_1 is a rational number, and sidebands of the carrier frequency may fall below the fundamental frequency, resulting in subharmonics. This is an issue when the ratio f_c/f_1 is small, which is typically the case in MV applications.

For synchronous PWM, a high degree of symmetry can be attained in the switching patterns. For odd ratios f_c/f_1, quarter-wave symmetry can be achieved in the single-phase switching pattern. To achieve symmetry between the three phases, triplen ratios f_c/f_1 are required. The phase of the modulating signal with regard to the carrier signals has a significant influence on the switching pattern, its symmetry, the number of pulses, and hence the switching frequency. When f_c/f_1 is small, the phase also has a considerable impact on the harmonic performance.

The *modulation cycle* is defined as the time interval during which the modulating signal is approximated by a sequence of three-phase switch positions. For asymmetric, regularly sampled CB-PWM, for example, this time interval is equal to $T_c/2$. An important property of CB-PWM and SVM is that their modulation cycle is symmetrical and of fixed length. This constrains the switching transitions to regular time intervals. Regardless of the sampling method, (at most) one switching transition per phase occurs within each carrier half-interval. Either in the overmodulation range or for certain phase shifts in the modulating signal, pulses might get dropped. The fact that the modulation cycle is symmetrical and of fixed length carries two advantages and one disadvantage.

First, note that the voltage harmonics that are due to the PWM stage are translated to current harmonics, which correspond to a ripple current. This ripple current is superimposed on the fundamental current component. The use of a symmetrical modulation cycle ensures that at the peaks of the carrier signals the ripple current is zero in all three phases. This facilitates the sampling of the current at regular and evenly spaced time instants. Specifically, when sampling the current at the carrier peaks, only the fundamental component of the current is measured. This important property of CB-PWM and SVM enables the use of linear control schemes that neglect the switching characteristic of the PWM through the notion of averaging.

Second, the sampling process in asymmetric or symmetric regularly sampled CB-PWM leads to a time delay (and thus to a phase shift) between the reference signal at the input of the PWM and the switched waveform at its output. Because the modulation cycle is of a fixed length, this phase shift is time-invariant, allowing one to compensate for it during the controller design. This enables the use of aggressively tuned linear controllers with a high bandwidth.

Third, however, the restriction to one switching transition per phase and carrier half-interval limits the achievable harmonic performance, particularly at low ratios between the carrier and fundamental frequency. When abolishing this restriction, as it is done for OPPs, additional degrees of freedom arise that can be exploited to either shape the spectrum or to reduce the harmonic distortions, as will be shown in the next section.

We conclude that CB-PWM is both easy to understand and implement. It works well in practise and provides an adequate harmonic performance, provided that the ratio between the carrier and fundamental frequency is not significantly below 20. CB-PWM also works well in connection with linear control schemes such as vector control (see Sect. 3.6.2). At low f_c/f_1 ratios, however, the performance of CB-PWM is rather poor. The harmonic distortions are high, sideband harmonics spread into the low-frequency range, the harmonic spectrum cannot be shaped, and the phase delay is significant. For such low f_c/f_1 ratios, more sophisticated control and modulation methods are required.

3.4 Optimized Pulse Patterns

Optimal PWM patterns can be calculated in an offline procedure by minimizing a cost function subject to constraints. The result of this optimization procedure is a set of optimal switching angles and switch positions over a fundamental period. Two different optimization criteria are commonly used: the selective harmonic elimination (SHE) of low-order harmonics, and the minimization of current distortions. We refer to the latter as optimized pulse patterns (or OPPs).

Starting with [21, 22] in the 1970s, the literature on SHE is extensive, indicating its popularity in academia and widespread adoption in industry. In SHE, a certain number of low-order voltage harmonics are eliminated. Given d switching angles over a quarter of the fundamental period, $d - 1$ harmonics can be eliminated, such as the 5th, 7th, 11th, and so on, harmonics. The dth degree of freedom is required to set the magnitude of the fundamental voltage according to the desired modulation index. To derive the optimal switching angles, an algebraic equation system consisting of d nonlinear equations needs to be solved. A cost function is not part of the SHE problem formulation. The main advantage of SHE is that the switching angles as a function of the modulation index are continuous. With regard to the current controller, this important property permits the use of the vector control principle (see also Sect. 3.6.2) albeit in a slow closed-loop setting. Significant results regarding SHE include [23–25] for two-level, [26] for three-level, [27, 28] for five-level, and [29] for general multilevel converters.

In the second variety of optimal PWM—OPPs—a cost function is minimized subject to constraints when computing the pulse pattern. This gives rise to a nonlinear optimization problem with multiple (local) minima. The cost function typically relates to the current distortions.

In general, with a few notable exceptions discussed next, optimal PWM based on the minimization of a cost function is rarely considered, reported, or used. This is mainly due to the difficulty in computing and using such OPPs in a closed-loop control setting. Discontinuities in the switching angles pose formidable challenges for linear control methods based on averaging. Dedicated control methods are required, such as closed-loop control based on trajectory tracking.

Early results of the computation of OPPs for two-level converters are provided in [30, 31]. Assuming an inductive load, the current TDD is proportional to the voltage TDD when scaling the voltage harmonics by their respective frequency. For a given number of switching angles, the current TDD is minimized in [31] using a gradient method. An algorithm to compute OPPs for multilevel converters is explained in [32] and [33].

In the early 1990s, the development of the notion of trajectory tracking enabled the application of OPPs to industrial drives. The initial results for stator current trajectory tracking [34–36] were adapted a decade later to the tracking of the stator flux trajectory [37, 38]. This modification made the trajectory controller independent from variations in the total leakage inductance. Interestingly, related ideas had already been proposed earlier [39]. OPPs for drive systems with four-level and five-level converters have also been investigated in [40, 41].

This section focuses on OPPs that minimize the current TDD. The mathematical equations describing the voltage and current harmonics of OPPs as a function of the switching angles and switching transitions are derived. The optimization problem is formulated and solved for three-level and five-level converters. The properties of OPPs and the inherent challenges when using them in a closed-loop setting are discussed at the end of this section. As the literature on OPPs is limited, this section is more detailed than the previous one on CB-PWM.

3.4.1 Pulse Pattern and Harmonic Analysis

Consider the three-level, single-phase switched waveform $u(\theta)$, with the angle θ as argument and $u(\theta) \in \mathcal{U}$ with $\mathcal{U} = \{-1, 0, 1\}$. Without loss of generality, we assume that $u(\theta)$ is periodic with the period 2π, that is, $u(\theta) = u(2\pi + \theta)$ for all θ. As is common practise, we impose quarter-wave symmetry on $u(\theta)$, that is, we require

$$u(\theta) = -u(\pi + \theta) \tag{3.20a}$$

$$u(\theta) = u(\pi - \theta). \tag{3.20b}$$

The first constraint ensures that when shifting the switched waveform by half its period, it is equal to the negative of the original waveform. This is known as half-wave symmetry. The second constraint imposes symmetry about the midpoints of the positive and negative half-waves. Waveforms that meet the two constraints (3.20) are quarter-wave symmetric. These constraints also imply that u is an odd function, that is, $u(\theta) = -u(-\theta)$ holds.

We define the *pulse number* d as the number of switching transitions in the single-phase switched waveform $u(\theta)$ between 0 and $\frac{\pi}{2}$, that is, within the first quarter of its fundamental period. The switching transitions are characterized by the switching angles α_i, $i \in \{1, 2, \dots, d\}$, which we refer to as the *primary* switching angles. We impose the order

$$0 \leq \alpha_1 \leq \alpha_2 \leq \cdots \leq \alpha_d \leq \frac{\pi}{2} \tag{3.21}$$

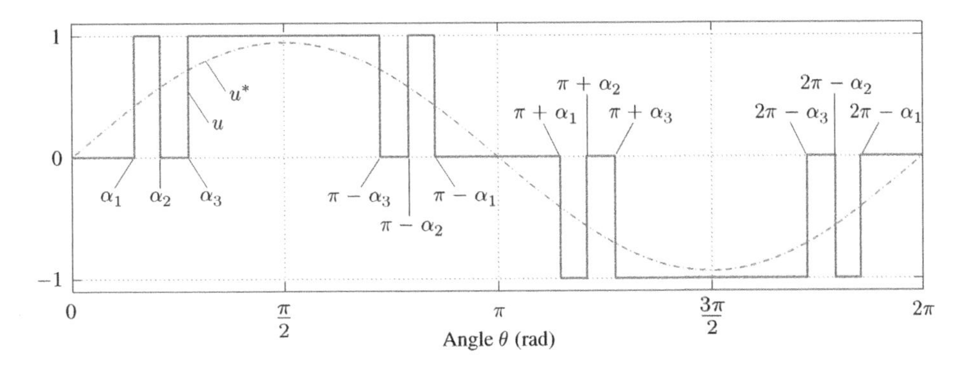

Figure 3.16 Single-phase pulse pattern u with quarter-wave symmetry and the $d = 3$ primary switching angles α_1, α_2, and α_3. The switched waveform u approximates the modulating signal u^*, which is shown as the dash-dotted sinusoidal line

on the primary switching angles. At the angle α_i, the waveform switches from u_{i-1} to u_i. The switch positions u_i, $i \in \{0, 1, \dots, d\}$, are restricted to the set $\mathcal{U}$. Half-wave symmetry (3.20a) implies that the switch position at the beginning of the quarter wave is zero, that is, $u_0 = 0$.

As a result of the symmetry properties (3.20), the switched waveform $u(\theta)$ is fully described by the d switching angles α_i and the $d + 1$ switch positions u_i. We often refer to $u(\theta)$ as the single-phase pulse pattern. An example of the latter is shown in Fig. 3.16 for the case of $d = 3$ primary switching angles and the modulation index $m = 0.95$.

As the single-phase pulse pattern $u(\theta)$ is a periodic signal, it can be described by the Fourier series

$$u(\theta) = \frac{a_0}{2} + \sum_{n=1}^{\infty} (a_n \cos(n\theta) + b_n \sin(n\theta)). \tag{3.22}$$

The Fourier series is an infinite sum of sine and cosine terms of harmonic order n. The Fourier coefficients a_n and b_n relate to the peak value of the nth harmonic component. These coefficients are given by

$$a_n = \frac{1}{\pi} \int_{-\pi}^{\pi} u(\theta) \cos(n\theta) \mathrm{d}\theta, \quad n \geq 0 \tag{3.23a}$$

$$b_n = \frac{1}{\pi} \int_{-\pi}^{\pi} u(\theta) \sin(n\theta) \mathrm{d}\theta, \quad n \geq 1. \tag{3.23b}$$

As the pulse pattern is an odd function with quarter-wave symmetry, the Fourier coefficients a_n are all zero and the b_n are only nonzero for odd n. This fact, which is shown in Appendix 3.A, leads to the compact representation

$$u(\theta) = \sum_{n=1}^{\infty} \hat{u}_n \sin(n\theta) \tag{3.24a}$$

$$\hat{u}_n = \begin{cases} \dfrac{4}{n\pi} \displaystyle\sum_{i=1}^{d} \Delta u_i \cos(n\alpha_i), & \text{if } n = 1, 3, 5, \dots \\ 0, & \text{if } n = 2, 4, 6, \dots, \end{cases} \tag{3.24b}$$

where we have replaced b_n by $\hat{u}_n$ to indicate that the Fourier coefficient relates to the peak value of the nth harmonic signal. The Fourier coefficient is a function of the changes in the switch positions (i.e., the switching transitions), which are defined as

$$\Delta u_i = u_i - u_{i-1} \quad \text{for } i = 1, 2, \ldots, d. \tag{3.25}$$

The detailed derivation of (3.24) is provided in Appendix 3.A.

It is clear from (3.24) that the harmonics of the single-phase pulse pattern are located solely at odd multiples of the fundamental frequency. In particular, harmonics of even order do not exist. The maximum achievable amplitude of the fundamental waveform can now be easily determined. It can be seen from Fig. 3.14 that this is achieved in the six-step operation. Setting $d = 1$, $\Delta u_1 = 1$, and $\alpha_1 = 0$ in (3.24b) leads to the amplitudes

$$\hat{u}_n = \begin{cases} \dfrac{4}{n\pi}, & \text{if } n = 1, 3, 5, \ldots \\ 0, & \text{if } n = 2, 4, 6, \ldots \end{cases} \tag{3.26}$$

of the harmonic components in the six-step operation. The fundamental component is given by $\hat{u}_1 = 4/\pi$.

So far, we have focused on the single-phase case. In a three-phase system, the phase a pulse pattern is used also for phases b and c, but its phase is shifted by 120° and 240°, respectively.

$$\boldsymbol{u}_{abc}(\theta) = \begin{bmatrix} u(\theta) \\ u(\theta - \frac{2\pi}{3}) \\ u(\theta - \frac{4\pi}{3}) \end{bmatrix}. \tag{3.27}$$

It can be shown that the Fourier representation of the pulse pattern with the phase shift ϕ is given by

$$u(\theta - \phi) = \sum_{n=1}^{\infty} \hat{u}_n \sin(n\theta - n\phi). \tag{3.28}$$

Note that for $\phi = 0$ and thus for phase a, all harmonics are in phase. For phase b, however, the harmonic of order n has the phase shift $n2\pi/3$. This implies that triplen harmonics, that is, harmonics of the orders $n = 3, 6, \ldots$, are again in phase. The same applies for phase c. As the triplen harmonics are all in phase, they constitute the so-called common-mode voltage harmonics. Assuming the star point in a wye-connected load to be floating, common-mode voltage harmonics do not drive a harmonic current. As a result, a three-phase pulse pattern has—apart from the fundamental component with the frequency f_1—only current harmonics at odd and non-triplen multiples of the fundamental frequency, that is, at nf_1 with $n = 5, 7, 11, 13, \ldots$. We refer to these harmonics as differential-mode harmonics.

For a three-level converter, the magnitude of the nth voltage harmonic is given by

$$\hat{v}_n = \frac{V_{dc}}{2} \hat{u}_n, \tag{3.29}$$

when neglecting the fluctuations of the neutral point potential and assuming a nominal dc-link voltage. These voltage harmonics drive a harmonic current. In the case of an induction machine, the impedance seen by the voltage harmonics is determined by the stator resistance

R_s and the total leakage reactance X_σ. These two quantities and the corresponding harmonic model of the induction machine are described in Sect. 2.2.5. The amplitudes of the current harmonics that result from a three-phase pulse pattern are given by

$$\hat{i}_n = \frac{\hat{v}_n}{n\omega_1 X_\sigma}, \tag{3.30}$$

where we have neglected the stator resistance and assumed that all quantities are given in the pu system. Recall that ω_1 denotes the angular fundamental frequency.

Inserting the current harmonics into the definition of the (stator) current TDD (3.2) leads to

$$I_{\text{TDD}} = \frac{1}{\sqrt{2} I_{s,\text{nom}} \omega_1 X_\sigma} \frac{V_{\text{dc}}}{2} \sqrt{\sum_{n \neq 1} \left(\frac{\hat{u}_n}{n} \right)^2}, \tag{3.31}$$

where we have also, based on (3.29), replaced the amplitude of the voltage harmonic $\hat{v}_n$ by the corresponding harmonic $\hat{u}_n$ of the pulse pattern.

Similarly, for a grid-connected converter, the impedance seen by the converter is the grid resistance R_g and the grid reactance X_g. As explained in Sect. 2.5.4, the latter dominates over the resistance, with the ratio X_g/R_g typically being around 10. Therefore, the current harmonics for a grid-side converter are given by

$$\hat{i}_n = \frac{\hat{v}_n}{n X_g}, \tag{3.32}$$

where we assumed that the grid frequency is equal to the base frequency.

As both on the machine side and the grid side the loads are of a predominantly inductive nature, the magnitude of the voltage harmonics is scaled by their order n to their corresponding current harmonic.

3.4.2 *Optimization Problem for Three-Level Converters*

By inserting the Fourier coefficient (3.24b) into (3.31), the current TDD

$$I_{\text{TDD}} = \frac{\sqrt{2}}{\pi} \frac{V_{\text{dc}}}{I_{s,\text{nom}} \omega_1 X_\sigma} \sqrt{\sum_{n=5,7,\ldots} \left(\frac{1}{n^2} \sum_{i=1}^{d} \Delta u_i \cos(n\alpha_i) \right)^2} \tag{3.33}$$

can be expressed as a function of the converter parameters, load and pulse pattern. More specifically, the TDD expression consists of two terms. The first term is a constant scaling factor, which includes the dc-link voltage, the nominal current, the (angular) fundamental frequency, and the total leakage reactance. This term depends on the converter and the load, but not on the pulse pattern. The second term is the square root of the sum of the squared differential-mode voltage harmonics scaled by their harmonic order. This pulse-pattern-dependent part is chosen as the cost function

$$J(\alpha_i) = \sum_{n=5,7,\ldots} \left(\frac{1}{n^2} \sum_{i=1}^{d} \Delta u_i \cos(n\alpha_i) \right)^2, \tag{3.34}$$

which represents the weighted sum of the squared differential-mode voltage harmonics. For inductive loads, the cost function J is proportional to the current TDD. Therefore, by minimizing J, the minimum current TDD can be obtained. Note that for a given pulse number d and set of switching transitions Δu_i, the cost function is a function of the primary switching angles α_i, with $i = 1, 2, \ldots, d$.

For three-level converters, switching is performed between 0 and 1 in the first quarter of the fundamental period. The switching transitions are given by $\Delta u_i = (-1)^{i+1}$, with $i = 1, 2, \ldots, d$. As the switching transitions are not a degree of freedom in the optimization problem, the latter is based only on real-valued variables. The same applies to two-level converters. However, as we will see later, for more complicated topologies such as five-level converters, the switching transitions form an integral part of the optimization problem and turn it into a mixed-integer problem, that is, an optimization problem involving real-valued variables and integer variables. This greatly complicates the process of solving the optimization problem.

Two sets of constraints are present in the optimization problem. First, the resulting amplitude of the fundamental voltage component of the pulse pattern must be equal to the desired one, that is, the modulation index m. Note that the former is given by $\hat{u}_1$ in (3.24b). Second, the ascending order (3.21) is imposed on the primary switching angles. This results in the following equality and $d + 1$ inequality constraints:

$$\frac{4}{\pi} \sum_{i=1}^{d} \Delta u_i \cos(\alpha_i) = m \tag{3.35a}$$

$$0 \leq \alpha_1 \leq \alpha_2 \leq \cdots \leq \alpha_d \leq \frac{\pi}{2}. \tag{3.35b}$$

The minimization of J subject to these constraints gives rise to the optimization problem

$$J_{\text{opt}} = \underset{\alpha_i}{\text{minimize}} \;\; J(\alpha_i) \tag{3.36a}$$

$$\text{subject to (3.35).} \tag{3.36b}$$

The set of constraints (3.35) defines a subset of the Euclidean space $\mathbb{R}^d$. This subset is often referred to as the *feasible region* or the *search space*. The cost function $J(\alpha_i) : \mathbb{R}^d \rightarrow \mathbb{R}$ is a function of the *optimization variable* α_i. The optimization variable that minimizes J is called the *optimal solution* or the *optimizer*. The minimum of the cost function is the *optimal value* J_{opt}.

As the optimization problem (3.36) includes trigonometric terms, it is not convex. In general, (3.36) has multiple local minima, making it difficult and often time consuming, particularly for high pulse numbers, to find the global minimum. For more details on mathematical optimization in general, and convexity, local and global minima, and solution methods in particular, the reader is referred to classic textbooks, such as [42] and [43]. The standard terminology used in connection with optimization problems is summarized in Appendix 3.B.

Example 3.7 *To visualize the characteristics of the optimization problem, consider a three-level pulse pattern with pulse number* $d = 2$. *This implies that the sequence of switch positions consists of the elements* $u_0 = 0$, $u_1 = 1$, *and* $u_2 = 0$. *It follows that the switching transitions are given by* $\Delta u_1 = 1$ *and* $\Delta u_2 = -1$.

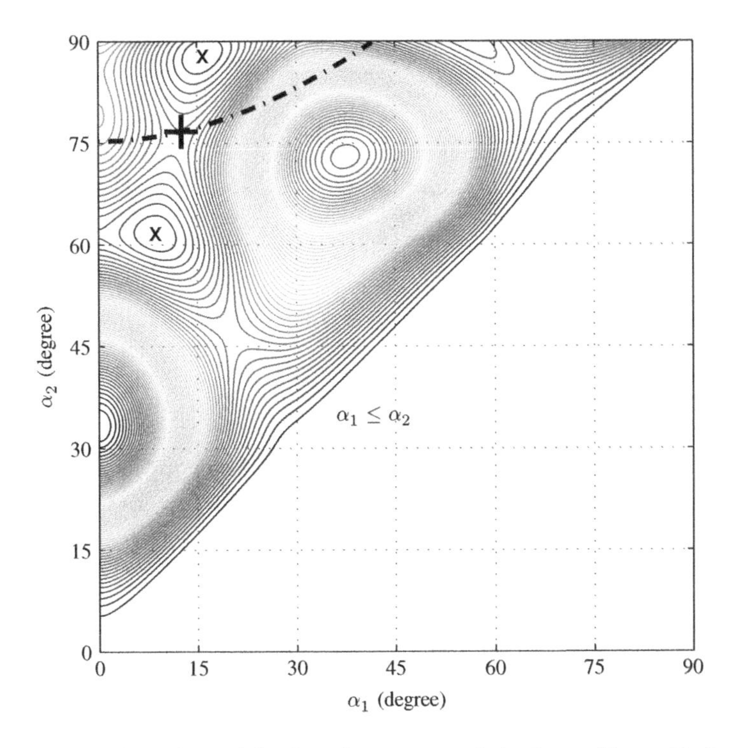

Figure 3.17 Cost function $J(\alpha_1, \alpha_2)$ for the pulse pattern optimization problem with $d = 2$ switching angles. The dash-dotted line refers to the desired modulation index $m = 0.95$. The optimal solution is indicated by the plus symbol

In a first step, we neglect the modulation index constraint (3.35a). The corresponding cost function J is shown in Fig. 3.17 with the two primary switching angles α_1 and α_2 as arguments. The cost function is smooth and exhibits three local minima. One minimum is along $\alpha_1 = \alpha_2$. The second and third minima are marked with crosses. The constraint $\alpha_1 \leq \alpha_2$ excludes the lower right-hand side of the domain.

When imposing the constraint (3.35a), the fundamental component of the resulting pulse pattern matches the desired modulation index m. This also reduces the optimization problem to a one-dimensional problem, limiting the set of admissible switching angles to a curved line. The latter is shown as the dash-dotted line in Fig. 3.17. The optimal set of primary switching angles is found by computing the minimum of the cost function along this line. The minimum is indicated by the plus symbol, and corresponds to the angles $\alpha_1 = 12.6°$ and $\alpha_2 = 76.7°$. The optimal solution is located at the saddle point between the two minima that disregard the constraint (3.35a).

It is clear from this example and from the optimization problem (3.36) that the equality constraint (3.35a) removes one dimension from the solution space. In the case of d primary switching angles, the optimization problem has $d - 1$ degrees of freedom.

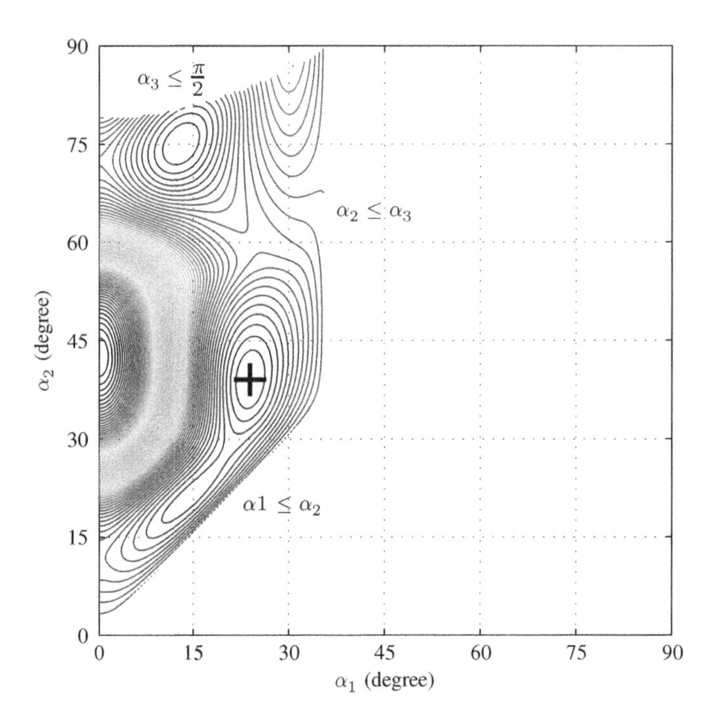

Figure 3.18 Cost function $J(\alpha_1, \alpha_2, \alpha_3)$ for the pulse pattern optimization problem with $d = 3$ switching angles. The third angle is determined by the desired modulation index $m = 0.95$. The optimal solution is indicated by the plus symbol

Example 3.8 *In a second example, we increase the pulse number to d* $= 3$. *In Fig. 3.18, the cost function J is shown as a function of the first two primary switching angles,* α_1 *and* α_2. *The third angle* α_3 *is determined by* α_1, α_2, *and the modulation index, provided that a solution exists. The angle domain is constrained in three ways. As previously, the constraint* $\alpha_1 \leq \alpha_2$ *excludes the lower right-hand side. The right-hand side is constrained by* $\alpha_2 \leq \alpha_3$, *and the top side by* $\alpha_3 \leq 90°$.

The cost function exhibits two local minima for $m = 0.95$. *The global minimum is indicated by the plus symbol, and corresponds to the switching angles* $\alpha_1 = 26.5°$, $\alpha_2 = 37.5°$, *and* $\alpha_3 = 49.9°$. *This set of angles corresponds to the pulse pattern shown in Fig. 3.16.*

Variations in the modulation index modify the location and the values of the minima. In general, the switching angles vary smoothly as a function of the modulation index. At about $m = 1.04$, *however, the upper local minimum turns into the global minimum. This leads to a stepwise change in the switching angles (see also Fig. 3.19(a)). Similar discontinuities can be observed in this figure at* $m = 0.51$, $m = 0.71$, *and* $m = 1.18$. *At these modulation indices, the optimal value* J_{opt} *of the cost function J is not smooth, as can be seen in Fig. 3.19(b). Note that the discontinuities in the switching angles are a result of the nonconvex nature of the pulse pattern optimization problem.*

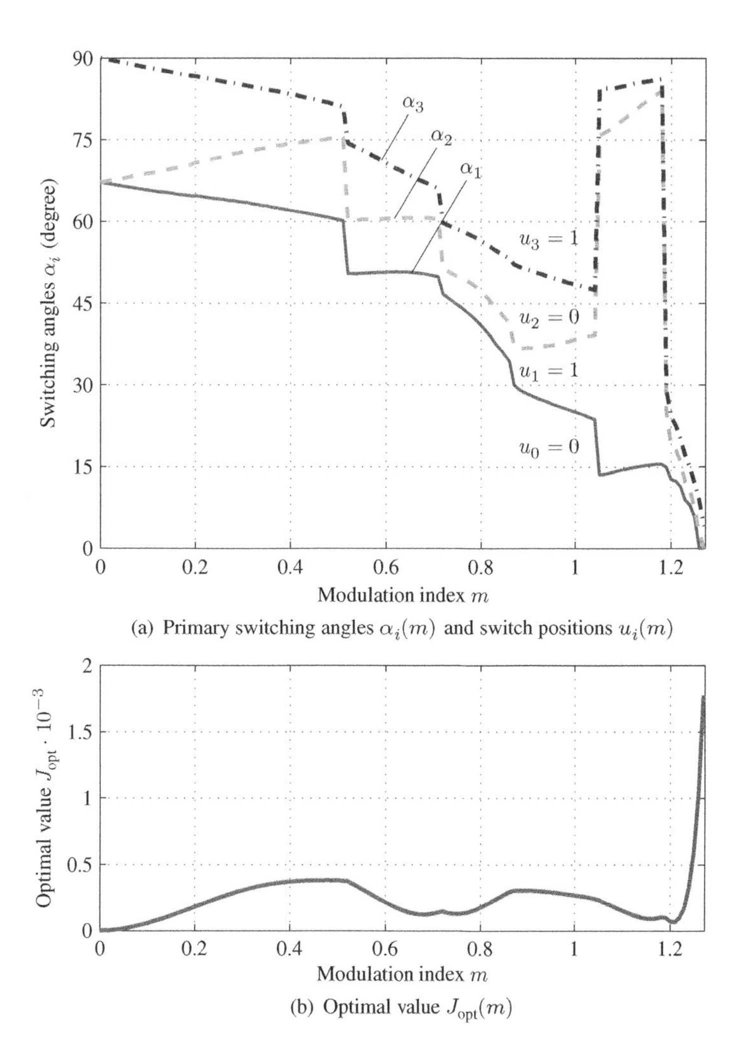

(a) Primary switching angles $\alpha_i(m)$ and switch positions $u_i(m)$

(b) Optimal value $J_{\text{opt}}(m)$

Figure 3.19 Primary switching angles α_i and optimal value J_{opt} of the cost function as a function of the modulation index m for a three-level OPP with $d = 3$ switching angles

The procedure to compute OPPs for three-level converters typically involves the following four steps [32, 33]:

Step 1. Equidistant gridding of the required modulation indices is performed by sampling the range $m \in [0, \frac{4}{\pi}]$ with a certain resolution. Typically, 256 discrete modulation indices suffice.

Step 2. For each modulation index, the corresponding single-phase pulse pattern is computed. Quarter-wave symmetry is imposed and d primary switching angles are considered.

The optimization problem (3.36) is solved by minimizing J over the set of admissible switching angles as defined by (3.35). A procedure to solve this optimization problem in an iterative manner is detailed in [32]. The method relies on a gradient solver to find a local minimum. To ensure a high success rate of finding the global minimum, the optimization stage is repeated several times with random initial sets of angles. The minimum with the smallest value is kept. In practise, this approach works sufficiently well for small pulse numbers of 10 or less.

Step 3 (optional). A post-processing stage is sometimes added to reduce the number of discontinuities in the switching angles, usually at the expense of a slight increase in the current TDD. As angle discontinuities are handled well by the controller proposed in Chap. 12, such discontinuities are removed only when the resulting deterioration in the current TDD is negligible.

These three computational steps lead to a set of primary switching angles as a function of the modulation index that fully characterizes the OPP. An example of such a set is shown in Fig. 3.19(a) for pulse number $d = 3$ and a three-level converter.

Step 4. For each modulation index, the single-phase pulse pattern over 360° is constructed. The primary switching angles of the quarter-wave pulse pattern are extended to the full fundamental waveform by applying quarter-wave symmetry. This procedure is exemplified in Fig. 3.16. The three-phase OPP is established by shifting the single-phase pulse pattern by 120° to obtain phase b and by another 120° to create phase c.

3.4.3 *Optimization Problem for Five-Level Converters*

When computing OPPs for converters with more than three voltage levels, the sequence of switching transitions is no longer predetermined, but constitutes an additional degree of freedom in the optimization problem. This fact is shown in this section, and the optimization problem (3.36) is generalized to account for it.

As explained in Sect. 2.4.2, the single-phase switch positions of a five-level converter are restricted to the integer set $\mathcal{U} = \{-2, -1, 0, 1, 2\}$. As the first quarter of the fundamental period corresponds to the first half of the positive fundamental half-wave, the single-phase switch positions are restricted in this quarter to the set $\mathcal{U}^+ = \{0, 1, 2\}$. As before, the initial switch position is zero, that is, $u_0 = 0$. This implies that the first switching transition is from 0 to 1, that is, $u_1 = 1$ and $\Delta u_1 = 1$. For the second transition, however, two options exist: switching one step up to $u_2 = 2$ or down to $u_2 = 0$. At the third switching transition, regardless of the previous choice, the switch position is always $u_3 = 1$. Therefore, for pulse number d,

$$2^{\text{floor}(\frac{d}{2})} \tag{3.37}$$

different sequences of switch positions exist, which we refer to as *switching sequences*. We denote them by $U = [u_1 \;\; u_2 \cdots u_d]^T$, where we omit u_0 from U.

For the case $d = 4$, four switching sequences can be identified, as shown in Fig. 3.20, namely $U = [1 \;\; 0 \;\; 1 \;\; 0]^T$, $U = [1 \;\; 0 \;\; 1 \;\; 2]^T$, $U = [1 \;\; 2 \;\; 1 \;\; 0]^T$, and $U = [1 \;\; 2 \;\; 1 \;\; 2]^T$. For the modulation index $m = 0.55$, the corresponding single-phase pulse patterns are shown over the first quarter of the fundamental period. Note that we impose the constraint $|\Delta u_i| = 1$ on the switching transitions. This implies that switching by more than one voltage level up or down is prohibited.

Two modifications are required to generalize the OPP optimization problem of a three-level converter to the case of a five-level converter. First, the switching sequences U are added.

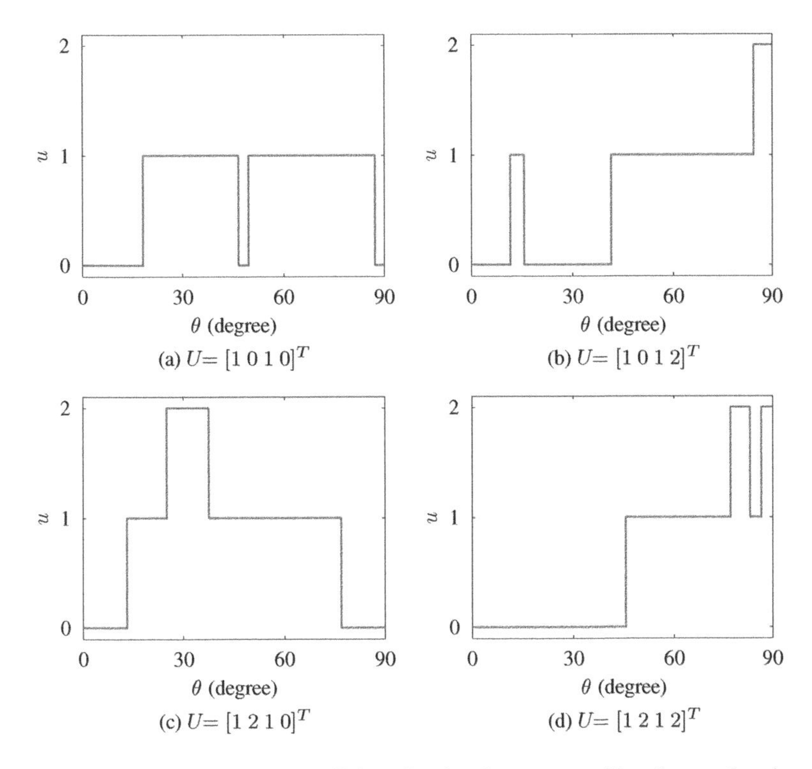

Figure 3.20 Switching sequences U for a five-level converter with pulse number $d = 4$

Second, the switching transitions in the cost function (3.34) and in the modulation index constraint (3.35a) need to be divided by 2. This scaling accounts for the fact that for a five-level converter, the voltage steps at its phase terminals are given by a quarter of the dc-link voltage rather than a half of it as for the three-level converter—compare (2.87) with (2.75). This leads to the revised cost function

$$J(\alpha_i, \Delta u_i) = \sum_{n=5,7,\ldots} \left(\frac{1}{n^2} \sum_{i=1}^{d} \frac{\Delta u_i}{2} \cos(n\alpha_i) \right)^2, \tag{3.38}$$

and to the revised optimization problem

$$J_{\text{opt}} = \underset{\alpha_i, \Delta u_i}{\text{minimize}} \;\; J(\alpha_i, \Delta u_i) \tag{3.39a}$$

$$\text{subject to} \quad \frac{4}{\pi} \sum_{i=1}^{d} \frac{\Delta u_i}{2} \cos(\alpha_i) = m \tag{3.39b}$$

$$0 \le \alpha_1 \le \alpha_2 \le \cdots \le \alpha_d \le \frac{\pi}{2} \tag{3.39c}$$

$$\Delta u_i \in \{-1, 1\}, \;\; \forall i = 1, 2, \ldots, d \tag{3.39d}$$

$$u_i = \Delta u_i + u_{i-1}, u_i \in \{0, 1, 2\}. \tag{3.39e}$$

The switching transitions Δu_i are now part of the optimization problem, leading to a mixed-integer program. This complicates the solution process, particularly for large d, because the number of switching sequences to be explored increases exponentially with d (see also (3.37)).

To compute OPPs for five-level converters, the procedure stated in the previous section is modified, as summarized in the following four steps [32, 33].

Step 1. The different switching sequences U are enumerated, with the number of sequences being given by (3.37). Equidistant gridding of the required modulation indices is performed, by sampling the range $m \in [0, \frac{4}{\pi}]$ with a given resolution.

Step 2. For each switching sequence U and modulation index m, the corresponding single-phase pulse pattern is computed by solving the optimization problem (3.39). By setting *a priori* the integer variables, the optimization problem is reduced to a nonlinear but real-valued program. This allows us to pursue the same optimization approach as for the three-level converter. Owing to the imposed quarter-wave symmetry, the result of this step are d primary switching angles.

Step 3. For each modulation index, several sets of switching angles exist, with each set relating to a different switching sequence. In a post-processing stage, the switching sequence as well as its corresponding set of angles is chosen that has the smallest value of the cost function. In this step, it is also possible to reduce the number of discontinuities in the switching angles.

The result of the first three steps is a set of primary switching angles and a switching sequence for each modulation index.

Step 4. For each modulation index, the three-phase OPP is constructed in the same way as before for the three-level inverter.

By enumerating the switching sequences, the complications of formulating and solving a mixed-integer program can be avoided, and an algorithm similar to the one for three-level OPPs can be used. In terms of its computational burden, this approach is viable only for small values of d, because enumeration compounds the combinatorial explosion inherent to mixed-integer programs, necessitating a mixed-integer optimization approach for large d.

Example 3.9 *In order to compute an OPP for a five-level converter, the algorithm stated earlier is used. For the pulse number* $d = 3$, *the two switching sequences* $U_1 = [1 \ \ 0 \ \ 1]^T$ *and* $U_2 = [1 \ \ 2 \ \ 1]^T$ *exist. Figure 3.21(a) and (b) shows the result of Step 2, that is, the primary switching angles for each of the two switching sequences. When selecting the switching sequence* U_1 *and avoiding the switch position* $u = 2$, *the achievable peak line-to-line voltage is limited to half the dc-link voltage. As a result, OPPs exist for this switching sequence only in the lower 50% of the modulation range. Note that the five-level switching angles in Fig. 3.21(a) are the same as the three-level switching angles in Fig. 3.19(a), but they are compressed to half the modulation range.*

On the other hand, when choosing the switching sequence U_2, *the full line-to-line voltage can be modulated, but the switch position* $u = 2$ *is a suboptimal choice for small modulation indices. This can be seen in Fig. 3.21(c) when comparing the values of the cost function for* U_1 *with the one for* U_2. *Indeed, for modulation indices below 0.35, it is optimal to remove the pulse with* $u = 2$ *from the OPP, by setting* $\alpha_3 = \alpha_2$. *The effective pulse number is then* $d = 1$ *with one switching transition from* $u_0 = 0$ *to* $u_1 = 1$.

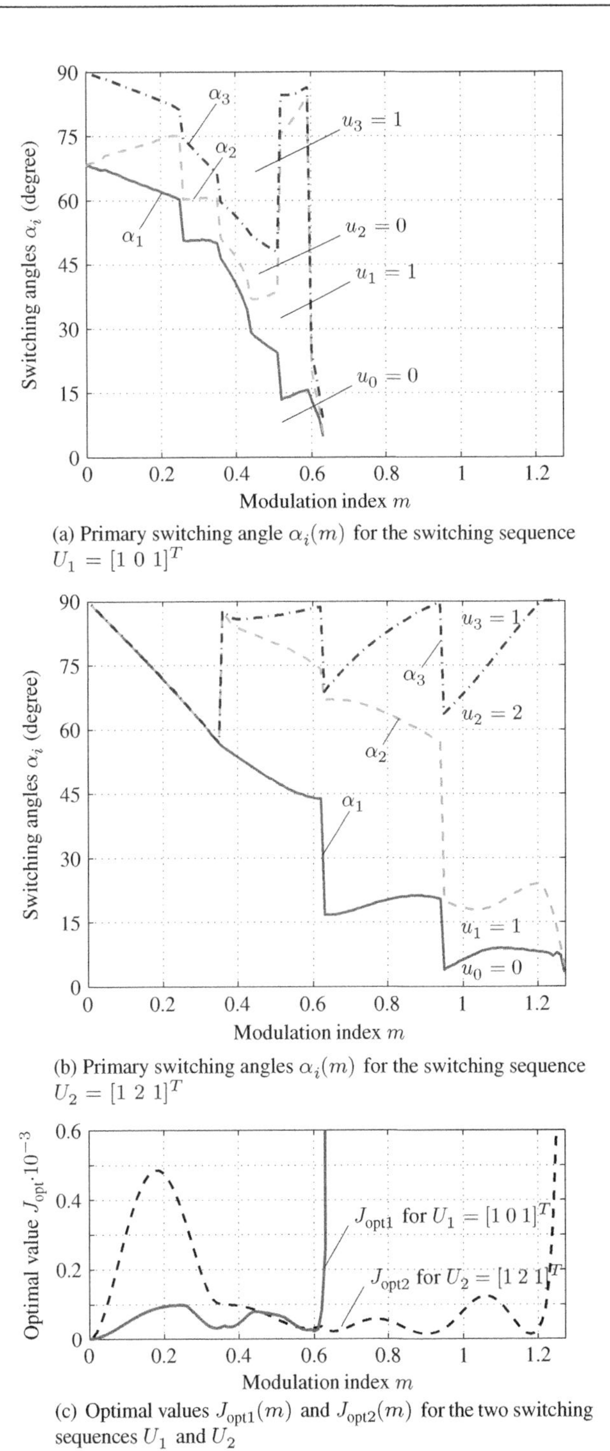

Figure 3.21 Primary switching angles and optimal cost function values as a function of the modulation index and the switching sequence for a five-level OPP with three switching angles

In Step 3 of the algorithm, the pulse patterns for the different switching sequences are consolidated into one. It is clear that U_1 is optimal in the lower modulation range, while U_2 is optimal in the upper range. In between, U_2 is optimal for $0.465 \leq m \leq 0.56$. For $0.56 < m \leq 0.612$, U_1 leads to a marginally lower value for the cost function than U_2, but as the difference is minor, and to avoid another discontinuity in the switching angles, we also use U_2 in this range. The resulting OPP is shown in Fig. 3.22 along with the switching sequence. As indicated by the vertical line, the different switching sequences available for the five-level topology lead to an additional angle discontinuity at $m = 0.465$.

In general, OPPs for five-level converters tend to exhibit more angle discontinuities than for three-level converters because of the different possible switching sequences. Nevertheless,

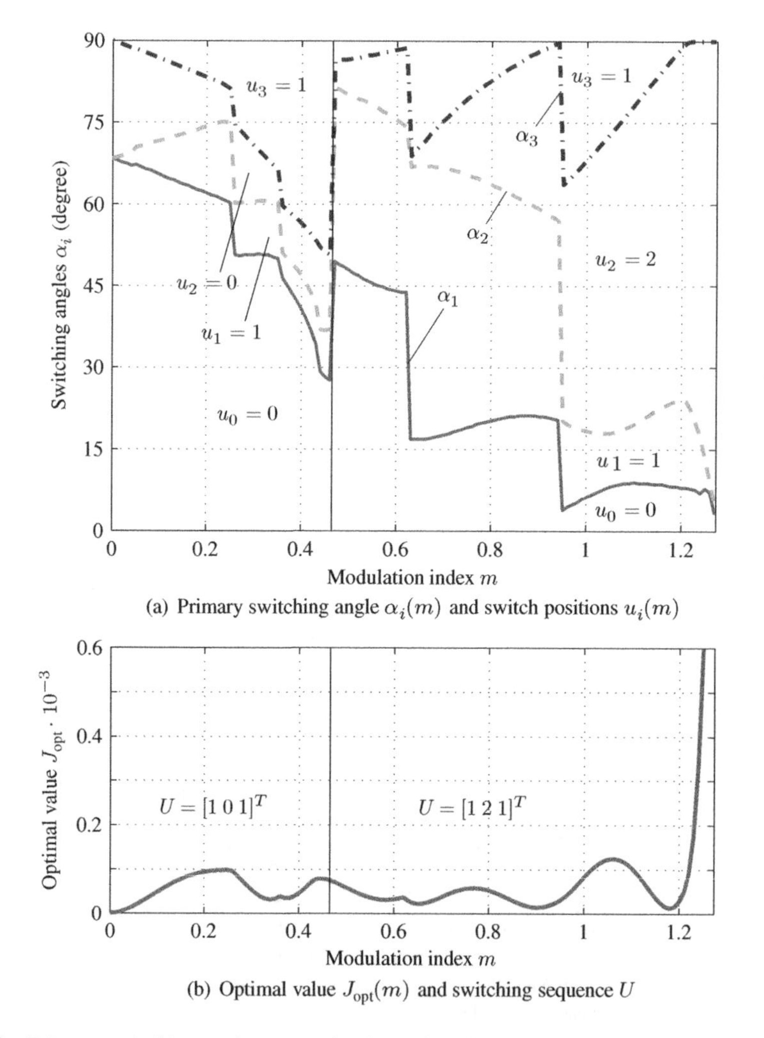

(a) Primary switching angle $\alpha_i(m)$ and switch positions $u_i(m)$

(b) Optimal value $J_{opt}(m)$ and switching sequence U

Figure 3.22 Primary switching angles α_i, optimal cost function value J_{opt}, and switching sequence U as a function of the modulation index m for a five-level OPP with three primary switching angles

not all switching sequences are applicable for all modulation indices. A switching sequence without the switch position $u = 2$, for example, cannot synthesize fundamental voltages with modulation indices exceeding $m = 0.637$.

3.4.4 Summary and Properties

OPPs are characterized by the integer pulse number d, which corresponds to the number of single-phase switching transitions within a quarter of the fundamental period. The pulse number and the fundamental frequency f_1 determine the switching frequency. For a three-level NPC converter, for example, the device switching frequency is given by $f_{\text{sw}} = df_1$. As d is an integer, the pulse pattern is synchronized to the fundamental voltage waveform at all operating points and load conditions. Therefore, modulation with OPPs belongs to the class of synchronous PWM methods.

The result of the offline OPP computation is a look-up table that holds the switching (firing) angles and the respective switch positions (or phase potential values). The content of this look-up table is a function of the pulse number and the modulation index, which is a normalized quantity that is proportional to the magnitude of the reference voltage in the linear operating range.

Owing to the synchronism between the pulse pattern and the fundamental waveform, subharmonic spectral components, that is, components below f_1, do not exist. Because the single-phase pulse pattern is an odd function with quarter-wave symmetry, all integer harmonics of even order are zero. Moreover, thanks to the 120° phase shift between the phases, all triplen harmonics are zero. As a result, three-phase OPPs exhibit a discrete amplitude spectrum that comprises only integer components of the orders 1, 5, 7, 11, 13, and so on.

OPPs do not feature a symmetric modulation cycle of fixed length. Regularly spaced time instants at which the ripple current is zero do not exist. This makes the direct sampling of the fundamental current component impossible, greatly complicating the task of establishing fast current control in systems operated with OPPs. Specifically, when feeding the current samples to a linear controller without adequate post-processing, the ripple current because of the OPP is interpreted by the controller as a current error. The current controller attempts to regulate the ripple current to zero, thereby worsening the harmonic performance of the OPP. To mitigate this issue, OPPs are typically used in control loops with a low bandwidth, such as V/f control or FOC that is tuned to act in a very slow manner.

To achieve fast closed-loop control with OPPs, the current controller and the modulator need to be formulated and solved in one computational stage. To this end, a model predictive pulse pattern controller is proposed in Chap. 12, which takes the OPP ripple current into account and achieves closed-loop current control by modifying the switching transitions of the OPP. A dynamic performance similar to DTC can be achieved by inserting additional switching transitions during transients.

3.5 Performance Trade-Off for Pulse Width Modulation

The TDD of the stator currents and the switching losses of the inverter pose a trade-off that is fundamental to power electronics. Specifically, for a given modulation method, it is well known in a qualitative manner that reducing the current TDD leads to higher switching losses, and vice versa. This trade-off can also be established in a quantitative way. Specifically, as will be shown in this section, the product of the two quantities is equal to a constant c. This implies that reducing the current TDD by a certain percentage point increases the switching losses by the same degree.

The constant c characterizes the steady-state performance of the considered modulation scheme. A similar figure of merit was introduced in [44], which is the product of the spectral amplitudes and the switching frequency. This section extends this concept by considering the switching losses rather than the switching frequency, because the former appears to be of a higher importance to the inverter operation than the latter.

Recall that for CB-PWM with the carrier frequency f_c and the fundamental frequency f_1, the voltage harmonics are located at the frequencies

$$f_{\mu\nu} = \mu f_c + \nu f_1, \quad \mu \in \mathbb{N}, \nu \in \mathbb{Z}. \tag{3.40}$$

The amplitudes of the voltage harmonics $\hat{v}_{\mu\nu}$ depend on the type of CB-PWM scheme used and the number of voltage levels in the inverter, as analyzed in detail in [15].

The amplitudes of the harmonic current components are equal to the voltage amplitudes $\hat{v}_{\mu\nu}$ divided by the total leakage impedance of the machine. In the pu system, the amplitudes of the current harmonics are given by

$$\hat{i}_{\mu\nu} = \hat{v}_{\mu\nu} \frac{f_B}{f_{\mu\nu}\, X_\sigma} = \hat{v}_{\mu\nu} \frac{f_B}{f_c(\mu + \nu\frac{f_1}{f_c})\, X_\sigma}, \tag{3.41}$$

where $f_B = \omega_B/(2\pi)$ denotes the base frequency of the pu system and X_σ is the total leakage reactance of the machine, as defined in Sect. 2.2.5.

3.5.1 Current TDD versus Switching Losses

Recall the definition of the current TDD (3.2), which we rewrite as

$$I_{\text{TDD}} = \frac{1}{\sqrt{2} I_{s,\text{nom}}} \sqrt{\sum_{\mu\in\mathbb{N},\nu\in\mathbb{Z}} (\hat{i}_{\mu\nu})^2} \tag{3.42}$$

in terms of the harmonic current components $\hat{i}_{\mu\nu}$. It should be clear that the fundamental component is excluded from the sum of squares; this includes $\mu = 0$ and $\nu = 1$ and any sideband of $\mu = 1$ that coincides with the fundamental component.

By inserting (3.41) into (3.42), the current TDD can be expressed as a function of the carrier frequency and the amplitudes of the voltage harmonics:

$$I_{\text{TDD}} = \frac{1}{\sqrt{2} I_{s,\text{nom}} X_\sigma} \frac{f_B}{f_c} \sqrt{\sum_{\mu\in\mathbb{N},\nu\in\mathbb{Z}} \left(\frac{\hat{v}_{\mu\nu}}{\mu + \nu\frac{f_1}{f_c}} \right)^2}. \tag{3.43}$$

Note that $\hat{v}_{\mu\nu}$ are independent of the carrier and fundamental frequencies. We have seen in Sect. 3.3 that the voltage harmonics are highly concentrated around the carrier multiples when the carrier-to-fundamental frequency ratio is not overly small. For the sidebands at νf_1, the amplitudes of the voltage harmonics quickly decay to zero as $|\nu|$ is increased. Specifically, for $|\nu| > 6$, the voltage harmonics can be usually neglected. Because of this, and assuming a reasonably high carrier-to-fundamental frequency ratio, such as $f_c/f_1 \geq 9$, the influence of the term $\nu f_1/f_c$ in (3.43) is small.

This allows us to conclude that the current TDD is (approximately) inversely proportional to the carrier frequency, that is,

$$I_{\text{TDD}} \propto \frac{1}{f_c}. \tag{3.44}$$

The average switching losses caused by CB-PWM can be derived analytically. The turn-off switching losses over one fundamental period $T_1 = 1/f_1$ for the ith semiconductor are given by

$$P_{i,\text{off}} = \frac{1}{T_1} \sum_{\ell=1}^{\ell_{i,\text{off}}} e_{i,\text{off}}(\ell), \tag{3.45}$$

where $e_{i,\text{off}}(\ell)$ denotes the energy losses of the ℓth switching transition, in which the semiconductor is turned off. The integer variable $\ell_{i,\text{off}}$ refers to the number of turn-off events for this semiconductor over one fundamental period.

According to (2.65), the turn-off energy losses are given by

$$e_{i,\text{off}}(\ell) = c_{\text{off}} v_T i_{\text{ph}}(\ell), \tag{3.46}$$

where c_{off} is a coefficient, v_T is the voltage across the semiconductor, and i_{ph} is the commutated phase current. The latter is equal to the anode current of the semiconductor device under commutation (see (2.65)). When operating in the linear modulation regime, the switching transitions generated by CB-PWM are evenly distributed over the fundamental period. Neglecting the ripple current, this implies

$$i_{\text{ph}}(\ell) \approx \hat{i}_{\text{ph}} \sin(2\pi \frac{\ell}{\ell_{i,\text{off}}}), \ell = 1, 2, \ldots, \ell_{i,\text{off}}, \tag{3.47}$$

with $\hat{i}_{\text{ph}}$ denoting the peak current of the fundamental waveform.

This leads to the (approximate) average turn-off switching losses

$$P_{i,\text{off}} \approx c_{\text{off}} v_T \frac{\hat{i}_{\text{ph}}}{T_1} \sum_{\ell=1}^{\ell_{i,\text{off}}} \sin(2\pi \frac{\ell}{\ell_{i,\text{off}}}). \tag{3.48}$$

It should be clear from Sect. 3.3 that the number of switching events is determined by the carrier-to-fundamental frequency ratio and is thus proportional to the carrier frequency.

It directly follows that $P_{i,\text{off}}$ is proportional to the carrier frequency. The same applies to the turn-on and reverse recovery losses, for which expressions similar to (3.48) can be derived. As a result, the total switching losses P_{sw} of all semiconductors in an inverter are also proportional to the carrier frequency, that is,

$$P_{\text{sw}} \propto f_c. \tag{3.49}$$

Multiplying (3.44) with (3.49) leads to the statement

$$I_{\text{TDD}} \cdot P_{\text{sw}} = \text{const}. \tag{3.50}$$

It is convenient to normalize (3.50). The switching losses can be normalized using the rated apparent power S_R (see Sect. 2.1.2). The current TDD already represents a normalized quantity. The relation (3.50) is then rewritten as

$$I_{\text{TDD}} \cdot \frac{P_{\text{sw}}}{S_R} = c_I, \tag{3.51}$$

where the constant c_I constitutes a performance metric that characterizes the modulation scheme under investigation. When reducing the carrier frequency by a certain factor, so as to reduce the switching losses accordingly, the current TDD is increased by the same factor, and vice versa. Note that, when expressing the current TDD in terms of the switching losses, a hyperbolic function results.

3.5.2 Torque TDD versus Switching Losses

A statement similar to (3.51) can be obtained for the electromagnetic torque. It is clear from (3.41) and the previous reasoning that the amplitudes of the current harmonics are inversely proportional to the carrier frequency, that is,

$$\hat{i}_{\mu\nu} \propto \frac{1}{f_c}. \tag{3.52}$$

It follows from the torque expression (2.61) that the amplitudes of the torque harmonics are also inversely proportional to the carrier frequency, that is, $\hat{T}_{\mu\nu} \propto 1/f_c$. Using (3.1), we conclude that

$$T_{\text{TDD}} \propto \frac{1}{f_c}, \tag{3.53}$$

which allows us to write $T_{\text{TDD}} \cdot P_{\text{sw}} = \text{const}$. The normalized trade-off for the electromagnetic torque is given by

$$T_{\text{TDD}} \cdot \frac{P_{\text{sw}}}{S_R} = c_T, \tag{3.54}$$

where c_T is the performance metric characterizing the trade-off between the torque TDD and the switching losses.

Owing to the equivalence between CB-PWM and SVM, which was shown in Sect. 3.3.2, the relations (3.51) and (3.54) also hold for SVM. For OPPs, however, particularly for low pulse numbers, (3.51) and (3.54) should be applied with some caution, as the switching transitions of OPPs are typically not evenly distributed over the fundamental period.

Example 3.10 *To highlight the trade-off between the current and torque TDD on one side and the switching losses on the other, consider a three-level NPC inverter driving an MV induction machine, as shown in Fig. 2.25. The detailed setup and the drive parameters are provided in Sect. 2.5.1.*

In simulations at 60% speed and at the rated torque, the carrier frequency was varied between 150 Hz and 1.2 kHz. Synchronous CB-PWM was used, for which the carrier frequency is an integer multiple of the fundamental frequency. After reaching steady-state operating conditions, the stator currents, stator voltages, and the electromagnetic torque were recorded.

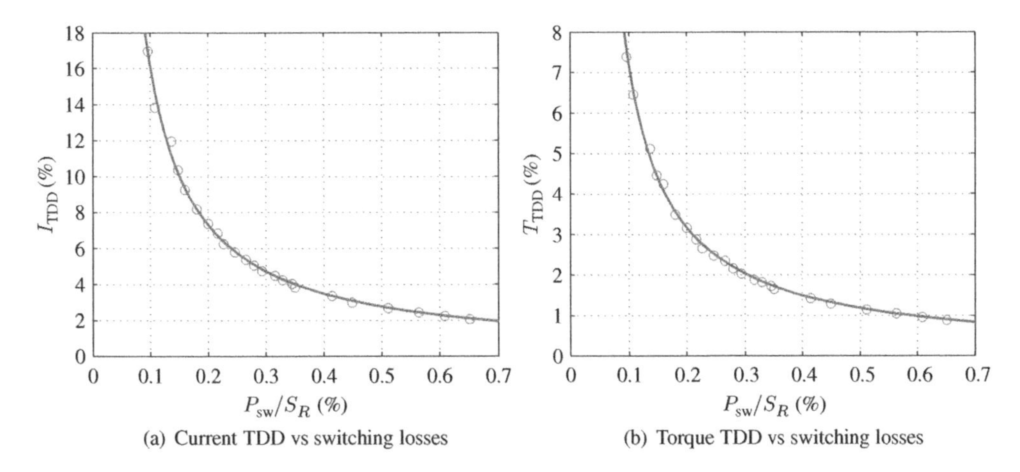

(a) Current TDD vs switching losses

(b) Torque TDD vs switching losses

Figure 3.23 Performance trade-off for synchronous CB-PWM, when applied to a three-level NPC inverter with an induction machine

Based on these quantities, the switching losses P_{sw} were computed according to Sect. 2.4.1. The current and torque TDDs were computed using Fourier series over a time window whose length is an integer multiple of the fundamental period.

Figure 3.23 shows the resulting harmonic distortions of the stator current and the electromagnetic torque, respectively, as a function of the normalized switching losses of the inverter. Both axes are given in percent. The individual simulations are denoted by circles. As anticipated by (3.51) *and* (3.54), *the data points can be approximately described by hyperbolic functions. Using a data-fitting tool, the constants $c_I = 1.3$ and $c_T = 0.55$ were obtained.*[6]

3.6 Control Schemes for Induction Machine Drives

Control schemes for induction machine drives can be broadly classified into scalar and vector control methods. Scalar control is based on the steady-state model of the machine and adjusts the magnitude and frequency of the applied stator voltage.

Vector control schemes, on the other hand, base their control actions on dynamic models of the induction machine. As a result, not only the magnitude and frequency of the applied stator voltage but also its instantaneous angular position is manipulated. This allows vector control schemes to control the instantaneous positions of the current and flux linkage vectors, thus achieving fast control of the electromagnetic torque and flux magnitude during load transients and reference changes.

A classification of control methods for induction machine drives adopted from [45] and slightly modified is shown in Fig. 3.24. The two most widely used vector control schemes are FOC and DTC. After outlining scalar control, these two control methods are explained in this section.

[6] A small offset in the switching losses of 0.02% is neglected here. This offset relates to the fact that the switching losses cannot be reduced to zero, as one switching transition per fundamental half-wave is always required to synthesize the fundamental component—see also the concept of six-step operation, which is displayed in Fig. 3.14(a).

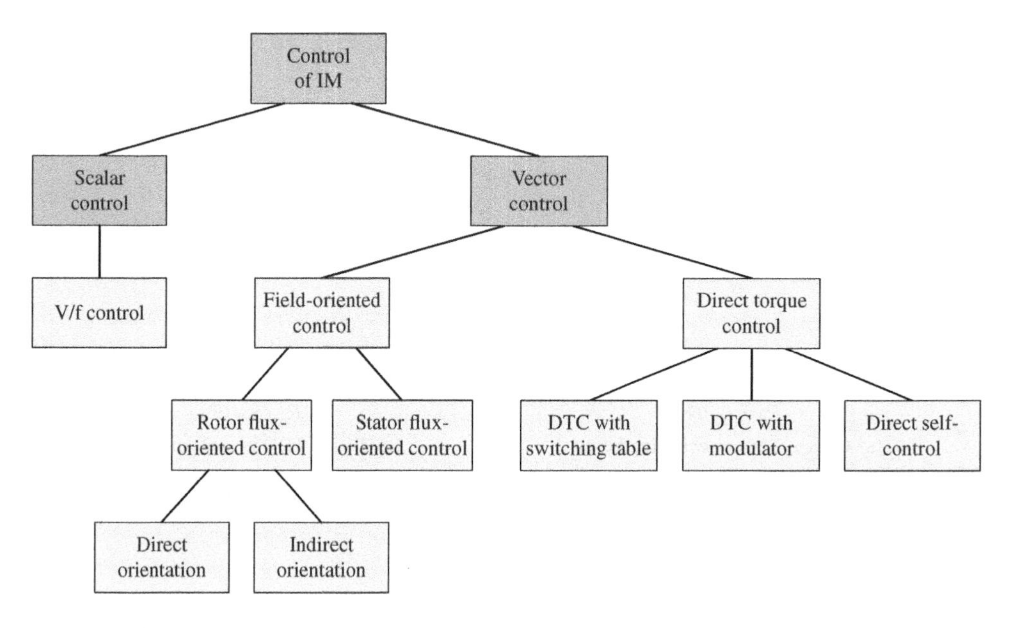

Figure 3.24 Classification of control methods for induction machine (IM) drives

3.6.1 Scalar Control

Scalar control aims at maintaining the stator flux magnitude close to its nominal value regardless of the speed and electromagnetic torque. The torque is not directly controlled but rather indirectly via the slip frequency.

Control of the stator flux magnitude is achieved by adjusting the magnitude of the stator voltage as a function of the electrical frequency. Consider a reference frame rotating with the stator frequency ω_s. Inserting $\omega_{\text{fr}} = \omega_s$ into the stator voltage equation (2.50a) leads to

$$\boldsymbol{v}_s = R_s \boldsymbol{i}_s + \frac{\mathrm{d}\boldsymbol{\psi}_s}{\mathrm{d}t} + \omega_s \begin{bmatrix} 0 & -1 \\ 1 & 0 \end{bmatrix} \boldsymbol{\psi}_s, \tag{3.55}$$

where we have dropped the superscript $'$ that indicates pu quantities. Recall that the variables in (3.55) are vectors in the rotating dq reference frame. The stator voltage, for example, is given by $\boldsymbol{v}_s = [v_{sd} \;\; v_{sq}]^T$. The stator current $\boldsymbol{i}_s$ and the stator flux $\boldsymbol{\psi}_s$ are defined accordingly.

By neglecting the stator resistance R_s and assuming steady-state operation, (3.55) simplifies to

$$\boldsymbol{v}_s = \omega_s \begin{bmatrix} 0 & -1 \\ 1 & 0 \end{bmatrix} \boldsymbol{\psi}_s. \tag{3.56}$$

We define the magnitude of the stator voltage as $v_s = \|\boldsymbol{v}_s\|$ and the magnitude of the stator flux as $\Psi_s = \|\boldsymbol{\psi}_s\|$. With this, we obtain

$$\frac{v_s}{\omega_s} = \Psi_s. \tag{3.57}$$

Because the ratio between the stator voltage magnitude and stator frequency is constant, scalar control schemes are often referred to as "voltage per frequency" or V/f control. Note that the stator resistance cannot be neglected at low-speed operation. To compensate for the corresponding voltage drop, a voltage offset is often added to the voltage–frequency relationship (3.57).

To achieve slip and thus torque control, the V/f term (3.57) is typically augmented by a speed controller that manipulates the slip frequency $\omega_{\text{sl}} = \omega_s - \omega_r$. The electrical angular speed of the rotor ω_r can be derived from the (measured) mechanical angular speed of the rotor ω_m through $\omega_r = p\omega_m$ (see also (2.35)), where p is the number of pole pairs. By adding the slip frequency to ω_r, the required stator frequency ω_s is directly obtained.

As scalar control schemes lack a current control loop, they have an open-loop characteristic and are thus inherently slow. This limits their scope of application to (quasi) steady-state operation and to drives that do not require fast torque and speed control. Many of the so-called general-purpose drives, such as pumps and fans, fall into this category. Thanks to their conceptual simplicity, scalar control schemes are an attractive choice for these drives. Few machine parameters are required, with the stator resistance being the most prominent one. This simplifies the commissioning and controller tuning.

For more details on scalar control schemes and their various extensions, the reader is referred to [46, Chap. 5], [47, Sect. 12.1], [48, Sect. 2.17], and [49] and the references therein.

3.6.2 Field-Oriented Control

The notion of FOC was proposed by Hasse [50, 51] and Blaschke [52–54] around 1970. Today, FOC constitutes the most widely used vector control method. Because of that, the terms FOC and vector control are often used synonymously.

A rotating reference frame is established in FOC, which rotates synchronously either with the stator, the air-gap or the rotor flux vector. In the rotating reference frame, the stator current vector $\boldsymbol{i}_s$ can be separated into a d-component and a q-component. These current components are, by definition, orthogonal. We will show that the d-component of the stator current can be used to control the flux magnitude or the degree of machine magnetization, while the q-component of the stator current directly relates to the electromagnetic torque. During steady-state operation, the machine quantities in the rotating reference frame are dc quantities and the two stator current components are effectively decoupled. This greatly simplifies the controller design.

3.6.2.1 Principle of Rotor Field Orientation

The principle of field orientation is illustrated in Fig. 3.25 for the case of rotor FOC. The d-axis of the rotating reference frame is aligned with the rotor flux vector $\boldsymbol{\psi}_r$. The reference frame rotates synchronously with the rotor flux at the angular speed $\omega_{\text{fr}} = \omega_s$. The reference frame is displaced by φ with respect to the stationary $\alpha\beta$ reference frame.

In Sect. 2.2.4, we derived the dynamic machine model (2.59), which uses the stator current and the rotor flux vectors as state variables. The stator and rotor equations are repeated here for

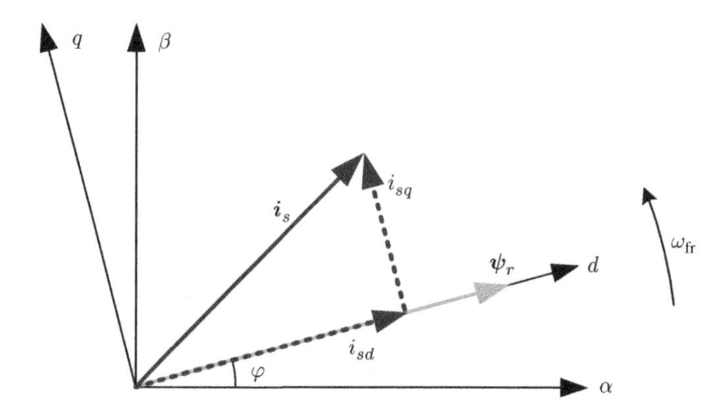

Figure 3.25 Illustration of the principle of rotor field orientation. The dq reference frame is aligned with the rotor flux vector ψ_r and rotates with the angular speed $\omega_{fr} = \omega_s$

the reader's convenience for the case of a squirrel-cage induction machine. The rotor voltage $\boldsymbol{v}_r$ is thus set to zero.

$$\frac{\mathrm{d}\boldsymbol{i}_s}{\mathrm{d}t} = -\frac{1}{\tau_s}\boldsymbol{i}_s - \omega_{fr}\begin{bmatrix} 0 & -1 \\ 1 & 0 \end{bmatrix}\boldsymbol{i}_s + \cdots + \left(\frac{1}{\tau_r}\boldsymbol{I}_2 - \omega_r\begin{bmatrix} 0 & -1 \\ 1 & 0 \end{bmatrix}\right)\frac{X_m}{D}\boldsymbol{\psi}_r + \frac{X_r}{D}\boldsymbol{v}_s \tag{3.58a}$$

$$\frac{\mathrm{d}\boldsymbol{\psi}_r}{\mathrm{d}t} = \frac{X_m}{\tau_r}\boldsymbol{i}_s - \frac{1}{\tau_r}\boldsymbol{\psi}_r - (\omega_{fr} - \omega_r)\begin{bmatrix} 0 & -1 \\ 1 & 0 \end{bmatrix}\boldsymbol{\psi}_r. \tag{3.58b}$$

The transient stator time and rotor time constants τ_s and τ_r, respectively, are defined in (2.60).

Assume the use of a high-bandwidth current controller that maintains the desired stator current by manipulating the stator voltage accordingly. In doing so, the current controller overrides the dynamics of the stator windings. This allows us to neglect the stator voltage equation (3.58a) for the time being.

Setting $\omega_{fr} = \omega_s$ in (3.58b) leads to the modified rotor equation

$$\frac{\mathrm{d}}{\mathrm{d}t}\begin{bmatrix} \psi_{rd} \\ \psi_{rq} \end{bmatrix} = \frac{X_m}{\tau_r}\begin{bmatrix} i_{sd} \\ i_{sq} \end{bmatrix} - \frac{1}{\tau_r}\begin{bmatrix} \psi_{rd} \\ \psi_{rq} \end{bmatrix} - (\omega_s - \omega_r)\begin{bmatrix} 0 & -1 \\ 1 & 0 \end{bmatrix}\begin{bmatrix} \psi_{rd} \\ \psi_{rq} \end{bmatrix}, \tag{3.59}$$

where we have replaced the stator current and rotor flux vectors by their dq-components. Owing to the rotor field orientation of the reference frame, the q-component of the rotor flux vector is, by definition, zero, that is, $\psi_{rq} = 0$. This also implies that the magnitude of the rotor flux vector is equal to its d-component, which allows us to write

$$\Psi_r = \|\boldsymbol{\psi}_r\| = \psi_{rd}. \tag{3.60}$$

To maintain the field orientation, the derivative of ψ_{rq} must be zero. This allows us to rewrite (3.59) in the form of the two scalar expressions

$$\frac{\mathrm{d}\Psi_r}{\mathrm{d}t} = \frac{X_m}{\tau_r} i_{sd} - \frac{1}{\tau_r}\Psi_r \tag{3.61a}$$

$$0 = \frac{X_m}{\tau_r} i_{sq} - (\omega_s - \omega_r)\Psi_r. \tag{3.61b}$$

A third equation is required, namely that of the electromagnetic torque. Setting $\psi_{rq} = 0$ in (2.61) leads to

$$T_e = \frac{1}{\mathrm{pf}}\frac{X_m}{X_r}\Psi_r i_{sq}. \tag{3.62}$$

The following three conclusions can be derived from (3.61) and (3.62). First, the differential equation (3.61a) of the rotor flux magnitude involves only i_{sd}. By manipulating the latter, the magnetization of the machine can be controlled. During steady-state operation, (3.61a) simplifies to

$$\Psi_r = X_m i_{sd}. \tag{3.63}$$

This highlights the linear steady-state relationship between the d-component of the stator current and the rotor flux magnitude. Second, provided that the rotor flux magnitude is constant, (3.62) implies that there is a linear relationship between i_{sq} and the electromagnetic torque, facilitating torque control. Third, given i_{sq}, Ψ_r, and ω_r, (3.61b) determines the (unique) stator frequency ω_s, which maintains the field orientation. This expression is often referred to as the *condition for rotor field orientation*. Note that $\omega_{\mathrm{sl}} = \omega_s - \omega_r$ is the slip frequency, which directly relates to the slip as defined in (2.37).

3.6.2.2 Indirect and Direct Field-Oriented Control

Field orientation can be achieved in two different ways. In the *indirect* method, the angular position of the rotor φ_r is measured with an incremental encoder. The slip frequency $\omega_{\mathrm{sl}} = \omega_s - \omega_r$ is reconstructed using (3.61b). Adding the integral of the slip frequency to φ_r yields the angular position of the rotor flux vector and thus of the rotating reference frame. As the slip frequency is determined in an open-loop fashion, it is vulnerable to machine parameter variations, such as operating-point-dependent changes in the rotor time constant.

In *direct* FOC, the angular position of the reference frame is determined with the help of an estimator. Based on the measured stator currents and stator voltages, the stator and rotor flux vectors can be estimated. The magnitude Ψ_r and angular position φ of the rotor flux vector directly follow. Instead of measuring the stator voltage, the latter is often reconstructed using the upper and lower dc-link voltages and the inverter switch positions. Obtaining accurate rotor flux positions at low speed is, however, inherently difficult because of machine parameter variations, measurement offsets, and drifts. For more details on (rotor) flux observer schemes, the reader is referred to [55, Sect. 4.5], [48, Sect. 5.3], and [56] and the references therein.

The block diagram of direct rotor FOC is summarized in Fig. 3.26. The reference for the rotor flux magnitude Ψ_r^* is translated into the d-component of the stator current reference i_{sd}^*

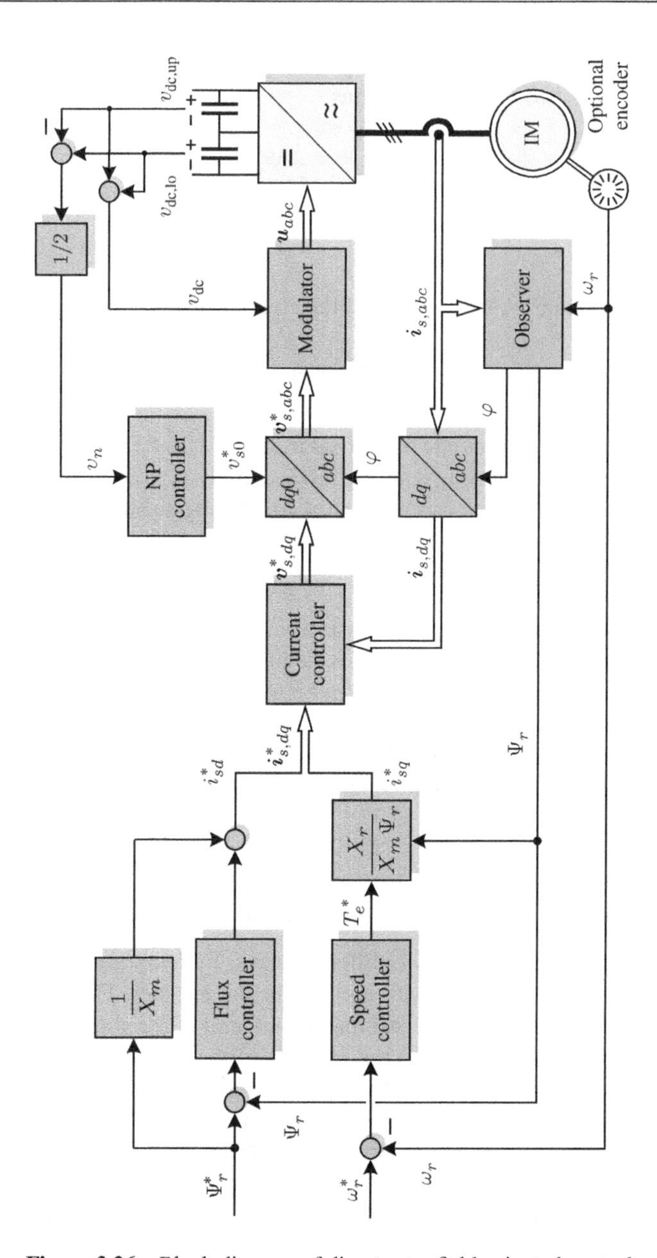

Figure 3.26 Block diagram of direct rotor field-oriented control

using (3.63) or (3.61a). A PI flux controller is typically added to achieve fast control of the rotor flux magnitude and to compensate for machine parameter variations. Given the error between the rotor speed ω_r and its reference, a speed controller manipulates the setpoint of the electromagnetic torque T_e^*. The latter is translated into the q-component of the stator current reference i_{sq}^* using the relationship (3.62). Note that this mapping depends on the actual magnitude of the rotor flux vector.

The two components of the stator current reference $\boldsymbol{i}^*_{s,dq}$ are fed to the current controller along with the measured stator currents. The latter are transformed from the three-phase abc system to the rotating dq reference frame with the angular position φ using the Park transformation (2.19). The current controller, which will be described next, manipulates the (differential-mode) stator voltage reference $\boldsymbol{v}^*_{s,dq}$. The common-mode voltage reference v^*_{s0} is set by another controller that maintains the neutral point (NP) potential v_n of an NPC inverter close to zero. Various methods are available to control the neutral point potential; the most prominent ones will be discussed at the end of this section.

The $dq0$ voltage reference is translated into the three-phase abc system using the inverse Park transformation (2.20). The resulting three-phase reference voltage is fed to the PWM stage, typically a CB-PWM or SVM, to generate the three-phase switching commands $\boldsymbol{u}_{abc}$, as explained in Sect. 3.3. To reduce the influence of dc-link voltage variations on the synthesized inverter voltage, the input voltage to the PWM is scaled by the instantaneous dc-link voltage v_{dc}. This is illustrated in Fig. 3.4.

3.6.2.3 Current Control

We assumed the use of a fast current controller when deriving the machine equations in rotor field orientation. This allowed us to neglect the dynamics of the stator windings and to focus exclusively on the rotor equation. This assumption will be justified in this section by designing a suitable current controller. Before doing so, however, it is expedient to point out two major advantages that are inherent to the combination of FOC with CB-PWM or SVM.

First, the use of CB-PWM ensures that the stator ripple current is zero at the peaks of the triangular carrier waveform, as discussed in Sect. 3.3.3. A similar statement holds true for SVM. By sampling the stator currents at the carrier peaks, only the fundamental component of the ac current waveform is captured. Directly related to this is the fact that CB-PWM and SVM use modulation cycles of fixed lengths. This facilitates sampling at equally spaced time instants. To this end, in the case of asymmetric, regularly sampled CB-PWM, the sampling interval is set to half that of the carrier interval, that is, $T_s = 0.5T_c$.

Second, by transforming the abc stator current into a dq reference frame, which rotates synchronously with the (stator or rotor) flux vector, the ac stator current is transformed into a dc quantity. The same applies to the orthogonal components of the flux vectors. As the bandwidth of the current controller is significantly higher than that of the outer flux and speed controllers, the stator current references are also effectively dc quantities. The fact that the current control loop involves (quasi) dc quantities facilitates the use of PI controllers. As shown in Fig. 3.27, two PI control loops are commonly used: one for the d-component of the stator current, and another one for its q-component.

The performance of the two PI loops is limited, however, by three fundamental issues. First, a digital implementation of the current controller leads to a computational delay of one sampling interval. For asymmetric, regularly sampled CB-PWM, the modulator incurs an additional delay of half a sampling interval (see Sect. 3.3.1). The overall delay in the current control loop thus amounts to $1.5T_s$. When operating at low switching frequencies with long carrier intervals and correspondingly long sampling intervals, this delay severely limits the achievable bandwidth of the current control loop. To compensate for the delay during steady-state operation, the angle $1.5\omega_s T_s$ can be added to the angular position of the reference frame when performing the inverse Park transformation (between the current controller and the modulator).

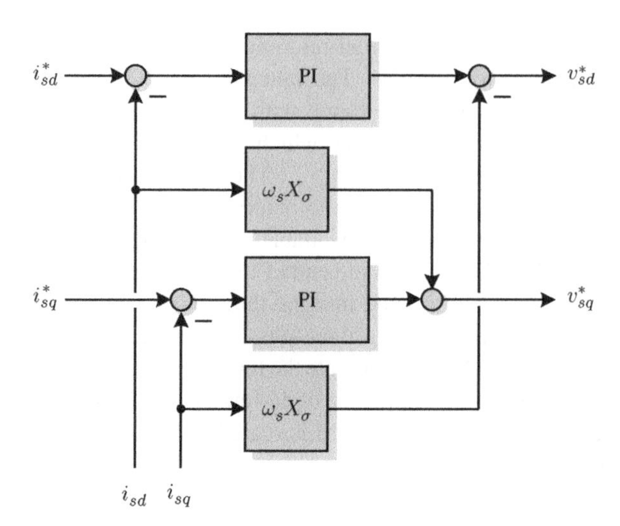

Figure 3.27 Current controller of the rotor FOC scheme. Decoupling of the d- and q-axes is achieved during steady-state operation with the feedforward terms

Second, the maximum voltage synthesizable by the modulator is limited, as discussed in Sect. 3.3.2. This imposes a physical constraint on the manipulated variable. Without appropriate countermeasures, the integrator of an aggressively tuned PI current controller operating close to its voltage limit might wind up. To prevent this, an anti-wind-up mechanism is often added. Such schemes typically monitor the difference between the commanded and the synthesized modulator voltage. If the difference is nonzero, the integrators in the PI controllers are turned off.

Third, the stator current dynamics in the d- and q-axes are not fully decoupled. This implies that the two PI loops are coupled to a certain degree, even when neglecting any delays. To highlight this fact, consider the system that is controlled, namely the stator equation (3.58a). Following the reasoning in [48, Sect. 4.2], we rearrange this equation in the rotating reference frame by expressing the stator voltage as a function of the stator current, its derivative, and the rotor flux vector.

$$\boldsymbol{v}_s = R_\sigma \boldsymbol{i}_s + X_\sigma \frac{\mathrm{d}\boldsymbol{i}_s}{\mathrm{d}t} + \omega_{\text{fr}} X_\sigma \begin{bmatrix} 0 & -1 \\ 1 & 0 \end{bmatrix} \boldsymbol{i}_s - R_r \frac{X_m}{X_r^2} \boldsymbol{\psi}_r + \omega_r \begin{bmatrix} 0 & -1 \\ 1 & 0 \end{bmatrix} \frac{X_m}{X_r} \boldsymbol{\psi}_r. \tag{3.64}$$

In (3.64), we have introduced the equivalent resistance

$$R_\sigma = R_s + R_r \frac{X_m^2}{X_r^2} \tag{3.65}$$

and used the total leakage reactance X_σ as defined in (2.63). Rotor field orientation implies that $\boldsymbol{\psi}_r = [\Psi_r \;\; 0]^T$. With this, the d- and q-components of the stator voltage can be written as

$$v_{sd} = R_\sigma i_{sd} + X_\sigma \frac{\mathrm{d}i_{sd}}{\mathrm{d}t} - \omega_{\text{fr}} X_\sigma i_{sq} - R_r \frac{X_m}{X_r^2} \Psi_r \tag{3.66a}$$

$$v_{sq} = R_\sigma i_{sq} + X_\sigma \frac{\mathrm{d}i_{sq}}{\mathrm{d}t} + \omega_{\text{fr}} X_\sigma i_{sd} + \omega_r \frac{X_m}{X_r} \Psi_r. \tag{3.66b}$$

Four distinct terms can be identified on the right-hand side of these two equations. The first and second terms involve only the current component of the respective axis. By manipulating the stator voltage, the dynamics of these two terms can be modified through closed-loop control, and their corresponding pole can be shifted. Most importantly, the two axes are decoupled when considering only the first two terms. This fact motivates the use of two independently operating PI control loops.

The third terms, however, feature the current component of the other axis. These so-called *motion-induced voltages* add coupling between the two axes. The fourth terms add a dependency on the rotor flux magnitude, particularly the rotor-induced voltage $\omega_r \frac{X_m}{X_r} \Psi_r$. As the latter is a function of i_{sd} (see (3.61a)), this adds further coupling from the d- to the q-axis.

A commonly used attempt to counteract these cross-coupling terms is to augment the control loops with feedforward terms. Specifically, the third terms in (3.66) are typically compensated for by setting $\omega_s = \omega_{\text{fr}}$ and adding

$$v_{sd,\text{ff}} = -\omega_s X_\sigma i_{sq} \tag{3.67a}$$

$$v_{sq,\text{ff}} = \omega_s X_\sigma i_{sd} \tag{3.67b}$$

to the outputs of the PI controllers, as shown in Fig. 3.27. Sometimes, the rotor-induced voltage $\omega_r \frac{X_m}{X_r} \Psi_r$ is added to (3.67b). The corresponding rotor flux term in (3.67a) is often neglected, because the rotor resistance R_r is small for high-power machines.

It should be clear that these feedforward terms achieve decoupling only during steady-state operation, while cross-coupling persists during transients. This leads to an adverse interaction between the two control loops and limits the control performance. To also achieve decoupling during transients, more sophisticated control methods are required that dynamically capture the cross-coupling. State feedback controllers [57, 58] and controllers with complex eigenvalues [59] are examples for such schemes, which achieve a high degree of decoupling and thus a high performance also during transients.

As a result of the ubiquitous use of FOC with current control and PWM, the related literature is vast, and numerous variations of the FOC concept have been proposed. Specifically, besides rotor field orientation, the dq reference frame can also be aligned with the stator flux or the air-gap flux. A good starting point to learn more about FOC and its variations are the well-known text books [47, 55, 60].

3.6.2.4 Control of the Neutral Point Potential

Besides the necessity to control the machine's stator current, the adoption of an NPC inverter gives rise to another control problem—that of balancing the inverter's neutral point potential. Despite the NPC inverter's natural balancing characteristic [61], active balancing techniques are commonly employed to avoid a lasting dc offset in the neutral point potential. The ac (or ripple) component of the neutral point potential is typically not targeted by closed-loop control.

Most control methods of the neutral point potential are based on the manipulation of the inverter's common-mode voltage. A positive common-mode voltage, for example, shifts the phase voltages to the upper inverter half. Depending on the sign of the phase current, this shift adds a positive or negative bias to the average current drawn from the neutral point [62, 63], which, in turn, modifies the neutral point potential.

Based on this principle, a dedicated neutral point controller can be designed that manipulates the common-mode voltage reference that is fed to the modulator. This control method was introduced in [64] and extended in [62, 63]. Common-mode voltage offsets can also be generated by manipulating the deadtime, which must be added between turn-*on* and turn-*off* switching transitions of the semiconductor switches [65].

On the other hand, instantaneous control of the common-mode voltage, and thus of the neutral point potential, can be achieved by exploiting the redundancy in the voltage vectors [66]. The inner voltage vectors form pairs, which generate the same differential-mode voltage but exhibit the opposite common-mode voltage. As a result, one of the two voltage vectors will always increase the neutral point potential while the other one will decrease it.

Accordingly in SVM, the existence of pairs of redundant voltage vectors can be exploited by varying the ratio of their *on* durations in the switching sequence [67]. This, in effect, also controls the neutral point potential via the common-mode voltage. Alternatively, at low output voltages, fast neutral point control can be achieved by shifting the entire pulse pattern either completely into the upper half or the lower half of the inverter [68].

Control of the neutral point potential through common-mode voltage manipulation is, however, not effective at high modulation indices, low power factors, or low phase currents [63]. To achieve neutral point potential balancing at no-load conditions, for example, a third-harmonic reactive current component can be injected, as proposed in [69].

3.6.3 Direct Torque Control

In FOC, as has been shown in the previous section, the electromagnetic torque and the machine magnetization are controlled *indirectly* through the stator currents. Alternatively, as was proposed by Takahashi and Noguchi in the mid-1980s, the torque and magnetization can also be controlled *directly*. This characteristic coined the term "direct torque control" (DTC) or "direct torque and flux control." Nowadays, DTC is a well-established, high-performance control method for motor drives and a viable alternative to FOC [45, 70–72].

The basic principle of DTC is to impose upper and lower bounds on the electromagnetic torque and the stator flux magnitude, and to use hysteresis controllers to enforce these bounds. The outputs of the hysteresis controllers are fed to a look-up table, which sets the inverter switch positions. Similar to FOC, flux and torque control are achieved independently of each other. The direct manipulation of the voltage vector applied to the stator windings exploits the fast stator flux dynamics of the machine. This leads to a control scheme that is conceptually simple and almost independent of the machine parameters, yet it achieves a very fast closed-loop torque and flux response.

By focusing on the stator flux vector, the notions of field orientation, rotating reference frames, and coordinate transformations are no longer required. Instead, DTC is formulated in the stationary and orthogonal $\alpha\beta$ coordinate system. The main characteristic that distinguishes DTC from other control methods is its use of closed-loop torque and flux magnitude controllers instead of current control loops [45]. Typically, DTC also lacks control and estimation loops that relate to rotor quantities [73].

3.6.3.1 Principle of Direct Flux Control

In order to accomplish torque and flux control, DTC bases the inner control loop on the stator *flux* instead of the stator *current* vector as in FOC. As stated in (2.53), the stator current can be expressed as a linear combination of the stator and rotor flux vectors. Specifically, we can write in the d-axis in rotor field orientation

$$i_{sd} = \frac{X_r}{D}\psi_{sd} - \frac{X_m}{D}\Psi_r, \tag{3.68}$$

where we have used $\Psi_r = \psi_{rd}$. This allows us to rewrite the scalar rotor flux magnitude equation (3.61a) as

$$\frac{\mathrm{d}\Psi_r}{\mathrm{d}t} = \frac{R_r}{D}(X_m\psi_{sd} - X_s\Psi_r). \tag{3.69}$$

This equation shows that the machine magnetization can be controlled through the d-component of the stator flux vector.

Recall that γ denotes the (load) angle between the stator and rotor flux vectors, as depicted, for example, in Fig. 3.28. Let $\Psi_s = \|\psi_s\|$ denote the magnitude of the stator flux vector. We can then write $\psi_{sd} = \cos(\gamma)\Psi_s$ in the rotor flux-oriented reference frame. At steady-state operating conditions, (3.69) reduces to

$$\Psi_r = \frac{X_m}{X_s}\cos(\gamma)\Psi_s. \tag{3.70}$$

To simplify the flux control loop in standard DTC, the stator flux magnitude is controlled instead of the rotor flux magnitude. As the load angle is typically limited to $\pm 15^\circ$, the error introduced in (3.70) by omitting the term $\cos(\gamma)$ is less than 3% and is thus in practice usually negligible. By maintaining the stator flux magnitude close to its desired level, which is

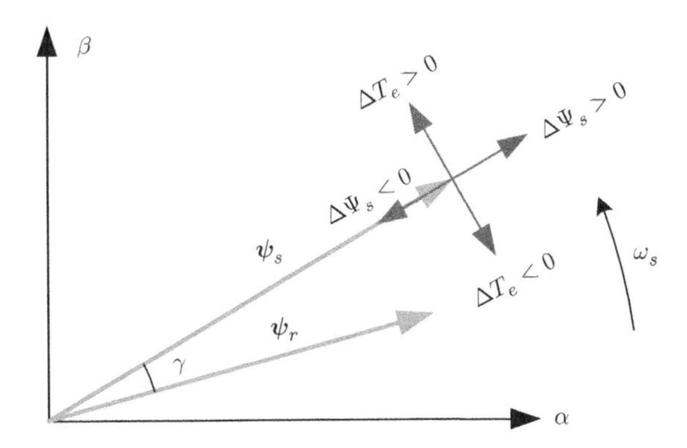

Figure 3.28 Illustration of the direct torque and flux control principle. The stator flux vector ψ_s is manipulated in the stationary orthogonal $\alpha\beta$ coordinate system by an appropriate voltage vector, which increases or decreases the stator flux magnitude Ψ_s and the electromagnetic torque T_e. The rate of change depends on the voltage vector's magnitude and direction

typically 1 unless field weakening is employed, the appropriate magnetization of the machine is achieved. As shown in Fig. 3.28, control of the stator flux magnitude is accomplished by manipulating the stator flux vector along its radial direction.

3.6.3.2 Principle of Direct Torque Control

FOC in rotor field orientation is based on the torque equation (3.62). Using (2.53), we can express the q-component of the stator current by the stator flux component in the q-axis as

$$i_{sq} = \frac{X_r}{D}\psi_{sq}. \tag{3.71}$$

Note that the q-component of the rotor flux vector is, by definition, zero. With the help of the expression $\psi_{sq} = \sin(\gamma)\Psi_s$, we can rewrite the torque equation (3.62) as

$$T_e = \frac{1}{\text{pf}}\frac{X_m}{D}\Psi_s\Psi_r\sin(\gamma). \tag{3.72}$$

This implies that the electromagnetic torque is the product of the sine of the load angle and the magnitudes of the stator and rotor flux vectors. Note that this statement is independent of the adopted reference frame. In particular, (3.72) holds also in the stationary coordinate system and can be derived also directly from (2.56).

Fast torque control can be achieved by manipulating the load angle in (3.72). To this end, the tangential component of the stator flux vector is manipulated by rotating the stator flux vector forward or backward. This principle is shown in Fig. 3.28.

To ensure that the torque controller is decoupled from the flux control loop, the magnitudes of the stator and rotor flux vectors must be constant. To achieve this, the stator flux magnitude is tightly controlled in DTC around its nominal value. The magnitude of the rotor flux vector can be considered to be constant within several milliseconds, thanks to the small rotor resistance R_r in (3.69), which causes a long rotor time constant.

3.6.3.3 Fast Stator Flux Control

As previously mentioned, DTC relies on fast stator flux control. To illustrate this principle, consider the stator equation (2.50a) in stationary coordinates. To this end, we set the angular speed of the reference frame to zero and drop the superscript $'$, which indicates pu quantities. This leads to

$$\frac{d\boldsymbol{\psi}_s}{dt} = \boldsymbol{v}_s - R_s\boldsymbol{i}_s, \tag{3.73}$$

where the stator flux vector is given by $\boldsymbol{\psi}_s = [\psi_{s\alpha} \;\; \psi_{s\beta}]^T$. The stator voltage $\boldsymbol{v}_s$ and the stator current $\boldsymbol{i}_s$ are defined accordingly.

As the stator resistance R_s is in many cases negligible, it is clear from (3.73) that the stator flux vector can be directly manipulated by choosing a suitable voltage vector, which is applied to the stator windings. In particular, within the sampling interval T_s, the stator flux vector is modified by

$$\Delta\boldsymbol{\psi}_s = \boldsymbol{v}_s T_s, \tag{3.74}$$

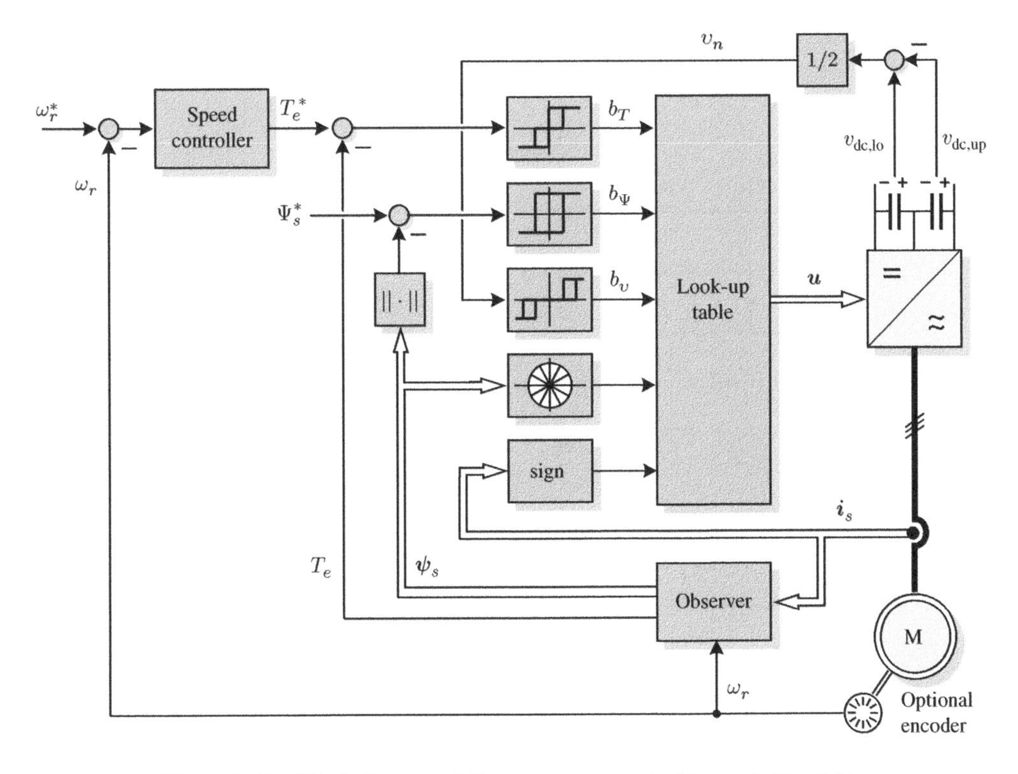

Figure 3.29 Block diagram of direct torque control with a switching table

where we have assumed that the voltage vector $\boldsymbol{v}_s$ is held constant for T_s. The stator flux vector is driven in the direction of the voltage vector. The rate of change corresponds to the length of the voltage vector. The available dc-link voltage imposes an upper bound on $\|\Delta\boldsymbol{\psi}_s\|$.

3.6.3.4 Direct Torque and Flux Control

The block diagram of standard DTC is shown in Fig. 3.29. A speed controller manipulates the torque reference T_e^*. The rotor flux magnitude is usually controlled indirectly via the stator flux magnitude reference Ψ_s^*. DTC requires only measurements of the stator current $\boldsymbol{i}_s$ and the upper and lower dc-link voltages. Based on the latter and the switch positions $\boldsymbol{u}$, the stator voltage $\boldsymbol{v}_s$ is reconstructed. Using $\boldsymbol{i}_s$, $\boldsymbol{v}_s$, and a machine model, an observer constructs the stator flux vector $\boldsymbol{\psi}_s$ and the electromagnetic torque T_e. The torque error is the difference between the torque reference and the estimated torque. The error of the stator flux magnitude is computed accordingly, by applying the Euclidean norm to the reconstructed stator flux vector.

The core of the DTC scheme is the hysteresis control unit and the look-up table, which contains the switching table. If either the torque or the flux error violates a hysteresis bound, a new voltage vector is selected that aims to drive the stator flux vector to a position such that the torque and flux errors both respect their corresponding hysteresis bounds. In the case of an

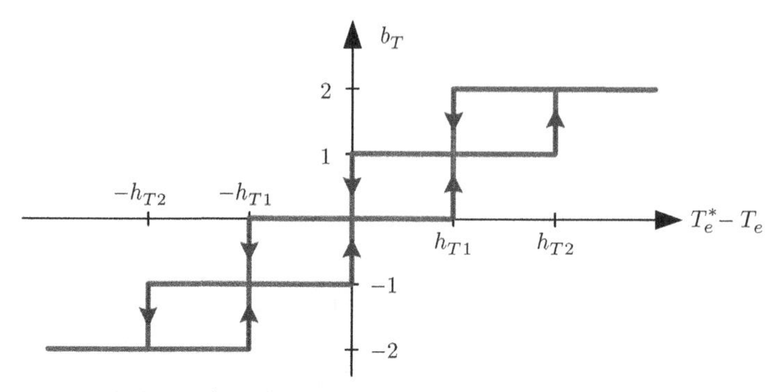

(a) Hysteresis on the error of the electromagnetic torque

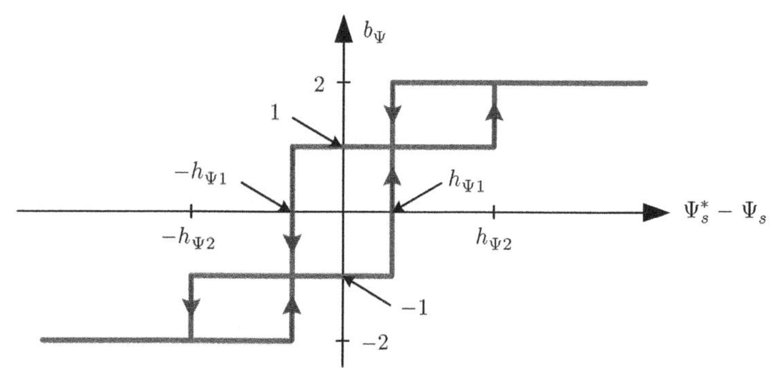

(b) Hysteresis on the error of the stator flux magnitude

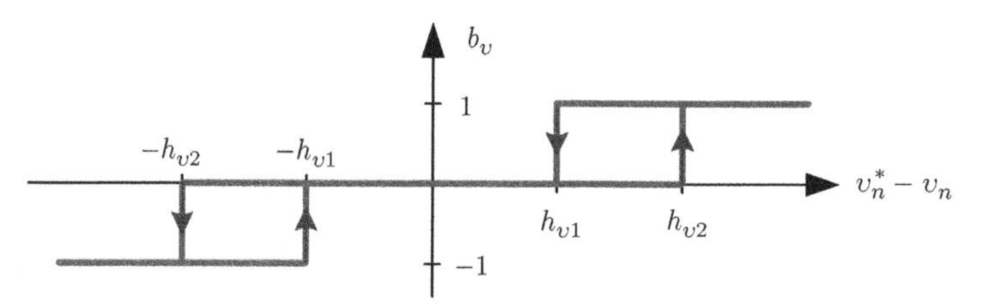

(c) Hysteresis on the error of the neutral point potential

Figure 3.30 DTC hysteresis controllers

NPC inverter, a third hysteresis control unit is added that maintains the neutral point potential υ_n within prespecified bounds around zero.

The torque hysteresis controller is shown in Fig. 3.30(a) for a three-level inverter. It uses hysteresis bands at zero, $\pm h_{T1}$, and $\pm h_{T2}$. Depending on the torque error and the state of the hysteresis, the output signal b_T is determined. The latter is an integer variable with the five possible values 0, ± 1, and ± 2. The flux hysteresis controller shown in Fig. 3.30(b) is similar to that of the torque controller, but uses only four hysteresis bands at $\pm h_{\Psi 1}$ and $\pm h_{\Psi 2}$. As a result, its output b_Ψ is equal to one of the four values ± 1 and ± 2.

The hysteresis controller for the neutral point potential is somewhat different in that it uses a large band around zero, in which the controller is inactive and the neutral point potential is allowed to float. Once the outer threshold at $\pm h_{v2}$ is violated, the controller is activated and its output is set to $b_v = \pm 1$ until the inner threshold at $\pm h_{v1}$ is activated. This principle is shown in Fig. 3.30(c).

In general, the bands of the hysteresis controllers are symmetric around zero. The motivation to use five bands for the torque controller and to omit the zero band for the flux controller will be expanded upon when explaining the switching table.

The outputs of the hysteresis controllers are fed into a switching table. For each combination of inputs, this look-up table holds a suitable voltage vector or, equivalently, a three-phase switch position. The voltage vectors are selected based on the sign and magnitude of the torque and flux errors along with the number of the sector that contains the stator flux vector. For the balancing of the neutral point potential, the sign of the phase currents is also required.

3.6.3.5 Switching Table

The (differential-mode components of the) voltage vectors produced by a two- or three-level inverter are the same when rotated by an integer multiple of $60°$. To exploit this symmetry, the $\alpha\beta$ coordinate system is divided into six pie-shaped sectors, which are enumerated from 0 to 5. Each sector covers $60°$. The design of the look-up table can be restricted to the zero-sector with angles between $-30°$ and $30°$. The look-up table for the other five sectors can be easily derived by rotating the voltage vectors in the zero-sector by multiples of $60°$.

For the three-level inverter, the zero-sector is further divided into two subsectors—a lower subsector between $-30°$ and $0°$, and an upper subsector between $0°$ and $30°$. To this end, the $360°$ circle is divided into 12 subsectors, each with an angular spread of $30°$. The white areas in Fig. 3.31 correspond to the lower subsectors, while the gray areas depict the upper subsectors.

As can be seen in Fig. 2.19, a three-level converter is capable of producing voltage vectors of approximately three different magnitudes: zero (the three zero vectors), short (the 12 vectors of the inner hexagon), and intermediate and long (the 12 vectors of the outer hexagon). The intermediate and long voltage vectors are of similar magnitude and can thus be lumped together. Therefore, we associate the integers 0, ± 1, and ± 2 with the voltage vector magnitudes.

A voltage vector that acts in a direction that is *orthogonal* to the stator flux vector increases or decreases the electromagnetic torque. Therefore, to accomplish torque control, the orthogonal component of the voltage vector is manipulated. To this end, we associate the five output levels of the torque controller with the orthogonal axis shown in Fig. 3.31(a). Note that we assume here a counterclockwise rotation of the flux vectors. The following switching logic can then be derived: $b_T = 0$ implies the use of a zero vector, $b_T = 1$ necessitates a short vector in the direction of the rotation, while an intermediate or long voltage vector is required in case of $b_T = 2$. Negative b_T values, on the other hand, trigger voltage vectors that point in the opposite direction.

Voltage vectors *parallel* to the stator flux vector either increase or decrease the stator flux magnitude. Therefore, the output of the flux controller determines the magnitude and sign of the voltage vector component in this axis, as shown in Fig. 3.31(b). Typically, only two levels are used for the flux controller, namely ± 1. The zero level is not used, because the zero vectors have only a minor influence on the stator flux magnitude. When the torque hysteresis controller outputs $b_T = 0$, however, control of the stator flux magnitude is temporarily lost. To avoid this, the flux hysteresis controller is augmented by the output levels ± 2. This enables DTC to switch to a short voltage vector when required to maintain flux control.

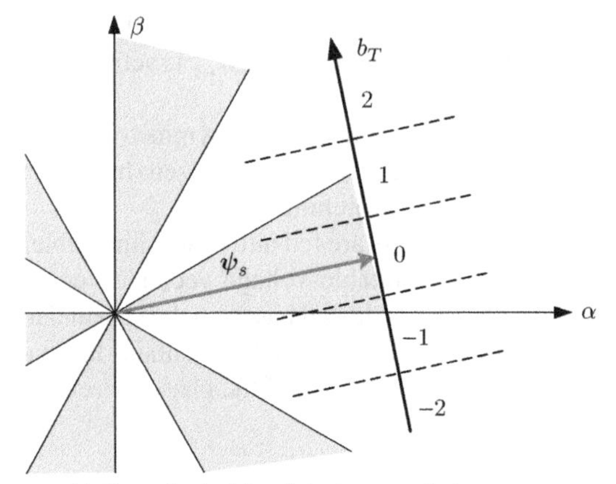

(a) Control principle of electromagnetic torque

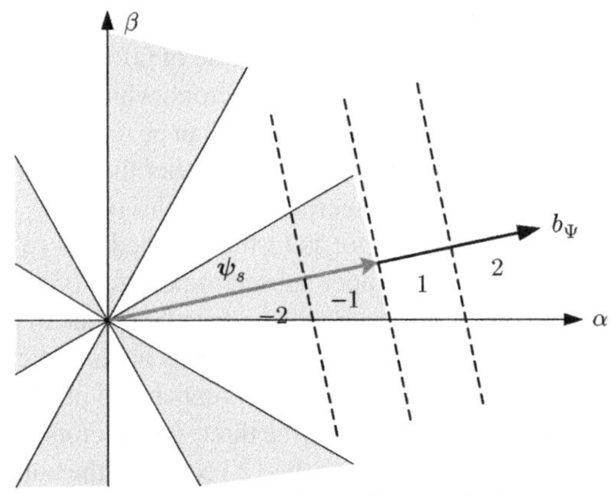

(b) Control principle of stator flux magnitude

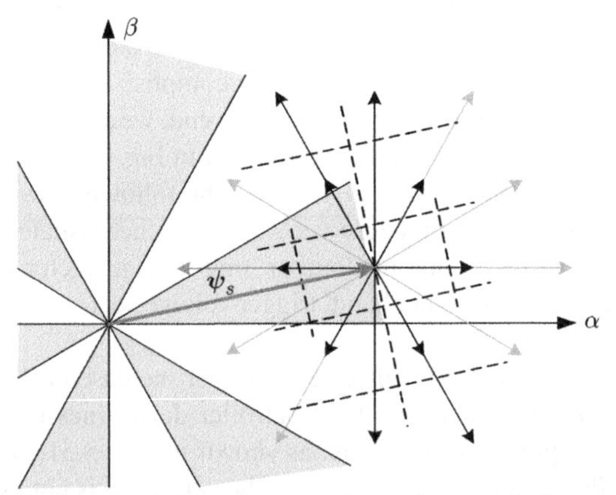

(c) Control principle of torque and stator flux magnitude

Figure 3.31 Voltage vector selection principle in DTC. The $\alpha\beta$-plane with the voltage vectors is divided into orthogonal bands that correspond to the integer outputs of the hysteresis controllers. The voltage vectors, and thus the three-phase switch positions, are chosen accordingly

The switching table can be derived by combining the reasoning for the torque and flux hysteresis controllers. The $\alpha\beta$-plane of the voltage vectors is divided into orthogonal bands that correspond to the integer outputs of the two hysteresis controllers. Each combination of the controller outputs corresponds to one (differential-mode) voltage vector. These voltage vectors are shown in Fig. 3.31(c) for the upper part of the zero-sector. It can be seen that all of the available short vectors are used. The voltage vectors are chosen such that they are applicable to any stator flux vector that lies within this subsector. It is then also apparent why DTC only requires knowledge of the subsector the stator flux lies in, rather than of its precise angular position.

To accomplish the balancing of the neutral point potential when $b_\upsilon = \pm 1$, the redundancy in the pairs of short voltage vectors is utilized. Each pair of vectors influences the neutral point potential by the same amount, but with opposite signs. To determine the desired redundant voltage vector, the sign of the three-phase stator currents is required. On the other hand, when the control of the neutral point potential is inactive for $b_\upsilon = 0$, the redundancy can be used to help reduce the switching frequency.

3.6.3.6 Characteristics and Discussion

In DTC, the torque, flux, and neutral point controllers are realized as hysteresis controllers. When compared to FOC, these controllers replace the inner current control loop along with the torque and flux feedforward terms. The DTC switching table supersedes the modulator. Instead of regulating the torque, flux, and neutral point potential to their respective references, DTC aims at keeping these quantities within certain bounds around their references.

The widths of the hysteresis bounds determine the switching frequency. For fixed-width hysteresis bands, the switching frequency is operating-point-dependent and varies with the fundamental frequency and torque setpoint. For an analysis and prediction of the switching frequency for two-level inverter drives, the reader is referred to [74]. The hysteresis controllers entail a harmonic spectrum of the stator currents and electromagnetic torque that is neither restricted to discrete frequencies nor is deterministic. Nevertheless, thanks to use of 12 subsectors in the switching table for the three-level inverter, the 11th and 13th harmonics are pronounced in the stator voltage and current.

To minimize the current and torque ripples, as many hysteresis levels as possible should be used. This is particularly important for the torque hysteresis controller, which controls the angular position of the stator flux vector. During steady-state operation, the magnitude of the orthogonal component of the voltage vector should nearly match the magnitude of the speed-dependent back-EMF $\omega_s \Psi_s$ (see also (3.57) and the discussion on scalar control in Sect. 3.6.1). This motivates the use of five levels for the torque hysteresis controller. Assuming a counterclockwise rotation of the flux vectors, the torque hysteresis switches between 0 and 1 at low-speed operation and thus applies zero and short active voltage vectors. At nominal speed, however, the torque hysteresis switches between 1 and 2, that is, between a short and a long vector.

When compared to FOC, DTC has the following benefits regarding implementation and performance:

- *Simplicity*. DTC requires neither a modulator nor coordinate transformations. The PI current control loops, the decoupling network, and parts of the torque and flux control loops are replaced by hysteresis controllers. The latter are conceptually and computationally simple. The tuning and commissioning effort of DTC schemes is minimal.

- *Robustness.* DTC requires the estimation of stator quantities, namely of the stator flux vector and the electromagnetic torque. To do so, only the stator resistance must be known. DTC is thus insensitive to variations in the rotor parameters. Neither the angular speed of the flux vectors nor the angular position or magnitude of the rotor flux vector is required. The use of hysteresis controllers further adds to the notable robustness of DTC.
- *Torque response.* The dynamic torque response achieved by DTC is very fast and limited only by the available dc-link voltage. Comparisons with FOC indicate that DTC tends to outperform FOC in this regard (see, e.g., [75]).

On the other hand, the following disadvantages are typically associated with DTC:

- *Harmonic distortions.* DTC produces pronounced ripples on the stator current and electromagnetic torque. Standard DTC tends to suffer from higher stator current and torque TDDs than FOC, at least in the case of two-level inverters [75]. For multilevel inverters, however, an advanced DTC scheme yields a lower torque ripple than FOC with PWM [76].
- *Switching frequency.* The use of hysteresis control loops leads to an operating-point-dependent switching frequency. To nevertheless achieve a sufficiently constant switching frequency, the widths of the hysteresis bounds require adjustment. This can be done, for example, by monitoring the switching frequency and by modifying the bound widths by a suitable closed-loop switching frequency control loop.
- *Sampling frequency.* To limit the violations of the hysteresis bounds, DTC requires a high sampling frequency. Typically, a sampling interval of $T_s = 25$ μs is adopted in DTC, which corresponds to a sampling frequency of 40 kHz. The latter is at least one order of magnitude higher than in FOC.
- *Torque error.* The use of hysteresis controllers with multiple output levels inevitably leads to pronounced steady-state errors, particularly in the torque control loop. To nevertheless achieve zero torque errors, an offset controller with an integrator is often added to the torque control loop.

3.6.3.7 Extensions of the DTC Concept and Related Control Methods

For a summary of the basic DTC concept and various extensions to improve the performance of DTC, the reader is referred to the survey paper [45], the book chapter [46, Chap. 8], and the text book [55]. Several notable extensions of the basic DTC concept for two-level inverters are briefly summarized in the following:

- To reduce the torque and flux ripple, additional (discrete) voltage vectors are introduced in [77], using the notion of SVM. The hysteresis controllers are kept in place.
- As a further step in the direction of SVM, the hysteresis control loops are replaced in [78] by a deadbeat controller. The stator voltage that is predicted to minimize the torque and flux errors at the next sampling instant is computed. This voltage is translated into switching signals using SVM. This ensures a deterministic harmonic spectrum and relatively small current distortions.

- Alternatively, a constant switching frequency can be achieved without the use of a modulator. As proposed in [79] for a two-level converter, the hysteresis control loops and look-up table determine the active voltage vector, which is complemented by a zero vector. Using a fixed-length switching interval, the duty cycle of the active voltage vector is adjusted such that it minimizes the torque ripple.

DTC was originally developed for two-level inverters with three hysteresis levels for the torque control loop and two levels for the flux loop [70]. Even today, most DTC-related research efforts focus on two-level inverters. On the other hand, the DTC literature on inverters with three or more levels is limited. Notable exceptions include [80, 81] for three-level inverters and [76, 82] for multilevel inverters.

A method closely related to DTC is direct self-control (DSC), which was proposed by Depenbrock [83, 84] in parallel to DTC. For two-level inverters, control of the stator flux magnitude is achieved by regulating the stator flux vector along a hexagonal path with the help of three hysteresis controllers—one for each of the three abc flux components. A fourth hysteresis controller accomplishes torque control by adding zero vectors to the hexagonal flux path.

As a result, standard DSC is inherently based on a six-step operation and allows the full utilization of the available dc-link voltage. Corner-folding of the flux trajectory was later introduced to reduce the torque ripple [85]. Extensions to three-level inverters are also possible [86]. The fact that DSC is very suitable for drives operating at very low switching frequencies and in the field weakening range makes DSC a popular control method for traction drive applications [87].

Appendix 3.A: Harmonic Analysis of Single-Phase Optimized Pulse Patterns

The Fourier series (3.24) for the single-phase pulse pattern $u(\theta)$ is derived in this appendix. Recall that $u(\theta)$ is a periodic signal with the period 2π. Therefore, $u(\theta)$ can be described by the Fourier series

$$u(\theta) = \frac{a_0}{2} + \sum_{n=1}^{\infty}(a_n \cos(n\theta) + b_n \sin(n\theta)) \tag{3.A.1}$$

with the Fourier coefficients

$$a_n = \frac{1}{\pi}\int_{-\pi}^{\pi} u(\theta)\cos(n\theta)\mathrm{d}\theta, \quad n \geq 0 \tag{3.A.2a}$$

$$b_n = \frac{1}{\pi}\int_{-\pi}^{\pi} u(\theta)\sin(n\theta)\mathrm{d}\theta, \quad n \geq 1. \tag{3.A.2b}$$

Owing to the imposed symmetry (3.20), the pulse pattern is an odd function, that is, $u(\theta) = -u(-\theta)$. The terms $\cos(n\theta)$ in the Fourier coefficient a_n are even functions. As the product of an odd function with an even function is an odd function, the integral over the interval $[-\pi,\ \pi]$ is zero. We conclude that all the coefficients a_n are zero.

Because the terms $\sin(n\theta)$ are odd functions, the integrand in b_n is an even function. The integral from $-\pi$ to 0 and the integral from 0 to π are thus the same, allowing us to simplify (3.A.2b) to

$$b_n = \frac{2}{\pi}\int_{0}^{\pi} u(\theta)\sin(n\theta)\mathrm{d}\theta. \tag{3.A.3}$$

We divide the integration interval $[0, \ \pi]$ into two parts:

$$b_n = \frac{2}{\pi}\int_0^{\frac{\pi}{2}} u(\theta)\sin(n\theta)\mathrm{d}\theta + \frac{2}{\pi}\int_{\frac{\pi}{2}}^{\pi} u(\pi-\theta)\sin(n\theta)\mathrm{d}\theta. \tag{3.A.4}$$

In the second integral, we have exploited the quarter-wave symmetry (3.20) of the single-phase pulse pattern. The change of variable $\theta' = \pi - \theta$ in the second integral leads to

$$b_n = \frac{2}{\pi}\int_0^{\frac{\pi}{2}} u(\theta)\sin(n\theta)\mathrm{d}\theta - \frac{2}{\pi}\int_{\frac{\pi}{2}}^{0} u(\theta')\sin(n\pi - n\theta')\mathrm{d}\theta'. \tag{3.A.5}$$

With the help of the identity

$$\sin(n\pi - n\theta') = -(-1)^n \sin(n\theta')$$

(3.A.5) can be rewritten as

$$b_n = \frac{2}{\pi}\int_0^{\frac{\pi}{2}} u(\theta)\sin(n\theta)\mathrm{d}\theta - \frac{2}{\pi}(-1)^n\int_0^{\frac{\pi}{2}} u(\theta)\sin(n\theta)\mathrm{d}\theta, \tag{3.A.6}$$

where we have replaced θ' by θ to simplify the notation. The Fourier coefficient can be further simplified to

$$b_n = \begin{cases} \frac{4}{\pi}\int_0^{\frac{\pi}{2}} u(\theta)\sin(n\theta)\mathrm{d}\theta, & \text{if } n = 1,3,5,\ldots \\ 0, & \text{if } n = 2,4,6,\ldots \end{cases} \tag{3.A.7}$$

Recall that the pulse pattern $u(\theta)$ is fully characterized by the primary switching angles α_i and the switch positions u_i, with $i \in \{1,2,\ldots,d\}$. The pulse pattern is a piecewise constant signal. In the interval $0 \le \theta \le \frac{\pi}{2}$, it can be described by the statement

$$u(\theta) = \begin{cases} 0, & \text{if } 0 \le \theta < \alpha_1 \\ u_1, & \text{if } \alpha_1 \le \theta < \alpha_2 \\ u_2, & \text{if } \alpha_2 \le \theta < \alpha_3 \\ \vdots & \vdots \\ u_{d-1}, & \text{if } \alpha_{d-1} \le \theta < \alpha_d \\ u_d, & \text{if } \alpha_d \le \theta \le \frac{\pi}{2} \end{cases} . \tag{3.A.8}$$

As an example for $u(\theta)$ with $d = 3$ primary switching angles, the reader is referred to Fig. 3.16.

We define the change in the switch position (or the switching transition) as

$$\Delta u_i = u_i - u_{i-1}. \tag{3.A.9}$$

As $u_0 = 0$, it follows that the ith switch position can be represented as the sum of i switching transitions

$$u_i = \sum_{j=1}^{i} \Delta u_j. \tag{3.A.10}$$

Inserting (3.A.10) and (3.A.8) into (3.A.7) for odd n leads to the sum of d integral terms

$$\begin{aligned} b_n &= \frac{4}{\pi}\Delta u_1 \int_{\alpha_1}^{\alpha_2} \sin(n\theta)\mathrm{d}\theta + \frac{4}{\pi}(\Delta u_1 + \Delta u_2) \int_{\alpha_2}^{\alpha_3} \sin(n\theta)\mathrm{d}\theta + \cdots \\ &+ \frac{4}{\pi}\sum_{j=1}^{d-1} \Delta u_j \int_{\alpha_{d-1}}^{\alpha_d} \sin(n\theta)\mathrm{d}\theta + \frac{4}{\pi}\sum_{j=1}^{d} \Delta u_j \int_{\alpha_d}^{\frac{\pi}{2}} \sin(n\theta)\mathrm{d}\theta. \end{aligned} \tag{3.A.11}$$

By rearranging and combining the integral terms for each Δu_i, the expression can be simplified to

$$\begin{aligned} b_n &= \frac{4}{\pi}\Delta u_1 \int_{\alpha_1}^{\frac{\pi}{2}} \sin(n\theta)\mathrm{d}\theta + \frac{4}{\pi}\Delta u_2 \int_{\alpha_2}^{\frac{\pi}{2}} \sin(n\theta)\mathrm{d}\theta + \cdots \\ &+ \frac{4}{\pi}\Delta u_{d-1} \int_{\alpha_{d-1}}^{\frac{\pi}{2}} \sin(n\theta)\mathrm{d}\theta + \frac{4}{\pi}\Delta u_d \int_{\alpha_d}^{\frac{\pi}{2}} \sin(n\theta)\mathrm{d}\theta. \end{aligned} \tag{3.A.12}$$

The integral terms can now be easily solved to

$$\int_{\alpha_i}^{\frac{\pi}{2}} \sin(n\theta)\mathrm{d}\theta = -\frac{1}{n}\left(\cos(n\frac{\pi}{2}) - \cos(n\alpha_i)\right) = \frac{1}{n}\cos(n\alpha_i), \tag{3.A.13}$$

where we have used the fact that n is odd. This leads to

$$b_n = \frac{4}{n\pi}\sum_{i=1}^{d} \Delta u_i \cos(n\alpha_i). \tag{3.A.14}$$

As all a_n are zero, this allows us to state the Fourier series (3.A.1) with its coefficients (3.A.2) in the compact representation

$$u(\theta) = \sum_{n=1}^{\infty} \hat{u}_n \sin(n\theta) \tag{3.A.15}$$

$$\hat{u}_n = \begin{cases} \dfrac{4}{n\pi}\displaystyle\sum_{i=1}^{d} \Delta u_i \cos(n\alpha_i), & \text{if } n = 1,3,5,\ldots \\ 0, & \text{if } n = 2,4,6,\ldots \end{cases} \tag{3.A.16}$$

where we have replaced b_n by $\hat{u}_n$ to indicate that the Fourier coefficient relates to the peak value of the nth harmonic signal.

Appendix 3.B: Mathematical Optimization

In this appendix, we review the basic terminology of mathematical programming, and introduce mixed-integer programs (MIP) and quadratic programs (QP). For more details on mathematical optimization and proofs of the statements made next, the reader is referred to one of the well-known text books, such as [42, 43, 88–90].

3.B.1 General Optimization Problems

We start by introducing basic terminology [43]. Consider the general form of an optimization problem

$$\underset{\boldsymbol{x}}{\text{minimize}} \quad J(\boldsymbol{x}) \tag{3.B.1a}$$

$$\text{subject to} \quad g_i(\boldsymbol{x}) \leq 0, \qquad i = 1, \ldots, n_g \tag{3.B.1b}$$

$$h_i(\boldsymbol{x}) = 0, \qquad i = 1, \ldots, n_h \tag{3.B.1c}$$

with the *optimization* or *decision variable* $\boldsymbol{x} = [\boldsymbol{x}_r^T \ \boldsymbol{x}_b^T]^T$, which contains, in general, both a real-valued part $\boldsymbol{x}_r \in \mathbb{R}^{n_r}$ and a binary part $\boldsymbol{x}_b \in \{0, 1\}^{n_b}$. The *objective function* or *cost function* $J(\boldsymbol{x}) : \mathbb{R}^{n_r} \times \{0, 1\}^{n_b} \rightarrow \mathbb{R}$ is a mapping from the vector $\boldsymbol{x}$ to the scalar J.

The cost function is to be minimized subject to the inequality constraints (3.B.1b) and the equality constraints (3.B.1c). These constraints define the subset $\boldsymbol{\mathcal{X}}$ of the space $\mathbb{R}^{n_r} \times \{0, 1\}^{n_b}$. This subset is often referred to as the *feasible set*, *feasible region*, or the *search space*. We refer to the functions $g_i(\boldsymbol{x})$ and $h_i(\boldsymbol{x})$ as the inequality and equality *constraint functions*. The point $\boldsymbol{x}$ is said to be *feasible* if it is part of the feasible set $\boldsymbol{\mathcal{X}}$ and thus satisfies all constraints. The problem (3.B.1) is *feasible* if at least one feasible point exists; or else it is *infeasible*.

The optimization problem (3.B.1) amounts to finding the vector $\boldsymbol{x} \in \boldsymbol{\mathcal{X}}$ that minimizes the cost function $J(\boldsymbol{x})$ over the feasible set $\boldsymbol{\mathcal{X}}$. The *optimal value* J_{opt} of the optimization problem (3.B.1) is defined as

$$J_{\text{opt}} = \inf\{J(\boldsymbol{x}) | \ \boldsymbol{x} \in \boldsymbol{\mathcal{X}}\}. \tag{3.B.2}$$

The solution $\boldsymbol{x}_{\text{opt}}$ of the (feasible) minimization problem (3.B.1) is referred to as the *optimal solution* or the *optimizer*. We can write $J(\boldsymbol{x}_{\text{opt}}) = J_{\text{opt}}$. If the problem is infeasible, we set $J_{\text{opt}} = \infty$, and if the problem is unbounded below, $J_{\text{opt}} = -\infty$ results. We say that a feasible point $\boldsymbol{x}$ is *locally optimal* if it minimizes J in a subset of the feasible set, whereas $\boldsymbol{x}$ is (*globally*) *optimal* if it minimizes J over the whole feasible set.

3.B.2 Mixed-Integer Optimization Problems

When the optimization variable contains integer variables ($n_b \geq 1$), (a subset of) the feasible set $\boldsymbol{\mathcal{X}}$ is discrete and the optimization problem (3.B.1) constitutes an MIP. In the worst case, all possible integer solutions need to be explored when solving MIPs. This implies that the solution time grows exponentially with the number of integer optimization variables [91]. In fact, MIPs are in general non-deterministic, polynomial-time hard (*NP*-hard) problems.

Nevertheless, optimization techniques have been proposed that often allow one to solve MIPs efficiently. One such technique is the so-called *branch-and-bound* method, which is explained next. Other techniques to solve MIPs include cutting plane, decomposition, and logic-based algorithms. For more details on these techniques, the reader is referred to [88] and [92].

The branch-and-bound concept was developed in the 1960s. It has since become of paramount importance in solving (mixed) integer optimization problems. Rather than enumerating all possible (candidate) solutions, branch-and-bound methods seek to reduce the number of investigated candidate solutions. By applying bounds, uninvestigated candidate

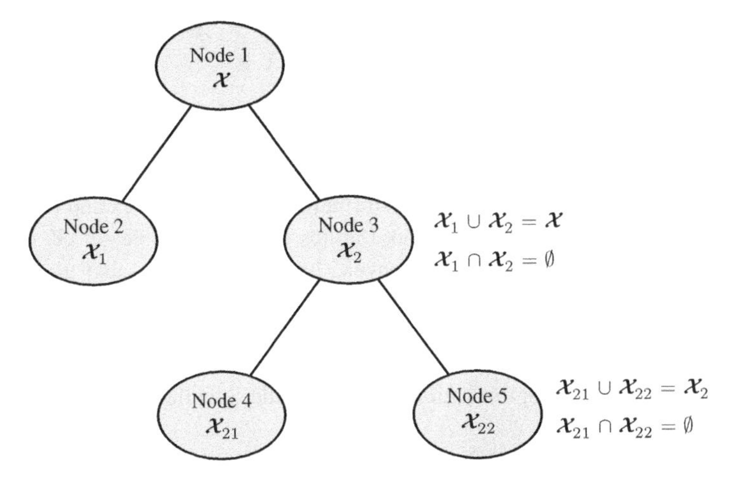

Figure 3.B.1 Illustration of the branch-and-bound concept. The feasible set $\mathcal{X}$ is split into disjoint subsets. Upper and lower bounds on the cost function are applied to identify and remove subproblems that contain only suboptimal candidate solutions

solutions can be removed from further consideration by proving that these solutions would be suboptimal. As a result, in general, only a small subset of the set of candidate solutions—and thus of the search tree—needs to be enumerated to find the optimal solution.

Specifically, as the name indicates, branch-and-bound methods are based on the following two operations.

1. *Branching*. The optimization problem is split into subproblems by dividing the feasible set $\mathcal{X}$ into two or more disjoint subsets, for example, into $\mathcal{X}_1$ and $\mathcal{X}_2$, such that $\mathcal{X}_1 \cup \mathcal{X}_2 = \mathcal{X}$ and $\mathcal{X}_1 \cap \mathcal{X}_2 = \emptyset$. Heuristics are typically employed to decide which subset is to be explored first.
2. *Bounding*. The smallest value of the cost function of all investigated candidate solutions serves as an upper bound. By definition, this value is equal to the cost function value of the best solution found so far. Lower bounds on the subsets' optimal solutions are usually provided by relaxations to ensure that they are quick to compute. If the lower bound of a subset exceeds the upper bound, then the optimal solution cannot be part of this subset and the corresponding subproblem can be removed from further consideration.

Branch-and-bound is a universal concept that is highlighted in Fig. 3.B.1, which was adopted from [93]. A good introduction and summary of the branch-and-bound methodology is provided in [93] and [94, Chaps. 12 and 13]. A more mathematical account is presented in [95] and [96, Chap. 8], while [97] provides a survey on branch-and-bound methods.

3.B.3 Convex Optimization Problems

An important subclass of (3.B.1) are *convex* optimization problems. The set $\mathcal{X} = \mathbb{R}^{n_r}$ is convex if the line segment between any of its two points $\boldsymbol{x}_1$ and $\boldsymbol{x}_2$ is also part of the set, that is,

$$\ell \boldsymbol{x}_1 + (1-\ell)\boldsymbol{x}_2 \in \mathcal{X} \tag{3.B.3}$$

for all $\boldsymbol{x}_1, \boldsymbol{x}_2 \in \boldsymbol{\mathcal{X}}$ and $\ell \in [0, 1]$. Consider the function $J : \boldsymbol{\mathcal{X}} \to \mathbb{R}$, where $\boldsymbol{\mathcal{X}} \in \mathbb{R}^{n_r}$ is a nonempty and convex set. The function J is convex on $\boldsymbol{\mathcal{X}}$ if

$$J(\ell \boldsymbol{x}_1 + (1-\ell)\boldsymbol{x}_2) \leq \ell J(\boldsymbol{x}_1) + (1-\ell)J(\boldsymbol{x}_2) \tag{3.B.4}$$

for all $\boldsymbol{x}_1, \boldsymbol{x}_2 \in \boldsymbol{\mathcal{X}}$ and $\ell \in [0, 1]$.

For convex optimization problems, the constraint functions are convex and the optimization variable is real-valued, that is, $n_b = 0$. This implies that the feasible set $\boldsymbol{\mathcal{X}}$ is convex. As the cost function is also required to be convex, local minima are also globally optimal. The importance of convex optimization problems stems from the fact that they can be solved in polynomial time, that is, the computation time scales well with the dimension of the problem.

One relevant type of convex optimization problems is the so-called *quadratic program* (QP)

$$\underset{\boldsymbol{x}}{\text{minimize}} \ \frac{1}{2}\boldsymbol{x}^T \boldsymbol{H} \boldsymbol{x} + \boldsymbol{c}^T \boldsymbol{x} \tag{3.B.5a}$$

$$\text{subject to} \ \boldsymbol{G}\boldsymbol{x} \leq \boldsymbol{g}. \tag{3.B.5b}$$

The cost function $J(\boldsymbol{x}) : \mathbb{R}^{n_r} \to \mathbb{R}$ is quadratic, the optimization variable $\boldsymbol{x} \in \mathbb{R}^{n_r}$ is real-valued, and the constraints are linear. The matrix $\boldsymbol{H}$, the so-called *Hessian*, is of the dimension $n_r \times n_r$ and the constraint matrix $\boldsymbol{G}$ is of the dimension $n_g \times n_r$, where n_g denotes the number of inequality constraints. The linear constraints (3.B.5b) define the feasible set $\boldsymbol{\mathcal{X}}$.

QPs are typically solved using the interior point method [98, 99]. Other solution approaches include the active set [100] and gradient methods [101]. Examples of QP solvers include SeDuMi [102], CPLEX [103], and IpOpt [104]. Recently, initial efforts have been reported in the literature to solve QPs in embedded systems, particularly when running on field-programmable gate arrays (FPGAs) (see, e.g., [105–108]).

We now require the notion of positive-definite matrices. A symmetric $n_r \times n_r$ matrix $\boldsymbol{H}$ is positive *semidefinite* if

$$\boldsymbol{x}^T \boldsymbol{H} \boldsymbol{x} \geq 0 \tag{3.B.6}$$

holds for all $\boldsymbol{x}$ in $\mathbb{R}^{n_r}$. Similarly, $\boldsymbol{H}$ is positive *definite* if

$$\boldsymbol{x}^T \boldsymbol{H} \boldsymbol{x} > 0 \tag{3.B.7}$$

for all nonzero $\boldsymbol{x}$ in $\mathbb{R}^{n_r}$.

One can show that the quadratic cost function $J(\boldsymbol{x})$ in (3.B.5a) is convex if and only if $\boldsymbol{H}$ is positive semidefinite. The expression *if and only if* signifies that this is a necessary and sufficient condition.

Consider the optimization problem (3.B.5) without the constraints (3.B.5b), that is, assume $\boldsymbol{\mathcal{X}} = \mathbb{R}^{n_r}$. We refer to this problem as the *unconstrained* QP. Assume that J is differentiable at $\boldsymbol{x}_{\text{opt}}$. If J is convex, then $\boldsymbol{x}_{\text{opt}}$ is a global optimal solution if and only if

$$\nabla J(\boldsymbol{x}_{\text{opt}}) = 0. \tag{3.B.8}$$

The gradient

$$\nabla J(\boldsymbol{x}) = \begin{bmatrix} \frac{\mathrm{d}J(\boldsymbol{x})}{\mathrm{d}x_1} \\ \frac{\mathrm{d}J(\boldsymbol{x})}{\mathrm{d}x_2} \\ \vdots \\ \frac{\mathrm{d}J(\boldsymbol{x})}{\mathrm{d}x_{n_r}} \end{bmatrix} \tag{3.B.9}$$

is a vector in $\mathbb{R}^{n_r}$, which consists of the partial derivatives of $J(\boldsymbol{x})$, where $\boldsymbol{x} = [x_1\ x_2 \ldots x_{n_r}]^T$. For the cost function (3.B.5a) of the QP, the gradient can easily be computed, and the optimality condition (3.B.8) can be stated as

$$\nabla J(\boldsymbol{x}_{\text{opt}}) = \boldsymbol{H}\boldsymbol{x}_{\text{opt}} + \boldsymbol{c} = 0. \tag{3.B.10}$$

Recall that the minimum of a function with a scalar argument can be found by setting the first derivative to zero and requiring the second derivative to be positive. For functions with a vector as an argument, this statement has been generalized previously. In particular, the point $\boldsymbol{x}_{\text{opt}}$ that minimizes J requires the gradient to be zero and the Hessian to be positive (semi)definite.

An alternative approach to deriving the optimality condition (3.B.10) is provided here. The use of derivatives is avoided—we only require the Hessian to be invertible and symmetric. Starting from the cost function $J(\boldsymbol{x}_{\text{opt}})$ in (3.B.5a), we use the technique of *completing the squares* to rewrite the cost function as

$$J(\boldsymbol{x}) = \frac{1}{2}(\boldsymbol{x} + \boldsymbol{H}^{-1}\boldsymbol{c})^T\boldsymbol{H}(\boldsymbol{x} + \boldsymbol{H}^{-1}\boldsymbol{c}) - \frac{1}{2}\boldsymbol{c}^T\boldsymbol{H}^{-1}\boldsymbol{c}. \tag{3.B.11}$$

Through algebraic manipulations and by using the fact that the Hessian is symmetric, that is, $\boldsymbol{H}^T = \boldsymbol{H}$, one can show that (3.B.11) is equal to (3.B.5a). By defining $\boldsymbol{y} = \boldsymbol{x} + \boldsymbol{H}^{-1}\boldsymbol{c}$, (3.B.11) can be further simplified to

$$J(\boldsymbol{y}) = \frac{1}{2}\boldsymbol{y}^T\boldsymbol{H}\boldsymbol{y} - \frac{1}{2}\boldsymbol{c}^T\boldsymbol{H}^{-1}\boldsymbol{c}. \tag{3.B.12}$$

The second term is a scalar offset that has no influence on the optimal solution. As $\boldsymbol{H}$ is positive semidefinite, the unconstrained minimum of $J(\boldsymbol{y})$ is attained at $\boldsymbol{y}_{\text{opt}} = 0$. Equivalently, the optimum of $J(\boldsymbol{x})$ is at

$$\boldsymbol{x}_{\text{opt}} = -\boldsymbol{H}^{-1}\boldsymbol{c}, \tag{3.B.13}$$

c.f. (3.B.10).

References

[1] J. Song-Manguelle, S. Schröder, T. Geyer, G. Ekemb, and J.-M. Nyobe-Yome, "Prediction of mechanical shaft failures due to pulsating torques of variable-frequency drives," *IEEE Trans. Ind. Appl.*, vol. 46, pp. 1979–1988, Sep./Oct. 2010.

[2] IEEE Std 519-1992, "IEEE recommended practices and requirements for harmonic control in electrical power systems," Apr. 1993.

[3] P. G. Cummings, "Estimating effect of system harmonics on losses and temperature rise of squirrel-cage motors," *IEEE Trans. Ind. Appl.*, vol. IA-22, pp. 1121–1126, Nov./Dec. 1986.
[4] M. Schweizer, *System-oriented efficiency optimization of variable speed drives*. PhD thesis, ETH Zurich, 2012.
[5] T. M. Blooming and D. J. Carnovale, "Application of IEEE Std 519-1992 harmonic limits," in *Annual Pulp and Paper Industry Technical Conference*, Jun. 2006.
[6] S. Chen, T. A. Lipo, and D. Fitzgerald, "Source of induction motor bearing currents caused by PWM inverters," *IEEE Trans. Energy Convers.*, vol. 11, pp. 25–32, Mar. 1996.
[7] S. Chen, T. A. Lipo, and D. Fitzgerald, "Modeling of motor bearing currents in PWM inverter drives," *IEEE Trans. Ind. Appl.*, vol. 32, pp. 1365–1370, Nov./Dec. 1996.
[8] T. Hoevenaars, K. LeDoux, and M. Colosino, "Interpreting IEEE std 519 and meeting its harmonic limits in VFD applications," in *Annual Petroleum and Chemical Industry Conference*, pp. 145–150, Sep. 2003.
[9] IEC 61000-2-4, "Electromagnetic compatibility (EMC)—part 2-4: Environment—compatibility levels in industrial plants for low-frequency conducted disturbances," Sep. 2002.
[10] P. Pillay and M. Manyage, "Definitions of voltage unbalance," *Power Eng. Rev.*, vol. 21, pp. 50–51, May 2001.
[11] H. S. Black, *Modulation theory*. D van Nostrand Co, NY & Toronto, 1953.
[12] G. Carrara, S. Gardella, M. Marchesoni, R. Salutari, and G. Sciutto, "A new multilevel PWM method: a theoretical analysis," *IEEE Trans. Power Electron.*, vol. 7, pp. 497–505, July 1992.
[13] J. Shen, S. Schröder, H. Stagge, and R. W. De Doncker, "Impact of modulation schemes on the power capability of high-power converters with low pulse ratios," *IEEE Trans. Power Electron.*, vol. 29, pp. 5696–5705, Nov. 2014.
[14] S. Bowes and B. Bird, "Novel approach to the analysis and synthesis of modulation processes in power converters," *IEE Proc.*, vol. 122, pp. 507–513, May 1975.
[15] D. G. Holmes and T. A. Lipo, *Pulse width modulation for power converters: Principles and practice*. IEEE Press, 2003.
[16] H. du Toit Mouton, B. McGrath, D. G. Holmes, and R. H. Wilkinson, "One-dimensional spectral analysis of complex PWM waveforms using superposition," *IEEE Trans. Power Electron.*, vol. 29, pp. 6762–6778, Dec. 2014.
[17] B. P. McGrath and D. G. Holmes, "An analytical technique for the determination of spectral components of multilevel carrier-based PWM methods," *IEEE Trans. Ind. Electron.*, vol. 49, pp. 847–857, Aug. 2002.
[18] N. Celanovic and D. Boroyevich, "A fast space-vector modulation algorithm for multilevel three-phase converters," *IEEE Trans. Ind. Appl.*, vol. 37, pp. 637–641, Mar./Apr. 2001.
[19] B. P. McGrath, D. G. Holmes, and T. Lipo, "Optimized space vector switching sequences for multilevel inverters," *IEEE Trans. Power Electron.*, vol. 18, pp. 1293–1301, Nov. 2003.
[20] D. G. Holmes, "The general relationship between regular-sampled pulse-width-modulation and space vector modulation for hard switched converters," in *Proceedings of the IEEE Industry Applications Society Annual Meeting*, pp. 1002–1009, 1992.
[21] H. S. Patel and R. G. Hoft, "Generalized techniques of harmonic elimination and voltage control in thyristor inverters: Part I–Harmonic elimination," *IEEE Trans. Ind. Appl.*, vol. IA-9, pp. 310–317, May/Jun. 1973.
[22] H. S. Patel and R. G. Hoft, "Generalized techniques of harmonic elimination and voltage control in thyristor inverters: Part II–Voltage control techniques," *IEEE Trans. Ind. Appl.*, vol. IA-10, pp. 666–673, Sep./Oct. 1974.
[23] P. N. Enjeti, P. D. Ziogas, and J. F. Lindsay, "Programmed PWM techniques to eliminate harmonics: A critical evaluation," *IEEE Trans. Ind. Appl.*, vol. 26, pp. 302–316, Mar./Apr. 1990.
[24] J. R. Espinoza, G. Joós, J. I. Guzmán, L. A. Morán, and R. P. Burgos, "Selective harmonic elimination and current/voltage control in current/voltage-source topologies: A unified approach," *IEEE Trans. Ind. Electron.*, vol. 48, pp. 71–81, Feb. 2001.
[25] J. N. Chiasson, L. M. Tolbert, K. J. McKenzie, and Z. Du, "A complete solution to the harmonic elimination problem," *IEEE Trans. Power Electron.*, vol. 19, pp. 491–499, Mar. 2004.
[26] V. G. Agelidis, A. I. Balouktsis, and C. Cossar, "On attaining the multiple solutions of selective harmonic elimination PWM three-level waveforms through function minimization," *IEEE Trans. Ind. Electron.*, vol. 55, pp. 996–1004, Mar. 2008.
[27] M. S. A. Dahidah, V. G. Agelidis, and M. V. Rao, "On abolishing symmetry requirements in the formulation of a five-level selective harmonic elimination pulse-width modulation technique," *IEEE Trans. Power Electron.*, vol. 21, pp. 1833–1837, Nov. 2006.
[28] V. G. Agelidis, A. I. Balouktsis, and M. S. A. Dahidah, "A five-level symmetrically defined selective harmonic elimination PWM strategy: Analysis and experimental validation," *IEEE Trans. Power Electron.*, vol. 23, pp. 19–26, Jan. 2008.

[29] J. N. Chiasson, L. M. Tolbert, K. J. McKenzie, and Z. Du, "Elimination of harmonics in a multilevel converter using the theory of symmetric polynomials and resultants," *IEEE Trans. Control. Syst. Technol.*, vol. 13, pp. 216–223, Mar. 2005.
[30] G. S. Buja and G. B. Indri, "Optimal pulse width modulation for feeding AC motors," *IEEE Trans. Ind. Appl.*, vol. IA-13, pp. 38–44, Jan./Feb. 1977.
[31] G. S. Buja, "Optimum output waveforms in PWM inverters," *IEEE Trans. Ind. Appl.*, vol. 16, pp. 830–836, Nov./Dec. 1980.
[32] A. K. Rathore, J. Holtz, and T. Boller, "Synchronous optimal pulse width modulation for low-switching-frequency control of medium-voltage multilevel inverters," *IEEE Trans. Ind. Electron.*, vol. 57, pp. 2374–2381, Jul. 2010.
[33] A. K. Rathore, J. Holtz, and T. Boller, "Generalized optimal pulse width modulation of multilevel inverters for low-switching-frequency control of medium-voltage high-power industrial AC drives," *IEEE Trans. Ind. Electron.*, vol. 60, pp. 4215–4224, Oct. 2013.
[34] J. Holtz and B. Beyer, "Off-line optimized synchronous pulse width modulation with on-line control during transients," *EPE Journal*, vol. 1, pp. 193–200, Dec. 1991.
[35] J. Holtz and B. Beyer, "The trajectory tracking approach—A new method for minimum distortion PWM in dynamic high-power drives," *IEEE Trans. Ind. Appl.*, vol. 30, pp. 1048–1057, Jul./Aug. 1994.
[36] J. Holtz and B. Beyer, "Fast current trajectory tracking control based on synchronous optimal pulse width modulation," *IEEE Trans. Ind. Appl.*, vol. 31, pp. 1110–1120, Sep./Oct. 1995.
[37] J. Holtz and N. Oikonomou, "Synchronous optimal pulse width modulation and stator flux trajectory control for medium-voltage drives," *IEEE Trans. Ind. Appl.*, vol. 43, pp. 600–608, Mar./Apr. 2007.
[38] J. Holtz and N. Oikonomou, "Fast dynamic control of medium voltage drives operating at very low switching frequency—An overview," *IEEE Trans. Ind. Electron.*, vol. 55, pp. 1005–1013, Mar. 2008.
[39] D. Kulka, W. Gens, and G. Berger, "Direct torque controlling technique with synchronous optimized pulse pattern," in *Proceedings of IEEE Power Electronics Specialists Conference*, pp. 245–250, Jun. 1993.
[40] J. Meili, S. Ponnaluri, L. Serpa, P. K. Steimer, and J. W. Kolar, "Optimized pulse patterns for the 5-level ANPC converter for high speed high power applications," in *Proceedings of IEEE Industrial Electronics Society, Annual Conference*, pp. 2587–2592, 2006.
[41] H. Weng, K. Chen, J. Zhang, R. Datta, X. Huang, L. J. Garces, R. Wagoner, A. M. Ritter, and P. Rotondo, "A four-level converter with optimized switching patterns for high-speed electric drives," in *Proceedings of IEEE Power Electronics Specialists Conference*, pp. 1585–1591, 2007.
[42] D. P. Bertsekas, *Nonlinear programming*. Athena Science, 2nd ed., 1999.
[43] S. Boyd and L. Vandenberghe, *Convex optimization*. Cambridge Univ. Press, 2004.
[44] J. Holtz, "Pulse width modulation—A survey," *IEEE Trans. Ind. Electron.*, vol. 32, pp. 410–420, Dec. 1992.
[45] G. S. Buja and M. P. Kazmierkowski, "Direct torque control of PWM inverter-fed AC motors—A survey," *IEEE Trans. Ind. Electron.*, vol. 51, pp. 744–757, Aug. 2004.
[46] A. Trzynadlowski, *Control of induction motors*. Academic Press, 2001.
[47] W. Leonhard, *Control of electrical drives*. Springer, 3rd ed., 2001.
[48] S.-K. Sul, *Control of electric machine drive systems*. IEEE Press, 2011.
[49] M. K. Kazmierkowski, "Control of PWM inverter-fed induction motors," in *Control in power electronics*, pp. 161–207. Academic Press, 2002.
[50] K. Hasse, "Zum dynamischen Verhalten der Asynchronmaschine bei Betrieb mit variabler Ständerfrequenz und Ständerspannung," *ETZ-A*, vol. 89, pp. 387–391, 1968.
[51] K. Hasse, *Zur Dynamik drehzahlgeregelter Antriebe mit stromrichtergespeisten Asynchron-Kurzschlußläufermaschinen*. PhD thesis, TH Darmstadt, 1969.
[52] F. Blaschke, "Das Prinzip der Feldorientierung, die Grundlage für die Transvector-Regelung von Drehfeldmaschinen," *Siemens Z.*, vol. 45, pp. 757–760, 1971.
[53] F. Blaschke, "Das Verfahren der Feldorientierung zur Regelung der Asynchronmaschine," *Siemens Forsch.- und Entw.-Ber.*, pp. 184–193, 1972.
[54] F. Blaschke, "The principle of field orientation applied to the new transvector closed-loop control system for rotating field machines," *Siemens Rev.*, vol. 39, pp. 217–220, 1972.
[55] P. Vas, *Sensorless vector and direct torque control*. Oxford Univ. Press, 1998.
[56] J. Holtz and J. Quan, "Drift- and parameter-compensated flux estimator for persistent zero-stator-frequency operation of sensorless-controlled induction motors," *IEEE Trans. Ind. Appl.*, vol. 39, pp. 1052–1060, Jul./Aug. 2003.

[57] T. Murata, T. Tsuchiya, and I. Takeda, "Vector control for induction machine on the application of optimal control theory," *IEEE Trans. Ind. Electron.*, vol. 37, pp. 283–290, Aug. 1990.
[58] P. Marion, M. Milano, and F. Vasca, "Linear quadratic state feedback and robust neural network estimator for field-oriented-controlled induction motors," *IEEE Trans. Ind. Electron.*, vol. 46, pp. 150–161, Feb. 1999.
[59] J. Holtz, J. Quan, J. Pontt, J. Rodríguez, P. Newman, and H. Miranda, "Design of fast and robust current regulators for high-power drives based on complex state variables," *IEEE Trans. Ind. Appl.*, vol. 40, pp. 1388–1397, Sep./Oct. 2004.
[60] D. W. Novotny and T. A. Lipo, *Vector control and dynamics of AC drives*. Oxford Univ. Press, 1996.
[61] H. du Toit Mouton, "Natural balancing of three-level neutral-point-clamped PWM inverters," *IEEE Trans. Ind. Electron.*, vol. 49, pp. 1017–1025, Oct. 2002.
[62] S. Ogasawara and H. Akagi, "Analysis of variation of neutral point potential in neutral-point-clamped voltage source PWM inverters," in *Proceedings of IEEE Industry Applications Society Annual Meeting*, Oct. 1993.
[63] C. Newton and M. Sumner, "Neutral point control for multi-level inverters: Theory, design and operational limitations," in *Proceedings of IEEE Industry Applications Society Annual Meeting*, Oct. 1997.
[64] J. K. Steinke, "Switching frequency optimal PWM control of a three-level inverter," *IEEE Trans. Power Electron.*, vol. 7, pp. 487–496, Jul. 1992.
[65] M. Sprenger, R. Alvarez, and S. Bernet, "Direct dead-time control—A novel DC-link neutral-point balancing method for the three-level neutral-point-clamped voltage source inverter," in *Proceedings of IEEE Energy Conversion Congress and Exposition* (Raleigh, NC, USA), pp. 1157–1163, Sep. 2012.
[66] J. Pou, R. Pinando, D. Borojevich, and P. Rodríguez, "Evaluation of the low-frequency neutral-point voltage oscillations in the three-level inverter," *IEEE Trans. Ind. Electron.*, vol. 52, pp. 1582–1588, Dec. 2005.
[67] T. Brückner and D. G. Holmes, "Optimal pulse-width modulation for three-level inverters," *IEEE Trans. Power Electron.*, vol. 20, pp. 82–89, Jan. 2005.
[68] J. Holtz and N. Oikonomou, "Neutral point potential balancing algorithm at low modulation index for three-level inverter medium-voltage drives," *IEEE Trans. Ind. Appl.*, vol. 43, pp. 761–768, May/Jun. 2007.
[69] M. Marchesoni, S. Segarich, and E. Sorressi, "A new control strategy for neutral-point-clamped active rectifiers," *IEEE Trans. Ind. Electron.*, vol. 52, pp. 462–470, Apr. 2005.
[70] I. Takahashi and T. Noguchi, "A new quick response and high efficiency control strategy for the induction motor," *IEEE Trans. Ind. Appl.*, vol. 22, pp. 820–827, Sep./Oct. 1986.
[71] I. Takahashi and Y. Ohmori, "High-performance direct torque control of an induction motor," *IEEE Trans. Ind. Appl.*, vol. 25, pp. 257–264, Mar./Apr. 1989.
[72] P. Pohjalainen, P. Tiitinen, and J. Lulu, "The next generation motor control method—Direct torque control, DTC," in *Proceedings of European on Power Electronics Chapter Symposium*, vol. 1 (Lausanne, Switzerland), pp. 115–120, 1994.
[73] Y.-S. Lai and J.-H. Chen, "A new approach to direct torque control of induction motor drives for constant inverter switching frequency and torque ripple reduction," *IEEE Trans. Energy Convers.*, vol. 16, pp. 220–227, Sep. 2001.
[74] J.-K. Kang and S.-K. Sul, "Analysis and prediction of inverter switching frequency in direct torque control of induction machine based on hysteresis bands and machine parameters," *IEEE Trans. Ind. Electron.*, vol. 48, pp. 545–553, Jun. 2001.
[75] D. Casadei, F. Profumo, G. Serra, and A. Tanni, "FOC and DTC: Two viable schemes for induction motors torque control," *IEEE Trans. Power Electron.*, vol. 17, pp. 779–787, Sep. 2002.
[76] C. A. Martins, X. Roboam, T. Meynard, and A. Carvalho, "Switching frequency imposition and ripple reduction in DTC drives by using a multilevel converter," *IEEE Trans. Power Electron.*, vol. 17, pp. 286–297, Mar. 2002.
[77] D. Casadei, G. Serra, and A. Tanni, "Implementation of a direct torque control algorithm for induction motors based on discrete space vector modulation," *IEEE Trans. Power Electron.*, vol. 15, pp. 769–777, Jul. 2000.
[78] T. G. Habetler, F. Profumo, M. Pastorelli, and L. M. Tolbert, "Direct torque control of induction machines using space vector modulation," *IEEE Trans. Ind. Appl.*, vol. 28, pp. 1045–1053, Sep./Oct. 1992.
[79] J.-K. Kang and S.-K. Sul, "New direct torque control of induction motor for minimum torque ripple and constant switching frequency," *IEEE Trans. Ind. Appl.*, vol. 35, pp. 1076–1082, Sep./Oct. 1999.
[80] K.-B. Lee, J.-H. Song, I. Choy, and J.-Y. Yoo, "Torque ripple reduction in DTC of induction motor driven by three-level inverter with low switching frequency," *IEEE Trans. Power Electron.*, vol. 17, pp. 255–264, Mar. 2002.
[81] A. Sapin, P. Steimer, and J.-J. Simond, "Modeling, simulation, and test of a three-level voltage-source inverter with output *LC* filter and direct torque control," in *Proceedings of the IEEE Industry Applications Society Annual Meeting*, vol. 43, pp. 469–475, Mar./Apr. 2007.

[82] J. Rodríguez, J. Pontt, S. Kouro, and P. Correa, "Direct torque control with imposed switching frequency in an 11-level cascaded inverter," *IEEE Trans. Ind. Electron.*, vol. 51, pp. 827–833, Aug. 2004.
[83] M. Depenbrock, "Direkte Selbstregelung (DSR) für hochdynamische Drehfeldantriebe mit Stromrichterschaltung," *ETZ-A*, vol. 7, pp. 211–218, 1985.
[84] M. Depenbrock, "Direct self control (DSC) of inverter fed induction machine," *IEEE Trans. Power Electron.*, vol. 3, pp. 420–429, Oct. 1988.
[85] A. Steimel and J. Wiesemann, "Further development of direct self control for application in electric traction," in *Proceedings of the IEEE International Symposium on Industrial Electronics* (Warsaw, Poland), 1996.
[86] M. Janßen and A. Steimel, "Direct self control with minimum torque ripple and high dynamics for a double three-level GTO inverter drive," *IEEE Trans. Ind. Electron.*, vol. 49, pp. 1065–1071, Oct. 2002.
[87] A. Steimel, "Direct self-control and synchronous pulse techniques for high-power traction inverters in comparison," *IEEE Trans. Ind. Electron.*, vol. 51, pp. 810–820, Aug. 2004.
[88] C. A. Floudas, *Nonlinear and mixed-integer optimization*. Oxford Univ. Press, 1995.
[89] D. P. Bertsekas, A. Nedic, and A. E. Ozdaglar, *Convex analysis and optimization*. Athena Science, 2003.
[90] M. S. Bazaraa, H. D. Sherali, and C. M. Shetty, *Nonlinear programming: Theory and algorithms*. John Wiley & Sons, Inc., 3rd ed., 2006.
[91] M. R. Garey and D. S. Johnson, *Computers and intractability: A guide to the theory of NP-completeness*. W. H. Freeman and Company, 1979.
[92] C. A. Floudas and P. M. Pardalos, *Encyclopedia of optimization*. Springer, 2nd ed., 2009.
[93] N. Agin, "Optimum seeking with branch and bound," *Manage. Sci.*, vol. 13, pp. 176–185, Dec. 1966.
[94] J. W. Chinneck, *Practical optimization: A gentle introduction*. Available online at www.sce.carleton.ca/faculty/chinneck/po.html, Draft, 2004.
[95] L. G. Mitten, "Branch-and-bound methods: General formulation and properties," *Oper. Res.*, vol. 18, pp. 24–34, Jan./Feb. 1970.
[96] W. Forst and D. Hoffmann, *Optimization—Theory and practice*. Springer, 2010.
[97] E. L. Lawler and D. E. Wood, "Branch and bound methods: A survey," *Oper. Res.*, vol. 14, pp. 699–719, Jul./Aug. 1966.
[98] N. Karmarkar, "A new polynomial-time algorithm for linear programming," in *Proceedings of ACM Symposium on Theory of Computing*, pp. 302–311, 1984.
[99] Y. Nesterov and A. Nemirovskii, *Interior-point polynomial algorithms in convex programming*. SIAM, 1994.
[100] R. Fletcher, "A general quadratic programming algorithm," *IMA J. Appl. Math.*, vol. 7, no. 1, pp. 76–91, 1971.
[101] Y. Nesterov, "A method of solving a convex programming problem with convergence rate $O(1/k^2)$," *Sov. Math. Dokl.*, vol. 27, no. 2, pp. 372–386, 1983.
[102] J. Sturm, *Using SeDuMi 1.02, a MATLAB toolbox for optimization over symmetric cones*, 1998.
[103] IBM ILOG, Inc., *CPLEX 12.5 user manual*. Somers, NY: IBM, USA, 2012.
[104] A. Wächter and L. Biegler, "On the implementation of a primal-dual interior point filter line search algorithm for large-scale nonlinear programming," *Math. Program.*, vol. 106, no. 1, pp. 25–57, 2006.
[105] S. Richter, C. Jones, and M. Morari, "Real-time input-constrained MPC using fast gradient methods," in *Proceedings of IEEE Conference on Decision and Control* (Shanghai, China), pp. 7387–7393, Dec. 2009.
[106] S. Richter, S. Mariéthoz, and M. Morari, "High-speed online MPC based on fast gradient method applied to power converter control," in *Proceedings of American Control Conference* (Baltimore, MD, USA), 2010.
[107] J. Jerez, G. Constantinides, and E. Kerrigan, "An FPGA implementation of a sparse quadratic programming solver for constrained predictive control," in *Proceedings of ACM/SIGDA International Symposium FPGA*, pp. 209–218, 2011.
[108] A. Domahidi, A. Zgraggen, M. N. Zeilinger, M. Morari, and C. N. Jones, "Efficient interior point methods for multistage problems arising in receding horizon control," in *Proceedings of the IEEE Conference on Decision and Control* (Maui, HI, USA), pp. 668–674, Dec. 2012.

Part Two

Direct Model Predictive Control with Reference Tracking

4

Predictive Control with Short Horizons

The most intuitively accessible and easy-to-implement variety of predictive controllers for power electronics uses a prediction horizon of one step and regulates one or more variables along their references. This chapter introduces this concept using a single-phase inverter setting with a resistive–inductive (RL) load and phase current regulation. In a second step, the controller is extended to a predictive current controller for a three-phase system with an induction machine. Alternatively, as shown in the last section of this chapter, the tracking of the electromagnetic torque and stator flux magnitude can be considered. The similarity between the current and torque controllers is shown by analyzing their corresponding cost functions. Moreover, the impact of the tracking error norm on stability is highlighted, and a method is reviewed to compensate for system delays.

4.1 Predictive Current Control of a Single-Phase RL Load

We start by introducing the notion of predictive control for a single-phase system. Specifically, one phase leg of a three-level inverter is considered, as shown in Fig. 4.1(a). An RL load is connected between the phase leg's terminal A and the neutral point N. Let $v_{\rm dc}$ denote the instantaneous dc-link voltage. The phase leg can produce the three phase voltages $-\frac{v_{\rm dc}}{2}$, 0, and $\frac{v_{\rm dc}}{2}$. We use the integer variable $u \in \{-1, 0, 1\}$ to denote the switch position in the phase leg. The values $-1, 0, 1$ correspond to the phase voltages $-\frac{v_{\rm dc}}{2}, 0, \frac{v_{\rm dc}}{2}$, respectively. Assuming the neutral point potential to be zero, the voltage applied to the RL load is given by $v = 0.5 v_{\rm dc} u$.

4.1.1 Control Problem

Let i^* denote the reference for the instantaneous current through the RL load, and i the actual current. The aim is to design a controller that manipulates the switches in the inverter phase

Model Predictive Control of High Power Converters and Industrial Drives, First Edition. Tobias Geyer.

Companion Website: www.wiley.com/go/geyermodelpredictivecontrol

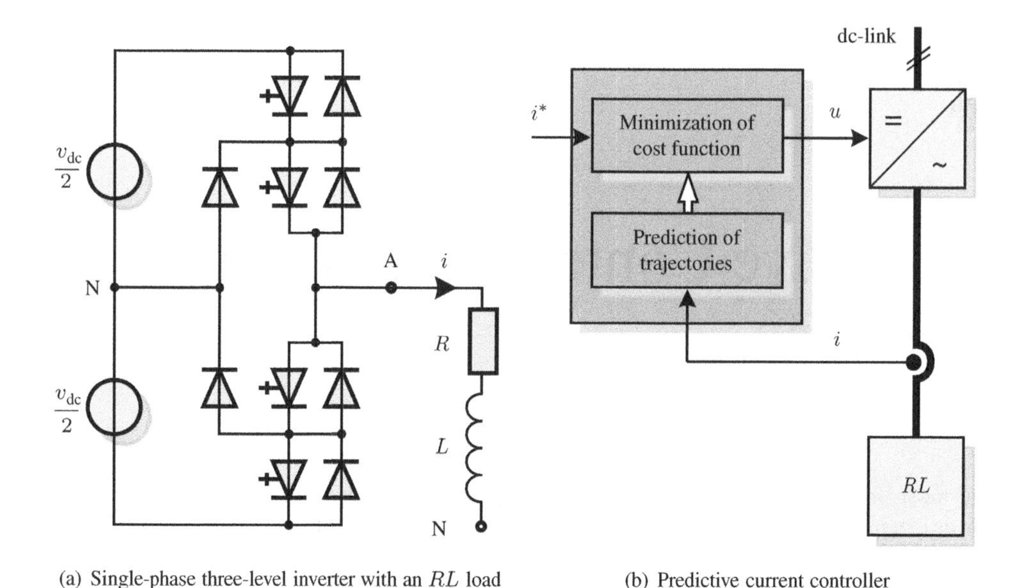

(a) Single-phase three-level inverter with an RL load

(b) Predictive current controller

Figure 4.1 Predictive current control of a single-phase RL load

leg such that the load current i closely tracks its reference i^*. Another requirement is to switch as little as possible. Switching directly between $u = 1$ and $u = -1$ is prohibited.

The block diagram of the controller to be designed is shown in Fig. 4.1(b). The controller consists of two parts. Starting from the current measurement, the first part predicts future current trajectories for different possible choices of control inputs. The second part is the optimization stage, in which the cost function capturing the control objectives is minimized. This yields the optimal control input u_{opt}, which is applied to the inverter. The controller operates in the discrete-time domain with the sampling interval T_s. Each of these blocks is explained in detail hereafter.

4.1.2 Prediction of Current Trajectories

To predict the future current trajectories, predictive control schemes require a model that captures the dynamics of the inverter system and its load. Starting from $v(t) = Ri(t) + L\frac{\mathrm{d}i(t)}{\mathrm{d}t}$, such a model in the continuous-time domain is directly obtained as

$$\frac{\mathrm{d}i(t)}{\mathrm{d}t} = Fi(t) + Gu(t) \tag{4.1}$$

with

$$F = -\frac{R}{L} \text{ and } G = \frac{1}{L}\frac{v_{\text{dc}}}{2}. \tag{4.2}$$

The current i is the state variable in the state-space model (4.1), and u is the input variable.

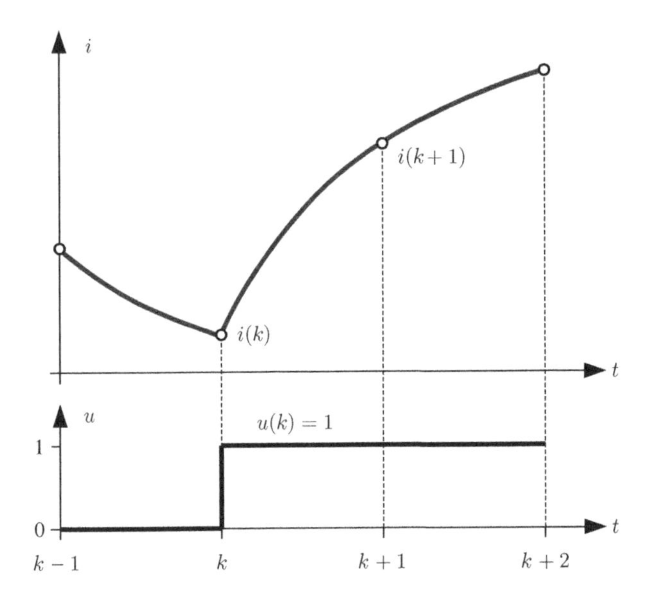

Figure 4.2 Evolution of the current i as a function of the switch position u, which is manipulated at discrete-time instants $t = kT_s$, with $k \in \mathbb{N}$

As the controller operates at time instants $t = kT_s$, with $k \in \mathbb{N}$, this model needs to be translated from the continuous- to the discrete-time domain. This is achieved by integrating (4.1) from $t = kT_s$ to $t = (k+1)T_s$. During this time interval, $u(t)$ is constant and equal to $u(k)$, where k refers to the kth sampling interval, as illustrated in Fig. 4.2. The integration yields

$$i(k+1) = Ai(k) + Bu(k) \tag{4.3}$$

with

$$A = e^{FT_s} \text{ and } B = \int_0^{T_s} e^{F\tau} d\tau\, G. \tag{4.4}$$

This provides the exact current evolution at the discrete-time instants $t = kT_s$. The detailed derivation of (4.3) is provided, for example, in [1, Sect. 4.3.3]. If F is nonzero, B in (4.4) can be further simplified to $B = (A-1)G/F$.

It is obvious from Fig. 4.2 that the evolution of the current between these time instants is described by exponentially shaped line segments. For short sampling intervals, these line segments can be approximated to be linear, and approximate discretization methods can be used, such as the forward Euler approximation or the bilinear method. Forward Euler, for example, yields (4.3) with

$$A = 1 + FT_s \text{ and } B = GT_s, \tag{4.5}$$

which uses the slope of the current at $t = kT_s$ to obtain its approximate value at $t = (k+1)T_s$. For short sampling intervals, the forward Euler approach is usually sufficiently accurate.

4.1.3 Optimization Problem

The control problem at time step k of tracking the current reference can be mapped into the cost function

$$J = (i^*(k+1) - i(k+1))^2 + \lambda_u |\Delta u(k)|. \tag{4.6}$$

The cost function consists of two terms. The first term penalizes the predicted current error at the next time step $k+1$ quadratically. The current error is given by the difference between the current reference and the predicted current in the phase leg. The second term penalizes the switching effort, which is the absolute value of $\Delta u(k) = u(k) - u(k-1)$. The penalty $\lambda_u \geq 0$ is a tuning parameter that adjusts the trade-off between the tracking accuracy (deviation of the current from its reference) and the switching effort, that is, whether switching is performed or not.[1] Note that J is always nonnegative.

The current at $k+1$ depends on the choice of $u(k)$. The discrete-time model of Sect. 4.1.2 can be used to predict the current at $k+1$ for all admissible $u(k)$. The admissible $u(k)$ are those that are either -1, 0, or 1 and differ by at most one step from the previous switch position, that is, $u(k-1)$. This optimization problem can formally be stated as

$$u_{\text{opt}}(k) = \arg \underset{u(k)}{\text{minimize}}\, J \tag{4.7a}$$

$$\text{subject to} \quad i(k+1) = Ai(k) + Bu(k) \tag{4.7b}$$

$$u(k) \in \{-1, 0, 1\}, |\Delta u(k)| \leq 1. \tag{4.7c}$$

The optimal control input, that is, the optimal switch position at time step k, $u_{\text{opt}}(k)$, is obtained by minimizing the cost function J. Note that the expression *minimize J* minimizes the cost function J subject to the constraints. This leads to the minimum (or optimum) value J_{opt}. On the other hand, *arg minimize J* yields the argument that minimizes the cost function. The argument is also referred to as the optimal solution or the optimizer, which is—in this case—the control input u_{opt}. For a short introduction to the notion of mathematical optimization, the reader is referred to Appendix 3.B.

The optimization problem and hence $u_{\text{opt}}(k)$ depend on the current $i(k)$, the previously chosen switch position $u(k-1)$, and the current reference $i^*(k+1)$. If the latter is not available, it can be approximated using the current reference at time steps $k-1$ and k.

4.1.4 Control Algorithm

The optimization problem (4.7) can be solved using tools from mathematical programming, as explained in Appendix 3.B. A simple alternative is to use the concept of enumeration. For each admissible switch position, the model response is predicted and the corresponding cost is computed by evaluating the cost function. The switch position with the smallest cost is, by definition, the optimal one and chosen as the control input. Depending on $u(k-1)$, the set of admissible switch positions $\mathcal{U}(k)$ consists of two or three elements, as shown in Table 4.1.

[1] As $\Delta u(k) \in \{-1, 0, 1\}$, the absolute value of the switching effort and its squared value give the same result, that is, $|\Delta u(k)| = (\Delta u(k))^2$. Even though we use $|\Delta u(k)|$ in (4.6), which is computationally simpler, the cost function effectively consists of two squared terms.

Table 4.1 Set of admissible switch positions $\mathcal{U}(k)$ that meets the switching constraint $|\Delta u(k)| \leq 1$

$u(k-1)$	$\mathcal{U}(k)$
1	$\{0, 1\}$
0	$\{-1, 0, 1\}$
-1	$\{-1, 0\}$

At time step k, the predictive current control algorithm computes $u_{\text{opt}}(k)$ according to the following procedure:

1. Given the previously applied switch position $u(k-1)$ and taking into account the constraints on the switching transitions (4.7c), the set of admissible switch positions at time step k, $\mathcal{U}(k)$, is determined.
2. For each switch position $u(k) \in \mathcal{U}(k)$, the current at time step $k+1$, $i(k+1)$, is predicted using the model (4.7b).
3. For each switch position $u(k) \in \mathcal{U}(k)$, the cost J is computed according to (4.6).
4. The switch position $u_{\text{opt}}(k)$ with the minimum cost is determined and applied to the inverter.

At the next time step, this procedure is repeated.

Example 4.1 *Consider the situation depicted in Fig. 4.3. The current at time step k is close to its reference. Assuming that $u(k-1)$ is zero, the set of admissible switch positions at time step k is $\mathcal{U}(k) = \{-1, 0, 1\}$. For each $u(k) \in \mathcal{U}(k)$, the predicted $i(k+1)$ is shown in the figure.*

The corresponding costs are summarized in Table 4.2. Switching to $u(k) = 1$ minimizes the predicted current error but incurs a switching penalty, which is given by λ_u. Refraining from

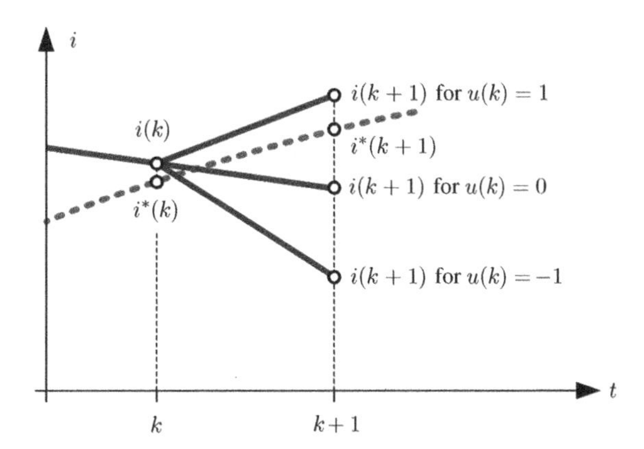

Figure 4.3 Prediction of the current at the next time step $i(k+1)$ as a function of the switch position $u(k)$. The current reference i^* is indicated by the dashed line

Table 4.2 Costs corresponding to the three admissible $u(k)$ in Example 4.1

$u(k)$	$(i^*(k+1) - i(k+1))^2$	$\lambda_u \lvert\Delta u(k)\rvert$
1	0.030^2	λ_u
0	0.052^2	0
−1	0.132^2	λ_u

switching, that is, using $u(k) = u(k-1) = 0$, leads to a slightly larger current error while incurring no switching penalty. The third choice, $u(k) = -1$, is clearly suboptimal because it entails a large current error and also requires switching. Whether $u(k) = 0$ or $u(k) = 1$ minimizes the cost and is thus optimal depends on the choice for λ_u. For large λ_u, the controller refrains from frequent switching and tolerates relatively large current errors. For small λ_u, the controller switches more often to minimize the deviations of the current from its reference. In this example, provided that λ_u is less than $0.052^2 - 0.030^2$, switching is avoided and $u_{\text{opt}}(k) = 1$ is selected; otherwise $u_{\text{opt}}(k) = 0$ is the optimal choice.

4.1.5 Performance Evaluation

In the following, simulation results are provided to demonstrate the performance of the predictive current controller with reference tracking. The performance is investigated during steady-state operating conditions as well as during transients, using the single-phase, three-level inverter with an RL load shown in Fig. 4.1(a). The time delay because of the controller computation is neglected. More specifically, the current $i(k)$ is sampled at time step k, the cost function is minimized, and the new control output $u(k)$ is applied to the inverter at the same time step.

The following parameters are assumed. The load has the ohmic resistance $R = 2\ \Omega$ and the inductance $L = 2$ mH. The rated root-mean-square (rms) phase voltage is $V_{\text{ph}R} = 3.3/\sqrt{3}$ kV, the rated fundamental frequency is $f_R = 50$ Hz, and the dc-link voltage is equal to its nominal value $V_{\text{dc}} = 5.2$ kV. The per unit (pu) system is established using the base quantities $V_B = \sqrt{2}V_{\text{ph}R} = 2694$ V, $\omega_B = \omega_R = 2\pi 50$ rad/s, and $Z_B = |R + \mathrm{j}\omega_R L| = 2.096\ \Omega$, with $\omega_R = 2\pi f_R$. For more details on the pu system, the reader is referred to Sect. 2.1.2. The sampling interval is set to $T_s = 25$ μs, which is equivalent to a sampling frequency of 40 kHz.

Under steady-state operating conditions, assume a current reference of a sinusoidal waveform with an amplitude of 0.8 pu and a fundamental frequency of 50 Hz. For very small λ_u, the current tracks its reference closely with a small current ripple, as shown in Fig. 4.4. Note that the switch positions have been (arbitrarily) scaled to half the peak current to simplify the illustration. The resulting total demand distortion (TDD) of the current, as defined in (3.2), is very small at $I_{\text{TDD}} = 1.66\%$, whereas the switching frequency $f_{\text{sw}} = 2650$ Hz is prohibitively high. The current spectrum was computed using a Fourier transformation. The amplitudes of the harmonics are effectively zero, except for the fundamental component at 50 Hz, which is equal to the reference amplitude of 0.8 pu.

Each switching transition incurs one *on* transition of an active switch (see Sect. 2.4.1). As the phase leg consists of four active switches, the average switching frequency of these active

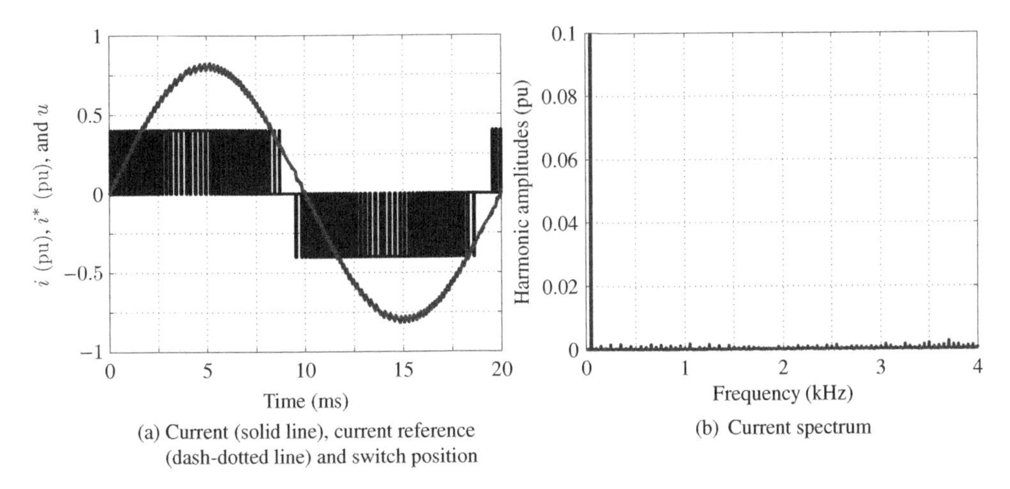

(a) Current (solid line), current reference (dash-dotted line) and switch position

(b) Current spectrum

Figure 4.4 Predictive current control with $\lambda_u = 0.0005$ during steady-state operation for a single-phase RL load, with $I_{\text{TDD}} = 1.66\%$ and $f_{\text{sw}} = 2650$ Hz

switches is upper-bounded by $0.25/T_s = 10$ kHz. Setting λ_u to zero leads to the classic deadbeat controller with $I_{\text{TDD}} = 1.03\%$ and $f_{\text{sw}} = 5475$ Hz, which is approximately half of the theoretical upper bound. To further lower the current TDD, the sampling interval has to be reduced. For $T_s = 5$ μs, for example, $I_{\text{TDD}} = 0.21\%$ and $f_{\text{sw}} = 27.3$ kHz are obtained.

Enlarging λ_u increases the current TDD and reduces the switching frequency accordingly, as exemplified in Fig. 4.5 for $\lambda_u = 0.005$ and in Fig. 4.6 for $\lambda_u = 0.0014$. Note that the amplitude spectra of the current are discrete and concentrated around $2f_{\text{sw}}$.

The relationship between the current TDD and the tuning parameter λ_u is effectively linear, particularly for small λ_u, as shown in Fig. 4.7(a). The switching frequency, however, depends in a nonlinear way on λ_u (see Fig. 4.7(b)). As λ_u is increased, particularly below 700 Hz, the

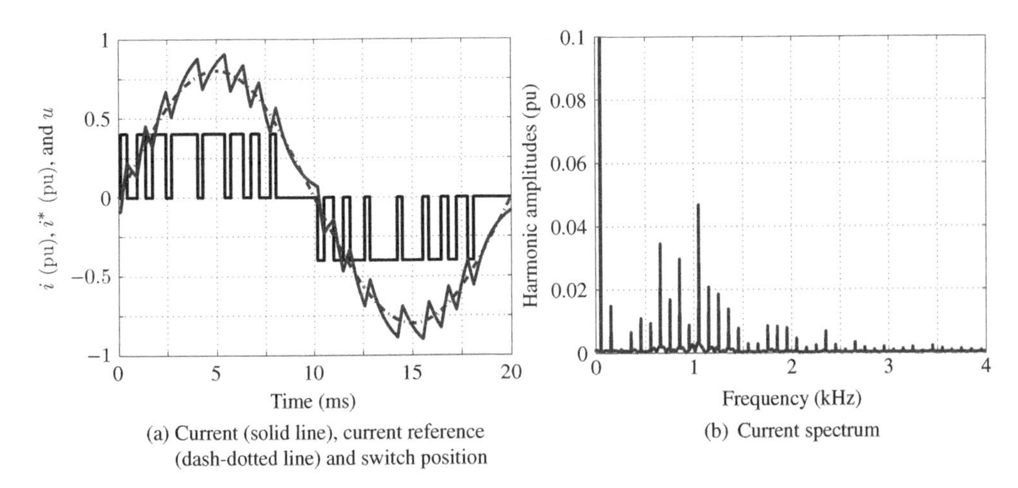

(a) Current (solid line), current reference (dash-dotted line) and switch position

(b) Current spectrum

Figure 4.5 Predictive current control as in Fig. 4.4, but with $\lambda_u = 0.005$, which results in $I_{\text{TDD}} = 8.47\%$ and $f_{\text{sw}} = 400$ Hz

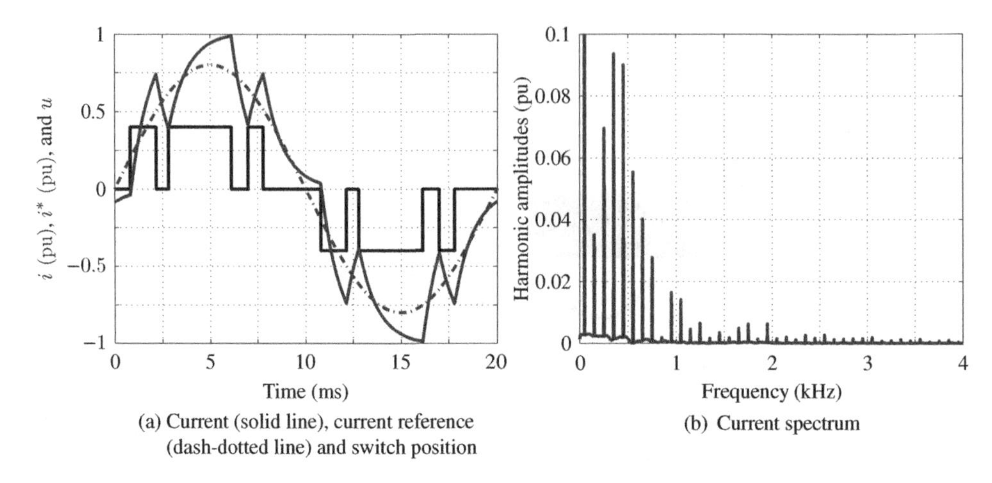

(a) Current (solid line), current reference (dash-dotted line) and switch position

(b) Current spectrum

Figure 4.6 Predictive current control as in Fig. 4.4, but with $\lambda_u = 0.0114$, which results in $I_{\text{TDD}} = 17.33\%$ and $f_{\text{sw}} = 150\,\text{Hz}$

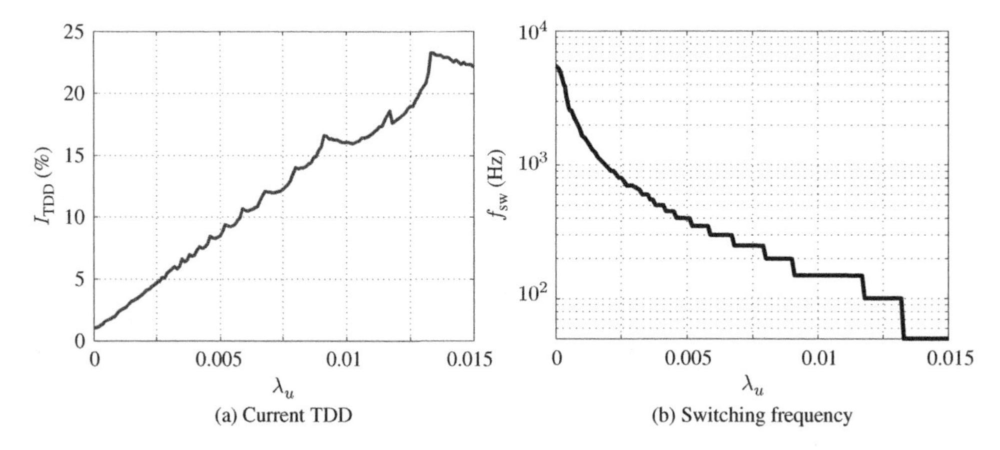

(a) Current TDD

(b) Switching frequency

Figure 4.7 Trade-off between the current TDD and the switching frequency for predictive current control as in Fig. 4.4

switching frequency drops in steps of 50 Hz, indicating a certain degree of periodicity in the switching pattern, even though this is not enforced in the cost function. This periodicity is also reflected in the shape of the current spectra, featuring harmonic amplitudes only at discrete frequencies.

Figure 4.8(a) highlights the performance of the predictive controller in the presence of step changes in the current reference. The switching penalty is set to $\lambda_u = 0.005$. At $t = 5$ ms, the current reference is reduced from 0.8 to 0.2 pu. When applying steps of significant magnitude, the current error suddenly increases and the corresponding penalty on the current error dominates over the switching penalty. As a result, switching is performed to reduce this error as quickly as possible—effectively regardless of the choice of λ_u. Specifically, the predictive controller inserts a negative pulse, which drives the current quickly to its new reference value within less than 0.5 ms. Note that the switching constraint (4.7c) is met, that is, switching is

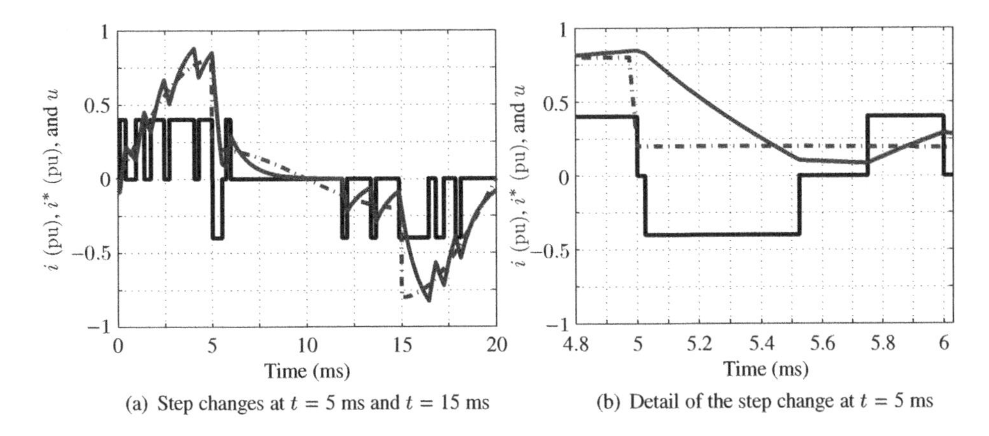

(a) Step changes at t = 5 ms and t = 15 ms

(b) Detail of the step change at t = 5 ms

Figure 4.8 Predictive current control for a single-phase RL load during step changes in the current reference from 0.8 pu to 0.2 pu and back to 0.8 pu at time instants $t = 5$ ms and $t = 15$ ms, respectively

performed from $u = 1$ to $u = -1$ via the intermediate switch position $u = 0$, as can be seen in Fig. 4.8(b).

At $t = 15$ ms, the current reference is stepped back up to 0.8 pu. The length of the resulting current transient at 1.2 ms is longer than in the step-down case, because less voltage is available to be applied across the RL load. In both cases, the predictive controller provides the fastest current response physically possible, effectively exhibiting a deadbeat behavior.

4.1.6 Prediction Horizons of more than 1 Step

The cost function (4.6) minimizes the current error at the next time step, based on the predicted current after one time step, which is a function of the to-be-selected switch position. As a result, the predictive current controller looks one step ahead, and the prediction horizon is $N_p = 1$. It is straightforward to generalize the controller to longer prediction horizons by rewriting (4.6) as

$$J = \sum_{\ell=k}^{k+N_p-1} (i^*(\ell+1) - i(\ell+1))^2 + \lambda_u |\Delta u(\ell)| . \tag{4.8}$$

The cost function is now a function of the sequence of switch positions, the so-called *switching sequence* $\boldsymbol{U}(k) = [u(k) \;\; u(k+1) \;\; \cdots \;\; u(k+N_p-1)]^T$. Accordingly, the generalized optimization problem is

$$\boldsymbol{U}_{\text{opt}}(k) = \arg \underset{\boldsymbol{U}(k)}{\text{minimize}} \; J \tag{4.9a}$$

$$\text{subject to} \;\; i(\ell+1) = Ai(\ell) + Bu(\ell) \tag{4.9b}$$

$$u(\ell) \in \{-1, 0, 1\}, \;\; |\Delta u(\ell)| \leq 1 \tag{4.9c}$$

$$\forall \ell = k, k+1, \ldots, k+N_p-1 . \tag{4.9d}$$

Out of the optimal switching sequence $\boldsymbol{U}_{\text{opt}}(k)$, only the first element $u_{\text{opt}}(k)$ is applied to the inverter. According to the receding horizon policy (see Sect. 1.3.2), new (current) measurements are obtained at the next time step $k+1$, and the optimization problem (4.9) is solved over the shifted time interval from $k+1$ until $k+1+N_p$.

Example 4.2 *Assume a prediction horizon of $N_p = 2$ steps and the previously applied switch position to be $u(k-1) = 0$. The set of admissible switching sequences is shown in Table 4.3. For each switching sequence, the current trajectory from k to $k + N_p$ can be predicted, as shown in Fig. 4.9. The numbering of the switching sequences is defined in Table 4.3.*

Table 4.3 Set $\mathcal{U}(k)$ of admissible switching sequences $\boldsymbol{U}(k) = [u(k)\ u(k+1)]^T$ for the prediction horizon $N_p = 2$, assuming the previously applied switch position to be $u(k-1) = 0$

Switching sequence	$u(k)$	$u(k+1)$
$\boldsymbol{U}_1$	1	1
$\boldsymbol{U}_2$	1	0
$\boldsymbol{U}_3$	0	1
$\boldsymbol{U}_4$	0	0
$\boldsymbol{U}_5$	0	−1
$\boldsymbol{U}_6$	−1	0
$\boldsymbol{U}_7$	−1	−1

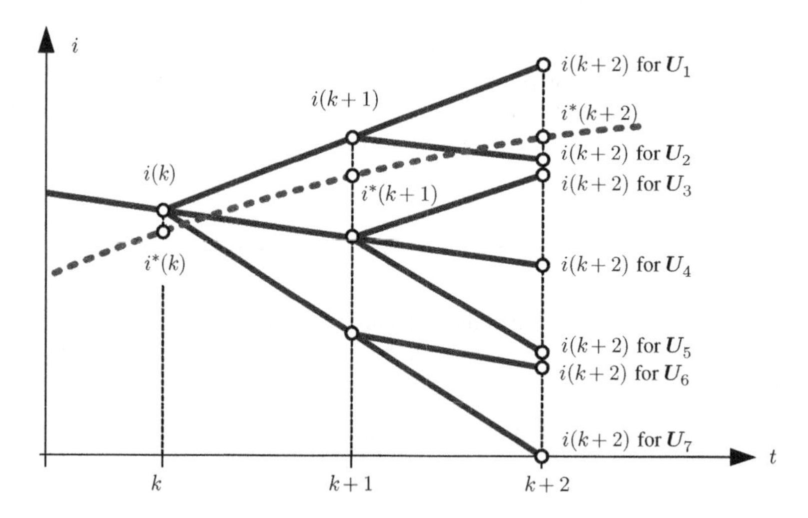

Figure 4.9 Prediction of the current trajectories $[i(k+1)\ i(k+2)]^T$ over the prediction horizon $N_p = 2$ as a function of the switching sequences $\boldsymbol{U}(k)$ as given in Table 4.3. The current reference i^* is indicated by the dashed line

As for a single-phase, three-level inverter there are three possible switch positions available at each step within the prediction horizon N_p, one would expect, in general, 3^{N_p} different switching sequences. However, because of the second constraint in (4.9c), this number merely constitutes an upper bound and the number of admissible switching sequences depends on $u(k-1)$. For $N_p = 2$, for example, $u(k-1) = 0$ entails 7 switching sequences, while $u(k-1) = \pm 1$ restricts this number to 5. Nevertheless, for long prediction horizons, the number of admissible switching sequences explodes. At the same time, the sampling intervals are typically in the range of several microseconds and thus very short, severely limiting the time available to perform the optimization step.

As a result, the concept of exhaustive enumeration is not suitable to solve the optimization problem for prediction horizons of more than a few steps. Instead, more sophisticated optimization methods are essential. One option is to use branch-and-bound techniques, as will be shown and explained in Chap. 5.

4.1.7 *Summary*

The predictive control concept introduced in this section is conceptually very simple and versatile. It consists of three main components, namely the cost function, the controller model, and the solution algorithm based on enumeration. Adapting the cost function to different control problems and the system model to a large variety of power electronic systems is straightforward. In the following, several examples are provided.

- Systems with more than one phase, such as three-phase and multiphase systems, can easily be controlled. The scalar switch position and current are replaced by vectors, with each component referring to one phase. Accordingly, the system model is given in matrix form, as shown in the next section.
- Multilevel inverter topologies can be addressed by expanding the discrete set for u. For a five-level inverter, for example, this leads to $u \in \{-2, -1, 0, 1, 2\}$. Additional switching constraints can be added to (4.9d).
- Different loads, such as induction or synchronous machines, can be considered by adapting the controller model accordingly. This adaptation will be exemplified in the next section for a squirrel-cage induction machine.
- Inverters with internal dynamics, such as a neutral point potential, can be addressed by adding these dynamics to the controller model.
- Instead of regulating the current along its reference trajectory, other quantities can be regulated, including the electromagnetic torque and the stator flux magnitude. This case will be investigated in Sect. 4.3.
- Bounds around the reference trajectory can be added, which can be interpreted as soft constraints. Within these bounds, the current error is not or only mildly penalized, while a large penalty comes into effect once a soft constraint is (or is predicted to be) violated.
- Different norms can be considered in the cost function, such as the 1-norm or the infinity-norm, rather than the 2-norm. The implications of using the 1-norm will be investigated in Sect. 4.2.7.

For a review of some of the related literature, the reader is referred to Sect. 4.4.

4.2 Predictive Current Control of a Three-Phase Induction Machine

The predictive current controller, which was introduced in the previous section for a single-phase RL load with a three-level inverter leg, is generalized in this section to a three-phase drive system.

4.2.1 Case Study

As a typically used medium-voltage (MV) drive system, consider a squirrel-cage induction machine driven by a three-level, three-phase, neutral-point-clamped (NPC) inverter, as shown in Fig. 4.10. To simplify the exposition in this section, the neutral point potential is fixed to zero. In each phase, the inverter produces the voltages $-\frac{v_{\text{dc}}}{2}$, 0, and $\frac{v_{\text{dc}}}{2}$, which correspond to the switch positions u_a, u_b, and $u_c \in \{-1, 0, 1\}$. The total dc-link voltage is denoted by v_{dc} and assumed to be constant. We use $\boldsymbol{u}_{abc} = [u_a \; u_b \; u_c]^T$ to denote the three-phase switch positions. The voltage applied to the machine terminals in orthogonal coordinates is given by

$$\boldsymbol{v}_s = \frac{v_{\text{dc}}}{2} \tilde{\boldsymbol{K}} \, \boldsymbol{u}_{abc} \tag{4.10}$$

with $\boldsymbol{v}_s = [v_{s\alpha} \; v_{s\beta}]^T$. The reduced Clarke transformation $\tilde{\boldsymbol{K}}$ is defined in (2.13). For more details on the NPC inverter, see Sect. 2.4.1.

State-space models of the squirrel-cage induction machine were derived in Sect. 2.2.4. In this section, we choose stationary orthogonal coordinates and set the angular speed of the reference frame to zero. For the current control problem at hand, it is convenient to represent the machine dynamics in terms of the stator current $\boldsymbol{i}_s = [i_{s\alpha} \; i_{s\beta}]^T$ and the rotor flux linkage vector $\boldsymbol{\psi}_r = [\psi_{r\alpha} \; \psi_{r\beta}]^T$. Recall that the rotor voltage $\boldsymbol{v}_r$ is zero in a squirrel-cage induction machine. This leads to the following set of continuous-time state-space equations:

$$\frac{\mathrm{d}\boldsymbol{i}_s}{\mathrm{d}t} = -\frac{1}{\tau_s}\boldsymbol{i}_s + \left(\frac{1}{\tau_r}\boldsymbol{I}_2 - \omega_r \begin{bmatrix} 0 & -1 \\ 1 & 0 \end{bmatrix}\right) \frac{X_m}{D}\boldsymbol{\psi}_r + \frac{X_r}{D}\boldsymbol{v}_s \tag{4.11a}$$

$$\frac{\mathrm{d}\boldsymbol{\psi}_r}{\mathrm{d}t} = \frac{X_m}{\tau_r}\boldsymbol{i}_s - \frac{1}{\tau_r}\boldsymbol{\psi}_r + \omega_r \begin{bmatrix} 0 & -1 \\ 1 & 0 \end{bmatrix} \boldsymbol{\psi}_r \tag{4.11b}$$

$$\frac{\mathrm{d}\omega_r}{\mathrm{d}t} = \frac{1}{M}(T_e - T_\ell), \tag{4.11c}$$

where $\boldsymbol{I}_2$ denotes the two-dimensional identity matrix. The model parameters are the stator and rotor resistances R_s and R_r, respectively, the stator, rotor, and mutual reactances X_{ls}, X_{lr}, and X_m, respectively, the moment of inertia M, and the mechanical load torque T_ℓ. The rotor quantities are referred to the stator circuit. Moreover, we had defined in Sect. 2.2

$$X_s = X_{ls} + X_m, \; X_r = X_{lr} + X_m, \; \text{and} \; D = X_s X_r - X_m^2. \tag{4.12}$$

The transient stator time constant and the rotor time constant are

$$\tau_s = \frac{X_r D}{R_s X_r^2 + R_r X_m^2} \; \text{and} \; \tau_r = \frac{X_r}{R_r}. \tag{4.13}$$

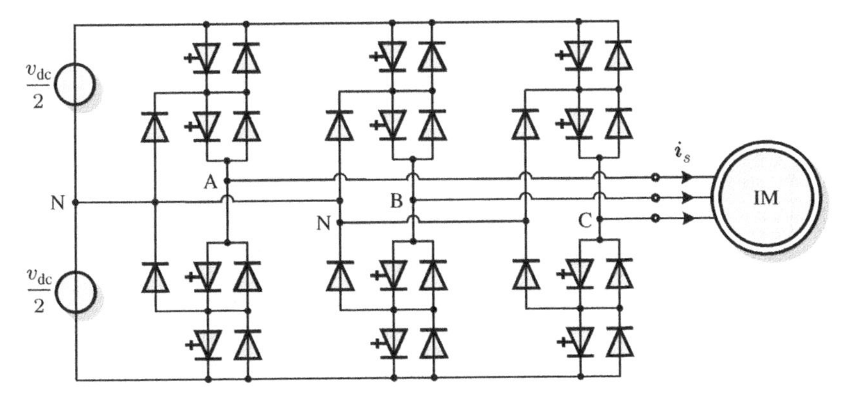

Figure 4.10 Three-level, three-phase, neutral-point-clamped voltage source inverter driving an induction motor with a fixed neutral point potential

The electromagnetic torque is given by

$$T_e = \frac{1}{\mathrm{pf}}\frac{X_m}{X_r}\boldsymbol{\psi}_r \times \boldsymbol{i}_s = \frac{1}{\mathrm{pf}}\frac{X_m}{X_r}(\psi_{r\alpha} i_{s\beta} - \psi_{r\beta} i_{s\alpha}). \tag{4.14}$$

4.2.2 *Control Problem*

The control problem of the inner current loop is formulated in the $\alpha\beta$ reference frame. Let $\boldsymbol{i}_s^*$ denote the reference of the instantaneous stator current, with $\boldsymbol{i}_s^* = [i_{s\alpha}^* \ i_{s\beta}^*]^T$. The objective of the current controller is to manipulate the three-phase switch position $\boldsymbol{u}_{abc}$ such that the stator current $\boldsymbol{i}_s$ closely tracks its reference. At the same time, the switching effort, that is, the switching frequency or the switching losses, is to be kept small. As previously, switching between 1 and -1 in a phase leg is prohibited.

The block diagram of the predictive current controller is shown in Fig. 4.11. The controller predicts the stator current at the next time step for all admissible switch positions. For the prediction, the measured stator current is required, along with the rotor flux vector. A flux observer estimates the rotor flux vector using the measured stator current and the stator voltage. The latter is typically not measured, but is rather reconstructed using the dc-link voltage and the three-phase switch position.

In a drive setting, outer control loops are added in a cascaded controller fashion. These outer loops manipulate $\boldsymbol{i}_s^*$ so as to keep the machine appropriately fluxed and to regulate the machine's rotational speed. An example is provided in Fig. 3.26, which shows the block diagram of a (direct) rotor field-oriented controller. The outer flux and speed control loops are adopted for the predictive current controller. These control loops are shown on the left-hand side of Fig. 4.11.

The rotor flux magnitude is defined as $\Psi_r = \|\boldsymbol{\psi}_r\|$. The outer control loops are formulated in the rotating dq reference frame with the angular position φ of the rotor flux vector. These control loops provide the stator current reference $\boldsymbol{i}_{s,dq}^*$ in dq, which serves, after being translated

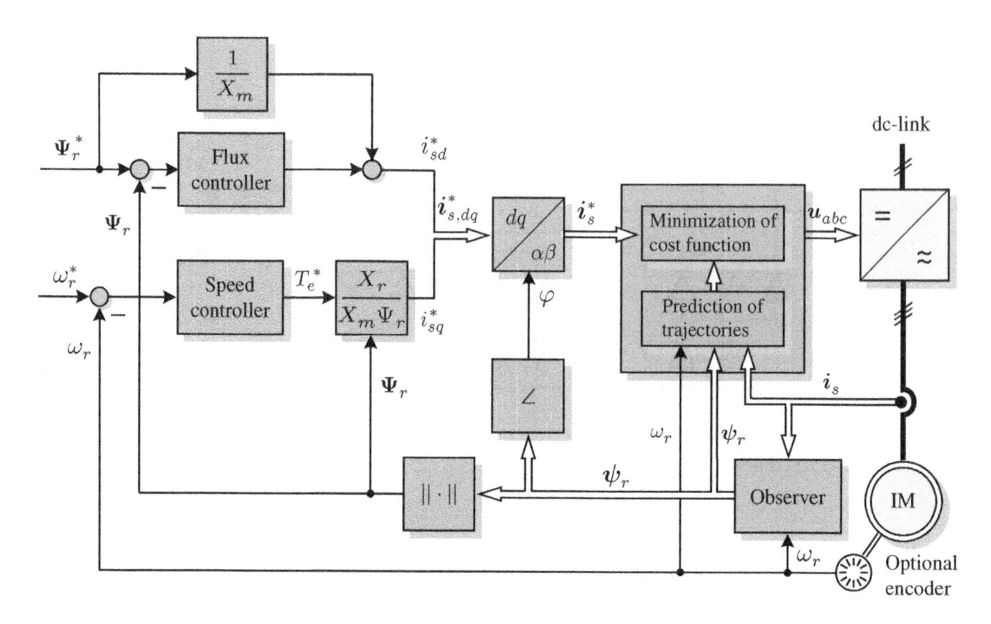

Figure 4.11 Predictive current controller for the three-phase, three-level NPC inverter with an induction machine, including the outer flux and speed control loops

into the stationary $\alpha\beta$ coordinate system, as the reference $\boldsymbol{i}_s^*$ to the predictive current controller. As the dynamics of the outer loops are slower than those of the inner loop—typically by an order of magnitude—we neglect them in the following to simplify the exposition. For more details on outer control loops, the reader is referred to Sect. 3.6.2.

4.2.3 Controller Model

The predictive current controller relies on an internal model of the physical drive system to predict future stator currents as a function of the three-phase switch position $\boldsymbol{u} = \boldsymbol{u}_{abc} = [u_a\ u_b\ u_c]^T \in \{-1, 0, 1\}^3$. By inserting (4.10) into (4.11a), the continuous-time state-space model of the stator current can be expressed in terms of the switch position $\boldsymbol{u}$ as

$$\frac{\mathrm{d}\boldsymbol{i}_s(t)}{\mathrm{d}t} = \boldsymbol{F}\boldsymbol{i}_s(t) + \boldsymbol{G}_1\boldsymbol{\psi}_r(t) + \boldsymbol{G}_2\boldsymbol{u}(t) \tag{4.15}$$

with

$$\boldsymbol{F} = -\frac{1}{\tau_s}\boldsymbol{I}_2,\ \boldsymbol{G}_1 = \frac{X_m}{D}\begin{bmatrix} \frac{1}{\tau_r} & \omega_r \\ -\omega_r & \frac{1}{\tau_r} \end{bmatrix} \text{ and } \boldsymbol{G}_2 = \frac{v_{\mathrm{dc}}}{3}\frac{X_r}{D}\begin{bmatrix} 1 & -\frac{1}{2} & -\frac{1}{2} \\ 0 & \frac{\sqrt{3}}{2} & -\frac{\sqrt{3}}{2} \end{bmatrix}. \tag{4.16}$$

Recall that the stator current $\boldsymbol{i}_s$ and the rotor flux $\boldsymbol{\psi}_r$ are represented in orthogonal coordinates with the components α and β.

The discrete-time state-space representation is obtained by integrating (4.15) from $t = kT_s$ to $t = (k+1)T_s$. We assume that the derivative of $\boldsymbol{i}_s$ at $t = kT_s$ is constant throughout the integration time interval. This approach is known as the *forward Euler method*, which is sufficiently accurate for short sampling intervals of several tens of microseconds. The discrete-time representation is then

$$\boldsymbol{i}_s(k+1) = \boldsymbol{A}\boldsymbol{i}_s(k) + \boldsymbol{B}_1\boldsymbol{\psi}_r(k) + \boldsymbol{B}_2\boldsymbol{u}(k) \tag{4.17}$$

with the system matrices

$$\boldsymbol{A} = \boldsymbol{I}_2 + \boldsymbol{F}T_s,\ \boldsymbol{B}_1 = \boldsymbol{G}_1 T_s \text{ and } \boldsymbol{B}_2 = \boldsymbol{G}_2 T_s\,. \tag{4.18}$$

Equation (4.17) allows one to predict the stator current at the next time step $k+1$. The forward Euler discretization method neglects the evolution of the rotor flux vector within the sampling interval, assuming that it is constant from $t = kT_s$ to $t = (k+1)T_s$. Therefore, the rotor flux equation (4.11b) can be neglected when adopting a prediction horizon of 1 and the forward Euler discretization method. As a result, the rotor flux vector can be considered to be a time-varying parameter rather than a system state. Similarly, the rotor speed ω_r is assumed to be constant within the prediction horizon, which turns the speed into a time-varying parameter.

It is clear that the discrete-time model (4.17) should be used only to predict the stator current at the next time step $k+1$. Predictions further ahead at time steps $k+\ell$, with $\ell \gg 1$, tend to be inaccurate, because the rotation of the rotor flux vector is not taken into account. For these predictions, the full state-space representation should be adopted, which includes the evolution of the rotor flux vector, as described in Chap. 5.

4.2.4 Optimization Problem

The previously defined cost function (4.6) is now generalized to vector quantities.

$$J = \|\boldsymbol{i}_{e,abc}(k+1)\|_2^2 + \lambda_u\|\Delta\boldsymbol{u}(k)\|_1\,. \tag{4.19}$$

The first term penalizes the predicted three-phase current error at the next time step $k+1$, using the squared 2-norm. The current error in abc is given by

$$\boldsymbol{i}_{e,abc} = \boldsymbol{i}^*_{s,abc} - \boldsymbol{i}_{s,abc}\,.$$

The second term penalizes switching at time step k. The latter is defined as

$$\Delta\boldsymbol{u}(k) = \boldsymbol{u}(k) - \boldsymbol{u}(k-1)\,,$$

referring to the switch positions in the three phases a, b, and c.[2] As previously, λ_u is a non-negative scalar weight.

As the state vector of the controller model is given by the stator currents in $\alpha\beta$ coordinates rather than in abc, it is convenient to express the first term in (4.19) in $\alpha\beta$. To this end, we define the current error in stationary orthogonal coordinates $\boldsymbol{i}_e = \boldsymbol{i}^*_s - \boldsymbol{i}_s$, and recall from Sect. 2.1.3 that

$$\boldsymbol{i}_{e,abc} = \tilde{\boldsymbol{K}}^{-1}\boldsymbol{i}_e\,.$$

[2] As in each phase switching is allowed only by one step up or down, the 1-norm and the (squared) 2-norm of the switching transition yield the same cost, that is, $\|\Delta\boldsymbol{u}(k)\|_1 = \|\Delta\boldsymbol{u}(k)\|_2^2$.

Noting that $\tilde{\boldsymbol{K}}^{-T}\tilde{\boldsymbol{K}}^{-1} = 1.5\boldsymbol{I}_2$, the first term in (4.19) can be rewritten as

$$||\boldsymbol{i}_{e,abc}||_2^2 = (\boldsymbol{i}_{e,abc})^T\,\boldsymbol{i}_{e,abc} = 1.5\;||\boldsymbol{i}_e||_2^2. \tag{4.20}$$

Omitting the factor 1.5 to simplify the expression, the equivalent cost function with the current error formulated in stationary orthogonal coordinates becomes

$$J = ||\boldsymbol{i}_e(k+1)||_2^2 + \lambda_u||\Delta\boldsymbol{u}(k)||_1. \tag{4.21}$$

The cost function requires the stator current reference $\boldsymbol{i}_s^*$ in orthogonal coordinates at the future time step $k+1$. Assuming steady-state operation with constant current references in the rotating dq reference frame, $\boldsymbol{i}_s^*(k+1)$ can be easily derived by modifying the angular position of the dq reference frame. Specifically, in the dq to $\alpha\beta$ transformation in Fig. 4.11, the angle $\varphi + \omega_s T_s$ is used as argument instead of φ. Recall that ω_s is the angular stator frequency, and $\omega_s T_s$ is the angular increment within the sampling interval T_s (see also Sect. 12.2.3).

The optimization problem underlying predictive current control with reference tracking for an induction machine can then be stated as

$$\boldsymbol{u}_{\text{opt}}(k) = \arg \underset{\boldsymbol{u}(k)}{\text{minimize}}\; J \tag{4.22a}$$

$$\text{subject to}\quad \boldsymbol{i}_s(k+1) = \boldsymbol{A}\,\boldsymbol{i}_s(k) + \boldsymbol{B}_1\,\boldsymbol{\psi}_r(k) + \boldsymbol{B}_2\,\boldsymbol{u}(k) \tag{4.22b}$$

$$\boldsymbol{u}(k) \in \{-1, 0, 1\}^3,\;\; ||\Delta\boldsymbol{u}(k)||_\infty \le 1\,. \tag{4.22c}$$

Note that $||\Delta\boldsymbol{u}||_\infty$ denotes the infinity-norm of the vector $\Delta\boldsymbol{u}$, which is defined as the component of $\Delta\boldsymbol{u}$ with the largest absolute value, that is, $||\Delta\boldsymbol{u}||_\infty = \max(|\Delta u_a|, |\Delta u_b|, |\Delta u_c|)$.

4.2.5 Control Algorithm

At time step k, the predictive current control algorithm computes $\boldsymbol{u}_{\text{opt}}(k)$ in effectively the same manner as in the single-phase case. The algorithm is repeated here for completeness.

1. Given the previously applied switch position $\boldsymbol{u}(k-1)$ and taking into account the constraints on the switching transitions (4.22c), the set of admissible switch positions at time step k, $\boldsymbol{\mathcal{U}}(k)$, is determined.
2. For each of these switch positions $\boldsymbol{u}(k) \in \boldsymbol{\mathcal{U}}(k)$, the stator current at time step $k+1$, $\boldsymbol{i}_s(k+1)$, is predicted using the model (4.22b).
3. For each switch position $\boldsymbol{u}(k) \in \boldsymbol{\mathcal{U}}(k)$, the cost J is computed according to (4.21).
4. The switch position $\boldsymbol{u}_{\text{opt}}(k)$ with the minimum cost is determined and applied to the inverter.

At the next time step, the procedure outlined here is repeated.

This predictive current control algorithm based on enumeration and a prediction horizon of $N_p = 1$ was originally introduced for a three-phase RL load with voltage sources; the case of the two-level inverter was considered in [2], whereas the predictive concept was extended to a three-level inverter in [3]. In both cases, instead of the squared 2-norm, the 1-norm was proposed for penalizing the predicted current error in the cost function. The implications of this will be analyzed and discussed in Sect. 4.2.7.

Example 4.3 *Consider the situation depicted in Fig. 4.12. The α-component of the stator current is shown in the top half, and the β-component is depicted in the bottom half. At time step k, both current components are close to their respective references. The previously applied switch position is $\mathbf{u}(k-1) = [1\ \ -1\ \ 0]^T$. The set of admissible switch positions $\mathcal{U}(k)$ contains 12 elements. The predicted stator currents are shown in the figure for three exemplary admissible switch positions $\mathbf{u}(k) \in \mathcal{U}(k)$.*

The corresponding costs are summarized in Table 4.4. Switching to $\mathbf{u}(k) = [1\ \ -1\ \ -1]^T$ minimizes the predicted current error but incurs a switching penalty, which is given by λ_u. Refraining from switching, that is, using $\mathbf{u}(k) = \mathbf{u}(k-1) = [1\ \ -1\ \ 0]^T$, leads to a more pronounced current error while incurring no switching penalty. The third choice, $\mathbf{u}(k) = [1\ \ 0\ \ 0]^T$, is clearly suboptimal, in that it entails a large current error and also requires two switching transitions.

If the weight λ_u is sufficiently small, prioritizing the tracking of the current, the first choice, that is, $\mathbf{u}(k) = [1\ \ -1\ \ -1]^T$, is selected as the optimal switch position $\mathbf{u}_{\text{opt}}(k)$. The corresponding stator current trajectory is shown as a solid line in Fig. 4.12. Large λ_u, on the other hand, accentuate the reduction of the switching effort at the expense of tracking accuracy. As a result, switching is avoided and $\mathbf{u}_{\text{opt}}(k) = \mathbf{u}(k-1)$ is chosen.

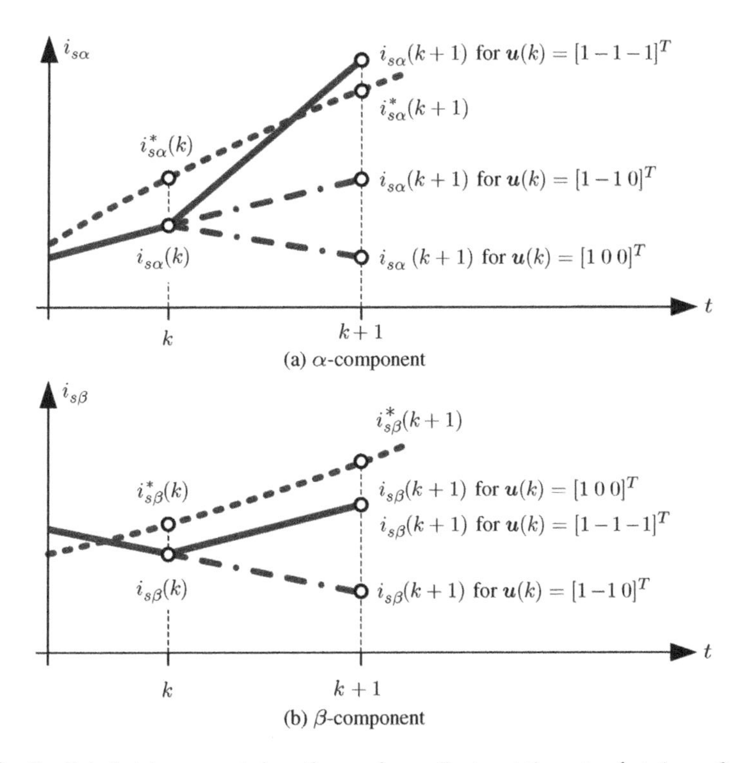

Figure 4.12 Predicted stator currents in orthogonal coordinates at time step $k+1$ as a function of the switch position $\mathbf{u}(k)$. The current reference is indicated by dashed lines

Table 4.4 Costs corresponding to the three switch positions $\boldsymbol{u}(k)$ shown in Fig. 4.12

$\boldsymbol{u}(k)$	$(i_{e\alpha}(k+1))^2 + (i_{e\beta}(k+1))^2$	$\lambda_u \|\Delta \boldsymbol{u}(k)\|_1$
$[1\ -1\ 0]^T$	$0.073^2 + 0.11^2$	0
$[1\ -1 -1]^T$	$0.025^2 + 0.03^2$	λ_u
$[1\ 0\ 0]^T$	$0.135^2 + 0.03^2$	$2\lambda_u$

4.2.6 Performance Evaluation

As a case study, consider a three-level NPC voltage source inverter driving an induction machine with a constant mechanical load. A 3.3 kV, 50 Hz squirrel-cage induction machine rated at 2 MVA with a total leakage reactance of 0.25 pu is used. The detailed parameters of the machine and inverter are provided in Sect. 2.5.1. Recall that, unlike in Sect. 2.5.1, the neutral point potential is fixed. The operating point is at nominal speed and rated torque. The sampling interval is 25 μs.

Simulation results at steady-state operation are shown in Fig. 4.13 for $\lambda_u = 0.003$. The three-phase stator currents, indicated by solid lines, along with their references, which are depicted by dash-dotted lines, are shown for one fundamental period, along with the electromagnetic torque and the three-phase switch positions. The current distortions are $I_{\text{TDD}} = 6.69\%$, and the average switching frequency per semiconductor device is $f_{\text{sw}} = 222$ Hz. Owing to the lack of periodicity, the harmonic current spectrum is spread over a large frequency range. The majority of the harmonic content is between 600 and 1500 Hz.

By varying the weight λ_u and running steady-state simulations, the trade-off between the current TDD and the average switching frequency per semiconductor device can be investigated. The corresponding analysis is shown in Fig. 4.14, where λ_u is varied between 0 and 0.02. Below $\lambda_u = 0.017$, the current TDD depends almost linearly on λ_u, while the switching frequency drops steeply from 3440 Hz at $\lambda_u = 0$ to about 70 Hz at $\lambda_u = 0.0175$. Note that the switching frequency is plotted using a logarithmic scale. A small discontinuity in the current TDD can be observed around $\lambda_u = 0.007$, where the switching frequency plateaus at 100 Hz. When increasing λ_u beyond 0.018, fundamental frequency switching (or six-step operation) is achieved with the switching frequency of 50 Hz. The current TDD settles at around 20%.

Of particular interest is the product of the current distortions and the device switching frequency

$$I_{\text{TDD}} \cdot f_{\text{sw}} = c_f\,.$$

We have seen in Sect. 3.5 that c_f is effectively a constant for pulse width modulation (PWM), regardless of the switching frequency. As can be seen in Fig. 4.14(c), for the predictive current controller under investigation, this statement needs to be refined. Specifically, five distinctive regions can be observed in Fig. 4.14(c).

1. $\lambda_u < 0.0005$: With the switching penalty effectively zero, a high c_f results, leading to an unfavorable ratio between current distortions and switching frequency. Correspondingly, when reducing λ_u below 0.0005, the switching frequency is increased significantly, while the current TDD, in relative terms, is reduced only mildly.

2. $0.0005 \leq \lambda_u < 0.0065$: For this range of weights, c_f is approximately equal to 1600. The switching frequency is bounded between 1 kHz and 100 Hz. The current TDD depends linearly on the weight λ_u, facilitating the tuning of the controller, which would typically be used in this operating region.
3. $\lambda_u = 0.0065$: A favorable ratio between current distortions and switching frequency is obtained for 100 Hz switching frequency with c_f being as low as 1100.
4. $0.0065 < \lambda_u < 0.018$: For relatively high switching penalties, low switching frequencies between 100 and 70 Hz result. The value of c_f is mostly very high, indicating an unfavorable ratio between current distortions and switching frequency.
5. $\lambda_u \geq 0.018$: For large switching penalties, fundamental frequency switching ($f_{\text{sw}} = 50$ Hz) results and a very favorable c_f of approximately 1000 is achieved.

This analysis shows that weights on the switching transition that are (effectively) zero lead to a poor steady-state performance. If very low current distortions are required, the sampling interval T_s should be reduced instead of setting λ_u to zero. This ensures operation in region 2, which is the recommended operating regime for the predictive controller. When very low

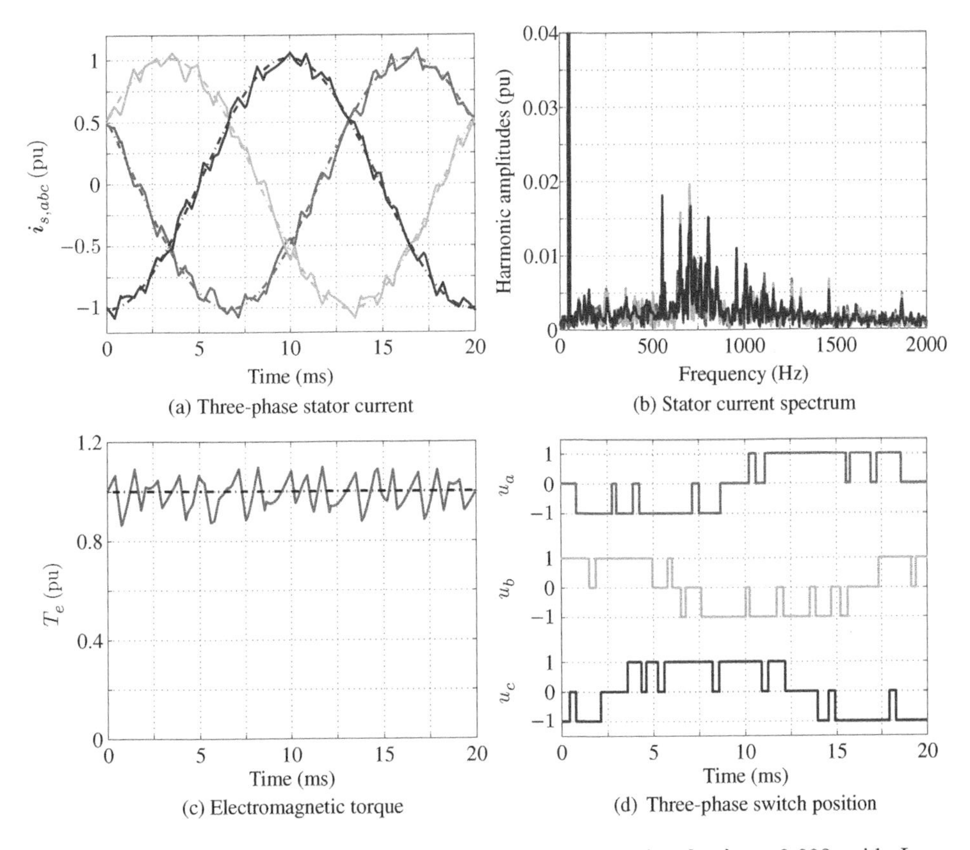

Figure 4.13 Predictive current control during steady-state operation for $\lambda_u = 0.003$, with $I_{\text{TDD}} = 6.69\%$ and $f_{\text{sw}} = 222$ Hz

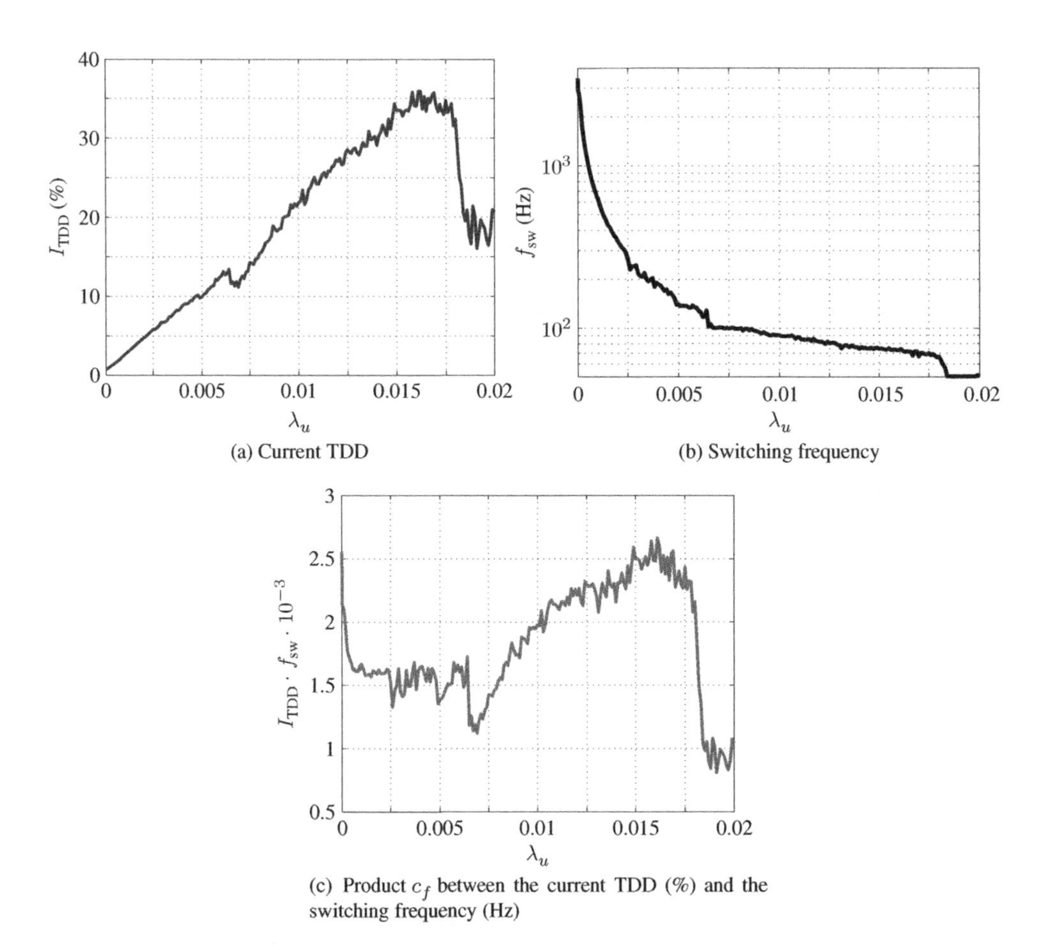

(a) Current TDD

(b) Switching frequency

(c) Product c_f between the current TDD (%) and the switching frequency (Hz)

Figure 4.14 Trade-off between the current TDD and the device switching frequency for predictive current control under steady-state operating conditions for an NPC inverter with an induction machine. As a function of the weight λ_u, the current TDD, the switching frequency, and the product of the two scaled by 1000 are shown. The sampling interval is kept constant at $T_s = 25$ μs

switching frequencies are required, regions 3 and 5 are also favorable. Interestingly, for large λ_u the switching frequency locks into integer multiples of the fundamental frequency. This phenomenon will be further investigated in Chap. 5 in the context of long prediction horizons. Region 4 should be avoided and the weight λ_u should be tuned accordingly.

A similar analysis can be performed for the sampling interval T_s. For the weight $\lambda = 0.003$, the sampling interval at which the controller is executed is varied between 5 and 500 μs.[3] Plotting the current TDD, the device switching frequency, and the product of the two scaled by 1000 as a function of the sampling interval using the logarithmic scale, Fig. 4.15 is obtained.

[3] One needs to distinguish between the controller and the simulation sampling interval. The controller is executed at the time instants defined by the controller sampling interval. The system sampling interval is used to simulate the response of the drive system in MATLAB—it is independent of the controller sampling interval and is in the range of a few microseconds and thus sufficiently small to ensure that the system evolution is accurately simulated. In this analysis, the controller sampling interval is varied while the system sampling interval is held constant.

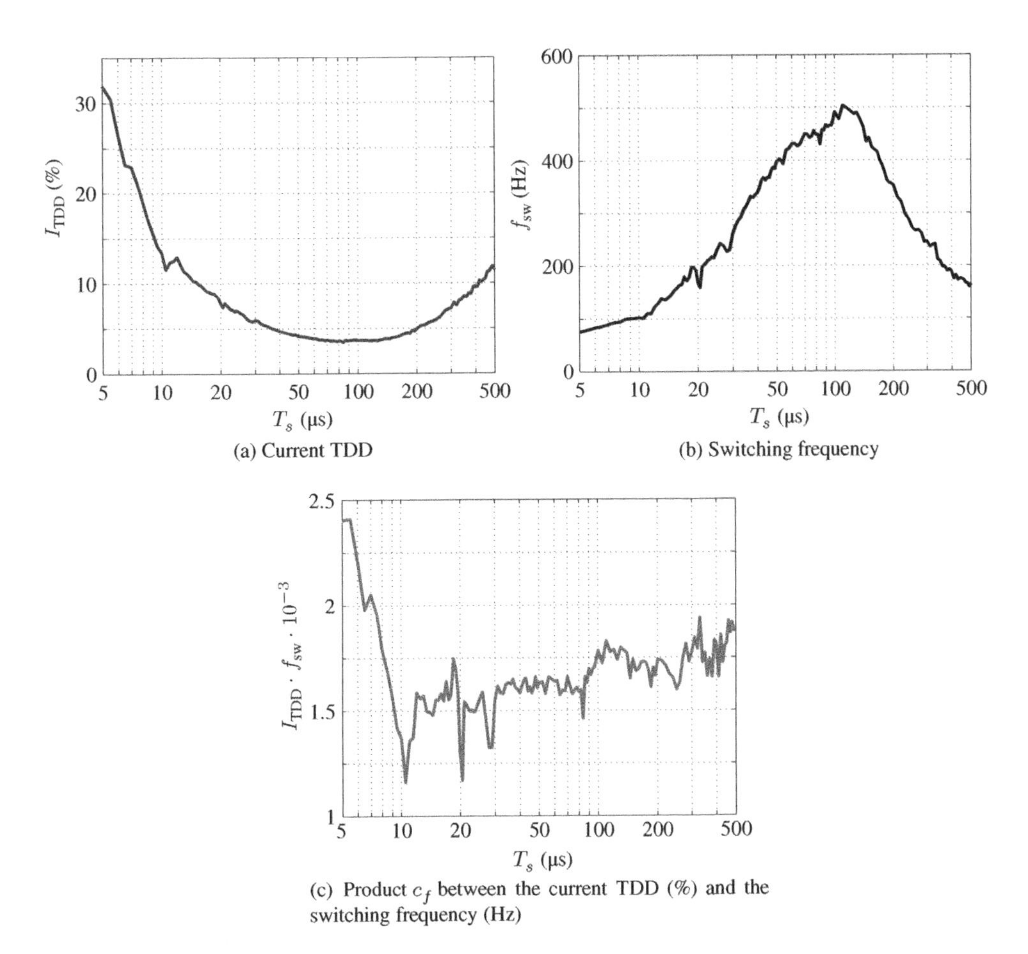

(a) Current TDD

(b) Switching frequency

(c) Product c_f between the current TDD (%) and the switching frequency (Hz)

Figure 4.15 Influence of the sampling interval on the current TDD, device switching frequency, and the product of the two at steady-state operation. The weight $\lambda = 0.003$ on the switching transitions is kept constant

As can be seen, for $\lambda = 0.003$, sampling intervals below 10 μs lead to high current distortions and to unfavorable products between the current distortion and the switching frequency, with c_f well exceeding 1600. The prediction interval (in time), that is, the prediction horizon (given as the number of steps) multiplied by the sampling interval, is very small for these sampling intervals, leading to a poor performance.

For sampling intervals in the range from 10 to 90 μs, low current distortions and relatively low switching frequencies are obtained, with c_f mostly below 1600. Very favorable c_f can be achieved for $T_s = 10, 20$, and 25 μs, for which the fundamental period (here 20 ms) is an integer multiple of the sampling interval. For sampling intervals exceeding 90 μs, c_f deteriorates and is above 1600. Even though the prediction interval is long, the granularity at which switching can be performed is overly coarse for long sampling intervals, impacting on the performance.

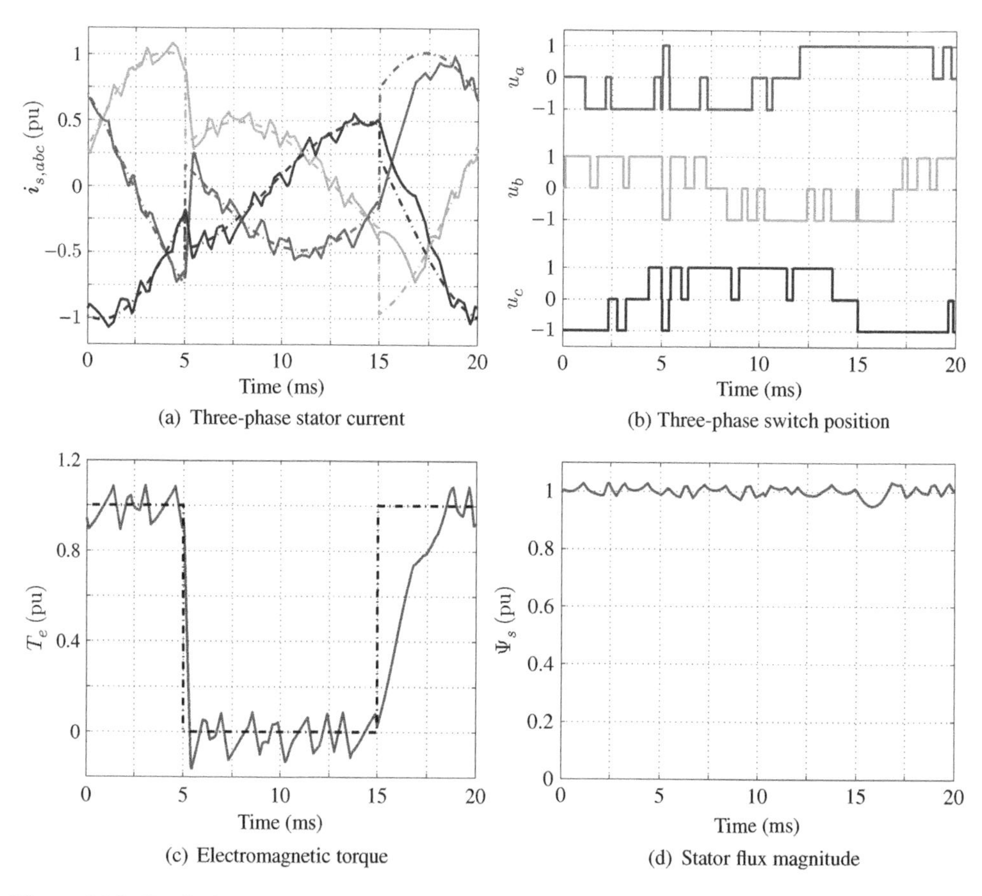

(a) Three-phase stator current

(b) Three-phase switch position

(c) Electromagnetic torque

(d) Stator flux magnitude

Figure 4.16 Predictive current control in the presence of step changes in the torque reference from 1 to 0 pu and back to 1 pu

We conclude that the sampling interval has almost as strong an influence on the steady-state performance as the weight λ_u. In effect, predictive current control uses two tuning parameters. This complicates the tuning process.

The performance of the predictive current controller during torque reference steps is examined next. The switching penalty is set to $\lambda_u = 0.003$, and the sampling interval $T_s = 25$ µs is again used. Operation is at nominal speed and initially at rated torque. At $t = 5$ ms, a torque reference step from 1 to 0 pu is applied, as shown in Fig. 4.16(c). The current references, which are shown as dash-dotted lines in Fig. 4.16(a), are stepped down accordingly. The predictive controller achieves a very fast current and thus torque settling time of about 0.3 ms, by inserting a switching pulse of the same duration, which effectively inverts the voltage applied to the machine (see Fig. 4.16(b)). The stator flux magnitude, which is shown in Fig. 4.16(d), remains constant and is not affected by the torque step.

At $t = 15$ ms, the torque reference is stepped back to 1 pu. The length of the resulting torque transient at 3.5 ms is significantly longer than in the step-down case. During the transient, a minor dip in the stator flux magnitude can be observed. The latter is due to the fact that the controller tracks the stator current references rather than the desired torque and stator flux magnitude. As in the single-phase case, the predictive controller provides the fastest torque response physically possible, effectively exhibiting the dynamic behavior of a deadbeat controller.

4.2.7 About the Choice of Norms

In the cost function, we have—so far—exclusively used the squared 2-norm, that is, the sums of squares of the current error components in the α- and β-axis. Predictive current control with reference tracking, however, was originally proposed with the 1-norm, or in other words, the sum of the absolute values of the current error components [2]. Even today, the vast majority of the literature on predictive control with reference tracking is based on the 1-norm, because it is computationally simpler. One might consider the choice of norm to be a subtlety, but this choice has indeed a profound impact on the controller performance, as will be shown in this section.

For this investigation, we replace the squared 2-norm by the 1-norm and repeat the steady-state simulations in the previous section for the predictive current controller. More specifically, the cost function (4.21) is replaced by

$$J = ||\boldsymbol{i}_e(k+1)||_1 + \lambda_u ||\Delta \boldsymbol{u}(k)||_1 . \tag{4.23}$$

Note that $||\boldsymbol{i}_e||_1 = |i_{e\alpha}| + |i_{e\beta}|$.

The drive system and operating point are the same as previously stated in Sect. 4.2.6. For $\lambda_u = 0.017$ and $T_s = 25$ μs, for example, the stator currents, three-phase switch positions, torque, and stator flux magnitude are shown in Fig. 4.17. Large deviations of the currents from their references occur, resulting in torque excursions of 20% and more from its reference, as well as fluctuations in the stator flux magnitude. This is despite the high switching frequency of 1180 Hz. The closed-loop system appears to become temporarily unstable.

Analogous to Fig. 4.14, but using the 1-norm in the cost function, λ_u is varied between 0 and 0.02. The current TDD and the device switching frequency as a function of λ_u are shown in Fig. 4.18. Current excursions manifest themselves for $\lambda_u > 0.0072$, which lead to a deterioration of the current TDD, causing a small, yet distinctive step in the current TDD in Fig. 4.18(a). For $\lambda_u > 0.0198$, the controller entirely fails to track the current references. The switching frequency drops close to zero and closed-loop stability is lost. For $\lambda_u > 0.027$, which is not shown in the figure, switching is avoided altogether and the switch position is frozen at its initial switching state $\boldsymbol{u} = [0\ 0\ 0]^T$.

Figures 4.14 and 4.18 use the same scaling to facilitate a direct comparison of the performance induced by the squared 2-norm and the 1-norm on the predicted current error, respectively. It is evident that the 1-norm is an unfortunate choice—it is effective only between $\lambda_u = 0$ and 0.0072. In this range, its tuning capability of setting the trade-off between the current distortion and the switching frequency is severely limited. In particular, closed-loop operation with current TDDs in the range of 5–10% with switching frequencies of a few hundred hertz is not possible. Beyond $\lambda_u = 0.0072$, the current excursions impact on the current TDD, leading to suboptimal results.

The root cause for the poor performance of the 1-norm is analyzed in the following. Consider a single-phase system with the current i, the reference current i^*, and the current error $i_{\text{err}} = i^* - i$. Assume that the current error at time step k is small, as depicted in Fig. 4.19(a). Starting

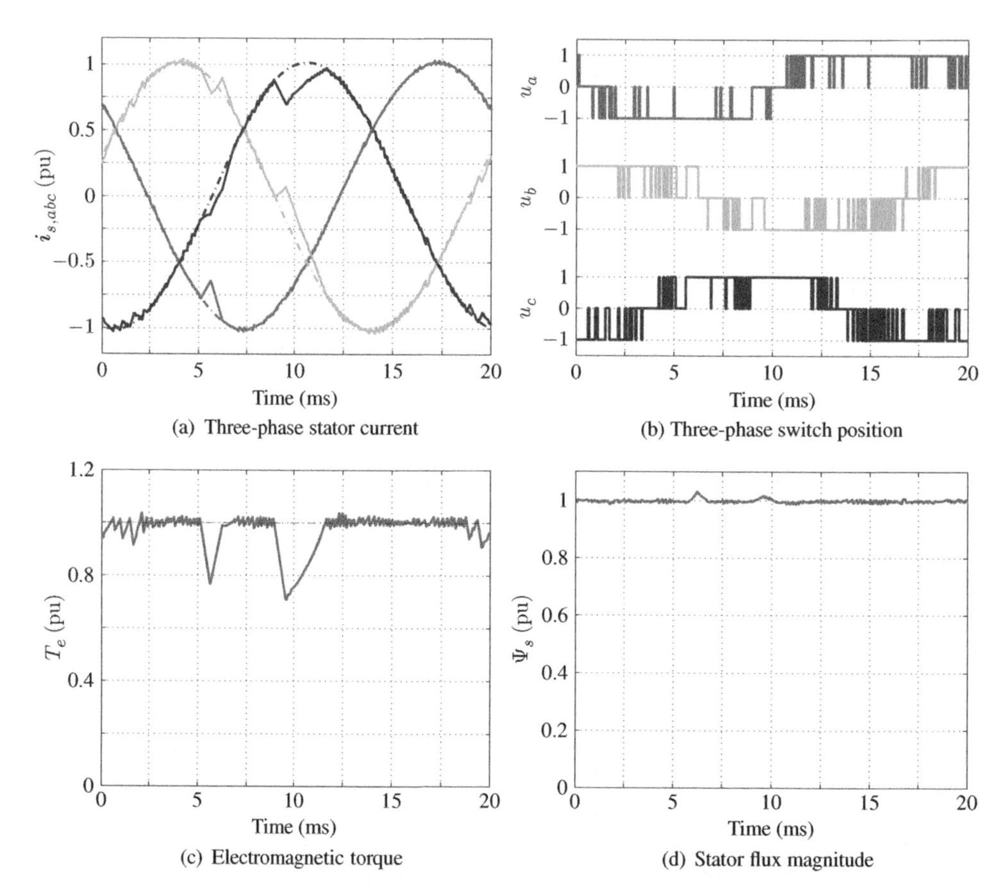

(a) Three-phase stator current

(b) Three-phase switch position

(c) Electromagnetic torque

(d) Stator flux magnitude

Figure 4.17 Predictive current control during steady-state operation with $\lambda_u = 0.017$ and the 1-norm penalizing the current error

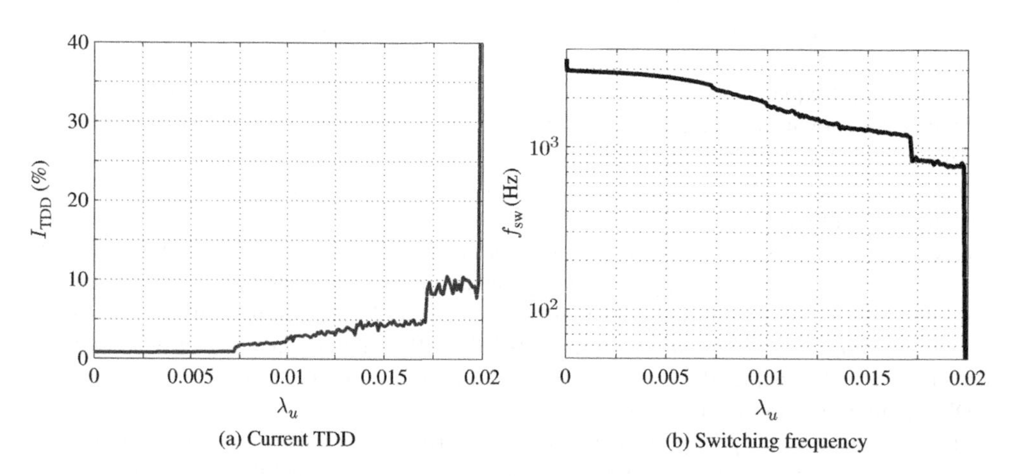

(a) Current TDD

(b) Switching frequency

Figure 4.18 Trade-off between the current TDD and the switching frequency when using the 1-norm in the cost function to penalize the predicted current error

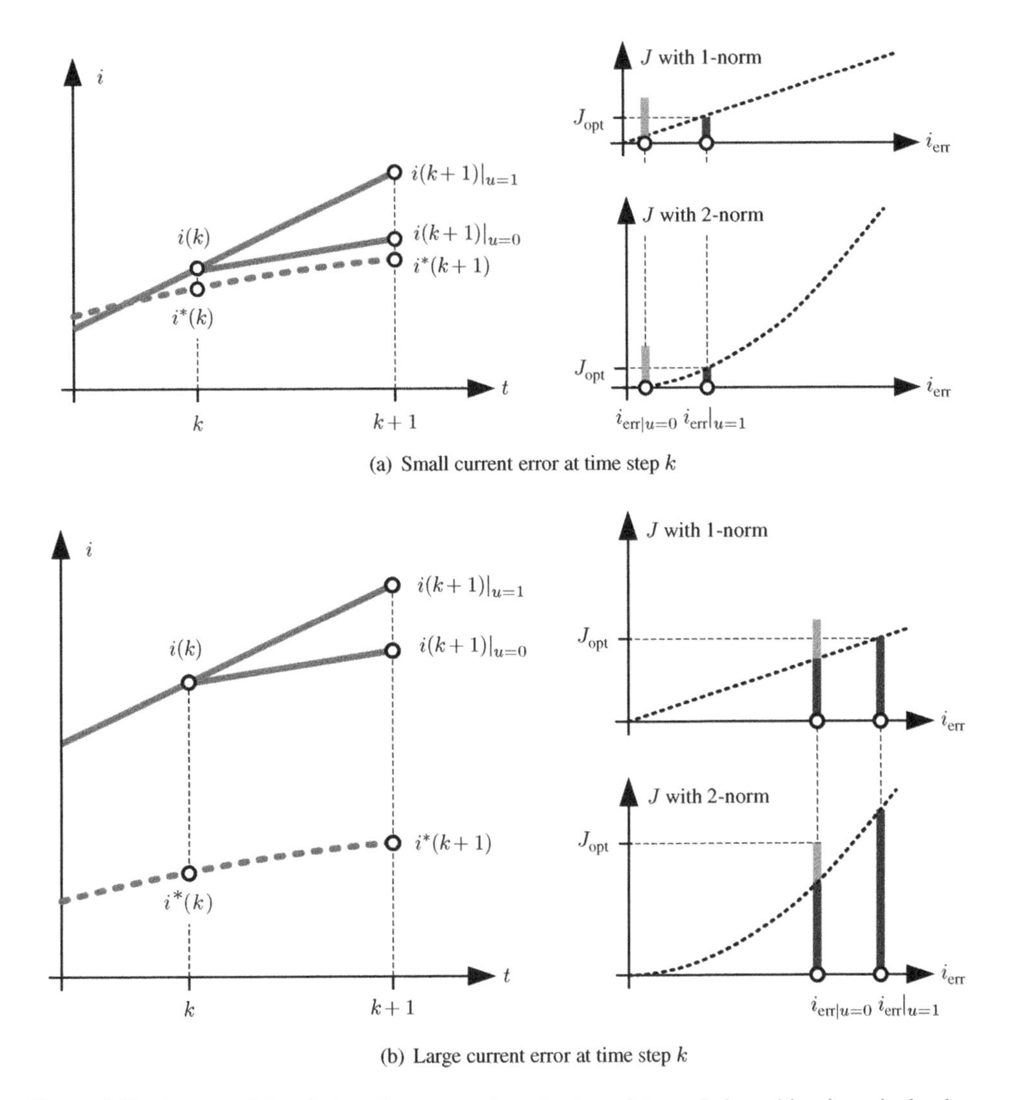

Figure 4.19 Impact of the choice of norm on the selection of the switch position in a single-phase setting, using the case of a small and a large current error as an example. For the 1-norm, when the switching cost (light gray bar) outweighs the *relative* reduction of the tracking error i_{err} (difference in the dark gray bars), switching is avoided regardless of the *absolute* tracking error. This leads to current excursions and ultimately to instability. When using the squared 2-norm, however, sufficiently large current errors will always dominate over the cost of switching, triggering a switching transition to reduce the current error. This ensures good tracking performance as well as stability of the closed-loop system

from the time step k and assuming $u(k-1) = 1$, the left plot shows the predicted currents $i(k+1)$ as a function of the switch position $u(k)$. To denote this, we introduce the notation $i(k+1)|_{u=u(k)}$.

The cost function[4] using the 1-norm is depicted on top. The tracking error cost depends linearly on the current error. This cost is shown as a dotted line and indicated by a dark gray bar. The switching cost is shown as a light gray bar. For the squared 2-norm, the penalty on the current error depends quadratically on the current error. For both cases, it is evident that refraining from switching yields the minimum cost and is thus optimal.

Figure 4.19(b) depicts the same situation as before, except for a large current error at time step k. The cost function based on the 1-norm selects $u(k) = u(k-1) = 1$ and thus fails to trigger a switching transition to $u(k) = 0$ to reduce the current error. Instead, the current error grows further. The reason for this is evident from the components in the cost function. The switching cost (light gray bar) outweighs the *relative* reduction in the tracking error, which is given by the difference in the predicted current error when using $u(k) = 1$ and $u(k) = 0$. As a result, switching is avoided regardless of the *absolute* tracking error. When using the squared 2-norm, however, the current error grows quadratically up to the point where the reduction in the tracking error cost exceeds the cost of switching, triggering a switching transition from $u(k-1) = 1$ to $u(k) = 0$. Therefore, when using the squared 2-norm, sufficiently large current errors will always trigger a switching transition to reduce this error, ensuring good tracking performance and closed-loop stability.

To simplify the exposition, a single-phase system was considered in this analysis, but the reasoning can directly be extended to three-phase systems. Even though the 1-norm might appear to be appealing in the sense that it is computationally simpler than the squared 2-norm, its inability to provide adequate tuning and to ensure closed-loop stability makes it a poor choice for predictive controllers with reference tracking.

4.2.8 Delay Compensation

So far we have assumed an ideal discrete-time setup without any time delay between the sampling of the measurements and the application of the new switch position. This case is exemplified in Fig. 4.20. The stator current $\boldsymbol{i}_s$ is sampled at time step k. The corresponding switch position $\boldsymbol{u}(k)$ is computed within an infinitesimally short time, and it is applied from time step k to $k+1$. The current is sampled again at time step $k+1$, based on which the switch position $\boldsymbol{u}(k+1)$ is computed, and so on.

In a practical inverter setting, however, physical limitations cause a time delay between the sampling of the measurements and the application of the new switch position. The most prominent and commonly encountered sources of delays can be summarized as follows:

- *Measurement delay.* The measured signals are typically sampled in the inverter at a fixed sampling frequency. The analog-to-digital (A/D) conversion incurs a time delay—albeit a small one—of about 1 μs.

[4] The cost J is the sum of the switching cost, $\lambda_u|\Delta u(k)|$, and the cost on the predicted tracking error. The latter is $(i_{\text{err}}(k+1))^2$ for the squared 2-norm and $|i_{\text{err}}(k+1)|$ for the 1-norm.

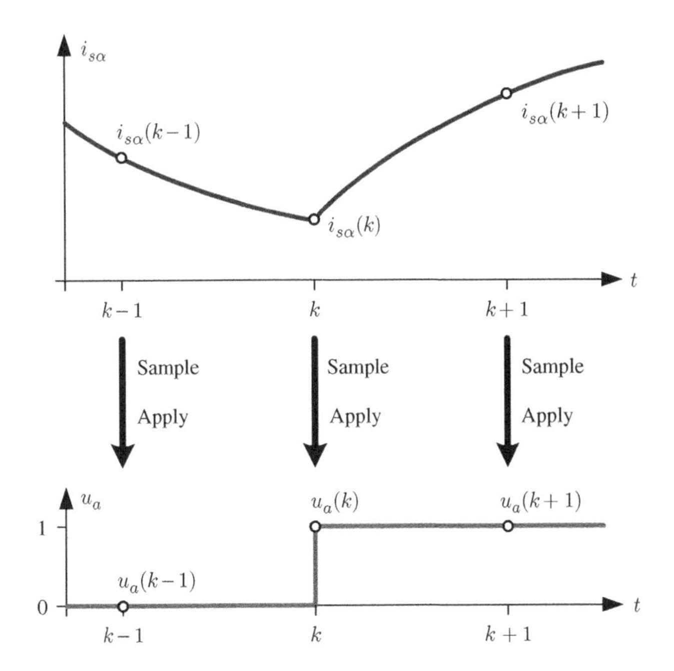

Figure 4.20 Idealized case of predictive current control without any delay. Only the α-component of the stator current and the phase a of the three-phase switch position are shown

- *Uplink communication delay.* The digital measurements are transmitted to a computational unit, such as a field-programmable gate array (FPGA), a digital signal processor (DSP), or a central processing unit (CPU). In the case of a serial link, the communication delay often amounts to 10 µs or more.
- *Computation delay.* The predictive control algorithm is executed on the computational unit. The computation of the new switch position requires a control-algorithm-specific number of clock cycles, ranging from several tens of cycles on a dedicated FPGA with a computationally simple algorithm to tens of thousands of cycles. The computational resources are often shared with other processes, such as outer control loops and monitoring tasks, which reduce the computational power available for the predictive controller and increase the computation delay. If some of the state or controlled variables cannot be measured, an observer needs to be run first to reconstruct them, further increasing the computation delay. As a result, delays of 10 µs are common even for simple predictive controllers.
- *Downlink communication delay.* The newly computed switch position is sent back to the inverter via a downlink, incurring another communication delay.
- *Actuation delay.* Before applying the switch position, a protection stage usually ensures that only admissible switch positions are applied to the semiconductors. Further delays often occur in the gate drivers and from the addition of interlocking times before the current commutates. For gate-commutated thyristors (GCTs), the interlocking time amounts to several microseconds.

The sum of all time delays typically amounts to several tens of microseconds in a high-power electronics setup.

Figure 4.21(a) depicts the communication and computation delays, which are generally the most significant delays. The symbols U and D refer to the uplink and downlink communication, respectively, whereas C denotes the computation. The sampling of the measurements and the application of the switch position are indicated by vertical arrows.

To simplify the exposition, we assume that the sum of the time delays T_d is equal to the sampling interval T_s. The switch position $\boldsymbol{u}(k)$ is computed based on the measurements at time step k. We use the notation $\boldsymbol{u}(k|k)$ to state that $\boldsymbol{u}(k)$ depends on measurements obtained at time step k. A delay of one sampling interval causes the application of $\boldsymbol{u}(k|k)$ at the next time step $k+1$. The delayed application of the switch position increases the current ripple, as indicated in Fig. 4.21(a), and adversely impacts on the closed-loop performance.

It is common practice to compensate for delays when implementing digital control schemes. To this end, an additional prediction step can be introduced, using the state-space model (4.17). Specifically, using the stator current sample $\boldsymbol{i}_s(k-1)$, the rotor flux estimate $\boldsymbol{\psi}_r(k-1|k-2)$, and the previously applied switch position $\boldsymbol{u}(k-1|k-2)$ at time step $k-1$, the stator current at time step k can be predicted with the help of

$$\boldsymbol{i}_s(k|k-1) = \boldsymbol{A}\boldsymbol{i}_s(k-1) + \boldsymbol{B}_1\boldsymbol{\psi}_r(k-1|k-2) + \boldsymbol{B}_2\boldsymbol{u}(k-1|k-2)\,. \tag{4.24}$$

This *initial state prediction* predicts the value of the stator current at time step k, at which the to-be-computed switch position will be applied to the inverter. The initial state prediction is based on information that is available at time step $k-1$. In particular, the estimate of the rotor flux vector $\boldsymbol{\psi}_r(k-1|k-2)$ and the switch position $\boldsymbol{u}(k-1|k-2)$ were computed at time step $k-2$. This situation is depicted in Fig. 4.21(b).

We are now ready to augment the predictive current controller in Sect. 4.2.5 with a delay compensation scheme. To this end, we add Step 0 to the control algorithm. Step 0 performs the initial state prediction and projects the stator current sample from time step $k-1$ to k. Also, the stator current reference needs to be projected one time step forward. Following the technique outlined in Sect. 4.2.4, we add $\omega_s T_s$ to the angular position of the dq reference frame, in which the outer control loops manipulate the stator current reference. At time step $k-1$, the predictive current control algorithm computes $\boldsymbol{u}_{\text{opt}}(k|k-1)$, which is applied at time step k.

0. The stator current $\boldsymbol{i}_s(k-1)$ is sampled, and the current at time step k is predicted using the initial state prediction (4.24).
1. Given the switch position $\boldsymbol{u}(k-1|k-2)$ and taking into account the constraints on the switching transitions (4.22c), the set of admissible switch positions at time step k, $\boldsymbol{\mathcal{U}}(k|k-1)$, is determined.
2. For each switch position $\boldsymbol{u}(k|k-1) \in \boldsymbol{\mathcal{U}}(k|k-1)$, the stator current at time step $k+1$, $\boldsymbol{i}_s(k+1)$, is predicted using the model (4.22b).
3. For each switch position $\boldsymbol{u}(k|k-1) \in \boldsymbol{\mathcal{U}}(k|k-1)$, the cost J is computed according to (4.21).
4. The switch position $\boldsymbol{u}_{\text{opt}}(k|k-1)$ with the minimum cost is determined and applied to the inverter.

This procedure is repeated at the next time step. Note that both the original and the revised algorithms compute an optimum switch position that is applied at time step k. To do so, the execution of the revised algorithm has already started at time step $k-1$ by sampling the current.

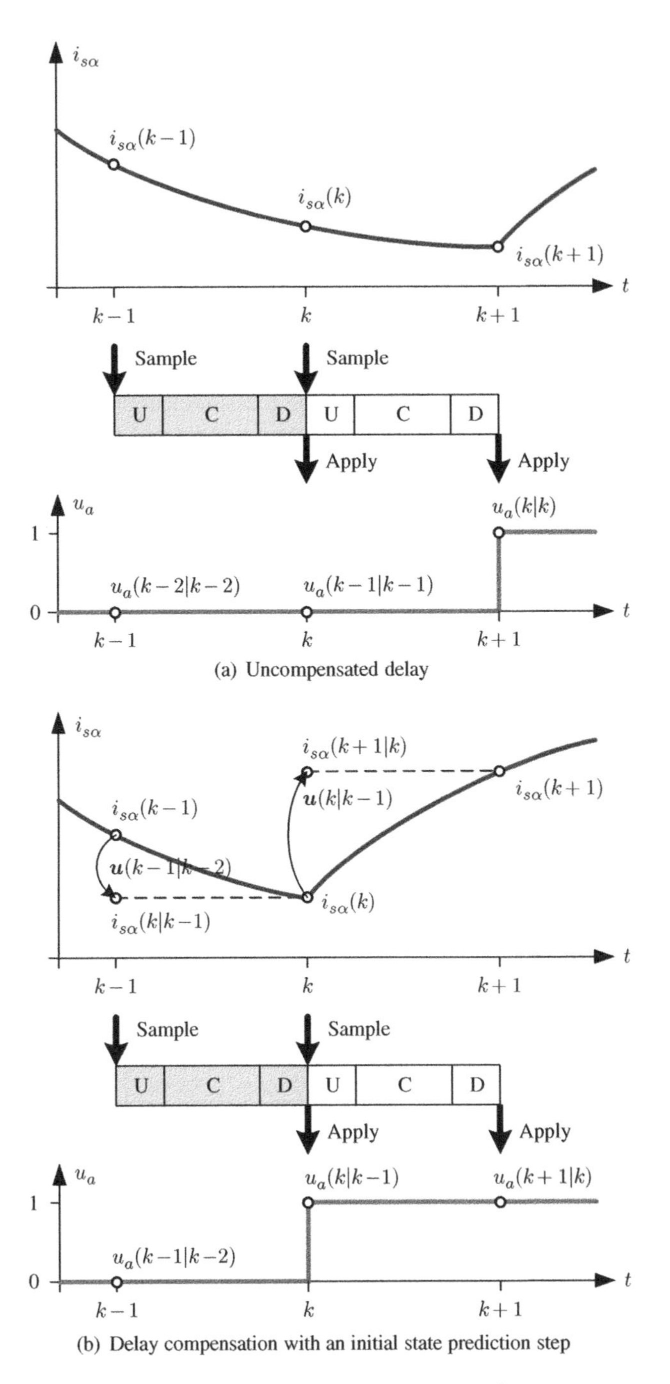

(a) Uncompensated delay

(b) Delay compensation with an initial state prediction step

Figure 4.21 Predictive current control without and with delay compensation, assuming a delay of one sampling interval. The symbols U, C, and D refer to the uplink communication, computation, and downlink communication, respectively. Only the α-component of the stator current and the phase a of the three-phase switch position are shown

If the delay is shorter than the sampling interval, the state-space model (4.24) for the initial state projection can be discretized with the length of the time delay T_d. Specifically, (4.15) is integrated from $t = kT_s$ to $t = kT_s + T_d$, and T_s is replaced by T_d in (4.18). For the prediction of the current at time step k in Step 2 of the algorithm, however, the state-space model is always discretized with the sampling interval T_s, regardless of the delay.

The delay compensation scheme is also applicable to cases involving time delays in excess of one sampling interval. Figure 4.22 depicts the timing diagram for a time delay of two sampling intervals. After sampling the current at time step $k-2$ and sending the measurements to the computational unit, two prediction steps are required to project the current sample from time step $k-2$ to k. To do so, the switch positions $\boldsymbol{u}(k-2|k-4)$ and $\boldsymbol{u}(k-1|k-3)$ are required. As indicated in Fig. 4.22, these switch positions are available at the time the initial state prediction is performed. Longer delays in the measurement and uplink communication chain can also be addressed [4]. Time delays that are non-integer multiples of the sampling interval can be compensated for by discretizing the state-space model accordingly, thus generalizing the technique described for the case $T_d < T_s$.

In the presence of (long) delays, the initial state prediction significantly improves the closed-loop performance, as shown in [4]. The performance of the delay compensation is, however, limited by the accuracy of the state-space model. Nondeterministic and time-varying

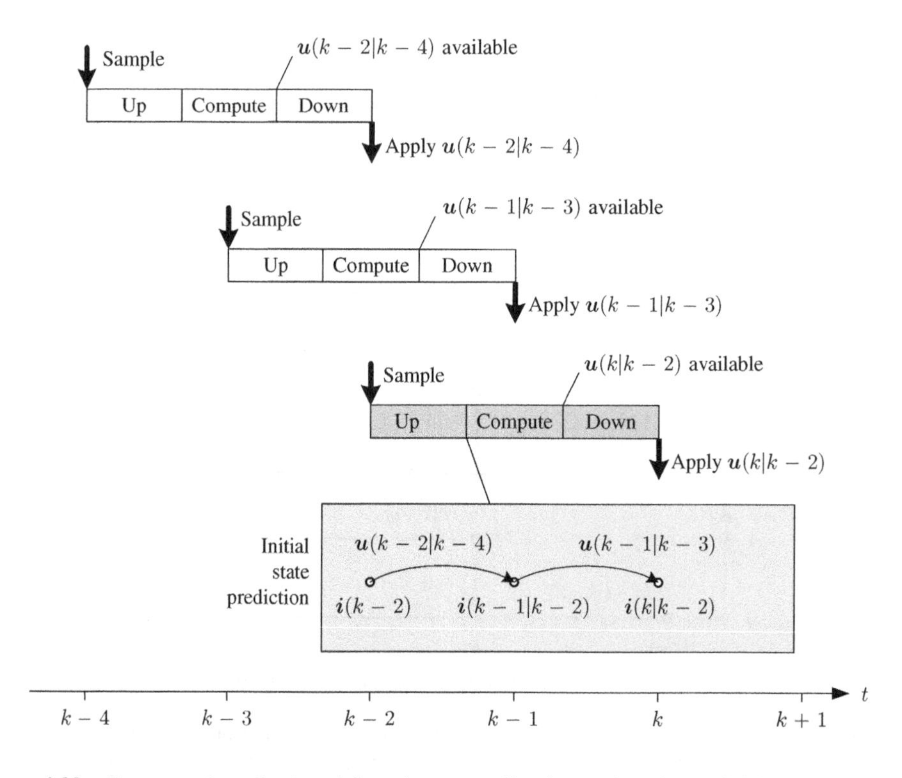

Figure 4.22 Compensation of a time delay of two sampling intervals, using an initial state prediction with two steps

delays also affect its performance. Furthermore, the delay compensation scheme is unable to avoid an initial delay in the controller response to reference changes and transients.

We have introduced in this section the delay compensation scheme for predictive current control with a prediction horizon of one step. It is straightforward to generalize the initial state prediction to other model predictive control (MPC) schemes that are formulated in the discrete-time domain. This includes MPC with reference tracking and long horizons (see Chaps. 5 and 6), MPC with bounds (as proposed in Chaps. 7–11), and indirect MPC with PWM (as detailed in Chap. 14). For more information on delays and their compensation in MPC for power electronics, the reader is referred to [5–7].

For model predictive pulse pattern control (MP^3C) (see Chaps. 12 and 13), a different delay compensation scheme is required. This will be discussed in Sect. 12.4. Throughout the book, all simulation results have been obtained for simulation setups without delays, unless otherwise noted. For the experimental results, the controllers were augmented with a delay compensation scheme similar to the one described in this section.

4.3 Predictive Torque Control of a Three-Phase Induction Machine

The notion of the predictive current controller with reference tracking has been illustrated in the previous section using an ac drive system setting. When controlling electrical drives, however, it is often convenient—similar to when using direct torque control (DTC)—to *directly* control the electromagnetic torque and the stator flux magnitude of the electrical machine, rather than to *indirectly* control these quantities via the stator currents. This can be easily accomplished by slightly modifying the predictive controller, as shown in this section. While the cost function and the internal controller model are adapted, the solution algorithm based on enumeration remains the same.

4.3.1 Case Study

Consider again the same drive system as in Sect. 4.2, which consists of an NPC inverter with a squirrel-cage induction machine. For predictive torque control, it is convenient to reformulate the machine model and to express its dynamics in terms of the stator and rotor flux linkage vectors $\boldsymbol{\psi}_s = [\psi_{s\alpha}\ \psi_{s\beta}]^T$ and $\boldsymbol{\psi}_r = [\psi_{r\alpha}\ \psi_{r\beta}]^T$, respectively. Setting the rotor voltage $\boldsymbol{v}_r$ and the angular speed of the reference frame to zero in (2.55) leads to the continuous-time machine equations in state-space representation:

$$\frac{\mathrm{d}\boldsymbol{\psi}_s}{\mathrm{d}t} = -R_s\frac{X_r}{D}\boldsymbol{\psi}_s + R_s\frac{X_m}{D}\boldsymbol{\psi}_r + \boldsymbol{v}_s \tag{4.25a}$$

$$\frac{\mathrm{d}\boldsymbol{\psi}_r}{\mathrm{d}t} = R_r\frac{X_m}{D}\boldsymbol{\psi}_s - R_r\frac{X_s}{D}\boldsymbol{\psi}_r + \omega_r\begin{bmatrix}0 & -1\\ 1 & 0\end{bmatrix}\boldsymbol{\psi}_r \tag{4.25b}$$

$$\frac{\mathrm{d}\omega_r}{\mathrm{d}t} = \frac{1}{M}(T_e - T_\ell)\,. \tag{4.25c}$$

By expanding the cross product in (2.56), the electromagnetic torque can be written in the form

$$T_e = \frac{1}{\mathrm{pf}}\frac{X_m}{D}\,\boldsymbol{\psi}_r \times \boldsymbol{\psi}_s = \frac{1}{\mathrm{pf}}\frac{X_m}{D}(\psi_{r\alpha}\psi_{s\beta} - \psi_{r\beta}\psi_{s\alpha}) \tag{4.26}$$

and the magnitude of the stator flux vector is

$$\Psi_s = \|\boldsymbol{\psi}_s\| = \sqrt{(\psi_{s\alpha})^2 + (\psi_{s\beta})^2}. \tag{4.27}$$

4.3.2 Control Problem

The control problem of predictive torque control is to track the references of these two quantities by manipulating the three-phase switch position $\boldsymbol{u}$ accordingly. As previously described, the switching frequency is to be minimized, and direct switching between 1 and -1 in a phase leg is prohibited.

The block diagram of the predictive torque controller is shown in Fig. 4.23. We introduce T_e^* to denote the torque reference and Ψ_s^* to denote the reference of the stator flux magnitude. The torque reference T_e^* is usually adjusted by an outer speed control loop as shown in Fig. 3.29. The predictive torque controller predicts the torque and stator flux magnitude at the next time step for all admissible switch positions. These predictions are based on the stator and rotor flux vectors, which are reconstructed by a flux observer.

4.3.3 Controller Model

The controller model predicts the electromagnetic torque and the magnitude of the stator flux vector at time step $k+1$ as a function of $\boldsymbol{u}(k)$. Using the stator and rotor flux vectors in orthogonal coordinates as the state vector, and treating the rotor speed ω_r as a parameter, the

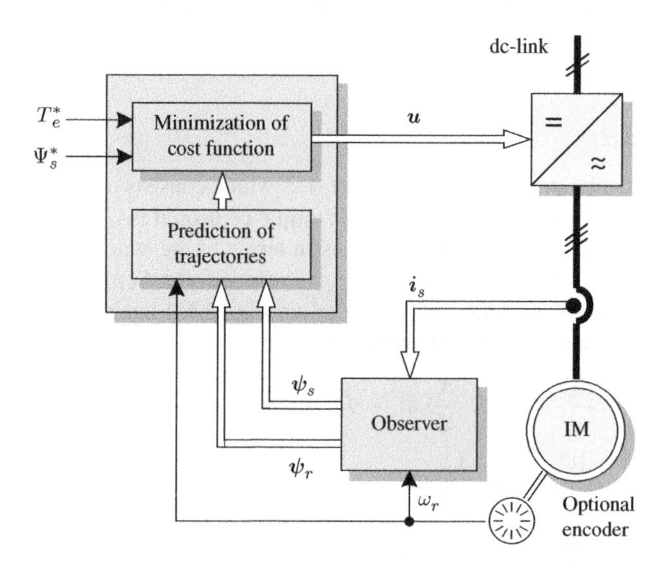

Figure 4.23 Predictive torque controller for the three-phase, three-level NPC inverter with an induction machine

simplified continuous-time state-space model

$$\frac{\mathrm{d}\boldsymbol{\psi}_s(t)}{\mathrm{d}t} = \boldsymbol{F}_1\boldsymbol{\psi}_s(t) + \boldsymbol{G}_1\boldsymbol{\psi}_r(t) + \boldsymbol{G}_2\boldsymbol{u}(t) \tag{4.28a}$$

$$\frac{\mathrm{d}\boldsymbol{\psi}_r(t)}{\mathrm{d}t} = \boldsymbol{F}_2\boldsymbol{\psi}_r(t) + \boldsymbol{G}_3\boldsymbol{\psi}_s(t) \tag{4.28b}$$

is obtained with the matrices

$$\boldsymbol{F}_1 = -R_s\frac{X_r}{D}\boldsymbol{I}_2,\ \boldsymbol{F}_2 = \omega_r\begin{bmatrix}0 & -1\\ 1 & 0\end{bmatrix} - R_r\frac{X_s}{D}\boldsymbol{I}_2, \tag{4.29a}$$

$$\boldsymbol{G}_1 = R_s\frac{X_m}{D}\boldsymbol{I}_2,\ \boldsymbol{G}_2 = \frac{v_{\mathrm{dc}}}{2}\tilde{\boldsymbol{K}},\ \boldsymbol{G}_3 = R_r\frac{X_m}{D}\boldsymbol{I}_2\,. \tag{4.29b}$$

Integrating (4.28) from $t = kT_s$ to $t = (k+1)T_s$ by applying the forward Euler approach, the discrete-time representation

$$\boldsymbol{\psi}_s(k+1) = \boldsymbol{A}_1\boldsymbol{\psi}_s(k) + \boldsymbol{B}_1\boldsymbol{\psi}_r(k) + \boldsymbol{B}_2\boldsymbol{u}(k) \tag{4.30a}$$

$$\boldsymbol{\psi}_r(k+1) = \boldsymbol{A}_2\boldsymbol{\psi}_r(k) + \boldsymbol{B}_3\boldsymbol{\psi}_s(k) \tag{4.30b}$$

results with the system matrices

$$\boldsymbol{A}_1 = \boldsymbol{I}_2 + \boldsymbol{F}_1T_s,\ \boldsymbol{A}_2 = \boldsymbol{I}_2 + \boldsymbol{F}_2T_s, \tag{4.31a}$$

$$\boldsymbol{B}_1 = \boldsymbol{G}_1T_s,\ \boldsymbol{B}_2 = \boldsymbol{G}_2T_s,\ \boldsymbol{B}_3 = \boldsymbol{G}_3T_s\,. \tag{4.31b}$$

4.3.4 Optimization Problem

The cost function

$$J = \lambda_T(T_e^*(k+1) - T_e(k+1))^2 + (1-\lambda_T)(\Psi_s^*(k+1) - \Psi_s(k+1))^2 + \lambda_u\|\Delta\boldsymbol{u}(k)\|_1 \tag{4.32}$$

consists of three terms. The first term penalizes the predicted deviation of the electromagnetic torque from its reference at time step $k+1$. Accordingly, the second term penalizes the predicted deviation of the stator flux magnitude from its reference. For both terms, the squared 2-norm is used, which is written as the sum of squares. The third term, which is adopted from the current controller, penalizes the switching effort at time step k, using the nonnegative scalar weight λ_u.

The weight λ_T is introduced to discount the torque ripple and to prioritize the flux magnitude ripple, without changing the balance between the cost on these two machine variables and the switching effort. In general, in order to obtain low current distortions, the stator flux ripple needs to be much smaller than the torque ripple, which can be achieved, for example, by setting λ_T to 0.1. The impact of λ_T on the current distortion will be analyzed in Sect. 4.3.8. To ensure that J is nonnegative, the weight λ_T is bounded between 0 and 1.

The optimization problem of the predictive torque controller with reference tracking is

$$\boldsymbol{u}_{\text{opt}}(k) = \arg \underset{\boldsymbol{u}(k)}{\text{minimize}}\ J \tag{4.33a}$$

$$\text{subject to}\ \boldsymbol{\psi}_s(k+1) = \boldsymbol{A}_1 \boldsymbol{\psi}_s(k) + \boldsymbol{B}_1 \boldsymbol{\psi}_r(k) + \boldsymbol{B}_2 \boldsymbol{u}(k) \tag{4.33b}$$

$$\boldsymbol{\psi}_r(k+1) = \boldsymbol{A}_2 \boldsymbol{\psi}_r(k) + \boldsymbol{B}_3 \boldsymbol{\psi}_s(k) \tag{4.33c}$$

$$T_e(k+1) = \frac{1}{\text{pf}} \frac{X_m}{D} \boldsymbol{\psi}_r(k+1) \times \boldsymbol{\psi}_s(k+1) \tag{4.33d}$$

$$\Psi_s(k+1) = \|\boldsymbol{\psi}_s(k+1)\| \tag{4.33e}$$

$$\boldsymbol{u}(k) \in \{-1, 0, 1\}^3, \quad \|\Delta \boldsymbol{u}(k)\|_\infty \leq 1 \tag{4.33f}$$

with the cost function as defined in (4.32).

One of the advantages of this formulation is that the torque and flux references are constant during steady-state operation. This is in contrast to the sinusoidally varying references that arise in the case of the predictive current controller. Hence we can usually assume that $T_e^*(k+1) = T_e^*(k)$ and $\Psi_s^*(k+1) = \Psi_s^*(k)$.

However, the predictive torque control formulation is computationally slightly more expensive than its current control counterpart. For the current controller, only one state-update equation with two components (see (4.22b)) needs to be computed. For the torque controller, the corresponding torque and stator flux magnitude also need to be calculated. Note that the second state-update equation (4.33c), which predicts the rotor flux vector at time step $k+1$, needs to be computed only once at time step k, because it is independent of $\boldsymbol{u}(k)$. This is due to the use of the forward Euler discretization method in (4.30). The exact Euler approach, however, would have led to a coupling—albeit a small one—between the rotor flux vector at time step $k+1$ and the switch position $\boldsymbol{u}(k)$.

4.3.5 Control Algorithm

This optimization problem (4.33) is solved and the optimal switch position at time step k, $\boldsymbol{u}_{\text{opt}}(k)$, is obtained through enumeration, by using a modified version of the predictive current control algorithm.

1. Given the previously applied switch position $\boldsymbol{u}(k-1)$ and taking into account the constraints on the switching transitions (4.33f), the set of admissible switch positions at time step $k, \boldsymbol{\mathcal{U}}(k)$, is determined.
2. The rotor flux vector at time step $k+1$, $\boldsymbol{\psi}_r(k+1)$, is computed using the state-update equation (4.33c).
3. For each of the switch positions $\boldsymbol{u}(k) \in \boldsymbol{\mathcal{U}}(k)$, the stator flux vector at time step $k+1$, $\boldsymbol{\psi}_s(k+1)$, is predicted using (4.33b). Based on this, the torque and stator flux magnitude at time step $k+1$ are predicted using (4.33d) and (4.33e), respectively.
4. For each switch position $\boldsymbol{u}(k) \in \boldsymbol{\mathcal{U}}(k)$, the cost J is computed according to (4.32).
5. The switch position $\boldsymbol{u}_{\text{opt}}(k)$ with the minimum cost is determined and applied to the inverter.

At the next time step, this procedure is repeated.

The predictive torque controller based on enumeration and a prediction horizon of one step was originally proposed for a two-level inverter in [5] with equal weights on the torque and flux magnitude error and no switching penalty. It has also been proposed to normalize the torque and flux error by the rated or nominal torque and flux. This normalization is implicitly accomplished in (4.32) by assuming that all quantities are provided in the pu system.

4.3.6 Analysis of the Cost Function

Before evaluating the closed-loop performance of the torque controller, it is beneficial to analyze its cost function (4.32) for the case of a prediction horizon of length 1. We start by rewriting (4.32) as

$$J = J_T + J_\Psi + \lambda_u \|\Delta \boldsymbol{u}(k)\|_1 \tag{4.34}$$

with the reference tracking error terms of the torque and stator flux magnitude

$$J_T = \lambda_T (T_e^* - T_e)^2 \tag{4.35a}$$

$$J_\Psi = (1 - \lambda_T)(\Psi_s^* - \Psi_s)^2 . \tag{4.35b}$$

We adopt the dq reference frame rotating in synchronism with the rotor flux. By aligning the rotor flux vector with the d-axis, the torque expression (4.26) can be simplified to

$$T_e = \frac{1}{\text{pf}} \frac{X_m}{D} \psi_{sq} \psi_{rd} . \tag{4.36}$$

With this, and recalling (4.27), the cost function terms in (4.35) can be expressed in terms of the d- and q-components of the stator flux vector as

$$J_T = \lambda_T \left(\frac{1}{\text{pf}} \frac{X_m}{D} \psi_{rd} \right)^2 (\psi_{sq}^* - \psi_{sq})^2 \tag{4.37a}$$

$$J_\Psi = (1 - \lambda_T)(\|\boldsymbol{\psi}_s^*\| - \|\boldsymbol{\psi}_s\|)^2 . \tag{4.37b}$$

The reference of the stator flux vector $\boldsymbol{\psi}_s^* = [\psi_{sd}^* \ \psi_{sq}^*]^T$ is obtained from (4.36) and (4.27) as

$$\psi_{sq}^* = \text{pf} \frac{D}{X_m} \frac{T_e^*}{\psi_{rd}} \tag{4.38}$$

and

$$\psi_{sd}^* = \sqrt{(\Psi_s^*)^2 - (\psi_{sq}^*)^2} . \tag{4.39}$$

Note that ψ_{rd} is equal to the magnitude of the rotor flux vector.

To visualize the cost function terms J_T and J_Ψ, consider again the MV drive system with the NPC inverter, as described in Sect. 4.2.6, operating at full speed and rated torque. A geometrical representation of the torque error term J_T is provided in Fig. 4.24(a). The rotor flux vector $\boldsymbol{\psi}_r$ is aligned with the d-axis. The reference of the stator flux vector $\boldsymbol{\psi}_s^*$ corresponds to nominal torque and a fully magnetized machine. The contour lines of the torque error term J_T with $\lambda_T = 0.052$ are shown as solid lines for the contour values $0.01, 0.02, \ldots, 0.08$.

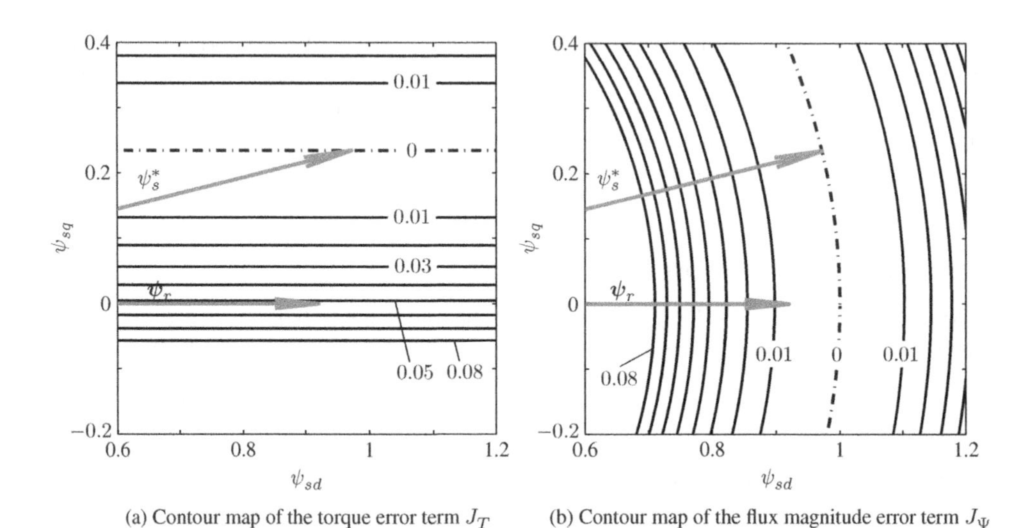

(a) Contour map of the torque error term J_T (b) Contour map of the flux magnitude error term J_Ψ

Figure 4.24 Geometrical representation of the cost function terms of the predictive torque controller in a plane spanned by the d- and q-components of the stator flux vector. The reference stator flux and the rotor flux vectors are shown by arrows. The references of the torque and stator flux magnitude are shown as dash-dotted lines in (a) and (b), respectively. The contour lines of the torque and flux magnitude error terms are solid lines

The dash-dotted line refers to $J_T = 0$. Owing to (4.36), the contour lines are straight lines that are parallel to the rotor flux vector.

Accordingly, the cost function term J_Ψ of the stator flux magnitude error is illustrated in Fig. 4.24(b). The contour lines of J_Ψ are depicted (as for the torque) for the values $0.01, 0.02, \ldots, 0.08$. These contour lines form concentric circles that are centered on the origin of the dq reference frame. The dash-dotted line refers to $J_\Psi = 0$. Adding the two cost functions terms to $J_T + J_\Psi$ leads to the contour map shown in Fig. 4.25(a).

4.3.7 Comparison of the Cost Functions for the Torque and Current Controllers

To provide further insight, we compare the contour plots of the predictive torque controller with those of the current controller for a prediction horizon of length 1. The control algorithm of the current controller was provided in Sect. 4.2.5. To this end, the current error term in the cost function (4.21),

$$J_I = \|\boldsymbol{i}_s^* - \boldsymbol{i}_s\|_2^2, \tag{4.40}$$

is rewritten in terms of the stator flux, by expressing the stator current as a linear combination of the stator and rotor flux vectors with d- and q-components. Using the upper row of (2.53), that is, $\boldsymbol{i}_s = \frac{1}{D}(X_r \boldsymbol{\psi}_s - X_m \boldsymbol{\psi}_r)$, the term

$$J_I = \left(\frac{X_r}{D}\right)^2 \|\boldsymbol{\psi}_s^* - \boldsymbol{\psi}_s\|_2^2 \tag{4.41}$$

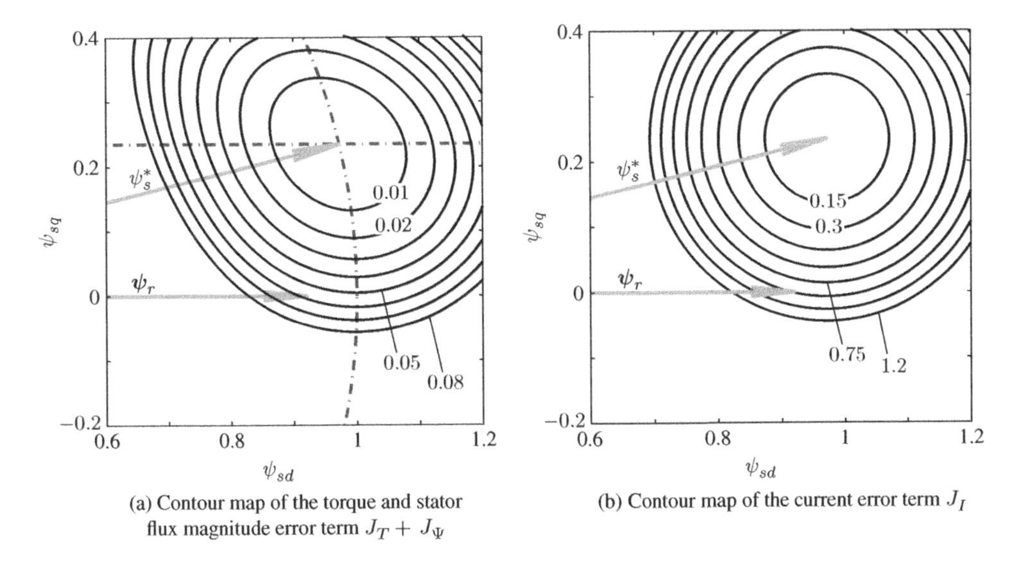

(a) Contour map of the torque and stator flux magnitude error term $J_T + J_\Psi$

(b) Contour map of the current error term J_I

Figure 4.25 Geometrical representations of the reference tracking error terms in the cost functions of the predictive torque and current controllers, in accordance with Fig. 4.24

is obtained. Note that J_I is independent of the rotor flux vector and its position. The contour lines of the current error term are shown in Fig. 4.25(b). These lines are concentric circles around the stator flux reference. Note that the contour lines are plotted for the values $0.15, 0.3, \ldots, 1.2$; that is, compared to the contour values used for the torque controller these values are multiplied by a factor of 15.

When comparing the reference tracking error term of the torque controller (4.37) with the current controller's term (4.41), it is obvious that the cost functions of the two controllers are not equivalent, despite the switching cost, $\lambda_u \|\Delta \boldsymbol{u}(k)\|_1$, being the same. This difference is illustrated by the different shapes of their corresponding contour lines in Fig. 4.25.

Nevertheless, by appropriately tuning the parameters in the cost function, a large degree of similarity between the two controllers can be achieved. Specifically, as shown in the following, λ_T can be chosen such that the contour lines of the torque controller approximate circles, particularly when the torque is close to zero.

To simplify the exposition in the following derivation, we set the torque reference to zero. Consider the stator flux vector

$$\boldsymbol{\psi}_s = \boldsymbol{\psi}_s^* + \begin{bmatrix} \psi_{\text{err}} \\ 0 \end{bmatrix} \tag{4.42}$$

with the flux error ψ_{err} in the d-axis. According to (4.37), the cost is

$$J_T + J_\Psi = (1 - \lambda_T)\psi_{\text{err}}^2 . \tag{4.43}$$

Similarly, for a stator flux vector with the flux error ψ_{err} in the q-axis, the cost is

$$J_T + J_\Psi = \lambda_T \left(\frac{1}{\text{pf}} \frac{X_m}{D} \psi_{rd} \right)^2 \psi_{\text{err}}^2 , \tag{4.44}$$

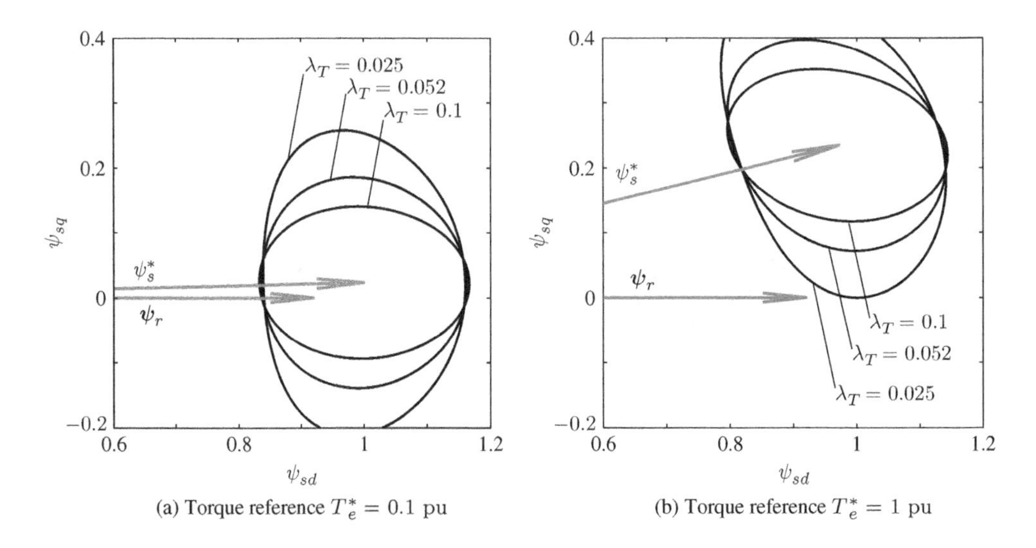

(a) Torque reference $T_e^* = 0.1$ pu

(b) Torque reference $T_e^* = 1$ pu

Figure 4.26 Tuning of the predictive torque controller using λ_T to achieve close-to-circular contour lines for the torque and stator flux magnitude error term $J_T + J_\Psi$

where we have neglected the minor contribution of J_Ψ. To achieve circular contour lines, both costs are required to be equal, which leads to

$$\lambda_T = \frac{(\text{pf}D)^2}{(\text{pf}D)^2 + (X_m \psi_{rd})^2}. \tag{4.45}$$

For the parameters of the drive system case study considered in this section, we obtain $\lambda_T = 0.052$.

The validity of this choice is confirmed by Fig. 4.26(a), which depicts for three different λ_T the contour lines with the same cost $J_T + J_\Psi = 0.025$. When the torque reference is close to zero, $\lambda_T = 0.052$ leads to an effectively circular contour line.

Variations in λ_T mostly affect the shape of the contour lines in the q-axis, which relates to the torque. Reducing λ_T, and hence the penalty on the torque error, widens the contour lines along the torque axis and increases the torque ripple. Conversely, when increasing λ_T and prioritizing the torque error, the torque ripple is reduced. In both cases, contour lines of elliptical shapes result. Note that variations of λ_T around 0.052 have only a minor effect on the contour lines along the d-axis, which relate to the stator flux magnitude and determine its ripple.

Increasing the torque reference from 0 to 1 pu distorts the contour lines along the circular reference of the stator flux magnitude, as can be seen in Fig. 4.26(b). In particular, the circular shape of the contour lines for $\lambda_T = 0.052$ becomes somewhat compromised. Nevertheless, as will be shown in the next section, the predictive torque and current control schemes provide similar performance results at all torque setpoints, provided that λ_T and the penalty on switching λ_u are appropriately chosen.

Tuning of the latter is required, because the diameters of the (almost circular) reference tracking contour lines of the two control schemes differ, as can be seen in Fig. 4.25. More specifically, errors in the stator flux vector are penalized more heavily for the current controller than for the torque controller. This implies that the switching penalty needs to be increased accordingly for the current controller to achieve the same switching frequency as the torque controller. Throughout this chapter, we have seen that the *ratio* between the cost values of the tracking error and the switching term determines the controller response. To achieve a similar closed-loop behavior for the torque and the current controllers, these ratios should be the same. We thus set

$$\frac{J_T + J_\Psi}{\lambda_{uT} \|\Delta \boldsymbol{u}(k)\|_1} = \frac{J_I}{\lambda_{uI} \|\Delta \boldsymbol{u}(k)\|_1}, \tag{4.46}$$

where we distinguish between the switching penalty of the torque controller, λ_{uT}, and the switching penalty of the current controller, λ_{uI}.

Consider again a zero torque reference, a stator flux error ψ_{err} in the d-axis, and zero flux error in the q-axis as in (4.42). Using (4.43) and (4.41), (4.46) can be simplified to

$$\frac{(1-\lambda_T)\psi_{\text{err}}^2}{\lambda_{uT}} = \left(\frac{X_r}{D}\right)^2 \frac{\psi_{\text{err}}^2}{\lambda_{uI}}. \tag{4.47}$$

This leads to

$$\lambda_{uI} = \left(\frac{X_r}{D}\right)^2 \frac{1}{1-\lambda_T} \lambda_{uT} = 16.25 \lambda_{uT}. \tag{4.48}$$

We conclude that both control schemes issue very similar switching commands when their penalties are selected according to the following rules:

- For the torque controller, set λ_T according to (4.45). Its penalty on switching, λ_{uT}, can be selected such that the desired switching frequency is achieved.
- For the current controller, scale its penalty on switching λ_{uI} according to (4.47).

As a result, the torque and current control schemes are expected to yield similar current and torque TDDs for a given switching frequency. This hypothesis will be substantiated in the next section through closed-loop simulations.

4.3.8 Performance Evaluation

For the performance evaluation of the predictive torque controller, we adopt the MV drive system case study that was previously considered in Sect. 4.2.6. The operating point is again at full speed, and the sampling interval is set to $T_s = 25$ µs. The penalty $\lambda_T = 0.052$ is chosen such that close-to-circular contour lines for the torque and flux error term are achieved, making the reference tracking error term in the cost function of the torque controller as similar as possible to that of the current controller. At rated torque, with the switching penalty $\lambda_{uT} = 0.198 \cdot 10^{-3}$, the predictive torque controller yields a current TDD of 7.74%, a torque TDD of 5.84%, and a device switching frequency of 221 Hz.

As shown in Table 4.5, these performance metrics are similar to those obtained by the predictive current controller when choosing a switching penalty of $\lambda_{uI} = 3 \cdot 10^{-3}$. The latter closely matches the design guideline (4.48) and achieves the same switching frequency as the torque controller with $\lambda_{uT} = 0.198 \cdot 10^{-3}$. At zero torque, both controllers achieve effectively

Table 4.5 Comparison of the predictive current and torque control schemes in terms of the current TDD I_{TDD} and torque TDD T_{TDD}

Torque reference	Control scheme	Controller settings	I_{TDD} (%)	T_{TDD} (%)	f_{sw} (Hz)
$T_e^* = 0$	Current	$\lambda_{uI} = 3 \cdot 10^{-3}$	6.38	5.57	220
$T_e^* = 0$	Torque	$\lambda_T = 0.052$, $\lambda_{uT} = 0.198 \cdot 10^{-3}$	6.45	5.76	219
$T_e^* = 1$	Current	$\lambda_{uI} = 3 \cdot 10^{-3}$	6.69	5.51	222
$T_e^* = 1$	Torque	$\lambda_T = 0.052$, $\lambda_{uT} = 0.198 \cdot 10^{-3}$	7.74	5.84	221

The switching penalties are chosen such that a switching frequency of approximately $f_{\text{sw}} = 220$ Hz results.

the same current and torque TDDs, while at rated torque the current TDD deteriorates by 16% when using the torque instead of the current controller. This worsening is due to the slightly noncircular shape of the contour lines for the stator flux error, which results in noncircular contour lines for the stator current error. The latter defines the current ripple and the current TDD. Nevertheless, at rated torque, the three-phase stator currents, electromagnetic torque, stator flux magnitude, switch positions, and stator current spectrum are similar to those of the current controller, which were shown in Fig. 4.13 and are not repeated here.

The similarity between the two control schemes is further underlined by Fig. 4.27, which shows the current and torque TDDs versus the switching frequency. At nominal speed and rated torque, and for switching frequencies in excess of 250 Hz, both schemes yield very similar current and torque TDDs for a given switching frequency, with the current controller slightly outperforming the torque controller. This small difference becomes more pronounced at low switching frequencies.

Figure 4.28(a) shows the influence λ_T has on the current TDD. The value of $\lambda_T = 0.052$ clearly achieves the lowest current TDD, confirming the cost function analysis provided in the previous section. For the torque TDD, however, the relatively small penalty of $\lambda_T = 0.052$

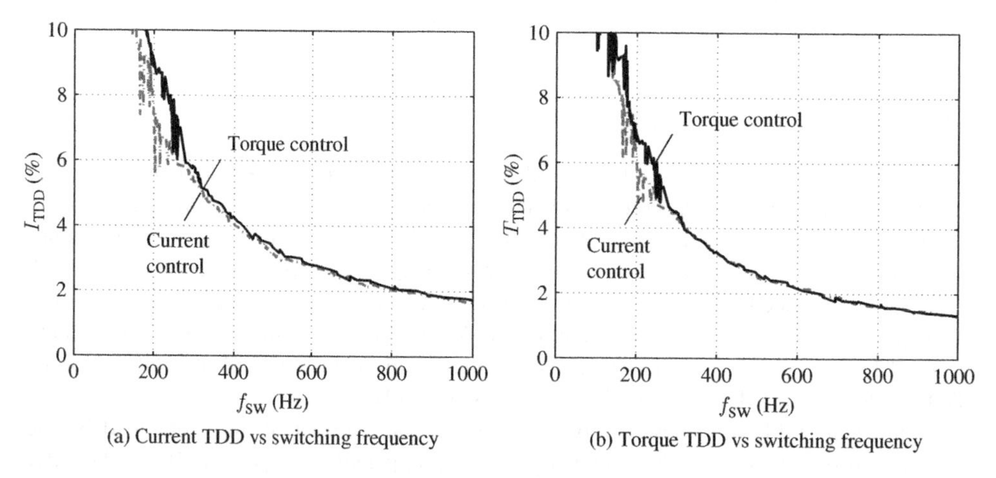

Figure 4.27 Trade-offs between the current and torque TDDs on one hand and the switching frequency on the other. Predictive torque control with $\lambda_T = 0.052$ is compared with predictive current control when operating at nominal speed and rated torque

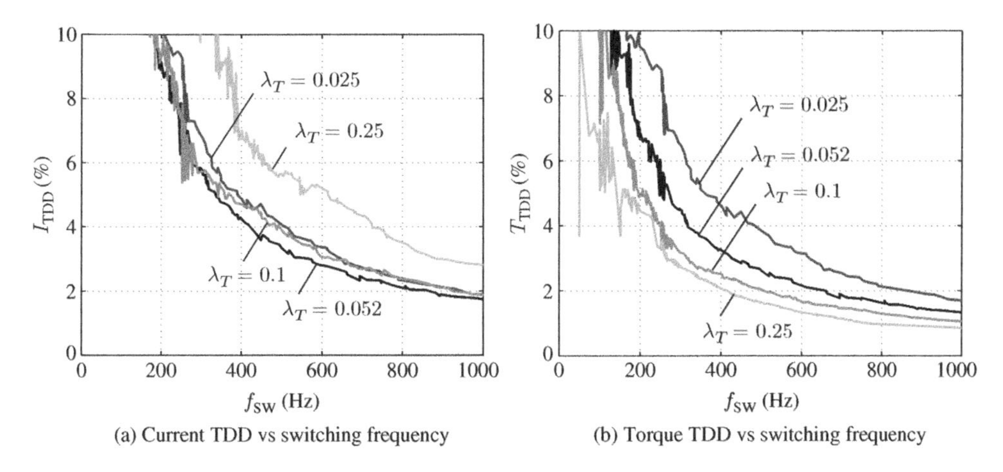

(a) Current TDD vs switching frequency

(b) Torque TDD vs switching frequency

Figure 4.28 Trade-offs between the current and torque TDDs on one hand and the switching frequency on the other for predictive torque control and different torque penalties λ_T, when operating at nominal speed and rated torque

leads to relatively large torque distortions, as shown in Fig. 4.28(b). Increasing the penalty fivefold to 0.25, for example, halves the torque TDD throughout the considered switching frequency range from 50 Hz to 1 kHz. This reduction in the torque TDD, however, comes at the price of pronounced current distortions (see Fig. 4.28(a)). Nevertheless, for some applications, very low torque TDDs might be beneficial. The parameter λ_T endows the torque controller with a degree of freedom to facilitate this.

When the predictive control schemes operate at low switching frequencies, the switching frequency tends to lock into integer multiples of the fundamental frequency, such as 50, 100, ..., 250 Hz, despite significant variations in the switching penalty. This phenomenon can be seen in Figs. 4.27 and 4.28, which leads to a certain degree of periodicity in the switching actions and a somewhat discrete current spectrum. This feature is more pronounced in the case of long prediction horizons, as shown and analyzed in Sect. 6.1.4.

We conclude that, in general, the predictive torque and current controllers yield similar performance metrics at steady-state operation, provided that the penalty λ_T is tuned appropriately. In particular, the influences of the switching penalty and the controller sampling interval T_s on the performance metrics are similar. The same applies to operation during transients, such as when current or torque reference steps are applied to the drive.

4.4 Summary

The notion of predictive controllers with horizon 1 and reference tracking was introduced in this section. Starting with a single-phase example in Sect. 4.1, the current of an RL load was regulated along its reference. In Sect. 4.2, the current controller was generalized to a three-phase drive system, and subsequently modified in Sect. 4.3 so as to track the references of the electromagnetic torque and stator flux magnitude of an electrical machine.

Owing to the versatility of the control concept, it can be applied to a wide range of power electronic topologies. Results have been reported for two-level inverters [2], three-level NPC inverters [3, 8], cascaded H-bridges [9–11], matrix converters [12], flying capacitor converters [13], and modular multilevel converters [14]. By including a corresponding term in the cost function, floating neutral point potentials in NPC inverters can be balanced around zero, as shown in [3]. On the grid side, active rectifiers can be controlled in a direct power control setting, as described in [15].

In light of the extensive literature, only a few references have been given. For further reading on the family of predictive controllers with horizon 1 and reference tracking, the reader is referred to the book [16] and to the survey papers [17–19], which include a detailed literature review.

References

[1] G. F. Franklin, J. D. Powell, and M. L. Workman, *Digital control of dynamic systems*. Addison-Wesley, 3rd ed., 1998.

[2] J. Rodríguez, J. Pontt, C. A. Silva, P. Correa, P. Lezana, P. Cortés, and U. Ammann, "Predictive current control of a voltage source inverter," *IEEE Trans. Ind. Electron.*, vol. 54, pp. 495–503, Feb. 2007.

[3] R. Vargas, P. Cortés, U. Ammann, J. Rodríguez, and J. Pontt, "Predictive control of a three-phase neutral-point-clamped inverter," *IEEE Trans. Ind. Electron.*, vol. 54, pp. 2697–2705, Oct. 2007.

[4] F. Kieferndorf, P. Karamanakos, P. Bader, N. Oikonomou, and T. Geyer, "Model predictive control of the internal voltages of a five-level active neutral point clamped converter," in *Proceedings of IEEE Energy Conversion Congress and Exposition* (Raleigh, NC, USA), Sep. 2012.

[5] H. Miranda, P. Cortés, J.-I. Yuz, and J. Rodríguez, "Predictive torque control of induction machines based on state-space models," *IEEE Trans. Ind. Electron.*, vol. 56, pp. 1916–1924, Jun. 2009.

[6] G. Papafotiou, J. Kley, K. G. Papadopoulos, P. Bohren, and M. Morari, "Model predictive direct torque control—Part II: Implementation and experimental evaluation," *IEEE Trans. Ind. Electron.*, vol. 56, pp. 1906–1915, Jun. 2009.

[7] P. Cortés, J. Rodríguez, C. Silva, and A. Flores, "Delay compensation in model predictive current control of a three-phase inverter," *IEEE Trans. Ind. Electron.*, vol. 59, pp. 1323–1325, Feb. 2012.

[8] G. Perantzakis, F. Xepapas, S. A. Papathanassiou, and S. Manias, "A predictive current control technique for three-level NPC voltage source inverters," in *Proceedings of IEEE Power Electronics Specialists Conference*, Jun. 2005.

[9] P. Correa, M. Pacas, and J. Rodríguez, "Predictive torque control for inverter-fed induction machine," *IEEE Trans. Ind. Electron.*, vol. 54, pp. 1073–1079, Apr. 2007.

[10] M. Pérez, P. Cortés, and J. Rodríguez., "Predictive control algorithm technique for multilevel asymmetric cascaded H-bridge inverters," *IEEE Trans. Ind. Electron.*, vol. 55, pp. 4354–4361, Dec. 2008.

[11] P. Cortés, A. Wilson, S. Kouro, J. Rodríguez, and H. Abu-Rub, "Model predictive control of multilevel cascaded H-bridge inverters," *IEEE Trans. Ind. Electron.*, vol. 57, pp. 2691–2699, Aug. 2010.

[12] S. Müller, U. Ammann, and S. Rees, "New time-discrete modulation scheme for matrix converters," *IEEE Trans. Ind. Electron.*, vol. 52, pp. 1607–1615, Dec. 2005.

[13] P. Lezana, R. P. Aguilera, and D. E. Quevedo, "Model predictive control of an asymmetric flying capacitor converter," *IEEE Trans. Ind. Electron.*, vol. 56, pp. 1839–1846, Jun. 2009.

[14] M. A. Pérez, J. Rodríguez, E. J. Fuentes, and F. Kammerer, "Predictive control of AC–AC modular multilevel converters," *IEEE Trans. Ind. Electron.*, vol. 59, pp. 2832–2839, Jul. 2012.

[15] P. Cortés, J. Rodríguez, P. Antoniewicz, and M. Kazmierkowski, "Direct power control of an AFE using predictive control," *IEEE Trans. Power Electron.*, vol. 23, pp. 2516–2523, Sep. 2008.

[16] J. Rodríguez and P. Cortés, *Predictive control of power converters and electrical drives*. Chichester, UK: John Wiley & Sons, Ltd, 2012.

[17] P. Cortés, M. P. Kazmierkowski, R. M. Kennel, D. E. Quevedo, and J. Rodríguez, "Predictive control in power electronics and drives," *IEEE Trans. Ind. Electron.*, vol. 55, pp. 4312–4324, Dec. 2008.

[18] S. Kouro, P. Cortés, R. Vargas, U. Ammann, and J. Rodríguez, "Model predictive control—a simple and powerful method to control power converters," *IEEE Trans. Ind. Electron.*, vol. 56, pp. 1826–1838, Jun. 2009.

[19] J. Rodríguez, M. P. Kazmierkowski, J. R. Espinoza, P. Zanchetta, H. Abu-Rub, H. A. Young, and C. A. Rojas, "State of the art of finite control set model predictive control in power electronics," *IEEE Trans. Ind. Inf.*, vol. 9, pp. 1003–1016, May 2013.

5

Predictive Control with Long Horizons

The optimization problem underlying direct model predictive control (MPC) with reference tracking is based on integer decision variables. This means that the number of possible solutions increases exponentially when extending the length of the prediction horizon. On one hand, integer decision variables facilitate the commonly used approach of enumerating all possible solutions (see Sect. 4.1.4 and the references therein). On the other hand, exhaustive enumeration quickly becomes computationally intractable when increasing the prediction horizon. This limitation is discussed in Sect. 4.1.6.

For direct MPC with reference tracking, the computational difficulty of solving the optimization problem has traditionally limited the length of the investigated prediction horizons to 1, except for a few attempts. These include [1], in which a horizon of 2 steps is used, and [2], in which a heuristic is proposed to reduce the number of switching sequences for longer horizons. In [3], a two-step prediction approach has been proposed; in a first step, the computational delay is compensated for, followed by the use of a standard predictive controller with a horizon of 1. The method in [3] can thus be considered to be equivalent to a one-step predictive controller.

We conclude that solving the optimization problem of direct MPC with long prediction horizons in an *efficient* manner has been—until recently—an unresolved problem. Consequently, the question of whether longer horizons lead to performance improvements or not remains largely unanswered.

This chapter examines the use of prediction horizons longer than 1 for direct MPC with reference tracking. For a linear system with a switched three-phase input vector with equal switching steps in all phases, the geometrical structure of the underlying optimization problem is exploited and an efficient optimization algorithm is derived. The algorithm uses elements of sphere decoding [4] to provide optimal switching sequences, requiring only modest computational resources. This enables the use of long prediction horizons in power electronics applications.

Model Predictive Control of High Power Converters and Industrial Drives, First Edition. Tobias Geyer.

Companion Website: www.wiley.com/go/geyermodelpredictivecontrol

5.1 Preliminaries

Direct MPC with reference tracking can be applied to general ac–dc, dc–dc, dc–ac, and ac–ac topologies with linear loads, including active front ends, inverters with RL loads, inverters with LC filters, and inverters with ac machines. Output quantities, such as voltages and currents, can be regulated along their time-varying references. We therefore opt to keep the problem formulation and the solution technique proposed in this chapter as general as possible.

5.1.1 Case Study

Nevertheless, we will often focus our description on the three-level inverter case study used frequently in this book. More specifically, we will consider the control problem of tracking the current reference of a medium-voltage (MV) variable speed drive system, which consists of a neutral-point-clamped (NPC) voltage source inverter (VSI) that drives an induction machine (IM). This case study is depicted in Fig. 5.1 and described in detail in Sect. 2.5.1 along with a summary of the parameters. To simplify the exposition, we assume the total dc-link voltage v_{dc} to be constant and the neutral point potential to be zero.

The mathematical model of the drive system was described in Sect. 4.2.1. As discussed there, the integer variables u_a, u_b, and $u_c \in \mathcal{U}$ denote the switch positions in the three phase legs. For a three-level inverter, the (single-phase) constraint set is given by

$$\mathcal{U} = \{-1, 0, 1\}. \tag{5.1}$$

The three switch positions are aggregated to the three-phase switch position $\boldsymbol{u} = [u_a \; u_b \; u_c]^T$.

The machine is modeled using the stationary $\alpha\beta$ reference frame. The stator currents $i_{s\alpha}$ and $i_{s\beta}$, and the rotor flux linkages $\psi_{r\alpha}$ and $\psi_{r\beta}$, are chosen as state variables. The rotor's angular velocity is treated as a (relatively slowly) varying parameter.

The objective of the current controller is to regulate the stator currents along their time-varying reference $\boldsymbol{i}_s^* = [i_{s\alpha}^* \; i_{s\beta}^*]^T$, by manipulating the switch positions $\boldsymbol{u}$, while minimizing the switching effort. Switching between 1 and -1 is prohibited. This control problem is described in detail in Sect. 4.2.2.

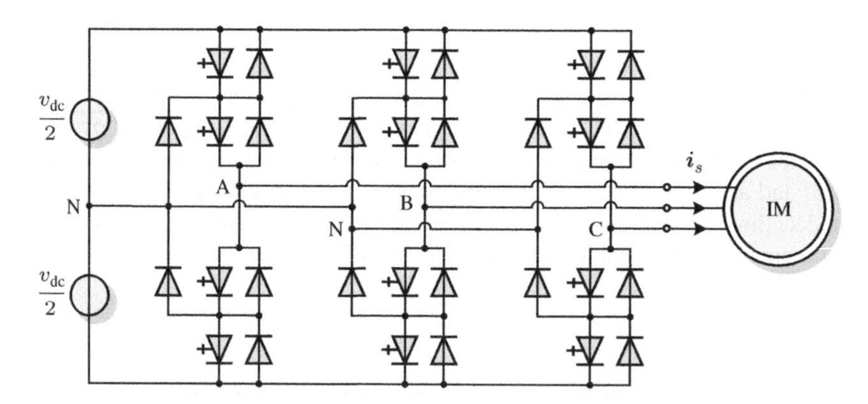

Figure 5.1 Three-level, three-phase, neutral-point-clamped voltage source inverter driving an induction machine with a fixed neutral point potential

5.1.2 Controller Model

For the derivation of the prediction model, it is convenient to introduce the state vector

$$\boldsymbol{x} = [i_{s\alpha}\ i_{s\beta}\ \psi_{r\alpha}\ \psi_{r\beta}]^T \tag{5.2}$$

of the drive model. The stator current is taken as the system output vector, that is,

$$\boldsymbol{y} = \boldsymbol{i}_s = [i_{s\alpha}\ i_{s\beta}]^T,$$

whereas the three-phase switch position $\boldsymbol{u}$ constitutes the input vector, which is provided by the controller.

Using the state vector $\boldsymbol{x}$, the machine model (4.11) described in Sect. 4.2.1 can be rewritten in state-space form as the continuous-time prediction model

$$\frac{\mathrm{d}\boldsymbol{x}(t)}{\mathrm{d}t} = \boldsymbol{F}\,\boldsymbol{x}(t) + \boldsymbol{G}\,\boldsymbol{u}(t), \tag{5.3a}$$

$$\boldsymbol{y}(t) = \boldsymbol{C}\,\boldsymbol{x}(t), \tag{5.3b}$$

where the matrices $\boldsymbol{F}$, $\boldsymbol{G}$, and $\boldsymbol{C}$ are provided in Appendix 5.A.[1]

By integrating (5.3a) from $t = kT_s$ to $t = (k+1)T_s$ and observing that $\boldsymbol{u}(t)$ is constant during this time interval and equal to $\boldsymbol{u}(k)$, we obtain the discrete-time representation

$$\boldsymbol{x}(k+1) = \boldsymbol{A}\,\boldsymbol{x}(k) + \boldsymbol{B}\,\boldsymbol{u}(k) \tag{5.4a}$$

$$\boldsymbol{y}(k) = \boldsymbol{C}\,\boldsymbol{x}(k) \tag{5.4b}$$

with $k \in \mathbb{N}$, where

$$\boldsymbol{A} = \boldsymbol{e}^{\boldsymbol{F}T_s} \text{ and } \boldsymbol{B} = \int_0^{T_s} \boldsymbol{e}^{\boldsymbol{F}\tau}\mathrm{d}\tau\boldsymbol{G}. \tag{5.5}$$

Note that $\boldsymbol{e}$ refers to the matrix exponential. For the derivation of the discrete-time matrices in (5.5), the reader is referred to [5, Sect. 4.3.3]. If $\boldsymbol{F}$ is nonsingular, the input matrix can be simplified to $\boldsymbol{B} = \boldsymbol{F}^{-1}(\boldsymbol{A} - \boldsymbol{I}_4)\boldsymbol{G}$, where $\boldsymbol{I}_4$ is the four-dimensional identity matrix.

If matrix exponentials were to pose computational difficulties, the forward Euler approximation is usually sufficiently accurate for short sampling intervals of up to several tens of microseconds and for short prediction horizons. In this case, the discrete-time system matrices are given by

$$\boldsymbol{A} = \boldsymbol{I}_4 + \boldsymbol{F}\,T_s \text{ and } \boldsymbol{B} = \boldsymbol{G}\,T_s. \tag{5.6}$$

5.1.3 Cost Function

The control problem of direct MPC with reference tracking over a finite prediction horizon of length N_p can be addressed through the minimization of the general cost function

$$J = \sum_{\ell=k}^{k+N_p-1} ||\boldsymbol{y}^*(\ell+1) - \boldsymbol{y}(\ell+1)||_{\boldsymbol{Q}}^2 + \lambda_u||\Delta\boldsymbol{u}(\ell)||_2^2. \tag{5.7}$$

[1] Note that $\boldsymbol{F}$ and $\boldsymbol{G}$ depend on the rotor speed ω_r and the dc-link voltage v_{dc}, respectively. Therefore, in a general setup, these two matrices need to be considered time-varying.

The first term in (5.7) penalizes the predicted tracking error, that is, the difference between the time-varying output reference $\boldsymbol{y}^*$ and the output vector $\boldsymbol{y}$. The tracking error is penalized at the future time steps $k+1, k+2, \dots, k+N_p$. For this, we use the compact representation

$$\|\boldsymbol{y}^* - \boldsymbol{y}\|_{\boldsymbol{Q}}^2 = (\boldsymbol{y}^* - \boldsymbol{y})^T \boldsymbol{Q} (\boldsymbol{y}^* - \boldsymbol{y}), \tag{5.8}$$

where $\boldsymbol{Q}$ denotes the penalty matrix on the tracking error. For technical reasons, we require $\boldsymbol{Q}$ to be positive semidefinite and symmetric, as will be explained in Sect. 5.1.4. $\boldsymbol{Q}$ is of the dimension $n_y \times n_y$, with n_y being the number of output variables, that is, $\boldsymbol{y} \in \mathbb{R}^{n_y}$.

To adapt this general formulation of the tracking error to the specific current control problem at hand, we set $\boldsymbol{y}^* = \boldsymbol{i}_s^*$, $\boldsymbol{y} = \boldsymbol{i}_s$ (as stated previously), and $\boldsymbol{Q} = \boldsymbol{I}_2$. The reference tracking problem in (5.7) is stated in stationary orthogonal $\alpha\beta$ coordinates. It can be easily shown (see Sect. 4.2.4) that the current reference tracking problem in the three-phase abc system is equivalent to the one formulated in the $\alpha\beta$ coordinate system.

The second term in (5.7) penalizes the switching effort

$$\Delta \boldsymbol{u}(k) = \boldsymbol{u}(k) - \boldsymbol{u}(k-1), \tag{5.9}$$

which refers to the switch positions in the three phases a, b, and c. The switching effort is penalized at the future time steps $k, k+1, \dots, k+N_p-1$. As switching is possible only by one step up or down in each phase, that is, we have $\|\Delta \boldsymbol{u}(k)\|_\infty \leq 1$, the 1-norm and the (squared) Euclidean norm of the switching effort yield the same cost:

$$\|\Delta \boldsymbol{u}(k)\|_1 = \|\Delta \boldsymbol{u}(k)\|_2^2.$$

The parameter $\lambda_u > 0$ in (5.7) is a tuning parameter that adjusts the trade-off between the tracking accuracy (deviation of the output from its reference) and the switching effort.

5.1.4 *Optimization Problem*

We introduce the switching sequence

$$\boldsymbol{U}(k) = [\boldsymbol{u}^T(k)\ \boldsymbol{u}^T(k+1)\ \dots\ \boldsymbol{u}^T(k+N_p-1)]^T,$$

which represents the sequence of inverter switch positions the controller has to decide upon. The optimization problem underlying direct MPC with reference tracking can then be stated as

$$\boldsymbol{U}_{\text{opt}}(k) = \arg \underset{\boldsymbol{U}(k)}{\text{minimize}}\ J \tag{5.10a}$$

$$\text{subject to}\ \boldsymbol{x}(\ell+1) = \boldsymbol{A}\,\boldsymbol{x}(\ell) + \boldsymbol{B}\,\boldsymbol{u}(\ell) \tag{5.10b}$$

$$\boldsymbol{y}(\ell+1) = \boldsymbol{C}\,\boldsymbol{x}(\ell+1) \tag{5.10c}$$

$$\Delta \boldsymbol{u}(\ell) = \boldsymbol{u}(\ell) - \boldsymbol{u}(\ell-1) \tag{5.10d}$$

$$\boldsymbol{U}(k) \in \mathbb{U} \tag{5.10e}$$

$$\|\Delta \boldsymbol{u}(\ell)\|_\infty \leq 1,\ \ \forall \ell = k, \dots, k+N_p-1. \tag{5.10f}$$

The cost function J depends on the state vector $\boldsymbol{x}(k)$, the previously chosen switch position $\boldsymbol{u}(k-1)$, and the switching sequence $\boldsymbol{U}(k)$. In (5.10e), $\mathbb{U} = \boldsymbol{\mathcal{U}} \times \cdots \times \boldsymbol{\mathcal{U}}$ is the N_p-times Cartesian product of the set $\boldsymbol{\mathcal{U}}$, where $\boldsymbol{\mathcal{U}}$ denotes the set of discrete three-phase switch positions. The latter is obtained from the single-phase set $\mathcal{U}$ via $\boldsymbol{\mathcal{U}} = \mathcal{U} \times \mathcal{U} \times \mathcal{U}$, as defined in (5.1). We refer to (5.10f) as switching constraints.

Following the principle of receding horizon control, only the first element of the optimal switching sequence $\boldsymbol{U}_{\text{opt}}(k)$ is applied to the semiconductor switches at time step k. At the next time step $k+1$, and given new information on $\boldsymbol{x}(k+1)$ and the output references, another optimization is performed, which provides the optimal switch positions at time step $k+1$. The optimization is repeated online and *ad infinitum*, as exemplified in Fig. 5.2.

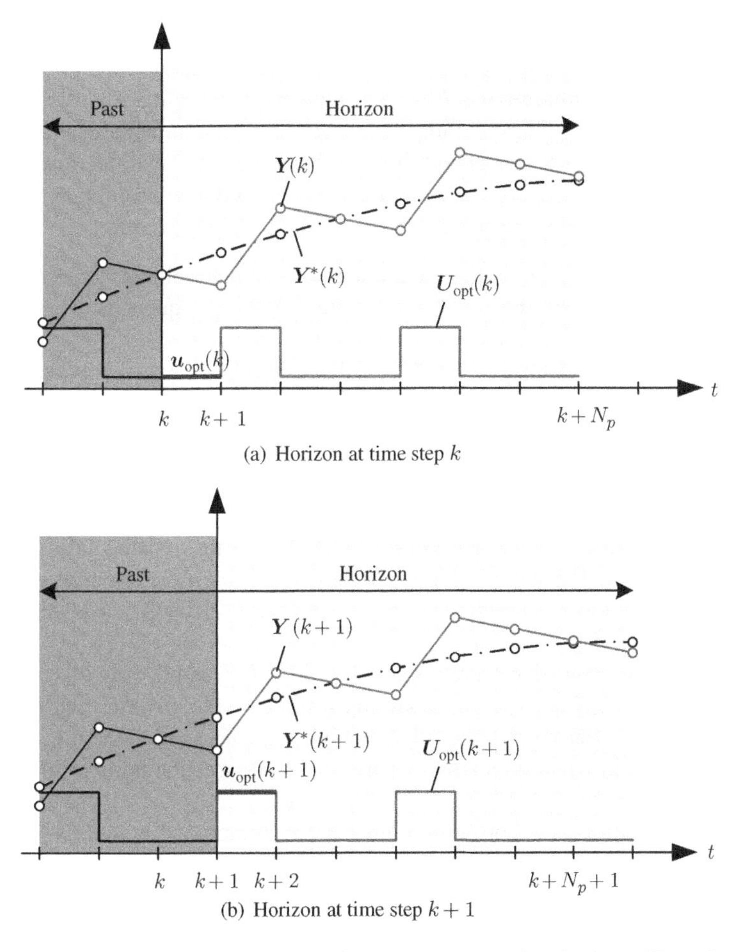

Figure 5.2 Receding horizon policy exemplified for the prediction horizon $N_p = 6$. The optimal switching sequence $\boldsymbol{U}_{\text{opt}}$ is chosen such that the predicted output trajectory $\boldsymbol{Y}$ tracks the output reference trajectory $\boldsymbol{Y}^*$. Out of the switching sequence $\boldsymbol{U}_{\text{opt}}$, only the first element $\boldsymbol{u}_{\text{opt}}$ is applied to the inverter

In this figure, $\boldsymbol{Y}(k)$ denotes the output trajectory over the prediction horizon from time step $k+1$ to $k+N_p$, that is

$$\boldsymbol{Y}(k) = [\boldsymbol{y}^T(k+1)\ \boldsymbol{y}^T(k+2)\ \ldots\ \boldsymbol{y}^T(k+N_p)]^T. \tag{5.11}$$

Correspondingly, $\boldsymbol{Y}^*(k)$ denotes the output reference trajectory.

5.1.5 Control Algorithm based on Exhaustive Search

Owing to the discrete nature of the decision variable $\boldsymbol{U}(k)$, the optimization problem (5.10) is difficult to solve except for problems with very short horizons. In fact, as the prediction horizon is enlarged and the number of decision variables is increased, the (worst case) computational complexity grows exponentially, thus it cannot be bounded by a polynomial. The difficulties associated with minimizing J become apparent when using exhaustive search. With this method, the set of admissible switching sequences $\boldsymbol{U}(k)$ is enumerated, and the cost function is evaluated for each such sequence. The switching sequence with the lowest cost is (by definition) the optimal one, and its first element is chosen as the control input.

At every time step k, exhaustive search entails the following procedure:

1. Given the previously applied switch position $\boldsymbol{u}(k-1)$, and taking into account the constraints (5.10e) and (5.10f), the set of admissible switching sequences over the prediction horizon is determined.
2. For each of these switching sequences $\boldsymbol{U}(k)$, the state trajectory is computed according to (5.10b) and the evolution of the output vector (5.10c) is predicted.
3. For each switching sequence, the cost J is computed according to (5.7).
4. The switching sequence $\boldsymbol{U}_{\text{opt}}(k)$ with the minimum cost is determined, and its first element, $\boldsymbol{u}_{\text{opt}}(k)$, is applied to the converter.

At the next time step $k+1$, this procedure is repeated using updated information on the state vector $\boldsymbol{x}(k+1)$ and the output reference trajectory $\boldsymbol{Y}^*(k+1)$. The corresponding algorithm for the current reference tracking problem with the horizon $N_p = 1$ was described in Sect. 4.2.5.

It is easy to see that exhaustive search is computationally feasible only for very short horizons N_p, such as 1 or 2. For $N_p = 5$, assuming a three-level converter and neglecting the switching constraint (5.10f), the number of switching sequences amounts to $1.4 \cdot 10^7$. This is clearly impractical, even when imposing (5.10f), which reduces the number of sequences by an order of magnitude.

Techniques from mathematical programming, such as branch-and-bound [6–8], can be used to reduce the computational burden of solving (5.10). In particular, off-the-shelf solvers such as CPLEX [9] include a wealth of smart heuristics and tailored optimization methods. However, none of the general methods takes advantage of the particular structure of the optimization problem (5.10) and the fact that in MPC the solution is implemented in a receding horizon manner. It will be shown in Sect. 5.3 how these distinguishing features of the problem at hand can be exploited to greatly reduce the computational burden, thereby enabling the use of long horizons in power electronics applications.

5.2 Integer Quadratic Programming Formulation

In this section, we reformulate the optimization problem (5.10) in vector form and state it as an integer quadratic program. The integer search space is truncated because of the finite number of inverter voltage levels.

5.2.1 Optimization Problem in Vector Form

The dynamic evolution of the prediction model (5.10b) and (5.10c) can be included in the cost function (5.7). After lengthy algebraic manipulations, which are provided in Appendix 5.B, the cost function can be written in the compact form

$$J = (\boldsymbol{U}(k))^T \boldsymbol{H} \boldsymbol{U}(k) + 2(\boldsymbol{\Theta}(k))^T \boldsymbol{U}(k) + \theta(k) \tag{5.12}$$

with

$$\boldsymbol{H} = \boldsymbol{\Upsilon}^T \tilde{\boldsymbol{Q}} \boldsymbol{\Upsilon} + \lambda_u \boldsymbol{S}^T \boldsymbol{S} \tag{5.13a}$$

$$(\boldsymbol{\Theta}(k))^T = -(\boldsymbol{Y}^*(k) - \boldsymbol{\Gamma} \boldsymbol{x}(k))^T \tilde{\boldsymbol{Q}} \boldsymbol{\Upsilon} - \lambda_u (\boldsymbol{E} \boldsymbol{u}(k-1))^T \boldsymbol{S} \tag{5.13b}$$

$$\theta(k) = \|\boldsymbol{Y}^*(k) - \boldsymbol{\Gamma} \boldsymbol{x}(k)\|_{\tilde{\boldsymbol{Q}}}^2 + \lambda_u \|\boldsymbol{E} \boldsymbol{u}(k-1)\|_2^2. \tag{5.13c}$$

The matrices $\boldsymbol{\Upsilon}$, $\boldsymbol{\Gamma}$, $\boldsymbol{S}$, and $\boldsymbol{E}$ are defined in Appendix 5.B. $\boldsymbol{Y}^*(k)$ denotes the output reference trajectory. The block diagonal penalty matrix on the tracking error is defined as $\tilde{\boldsymbol{Q}} = \text{diag}(\boldsymbol{Q}, \ldots, \boldsymbol{Q})$.

The cost function (5.12) consists of three terms. The first term is quadratic in the switching sequence $\boldsymbol{U}(k)$. The Hessian[2] matrix $\boldsymbol{H}$ is a function of the system matrices $\boldsymbol{A}$, $\boldsymbol{B}$, and $\boldsymbol{C}$, the penalty matrix $\tilde{\boldsymbol{Q}}$, the penalty λ_u on switching transitions, and the matrix $\boldsymbol{S}$. Provided that the system parameters are time-invariant, the Hessian is time-invariant, too. The Hessian is symmetric and positive definite for $\lambda_u > 0$. This is due to the fact that $\boldsymbol{S}^T \boldsymbol{S}$ in (5.13a) is positive definite and $\boldsymbol{\Upsilon}^T \tilde{\boldsymbol{Q}} \boldsymbol{\Upsilon}$ is positive semidefinite, because we require $\boldsymbol{Q}$ to be positive semidefinite. For $\lambda_u \geq 0$, the Hessian would be positive semidefinite. For the definition of positive (semi)definite matrices, the reader is referred to Sect. 3.8.

The second term in (5.12) is linear in the switching sequence $\boldsymbol{U}(k)$. The time-varying vector $\boldsymbol{\Theta}(k)$ is a function of the state vector at time step k, the output reference trajectory $\boldsymbol{Y}^*(k)$, and the previously chosen switch position $\boldsymbol{u}(k-1)$. The third term in (5.12) is a time-varying scalar that has the same arguments as $\boldsymbol{\Theta}(k)$.

By completing the squares, (5.12) can be rewritten as

$$J = (\boldsymbol{U}(k) + \boldsymbol{H}^{-1} \boldsymbol{\Theta}(k))^T \boldsymbol{H} (\boldsymbol{U}(k) + \boldsymbol{H}^{-1} \boldsymbol{\Theta}(k)) + \text{const}(k). \tag{5.14}$$

[2] Strictly speaking, $2\boldsymbol{H}$ is the Hessian matrix according to the commonly used definition (see also (3.94)).

The constant term in (5.14) is independent of $U(k)$ and thus does not influence the optimal solution. This allows us to omit the constant term from the cost function and to state the reformulated optimization problem as

$$U_{\text{opt}}(k) = \arg \underset{U(k)}{\text{minimize}}\, (U(k) + H^{-1}\Theta(k))^T H(U(k) + H^{-1}\Theta(k)) \tag{5.15a}$$

$$\text{subject to}\;\; U(k) \in \mathbb{U} \tag{5.15b}$$

$$\|\Delta u(\ell)\|_\infty \leq 1,\;\; \forall \ell = k, \ldots, k + N_p - 1. \tag{5.15c}$$

5.2.2 Solution in Terms of the Unconstrained Minimum

The *unconstrained* optimum of (5.15) is obtained by minimization, *neglecting* the constraints (5.15b) and (5.15c), thus allowing $U(k) \in \mathbb{R}^{3N_p}$. As H is positive definite, it follows directly from (5.15a) that the unconstrained solution at time step k is unique and given by

$$U_{\text{unc}}(k) = -H^{-1}\Theta(k). \tag{5.16}$$

For more details on the unconstrained minimum of a quadratic cost function, the reader is referred to Sect. 3.8.

As the first element of the unconstrained switching sequence $U_{\text{unc}}(k)$ does not meet the constraints (5.15b) and (5.15c), it cannot be directly used as gating signals to the semiconductor switches. However, $U_{\text{unc}}(k)$ can be used to state the solution to the *constrained* optimization problem (5.15)—including the constraints (5.15b) and (5.15c)—as shown next.

The cost function (5.15a) can be rewritten by inserting (5.16) as follows:

$$J = (U(k) - U_{\text{unc}}(k))^T H(U(k) - U_{\text{unc}}(k)). \tag{5.17}$$

As H is (by definition) symmetric and positive definite for $\lambda_u > 0$, a unique *invertible* and *lower triangular* matrix $V \in \mathbb{R}^{3N_p \times 3N_p}$ exists, which satisfies

$$V^T V = H. \tag{5.18}$$

The matrix V is the so-called *generator* matrix. It can be calculated by noting that its inverse V^{-1} is also lower triangular and is provided by the following Cholesky decomposition of H^{-1} (see, e.g., [10]):

$$V^{-1}V^{-T} = H^{-1}. \tag{5.19}$$

In terms of V and

$$\bar{U}_{\text{unc}}(k) = VU_{\text{unc}}(k), \tag{5.20}$$

the cost in (5.17) can be written as

$$J = (VU(k) - \bar{U}_{\text{unc}}(k))^T (VU(k) - \bar{U}_{\text{unc}}(k)). \tag{5.21}$$

5.2.3 Integer Quadratic Program

The optimization problem (5.15) underlying direct MPC with output reference tracking can now be stated as a (truncated) *integer quadratic* program. The optimal switching sequence

$\boldsymbol{U}_{\text{opt}}(k)$ is obtained by minimizing the cost function (5.21) subject to (5.15b) and (5.15c), that is,

$$\boldsymbol{U}_{\text{opt}}(k) = \arg \underset{\boldsymbol{U}(k)}{\text{minimize}} \, \|\boldsymbol{V}\boldsymbol{U}(k) - \bar{\boldsymbol{U}}_{\text{unc}}(k)\|_2^2 \tag{5.22a}$$

$$\text{subject to } \boldsymbol{U}(k) \in \mathbb{U} \tag{5.22b}$$

$$\|\Delta \boldsymbol{u}(\ell)\|_\infty \leq 1, \; \forall \ell = k, \dots, k + N_p - 1 \tag{5.22c}$$

with

$$\tilde{\boldsymbol{Q}} = \text{diag}(\boldsymbol{Q}, \dots, \boldsymbol{Q}) \tag{5.23a}$$

$$\boldsymbol{H} = \boldsymbol{\Upsilon}^T \tilde{\boldsymbol{Q}} \boldsymbol{\Upsilon} + \lambda_u \boldsymbol{S}^T \boldsymbol{S} \tag{5.23b}$$

$$\boldsymbol{V}^T \boldsymbol{V} = \boldsymbol{H} \tag{5.23c}$$

$$(\boldsymbol{\Theta}(k))^T = -(\boldsymbol{Y}^*(k) - \boldsymbol{\Gamma}\boldsymbol{x}(k))^T \tilde{\boldsymbol{Q}} \boldsymbol{\Upsilon} - \lambda_u (\boldsymbol{E}\boldsymbol{u}(k-1))^T \boldsymbol{S} \tag{5.23d}$$

$$\bar{\boldsymbol{U}}_{\text{unc}}(k) = -\boldsymbol{V}\boldsymbol{H}^{-1}\boldsymbol{\Theta}(k). \tag{5.23e}$$

To obtain (5.23e), we have inserted (5.16) in (5.20).

Recall that we require $\boldsymbol{Q}$ to be positive semidefinite and symmetric. Therefore, $\tilde{\boldsymbol{Q}}$ is also positive semidefinite and symmetric. As a result, the Hessian $\boldsymbol{H}$ is positive definite and symmetric, provided that λ_u is positive. The generator matrix $\boldsymbol{V}$ is lower triangular.

The dimensions of the matrices and vectors in (5.23) are summarized here. Let n_x (n_y) denote the number of state (output) variables, and assume the use of a three-phase system with three switch positions. It follows that the dimension of $\boldsymbol{Q}$ is $n_y \times n_y$ and $\tilde{\boldsymbol{Q}}$ and is of dimension $n_y N_p \times n_y N_p$. The Hessian $\boldsymbol{H}$ and the generator matrix $\boldsymbol{V}$ are of dimension $3N_p \times 3N_p$. The vector $\boldsymbol{\Theta}(k)$ has the dimension $3N_p \times 1$. The output trajectory $\boldsymbol{Y}(k)$ and its reference $\boldsymbol{Y}^*(k)$ are of dimension $n_y N_p \times 1$. The switching sequence $\boldsymbol{U}(k)$ and the transformed unconstrained solution $\bar{\boldsymbol{U}}_{\text{unc}}(k)$ are of dimension $3N_p \times 1$.

The matrices $\boldsymbol{\Gamma}$, $\boldsymbol{\Upsilon}$, $\boldsymbol{S}$, and $\boldsymbol{E}$, which are required in (5.23b) and (5.23c), are provided in Appendix 5.B. These matrices are of dimensions $n_y N_p \times n_x$, $n_y N_p \times 3N_p$, $3N_p \times 3N_p$, and $3N_p \times 3$, respectively.

In recent years, various efficient solution algorithms for (5.22a) subject to (5.22b)—but not taking into account (5.22c)—have been developed (see, e.g., [4, 11] and the references therein). In Sect. 5.3, we will tailor one such algorithm to the optimization problem of interest.

5.2.4 Direct MPC with a Prediction Horizon of 1

Next, we focus on the particular case where the prediction horizon is taken to be equal to 1. As in virtually all the literature on direct MPC with reference tracking a prediction horizon of 1 is considered, this case is of particular importance and deserves some additional attention. The low dimensionality of the problem at hand also allows for an intuitively accessible visualization.

Setting the horizon N_p to 1, we have $\boldsymbol{U}(k) = \boldsymbol{u}(k)$, $\boldsymbol{\Gamma} = \boldsymbol{C}\boldsymbol{A}$, $\boldsymbol{\Upsilon} = \boldsymbol{C}\boldsymbol{B}$, $\boldsymbol{S} = \boldsymbol{E} = \boldsymbol{I}_3$, and $\tilde{\boldsymbol{Q}} = \boldsymbol{Q}$. This simplifies the integer quadratic program (5.22) to

$$\boldsymbol{u}_{\text{opt}}(k) = \arg \underset{\boldsymbol{U}(k)}{\text{minimize}} \, \|\boldsymbol{V}\boldsymbol{u}(k) - \bar{\boldsymbol{u}}_{\text{unc}}(k)\|_2^2 \tag{5.24a}$$

$$\text{subject to } \boldsymbol{u}(k) \in \boldsymbol{\mathcal{U}} \tag{5.24b}$$

$$\|\Delta \boldsymbol{u}(k)\|_\infty \leq 1 \tag{5.24c}$$

with

$$\boldsymbol{H} = (\boldsymbol{C}\boldsymbol{B})^T \boldsymbol{Q}\boldsymbol{C}\boldsymbol{B} + \lambda_u \boldsymbol{I}_3 \tag{5.25a}$$

$$\boldsymbol{V}^T\boldsymbol{V} = \boldsymbol{H} \tag{5.25b}$$

$$(\boldsymbol{\Theta}(k))^T = -(\boldsymbol{y}^*(k+1) - \boldsymbol{C}\boldsymbol{A}\boldsymbol{x}(k))^T \boldsymbol{Q}\boldsymbol{C}\boldsymbol{B} - \lambda_u(\boldsymbol{u}(k-1))^T \tag{5.25c}$$

$$\bar{\boldsymbol{u}}_{\text{unc}}(k) = -\boldsymbol{V}\boldsymbol{H}^{-1}\boldsymbol{\Theta}(k). \tag{5.25d}$$

To further illustrate this case, we consider the drive system case study in Sect. 5.1.1 with current reference tracking and set $\boldsymbol{Q} = \boldsymbol{I}_2$. It is convenient to use the forward Euler approximation (5.6) for the prediction model to obtain

$$\boldsymbol{C}\boldsymbol{B} = \frac{v_{\text{dc}} X_r}{3D} T_s \begin{bmatrix} 1 & -\frac{1}{2} & -\frac{1}{2} \\ 0 & \frac{\sqrt{3}}{2} & -\frac{\sqrt{3}}{2} \end{bmatrix}, \tag{5.26}$$

so that

$$\boldsymbol{H} = \left(\frac{v_{\text{dc}} X_r T_s}{3D}\right)^2 \begin{bmatrix} 1 & -\frac{1}{2} & -\frac{1}{2} \\ -\frac{1}{2} & 1 & -\frac{1}{2} \\ -\frac{1}{2} & -\frac{1}{2} & 1 \end{bmatrix} + \lambda_u \begin{bmatrix} 1 & 0 & 0 \\ 0 & 1 & 0 \\ 0 & 0 & 1 \end{bmatrix}. \tag{5.27}$$

As in the $N_p > 1$ case, $\boldsymbol{H}$ is always symmetric and positive definite for $\lambda_u > 0$.

If the design parameter λ_u is chosen to be much larger than $(v_{\text{dc}} X_r T_s/3D)^2$, then the diagonal terms of $\boldsymbol{H}$ in (5.27) become dominant, that is, $\boldsymbol{H} \approx \lambda_u \boldsymbol{I}_3$. This turns $\boldsymbol{V}$ effectively into a diagonal matrix with $\boldsymbol{V} \approx \sqrt{\lambda_u}\boldsymbol{I}_3$ (see (5.18)). As a result, for sufficiently large values of λ_u, direct component-wise rounding of $\bar{\boldsymbol{u}}_{\text{unc}}(k)$ to the integer set will often give the optimal solution (see also [12]).

On the other hand, if λ_u is much smaller than $(v_{\text{dc}} X_r T_s/3D)^2$, then

$$\boldsymbol{V} \approx \frac{v_{\text{dc}} X_r T_s}{3D} \begin{bmatrix} 0 & 0 & 0 \\ -\frac{\sqrt{3}}{2} & \frac{\sqrt{3}}{2} & 0 \\ -\frac{1}{2} & -\frac{1}{2} & 1 \end{bmatrix}. \tag{5.28}$$

In particular, for $\lambda_u \approx 0$, direct component-wise rounding of $\bar{\boldsymbol{u}}_{\text{unc}}(k)$ will—in general—provide only suboptimal results. This conclusion stands in contrast to the proposition made in [13]. In Sect. 6.2, we will evaluate direct component-wise rounding for horizons larger than 1.

5.3 An Efficient Method for Solving the Optimization Problem

In this section, we will show how to adapt the sphere decoding algorithm [4, 14] to the task of finding the optimal switching sequence $\boldsymbol{U}_{\text{opt}}(k)$. The algorithm is based on branch-and-bound techniques and is—as will be illustrated in Sect. 5.4—by far more efficient than the exhaustive enumeration method described in Sect. 5.1.5.

5.3.1 Preliminaries and Key Properties

The basic idea of the algorithm is to iteratively consider candidate sequences, say $\boldsymbol{U}(k) \in \mathbb{U}$, that belong to a sphere of radius $\rho(k) > 0$ centered on $\bar{\boldsymbol{U}}_{\text{unc}}(k)$

$$\|\bar{\boldsymbol{U}}_{\text{unc}}(k) - \boldsymbol{V}\boldsymbol{U}(k)\|_2 \leq \rho(k), \tag{5.29}$$

and satisfy the switching constraint (5.22c).

A key property used in sphere decoding is that, because $\boldsymbol{V}$ is triangular, identifying candidate sequences that satisfy (5.29) is very simple. In our case, $\boldsymbol{V}$ is lower triangular, and (5.29) can be rewritten as

$$\rho^2(k) \geq (\bar{u}_{\text{unc},1}(k) - v_{(1,1)}u_1(k))^2 + (\bar{u}_{\text{unc},2}(k) - v_{(2,1)}u_1(k) - v_{(2,2)}u_2(k))^2 + \cdots \tag{5.30}$$

where $\bar{u}_{\text{unc},i}(k)$ denotes the ith element of $\bar{\boldsymbol{U}}_{\text{unc}}(k)$, $u_i(k)$ is the ith element of $\boldsymbol{U}(k)$, and $v_{(i,j)}$ refers to the (i,j)th entry of $\boldsymbol{V}$. Therefore, the solution set of (5.29) can be found by proceeding in a sequential manner somewhat akin to Gaussian elimination, in the sense that at each step only a one-dimensional problem needs to be solved (for details, see [4]).

To determine $\boldsymbol{U}(k)$, the algorithm requires an initial value for the radius used at time step k. On one hand, the radius $\rho(k)$ should be as small as possible, enabling us to remove as many candidate switching sequences *a priori* as possible. On the other hand, $\rho(k)$ must not be too small, to ensure that the solution set is not empty. We propose to choose the initial radius based on the following *educated guess*:

$$\boldsymbol{U}_{\text{ini}}(k) = \begin{bmatrix} \boldsymbol{0}_{3\times3} & \boldsymbol{I}_3 & \boldsymbol{0}_{3\times3} & \cdots & \boldsymbol{0}_{3\times3} \\ \boldsymbol{0}_{3\times3} & \boldsymbol{0}_{3\times3} & \boldsymbol{I}_3 & \ddots & \vdots \\ \vdots & & \ddots & \ddots & \boldsymbol{0}_{3\times3} \\ \boldsymbol{0}_{3\times3} & \cdots & \cdots & \boldsymbol{0}_{3\times3} & \boldsymbol{I}_3 \\ \boldsymbol{0}_{3\times3} & \cdots & \cdots & \boldsymbol{0}_{3\times3} & \boldsymbol{I}_3 \end{bmatrix} \boldsymbol{U}_{\text{opt}}(k-1). \tag{5.31}$$

The guessed switching sequence $\boldsymbol{U}_{\text{ini}}(k)$ is obtained by shifting the previous optimal switching sequence by one time step and repeating the last switch position. This is in accordance with the receding horizon paradigm used in MPC (see also Fig. 5.2). As the optimal switching sequence at the previous time step satisfies both constraints (5.22b) and (5.22c), the shifted sequence automatically meets these constraints, too. This statement holds true in all circumstances, including transients. Thus, $\boldsymbol{U}_{\text{ini}}(k)$ is a feasible candidate solution for the optimization problem (5.22). Given (5.31), the initial value of $\rho(k)$ is then set to

$$\rho(k) = \|\bar{\boldsymbol{U}}_{\text{unc}}(k) - \boldsymbol{V}\boldsymbol{U}_{\text{ini}}(k)\|_2. \tag{5.32}$$

5.3.2 Modified Sphere Decoding Algorithm

At each time step k, the controller first uses the current state $\boldsymbol{x}(k)$, the future reference values $\boldsymbol{Y}^*(k)$, the previous switch position $\boldsymbol{u}(k-1)$, and the previous optimizer $\boldsymbol{U}_{\text{opt}}(k-1)$ to calculate $\boldsymbol{U}_{\text{ini}}(k)$, $\rho(k)$ and $\bar{\boldsymbol{U}}_{\text{unc}}(k)$; see (5.31), (5.32), (5.23e), and (5.23d). The optimal switching sequence $\boldsymbol{U}_{\rm opt}(k)$ is then obtained by invoking Algorithm 1[3]:

$$\boldsymbol{U}_{\rm opt}(k) = \text{MSPHDEC}([\ ], 0, 1, \rho^2(k), \bar{\boldsymbol{U}}_{\rm unc}(k)), \tag{5.33}$$

where [ ] denotes an empty vector.

Algorithm 1 Modified sphere decoding algorithm

function $\boldsymbol{U}_{\rm opt}$ = MSPHDEC($\boldsymbol{U}$, d^2, i, ρ^2, $\bar{\boldsymbol{U}}_{\rm unc}$)
  for each $u \in \mathcal{U}$ **do**
    $u_i = u$
    $d'^2 = ||\bar{u}_{{\rm unc},i} - \boldsymbol{v}_{(i,1:i)}\boldsymbol{u}_{1:i}||_2^2 + d^2$
    if $d'^2 \leq \rho^2$ **then**
      if $i < 3N_p$ **then**
        MSPHDEC($\boldsymbol{U}, d'^2, i+1, \rho^2, \bar{\boldsymbol{U}}_{\rm unc}$)
      else
        if $\boldsymbol{U}$ meets (5.22c) **then**
          $\boldsymbol{U}_{\rm opt} = \boldsymbol{U}$
          $\rho^2 = d'^2$
        end if
      end if
    end if
  end for
end function

As can be seen in Algorithm 1, the proposed sphere decoder operates in a recursive manner. Starting with the first component, the switching sequence $\boldsymbol{U}(k)$ is built component by component, by considering the admissible single-phase switch positions in the set $\mathcal{U}$. If the associated squared distance is smaller than the current value of $\rho^2(k)$, then we proceed to the next component. Once the last component, that is, $u_{3N_p}(k)$, has been reached, meaning that $\boldsymbol{U}(k)$ is of full dimension, then $\boldsymbol{U}(k)$ is a candidate solution. If $\boldsymbol{U}(k)$ meets the switching constraint (5.22c) and if the distance is smaller than the current optimum, then we update the incumbent optimal solution $\boldsymbol{U}_{\rm opt}(k)$ and also the radius $\rho(k)$.

The computational advantages of this algorithm stem from adopting the notion of branch-and-bound [6, 7]. Branching is done over the set of single-phase switch positions $\mathcal{U}$; bounding is achieved by considering solutions only within the sphere with the radius $\rho(k)$ (see (5.29)). If the distance d' exceeds the radius, a so-called certificate (or proof) has been found that the branch (and all its associated switching sequences) provides only suboptimal solutions, that is, solutions that are worse than the incumbent optimum. Therefore, without exploring this branch, it can be pruned and removed from further consideration. During the optimization procedure, whenever a better incumbent solution is found, the radius is reduced and the sphere thus tightened, so that the set of candidate sequences is as small as possible, but nonempty.

[3] The notation $\boldsymbol{v}_{(i,1:i)}$ refers to the first i entries of the ith row of $\boldsymbol{V}$; similarly, $\boldsymbol{u}_{1:i}$ are the first i elements of the vector $\boldsymbol{U}$.

The majority of the computational burden relates to the computation of d' via the evaluation of the terms $\boldsymbol{v}_{(i,1:i)}\boldsymbol{u}_{1:i}(k)$. Thanks to (5.30), d' can be computed sequentially, by adding only the squared term involving the ith component of $\bar{U}(k)$. In particular, the sum of squares in d, accumulated over the layers 1 to $i-1$, need not be recomputed.

It is worth emphasizing that the computational advantages of the proposed algorithm do not come at the expense of optimality: the algorithm always provides the optimal switch position. This can easily be verified by recalling that the optimal constrained solution minimizes the Euclidean distance d to the unconstrained solution. Moreover, the use of the initial radius in (5.32) guarantees that a feasible switching sequence (which satisfies the constraints) will be returned. Successive values of $\rho^2(k)$ in the iterations are always associated with, and allow for, feasible sequences. The algorithm stops when the sphere centered on $\bar{U}_{\text{unc}}(k)$ only contains a single element. The latter amounts to the optimal (integer) solution.

5.3.3 Illustrative Example with a Prediction Horizon of 1

To provide additional insight into the operation of the algorithm, we provide an illustrative example of one problem instance. Consider the horizon $N_p = 1$ case with the sampling interval $T_s = 25$ μs and the penalties $\boldsymbol{Q} = \boldsymbol{I}_2$ and $\lambda_u = 1 \cdot 10^{-3}$. As before, we consider the current regulation problem for a drive system with a three-level inverter as in Fig. 5.1 with the parameters as in Sect. 2.5.1. The set of single-phase switch positions is $\mathcal{U} = \{-1, 0, 1\}$.

The set of admissible three-phase switch positions $\boldsymbol{u}(k) \in \boldsymbol{\mathcal{U}}$ is shown in Fig. 5.3(a) as black circles. To simplify the exposition, only the ab-plane is shown in this figure, neglecting the c-axis. Suppose that $\boldsymbol{u}(k-1) = [1\ 0\ 1]^T$ and that the problem instance at time step k yields the unconstrained solution $\boldsymbol{u}_{\text{unc}}(k) = [0.647\ \ -0.533\ \ -0.114]^T$, shown as a triangle in the figure. Rounding $\boldsymbol{u}_{\text{unc}}(k)$ to the next integer values leads to the possible feasible solution $\boldsymbol{u}_{\text{sub}}(k) = [1\ \ -1\ \ 0]^T$, which corresponds to the square. It turns out, however—as shown next—that the optimal solution is $\boldsymbol{u}_{\text{opt}}(k) = [1\ 0\ 0]^T$, which is indicated by the diamond.

The modified sphere decoding problem is solved in the transformed coordinate system, which is created by the generator matrix

$$\boldsymbol{V} = \begin{bmatrix} 36.45 & 0 & 0 \\ -6.068 & 36.95 & 0 \\ -5.265 & -5.265 & 37.32 \end{bmatrix} \cdot 10^{-3},$$

see (5.23c). Using $\boldsymbol{V}$, the integer solutions $\boldsymbol{u}(k) \in \boldsymbol{\mathcal{U}}$ in the orthogonal coordinate system can be transformed to $\boldsymbol{V}\boldsymbol{u}(k)$, which are shown as black circles in Fig. 5.3(b). The coordinate system created by $\boldsymbol{V}$ is slightly skewed, but almost orthogonal, with the angle between the axes being 98.2° for the chosen parameters. As discussed in Sect. 5.2.4, increasing λ_u results in this angle converging toward 90°.

The optimal solution $\boldsymbol{u}_{\text{opt}}(k)$ is obtained by minimizing the distance between the unconstrained solution and the integer switch positions in the transformed coordinate system. The initial value of $\rho(k)$ results from (5.32) and is equal to 0.638. This defines a ball of radius $\rho(k)$ around $\bar{\boldsymbol{u}}_{\text{unc}}(k) = \boldsymbol{V}\boldsymbol{u}_{\text{unc}}(k)$, which is shown in the ab-plane in Fig. 5.3(b) as the circle. This ball reduces the set of possible solutions from $3^3 = 27$ elements to 2, because only two transformed integer solutions $\boldsymbol{V}\boldsymbol{u}(k)$ lie within the ball—these are $\boldsymbol{V}\boldsymbol{u}_{\text{opt}}(k)$ (the diamond) and $\boldsymbol{V}\boldsymbol{u}_{\text{sub}}(k)$ (the square). The algorithm sequentially computes the distances between

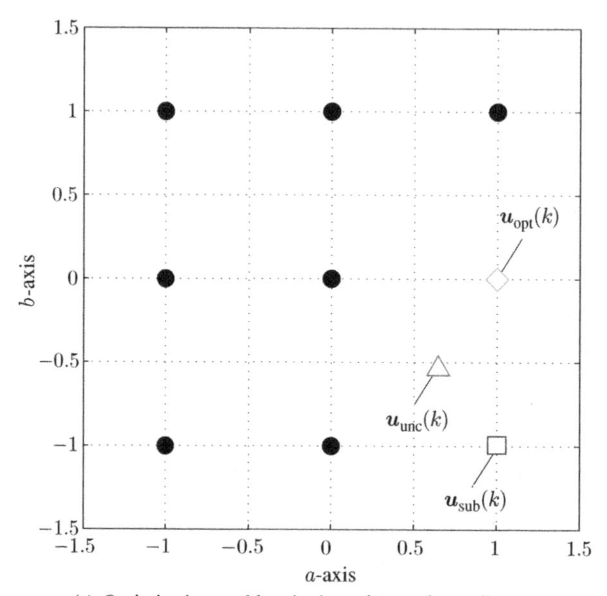

(a) Optimization problem in the orthogonal coordinate system

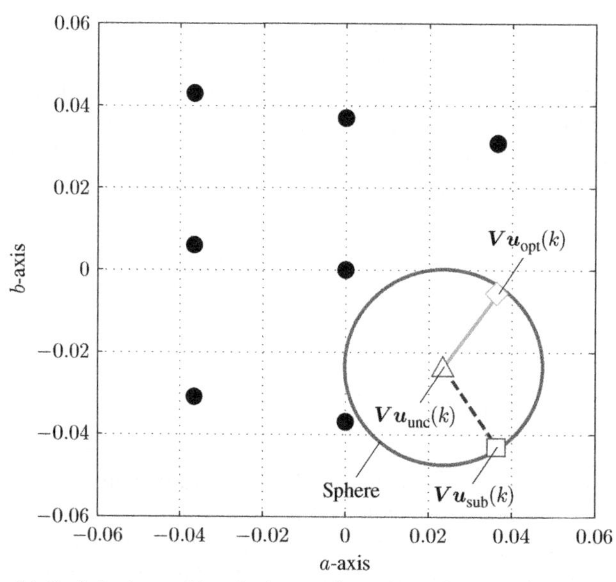

(b) Optimization problem in the transformed coordinate system created by the generator matrix V

Figure 5.3 Visualization of the sphere decoding algorithm in the ab-plane for the horizon $N_p = 1$

$\bar{\boldsymbol{u}}_{\text{unc}}(k)$ and each of these two points. These distances are indicated by the solid and dashed lines, respectively. The solid line is slightly shorter than the dashed one. Therefore, minimizing the distance yields the optimal solution $\boldsymbol{u}_{\text{opt}}(k) = [1\ 0\ 0]^T$ and not the (suboptimal) rounded switch position $\boldsymbol{u}_{\text{sub}}(k) = [1\ -1\ 0]^T$.

5.3.4 Illustrative Example with a Prediction Horizon of 2

A second example is provided in the following, which introduces the notion of the search tree and illustrates the traversal of the tree by the sphere decoding algorithm. Recall that the optimization problem solved by the algorithm is to find the optimal switching sequence U_{opt} of length $3N_p$ out of the set of possible switching sequences $\mathbb{U} = \mathcal{U} \times \cdots \times \mathcal{U}$, which is the $3N_p$-times Cartesian product of the set of single-phase switch positions $\mathcal{U}$. For the three-level converter, we have $\mathcal{U} = \{-1, 0, 1\}$.

The set $\mathbb{U}$ spans a tree of the depth $3N_p$ with nodes at the levels $i \in \{1, 2, \ldots, 3N_p\}$. Each node has three subsequent child nodes, except for the nodes at level $3N_p$, which are leaf nodes. The nodes at level i correspond to the switching decisions to be made with regard to the ith element in $\boldsymbol{U}$, u_i. Specifically, the branches starting from nodes at level i relate to u_i. Traversing the tree from the root node to one of its leaves corresponds to a unique switching sequence $\boldsymbol{U} = [u_1\ u_2 \ldots u_{3N_p}]^T$.

The exploration of the search tree is exemplified in Fig. 5.4 for the horizon $N_p = 2$ case, assuming the previously chosen switch position to be $\boldsymbol{u}(k-1) = [0\ 0\ 0]^T$. Starting from the

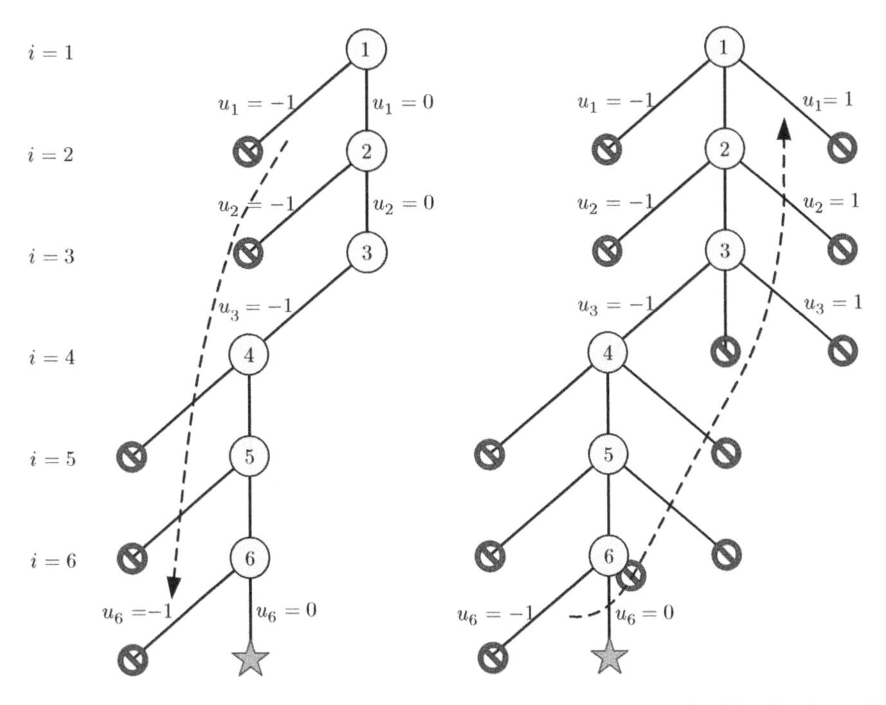

Figure 5.4 Visualization of the search tree traversal by the sphere decoding algorithm for the prediction horizon $N_p = 2$ and a three-level inverter. The optimal switching sequence is found after the exploration of 6 nodes out of 364 (left-hand side). These six nodes need to be fully explored to prove that this is the optimal switching sequence (right-hand side)

root node at $i = 1$, the algorithm evaluates $u_1 \in \mathcal{U}$. The choice $u_1 = -1$ exceeds the radius of the sphere, and the subtree starting at $u_1 = -1$ is pruned. This is indicated by the crossed circle. The choice $u_1 = 0$ is within the sphere, and the algorithm proceeds to the next level $i = 2$, without exploring $u_1 = 1$. The search direction of the algorithm is indicated by the dashed line.

After the exploration of six nodes, the first candidate solution $\boldsymbol{U} = [0\ 0\ -1\ 0\ 0\ 0]^T$ is found, which corresponds to the tentative switch positions $\boldsymbol{u}(k) = [0\ 0\ -1]^T$ and $\boldsymbol{u}(k+1) = [0\ 0\ 0]^T$. The leaf node corresponding to the candidate solution is indicated by a star. The radius of the sphere is tightened, before continuing the exploration, as illustrated on the right-hand side of Fig. 5.4. The remaining candidate switch positions are explored at the nodes visited so far, while moving up in the search tree toward the root node. Fully exploring these nodes provides the algorithm with a certificate that the candidate switching sequence is indeed the optimal solution. Note that some branches are discarded before exploration. As $u_3 = -1$ was chosen, switching to $u_6 = 1$ would violate the switching constraint (5.22c).

Owing to the tight sphere and the depth-first search of the algorithm, the optimal switching sequence is found in the majority of the cases after the exploration of the minimum number of nodes, which is given by $3N_p$. In some cases, however, additional candidate sequence are explored, as exemplified in Fig. 5.5. As shown on the left-hand side, after having found the

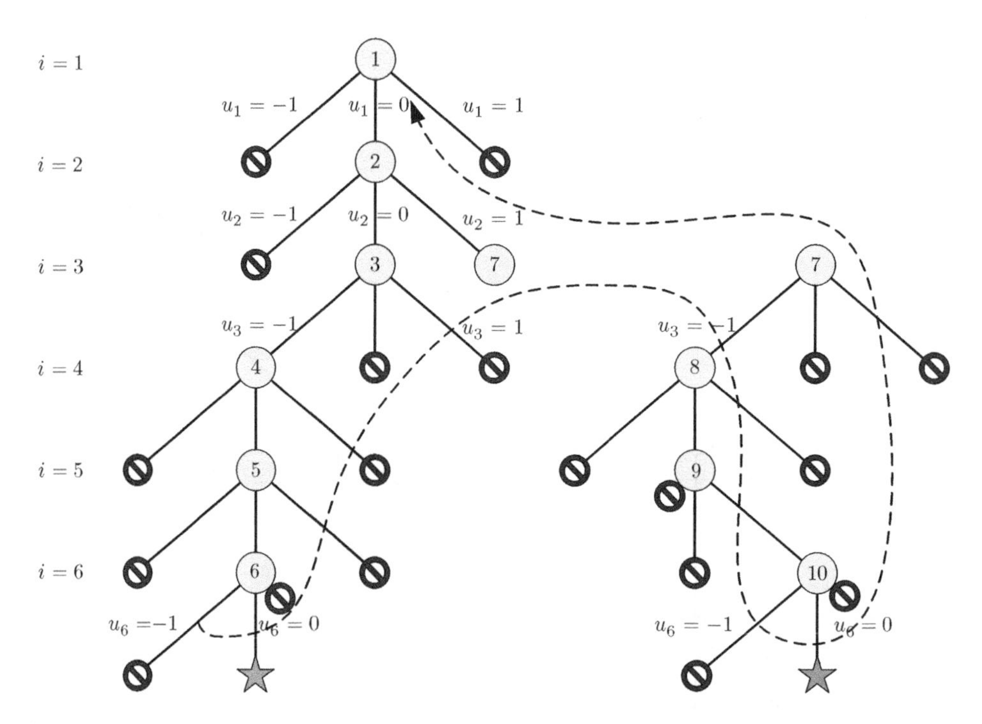

Figure 5.5 Visualization of the search tree traversal for another problem instance. The optimal switching sequence is found after the enumeration of two candidate switching sequences and the exploration of 10 nodes

first candidate switching sequence $\boldsymbol{U} = [0\ 0\ -1\ 0\ 0\ 0]^T$, the branch $u_2 = 1$ of node 2 turns out to be within the sphere, triggering the exploration of another part of the search tree with the nodes 7–10, which are shown on the right-hand side of Fig. 5.5. The second candidate solution is $\boldsymbol{U} = [0\ 1\ -1\ 0\ 1\ 0]^T$, which is, in this example, the optimal switching sequence.

For a three-level converter, when neglecting the switching constraint (5.22c), the search tree has

$$\sum_{i=0}^{3N_p-1} 3^i \tag{5.34}$$

nodes, where the base relates to the three levels of the inverter. For the horizon $N_p = 2$ considered previously, the search tree has 364 nodes. Of these, the sphere decoding algorithm visits only a very small subset, as highlighted by the examples shown in Figs. 5.4 and 5.5. Statistical data are provided in the next section.

5.4 Computational Burden

In this section, the computational burden of the modified sphere decoding algorithm is analyzed and compared with the exhaustive search method. In this analysis, a distinction is made between computations that can be made offline once and the computations that are performed online and in real time. For the latter, we distinguish between the preprocessing and the sphere decoding stage.

5.4.1 *Offline Computations*

The generator matrix $\boldsymbol{V}$ can be computed offline, provided that the rotor speed ω_r and the dc-link voltage v_{dc} are time-invariant. If these two quantities were time-varying, $\boldsymbol{V}$ could be precomputed for different combinations of rotor speeds and dc-link voltages, and the result could be stored in a look-up table. The steps required to compute $\boldsymbol{V}$ involve the following: First, the discrete-time state-space matrices $\boldsymbol{A}$, $\boldsymbol{B}$, and $\boldsymbol{C}$ are computed according to (5.5) or (5.6). Second, $\boldsymbol{\Upsilon}$ and $\boldsymbol{S}$ are calculated (see Appendix 5.B), based on which $\boldsymbol{H}$ is derived according to (5.23b). In the last step, the generator matrix $\boldsymbol{V}$ results from the Cholesky factorization (see (5.19)).

5.4.2 *Online Preprocessing*

The preprocessing stage involves four computational steps, which need to be performed in real time. First, the output reference trajectory $\boldsymbol{Y}^*(k)$ needs to be derived over the prediction horizon N_p. Second, $\boldsymbol{\Theta}(k)$ in (5.23d) is computed as a function of the state vector $\boldsymbol{x}(k)$, output reference trajectory $\boldsymbol{Y}^*(k)$, and the previously applied switch position $\boldsymbol{u}(k-1)$. Third, the transformed unconstrained solution (5.23e) is derived. Note that the term $\boldsymbol{V}\boldsymbol{H}^{-1}$ can be precomputed offline. In the last step, a feasible candidate switching sequence $\boldsymbol{U}_{\text{ini}}(k)$ is obtained from the optimal switching sequence $\boldsymbol{U}_{\text{opt}}(k-1)$ at the previous time step (see (5.31)). Based on this candidate sequence, the initial radius $\rho(k)$ of the sphere can be derived according to (5.32).

5.4.3 Sphere Decoding

The sphere decoding algorithm takes up the majority of the real-time computations. As a measure of the computational burden, we count the number of nodes that are visited by the algorithm in the search tree. For this, the drive system case study with current reference tracking is considered again. We assume operation at nominal speed and rated torque, and chose $Q = I_2$ and the sampling interval $T_s = 25$ μs. Different prediction horizons are investigated. The weight λ_u is chosen such that a switching frequency of approximately 300 Hz is obtained, regardless of the prediction horizon.

Over multiple fundamental periods, the number of nodes that the algorithm investigates at each time step is recorded. The average and the maximum number of nodes are stated in Table 5.1 as a function of the prediction horizon. By definition, at least $3N_p$ nodes need to be explored before the optimal switching sequence can be found. This constitutes the lower bound. The theoretical upper bound is given by (5.34). Even though this bound neglects the switching constraint, the computational burden of exhaustive search is similar to this upper bound.

When using sphere decoding for the horizon 1 case, the table shows that on an average 3.18 nodes need to be considered by the algorithm. This implies that, by choosing the initial radius of the sphere according to the educated guess (5.32), the sphere is sufficiently tight. Specifically, in the vast majority of cases, the sphere is perfectly tight, in the sense that, out of all the admissible switching sequences, only *one* is located within the sphere, which implies that only one leaf node in the search tree is visited. The same holds true for prediction horizons of lengths 2 and 3.

This is in stark contrast to the exhaustive search method. For the horizon 1 case, depending on the optimal switch position obtained at the previous time step, $u(k-1)$, and in accordance with the switching constraint, up to 13 nodes need to be investigated. As the prediction horizon is increased, the computational burden associated with sphere decoding initially grows slowly, while exhaustive search becomes computationally intractable for horizons of 5 or more. Using sphere decoding, the optimization problem for direct MPC with long prediction horizons such

Table 5.1 Number of nodes explored in the search tree as a function of the length of the prediction horizon

Prediction horizon N_p	Lower bound	Sphere decoding		Upper bound
		average	maximum	
1	3	3.18	7	13
2	6	6.39	13	364
3	9	9.72	22	9841
5	15	16.54	49	$7.17 \cdot 10^6$
10	30	37.10	249	$1.03 \cdot 10^{14}$

The average and maximum number of nodes explored by the sphere decoding algorithm are provided, along with lower and upper bounds.

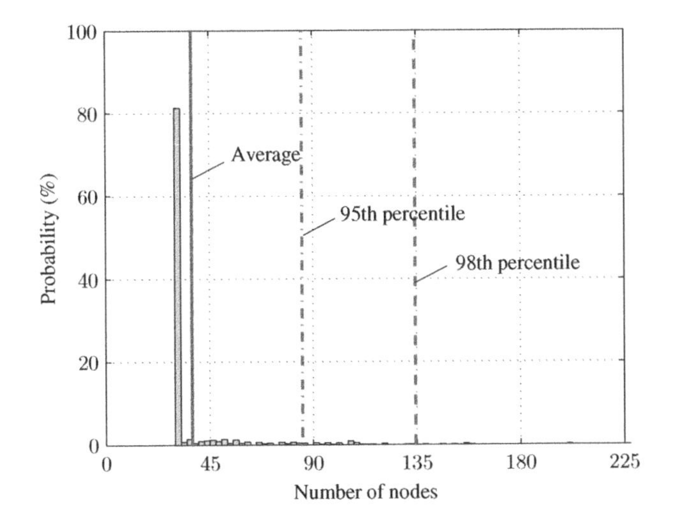

Figure 5.6 Histogram of the number of nodes investigated by the modified sphere decoding algorithm, when considering the prediction horizon $N_p = 10$

as 10 can be solved relatively quickly, with the empirical maximum number of nodes to be investigated being 249.

Figure 5.6 depicts the histogram of the average number of nodes that need to be explored at each time step when using a horizon of 10 steps. The histogram is highly concentrated at the lower bound of 30, but it exhibits a long, albeit very flat tail. It can be seen that, with sphere decoding, in 80% of the cases the optimization problem can be solved by exploring only one candidate switching sequence. The 95th and 98th percentiles are shown as dash-dotted and dashed lines, respectively. They indicate that in 95% of the cases fewer than 85 nodes need to be explored.

Appendix 5.A: State-Space Model

The matrices corresponding to the continuous-time prediction model (5.3) are

$$\boldsymbol{F} = \begin{bmatrix} -\frac{1}{\tau_s} & 0 & \frac{X_m}{\tau_r D} & \omega_r \frac{X_m}{D} \\ 0 & -\frac{1}{\tau_s} & -\omega_r \frac{X_m}{D} & \frac{X_m}{\tau_r D} \\ \frac{X_m}{\tau_r} & 0 & -\frac{1}{\tau_r} & -\omega_r \\ 0 & \frac{X_m}{\tau_r} & \omega_r & -\frac{1}{\tau_r} \end{bmatrix}, \quad \boldsymbol{G} = \frac{v_{\text{dc}}}{2} \frac{X_r}{D} \begin{bmatrix} 1 & 0 \\ 0 & 1 \\ 0 & 0 \\ 0 & 0 \end{bmatrix} \tilde{\boldsymbol{K}}, \tag{5.A.1a}$$

$$\boldsymbol{C} = \begin{bmatrix} 1 & 0 & 0 & 0 \\ 0 & 1 & 0 & 0 \end{bmatrix}. \tag{5.A.1b}$$

The transient stator and rotor time constants τ_s and τ_r were defined in (2.60). The determinant D is stated in (2.54).

Appendix 5.B: Derivation of the Cost Function in Vector Form

By successively using (5.10b), the state vector at time step $\ell + 1$ can be represented as a function of the state vector at time step k and the switching sequence comprising the switch positions $\boldsymbol{u}(k)$ to $\boldsymbol{u}(\ell)$ as

$$\boldsymbol{x}(\ell+1) = \boldsymbol{A}^{\ell-k+1}\,\boldsymbol{x}(k) + \boldsymbol{A}^{\ell-k}\boldsymbol{B}\,\boldsymbol{u}(k) + \cdots + \boldsymbol{A}^{0}\boldsymbol{B}\,\boldsymbol{u}(\ell) \tag{5.B.1}$$

with $\ell = k, \ldots, k + N_p - 1$.

Recall the definition of the output reference trajectory

$$\boldsymbol{Y}(k) = [\boldsymbol{y}^T(k+1)\ \boldsymbol{y}^T(k+2)\ \ldots\ \boldsymbol{y}^T(k+N_p)]^T \tag{5.B.2}$$

over the prediction horizon N_p. Substituting (5.B.1) into (5.10c) yields

$$\boldsymbol{Y}(k) = \boldsymbol{\Gamma}\,\boldsymbol{x}(k) + \boldsymbol{\Upsilon}\,\boldsymbol{U}(k), \tag{5.B.3}$$

where the matrices $\boldsymbol{\Gamma}$ and $\boldsymbol{\Upsilon}$ are defined as

$$\boldsymbol{\Gamma} = \begin{bmatrix} \boldsymbol{C}\boldsymbol{A} \\ \boldsymbol{C}\boldsymbol{A}^2 \\ \vdots \\ \boldsymbol{C}\boldsymbol{A}^{N_p} \end{bmatrix} \quad \text{and} \quad \boldsymbol{\Upsilon} = \begin{bmatrix} \boldsymbol{C}\boldsymbol{B} & \boldsymbol{0}_{n_y\times 3} & \cdots & \boldsymbol{0}_{n_y\times 3} \\ \boldsymbol{C}\boldsymbol{A}\boldsymbol{B} & \boldsymbol{C}\boldsymbol{B} & \cdots & \boldsymbol{0}_{n_y\times 3} \\ \vdots & \vdots & & \vdots \\ \boldsymbol{C}\boldsymbol{A}^{N_p-1}\boldsymbol{B} & \boldsymbol{C}\boldsymbol{A}^{N_p-2}\boldsymbol{B} & \cdots & \boldsymbol{C}\boldsymbol{B} \end{bmatrix}. \tag{5.B.4}$$

Note that n_y denotes the number of output variables, that is, $\boldsymbol{y} \in \mathbb{R}^{n_y}$.

We define the output tracking error $\boldsymbol{\xi} = \boldsymbol{y}^* - \boldsymbol{y}$. With this definition, the first term in the cost function (5.7) can be written as

$$J_1 = \sum_{\ell=k}^{k+N_p-1} ||\boldsymbol{\xi}(\ell+1)||^2_{\boldsymbol{Q}} = \sum_{\ell=k}^{k+N_p-1} (\boldsymbol{\xi}(\ell+1))^T\boldsymbol{Q}\,\boldsymbol{\xi}(\ell+1) \tag{5.B.5a}$$

$$= [\boldsymbol{\xi}^T(k+1)\ldots\boldsymbol{\xi}^T(k+N_p)]\ \tilde{\boldsymbol{Q}}\ [\boldsymbol{\xi}^T(k+1)\ldots\boldsymbol{\xi}^T(k+N_p)]^T \tag{5.B.5b}$$

$$= (\boldsymbol{\Xi}(k))^T\tilde{\boldsymbol{Q}}\,\boldsymbol{\Xi}(k) = ||\boldsymbol{\Xi}(k)||^2_{\tilde{\boldsymbol{Q}}}, \tag{5.B.5c}$$

where we have introduced the block diagonal matrix $\tilde{\boldsymbol{Q}} = \text{diag}(\boldsymbol{Q}, \ldots, \boldsymbol{Q})$ and the output error trajectory $\boldsymbol{\Xi}(k) = [\boldsymbol{\xi}^T(k+1)\ \boldsymbol{\xi}^T(k+2)\ \ldots\ \boldsymbol{\xi}^T(k+N_p-1)]^T$. By inserting (5.B.3) into $\boldsymbol{\Xi}(k) = \boldsymbol{Y}^*(k) - \boldsymbol{Y}(k)$ in (5.B.5c), the cost function term

$$J_1 = ||\boldsymbol{Y}^*(k) - \boldsymbol{\Gamma}\boldsymbol{x}(k) - \boldsymbol{\Upsilon}\boldsymbol{U}(k)||^2_{\tilde{\boldsymbol{Q}}} \tag{5.B.6}$$

directly follows.

Similarly, the second term in the cost function (5.7) can be rewritten as

$$J_2 = \sum_{\ell=k}^{k+N_p-1} \lambda_u||\Delta\boldsymbol{u}(\ell)||^2_2 \tag{5.B.7a}$$

$$= \lambda_u \sum_{\ell=k}^{k+N_p-1} (\boldsymbol{u}(\ell) - \boldsymbol{u}(\ell-1))^T(\boldsymbol{u}(\ell) - \boldsymbol{u}(\ell-1)) \tag{5.B.7b}$$

$$= \lambda_u (\boldsymbol{SU}(k) - \boldsymbol{Eu}(k-1))^T (\boldsymbol{SU}(k) - \boldsymbol{Eu}(k-1)) \tag{5.B.7c}$$

$$= \lambda_u \|\boldsymbol{SU}(k) - \boldsymbol{Eu}(k-1)\|_2^2, \tag{5.B.7d}$$

with the auxiliary matrices

$$\boldsymbol{S} = \begin{bmatrix} \boldsymbol{I}_3 & \boldsymbol{0}_{3\times3} & \cdots & \boldsymbol{0}_{3\times3} \\ -\boldsymbol{I}_3 & \boldsymbol{I}_3 & \cdots & \boldsymbol{0}_{3\times3} \\ \boldsymbol{0}_{3\times3} & -\boldsymbol{I}_3 & \cdots & \boldsymbol{0}_{3\times3} \\ \vdots & \vdots & & \vdots \\ \boldsymbol{0}_{3\times3} & \boldsymbol{0}_{3\times3} & \cdots & \boldsymbol{I}_3 \end{bmatrix} \text{ and } \boldsymbol{E} = \begin{bmatrix} \boldsymbol{I}_3 \\ \boldsymbol{0}_{3\times3} \\ \boldsymbol{0}_{3\times3} \\ \vdots \\ \boldsymbol{0}_{3\times3} \end{bmatrix}. \tag{5.B.8}$$

Summing up (5.B.6) and (5.B.7d) yields the cost function in vector notation:

$$J = \|\boldsymbol{Y}^*(k) - \boldsymbol{\Gamma x}(k) - \boldsymbol{\Upsilon U}(k)\|_{\tilde{\boldsymbol{Q}}}^2 + \lambda_u \|\boldsymbol{SU}(k) - \boldsymbol{Eu}(k-1)\|_2^2. \tag{5.B.9}$$

The first term in (5.B.9) penalizes the predicted tracking error, while the second term penalizes the switching effort.

To obtain the representation (5.12) of the cost function, some additional algebraic manipulations are required. In this last step, to improve the readability, we simplify the notation and drop the time dependence of the vectors and matrices in (5.B.9) and write

$$J = \|\boldsymbol{Y}^* - \boldsymbol{\Gamma x} - \boldsymbol{\Upsilon U}\|_{\tilde{\boldsymbol{Q}}}^2 + \lambda_u \|\boldsymbol{SU} - \boldsymbol{Eu}\|_2^2 \tag{5.B.10a}$$

$$\begin{aligned} &= (\boldsymbol{\Upsilon U})^T \tilde{\boldsymbol{Q}} \boldsymbol{\Upsilon U} + \lambda_u (\boldsymbol{SU})^T \boldsymbol{SU} - (\boldsymbol{Y}^* - \boldsymbol{\Gamma x})^T \tilde{\boldsymbol{Q}} \boldsymbol{\Upsilon U} \\ &\quad - (\boldsymbol{\Upsilon U})^T \tilde{\boldsymbol{Q}} (\boldsymbol{Y}^* - \boldsymbol{\Gamma x}) - \lambda_u (\boldsymbol{Eu})^T \boldsymbol{SU} - \lambda_u (\boldsymbol{SU})^T \boldsymbol{Eu} \\ &\quad + (\boldsymbol{Y}^* - \boldsymbol{\Gamma x})^T \tilde{\boldsymbol{Q}} (\boldsymbol{Y}^* - \boldsymbol{\Gamma x}) + \lambda_u (\boldsymbol{Eu})^T \boldsymbol{Eu}. \end{aligned} \tag{5.B.10b}$$

By exploiting the fact that $\tilde{\boldsymbol{Q}}$ is symmetric, that is, $\tilde{\boldsymbol{Q}}^T = \tilde{\boldsymbol{Q}}$, and that for a scalar ξ the statement $\xi^T = \xi$ holds, the terms in the cost function (5.B.10b) can be rearranged as

$$\begin{aligned} J &= \boldsymbol{U}^T (\boldsymbol{\Upsilon}^T \tilde{\boldsymbol{Q}} \boldsymbol{\Upsilon} + \lambda_u \boldsymbol{S}^T \boldsymbol{S}) \boldsymbol{U} - 2(\boldsymbol{Y}^* - \boldsymbol{\Gamma x})^T \tilde{\boldsymbol{Q}} \boldsymbol{\Upsilon U} \\ &\quad - 2\lambda_u (\boldsymbol{Eu})^T \boldsymbol{SU} + \|\boldsymbol{Y}^* - \boldsymbol{\Gamma x}\|_{\tilde{\boldsymbol{Q}}}^2 + \lambda_u \|\boldsymbol{Eu}\|_2^2. \end{aligned} \tag{5.B.11}$$

By defining

$$\boldsymbol{H} = \boldsymbol{\Upsilon}^T \tilde{\boldsymbol{Q}} \boldsymbol{\Upsilon} + \lambda_u \boldsymbol{S}^T \boldsymbol{S} \tag{5.B.12a}$$

$$(\boldsymbol{\Theta}(k))^T = -(\boldsymbol{Y}^*(k) - \boldsymbol{\Gamma x}(k))^T \tilde{\boldsymbol{Q}} \boldsymbol{\Upsilon} - \lambda_u (\boldsymbol{Eu}(k-1))^T \boldsymbol{S} \tag{5.B.12b}$$

$$\theta(k) = \|\boldsymbol{Y}^*(k) - \boldsymbol{\Gamma x}(k)\|_{\tilde{\boldsymbol{Q}}}^2 + \lambda_u \|\boldsymbol{Eu}(k-1)\|_2^2, \tag{5.B.12c}$$

(5.B.11) can be written in the final form

$$J = (\boldsymbol{U}(k))^T \boldsymbol{HU}(k) + 2(\boldsymbol{\Theta}(k))^T \boldsymbol{U}(k) + \theta(k), \tag{5.B.13}$$

where we have reintroduced the time dependency.

References

[1] P. Cortés, J. Rodríguez, S. Vazquez, and L. Franquelo, "Predictive control of a three-phase UPS inverter using two steps prediction horizon," in *Proceedings of the IEEE International Conference on Industrial Technology (Viña del Mar, Chile)*, pp. 1283–1288, Mar. 2010.

[2] P. Stolze, P. Landsmann, R. Kennel, and T. Mouton, "Finite-set model predictive control of a flying capacitor converter with heuristic voltage vector preselection," in *Proceedings of IEEE International Conference on Power Electronics and ECCE Asia*, Jun. 2011.

[3] S. Kouro, P. Cortés, R. Vargas, U. Ammann, and J. Rodríguez, "Model predictive control—a simple and powerful method to control power converters," *IEEE Trans. Ind. Electron.*, vol. 56, pp. 1826–1838, Jun. 2009.

[4] B. Hassibi and H. Vikalo, "On the sphere-decoding algorithm I. Expected complexity," *IEEE Trans. Signal Process.*, vol. 53, pp. 2806–2818, Aug. 2005.

[5] G. F. Franklin, J. D. Powell, and M. L. Workman, *Digital control of dynamic systems*. Addison-Wesley, 3rd ed., 1998.

[6] E. L. Lawler and D. E. Wood, "Branch and bound methods: A survey," *Oper. Res.*, vol. 14, pp. 699–719, Jul./Aug. 1966.

[7] L. G. Mitten, "Branch–and–bound methods: General formulation and properties," *Oper. Res.*, vol. 18, pp. 24–34, Jan./Feb. 1970.

[8] T. Geyer, "Computationally efficient model predictive direct torque control," *IEEE Trans. Power Electron.*, vol. 26, pp. 2804–2816, Oct. 2011.

[9] IBM ILOG, Inc., *CPLEX 12.5 user manual*. Somers, NY, USA, 2012.

[10] R. A. Horn and C. R. Johnson, *Matrix analysis*. Cambridge, UK: Cambridge Univ. Press, 1985.

[11] E. Agrell, T. Eriksson, A. Vardy, and K. Zeger, "Closest point search in lattices," *IEEE Trans. Inf. Theory*, vol. 48, pp. 2201–2214, Aug. 2002.

[12] D. E. Quevedo, G. C. Goodwin, and J. A. D. Doná, "Finite constraint set receding horizon quadratic control," *Int. J. Robust Nonlinear Control*, vol. 14, pp. 355–377, Mar. 2004.

[13] M. Pérez, P. Cortés, and J. Rodríguez., "Predictive control algorithm technique for multilevel asymmetric cascaded H-bridge inverters," *IEEE Trans. Ind. Electron.*, vol. 55, pp. 4354–4361, Dec. 2008.

[14] U. Fincke and M. Pohst, "Improved methods for calculating vectors of short length in a lattice, including a complexity analysis," *Math. Comput.*, vol. 44, pp. 463–471, Apr. 1985.

6

Performance Evaluation of Predictive Control with Long Horizons

In direct model predictive control (MPC) schemes with reference tracking, the prediction horizon has traditionally been set to 1 [1]. Indeed, it is often believed that a horizon of 1 suffices and that the use of longer horizons carries no performance benefits. This common belief might be a result of the fact that, because of the combinatorial explosion of the number of possible solutions, investigating the potential benefits of long horizons is intrinsically challenging, and horizons of 2 or 3 often offer only an incremental benefit, as will be shown in this chapter. Another reason might be that researchers have so far mostly focused on inverters directly connected to the load, such as an RL load [2]. In an orthogonal coordinate system, the fast (current) dynamic of such a setup constitutes a first-order system. In each coordinate axis, the transfer function from the manipulated variable (the inverter voltage) to the load current (the controlled variable) is of first order, implying that these power electronic systems can be controlled with ease.

Adopting the sphere decoding algorithm proposed in the previous chapter, the performance benefits of long horizons for inverter drive systems are investigated in this chapter. As will be shown for a three-level converter connected to an induction machine, long prediction horizons improve the converter performance at steady-state operating conditions, by either reducing the switching frequency or the total demand distortion (TDD) of the current, or both. Specifically, direct MPC with horizon $N_p = 10$ reduces the current TDD by approximately 20%, when compared to the $N_p = 1$ case. As a result, direct MPC can outperform space vector modulation (SVM) and carrier-based pulse width modulation (CB-PWM). In some cases, the performance of direct MPC may even approach that of optimized pulse patterns (OPPs), which are generally considered to provide the upper bound of the attainable steady-state performance. During transients, MPC with long horizons provides as short a transient response time as MPC with short horizons, often outperforming classic control arrangements such as field-oriented control.

Model Predictive Control of High Power Converters and Industrial Drives, First Edition. Tobias Geyer.

Companion Website: www.wiley.com/go/geyermodelpredictivecontrol

The main benefit of long horizons, however, becomes evident when considering power electronic systems of so-called *higher order*. In an orthogonal coordinate system, the fast dynamics of such systems feature more than one state variable per coordinate axis. In the second case study, an LC filter is added to the drive system, resulting in a third-order system. The direct MPC scheme can easily be extended to control all three state variables, without requiring an additional active damping loop. For such systems, the horizon 1 case leads to poor performance results with large current TDDs. Adopting long horizons, however, reduces the current TDD by about fivefold for the same switching frequency. Interestingly, device switching frequencies significantly below the filter resonance frequency can be used with long-horizon direct MPC.

6.1 Performance Evaluation for the NPC Inverter Drive System

In this section, the performance of direct MPC with long prediction horizons is investigated at steady-state operating conditions and during torque transients. An NPC inverter with an induction machine is used as a case study. We use the modified sphere decoding algorithm described in Sect. 5.3.2 to solve the optimization problem (5.22). To assess the performance of MPC, we need to first define a suitable framework.

6.1.1 Framework for Performance Evaluation

6.1.1.1 Simulation Setup

As stated in Sect. 5.1.1, we consider as a case study the current reference tracking problem of an NPC voltage source inverter. The inverter is connected to a medium-voltage (MV) induction machine with a constant mechanical load, as shown in Fig. 5.1. All simulations are performed in MATLAB, using an idealized setup. As such, second-order effects such as deadtimes, measurement noise, observer errors, saturation of the machine's magnetic material, parameter variations, and so on, are neglected. We assume that the manipulated variable is computed instantly after measuring the state variables, thus neglecting implementation-related delays. For more details on delays and a method to compensate them, the reader is referred to Sect. 4.2.8. Furthermore, the penalty matrix $\boldsymbol{Q}$ is set to the identity matrix $\boldsymbol{I}_2$.

Throughout this chapter, if not otherwise stated, all simulations were performed at nominal speed and rated torque, implying a fundamental frequency of 50 Hz and rated currents. To ensure that the drive system has settled at steady-state operating conditions, the system was first simulated over several fundamental periods without recording the results.

6.1.1.2 Modulation Methods Used for Benchmarking

To evaluate the steady-state performance of direct MPC with long horizons, we will benchmark this strategy with SVM and OPPs. Synchronous modulation is used, that is, the carrier frequency is an integer multiple of the fundamental frequency.

The OPPs were calculated offline for pulse numbers (the ratio between the switching frequency and the fundamental frequency) of up to 20. The switching angles were computed by minimizing the squared differential-mode voltage harmonics divided by the order of the harmonic. For an inductive load such as a machine, this approach is effectively equivalent to minimizing the current TDD. This is explained in detail in Sect. 3.4.

6.1.1.3 Performance Criteria

The key criteria related to the control performance are the device switching frequency f_{sw} and the current TDD I_{TDD}. In addition, we will also investigate the *a posteriori closed-loop* cost. It is obtained after the simulations, by evaluating the cost function J (see (5.7)), over all simulated time steps and dividing it by the total number of time steps k_{tot}. For the current reference tracking problem at hand, the closed-loop cost is given by

$$J_{cl} = \frac{1}{k_{tot}} \sum_{\ell=0}^{k_{tot}-1} \|\boldsymbol{i}_s^*(\ell+1) - \boldsymbol{i}_s(\ell+1)\|_2^2 + \lambda_u \|\Delta \boldsymbol{u}(\ell)\|_2^2. \tag{6.1}$$

The closed-loop cost (6.1) captures the squared root-mean-square (rms) current error (in stationary $\alpha\beta$ coordinates) and the averaged and squared switching effort (which is weighted with λ_u) over the closed-loop simulation.

6.1.1.4 Trade-Off between the Current TDD and Switching Frequency

Unavoidably, with switching power converters, a trade-off exists between the current TDD I_{TDD} and the switching frequency f_{sw}. It is convenient to plot these two quantities along two orthogonal axes. Figure 6.1 illustrates this performance trade-off for SVM and OPPs. In this

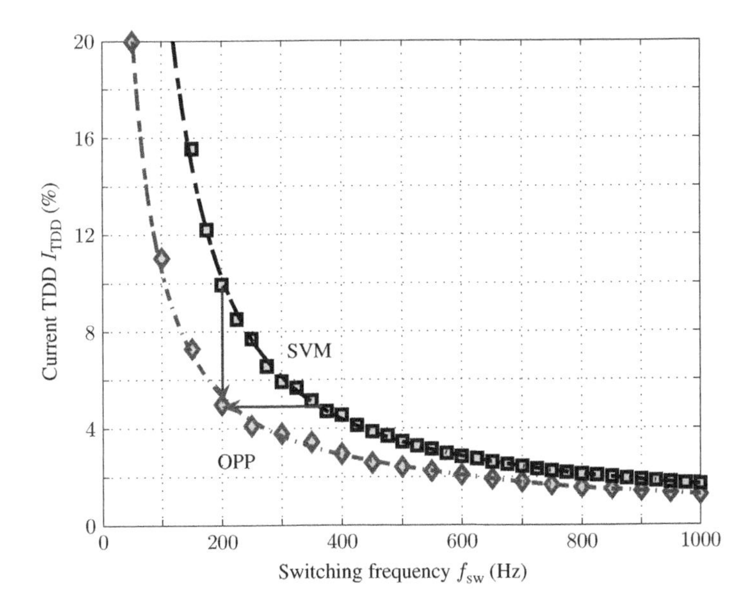

Figure 6.1 Trade-off between the current TDD and the switching frequency for synchronous SVM and OPPs

figure, each square corresponds to a unique simulation with synchronous SVM. The squares are approximated using a polynomial, indicated by the dash-dotted line. Accordingly, the diamonds correspond to volts per frequency (V/f) simulations with OPPs.

The switching frequency range between 200 and 350 Hz is of particular importance for MV power converters. As can be seen in Fig. 6.1, in this range there is scope for a significant reduction of the current TDD while maintaining the same switching frequency. For example, at $f_{sw} = 200$ Hz, the current TDD can be almost halved when replacing SVM by OPPs. Conversely, the switching frequency can be drastically reduced for the same current TDD. For $I_{\mathrm{TDD}} = 5\%$, for example, the switching frequency can be lowered from 350 to 200 Hz by adopting OPPs instead of SVM. This is a reduction of 42%. Both examples are indicated by arrows, which are shown in Fig. 6.1.

At higher switching frequencies, however, the performance benefit of OPPs compared to SVM becomes smaller. For $f_{sw} > 600$ Hz and pulse numbers greater than 12, the OPPs reduce the current distortions only by 15–20% when compared to SVM. Moreover, the optimization process to compute OPPs with high pulse numbers becomes computationally demanding. As a result, previous results shown in [3] were based on OPPs with slightly higher current distortions for pulse numbers exceeding 9.

With this as a background and recalling that OPPs exhibit—to a large extent—optimal steady-state behavior, we will quantify in this chapter the *relative* merits of MPC by normalizing the current TDD to the one obtained by OPPs. Specifically, we introduce

$$I_{\mathrm{TDD}}^{\mathrm{rel}} = \frac{I_{\mathrm{TDD}} - I_{\mathrm{TDD,OPP}}}{I_{\mathrm{TDD,OPP}}}, \tag{6.2}$$

which is the *relative* current TDD degradation, normalized to the current TDD of OPPs and given in percent. The normalization is performed with regard to the polynomial approximation of the OPPs shown in Fig. 6.1.

6.1.2 Comparison at the Switching Frequency 250 Hz

Consider direct MPC with the horizon $N_p = 1$, sampling interval $T_s = 125$ μs, and cost function (5.7) with the weighting factor $\lambda_u = 8.4 \cdot 10^{-3}$. This results in an average device switching frequency of $f_{sw} = 250$ Hz, which is typical for MV applications, and a current TDD of $I_{\mathrm{TDD}} = 5.96\%$.

Figure 6.2(a) illustrates three-phase stator current waveforms over one fundamental period. The colors black, light gray, and dark gray correspond to the phases a, b, and c, respectively. The currents are provided in a per unit (pu) system. The evolution of the stator current is simulated with a time resolution of 25 μs, based on which the spectrum of the stator current is computed with a Fourier transformation. The resulting current spectrum is shown in Fig. 6.2(b) and the three-phase switching sequence is depicted in Fig. 6.2(c). For direct MPC—unlike PWM—a repetitive switching pattern is not enforced. As a result, the current spectrum is predominantly flat without characteristic harmonics, despite a pronounced 11th harmonic.

Extending the prediction horizon to $N_p = 10$ reduces the current TDD by about one percentage point, as stated in Table 6.1. This initial result indicates that long prediction horizons do indeed reduce the current TDD, in this case by about 15%. The corresponding waveforms for the $N_p = 10$ case are shown in Fig. 6.3. It can be seen that the long horizon leads to a certain

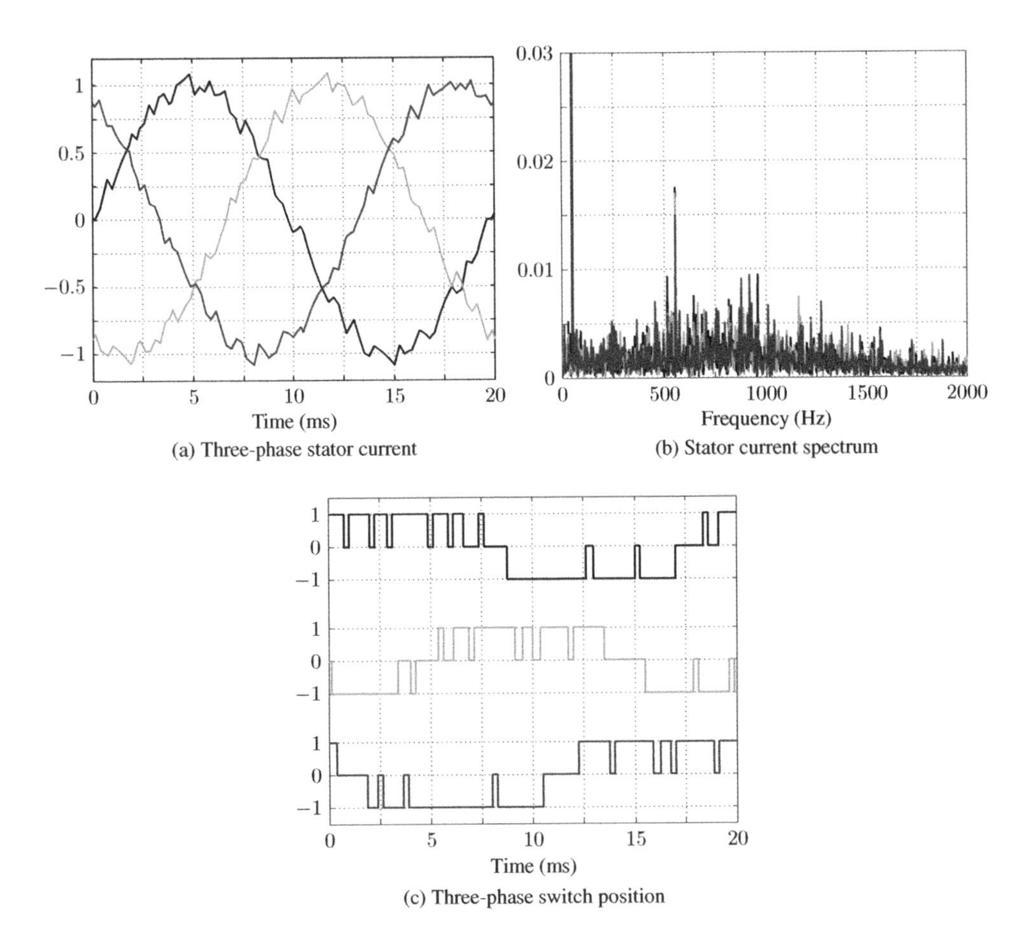

Figure 6.2 Simulated waveforms for direct MPC with the horizon $N_p = 1$, sampling interval $T_s = 125$ μs, and weighting $\lambda_u = 8.4 \cdot 10^{-3}$, at full speed and rated torque. The switching frequency is approximately 250 Hz

Table 6.1 Comparison of direct MPC with SVM and an OPP in terms of the current TDD I_{TDD}, torque TDD T_{TDD}, and switching frequency f_{sw}

Controller	Controller settings	I_{TDD} (%)	T_{TDD} (%)	f_{sw} (Hz)
MPC	$N_p = 1, \lambda_u = 8.4 \cdot 10^{-3}$	5.96	4.65	250
MPC	$N_p = 10, \lambda_u = 8.3 \cdot 10^{-3}$	5.05	4.03	254
SVM	$f_c = 450$ Hz	7.71	5.35	250
OPP	$d = 5$	4.12	3.40	250

The penalty λ_u, carrier frequency f_c, and pulse number d are chosen such that a switching frequency of approximately 250 Hz results.

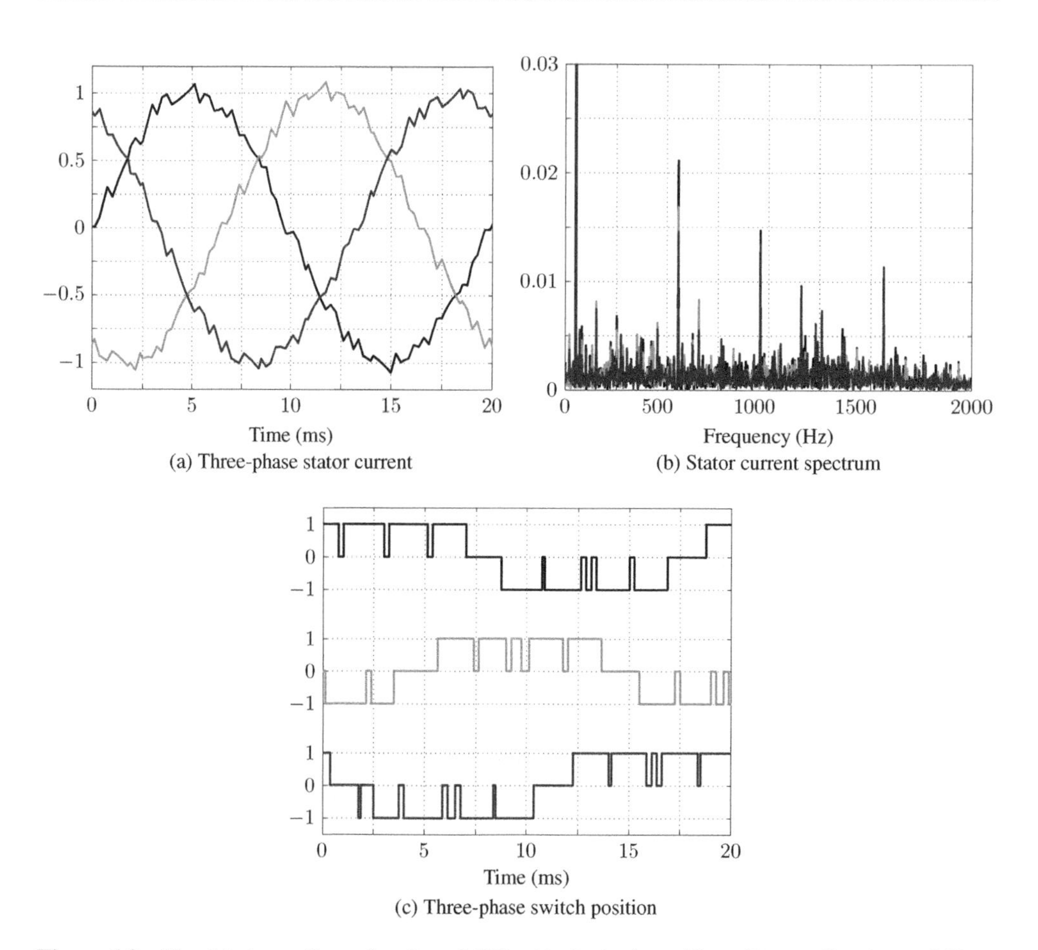

Figure 6.3 Simulated waveforms for direct MPC with the horizon $N_p = 10$, sampling interval $T_s = 125\ \mu\text{s}$, and weighting $\lambda_u = 8.3 \cdot 10^{-3}$. The operating point and the switching frequency are the same as in Fig. 6.2

degree of repetitiveness in the switching pattern. Accordingly, non-triplen odd-order harmonics are clearly identifiable in the current spectrum, such as the 11th, 13th, and 19th harmonics. Indeed, the degree of repetitiveness in the switching pattern and—correspondingly—the magnitude of the discrete harmonics in the current spectrum are remarkable. It can be observed that, in this case, long prediction horizons foster a discrete current spectrum by concentrating the harmonic power in harmonics of odd order. An analysis shows that the same applies to the triplen (common-mode) voltage harmonics. The shift of some of the harmonic ripple power into common-mode harmonics is one of the reasons why direct MPC with long prediction horizons leads—in general—to a lower current TDD than the horizon 1 case. Moreover, longer horizons tend to shift some of the differential-mode voltage harmonics from the low-frequency range to higher frequencies, resulting in a lower current TDD.

To facilitate a comparison with SVM, the corresponding waveforms of SVM are shown in Fig. 6.4. The equivalent carrier frequency $f_c = 450$ Hz results in the same switching frequency

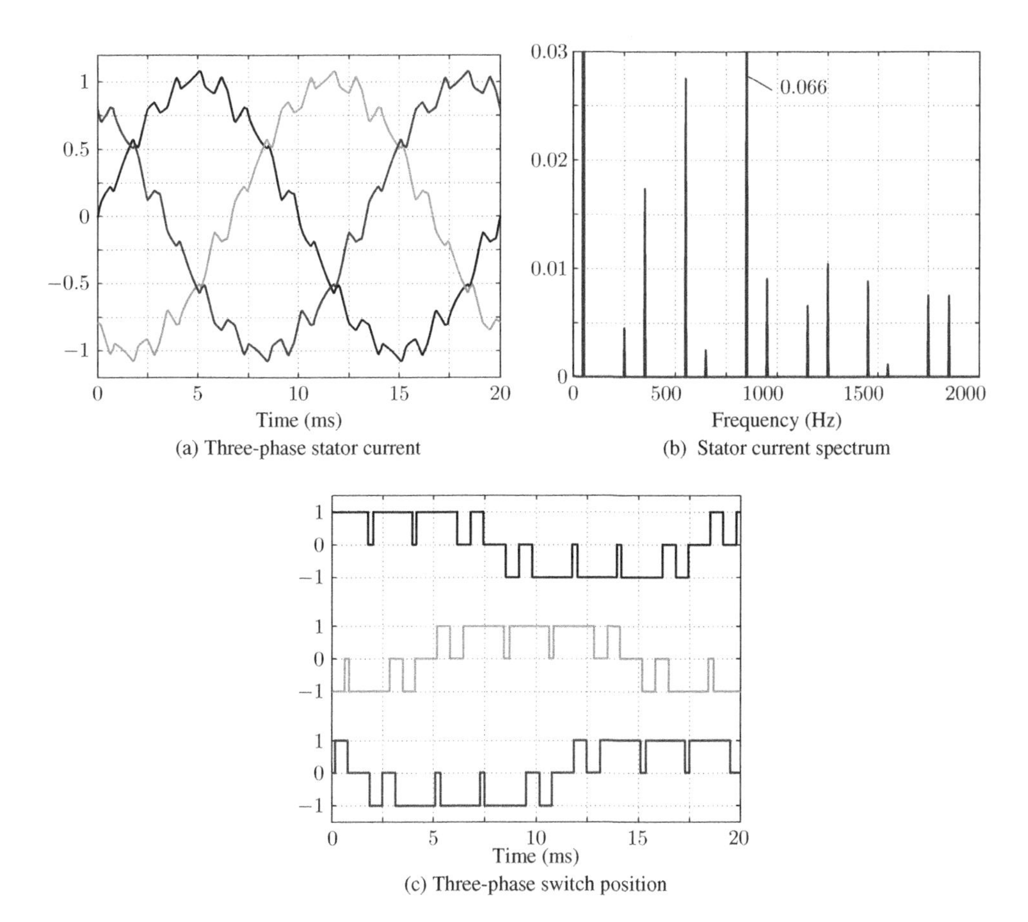

(a) Three-phase stator current

(b) Stator current spectrum

(c) Three-phase switch position

Figure 6.4 Simulated waveforms for volts per frequency (V/f) control and SVM with the equivalent carrier frequency $f_c = 450$ Hz. The operating point and the switching frequency are the same as in Fig. 6.2

as for MPC, that is, $f_{\rm sw} = 250$ Hz. The current TDD is at 7.71%, which is significantly higher than with direct MPC (see Table 6.1). As expected, because of the symmetry and repetitiveness of the switching pattern, SVM features a discrete current spectrum with distinctive harmonics at non-triplen and odd multiples of the fundamental frequency. Note that the 17th current harmonic has an amplitude of 0.066 pu.

On the other hand, for the same switching frequency and the pulse number $d = 5$, an OPP leads to a current TDD of 4.12%, which is approximately one percentage point lower than for direct MPC with $N_p = 10$. The corresponding waveforms of the OPP are shown in Fig. 6.5.

6.1.3 Closed-Loop Cost

Next, the influence of λ_u on the switching frequency and the current TDD is investigated. For each of the horizons $N_p = 1$, 3, 5, and 10, and for more than 1000 different values of λ_u,

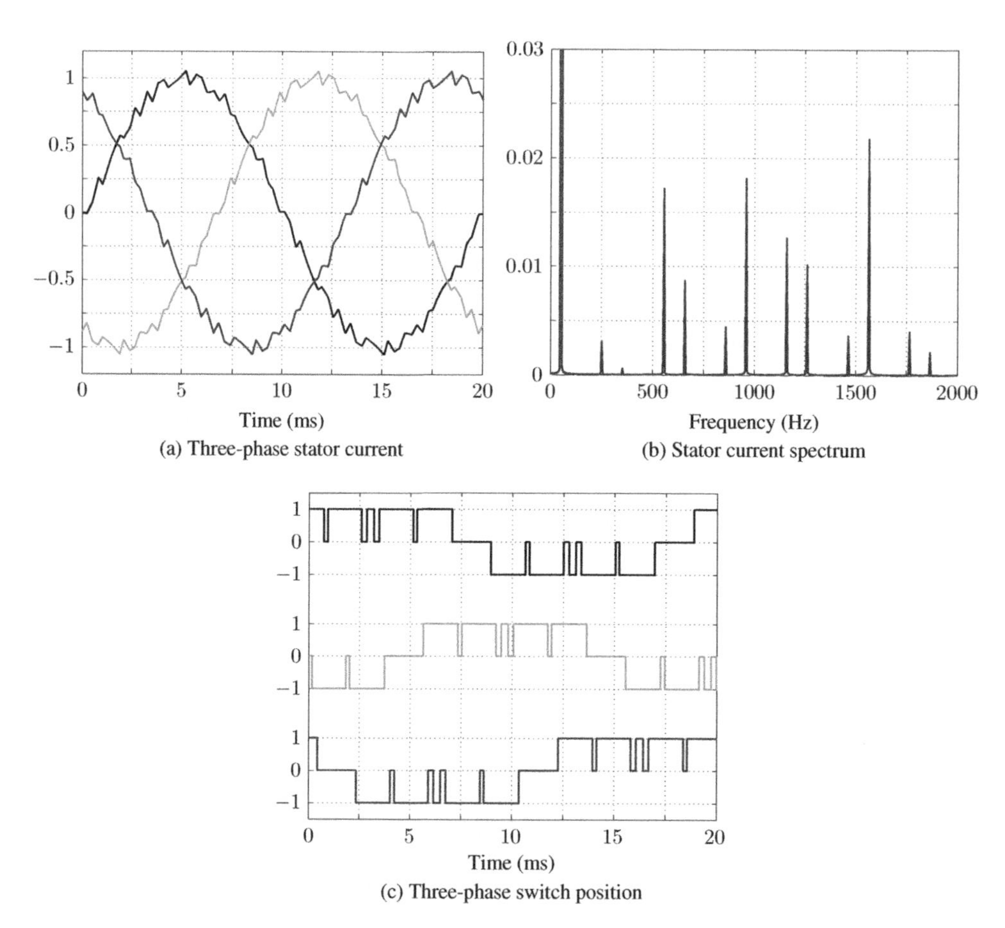

(a) Three-phase stator current
(b) Stator current spectrum
(c) Three-phase switch position

Figure 6.5 Simulated waveforms for V/f control and an OPP with pulse number $d = 5$. The operating point and the switching frequency are the same as in Fig. 6.2

ranging between 0 and 0.5, steady-state simulations were run. Focusing on switching frequencies between 100 Hz and 1 kHz, and current TDDs below 20%, the results are shown in Fig. 6.6 using a double logarithmic scale. Each simulation corresponds to a single data point. Polynomial functions, which approximate the individual data points, are overlaid. Figures 6.6(a) and (b) suggest that, for small prediction horizons, the relationship between λ_u and the performance variables is approximately linear in the double logarithmic scale; for larger values of N_p, the relationship is more complicated, but still monotonic.

When extending the horizon for a given λ_u, the switching frequency is increased while the current TDD is significantly reduced. This waterbed effect makes it difficult to assess from Fig. 6.6(a) and (b) the benefit long prediction horizons might have on these two key performance metrics. A more suitable measure is the *a posteriori* closed-loop cost (see (6.1)), which is illustrated in Fig. 6.6(c). As the prediction horizon is increased, the cost is clearly reduced, suggesting that horizons larger than 1 are beneficial. For example, with $\lambda_u = 0.01$ and $N_p = 1$, we have $J_{\text{cl}} \approx 0.05$, whereas with the horizon $N_p = 3$, the closed-loop cost can be reduced to

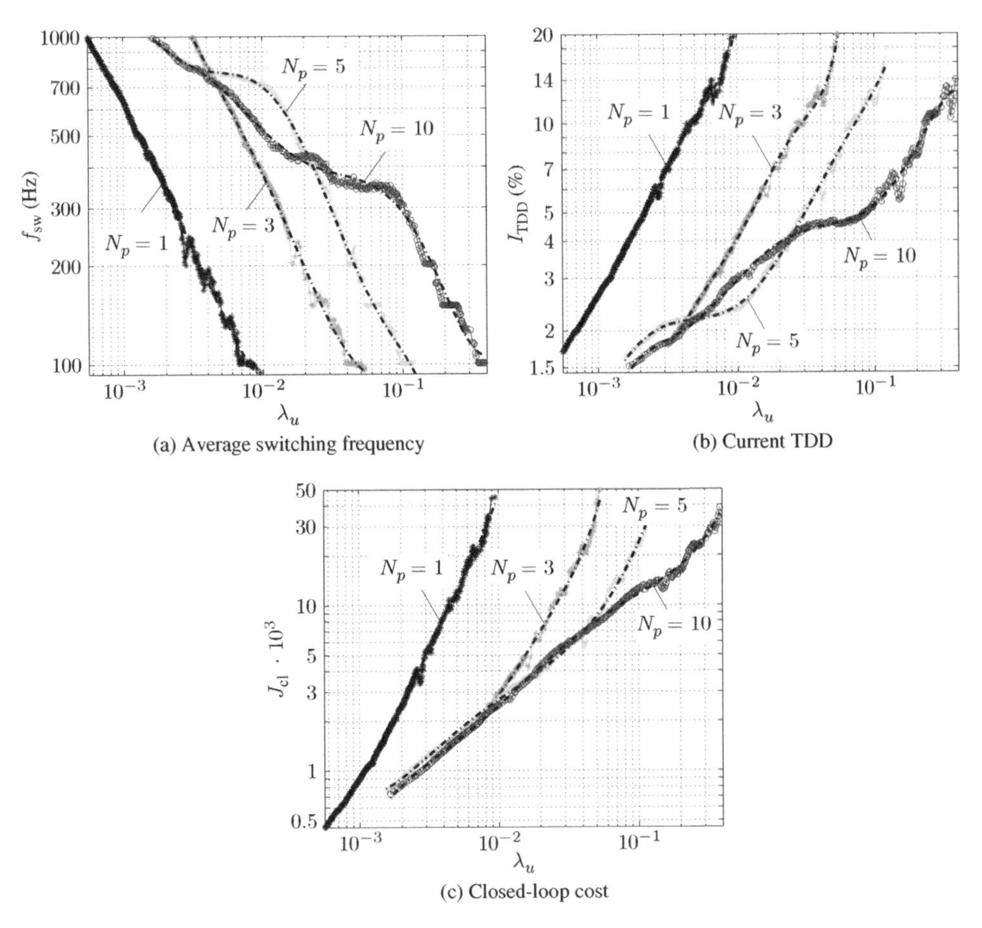

(a) Average switching frequency

(b) Current TDD

(c) Closed-loop cost

Figure 6.6 Key performance criteria of MPC for the prediction horizons $N_p = 1$, 3, 5, and 10 and sampling interval $T_s = 25$ µs. The switching frequency, current TDD, and closed-loop cost are shown as a function of the tuning parameter λ_u, using a double logarithmic scaling. The individual simulations are indicated using dots, and their overall trend is approximated using dash-dotted polynomials

$J_{cl} \approx 0.003$. This is a reduction by a factor of 17! We also note that, for this value of λ_u, the achieved *a posteriori* closed-loop cost is almost optimal. The benefit of long horizons on the current TDD and the switching frequency is investigated in the subsequent sections.

6.1.4 Relative Current TDD

6.1.4.1 Sampling Interval $T_s = 25$ µs

Figure 6.7 shows the *relative* current TDDs of SVM and of MPC, as defined in (6.2). In this figure, the simulations referring to SVM are indicated by squares, and those of OPPs are indicated by diamonds. Using the sampling interval $T_s = 25$ µs, hundreds of individual

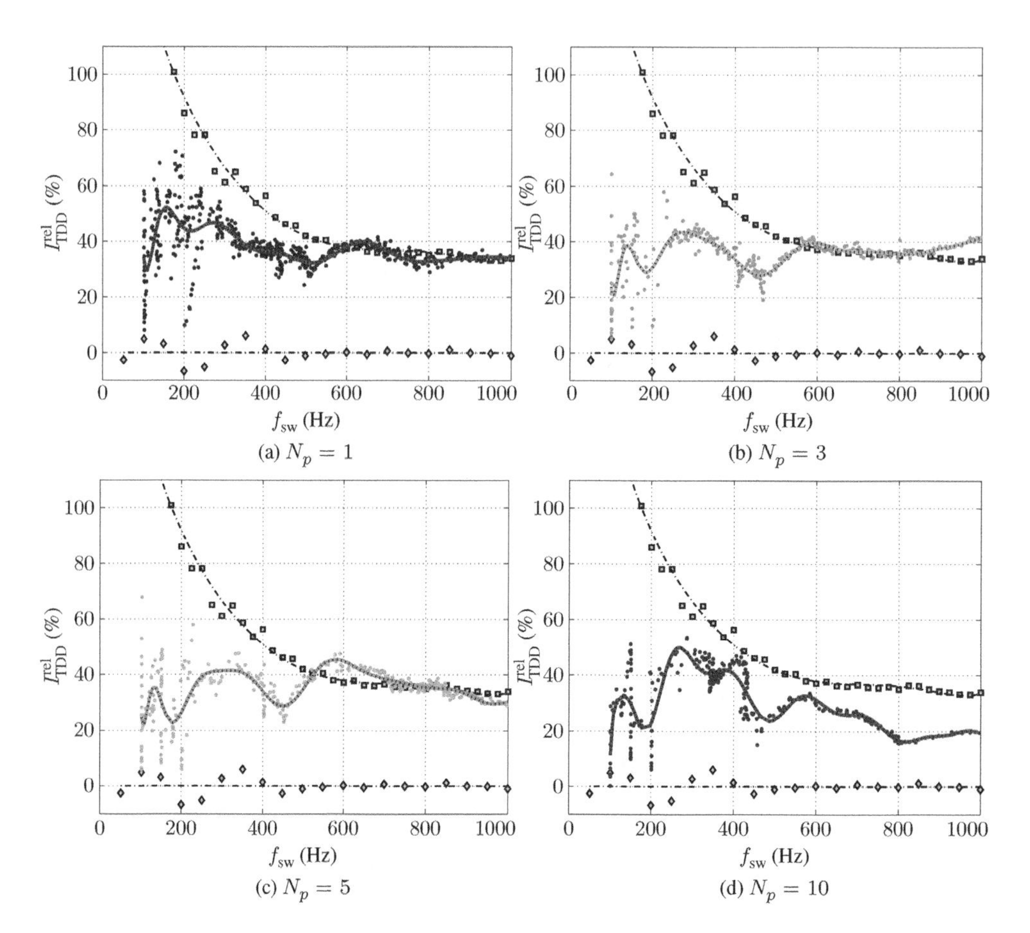

Figure 6.7 Trade-off between the *relative* current TDD and the switching frequency for MPC with the prediction horizons $N_p = 1, 3, 5,$ and 10, and sampling interval $T_s = 25$ μs. The relative current TDD of SVM and OPP modulation is indicated by squares and diamonds, respectively

simulations of MPC with prediction horizons $N_p = 1, 3, 5,$ and 10 were performed, using different weights λ_u. Specifically, λ_u was varied between 0 and 1. Each simulation corresponds to a dot in the figure. The individual simulation results were approximated by polynomials in a least-squares sense, which are shown in Fig. 6.7 as solid lines. The trend lines for the different prediction horizons are summarized in Fig. 6.9(a).

It can be clearly seen that long prediction horizons reduce the current TDD. In fact, for high switching frequencies above 600 Hz, the horizon 1 case resembles the performance of SVM. Increasing the horizon to 10 steps reduces the current TDD by about 15% compared to SVM. For even higher switching frequencies above 800 Hz, the current distortions are half way between that of SVM and OPPs. In terms of the absolute current TDD, however, the differences are small and in the range of a fraction of 1% (see also Fig. 6.1).

For low switching frequencies between 100 and 250 Hz, the performance results are somewhat scattered. The trend lines suggest that around $f_{\mathrm{sw}} = 200$ Hz the current TDD can be

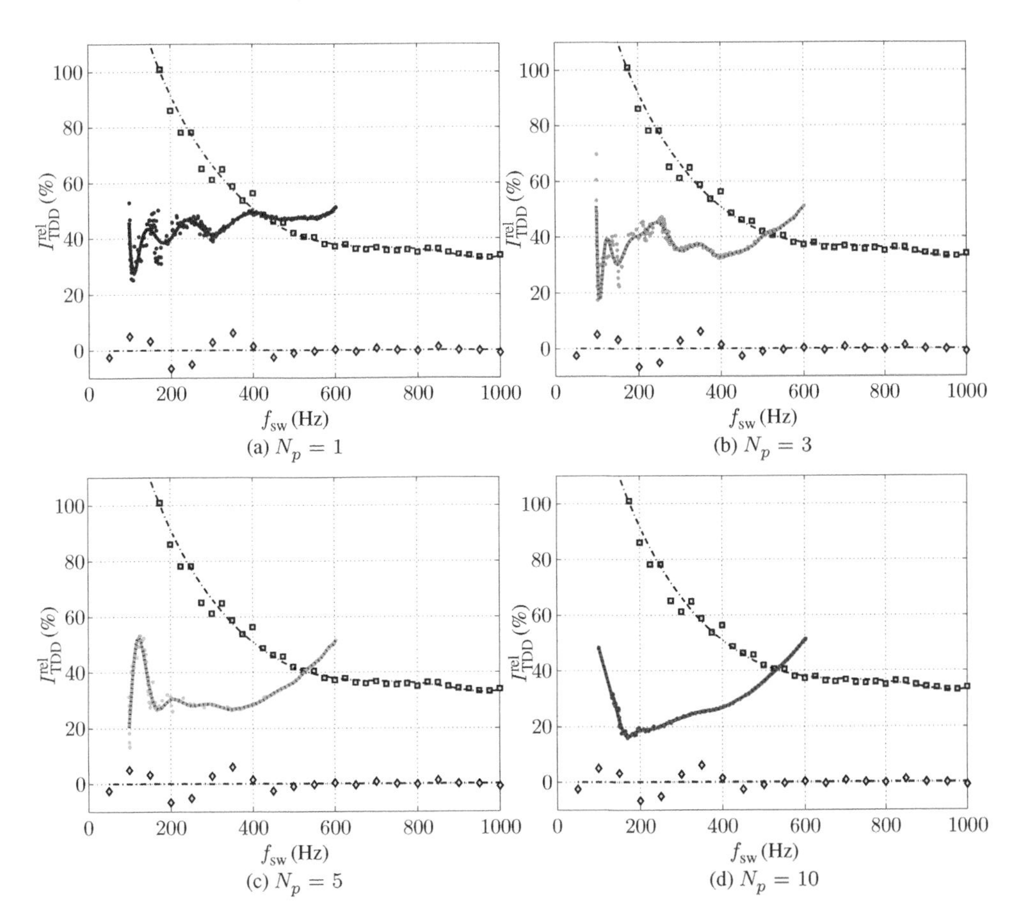

Figure 6.8 Trade-off between the *relative* current TDD and the switching frequency for MPC with the prediction horizons $N_p = 1, 3, 5$, and 10, and sampling interval $T_s = 125$ μs. The relative current TDD of SVM and OPP modulation is indicated by squares and diamonds, respectively

reduced by about 30% when increasing the prediction horizon from $N_p = 1$ to 5. Longer horizons do not appear to carry any additional performance benefit. Interestingly, for long horizons such as $N_p = 10$ and low switching frequencies, the switching frequency appears to lock into integer multiples of the fundamental frequency. This is apparent for $f_{\mathrm{sw}} = 100$, 150, and 200 Hz. For these switching frequencies and for particular choices of λ_u, MPC almost reproduces the steady-state performance of OPPs, in terms of the current TDD achieved for a given switching frequency.

6.1.4.2 Sampling Interval $T_s = 125$ μs

The simulations are repeated for a fivefold longer sampling interval, that is, $T_s = 125$ μs. Figure 6.8 shows the resulting trade-off relationships, analog to those in Fig. 6.7. The summary plot is provided in Fig. 6.9(b). As in the $T_s = 25$ μs case, longer prediction horizons improve

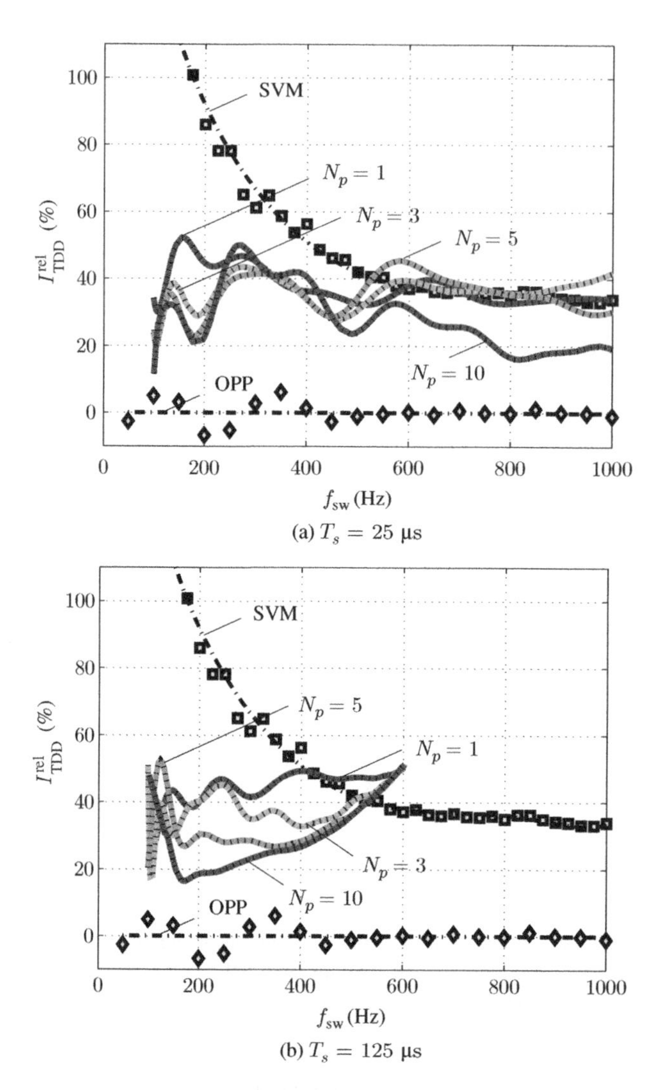

Figure 6.9 Trade-off between the *relative* current TDD and the switching frequency for MPC with the prediction horizons $N_p = 1, 3, 5$, and 10, and sampling interval $T_s = 25$ and 125 μs, respectively. The relative current TDD of SVM and OPP modulation is indicated by squares and diamonds, respectively

the performance of MPC by lowering the current TDD for a given switching frequency. This becomes particularly evident for switching frequencies between 150 and 450 Hz. In this range, MPC with $N_p = 10$ exhibits a relative current TDD that is approximately 20% lower than that for the $N_p = 1$ case. When comparing Figs. 6.9(a) and (b), we note that, in addition to the weight λ_u, the choice of sampling interval has a significant impact on the closed-loop performance. This somewhat complicates the tuning procedure for direct MPC.

All trade-off curves converge at $f_{sw} = 600$ Hz, which corresponds to an effectively zero penalty on the switching effort, that is, $\lambda_u \approx 0$. When not penalizing the switching transitions

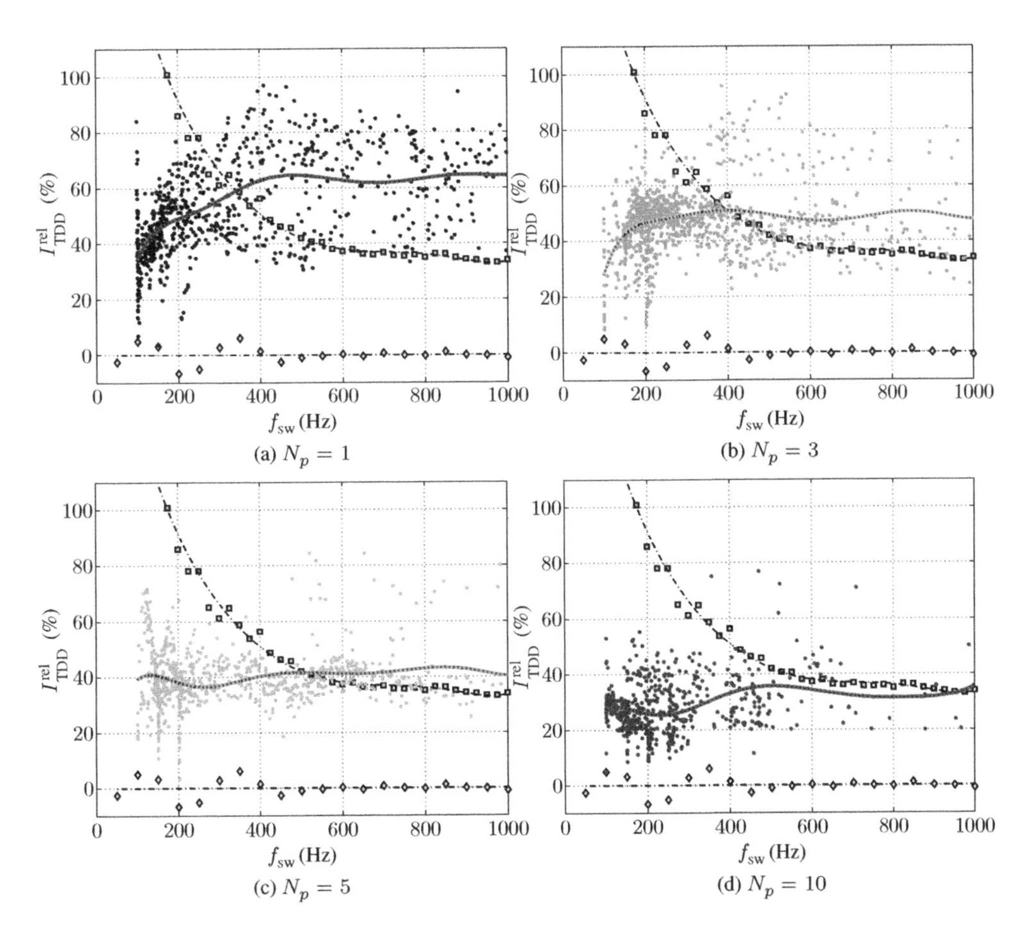

Figure 6.10 Trade-off between the *relative* current TDD and the switching frequency for MPC with the prediction horizons $N_p = 1, 3, 5$, and 10, using Monte Carlo simulations. The relative current TDD of SVM and OPP modulation is indicated by squares and diamonds, respectively

and thus only penalizing the predicted deviation of the current from its sinusoidal reference, MPC turns into a deadbeat controller. Here, the current loop effectively constitutes two first-order systems (one in the α-axis and the other one in the β-axis) and the length of the prediction horizon ceases to have an impact on the performance of MPC. In this situation, MPC with $N_p = 1$ yields the same control action as MPC with $N_p > 1$.

6.1.4.3 Monte Carlo Simulations

We have seen that, in addition to the tuning parameter λ_u and the horizon N_p, the sampling interval T_s has a profound influence on the MPC performance. The reason for this is that the MPC cost function in (5.7) evaluates system predictions over a prediction horizon of length N_pT_s *in time*.

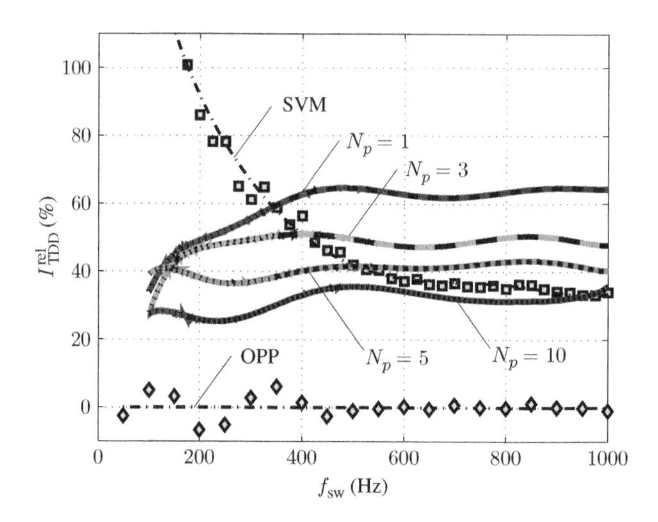

Figure 6.11 Trade-off between the *relative* current TDD and the switching frequency for MPC with the prediction horizons $N_p = 1, 3, 5,$ and 10, using Monte Carlo simulations. The relative current TDD of SVM and OPP modulation is indicated by squares and diamonds, respectively

To derive results that take into account a variety of sampling intervals, we carried out Monte Carlo simulations with random sampling intervals and random weights. Specifically, the sampling interval was randomly chosen from the interval $T_s \in [5, 200]$ μs, and the weight was chosen from $\lambda_u \in [0, 5]$. Moreover, the initial conditions of the drive system were random, including random initial stator currents and rotor fluxes for the induction machine, and random initial switch positions for the inverter. As previously, to ensure that the simulations were captured at steady state, presimulations were run, which were not recorded.

MPC with prediction horizons $N_p \in \{1, 3, 5, 10\}$ was considered, and approximately 10^4 simulations were performed. The results are depicted in Fig. 6.10, where each data point corresponds to one closed-loop simulation. To determine the trend lines, polynomials were fitted to the data points using a least-squares approach. The resulting curves are shown as solid lines in Fig. 6.10, and they are summarized in Fig. 6.11.

It can be clearly observed that, as the prediction horizon is extended, the performance of MPC is improved by reducing the current TDD for a given switching frequency. For switching frequencies above 400 Hz, MPC with the horizon $N_p = 10$ reduces the relative current TDD by about 30%, when compared to the popular $N_p = 1$ case. Even the relatively short horizon of $N_p = 3$ shows an improvement on the horizon 1 case by 15%. One can also observe that the performance improvement is most significant when enlarging the horizon from 1 to 3, whereas the performance gains level off when further increasing N_p to 5 and 10. For very low switching frequencies, the trade-off curves for different prediction horizons converge, getting close to the point of six-step operation, that is, fundamental frequency modulation.

For switching frequencies above 340 Hz, MPC with the horizon $N_p = 1$ performs worse than SVM. In terms of the average current TDD, MPC with the horizon $N_p = 10$ always outperforms SVM and achieves a steady-state performance that is not dissimilar to that of OPPs when operating at low switching frequencies of 250 Hz and below.

6.1.5 Operation during Transients

One of the major benefits of direct MPC is its very fast dynamic behavior during transients. Consider the MPC scheme with the horizon 1, sampling interval $T_s = 25$ µs, and weight $\lambda_u = 2.55 \cdot 10^{-3}$. At nominal speed, reference torque steps of magnitude 1 pu are imposed (see Fig. 6.12(a)). The steps on the torque reference are translated into steps in the current references, shown as dash-dotted lines in Fig. 6.13(a). The corresponding switching pattern is shown in Fig. 6.14(a), with the switching frequency being $f_{\rm sw} = 252$ Hz.

When switching from the rated torque to zero, the voltage applied to the machine is momentarily inverted, leading to an extremely short settling time of 0.35 ms. On the other hand, the torque step from 0 to 1 pu at 4 ms is significantly slower. This is due to the small voltage

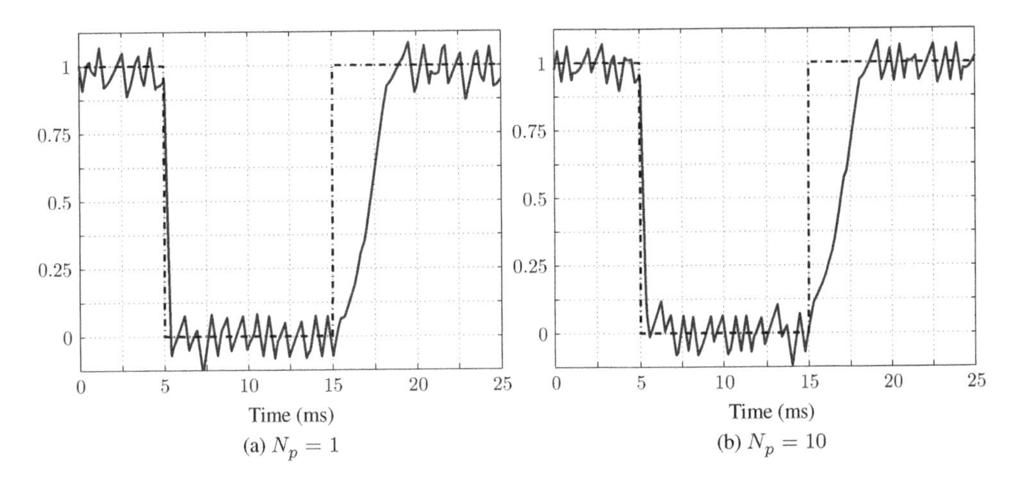

Figure 6.12 Torque for the horizons $N_p = 1$ and $N_p = 10$ during torque steps

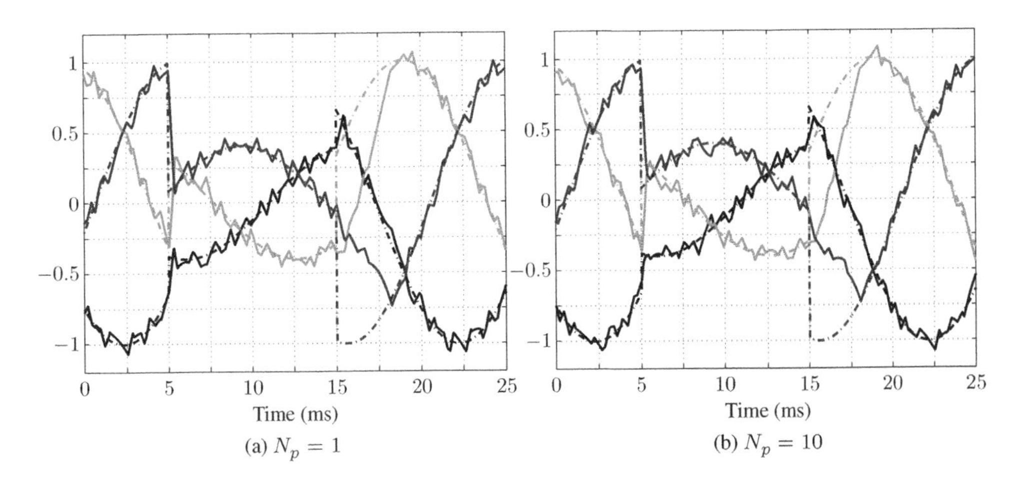

Figure 6.13 Three-phase stator currents for the horizons $N_p = 1$ and $N_p = 10$ during torque steps

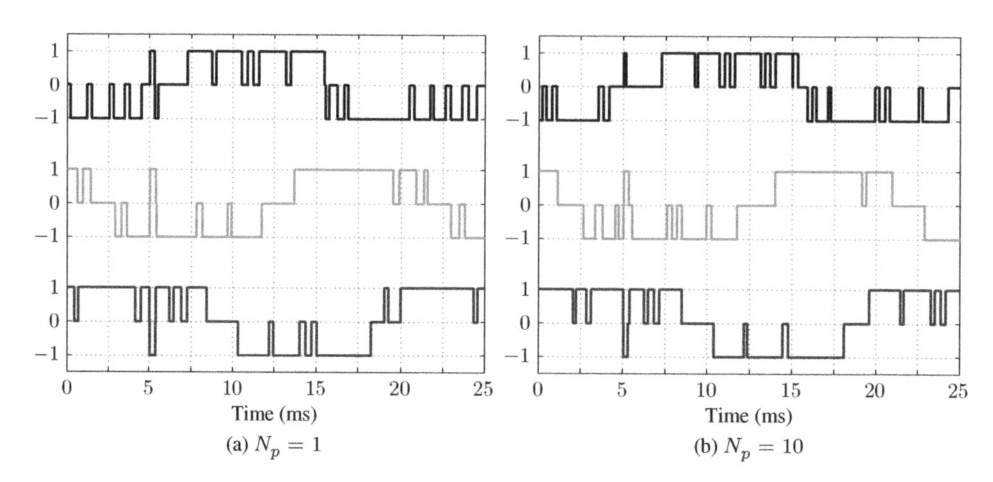

Figure 6.14 Three-phase switch positions for the horizons $N_p = 1$ and $N_p = 10$ during torque steps

margin available, which results from the machine operating at nominal speed. Nevertheless, as can be seen in Fig. 6.13(a), the currents are regulated as quickly as possible to their new references. Note that, because of the constraint (5.22c), switching between −1 and 1 is inhibited, and switching is performed via an intermediate zero switch position, which is applied for one sampling interval T_s.

Figures 6.12(b), 6.13(b), and 6.14(b) show the corresponding step responses for MPC with horizon 10. The settling times are nearly identical to those of the horizon 1 case. The weight $\lambda_u = 120 \cdot 10^{-3}$ was chosen, which results in the same switching frequency as previously, that is, $f_{\text{sw}} = 250\,\text{Hz}$.

When operating at 50% speed and applying the same torque steps as before, the torque settling times are 0.5 ms for the step-down and 1.1 ms for the step-up case, for MPC with both horizon 1 and horizon 10. We conclude that during transients, the dynamic performance of direct MPC is effectively limited only by the available voltage, regardless of the length of the prediction horizon. In particular, long horizons do not slow down the dynamic response of MPC. This is also due to the fact that the computation of the controller output is assumed to take the same time regardless of the length of the prediction horizon. Indeed, we assume in this chapter that the computation time is zero.

6.2 Suboptimal MPC via Direct Rounding

We have seen in Sects. 5.2.4 and 5.3.3 that, in general, direct rounding of the unconstrained solution provides suboptimal solutions. However, in some cases, the generator matrix $\boldsymbol{V}$ is almost diagonal and its basis vectors are almost orthogonal. This motivates the investigation of an approximate solution, based on trivial quantization (rounding). This approach yields suboptimal solutions but is computationally very fast, because it requires only basic matrix manipulations. Sphere decoding and branching is not required.

Specifically, instead of invoking Algorithm 1 in Sect. 5.3.2, the (suboptimal) sequence of switch positions is obtained by rounding (quantizing) the unconstrained solution component-wise to the nearest integer in the set $\mathcal{U}$:

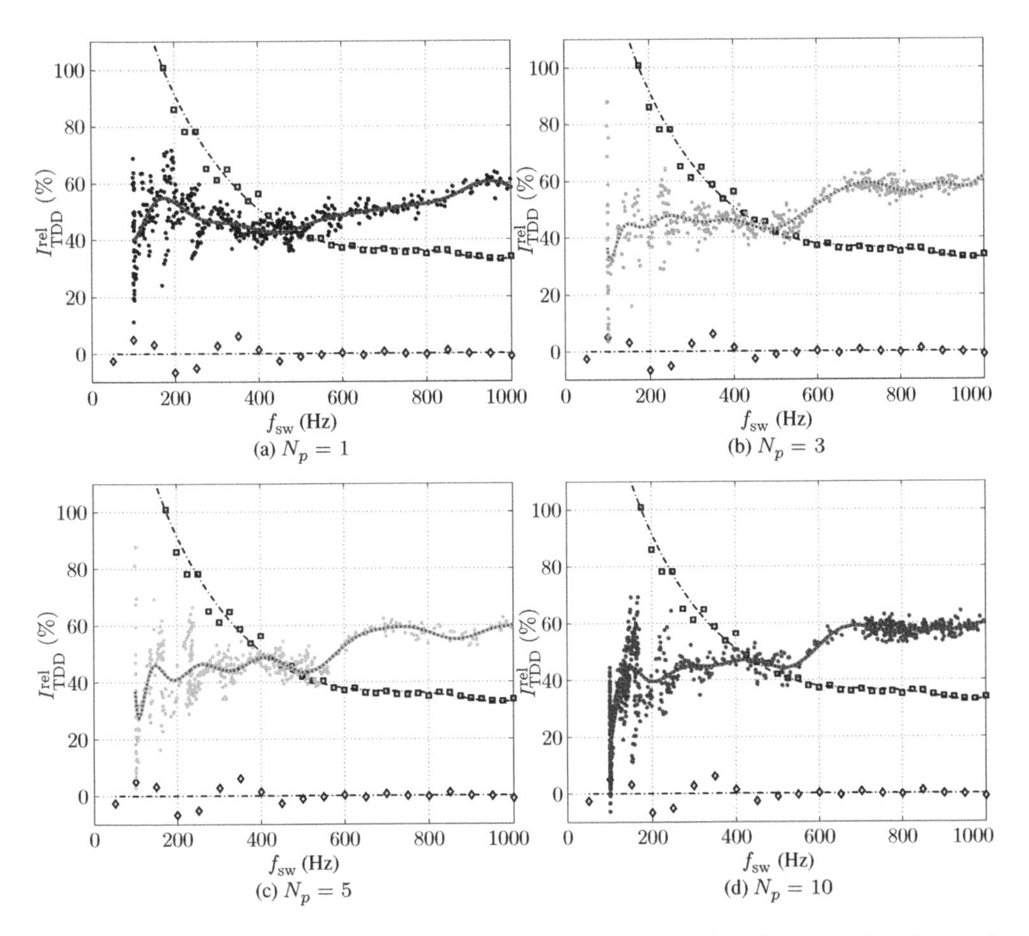

Figure 6.15 Trade-off between the *relative* current TDD and the switching frequency for *suboptimal* MPC with the prediction horizons $N_p = 1, 3, 5$, and 10, and sampling interval $T_s = 25$ μs, when rounding the unconstrained solution

$$\boldsymbol{U}_{\text{sub}}(k) = \text{round}_{\mathcal{U}}(\boldsymbol{U}_{\text{unc}}(k)). \tag{6.3}$$

Recall that, as defined in (5.23e), $\boldsymbol{U}_{\text{unc}}(k)$ denotes the unconstrained solution to the optimization problem (5.22) at time step k.

The simulations in Sect. 6.1.4 for (optimal) direct MPC were repeated for this suboptimal design. Using the sampling interval $T_s = 25$ μs, the resulting trade-off curves for suboptimal MPC are depicted in Fig. 6.15 for the horizons 1, 3, 5, and 10. The trend lines were fitted as before and are shown separately in Fig. 6.16.

For switching frequencies below 300 Hz, the weight λ_u is large and the diagonal terms dominate over the off-diagonal terms in the generator matrix $\boldsymbol{V}$. As a result, the component-wise quantization in (6.3) yields solutions close to the optimal one. This can be seen when comparing Figs. 6.9(a) and 6.16 with each other. For $N_p = 1$, suboptimal MPC exhibits a performance

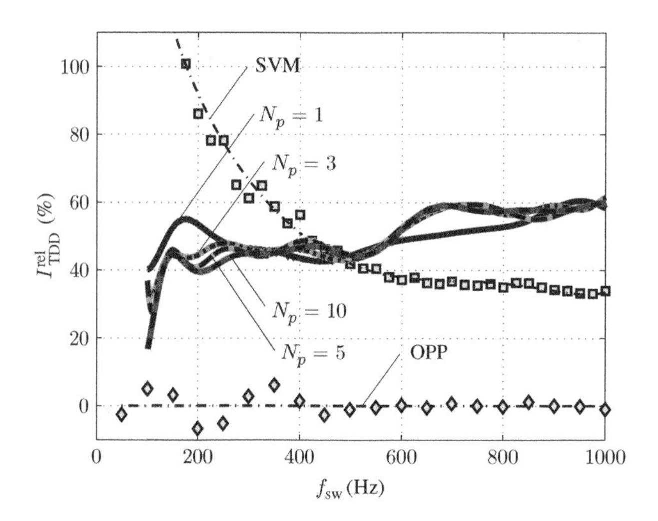

Figure 6.16 Trade-off between the *relative* current TDD and the switching frequency for *suboptimal* MPC with the prediction horizons $N_p = 1, 3, 5$, and 10, and sampling interval $T_s = 25$ μs, when rounding the unconstrained solution

that is very similar to that of optimal MPC. Longer horizons improve the performance of suboptimal MPC (6.3), but to a lesser degree than for the optimal case (5.22).

High switching frequencies are the result of small λ_u and generator matrices that are correspondingly less orthogonal. Using the trivial quantization in (6.3) for switching frequencies above 300 Hz leads to suboptimal solutions that are clearly inferior, with the $N_p = 1$ case being about 15% worse than the optimal solution. Moreover, extending the horizon appears to be of very little benefit, if any at all. As λ_u is decreased and the switching frequency is increased, the relative deterioration of suboptimal MPC becomes more prominent. The absolute performance loss, in terms of current TDD for a given switching frequency, is, however, very small.

6.3 Performance Evaluation for the NPC Inverter Drive System with an *LC* Filter

For direct MPC schemes, the performance benefits of long prediction horizons become more pronounced when increasing the complexity of the power electronic system to be controlled. As an example, consider an inverter driving an induction machine via an intermediate *LC* filter. Ignoring the rotor dynamics, which are slow compared to the dynamics of the stator and *LC* filter, this system consists of two third-order systems in an orthogonal coordinate system. To control such a system, it is common practice to design two single-input single-output (SISO) proportional integral (PI) controllers for the inverter current. To rein in the filter resonance, an additional active damping loop is typically added [4].

A more elegant—and in the end more promising—approach is to treat the higher order system as a multiple-input multiple-output (MIMO) system and to design a single MIMO

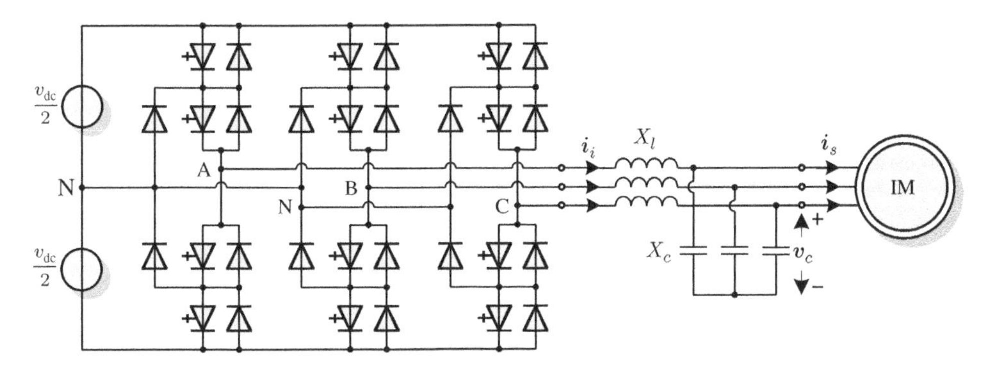

Figure 6.17 Three-level, three-phase, neutral-point-clamped voltage source inverter with an LC filter driving an induction machine. The neutral point potential is fixed at zero

controller for it. This can easily be accomplished with MPC, by regulating the inverter current, capacitor voltage, and stator current simultaneously. For the MPC scheme to perform well, however, a long prediction horizon is required that covers a significant fraction of the oscillation period of the filter resonance. To solve the underlying integer optimization problem, the sphere decoding algorithm proposed in Chap. 5 is used.

6.3.1 Case Study

In the following, the previously used case study consisting of an NPC inverter and an MV induction machine is augmented by an LC filter. As shown in Fig. 6.17, a symmetrical three-phase LC filter is placed between the inverter and the machine to reduce the harmonic distortions at the stator windings. Adopting again the pu system, we introduce the reactances[1] X_l and X_c for the filter inductor and capacitor, respectively. The internal resistors of the inductor and capacitor are denoted by R_l and R_c, respectively.

The total dc-link voltage v_{dc} is assumed to be constant, and the neutral point potential is clamped to zero. We use $\boldsymbol{u} = [u_a \; u_b \; u_c]^T$ to denote the three-phase switch position. The inverter voltage $\boldsymbol{v}_i = [v_{i\alpha} \; v_{i\beta}]^T$ in stationary orthogonal coordinates is given by

$$\boldsymbol{v}_i = \frac{1}{2} v_{\mathrm{dc}} \, \tilde{\boldsymbol{K}} \, \boldsymbol{u}. \tag{6.4}$$

The reduced Clarke transformation $\tilde{\boldsymbol{K}}$ has been defined in (2.13).

The state-space equations of the filter in the $\alpha\beta$ coordinate system are

$$\frac{\mathrm{d}\boldsymbol{i}_i}{\mathrm{d}t} = \frac{1}{X_l}(\boldsymbol{v}_i - \boldsymbol{v}_c - R_l \boldsymbol{i}_i - R_c(\boldsymbol{i}_i - \boldsymbol{i}_s)) \tag{6.5a}$$

$$\frac{\mathrm{d}\boldsymbol{v}_c}{\mathrm{d}t} = \frac{1}{X_c}(\boldsymbol{i}_i - \boldsymbol{i}_s), \tag{6.5b}$$

where $\boldsymbol{i}_i$ denotes the inverter current, $\boldsymbol{v}_c$ is the capacitor voltage, and $\boldsymbol{i}_s$ is the stator current.

[1] Note that the notation of X_c is slightly misleading, because X_c is not a reactance, but rather its inverse. Specifically, we have $\frac{1}{X_c} = \frac{1}{\omega_B C_c} \frac{1}{Z_B}$, see also Footnote 1 in Sect. 2.4.1.

Table 6.2 Parameters in the SI (left) and per unit system (right) of the three-level NPC inverter drive system with an LC filter

Parameter	SI symbol	SI value	pu symbol	pu value
Stator resistance	R_s	57.61 mΩ	R_s	0.0108 pu
Rotor resistance	R_r	48.89 mΩ	R_r	0.0091 pu
Stator leakage inductance	L_{ls}	2.544 mH	X_{ls}	0.1493 pu
Rotor leakage inductance	L_{lr}	1.881 mH	X_{lr}	0.1104 pu
Main inductance	L_m	40.01 mH	X_m	2.349 pu
Number of pole pairs	p	5		
dc-link voltage	V_{dc}	5.2 kV	V_{dc}	1.930 pu
Filter inductor	L	2 mH	X_l	0.1174 pu
Filter capacitor	C	200 μF	X_c	0.3363 pu
Filter inductor resistance	R_l	2 mΩ	R_l	0.0004 pu
Filter capacitor resistance	R_c	2 mΩ	R_c	0.0004 pu

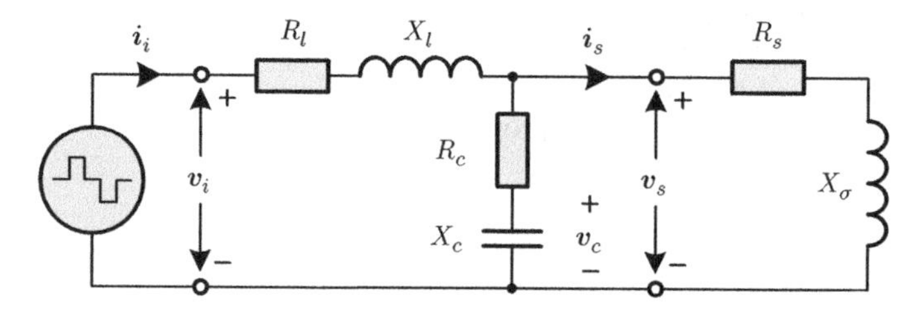

Figure 6.18 Harmonic model of the drive system with an LC filter given in stationary orthogonal coordinates. The model is valid for frequencies above the fundamental frequency

The voltage applied to the stator windings of the machine is

$$\boldsymbol{v}_s = \boldsymbol{v}_c + R_c(\boldsymbol{i}_i - \boldsymbol{i}_s). \tag{6.6}$$

The state-space model of the induction machine remains the same as in (4.11)–(4.14), using the stator current $\boldsymbol{i}_s = [i_{s\alpha}\ i_{s\beta}]^T$ and the rotor flux linkage vector $\boldsymbol{\psi}_r = [\psi_{r\alpha}\ \psi_{r\beta}]^T$ as state variables.

The previously considered 3.3 kV, 50 Hz squirrel-cage induction machine rated at 2 MVA with 356 A rated current is used again. The machine's total leakage reactance is $X_\sigma = 0.25$ pu. The values of the LC filter are given by $L = 2$ mH and $C = 200$ μF. The detailed parameters of the machine, inverter, and LC filter are provided in Table 6.2. The pu system is established using the rated machine values, which are stated in Table 2.9.

For frequencies beyond the fundamental frequency, a simplified model can be derived. In particular, the model of the induction machine (4.11) is replaced by its harmonic model. As explained in Sect. 2.2.5, the machine can be represented by the series connection of the stator resistor R_s and the total leakage reactance X_σ. This leads to the harmonic model shown in Fig. 6.18. The voltage source on the left-hand side of the circuit represents the voltage harmonics injected by the inverter.

Two conclusions can be drawn. First, the three resistors in the system are almost negligible and provide effectively no passive damping. Second, the dominant resonance of the system is constituted by the filter capacitance oscillating against the two inductances, namely the filter inductance and the total leakage inductance of the machine. Therefore, the resonance frequency in hertz can be expressed in SI units and in the pu system, respectively, as

$$f_{\rm res} = \frac{1}{2\pi\sqrt{C\frac{LL_\sigma}{L+L_\sigma}}} = f_B\frac{1}{\sqrt{X_c\frac{X_l X_\sigma}{X_l+X_\sigma}}}, \tag{6.7}$$

where $f_B = \omega_B/(2\pi) = 50$ Hz is the base frequency. For the given parameters, $f_{\rm res} = 304$ Hz results.

6.3.2 Controller Model

As the state vector for the prediction model, we chose the eight-dimensional vector

$$\boldsymbol{x} = [\boldsymbol{i}_i^T\ \boldsymbol{v}_c^T\ \boldsymbol{i}_s^T\ \boldsymbol{\psi}_r^T]^T \tag{6.8}$$

in the stationary orthogonal coordinate system. The three-phase switch position $\boldsymbol{u}$ constitutes the input vector, whereas the inverter current, the capacitor voltage, and the stator current are the output variables, which form the output vector

$$\boldsymbol{y} = [\boldsymbol{i}_i^T\ \boldsymbol{v}_c^T\ \boldsymbol{i}_s^T]^T. \tag{6.9}$$

As discussed in the previous chapters, the dynamic of the rotor is neglected and the rotor speed ω_r is considered to be a time-varying parameter. By inserting the inverter voltage (6.4) into the filter model (6.5) and the filter output voltage (6.6) into the machine model (4.11), the continuous-time prediction model

$$\frac{{\rm d}\boldsymbol{x}(t)}{{\rm d}t} = \boldsymbol{F}\boldsymbol{x}(t) + \boldsymbol{G}\boldsymbol{u}(t) \tag{6.10a}$$

$$\boldsymbol{y}(t) = \boldsymbol{C}\boldsymbol{x}(t) \tag{6.10b}$$

can be obtained, where the matrices $\boldsymbol{F}$, $\boldsymbol{G}$, and $\boldsymbol{C}$ are provided in Appendix 6.A. Using the exact Euler approach—see (5.5)—with the sampling interval T_s results in the equivalent discrete-time state-space model

$$\boldsymbol{x}(k+1) = \boldsymbol{A}\boldsymbol{x}(k) + \boldsymbol{B}\boldsymbol{u}(k) \tag{6.11a}$$

$$\boldsymbol{y}(k) = \boldsymbol{C}\boldsymbol{x}(k). \tag{6.11b}$$

6.3.3 Optimization Problem

For the drive system with an *LC* filter, the three output variables in $\alpha\beta$ coordinates need to be regulated along their trajectories. Specifically, the inverter current $\boldsymbol{i}_i$, the capacitor voltage $\boldsymbol{v}_c$, and the stator current $\boldsymbol{i}_s$ should track their references $\boldsymbol{i}_i^*$, $\boldsymbol{v}_c^*$, and $\boldsymbol{i}_s^*$, respectively. The latter are joined together in the output reference vector $\boldsymbol{y}^*$. Moreover, the switching frequency needs to be minimized. The block diagram of the proposed MPC scheme with reference tracking of $\boldsymbol{y}^*$ is depicted in Fig. 6.19.

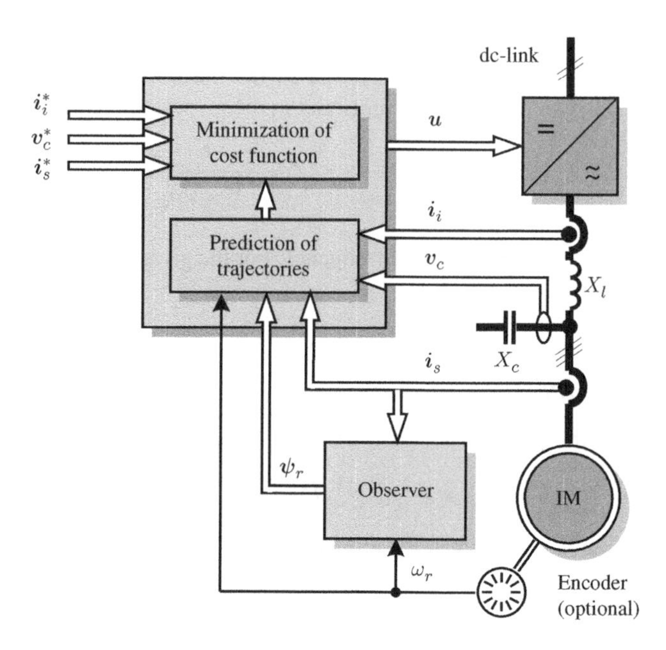

Figure 6.19 Direct MPC with reference tracking for the drive system with an LC filter

The general cost function (5.7) introduced in Sect. 5.1.3 is repeated here:

$$J = \sum_{\ell=k}^{k+N_p-1} \|\boldsymbol{y}^*(\ell+1) - \boldsymbol{y}(\ell+1)\|_{\boldsymbol{Q}}^2 + \lambda_u \|\Delta \boldsymbol{u}(\ell)\|_2^2, \tag{6.12}$$

which addresses all three output variables. As before, we require the penalty matrix $\boldsymbol{Q} \in \mathbb{R}^{6\times 6}$ to be positive semidefinite and symmetric. To achieve a low stator current TDD, the stator current ripple needs to be small. This can be achieved by prioritizing the tracking of the stator current, by choosing large values for the weighting factors of the α- and β-components of the stator current. These weighting factors correspond to the last two diagonal entries in $\boldsymbol{Q}$. The scalar penalty on the switching effort λ_u is required to be positive. The ratio between $\boldsymbol{Q}$ and λ_u decides on the trade-off between the overall tracking accuracy and the switching effort. When prioritizing the tracking of the stator current, this trade-off is equivalent to that between the stator current TDD and the switching frequency of the inverter.

Note that the output reference vector $\boldsymbol{y}^*$ in (6.12) is time-varying. Its evolution within the prediction horizon can be computed by transforming the drive system model (6.10) from the $\alpha\beta$ frame into the rotating dq reference frame. The electrical angular speed of the rotor ω_r, the torque reference T_e^*, and the reference of the stator flux magnitude Ψ_s^* define the operating point. Based on these, the relationship between the components of the output vector $\boldsymbol{y}$, the stator and rotor flux vectors $\boldsymbol{\psi}_s$ and $\boldsymbol{\psi}_r$, and the inverter voltage $\boldsymbol{v}_s$ can be computed at steady-state operating conditions. This is shown in detail in Appendix 6.B. When operating at nominal speed and rated torque, the relationship between these quantities in the rotating dq

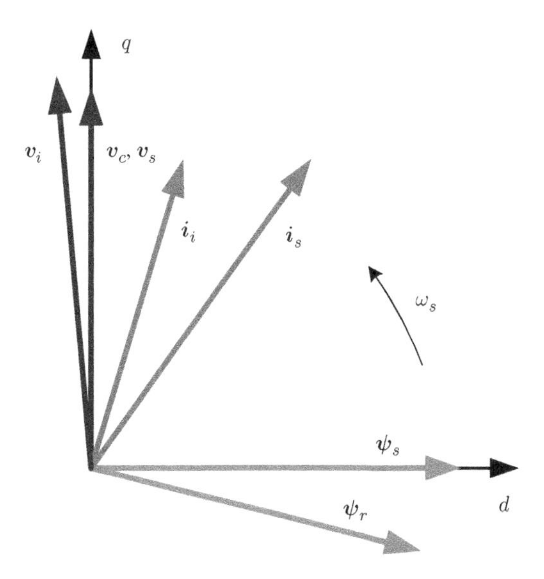

Figure 6.20 Drive system quantities in the rotating dq reference frame at steady-state operation. The reference frame rotates with the stator frequency ω_s

frame is exemplified in Fig. 6.20. As the filter capacitor resistance R_c is effectively zero, the capacitor and stator voltage vectors are almost equal.

The integer quadratic program underlying direct MPC with output reference tracking was stated in (5.22). Even though the dimension of the output reference trajectory $\boldsymbol{Y}^*(k)$ is increased from $2N_p$ to $6N_p$ and the dimension of the state vector $\boldsymbol{x}(k)$ is doubled from 4 to 8—when compared to the case study without an LC filter—the dimensions of $\boldsymbol{\Theta}(k)$ and the Hessian $\boldsymbol{H}$ remain the same, that is, $\boldsymbol{\Theta}(k) \in \mathbb{R}^{3N_p}$ and $\boldsymbol{H} \in \mathbb{R}^{3N_p \times 3N_p}$. This is due to the fact that the dimension of the optimizer, that is, of the switching sequence $\boldsymbol{U}(k)$, is the same. This implies that the computational burden required to solve the integer quadratic program remains unchanged. Nevertheless, the effort to compute $\boldsymbol{\Theta}(k)$ and the unconstrained solution $\bar{\boldsymbol{U}}_{\text{unc}}(k)$ is somewhat increased because of the higher dimensionality of $\boldsymbol{Y}^*(k)$ and $\boldsymbol{x}(k)$.

6.3.4 *Steady-State Operation*

Simulation results are presented in this section to highlight the benefits long prediction horizons bring when using direct MPC for higher order systems. The MV drive with an LC filter, as shown in Fig. 6.17, is used for this purpose. The NPC inverter is fed by a constant dc-link voltage and a fixed neutral point potential. The parameters of the drive system are provided in Table 6.2.

The controller sampling interval $T_s = 125$ µs is chosen to facilitate long prediction intervals *in time*. Even though such a relatively long sampling interval reduces the granularity of switching, it is beneficial when operating at low switching frequencies. The penalty on the tracking error is set to $\boldsymbol{Q} = \text{diag}(1, 1, 5, 5, 150, 150)$ in the cost function (6.12). The weight λ_u is tuned such that the desired switching frequency is achieved.

In the first step, the performance of direct MPC with reference tracking is investigated at steady-state operating conditions. Operation at nominal speed and rated torque is considered. To ensure that the drive system has settled at steady-state operation, the system is first simulated over several fundamental periods without recording the results. The prediction horizon of $N_p = 15$ is investigated and $\lambda_u = 0.28$ is chosen, resulting in an average device switching frequency of 303 Hz, which is a typical value for MV applications. A very low stator current TDD of 1.156% is achieved.

The steady-state waveforms of the electromagnetic torque and of the three-phase inverter current are shown in Figs. 6.21(a) and (b) over one fundamental period. Figure 6.21(c) displays the three-phase voltage across the filter capacitors, and Fig. 6.21(d) shows the three-phase stator current waveforms. The latter are effectively sinusoidal, despite operation at a low switching frequency. Figure 6.21(e) depicts the three-phase switch position. For each phase of the stator current, the spectrum was computed using a Fourier transformation of the current waveform, which was recorded over 15 fundamental periods. The spectrum of each phase is shown separately in Fig. 6.21(f). The amplitude of the fundamental component is 1 pu. To ensure a high resolution, the drive system was simulated with a sampling interval of 25 μs, despite the control algorithm being executed every 125 μs.

In the second step, the influence of the prediction horizon on the TDD of the stator current is investigated for a given switching frequency. Specifically, for different prediction horizons, the penalty λ_u was tuned such that an almost constant switching frequency of 300 Hz results. As shown in Fig. 6.22, when using a prediction horizon of one step, the stator current TDD at 7.43% is high, making the direct MPC scheme unsuitable for industrial applications. Increasing the prediction horizon to three steps, however, drastically reduces the TDD to 2.17%. This is a reduction by a factor of three. Further increases in the prediction horizon lead to further decreases in the TDD. For the prediction horizon $N_p = 20$, for example, a stator current TDD of 1.01% is achieved when operating at a switching frequency of 303 Hz.

This result is remarkable in that a very low stator current TDD can be achieved with a direct MPC scheme without the addition of an outer damping loop. Note that the drive system does not provide passive damping; the resistances of the filter inductor and capacitor are almost zero, and the stator resistance at 0.01 pu is very small. Moreover, it is remarkable that the optimization problem with such a long prediction horizon can be solved. Without the sphere decoder and when resorting to full enumeration, about 10^{N_p} switching sequences would have to be computed every 125 μs, which is computationally intractable for long prediction horizons N_p[2].

When operating at 200 Hz and extending the prediction horizon, the stator current TDD slowly drops from its peak value of 10.2% at $N_p = 1$. For the horizon $N_p = 4$, the TDD is halved to 5.03%. To halve it again, the prediction horizon needs to be extended to 15 steps, resulting in a TDD of 2.43%. We conclude that long prediction horizons tend to be of greater benefit when operating at low device switching frequencies. This observation is in line with the results in Sect. 6.1.4.

The trade-off between the stator current TDD and the device switching frequency is investigated in Fig. 6.23. Each data point corresponds to a steady-state simulation at nominal speed and rated torque. The switching penalty λ_u is varied as follows: for $N_p = 1$ between 0.025 and

[2] Note that the number of switching sequences is less than the theoretical upper bound of 27^{N_p} thanks to the switching constraint (5.22c).

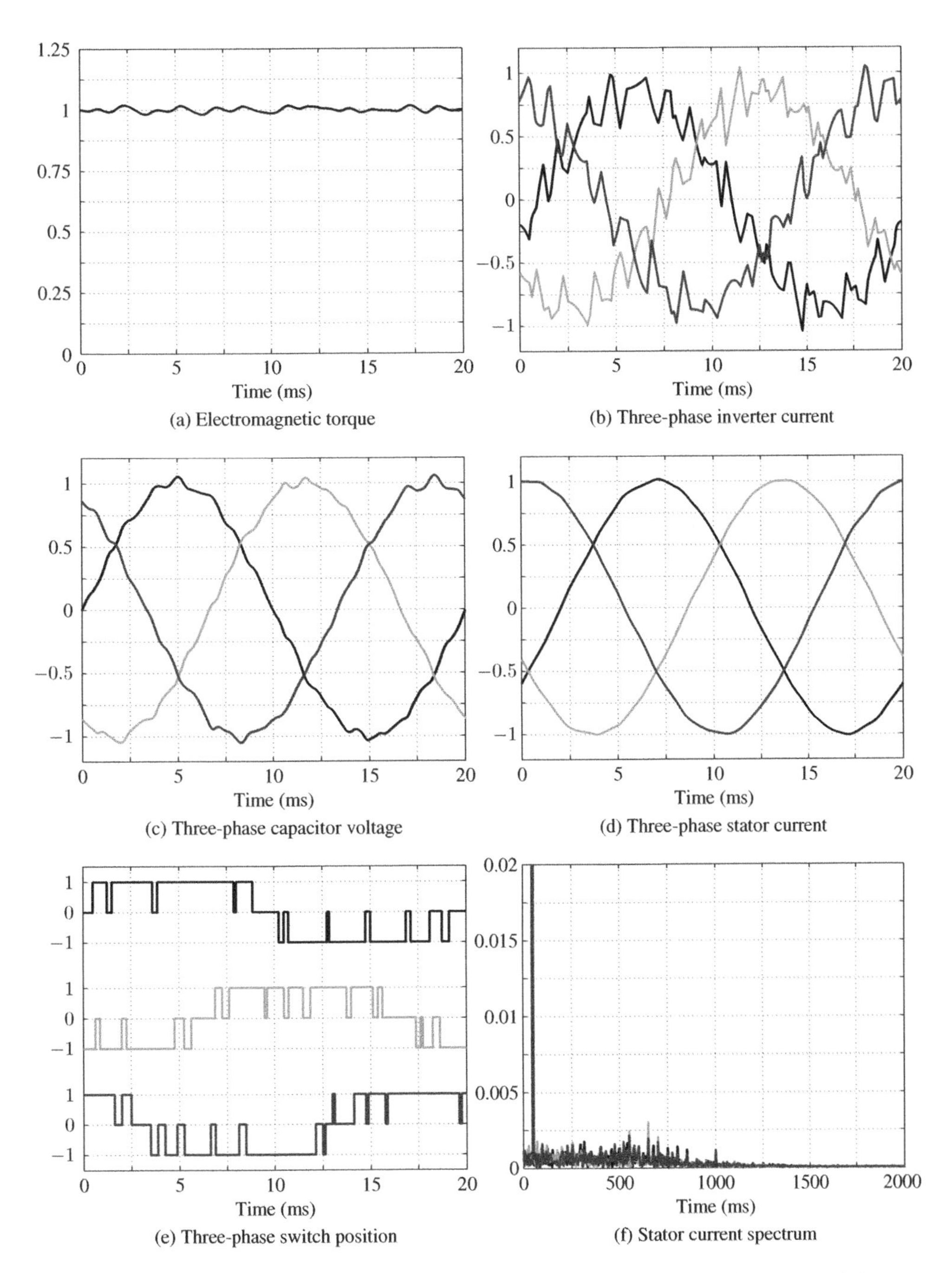

Figure 6.21 Simulated waveforms for direct MPC with the prediction horizon $N_p = 15$ and the controller sampling interval $T_s = 125$ μs. For the device switching frequency of 303 Hz, a stator current TDD of 1.156% is achieved

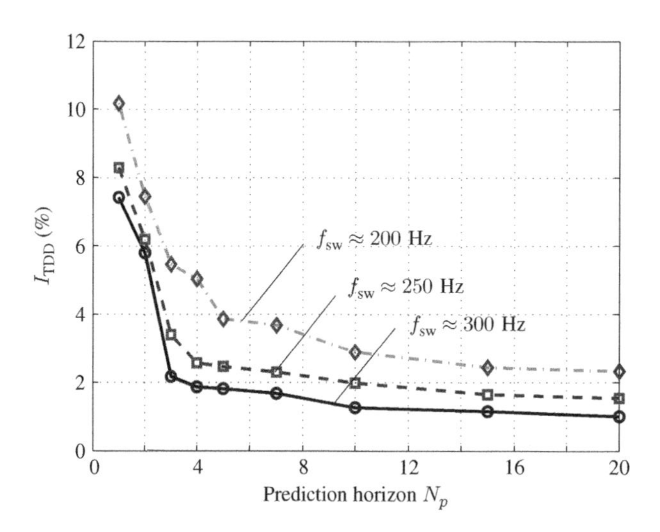

Figure 6.22 Stator current TDD as a function of the prediction horizon for the device switching frequency 200, 250, and 300 Hz, respectively. The data points refer to individual simulation results

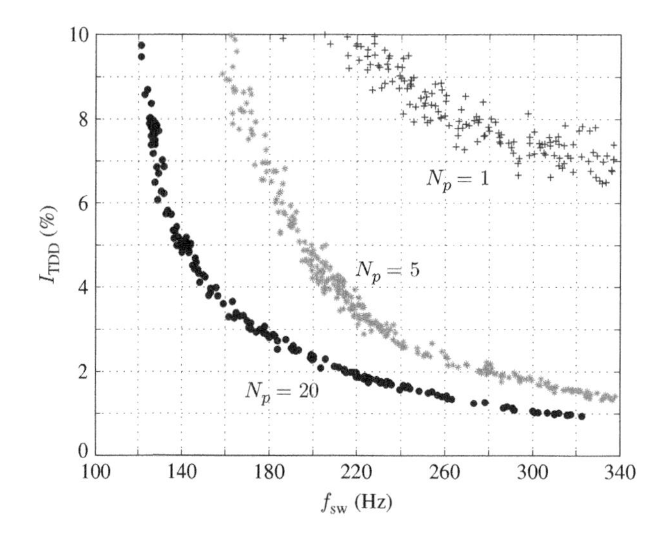

Figure 6.23 Trade-off between the stator current TDD and the device switching frequency, when using the prediction horizons $N_p = 1$, 5, and 20

0.065, for $N_p = 5$ between 0.15 and 2.5, and for $N_p = 20$ between 0.18 and 85. The horizon $N_p = 1$ case is clearly not suitable for the drive system with an LC filter. Extending the prediction horizon to $N_p = 5$ greatly reduces the current TDD for the same switching frequency, or vice versa. Long prediction horizons such as $N_p = 20$ lead to superior results.

In particular, the drive system can be successfully operated at switching frequencies significantly below the resonance frequency of the system. For the prediction horizon $N_p = 20$ and the penalty $\lambda_u = 9.6$, for example, the converter operates at the device switching frequency 138 Hz, which is less than half the 304 Hz of the LC filter resonance. The resulting stator current TDD is 4.99%. To achieve this, the direct MPC scheme shapes the stator current spectrum based on information extracted from the internal prediction model of the drive system, which captures the filter resonance and the effect the switching actions have on it. To successfully shape the current spectrum at low switching frequencies, long prediction horizons are required.

6.3.5 Operation during Transients

In the last step, the performance of direct MPC is examined during torque transients to highlight its very fast dynamic behavior. For the examined scenario, a 15-step prediction horizon is considered with the penalty $\lambda_u = 0.28$ on the switching transitions. While operating at the rated speed, reference torque steps of magnitude 1 pu are imposed. During the transients, a switching frequency of 280 Hz is measured, which is slightly below that for $\lambda_u = 0.28$ at steady-state operation.

The drive system response to torque steps at $t = 10$ ms and $t = 30$ ms is shown in Fig. 6.24. The same subfigures as in Fig. 6.21 are used, except for the current spectrum, which is not shown because it is of no relevance during transients. The steps on the torque reference are translated into the corresponding steady-state references $\boldsymbol{y}^*$ on the inverter current, capacitor voltage, and stator current. The latter are shown as dash-dotted lines in Fig. 6.24(d). Note that the torque steps cannot be anticipated by the controller and are thus not captured in the output reference trajectory $\boldsymbol{Y}^*$.

As can be observed in Fig. 6.24(e), when the torque reference is stepped down from the rated torque to zero, the voltage applied to the LC filter is instantly inverted, as shown in Fig. 6.24(f). Because of this, a short torque settling time of 2.5 ms is achieved. When the torque reference is stepped up from 0 to 1 pu, the transient lasts significantly longer and the steady-state operating point is reached within 6 ms. This is due to the small voltage margin available, which results from the machine operating at nominal speed. Nevertheless, the stator currents are quickly regulated to their new reference values, as can be seen in Fig. 6.24(d).

During torque transients, significant energy is to be moved between the inverter, filter, and the machine. The magnitude and phase of the inverter current through the filter inductor is changed, as is the phase of the capacitor voltage as well as the magnitude and phase of the stator current.

When large torque steps are imposed, the direct MPC scheme acts effectively like a deadbeat controller. To move the third-order system to its new references as quickly as possible, notches in the switching sequence are created, which lead to notches in the inverter current. These, in turn, lead to one distinctive notch in the capacitor voltage. When voltage margin is available, that is, when stepping the torque down from 1 to 0 pu, these notches can easily be identified in Fig. 6.24(f). When stepping the torque up, very little voltage margin is available, limiting the magnitude of the notches and the speed of the torque response.

During transients, aggressively tuned controllers such as this one are prone to producing overshoots in the controlled variables. When using the penalty matrix $\boldsymbol{Q} = \text{diag}(1, 1, 5, 5, 150, 150)$, overshoots in the torque by 25% and in the stator current by 60% occur during the negative torque step. To avoid this, we switch during transients to a different penalty matrix, which provides a more equal weighting of the tracking errors. In

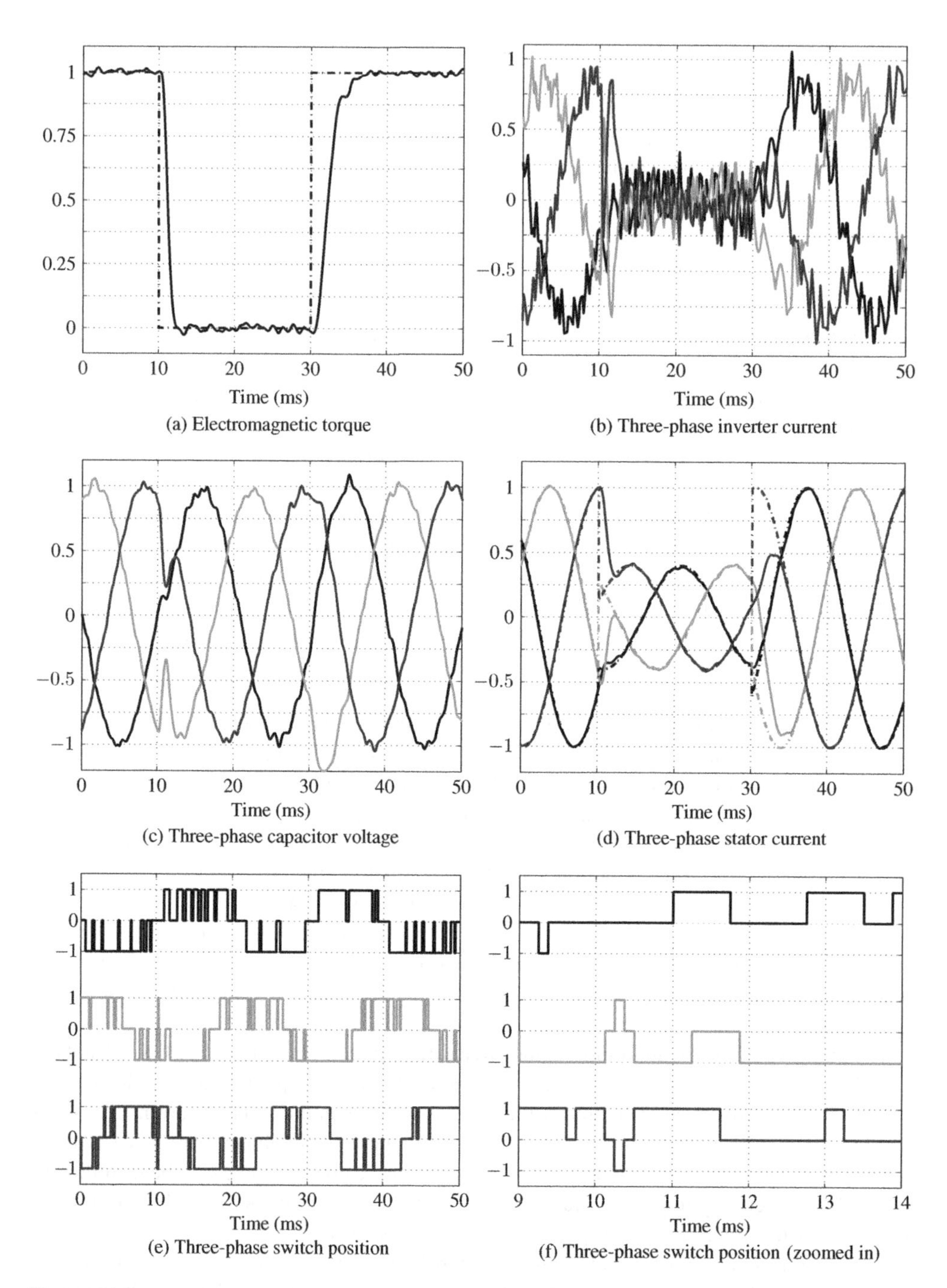

Figure 6.24 Steps in the torque reference for direct MPC with the horizon $N_p = 15$ at nominal speed. The references are shown as dash-dotted lines

particular, by choosing $Q_{\text{trans}} = \text{diag}(1, 1, 5, 5, 15, 15)$, the penalty on the stator current error is reduced and the tracking of the inverter current reference is improved. When reaching the desired torque, the original penalty matrix Q is applied again. As can be seen in Fig. 6.24, this approach avoids any overshoot in the torque and in the inverter current. Small overshoots are observable in the capacitor voltage, which are, in general, less critical than current overshoots.

The effect of the switching constraint (5.22c) is visible in Fig. 6.24(f). Whenever a switching transition from -1 to 1, and vice versa, is required, a mandatory intermediate zero switch position is added for the duration of the controller sampling interval. This slightly slows down the transient performance of the controller. Nevertheless, the torque settling times are similar to those for the drive system without an LC filter (see Sect. 6.1.5).

6.4 Summary and Discussion

In this final section, the proposed MPC algorithm, its performance during steady-state and transient operation, the choice of the cost function, and its computational complexity are discussed and conclusions are provided.

6.4.1 *Performance at Steady-State Operating Conditions*

When assessing the steady-state performance of a current controller, the two key performance metrics are arguably the current TDD and the switching effort. As the switching frequency is easy to quantify, it is usually used as a measure for the switching effort, rather than the switching losses, which might be more meaningful. OPPs are typically considered to yield the lowest achievable current TDD for a given switching frequency, while SVM—particularly for low switching frequencies—entails a significantly higher current TDD.

When tracking the current reference in MPC and directly setting the converter switch positions without the use of a modulator, a horizon of $N_p = 1$ is almost universally used in the literature [5, 6]. Alas, the penalty on the switching effort is often omitted in the literature, resulting in a deadbeat control scheme. Such schemes are well known to be highly sensitive to noise in the measurements and estimates. Adding a penalty on the switching effort not only reduces the switching frequency but also lessens the sensitivity to such noise. By enlarging the prediction horizon, this sensitivity is further reduced, as shown, for example, in Sect. 13.1.2 for the model predictive pulse pattern controller.

For the low switching frequencies typically used in MV applications, the horizon 1 case tends to improve on SVM, by reducing the current TDD for a given switching frequency, or vice versa. For higher switching frequencies, however, MPC with $N_p = 1$ performs either similarly to SVM or worse. The use of long prediction horizons significantly reduces the current TDD. For a three-level inverter with an induction machine, for example, when compared to the horizon 1 case and at the same switching frequency, direct MPC with the horizon $N_p = 10$ leads to a 20% reduction in the current TDD. Indeed, for long prediction horizons, the resulting steady-state performance in terms of current TDD per switching frequency gets close to that of OPPs. When considering multilevel inverters with a higher number of voltage levels, the benefit of long horizons would be expected to be even more pronounced.

For higher order systems, such as drive systems with an *LC* filter, the performance benefits of long prediction horizons are even more prominent. A minimum horizon length of a few steps is required to provide the system with a sufficient degree of active damping. For the case study investigated in this chapter with the relatively long sampling interval of $T_s = 125$ μs, a prediction horizon of at least three steps is mandatory. Longer prediction horizons further improve the performance. Overall, when extending the prediction horizon from 1 to 20, the stator current TDD can be reduced by fourfold to sevenfold for the same switching frequency.

Not only the weight λ_u but also the sampling interval T_s has a profound impact on the closed-loop characteristic of MPC. Even though this second degree of freedom complicates the tuning procedure, it can be exploited to one's advantage. Specifically, it is important to achieve a long prediction *interval* in time. If a low switching frequency is desired, it is beneficial to use a fairly long sampling interval such as $T_s = 125$ μs, even though this reduces the granularity of switching. For high switching frequencies, a high granularity is important, requiring a high sampling frequency and thus a short sampling interval, such as $T_s = 25$ μs.

6.4.2 *Performance during Transients*

During transients, MPC achieves an excellent dynamic performance, similar to that of deadbeat control (see also Sect. 8.1.3 and [5]). When applying torque steps, the settling time is limited in effect only by the available dc-link voltage. If required, MPC temporarily inverts the voltage applied to the load, in order to achieve as short a transient as possible. For a drive system without an *LC* filter, the horizon length has no impact on the settling time, with long horizons resulting in the same transient performance as the short ones. When adding an *LC* filter to the drive, long horizons are also required during transients to provide active damping of the filter resonance.

The transient performance of direct MPC is by far superior to that typically achieved with OPPs, because traditionally it has only been possible to use OPPs in a modulator driven by a very slow control loop (see, e.g., [7]). Notable exceptions to this include the stator flux trajectory controller with pulse insertion, as summarized in Chap. 12 and [8].

6.4.3 *Cost Function*

Horizons longer than 1 significantly reduce the closed-loop cost (6.1), when compared to the $N_p = 1$ case. For very long prediction horizons, however, when further increasing the horizon, the incremental cost reduction becomes very small and ceases at some point. This is a general characteristic of MPC (see, e.g., [9, 10]) and can be seen in Fig. 6.6(c) for the drive system case study without an *LC* filter. For the case study with a filter, the same characteristic can be observed in Fig. 6.22. The larger the weight on the switching effort, the later this leveling off occurs, indicating that long horizons are particularly beneficial when switching is expensive and the switching frequency is low.

The cost function consists of two terms. The first term captures the rms output tracking error, which corresponds to the stator current TDD in both case studies. To prioritize the minimization of the stator current ripple for the higher order system, large penalties are imposed in Q on the stator current tracking error and smaller penalties are used for the inverter current and the capacitor voltage. The second term in the cost function represents the switching effort, which is a direct measure of the switching frequency. Both terms are penalized in the cost function, and the trade-off between the two is adjusted by the weight λ_u. When increasing the length of the prediction horizon for a given λ_u, a drastic reduction of the closed-loop cost can be observed, but only comparatively minor reductions in the current TDD and switching frequency are achieved. In particular, long horizons shift the trade-off point along the trade-off curve while only marginally improving it.

As an alternative, in model predictive direct current control, this shift is avoided by fixing one of the two quantities (see Sect. 11.1). More specifically, the width of the current bounds determines the current TDD, and the cost function captures the switching effort, which is to be minimized. Fixing one of the two performance metrics, while minimizing the other one—rather than aiming at minimizing both—merits further investigations. In addition, the effect of final state weighting is worth exploring, as it allows one to approximate infinite horizon problems (see [9, 11]).

6.4.4 Control Objectives

It is conceivable that one can directly minimize the switching losses rather than the switching frequency. To achieve this, one might replace the constant scalar weight λ_u by a time-varying 3×3 matrix $\boldsymbol{Q}_u$. Specifically, the switching effort term $\lambda_u \|\Delta \boldsymbol{u}(\ell)\|_2^2$ could be replaced by $\Delta \boldsymbol{u}^T \boldsymbol{Q}_u \Delta \boldsymbol{u}$. The penalty matrix $\boldsymbol{Q}_u$ would be diagonal with three phase-specific weights. These weights could be adjusted online according to the phase current. Specifically, the phases with high instantaneous currents would feature large weights, while phases with low instantaneous currents would have a correspondingly small weight. As a result, it is expected that the switching transitions are shifted from phases with high currents to phases with lower currents, thus reducing the average switching losses, albeit at the computational expense of a time-varying Hessian matrix.

6.4.5 Computational Complexity

When considering multilevel converters with more than three levels, the computational complexity increases—in the worst case—exponentially with a large base. Nevertheless, our empirical results for the modified sphere decoding algorithm suggest that the average computational burden is effectively independent of the number of inverter levels, because the search for the optimal switching sequence is restricted to a sphere centered on the unconstrained solution. The size of the sphere is independent of the number of levels. The same holds true when considering higher order systems.

Therefore, this algorithm appears to be particularly well suited to multilevel converter topologies with a very large number of levels. Nevertheless, the modified sphere decoding algorithm provides significant computational savings also for three-level converters, when compared to an exhaustive search. Notably, even for the horizon 1 case, the computational burden can be reduced by one order of magnitude on average, which makes sphere decoding an attractive alternative to solve the optimization problem of direct MPC also in cases where long horizons are not strictly required.

The real-time computational complexity of the proposed sphere decoding algorithm can be further reduced, as shown in [12]. In a preprocessing stage, a lattice reduction algorithm [13] transforms the integer optimization problem into an equivalent problem that is well conditioned and can be solved more efficiently. During the optimization stage, the initial radius of the sphere can be calculated in a more effective way, by adopting the notion of the Babai estimate [14]. The computational complexity can be further reduced by allowing suboptimal solutions [15]. In particular, an upper limit can be imposed on the operations performed in real time while guaranteeing that a feasible solution is always found.

Appendix 6.A: State-Space Model

The matrices of the state-space model (6.10) of the drive system in the continuous-time domain are given by

$$F = \begin{bmatrix} -\frac{R_l+R_c}{X_l}\boldsymbol{I}_2 & -\frac{1}{X_l}\boldsymbol{I}_2 & \frac{R_c}{X_l}\boldsymbol{I}_2 & \boldsymbol{0}_{2\times2} \\ \frac{1}{X_c}\boldsymbol{I}_2 & \boldsymbol{0}_{2\times2} & -\frac{1}{X_c}\boldsymbol{I}_2 & \boldsymbol{0}_{2\times2} \\ \frac{X_r}{D}R_c\boldsymbol{I}_2 & \frac{X_r}{D}\boldsymbol{I}_2 & -(\frac{1}{\tau_s}+\frac{X_r}{D}R_c)\boldsymbol{I}_2 & \left(\frac{1}{\tau_r}\boldsymbol{I}_2 - \omega_r\begin{bmatrix}0 & -1\\ 1 & 0\end{bmatrix}\right)\frac{X_m}{D} \\ \boldsymbol{0}_{2\times2} & \boldsymbol{0}_{2\times2} & \frac{X_m}{\tau_r}\boldsymbol{I}_2 & -\frac{1}{\tau_r}\boldsymbol{I}_2 + \omega_r\begin{bmatrix}0 & -1\\ 1 & 0\end{bmatrix} \end{bmatrix},$$

$$\boldsymbol{G} = \frac{1}{2}v_{dc}\left[\frac{1}{X_l}\boldsymbol{I}_2 \;\; \boldsymbol{0}_{2\times6}\right]^T\tilde{\boldsymbol{K}} \quad\text{and}\quad \boldsymbol{C} = \left[\boldsymbol{I}_6 \;\; \boldsymbol{0}_{6\times2}\right].$$

The transient stator and rotor time constants τ_s and τ_r were defined in (2.60). The determinant D is stated in (2.54).

Appendix 6.B: Computation of the Output Reference Vector

The output reference vector $\boldsymbol{y}^*$ is required at each time step within the prediction horizon. For the operating point, which is determined by the angular rotor speed ω_r, the torque reference T_e^*, and the reference of the stator flux magnitude Ψ_s^*, the steady-state relationship between the output variables can be computed. To this end, only fundamental components are considered and switching is neglected. The computation of the output reference vector is performed in three steps.

6.B.1 Step 1: Stator Frequency

Consider the machine model (2.55), which uses as state variables the stator and rotor flux vectors in the dq reference frame, which rotates synchronously at the flux vectors with the angular speed $\omega_{\text{fr}} = \omega_s$. We chose to align the stator flux vector with the d-axis of the reference frame, that is

$$\boldsymbol{\psi}_{s,dq} = [\Psi_s^* \; 0]^T. \tag{6.B.1}$$

The magnitude of the stator flux is set equal to its reference.

The q-component of the rotor flux

$$\psi_{rq} = -\text{pf} \frac{T_e^*}{\Psi_s^*} \frac{D}{X_m} \tag{6.B.2}$$

can be computed by inserting (6.B.1) into the torque equation (2.56). At stationary operation, the flux derivatives in the dq reference frame are zero. Inserting (6.B.1) into (2.55b) and assuming the use of a squirrel-cage induction machine, the d-component of the rotor flux results in

$$\psi_{rd} = -R_r \frac{X_s}{D} \frac{\psi_{rq}}{\omega_s - \omega_r}. \tag{6.B.3}$$

By inserting (6.B.1) and (6.B.3) into (2.55a), and after some algebraic manipulations, we obtain two possible solutions for the d-component of the rotor flux.

$$\psi_{rd} = \frac{X_m}{2X_s} \Psi_s^* \pm \sqrt{\frac{X_m^2}{4X_s^2} (\Psi_s^*)^2 - \psi_{rq}^2} \tag{6.B.4}$$

It is easy to show that the solution with the plus operation is the correct one.

Having computed the rotor flux vector, the stator frequency can be derived as a function of the electrical angular speed of the rotor ω_r. This also implicitly defines the slip. To this end, (6.B.3) is solved for the stator frequency

$$\omega_s = \omega_r - R_r \frac{X_s}{D} \frac{\psi_{rq}}{\psi_{rd}}. \tag{6.B.5}$$

Note that, in motoring operation, the rotor flux lags behind the stator flux. As we have aligned the stator flux with the reference frame's d-axis, the q-component of the rotor flux is negative and, according to (6.B.5), $\omega_s > \omega_r$ holds.

6.B.2 Step 2: Inverter Voltage

Here, the continuous-time model (6.10) of the drive system is transformed from the stationary orthogonal $\alpha\beta$ reference frame to the rotating dq reference frame with the angular position φ. The dq frame rotates synchronously with the stator quantities at the angular speed ω_s.

Consider the state-space model (6.5) of the LC filter, replace all quantities in $\alpha\beta$ by $\boldsymbol{R}^{-1}(\varphi)\, \boldsymbol{\xi}_{dq}$, and multiply the differential equations on the left with $\boldsymbol{R}(\varphi)$. Note that $\boldsymbol{R}(\varphi)$ and $\boldsymbol{R}^{-1}(\varphi)$ denote the rotation matrices from $\alpha\beta$ to dq, and vice versa, as defined in (2.25). Because

$$\boldsymbol{R}(\varphi) \frac{\text{d}}{\text{d}t}(\boldsymbol{R}^{-1}(\varphi)\boldsymbol{\xi}_{dq}) = \frac{\text{d}\boldsymbol{\xi}_{dq}}{\text{d}t} + \omega_s \begin{bmatrix} 0 & -1 \\ 1 & 0 \end{bmatrix} \boldsymbol{\xi}_{dq}, \tag{6.B.6}$$

the two LC filter state-space equations in the rotating reference frame are

$$\frac{d\boldsymbol{i}_{i,dq}}{dt} + \omega_s \begin{bmatrix} 0 & -1 \\ 1 & 0 \end{bmatrix} \boldsymbol{i}_{i,dq} = \frac{1}{X_l}(\boldsymbol{v}_{i,dq} - \boldsymbol{v}_{c,dq} - R_l\boldsymbol{i}_{i,dq} - R_c(\boldsymbol{i}_{i,dq} - \boldsymbol{i}_{s,dq})) \tag{6.B.7a}$$

$$\frac{d\boldsymbol{v}_{c,dq}}{dt} + \omega_s \begin{bmatrix} 0 & -1 \\ 1 & 0 \end{bmatrix} \boldsymbol{v}_{c,dq} = \frac{1}{X_c}(\boldsymbol{i}_{i,dq} - \boldsymbol{i}_{s,dq}). \tag{6.B.7b}$$

The filter equations are augmented by the state-space equations of a machine model in the rotating reference frame. Using the stator current and the rotor flux as state variables, the machine model (2.59) is adopted. We set $\omega_{\text{fr}} = \omega_s$ and the rotor voltage $\boldsymbol{v}_r$ to zero.

Combining (6.B.7) with (2.59) leads to the reformulated drive system model

$$\frac{d\boldsymbol{x}_{dq}(t)}{dt} = \boldsymbol{F}_{dq}\boldsymbol{x}_{dq}(t) + \boldsymbol{G}_{dq}\boldsymbol{v}_{i,dq}(t) \tag{6.B.8a}$$

$$\boldsymbol{\psi}_{s,dq}(t) = \boldsymbol{C}_{dq}\boldsymbol{x}_{dq}(t) \tag{6.B.8b}$$

in the rotating dq reference frame, where we have also used (6.6). The inverter voltage $\boldsymbol{v}_{i,dq}$ in the rotating reference frame is used as the input vector rather than the three-phase switch position $\boldsymbol{u}$. The stator flux $\boldsymbol{\psi}_{s,dq}$ is defined as the output, which is, according to (2.58), a linear combination of the stator current and the rotor flux. The input and output matrices are

$$\boldsymbol{G}_{dq} = \left[\tfrac{1}{X_l}\boldsymbol{I}_2 \;\; \boldsymbol{0}_{2\times 6}\right]^T \text{ and } \boldsymbol{C}_{dq} = \left[\boldsymbol{0}_{2\times 4} \;\; \tfrac{D}{X_r}\boldsymbol{I}_2 \;\; \tfrac{X_m}{X_r}\boldsymbol{I}_2\right]. \tag{6.B.9}$$

The system matrix $\boldsymbol{F}_{dq}$ can easily be derived.

At steady-state operation, in the rotating orthogonal reference frame, the derivative of the state vector in (6.B.8a) is equal to zero. This allows us to express the state vector

$$\boldsymbol{x}_{dq} = -(\boldsymbol{F}_{dq})^{-1}\,\boldsymbol{G}_{dq}\,\boldsymbol{v}_{i,dq} \tag{6.B.10}$$

as a function of the inverter voltage. Note that the operating point (rotor speed, torque, and stator flux magnitude) is determined by the system matrix $\boldsymbol{F}_{dq}$.

As stated in (6.B.1), we choose to align the stator flux vector with the reference frame's d-axis. By inserting (6.B.10) into (6.B.8b) and by inverting the equation, the inverter voltage

$$\boldsymbol{v}_{i,dq} = -(\boldsymbol{C}_{dq}\,\boldsymbol{F}_{dq}^{-1}\,\boldsymbol{G}_{dq})^{-1}\,\boldsymbol{\psi}_{s,dq} \tag{6.B.11}$$

can be computed as a function of the stator flux. Note that the inverted matrix in (6.B.11) is of the dimension 2×2.

6.B.3 Step 3: Output Reference Vector

The output reference vector in the rotating reference frame is given by

$$\boldsymbol{y}_{dq}^* = \boldsymbol{C}\boldsymbol{x}_{dq} = -\boldsymbol{C}\,(\boldsymbol{F}_{dq})^{-1}\,\boldsymbol{G}_{dq}\,\boldsymbol{v}_{i,dq} \tag{6.B.12}$$

with $\boldsymbol{C}$ as defined in Appendix 6.A. The evolution of $\boldsymbol{y}^*$ in stationary coordinates is obtained by rotating each one of the three components from dq to $\alpha\beta$ according to (2.24), using $\boldsymbol{R}^{-1}(\varphi)$

as stated in (2.25). This operation needs to be performed at each time step within the prediction horizon.

References

[1] J. Rodríguez, M. P. Kazmierkowski, J. R. Espinoza, P. Zanchetta, H. Abu-Rub, H. A. Young, and C. A. Rojas, "State of the art of finite control set model predictive control in power electronics," *IEEE Trans. Ind. Informatics*, vol. 9, pp. 1003–1016, May 2013.

[2] J. Rodríguez, J. Pontt, C. A. Silva, P. Correa, P. Lezana, P. Cortés, and U. Ammann, "Predictive current control of a voltage source inverter," *IEEE Trans. Ind. Electron.*, vol. 54, pp. 495–503, Feb. 2007.

[3] T. Geyer and D. E. Quevedo, "Performance of multistep finite control set model predictive control for power performance," *IEEE Trans. Power Electron.*, vol. 30, pp. 1633–1644, Mar. 2015.

[4] Y. W. Li, "Control and resonance damping of voltage-source and current-source converters with *LC* filters," *IEEE Trans. Ind. Electron.*, vol. 56, pp. 1511–1521, May 2009.

[5] P. Cortés, M. P. Kazmierkowski, R. M. Kennel, D. E. Quevedo, and J. Rodríguez, "Predictive control in power electronics and drives," *IEEE Trans. Ind. Electron.*, vol. 55, pp. 4312–4324, Dec. 2008.

[6] J. Rodríguez and P. Cortés, *Predictive control of power converters and electrical drives*. Chichester, UK John Wiley & Sons, Ltd, 2012.

[7] J. Holtz and B. Beyer, "Fast current trajectory tracking control based on synchronous optimal pulsewidth modulation," *IEEE Trans. Ind. Appl.*, vol. 31, pp. 1110–1120, Sep./Oct. 1995.

[8] J. Holtz and N. Oikonomou, "Synchronous optimal pulsewidth modulation and stator flux trajectory control for medium-voltage drives," *IEEE Trans. Ind. Appl.*, vol. 43, pp. 600–608, Mar./Apr. 2007.

[9] D. E. Quevedo and G. C. Goodwin, "Multistep optimal analog-to-digital conversion," *IEEE Trans. Circuits Syst. I*, vol. 52, pp. 503–515, Mar. 2005.

[10] L. Grüne and A. Rantzer, "On the infinite horizon performance of receding horizon controllers," *IEEE Trans. Automat. Control*, vol. 53, pp. 2100–2111, Oct. 2008.

[11] D. E. Quevedo, G. C. Goodwin, and J. A. D. Doná, "Finite constraint set receding horizon quadratic control," *Int. J. Robust Nonlinear Control*, vol. 14, pp. 355–377, Mar. 2004.

[12] P. Karamanakos, T. Geyer, and R. Kennel, "Reformulation of the long-horizon direct model predictive control problem to reduce the computational effort," in *Proceedings of IEEE Energy Conversion Congress and Exposition* (Pittsburgh, PA, USA), Sep. 2014.

[13] A. K. Lenstra, H. W. Lenstra Jr., and L. Lovász, "Factoring polynomials with rational coefficients," *Math. Ann.*, vol. 261, no. 4, pp. 515–534, Dec. 1982.

[14] L. Babai, "On Lovász' lattice reduction and the nearest lattice point problem," *Combinatorica*, vol. 6, no. 1, pp. 1–13, 1986.

[15] P. Karamanakos, T. Geyer, and R. Kennel, "Suboptimal search strategies with bounded computational complexity to solve long-horizon direct model predictive control problems," in *Proceedings of IEEE Energy Conversion Congress and Exposition* (Montreal, Canada), Sep. 2015.

Part Three

Direct Model Predictive Control with Bounds

7

Model Predictive Direct Torque Control

7.1 Introduction

Direct torque control (DTC) is a machine-side control concept that imposes hysteresis bounds on the electromagnetic torque and the stator flux magnitude of an electrical machine, as was shown in Sect. 3.6.3. The outputs of the hysteresis controllers are fed into a look-up table, which chooses the discrete three-phase switch position for the inverter. As a result, DTC controls the torque directly (rather than indirectly via the stator currents) and does not require a modulator. DTC is sometimes interpreted as a predictive control strategy, but it predicts only one switching transition ahead, and lacks an internal controller model, a cost function, the notion of optimality, and the receding horizon policy, which are all fundamental elements of model predictive control (MPC) schemes, as explained in Sect. 1.3.

The DTC look-up table can be replaced by an MPC method with the aim of reducing the switching losses while maintaining the very fast torque response that is inherent to DTC. To this end, three preliminary MPC methods were proposed. Starting with a closed-form, mixed-integer linear optimization problem that was solved online [1, 2], the optimization problem was subsequently pre-solved offline and a look-up table was computed [3]. An alternative open-form problem formulation, which was also solved offline, reduced the computational burden by an order of magnitude [4].

The concept of model predictive direct torque control (MPDTC) was proposed in 2004 [5]. It was described and analyzed in detail in 2005 [6] and later in [7]. MPDTC keeps the electromagnetic torque T_e and the stator flux magnitude Ψ_s within upper and lower bounds. The three-phase switch position $\boldsymbol{u}$ is directly set by MPDTC, thus not requiring the use of a modulator. To achieve this, MPDTC enumerates the set of admissible switch positions, predicts trajectories of the torque and stator flux magnitude as a function of these switch positions using the notion of extrapolation, and minimizes a cost function that captures the switching frequency or the switching losses. For multilevel inverters, MPDTC also imposes bounds on internal inverter voltages, such as the neutral point potential in a neutral-point-clamped (NPC) inverter.

Model Predictive Control of High Power Converters and Industrial Drives, First Edition. Tobias Geyer.

Companion Website: www.wiley.com/go/geyermodelpredictivecontrol

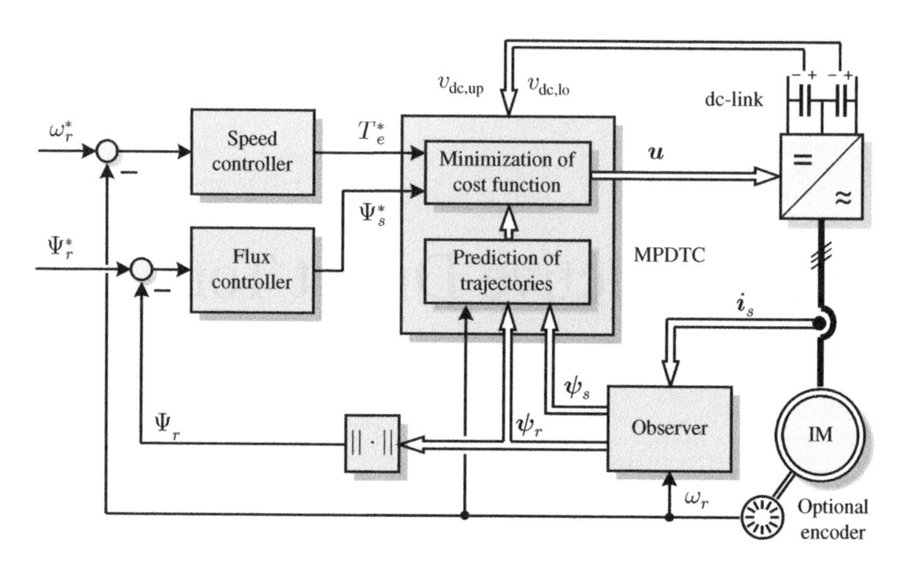

Figure 7.1 Block diagram of MPDTC for an NPC inverter with an induction machine (IM)

As shown in Fig. 7.1, MPDTC constitutes the inner control loop that is augmented by a speed controller and an (optional) rotor flux controller. An additional loop adjusts the widths of the bounds that are imposed on the torque and stator flux magnitude. In doing so, the average switching frequency can be controlled. A flux observer is needed to reconstruct the stator and rotor flux linkage vectors. It is clear that MPDTC inherits the basic structure of the control loops from DTC (see also the DTC block diagram shown in Fig. 3.29). Note also that MPDTC is applicable both to drives with position encoders and to sensorless drives.

Using extrapolation and relying on a tailored open-form formulation of the optimization problem, MPDTC in its simplest form has a modest computational burden. This facilitated the implementation of MPDTC on the existing control platform of an industrial medium-voltage (MV) drive system, which is based on an NPC inverter. More specifically, the successful test runs on ABB's ACS6000 drive with power levels exceeding 1 MW [8, 9] can be considered as a milestone in the development of MPC for industrial drives.

The initial MPDTC algorithm was generalized in 2009, allowing for longer prediction horizons that include multiple hinges (groups of switching transitions) and extrapolation segments [10]. Prediction horizons in excess of 100 time steps are typically achieved, greatly reducing the switching frequency per harmonic distortions. Moreover, instead of indirectly reducing the switching losses through the minimization of the switching frequency, the switching losses can be directly minimized in MPDTC by predicting and minimizing them in the cost function, as proposed in [11]. Since 2009, a large number of publications have appeared that further extend and improve the MPDTC concept. Some of these extensions, such as the analysis, resolution, and avoidance of deadlocks, branch-and-bound methods to reduce the computational burden, and smart extrapolation techniques to achieve more accurate predictions, will be detailed in this and subsequent chapters.

Precursors to MPDTC had already been proposed in the 1980s [12, 13] and early 1990s [14]. Even though these control methods control the current (instead of the torque and the stator flux magnitude) for a two-level converter using a very simplistic model of the machine, the controlled quantities are kept within bounds, the set of switch positions is enumerated, linear extrapolation is used to predict future current trajectories, and a cost function is minimized that captures the switching frequency. MPDTC reinvented many of these concepts, albeit in a more general and formal MPC framework.

Among others, MPDTC adds the notion of the receding horizon policy, achieves long prediction horizons with multiple hinges, minimizes the switching losses, avoids deadlocks, uses branch-and-bound methods, and predicts future trajectories using smart extrapolation techniques. Moreover, MPDTC targets multilevel converters. Owing to the different shapes of the bounds, MPDTC with short horizons tends to lead to slightly lower current and torque distortions for the same switching frequency even when using the same cost function and the same extrapolation technique. For MPDTC with long horizons, the performance difference is more pronounced. For an in-depth comparison of MPDTC with these early predictive methods, the reader is referred to [15].

7.2 Preliminaries

7.2.1 Case Study

As a case study, consider an MV squirrel-cage induction machine fed by a three-level NPC inverter with the floating neutral point potential N, as shown in Fig. 7.2. The total (instantaneous) dc-link voltage is $v_{\text{dc}} = v_{\text{dc,up}} + v_{\text{dc,lo}}$, with $v_{\text{dc,up}}$ and $v_{\text{dc,lo}}$ denoting the voltages over the upper and lower dc-link capacitors, respectively. The total dc-link voltage is assumed to be constant and equal to its nominal value V_{dc}.

At each of the three phase terminals A, B, and C, the inverter produces the voltages $-\frac{v_{\text{dc}}}{2}$, 0, and $\frac{v_{\text{dc}}}{2}$ with respect to the dc-link midpoint. These phase voltages correspond to the switch positions $u_a, u_b, u_c \in \{-1, 0, 1\}$. We use $\boldsymbol{u}_{abc} = [u_a\ u_b\ u_c]^T$ to denote the three-phase switch position. The voltage applied to the machine terminals in orthogonal coordinates is given by

$$\boldsymbol{v}_s = \frac{1}{2} v_{\text{dc}}\, \tilde{\boldsymbol{K}} \boldsymbol{u}_{abc}, \tag{7.1}$$

where $\boldsymbol{v}_s = [v_{s\alpha}\ v_{s\beta}]^T$. $\tilde{\boldsymbol{K}}$ denotes the reduced Clarke transformation (2.12).

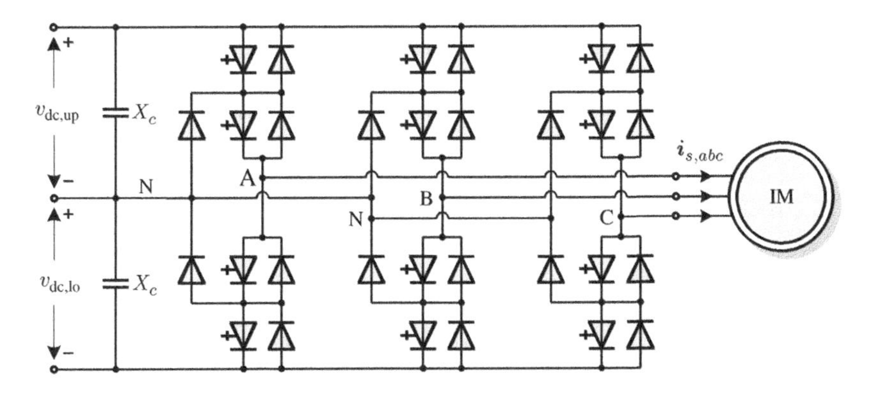

Figure 7.2 Three-level NPC voltage source inverter driving an induction machine (IM)

The neutral point potential is defined as

$$v_n = \frac{1}{2}(v_{\mathrm{dc,lo}} - v_{\mathrm{dc,up}}), \tag{7.2}$$

and its dynamic evolution in the per unit (pu) system is given by

$$\frac{\mathrm{d}v_n}{\mathrm{d}t} = \frac{1}{2X_c}|\boldsymbol{u}_{abc}|^T \boldsymbol{i}_{s,abc} \tag{7.3}$$

according to (2.83). Recall that X_c denotes the pu equivalence of one of the two dc-link capacitors. The differential equation (7.3) is a function of the component-wise absolute value of the inverter switch position $|\boldsymbol{u}_{abc}| = [|u_a|\ |u_b|\ |u_c|]^T$ and the three-phase stator current $\boldsymbol{i}_{s,abc} = [i_{sa}\ i_{sb}\ i_{sc}]^T$. For more details on the NPC inverter, the reader is referred to Sect. 2.4.1. The details of the drive system and its parameters are provided in Sect. 2.5.1.

Next, we state the differential equations of the induction machine using normalized quantities and a normalized time axis. As in Sect. 4.3.1, we adopt the stationary $\alpha\beta$ coordinate system and express the machine dynamics in terms of the stator and rotor flux vectors $\boldsymbol{\psi}_s = [\psi_{s\alpha}\ \psi_{s\beta}]^T$ and $\boldsymbol{\psi}_r = [\psi_{r\alpha}\ \psi_{r\beta}]^T$, respectively. Setting the rotor voltage $\boldsymbol{v}_r$ and the angular speed of the reference frame to zero in (2.55) leads to the continuous-time machine equations in state-space representation

$$\frac{\mathrm{d}\boldsymbol{\psi}_s}{\mathrm{d}t} = -R_s\frac{X_r}{D}\boldsymbol{\psi}_s + R_s\frac{X_m}{D}\boldsymbol{\psi}_r + \frac{1}{2}v_{\mathrm{dc}}\,\tilde{\boldsymbol{K}}\boldsymbol{u}_{abc} \tag{7.4a}$$

$$\frac{\mathrm{d}\boldsymbol{\psi}_r}{\mathrm{d}t} = R_r\frac{X_m}{D}\boldsymbol{\psi}_s - R_r\frac{X_s}{D}\boldsymbol{\psi}_r + \omega_r\begin{bmatrix}0 & -1\\ 1 & 0\end{bmatrix}\boldsymbol{\psi}_r \tag{7.4b}$$

$$\frac{\mathrm{d}\omega_r}{\mathrm{d}t} = \frac{1}{M}(T_e - T_\ell), \tag{7.4c}$$

where we have used (7.1).

The machine parameters are the stator and rotor resistances R_s and R_r, and the stator, rotor, and mutual reactances X_{ls}, X_{lr}, and X_m, respectively. In (2.33) and (2.54), we had also defined

$$X_s = X_{ls} + X_m,\ X_r = X_{lr} + X_m,\ \text{and}\ D = X_sX_r - X_m^2 \tag{7.5}$$

as the stator self-reactance, rotor self-reactance, and determinant, respectively. All rotor quantities are referred to the stator circuit.

Recall that ω_r denotes the electrical angular rotor speed, M is the moment of inertia, and T_ℓ is the load torque. According to (2.56), the electromagnetic torque is given by

$$T_e = \frac{1}{\mathrm{pf}}\frac{X_m}{D}\,\boldsymbol{\psi}_r \times \boldsymbol{\psi}_s = \frac{1}{\mathrm{pf}}\frac{X_m}{D}(\psi_{r\alpha}\psi_{s\beta} - \psi_{r\beta}\psi_{s\alpha}) \tag{7.6}$$

and the magnitude of the stator flux vector is

$$\Psi_s = \|\boldsymbol{\psi}_s\| = \sqrt{(\psi_{s\alpha})^2 + (\psi_{s\beta})^2}. \tag{7.7}$$

7.2.2 Control Problem

The control problem of a high-performance, variable-speed drive (VSD) system has a high degree of complexity with multiple conflicting objectives. Regarding the machine, the electromagnetic torque and the degree of the machine's magnetization are the key quantities to be controlled. In DTC and MPDTC, these two quantities are directly controlled. Rather than forcing them along their references, a certain ripple is tolerated. To this end, upper and lower bounds around the references are introduced, and the torque and stator flux magnitude are kept within these bounds.

At steady-state operating conditions, the total demand distortion (TDD) of the stator current (3.2) is to be minimized. This reduces the iron and copper losses in the machine, which, in turn, lowers the thermal losses. In addition, low torque TDDs (3.1) are required to avoid problems with the mechanical drive systems, such as accelerated wear of the shaft and the excitation of eigenmodes of the shaft and mechanical load. During transients, a high dynamic performance, that is, a short torque settling time in the range of a few milliseconds, is demanded in special-purpose drive applications.

Regarding the inverter, the neutral point potential needs to be balanced around zero to limit the blocking voltages of the semiconductor switches. The temperature of the semiconductor devices has to be kept below a maximum value to ensure their safe operation. As the inverter has limited cooling capability, the temperature requirement can be translated into a limit on the total semiconductor losses that can be tolerated. As the conduction losses are given by the fundamental current and (half) the dc-link voltage, only the switching losses can be influenced by the control and modulation scheme. Reducing the switching losses increases the efficiency and the power capability of the inverter. Lowering the switching losses also tends to reduce the failure rate of the semiconductor devices. An indirect, and less effective, way of minimizing the switching losses is to minimize the number of commutations, that is, the switching frequency. Unlike the switching losses, however, the switching frequency can be easily measured and monitored.

7.2.3 Controller Model

In the following, we derive a discrete-time model of the drive system that is suitable to serve as an internal prediction model for MPDTC. The model's purpose is to predict the trajectories of the electromagnetic torque, stator flux magnitude, and neutral point potential over the prediction horizon with sufficient accuracy. The length of this time interval is typically between 1 and 3 ms.

As the time constant of the rotor speed dynamics exceeds the length of the prediction interval by at least an order of magnitude, we neglect the speed dynamics and assume the rotor's angular speed ω_r to be constant within the prediction horizon.[1] This allows us to treat the speed as a model parameter rather than as a state variable. The saturation of the machine flux and the skin effect in the rotor are neglected, even though these variations could also be incorporated in the model.

[1] For highly dynamic drives and/or drives with a small inertia, it might be necessary to include the speed as an additional state variable in the model.

Modeling the squirrel-cage induction machine in orthogonal coordinates, it is convenient to select the stator and the rotor flux linkages as the machine state vector

$$\boldsymbol{x}_m = [\psi_{s\alpha}\ \psi_{s\beta}\ \psi_{r\alpha}\ \psi_{r\beta}]^T. \tag{7.8}$$

The three-phase switch position constitutes the input vector

$$\boldsymbol{u} = \boldsymbol{u}_{abc} = [u_a\ u_b\ u_c]^T \in \{-1, 0, 1\}^3. \tag{7.9}$$

7.2.3.1 Continuous-Time Model

We write the differential equations (7.4a) and (7.4b) in the state-space representation as

$$\frac{\mathrm{d}\boldsymbol{x}_m(t)}{\mathrm{d}t} = \boldsymbol{F}_m\, \boldsymbol{x}_m(t) + \boldsymbol{G}_m\, \boldsymbol{u}(t), \tag{7.10}$$

where $\boldsymbol{F}_m$ and $\boldsymbol{G}_m$ are provided in Appendix 7.A.

In order to represent the differential equation of the neutral point potential as a function of the machine state vector, we take advantage of the fact that the α- and β-components of the stator current $\boldsymbol{i}_s = [i_{s\alpha}\ i_{s\beta}]^T$ are linear combinations of the stator and rotor flux components. Specifically, repeating (2.53), we write

$$\boldsymbol{i}_s = \frac{1}{D}\left[X_r \boldsymbol{I}_2\ \ -X_m \boldsymbol{I}_2\right] \begin{bmatrix} \boldsymbol{\psi}_s \\ \boldsymbol{\psi}_r \end{bmatrix}, \tag{7.11}$$

where $\boldsymbol{I}_2$ denotes the 2×2 identity matrix. Transforming the stator current from orthogonal coordinates to the three-phase system using $\tilde{\boldsymbol{K}}^{-1}$ and inserting the result into (7.3) leads to

$$\frac{\mathrm{d}v_n(t)}{\mathrm{d}t} = \frac{1}{2X_c D} |\boldsymbol{u}(t)|^T\, \tilde{\boldsymbol{K}}^{-1} \left[X_r \boldsymbol{I}_2\ \ -X_m \boldsymbol{I}_2\right] \boldsymbol{x}_m(t). \tag{7.12}$$

We define the neutral point potential as the inverter state variable $x_i = v_n$ and write

$$\frac{\mathrm{d}x_i(t)}{\mathrm{d}t} = \boldsymbol{f}_i(\boldsymbol{u}(t))\, \boldsymbol{x}_m(t), \tag{7.13}$$

where $\boldsymbol{f}_i$ is provided in Appendix 7.A. Note that $\boldsymbol{f}_i$ is a function of the component-wise absolute value of the three-phase switch position $\boldsymbol{u}$.

The state-space representation of the drive system is obtained by combining the state-space models of the machine (7.10) and of the inverter (7.13) to

$$\frac{\mathrm{d}\boldsymbol{x}(t)}{\mathrm{d}t} = \boldsymbol{F}(\boldsymbol{u}(t))\, \boldsymbol{x}(t) + \boldsymbol{G}\, \boldsymbol{u}(t) \tag{7.14a}$$

$$\boldsymbol{y}(t) = \boldsymbol{h}(\boldsymbol{x}(t)). \tag{7.14b}$$

The drive state vector

$$\boldsymbol{x} = [\boldsymbol{x}_m^T\ x_i]^T \tag{7.15}$$

is the concatenation of the machine and inverter state vectors. The system and input matrices are

$$\boldsymbol{F}(\boldsymbol{u}) = \begin{bmatrix} \boldsymbol{F}_m & \boldsymbol{0}_{4\times 1} \\ \boldsymbol{f}_i(\boldsymbol{u}) & 0 \end{bmatrix}, \quad \boldsymbol{G} = \begin{bmatrix} \boldsymbol{G}_m \\ \boldsymbol{0}_{1\times 3} \end{bmatrix}. \tag{7.16}$$

The output vector

$$\boldsymbol{y} = [T_e \; \Psi_s \; v_n]^T \tag{7.17}$$

comprises the electromagnetic torque, the stator flux magnitude, and the neutral point potential. The output function $\boldsymbol{h}(\boldsymbol{x})$, which is provided in Appendix 7.A, maps the state vector to the output vector.

The drive model (7.14) is nonlinear because of two terms. The dynamic equation of the neutral point potential entails a multiplication of the input vector with the state vector. In the output equation $\boldsymbol{h}(\boldsymbol{x})$, the electromagnetic torque and the stator flux magnitude are nonlinear functions of the state vector.

7.2.3.2 Discrete-Time Model

In the following, we derive the discrete-time representation of the continuous-time drive model (7.14). The discrete-time state-space model is valid at the discrete-time steps $t = kT_s$, where $k \in \mathbb{N}$ denotes the time step and T_s is the sampling interval.

We compute the discrete-time state-space model in three steps. First, the differential equation of the machine (7.10) is discretized. To achieve high accuracy and to take into account the coupling between the stator and the rotor fluxes, we resort to the exact Euler discretization method and integrate (7.10) from $t = kT_s$ to $t = (k+1)T_s$. Recall that the three-phase switch position $\boldsymbol{u}(t)$ is—by definition—constant between $t = kT_s$ and $t = (k+1)T_s$ and thus equal to $\boldsymbol{u}(k)$. This leads to the discrete-time representation of the machine model

$$\boldsymbol{x}_m(k+1) = \boldsymbol{A}_m \, \boldsymbol{x}(k) + \boldsymbol{B}_m \boldsymbol{u}(k) \tag{7.18}$$

with the matrices

$$\boldsymbol{A}_m = \boldsymbol{e}^{\boldsymbol{F}_m T_s} \text{ and } \boldsymbol{B}_m = -\boldsymbol{F}_m^{-1}(\boldsymbol{I}_4 - \boldsymbol{A}_m)\boldsymbol{G}_m . \tag{7.19}$$

Note that $\boldsymbol{e}$ refers to the matrix exponential.

Second, we discretize the differential equation of the neutral point potential (7.13). For this, we choose the forward Euler discretization method and assume the derivative of x_i at $t = kT_s$ to be constant throughout the integration step. This leads to the discrete-time representation

$$x_i(k+1) = x_i(k) + \boldsymbol{f}_i(\boldsymbol{u}(k)) \, T_s \, \boldsymbol{x}_m(k) . \tag{7.20}$$

Third, we combine the discrete-time models of the machine (7.18) and of the inverter (7.20) to the discrete-time drive model:

$$\boldsymbol{x}(k+1) = \boldsymbol{A}(\boldsymbol{u}(k)) \, \boldsymbol{x}(k) + \boldsymbol{B}\boldsymbol{u}(k) \tag{7.21a}$$

$$\boldsymbol{y}(k) = \boldsymbol{h}(\boldsymbol{x}(k)) \tag{7.21b}$$

with the matrices

$$\boldsymbol{A}(\boldsymbol{u}) = \begin{bmatrix} \boldsymbol{A}_m & \boldsymbol{0}_{4\times 1} \\ \boldsymbol{f}_i(\boldsymbol{u}) \, T_s & 1 \end{bmatrix}, \quad \boldsymbol{B} = \begin{bmatrix} \boldsymbol{B}_m \\ \boldsymbol{0}_{1\times 3} \end{bmatrix} . \tag{7.22}$$

Using the exact Euler method for the machine model and forward Euler for the neutral point potential is facilitated by the fact that the machine states do not depend on the inverter state.

The proposed discretization approach has two advantages. First, the discrete-time model of the machine captures the coupling between the stator and rotor fluxes, which enhances the accuracy of the prediction model and is beneficial when considering long prediction horizons. Second, only the time-varying part of the system matrix $\boldsymbol{A}$, that is, the vector $\boldsymbol{f}_i(\boldsymbol{u})$ that relates to the neutral point potential, needs to be updated. This reduces the computation time.

7.2.4 Switching Effort

Throughout this book, we use the term *switching effort* as a generic term that refers to both the switching frequency and the switching losses of the inverter. The switching frequency of each active semiconductor device can be computed by identifying the corresponding *on* transitions, as summarized in Table 2.4. By counting the number of *on* transitions over a time interval and by dividing this number by the interval's length yields the switching frequency for each of the 12 active switches. The *average* switching frequency is obtained by averaging over the individual switching frequencies of the 12 active switches. As each switching transition in $\boldsymbol{u}$ corresponds to one *on* transition (see Table 2.4), the average switching frequency per semiconductor can be conveniently computed via

$$f_{\text{sw}} = \lim_{N\to\infty} \frac{1}{12NT_s} \sum_{\ell=0}^{N-1} ||\boldsymbol{u}(\ell) - \boldsymbol{u}(\ell-1)||_1, \tag{7.23}$$

where $||\cdot||_1$ denotes the 1-norm. We often refer to (7.23) as the (average) *device* switching frequency.

The procedure to compute the switching losses is more involved. When assuming the nominal voltage $V_{\text{dc}}/2$ over each half of the dc-link and the use of integrated-gate-commutated thyristors (IGCTs) as active semiconductor devices, the turn-off and turn-on losses of the gate-commutated thyristors (GCTs), e_{off} and e_{on}, are proportional to the commutated phase current. On the other hand, the diode's reverse recovery losses e_{rr} are nonlinear in the commutated current. In the pu system, the switching (power) *losses* can be summarized by

$$e_{\text{off}} = c_{\text{off}} \frac{V_{\text{dc}}}{2} i_x \tag{7.24a}$$

$$e_{\text{on}} = c_{\text{on}} \frac{V_{\text{dc}}}{2} i_x \tag{7.24b}$$

$$e_{\text{rr}} = c_{\text{rr}} \frac{V_{\text{dc}}}{2} f_{\text{rr}}(i_x), \tag{7.24c}$$

as has been previously stated in (2.95) and (2.96). The loss coefficients c_{off}, c_{on}, and c_{rr} are provided in Table 2.11. The quantity i_x, with $x \in \{a, b, c\}$, is the commutated phase current, which is—by definition—always nonnegative. Furthermore, according to (7.11), the phase current depends linearly on the state vector $\boldsymbol{x}$. For more details on semiconductors and switching losses in general, the reader is referred to Sect. 2.3. The specific switching losses of NPC inverters are explained in Sects. 2.4.1 and 2.5.1.

For a given switching transition from $\boldsymbol{u}(\ell-1)$ to $\boldsymbol{u}(\ell)$ and polarity of the commutated phase current, the semiconductor devices that are turned on and off can be identified. These cases are summarized in Table 2.5 and can be easily translated into a small look-up table. By using (7.24) and Table 2.5, the switching loss $e_{\rm sw}(\boldsymbol{x}(\ell), \boldsymbol{u}(\ell), \boldsymbol{u}(\ell-1))$ can be computed. By summing up the instantaneous switching (energy) losses over a time interval and dividing the sum by the length of this time interval, the switching (power) losses

$$P_{\rm sw} = \lim_{N\to\infty} \frac{1}{NT_s} \sum_{\ell=0}^{N-1} e_{\rm sw}(\boldsymbol{x}(\ell), \boldsymbol{u}(\ell), \boldsymbol{u}(\ell-1)) \tag{7.25}$$

of the inverter are obtained. Note that the switching energy losses $e_{\rm sw}$ are always positive.

7.3 Control Problem Formulation

As stated in Sect. 7.2.2, the control objective is to keep the output (or controlled) variables within given bounds around their respective references while minimizing the switching effort. For the drive system based on an NPC inverter, the output variables are the electromagnetic torque, the stator flux magnitude, and the neutral point potential. If a bound has been violated, the violation has to be eliminated as quickly as possible.

The MPDTC controller is endowed with the discrete-time controller model of the drive (7.21). This internal model enables the controller to anticipate and predict the impact of its decisions. The control objectives are mapped to a cost function that yields a scalar cost value. This cost function is minimized subject both to the dynamic evolution of the controller model and to the constraints. The latter include switching constraints and integer constraints on the switch positions. Upper and lower constraints on the phase currents are also conceivable.

At each time step, the controller computes a sequence of three-phase switch positions over the time interval of the prediction horizon. This sequence of switch positions keeps the controlled variables within their bounds and minimizes the switching effort over the prediction horizon. From this sequence, only the first switch position at the current time step is applied to the inverter. The predictions are recomputed at the next time step using new measurements and flux estimates, and a shifted—and if necessary, revised—sequence of switch positions is derived. This is referred to as the *receding horizon policy*, which provides feedback and makes MPDTC robust to parameter uncertainties in the underlying prediction model (see also [16]).

7.3.1 *Naive Optimization Problem*

Writing the aforementioned control problem as a closed-form optimization problem leads to

$$\boldsymbol{U}_{\rm opt}(k) = \arg \underset{\boldsymbol{U}(k)}{\text{minimize}}\ J \tag{7.26a}$$

$$\text{subject to } \boldsymbol{x}(\ell+1) = \boldsymbol{A}(\boldsymbol{u}(\ell))\boldsymbol{x}(\ell) + \boldsymbol{B}\boldsymbol{u}(\ell) \tag{7.26b}$$

$$\boldsymbol{y}(\ell+1) = \boldsymbol{h}(\boldsymbol{x}(\ell+1)) \tag{7.26c}$$

$$\begin{cases} \varepsilon_j(\ell+1) = 0, & \text{if } \varepsilon_j(\ell) = 0 \\ \varepsilon_j(\ell+1) < \varepsilon_j(\ell), & \text{if } \varepsilon_j(\ell) > 0 \end{cases} \tag{7.26d}$$

$$\boldsymbol{u}(\ell) \in \boldsymbol{\mathcal{U}},\ \|\Delta \boldsymbol{u}(\ell)\|_\infty \le 1 \tag{7.26e}$$

$$\forall \ell = k, \ldots, k+N_p-1,\ \forall j = 1,2,3. \tag{7.26f}$$

The sequence of control inputs

$$U(k) = [u^T(k)\ u^T(k+1)\ \dots\ u^T(k+N_p-1)]^T \tag{7.27}$$

over the prediction horizon N_p represents the sequence of three-phase switch positions the controller decides upon. The cost function J is minimized over the set of all admissible $U(k)$ subject to the dynamic evolution of the drive system (7.26b), its output variables (7.26c), and the constraints (7.26d) and (7.26e). The argument of this minimization is the optimal switching sequence $U_{\mathrm{opt}}(k)$. The variables in (7.26) are defined in the following subsections, and the respective equations are explained in detail.

7.3.2 Constraints

To quantify the degree of a bound violation, we introduce the nonnegative auxiliary variable

$$\varepsilon_T(\ell) = \frac{1}{T_{e,\max} - T_{e,\min}} \begin{cases} T_e(\ell) - T_{e,\max} & \text{if } T_e(\ell) > T_{e,\max} \\ T_{e,\min} - T_e(\ell) & \text{if } T_e(\ell) < T_{e,\min} \\ 0 & \text{else} \end{cases} \tag{7.28}$$

for the torque, where $T_{e,\min}$ ($T_{e,\max}$) denotes the lower (upper) bound on the torque. Note that we normalize the degree of the bound violation by the width of the torque bounds. This facilitates the deadlock resolution mechanism, which will be proposed in Sect. 9.4, but the normalization can be omitted here when imposing the constraint (7.26d).

For the stator flux magnitude and the neutral point potential, the auxiliary variables ε_Ψ and ε_v are defined accordingly. Aggregating these variables, we define the degree of the bound violation as

$$\boldsymbol{\varepsilon} = [\varepsilon_T\ \varepsilon_\Psi\ \varepsilon_v]^T. \tag{7.29}$$

In (7.26d), we use the index j, with $j \in \{1, 2, 3\}$, to refer to the jth component of $\boldsymbol{\varepsilon}$.

The constraint (7.26d) is imposed component-wise, that is, separately for each output variable. If at time step ℓ_0 an output variable is within its bounds, then it has to remain within them for the future time steps $\ell > \ell_0$ until the end of the prediction horizon. We say that the output variable is *feasible* or *within its feasible region*. The latter is defined as the set between the respective upper and lower bounds. During steady-state operation, the outputs usually remain within their bounds.

If, however, at time step ℓ_0 an output variable violates a bound, then it has to move closer to the bound at every time step within the prediction horizon. In doing so, the bound violation is reduced until the output variable becomes feasible. This scenario typically occurs during transients, such as torque reference steps. The concept of feasibility and diminishing bound violations is illustrated in Fig. 7.3 for torque trajectories. Note that the trajectory on the lower right-hand side of this figure violates the constraint (7.26d) and must be excluded. The switching sequence corresponding to this torque trajectory would allow MPDTC to postpone any switching transition from time step k to $k+1$. Owing to the receding horizon policy, MPDTC would be able to postpone switching again at time step $k+1$, and so on.

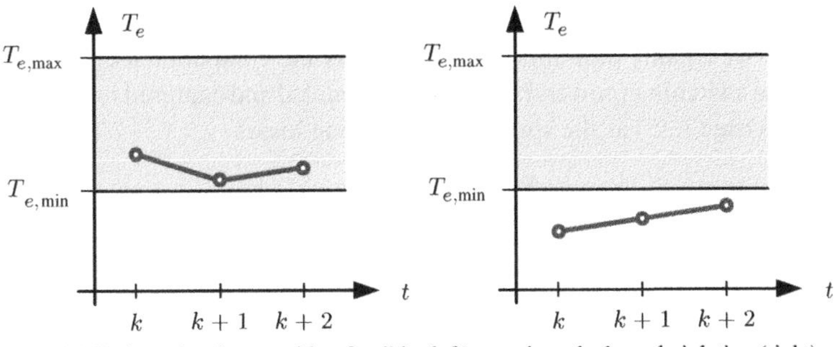

(a) Trajectories that are either feasible (left) or reduce the bound violation (right)

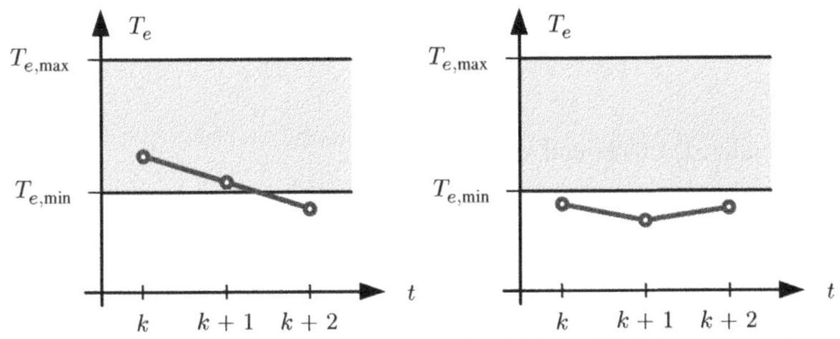

(b) Trajectories that are neither feasible nor reduce the bound violation at every time step

Figure 7.3 Example torque trajectories that meet the constraint (7.26d) (a) and torque trajectories that violate it (b). The feasible region between the upper and lower torque bounds is shaded

The constraint (7.26e) limits the switch position $\boldsymbol{u}$ to the integer values $\mathcal{U} = \{-1, 0, 1\}^3$, which are available for the three-level inverter. Switching in a phase by more than one step up or down is not allowed. This is enforced by the second constraint in (7.26e), $\|\Delta \boldsymbol{u}(\ell)\|_\infty \leq 1$, which limits the elements in $\Delta \boldsymbol{u}(\ell) = \boldsymbol{u}(\ell) - \boldsymbol{u}(\ell - 1)$ to ± 1. These constraints have to be met at every time step ℓ within the prediction horizon.

7.3.3 Cost Function

The cost function

$$J = J_{\text{sw}} + J_{\text{bnd}} + J_t \tag{7.30}$$

in (7.26a) consists of three terms. The first term captures the switching effort. Specifically,

$$J_f = \frac{1}{N_p} \sum_{\ell=k}^{k+N_p-1} \|\Delta \boldsymbol{u}(\ell)\|_1 \tag{7.31}$$

represents the sum of the switching transitions (i.e., the number of commutations) over the prediction horizon divided by the length of the horizon. As a result, J_f approximates the short-term switching *frequency*. Strictly speaking, (7.31) should be multiplied with $1/(12T_s)$, as stated in (7.23). We usually omit this term to simplify the computations.

Alternatively, the switching (power) *losses* can be modeled and captured in the cost function. To this end, we rewrite (7.25) as the short-term switching losses

$$J_P = \frac{1}{N_p} \sum_{\ell=k}^{k+N_p-1} e_{\text{sw}}(\boldsymbol{x}(\ell), \boldsymbol{u}(\ell), \boldsymbol{u}(\ell-1)) \tag{7.32}$$

over the prediction horizon. As for the switching frequency, we drop the factor $1/T_s$ from (7.32). In the cost function (7.30), we either set the switching effort term to $J_{\text{sw}} = J_f$ or to $J_{\text{sw}} = J_P$.

The second term in the cost function

$$J_{\text{bnd}} = \boldsymbol{q}^T \boldsymbol{\epsilon}(k) \tag{7.33}$$

penalizes the (normalized) rms bound violation

$$\boldsymbol{\epsilon}(k) = [\epsilon_T(k)\ \epsilon_\Psi(k)\ \epsilon_\upsilon(k)]^T \tag{7.34}$$

of the output vector $\boldsymbol{y}$ over the prediction horizon. The rms bound violation of the torque, for example, is defined as

$$\epsilon_T(k) = \sqrt{\frac{1}{N_p} \sum_{\ell=k}^{k+N_p-1} (\varepsilon_T(\ell))^2}, \tag{7.35}$$

with $\varepsilon_T(\ell)$ as in (7.28). Alternatively, to reduce the computational burden, the square root can be omitted from (7.35). For the stator flux magnitude and the neutral point potential, the bound violations ϵ_Ψ and ϵ_υ are defined accordingly. The weighting vector $\boldsymbol{q}$ has nonnegative components.

The third term J_t is an optional penalty on the predicted output quantities at the end of the prediction horizon, in the form of either a terminal soft constraint or a terminal weight. The rationale for such a penalty and its benefits will be explained in detail in Sect. 9.4.

7.4 Model Predictive Direct Torque Control

To assess the length of the prediction horizon that is required to provide a good steady-state performance, recall that one of the controller's main objectives is to minimize the switching effort. For the model predictive controller to be able to address this objective sufficiently well, the prediction horizon should comprise several switching transitions. In an MV power electronics setting, the switching frequency is in the range of a few hundred hertz.

Example 7.1 *For an NPC inverter and a device switching frequency of 250 Hz, a switching transition occurs on average every 2 ms per phase and every 0.67 ms in the three-phase inverter. If, on average, the prediction horizon was to capture six transitions in the inverter, a prediction horizon amounting to 4 ms would be required.*

On the other hand, with a direct control scheme like MPDTC, switching is restricted to discrete-time steps and thus to the sampling instants. To avoid that the discretization of the time axis restricts the controller in its switching decisions, potentially worsening the current distortions, a short sampling interval is required. This ensures a fine granularity of the time axis and thus of possible switching events. If the ratio between the sampling frequency and the switching frequency is large, say more than 100, the restriction of the switching transitions to discrete-time instants becomes negligible and the time axis can be deemed to be continuous. To achieve this, we typically chose a sampling interval of 25 μs.

The combination of low switching frequencies, the need to capture several switching transitions within the prediction horizon, and the necessity of short sampling intervals leads to prediction horizons that exceed 100 time steps. For example, when adopting a sampling interval of 25 μs, Example 7.1 implies a horizon of $N_p = 160$ steps.

The optimization problem (7.26) is a nonlinear, mixed-integer program. When solving (7.26) to find the optimal solution, it is well known that—in the worst case—all possible solutions need to be enumerated. This makes the solution process computationally challenging even for very short prediction horizons. Solving it for reasonably long horizons is not feasible, as the next example will show, unless simplifications or approximations are made.

Example 7.2 *Using a three-level inverter as an example, the number of admissible switching transitions from* $\mathbf{u}(\ell)$ *to* $\mathbf{u}(\ell + 1)$ *is on average 12. Owing to the switching constraint (7.26e), this number is smaller than the theoretically possible 27 transitions. Nevertheless, for the horizon* $N_p = 75$*, for example, the number of possible switching sequences amounts to* $12^{N_p} \approx 10^{80}$*, which is equal to the estimated number of atoms in the observable universe.*

7.4.1 Definitions

To facilitate the exposition of the MPDTC algorithm, we first define the following terms:

- The *switching sequence* $\boldsymbol{U}(k)$ is the sequence of three-phase switch positions $\boldsymbol{u}$. Generalizing (7.27), we define the switching sequence

$$\boldsymbol{U}(k) = [\boldsymbol{u}^T(k)\ \boldsymbol{u}^T(k+1)\ \dots\ \boldsymbol{u}^T(k+N-1)]^T \tag{7.36}$$

  of length N. Its first element is the switch position at the current time step k.
- Associated with a switching sequence is the *state trajectory* $\boldsymbol{X}(k) = [\boldsymbol{x}^T(k+1) \dots \boldsymbol{x}^T(k+N)]^T$, which is the sequence of state vectors $\boldsymbol{x}$ that fully describes the evolution of the drive from time step $k+1$ until time step $k+N$, when applying the switching sequence $\boldsymbol{U}(k)$. Note that the initial state $\boldsymbol{x}(k)$ is not included in $\boldsymbol{X}(k)$, because it does not depend on $\boldsymbol{U}(k)$. The state vector encompasses the four components of the machine fluxes and the neutral point potential (see also (7.15)).
- Similarly, the evolution of the drive outputs is described by the *output trajectory* $\boldsymbol{Y}(k) = [\boldsymbol{y}^T(k+1)\ \dots\ \boldsymbol{y}^T(k+N)]^T$, where $\boldsymbol{y}$ is composed of the electromagnetic torque, the stator flux magnitude, and the neutral point potential as defined in (7.17).

- An *admissible* switching sequence meets the switching constraint (7.26e) at every time step.
- A *candidate* switching sequence is an admissible switching sequence that yields output trajectories that meet the constraint (7.26d) at every time step. These output trajectories are either feasible or reduce the bound violation. Feasibility means that the output variables lie within their corresponding bounds; reducing the bound violation refers to the case in which an output variable is not feasible, but the degree of the bound violation decreases at every time step. These conditions must hold component-wise for all output variables.[2] The notion of candidate switching sequences is exemplified in Fig. 7.3 for torque trajectories.

7.4.2 Simplified Optimization Problem

It is important to recognize that optimal solutions to the optimization problem (7.26) correspond to switching sequences $\boldsymbol{U}$ that feature switching transitions almost exclusively in the vicinity of the bounds or when bounds have been violated. This is due to the combination of the cost function (7.26a) with the constraint (7.26d). The former penalizes the switching effort and bound violations, while the latter requires the output variables to remain within their bounds, or, when a bound has been violated, to reduce the bound violation. These objectives are best met by switching effectively only when an output variable is about to violate its bound, or, in case a bound has been violated, when the rate of convergence toward the bound is insufficient.

Therefore, an attractive way to simplify the solution procedure of (7.26) is to consider switching transitions only when at least one of the output variables is close to its respective bound, that is, when switching is imminently required to keep the output variables within their bounds. When the outputs are well within their bounds, it is not necessary to consider switching transitions, and the three-phase switch position can be frozen.

This approach meets the constraints (7.26d) and (7.26e), minimizes the switching effort in the cost function (7.26a), and greatly reduces the number of switching sequences to be considered and thus the computational complexity of the problem at hand. However, the switching sequences obtained with this approach are in general suboptimal solutions to the original optimization problem (7.26), because only a subset of the admissible switching sequences is considered.

7.4.3 Concept of the Switching Horizon

To describe the switching sequences that are considered in the simplified optimization problem, we introduce the so-called *switching horizon*. The switching horizon N_s is a string that is composed of the characters *S*, *E*, and *e*.

The character *S* stands for *switch*, that is, a switching transition from $\boldsymbol{u}(\ell)$ to $\boldsymbol{u}(\ell+1)$. This involves considering at time step ℓ all admissible switch positions

$$\boldsymbol{u}(\ell) \in \boldsymbol{\mathcal{U}} \text{ such that } ||\Delta \boldsymbol{u}(\ell)||_\infty \leq 1, \tag{7.37}$$

[2] As an example, consider the case where the torque is feasible, the stator flux violates its bounds but the bound violation is reduced at each time step, and the neutral point potential is feasible. The underlying switching sequence is a candidate switching sequence, provided that it is admissible.

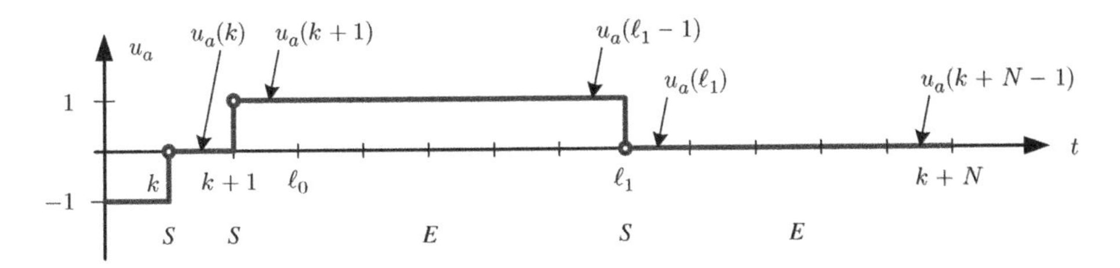

Figure 7.4 Concept of the switching horizon with the elements S and E, which refer to *switch* and *extend*, respectively, exemplified for a single-phase switching sequence with the switching horizon $N_s = SSESE$. Note that $u_a(\ell_0-1) = u_a(k+1)$ and $u_a(\ell_0-2) = u_a(k)$

see (7.26e), and performing the prediction step

$$\boldsymbol{x}(\ell+1) = \boldsymbol{A}(\boldsymbol{u}(\ell))\boldsymbol{x}(\ell) + \boldsymbol{B}\boldsymbol{u}(\ell) \tag{7.38a}$$

$$\boldsymbol{y}(\ell+1) = \boldsymbol{h}(\boldsymbol{x}(\ell+1)) \tag{7.38b}$$

as in (7.26b) and (7.26c). At time step $\ell+1$, the constraint

$$\begin{cases} \varepsilon_j(\ell+1) = 0, & \text{if } \varepsilon_j(\ell) = 0 \\ \varepsilon_j(\ell+1) < \varepsilon_j(\ell), & \text{if } \varepsilon_j(\ell) > 0 \end{cases} \tag{7.39}$$

must be met for each output variable y_j, with $j \in \{1,2,3\}$. The degree of the jth bound violation ε_j was defined in (7.28), using the torque as an example. Transitions to switch positions $\boldsymbol{u}(\ell+1)$ that violate (7.39) are discarded.

The character E refers to *extend*, that is, an extension step, which is preceded by a switching transition. Assume that the last switching transition (was predicted to have) occurred at time step ℓ_0-1 from the old switch position $\boldsymbol{u}(\ell_0-2)$ to the new switch position $\boldsymbol{u}(\ell_0-1)$ (see also Fig. 7.4). By definition, the output variables meet the constraint (7.39) at time step ℓ_0, at which the extension step starts. The switch position is frozen from time step ℓ_0-1 until time step ℓ_1, that is, $\boldsymbol{u}(\ell_0-1) = \boldsymbol{u}(\ell_0) = \cdots = \boldsymbol{u}(\ell_1-1)$. The time step ℓ_1 is defined as the last time step at which the constraint (7.39) is met. Specifically, at time step ℓ_1+1, one of the output variables will either violate its bounds or the requirement to reduce the bound violations will no longer be met.

This extension mechanism can be formally stated as the following maximization problem:

$$\ell_1 = \arg\underset{\ell}{\text{maximize}}\ \ell \tag{7.40a}$$

$$\text{subject to } \boldsymbol{x}(\ell) = \boldsymbol{A}(\boldsymbol{u}(\ell-1))\boldsymbol{x}(\ell-1) + \boldsymbol{B}\boldsymbol{u}(\ell-1) \tag{7.40b}$$

$$\boldsymbol{y}(\ell) = \boldsymbol{h}(\boldsymbol{x}(\ell)) \tag{7.40c}$$

$$\begin{cases} \varepsilon_j(\ell) = 0, & \text{if } \varepsilon_j(\ell-1) = 0 \\ \varepsilon_j(\ell) < \varepsilon_j(\ell-1), & \text{if } \varepsilon_j(\ell-1) > 0 \end{cases} \tag{7.40d}$$

$$\boldsymbol{u}(\ell-1) = \boldsymbol{u}(\ell_0-1) \tag{7.40e}$$

$$\forall \ell = \ell_0+1, \ell_0+2, \ldots, \forall j = 1,2,3. \tag{7.40f}$$

The extension leg commences at time step ℓ_0 and ends at $\ell_1 > \ell_0$. It uses the state vector $\boldsymbol{x}(\ell_0)$ and the switch position $\boldsymbol{u}(\ell_0 - 1)$ as initial conditions.

The third possible element *e* denotes an optional extension step. The switching horizon $N_s = eSE$, for example, implies two cases to be investigated: one with the switching horizon *ESE*, and the other with *SE*.

Example 7.3 *The switching horizon* $N_s =$ SSESE *entails two switching transitions at time steps* k *and* $k+1$*, followed by an extension step and a switching transition at time step* $\ell_1 > k+1$*, which is followed by another extension step. Thus, switching is considered only at time steps* k*,* $k+1$*, and* ℓ_1*, while the switch position is frozen during the extension steps. As a result, the switch positions in the general switching sequence* (7.36) *meet the constraints*

$$\boldsymbol{u}(k+1) = \boldsymbol{u}(k+2) = \cdots = \boldsymbol{u}(\ell_1 - 1) \tag{7.41a}$$

$$\boldsymbol{u}(\ell_1) = \boldsymbol{u}(\ell_1 + 1) = \cdots = \boldsymbol{u}(k + N - 1). \tag{7.41b}$$

A corresponding single-phase switching sequence is depicted in Fig. 7.4.

We have seen in Example 7.2 that the number of possible switching sequences for a three-level inverter amounts to approximately 12^{N_p}, where N_p is the length of the prediction horizon. For the switching horizon $N_s = SSESE$, prediction horizons of similar lengths can be achieved while considering only $12^3 = 1728$ switching sequences. This compelling reduction motivates the simplification of the optimization problem and the adoption of the switching horizon.

We define the *prediction horizon* as the length (in terms of the number of time steps) of the *longest* switching sequence. Specifically, let $\mathcal{I}$ denote the index set of candidate switching sequences and $i \in \mathcal{I}$ the ith switching sequence, which is of length N_i. The prediction horizon is then defined as

$$N_p = \max_{i \in \mathcal{I}} N_i. \tag{7.42}$$

Note that the prediction horizon is time-varying. There is a weak correlation between the switching horizon and the prediction horizon, in the sense that adding *S* and *E* elements to the switching horizon increases, in general, the length of the prediction horizon.

Example 7.4 *Figure 7.5 depicts the concept of the switching horizon for a three-level inverter drive system using the switching horizon* $N_s =$ eSSESE*. Not utilizing the optional extension leg, a first candidate switching sequence is shown in Fig. 7.5(c). Its corresponding output trajectories are shown in Figs. 7.5(a) and (b). Switching in phases* b *and* a *is performed at time steps* k *and* $k+1$*, respectively, from* $\boldsymbol{u}(k-1) = [0\ 1\ -1]^T$ *via* $\boldsymbol{u}(k) = [0\ 0\ -1]^T$ *to* $\boldsymbol{u}(k+1) = [1\ 0\ -1]^T$.

After freezing the switch position at time step $k+1$ *at* $\boldsymbol{u} = [1\ 0\ -1]^T$*, the extension step* (7.40) *predicts that the torque will hit its lower bound between* $k+5$ *and* $k+6$*. This triggers a predicted switching transition at time step* $k+5$ *in phases* a *and* b *to* $\boldsymbol{u}(k+5) = [0\ 1\ -1]^T$ *with the aim of keeping the torque within its bounds. Using this switch position and another extension leg, the stator flux magnitude is predicted to violate its upper bound shortly after time step* $k+24$*. As a result, the switching sequence shown in Fig. 7.5(c) switches at three time steps and is of length* $N_1 = 24$ *steps. Switching will be required at time step* $k+24$.

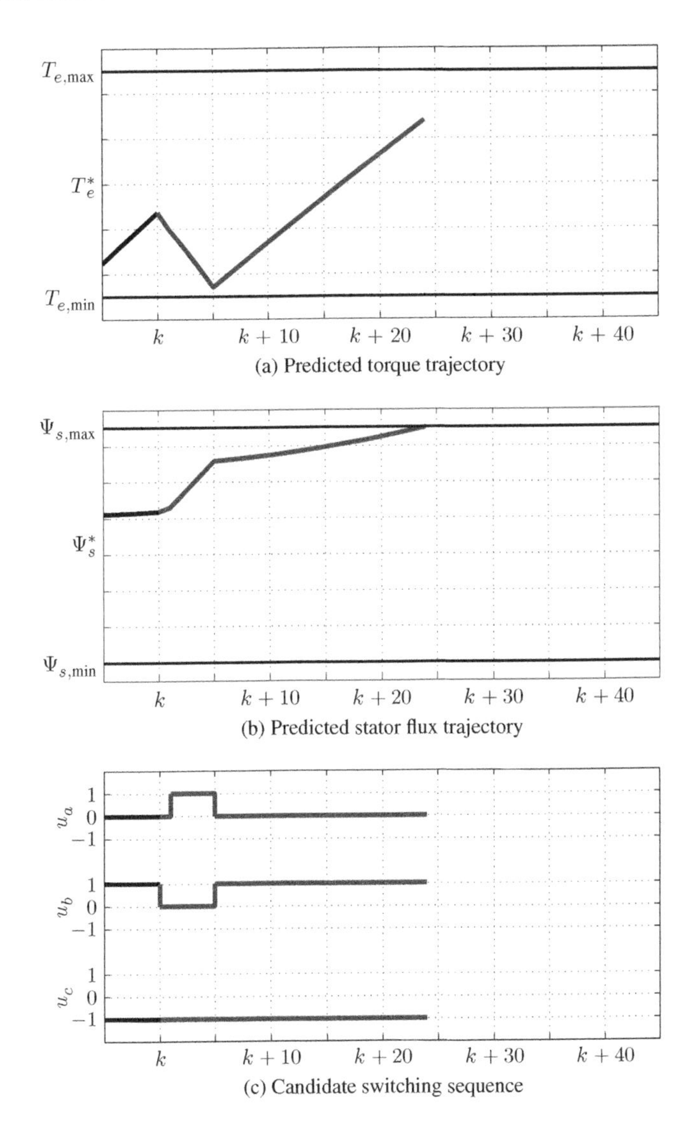

Figure 7.5 The first candidate switching sequence of Example 7.4 with the switching horizon *eSSESE*. The corresponding torque and stator flux trajectories are shown in (a) and (b) within their respective upper and lower bounds. The neutral point potential is not shown, but it is treated in the same way as the torque and stator flux magnitude

Another candidate switching sequence along with its output trajectories is indicated by the solid lines in Fig. 7.6. The optional extension step predicts that switching is not required until the time step $k + 13$, when the torque is about to violate its upper bound. One way to avoid this violation is to switch in phase c from -1 to 0. The degree of freedom of the second switching transition (in the switching horizon $N_s =$ eSSESE) at time step $k + 14$ is not taken up and the switch positions are kept constant. Another extension leg predicts a bound violation of the

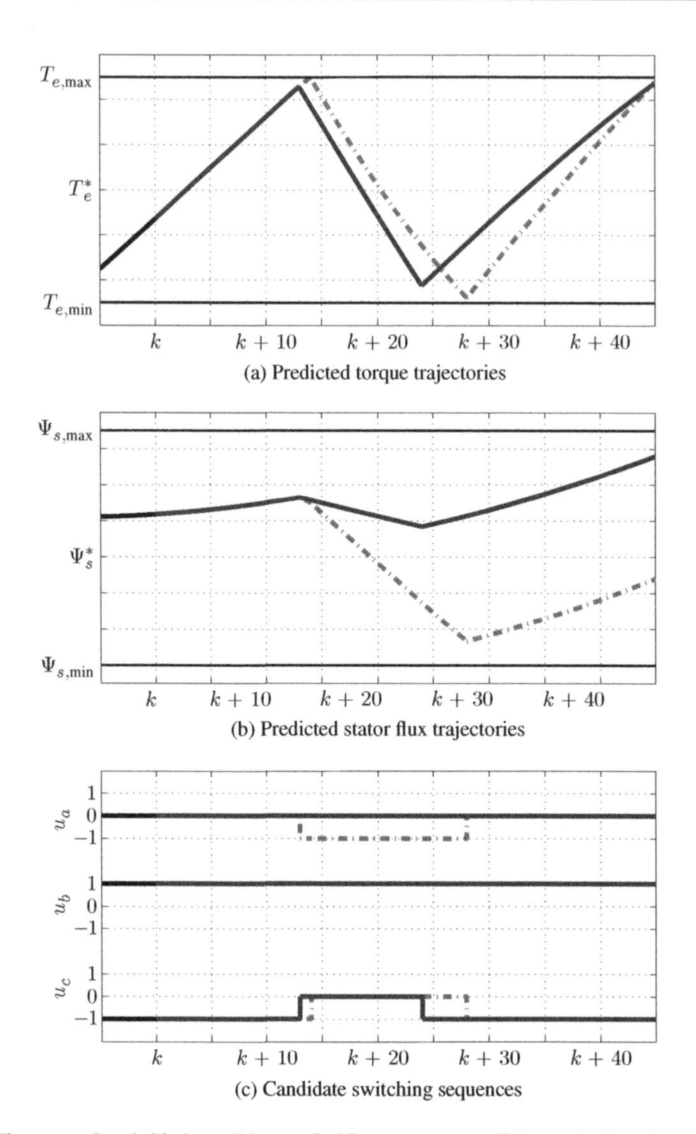

Figure 7.6 The second and third candidate switching sequences of Example 7.4 along with their corresponding torque and stator flux trajectories. Solid lines correspond to the second switching sequence, while dash-dotted lines refer to the third one

torque just before $k+25$, triggering a switching transition at time step $k+24$ in phase c. The last extension step indicates that this switching sequence can be applied for $N_2 = 45$ sampling intervals before an output variable violates its bound.

A third candidate switching sequence is shown in Fig. 7.6 using dash-dotted lines. Switching is predicted to occur at time steps $k+13$, $k+14$, and $k+33$. The length of the switching sequence is again $N_3 = 45$ time steps.

Table 7.1 Summary of the three candidate switching sequences of Example 7.4 when minimizing the switching frequency

Index i of the candidate switching sequence	Number of switching transitions	Length N_i of the switching sequence	Cost J_i
1	4	24	0.167
2	2	45	0.044
3	4	45	0.089

In order to assess which candidate switching sequence is the optimal one, the cost function J *stated in* (7.30) *needs to be evaluated for each candidate sequence. As the bounds are not violated in this example,* J_{bnd} *in* (7.30) *is always zero. When penalizing the switching frequency with the cost function term* J_f *as defined in* (7.31)*, the second candidate switching sequence is optimal, because it requires the least number of switching transitions and features a long switching sequence. The corresponding values that yield this result are shown in Table 7.1.*

When minimizing the switching losses J_P *as in* (7.32)*, the first switching sequence could be the optimal one in case the sum of the absolute values of the commutated currents is less than half that of the second switching sequence. Specifically, in case the instantaneous currents in phases* a *and* b *were close to zero, the first sequence would be the optimal one because the phase* c *current would be high. The third switching sequence will always remain suboptimal because for the same length as the second one it requires two more switching transitions in phase* a*.*

In summary, considering switching transitions only when the constraint (7.26d) is triggered gives rise to two different prediction horizons—the *switching* horizon N_s (consisting of switching events S, extension legs E, and optional extension legs e), and the *prediction* horizon N_p (the number of time steps the MPDTC algorithm looks into the future). By inserting extension legs between the switching instants, the switch positions are frozen until the constraint (7.26d) is about to be violated. The switching elements act like hinges that are connected by extension legs. Multiple switching events may form clusters of switching transitions, such as *SSS*. The concept of extension legs leads to very long prediction horizons (typically 30–200 time steps) while the switching horizon is short (usually 2–5 elements).

It will later become apparent that the computational complexity of the MPDTC algorithm strongly depends on the switching horizon, that is, the controller's degree of freedom, while the closed-loop performance is a function of the prediction horizon. We have seen in Example 7.1 that a prediction horizon of $N_p = 160$ steps was required to ensure that the controller is capable of adequately anticipating the impact of its decisions, by including on average two switching transitions per phase within the prediction horizon. For the NPC inverter, 12^{160} possible switching sequences result. The notion of the switching horizon reduces this number to a few thousands. Branch-and-bound techniques will be described in Chap. 10, which further reduce the number of switching sequences to be explored to a few hundreds. This results in a computational complexity that is amenable for implementation on a digital signal processor (DSP) or a field-programmable gate array (FPGA).

7.4.4 Search Tree

The switching horizon induces a search tree. A node in the search tree at time step ℓ_0, $\ell_0 = k, k+1, \ldots$, is specified by the 9-tuple[3] $(\boldsymbol{u}(\ell_0 - 1), \boldsymbol{x}(\ell_0), \boldsymbol{y}(\ell_0 - 1), \boldsymbol{y}(\ell_0), E_{sw}, S_{sw}, \ell_0, A, \boldsymbol{u}(k))$, which is defined as follows.

- As before, $\boldsymbol{u}(\ell_0 - 1)$ and $\boldsymbol{x}(\ell_0)$ denote the previously applied three-phase switch position and the current state vector of the drive, respectively. The pair $\boldsymbol{u}(\ell_0 - 1)$ and $\boldsymbol{x}(\ell_0)$ thus fully defines the state of the drive including its switching state.
- The output vector $\boldsymbol{y}$ at time steps $\ell_0 - 1$ and ℓ_0 is required to assess whether the output trajectory at time step ℓ_0 is feasible, reduces the bound violation, or fulfills neither of the two conditions.
- $E_{sw} = \sum_{\ell=k}^{\ell_0} e_{sw}(\ell)$ is the sum of the predicted individual switching losses $e_{sw}(\ell)$ up to time step ℓ_0. The unit of the switching energy losses is watt-second.
- $S_{sw} = \sum_{\ell=k}^{\ell_0} ||\Delta \boldsymbol{u}(\ell)||_1$ is the sum of the predicted number of commutations up to time step ℓ_0.
- A denotes the sequence of actions to be performed on the node. A is a string that consists of elements from the set $\{S, E, e\}$.
- Instead of storing the complete switching sequences $\boldsymbol{U}(k)$ in the 9-tuple, it suffices to store only its first element, that is, the switch position $\boldsymbol{u}(k)$. Only the latter is applied to the inverter at time step k.

An example of a search tree is provided in Fig. 7.7 for the switching horizon *SSESE*. We distinguish between the following nodes in the search tree.

- The *root node* is the initial node at time step k. It is initialized with $(\boldsymbol{u}(k-1), \boldsymbol{x}(k), \boldsymbol{y}(k-1), \boldsymbol{y}(k), 0, 0, k, N_s, [\,])$. The symbol [ ] denotes an empty switch position. The root node is depicted as the gray circle at the top of Fig. 7.7.
- *Bud nodes* correspond to *incomplete candidate* switching sequences with actions remaining that induce child nodes. The corresponding output trajectories fulfill the candidacy requirement (so far). Bud nodes are shown as gray circles.
- *Leaf nodes* come in two varieties. (i) Nodes corresponding to *complete candidate* switching sequences that have been fully computed with no actions remaining and candidacy fulfilled at every time step. These are shown as stars in Fig. 7.7. (ii) Nodes that correspond to *non-candidate* switching sequences, which are not further considered, are marked with a stop sign.

As the path from the root node to a subsequent node is unique, the switching sequence to each node is also unique. As a result, there is a direct correspondence between nodes and switching sequences, allowing us to subsequently use both terms interchangeably. Note also that leaf nodes with complete candidate switching sequences correspond to complete solutions (switching sequences) of the optimization problem.

[3] Note that in the node either the switching losses E_{sw} or the number of commutations S_{sw} is required, allowing one to reduce the node to an 8-tuple.

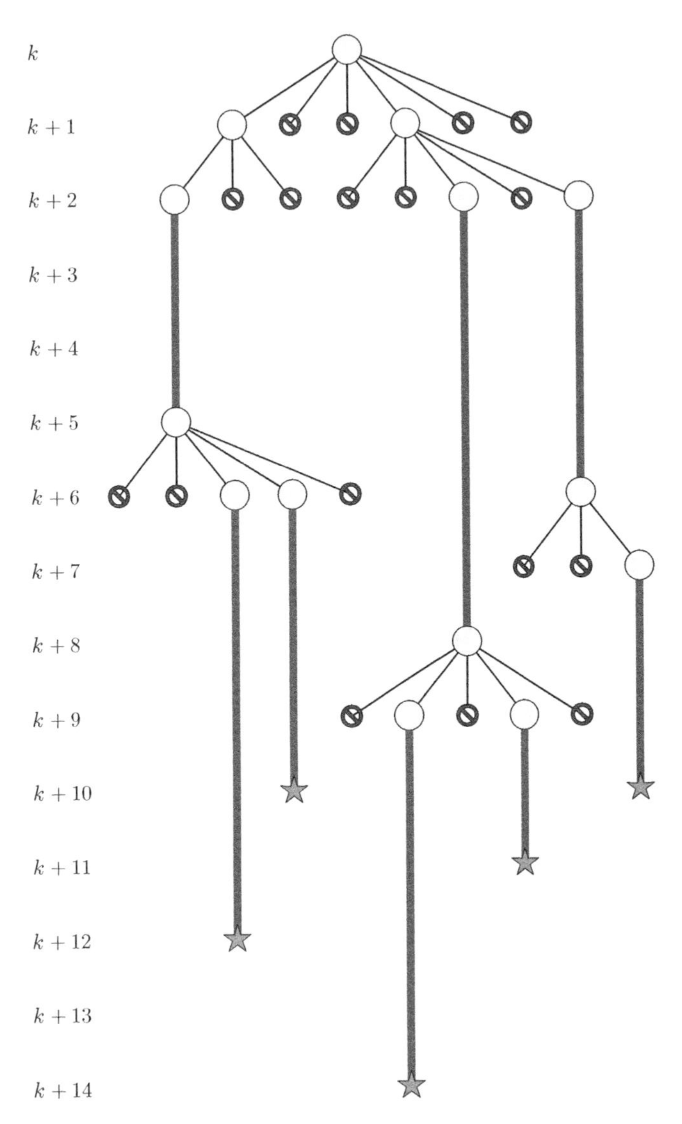

Figure 7.7 Example of a search tree that is induced by the switching horizon *SSESE*. The gray circles denote the root and bud nodes, stars refer to leaf nodes that correspond to complete candidate switching sequences, and stop signs mark leaf nodes that relate to non-candidate switching sequences. Switching transitions *S* are shown as thin lines, while extension steps *E* are thick vertical lines. The discrete-time axis is shown on the left, where the prediction horizon has 14 steps

7.4.5 MPDTC Algorithm with Full Enumeration

We are now ready to specify the MPDTC algorithm. The MPDTC algorithm relies on the notions of the switching horizon, enumeration, and extension steps. Specifically, in its basic form, the MPDTC algorithm enumerates all admissible switching sequences that are candidate sequences and computes their corresponding output trajectories and their cost. Starting at the current time step k, the MPDTC algorithm iteratively explores the tree of admissible switching sequences moving forward in time. Hence all nodes in the search tree are visited that belong to candidate switching sequences. The algorithm is based on a stack model, which is commonly used in the field of computer science.

At time step k, the drive's system state is fully described by the pair $\boldsymbol{x}(k)$ and $\boldsymbol{u}(k-1)$, that is, the state vector and the previously chosen inverter switch position. Based on these, the MPDTC algorithm computes the optimal control input $\boldsymbol{u}_{\text{opt}}(k)$ according to the following procedure in which the switching losses are minimized. The modifications required to minimize the switching frequency will be stated at the end of this section.

1a. The root node is initialized and pushed onto the stack.

1b. Optional step: The output trajectories are extended using (7.40). If the length of the extended trajectory exceeds a given threshold, $\boldsymbol{u}_{\text{opt}}(k) = \boldsymbol{u}(k-1)$ is set and the algorithm proceeds with Step 4. Otherwise, the extension leg is discarded and the algorithm proceeds with Step 2a.

2a. The top node i with a nonempty sequence of actions, $A_i \neq \emptyset$, is taken from the stack.

2b. The first element is read out from A_i and removed.

- For S, all admissible switching transitions are enumerated according to (7.37). The state and output vectors at the next time step are predicted for each admissible switching transition using (7.38). If the output constraint (7.39) is met, the new node j is created. Assuming that the switching transition occurred at time step ℓ, the switching losses of the semiconductor switches are predicted with the help of (7.24) and Table 2.5. This yields the losses $e_{\text{sw}}(\ell)$, which are added to the sum of the switching losses $E_{\text{sw},i}$ incurred so far for this switching sequence by setting $E_{\text{sw},j} = E_{\text{sw},i} + e_{\text{sw}}(\ell)$. The node i is removed, but multiple child nodes have been created.
- For E, the output trajectories are extended using (7.40) and the node i is updated. A new node is not created.
- For e, the node i is kept and the optional extension leg is ignored. The new node j is created as a copy of the node i, and its trajectories are extended using (7.40).

2c. The newly created and updated nodes are pushed onto the stack. By definition, these nodes relate to candidate switching sequences.

2d. If at least one node with a nonempty set of actions A remains, the algorithm proceeds with Step 2a; or else it proceeds with Step 3a.

The results of Step 2 are the leaf nodes $i \in \mathcal{I}$, where $\mathcal{I}$ is an index set. These nodes correspond to the candidate switching sequences $\boldsymbol{U}_i(k)$.

3a. For each leaf node $i \in \mathcal{I}$, the associated cost $J_i = E_{\text{sw},i}/N_i + J_{\text{bnd},i} + J_{ti}$ is computed, as defined in (7.30) and (7.32). Note that N_i is the length of the switching sequence $\boldsymbol{U}_i$.

3b. The leaf node with the index

$$i = \arg \min_{i \in \mathcal{I}} J_i$$

is chosen that has the minimum cost value. The associated switch position at time step k is read out and set as the optimal one, $\boldsymbol{u}_{\text{opt}}(k) = \boldsymbol{u}_i(k)$.

4. The switch position $\boldsymbol{u}_{\text{opt}}(k)$ is applied to the inverter, and this procedure is executed again at the next time step $k+1$.

Step 1b is an optional preprocessing step, which greatly reduces the average computation time. For long switching horizons, it also tends to slightly improve the closed-loop performance of MPDTC, as will be explained at the end of Sect. 9.2.2. The threshold for the length of the output trajectories is usually set to two time steps.

Steps 2a–2d are executed until all nodes of the search tree have been enumerated. Subsequently, Steps 3a and 3b are run on the candidate leaf nodes, which form a subset of all admissible nodes.

If the switching frequency is to be minimized, $E_{\text{sw},i}$ is replaced by the number of commutations $S_{\text{sw},i}$. When switching in Step 2b, the incremental number of commutations $\|\Delta \boldsymbol{u}(\ell)\|_1$ is added to the sum of the number of commutations $S_{\text{sw},j} = S_{\text{sw},i} + \|\Delta \boldsymbol{u}(\ell)\|_1$. The cost function $J_i = S_{\text{sw},i}/N_i + J_{\text{bnd},i} + J_{ti}$ is adopted in Step 3a, which approximates the switching frequency over the length of the switching sequence.

In summary, the MPDTC algorithm derives a long sequence of switch positions that minimizes the predicted switching effort and ensures that the output variables are maintained within their bounds. From this sequence, only the first gating signal at the current time step k is applied to the inverter. New measurements are obtained at the next sampling instant $k+1$, the optimization procedure is repeated, and a new switching sequence is computed.

At steady-state operating conditions, the new switching sequence is in many cases equal to the one previously computed, albeit being shifted by one step in time. In some cases, however, the new switching sequence is slightly modified in order to account for model mismatches, dc-link voltage fluctuations, measurement noise, observer errors, and so on. This strategy, which is referred to as the *receding horizon policy*, provides feedback and makes MPDTC robust. During step changes of the torque reference, however, the new switching sequence differs significantly from the previously computed one.

7.5 Extension Methods

One of the main characteristics of the MPDTC algorithm is the extension mechanism in which the number of time steps is predicted for which a switch position can be applied before one of the output variables either violates a bound, or, in case of a bound violation, ceases to reduce the violation. As stated in (7.40), the extension step can be formulated as an optimization problem. The fact that the controller model (7.21) is nonlinear implies that the optimization problem is also nonlinear with the nonlinear constraints (7.40b) and (7.40c). The solution to this optimization problem constitutes arguably the most difficult step in the MPDTC algorithm from a conceptual and a computational point of view.

This section analyzes the nonlinear characteristic of the state and output trajectories, and proposes methods to solve the extension problem (7.40) in an approximate manner. The most straightforward and commonly used approach is linear extrapolation, as proposed in [6, 9]. At low-speed operation, when the machine's back electromotive force (back-EMF) is small, linear extrapolation is usually sufficiently accurate. An alternative and more accurate approach is to adopt quadratic extrapolation. This might be required during high-speed operation, when the machine's back-EMF is consequently high [6]. The concept of prediction with interpolation is even more accurate. In this approach, the controller model is discretized with an integer multiple of the sampling interval, and the state (or output) vector is predicted at time steps that lie multiple sampling intervals in the future. When this operation is performed twice, quadratic interpolation can be achieved.

7.5.1 Analysis of the State and Output Trajectories

We focus again on the NPC inverter drive system with an induction machine. The stator and rotor flux vectors and the neutral point potential form the drive's state vector; the electromagnetic torque, the stator flux magnitude, and the neutral point potential are its three output variables. When applying a constant three-phase switch position $\boldsymbol{u}$ from time instant t_0 until t, with $t \geq t_0$, the trajectories of the state and output variables are analyzed in this section.

To simplify the exposition, we assume that the angular rotor speed ω_r and the total dc-link voltage are constant. Recall that the stator resistance R_s is very small in MV machines. Moreover, the magnetizing reactance dominates over the leakage reactances (see (7.5)), which allows us to write $X_m \approx X_s \approx X_r$. As the stator and rotor flux vectors of an induction machine are close to each other even at full torque, the first two terms in (7.4a) are not only small but almost cancel each other out. The same applies to the first two terms in the rotor flux equation (7.4b). This allows us to approximate the machine model (7.4) for $t \geq t_0$ by the differential equations

$$\frac{\mathrm{d}\boldsymbol{\psi}_s(t)}{\mathrm{d}t} = \frac{1}{2} v_{\mathrm{dc}} \tilde{\boldsymbol{K}} \boldsymbol{u}(t_0) \tag{7.43a}$$

$$\frac{\mathrm{d}\boldsymbol{\psi}_r(t)}{\mathrm{d}t} = \omega_r \begin{bmatrix} 0 & -1 \\ 1 & 0 \end{bmatrix} \boldsymbol{\psi}_r(t). \tag{7.43b}$$

Integrating (7.43a) from time instant t_0 to t provides the stator flux vector at time t in orthogonal coordinates:

$$\boldsymbol{\psi}_s(t) = \boldsymbol{\psi}_s(t_0) + \frac{1}{2} v_{\mathrm{dc}} \tilde{\boldsymbol{K}} \boldsymbol{u}(t_0)\,(t - t_0). \tag{7.44}$$

The α- and β-components of the stator flux vector are linear functions of time. The squared magnitude of the stator flux vector $\Psi_s^2 = ||\boldsymbol{\psi}_s||^2$ is quadratic in time. As the stator flux magnitude Ψ_s is kept close to 1, the magnitude is effectively also quadratic in time. As a result, the trajectory of the stator flux magnitude can be accurately described by a quadratic function.

The trajectory of the rotor flux vector in stationary coordinates can be obtained from (7.43b), which states that the rotor flux vector rotates with the constant magnitude $\Psi_r = ||\boldsymbol{\psi}_r||$ at the angular speed ω_r. The rotor flux vector at time t is given by

$$\boldsymbol{\psi}_r(t) = \begin{bmatrix} \cos(\omega_r(t - t_0) + \varphi_0) \\ \sin(\omega_r(t - t_0) + \varphi_0) \end{bmatrix} \Psi_r, \tag{7.45}$$

where $\varphi_0 = \varphi(t_0)$ denotes the angular position of the rotor flux vector at time t_0. The α- and β-components of the rotor flux vector are trigonometric functions of time.

The electromagnetic torque is the cross product between the stator and rotor flux vectors (see (7.6)). Therefore, the evolution of the torque is described by terms that include $t\cos(\omega_r t)$, $t\sin(\omega_r t)$, $\cos(\omega_r t)$, $\sin(\omega_r t)$, and t. At low-speed operation, ω_r is small and the sine and cosine terms in (7.45) can be accurately approximated by linear functions. The torque is then a quadratic function of time. Close to standstill, the rotor flux vector is effectively stationary, turning the torque into a linear function of time.

The trajectory of the neutral point potential (7.3) is the integral of the stator currents $\boldsymbol{i}_{s,abc}$ over time, weighted with the component-wise absolute value of the inverter switch positions. More specifically, by integrating (7.3), the neutral point potential at time t follows as

$$\upsilon_n(t) = \upsilon_n(t_0) + \frac{1}{2X_c}|\boldsymbol{u}(t_0)|^T \int_{t_0}^{t} \boldsymbol{i}_{s,abc}(\tau)\mathrm{d}\tau. \tag{7.46}$$

The integral acts like a low-pass filter, allowing us to approximate the stator currents by linear functions in time. The trajectory of the neutral point potential is then approximately quadratic in time.

7.5.2 Linear Extrapolation

Let the extension leg commence at time step $\ell_0 \in \mathbb{N}$. An extension leg is typically preceded by a switching transition, which decides the switch position $\boldsymbol{u}(\ell_0 - 1)$. The output vector $\boldsymbol{y}(\ell_0)$ is predicted during the switching step *S*. This case is exemplified in Fig. 7.8 for the switching horizon *SE*, which implies $\ell_0 = k + 1$. Alternatively, the extension leg may constitute the first element in the switching horizon and thus start at the current time step $\ell_0 = k$. In both cases, the output vector $\boldsymbol{y}$ is available at time steps $\ell_0 - 1$ and ℓ_0 along with the previously selected three-phase switch position $\boldsymbol{u}(\ell_0 - 1)$, which is to be kept constant during the extension leg.

Let the index $j \in \{1, 2, 3\}$ denote the jth output variable y_j. In the following, we consider each one of the three output variables separately. Based on y_j at time steps $\ell_0 - 1$ and ℓ_0, the future output trajectory can be predicted. Using linear extrapolation, future values of y_j are given by

$$y_j(\ell_0 + n_j) = y_j(\ell_0) + (y_j(\ell_0) - y_j(\ell_0 - 1))n_j, \tag{7.47}$$

where $n_j \in \mathbb{N}$ denotes the discrete-time step within the extension leg.

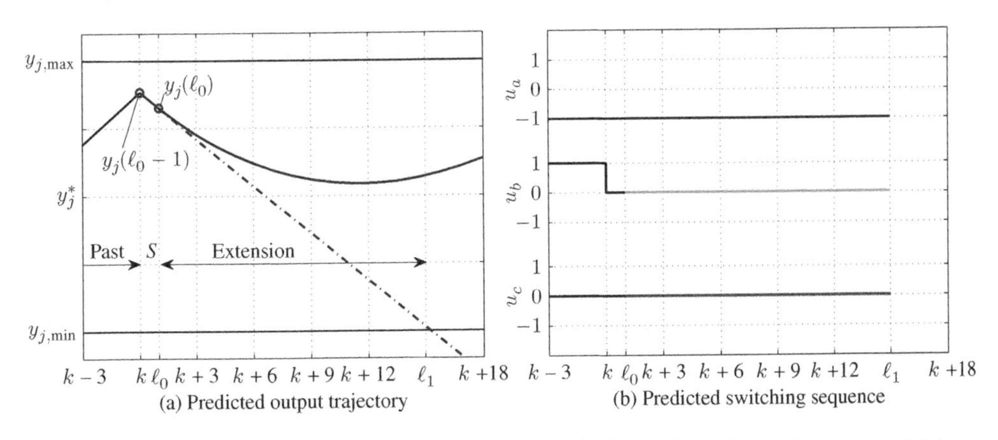

Figure 7.8 Linear extrapolation: Predicted trajectory (dash-dotted line) of the jth output variable y_j starting at time step ℓ_0, where $\ell_0 = k + 1$. The output variable is predicted to violate its lower bound between time steps $\ell_1 = \ell_0 + n_j$ and $\ell_1 + 1$, where $\ell_1 = k + 15$. The nonlinear trajectory of y_j is indicated by a solid line

For $y_j(\ell_0) \neq y_j(\ell_0 - 1)$, the linearly extrapolated trajectory has exactly one intersection point with each bound. By setting (7.47) equal to the upper and lower bound, $y_{j,\max}$ and $y_{j,\min}$, respectively, the two corresponding intersection points can be easily derived. Regardless of the location of $y_j(\ell_0)$ with respect to its bounds (it is either within the bounds or it violates one of them), we always require the second intersection point. We therefore take the maximum of the two points in time and define the number of time steps:

$$n_j = \left\lfloor \max\left(\frac{y_{j,\max} - y_j(\ell_0)}{y_j(\ell_0) - y_j(\ell_0 - 1)}, \frac{y_{j,\min} - y_j(\ell_0)}{y_j(\ell_0) - y_j(\ell_0 - 1)} \right) \right\rfloor . \tag{7.48}$$

In the degenerate case, when the extrapolated trajectory is parallel to the bounds, we set $n_j = \infty$.

Our objective is to maintain the output variables within their bounds and to switch *before* they violate them. To facilitate this, we apply the floor operator $\lfloor \cdot \rfloor$ to the expression in (7.48). As a result, the output variable y_j is predicted to remain within its bounds for n_j time steps before violating a bound between time steps $\ell_1 = \ell_0 + n_j$ and $\ell_1 + 1$.

The length of the extension leg is determined by the output variable that is predicted to *first* violate a bound. Therefore, when considering all three outputs, the length of the extension leg is given by

$$n = \min_j \; n_j . \tag{7.49}$$

Example 7.5 *An example is provided in Fig. 7.8 that illustrates the concept of linear extrapolation of output trajectories for the switching horizon* SE. *The output variable relates to the electromagnetic torque when operating the machine at nominal speed. Following the switching transition at time step* $k = \ell_0 - 1$ *to* $\mathbf{u}(k) = [-1\ 0\ 0]^T$*, the linearly extrapolated output trajectory, which is shown as the dash-dotted line, is predicted to remain within its bounds for* $n_j = 14$ *steps, that is, until time step* $\ell_1 = k + 15$. *To avoid a violation of the bounds between* ℓ_1 *and* $\ell_1 + 1$*, a switching transition will be required at time step* ℓ_1. *Using the nonlinear controller model, the exact trajectory of* y_j *can be predicted, which is shown as the solid line in Fig. 7.8(a).*

The approximation error is significant in this example, motivating the investigation of more accurate techniques to perform the extension step. Linear extrapolation is nevertheless often sufficiently accurate, resulting only in a minor degradation in performance, particularly at low speed. This is evidenced by the successful experimental tests of an early version of MPDTC with linear extrapolation for an MV NPC inverter drive system (see [9]).

7.5.3 *Quadratic Extrapolation*

A more accurate method to predict the future evolution of output trajectories is based on quadratic extrapolation. To this end, the output values at time steps $\ell_0 - 1$, ℓ_0, and $\ell_0 + 1$ are required. The first two values are generally available, but the third one at time step $\ell_0 + 1$ needs to be computed with the help of the prediction step (7.38), in which we set $\ell = \ell_0$.

With quadratic extrapolation, the trajectory of the output variable y_j beginning at time step ℓ_0 is given by

$$y_j(\ell_0 + n_j) = a_j n_j^2 + b_j n_j + c_j . \tag{7.50}$$

The coefficients a_j, b_j, and c_j are obtained from the equation system

$$y_j(\ell_0 - 1) = a_j - b_j + c_j \tag{7.51a}$$

$$y_j(\ell_0) = c_j \tag{7.51b}$$

$$y_j(\ell_0 + 1) = a_j + b_j + c_j\,, \tag{7.51c}$$

which can be solved to give

$$a_j = \frac{1}{2}(y_j(\ell_0 - 1) - 2y_j(\ell_0) + y_j(\ell_0 + 1)) \tag{7.52a}$$

$$b_j = \frac{1}{2}(-y_j(\ell_0 - 1) + y_j(\ell_0 + 1)) \tag{7.52b}$$

$$c_j = y_j(\ell_0)\,. \tag{7.52c}$$

In general, up to four intersection points of the extrapolated output trajectory (7.50) and the bounds exist. To determine the sought-after intersection point, we introduce three criteria: (i) value of $y_j(\ell_0)$ with respect to the bounds, that is, it is either above, within, or below the bounds; (ii) sign of a_j, that is, when it is positive (negative), the extrapolated trajectory is convex (concave), whereas in the event of the degenerative case with $a_j = 0$, (7.50) turns into a linear equation and linear extrapolation as explained in the previous section should be used; and (iii) existence of intersection points with the upper or the lower bound or with both.

Based on these criteria, one can derive statements based on which the number of time steps n_j can be derived, for which the extrapolated output trajectory is predicted to be applicable before the output variable leaves a bound or ceases to reduce the bound violation. For the case shown in Fig. 7.9, for example, $y_j(\ell_0)$ is within the bounds, the approximated output trajectory is convex (a_j is positive), and intersection points exist only with the upper bound. Therefore, the second intersection with the upper bound

$$n_j = \left\lfloor \frac{1}{2a_j}\left(-b_j + \sqrt{b_j^2 - 4a_j(c_j - y_{j,\max})}\right)\right\rfloor \tag{7.53}$$

is the desired solution. If the quadratically extrapolated output trajectory also intersected its lower bound, the first intersection with the lower bound

$$n_j = \left\lfloor \frac{1}{2a_j}\left(-b_j - \sqrt{b_j^2 - 4a_j(c_j - y_{j,\min})}\right)\right\rfloor \tag{7.54}$$

would be the solution.

Example 7.6 *Reconsider the extension problem of Example 7.5. Using the values of the output variable y_j at time steps $\ell_0 - 1$, ℓ_0, and $\ell_0 + 1$, the output trajectory is predicted using quadratic extrapolation. The latter is shown as the dash-dotted line in Fig. 7.9(a), while the solid line refers to the exact output trajectory. The quadratically extrapolated output trajectory is predicted to remain within its bounds for $n_j = 22$ steps, that is, until time step $\ell_1 = k + 23$.*

In a last step, the length of all three output trajectories is considered and their minimum is taken, as was stated in (7.49). Quadratic extrapolation is more complex to implement than linear extrapolation. In particular, the decision table adds to its complexity. Yet, it is significantly

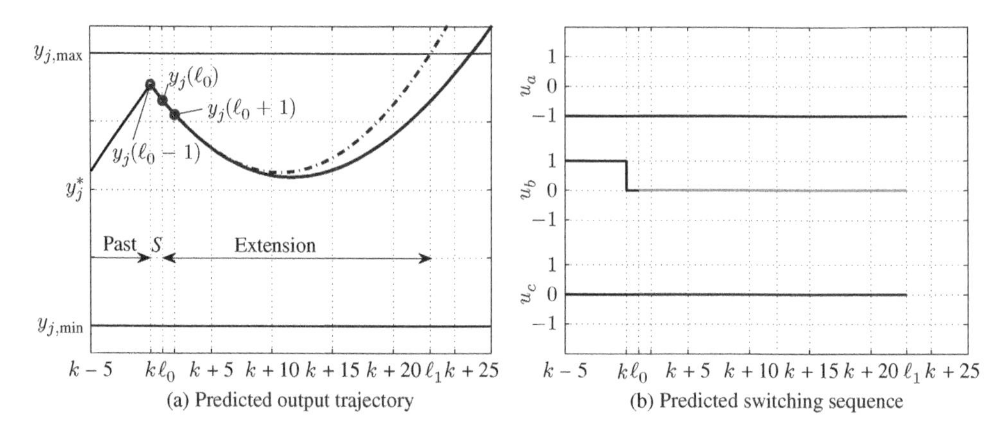

(a) Predicted output trajectory (b) Predicted switching sequence

Figure 7.9 Quadratic extrapolation: Predicted trajectory (dash-dotted line) of the jth output variable y_j starting at time step ℓ_0, where $\ell_0 = k+1$. The output variable is predicted to violate its upper bound between time steps $\ell_1 = \ell_0 + n_j$ and $\ell_1 + 1$, where $\ell_1 = k + 23$. The nonlinear trajectory of y_j is indicated by a solid line

more accurate and causes only minor approximation errors. Nevertheless, quadratic extrapolation is somewhat vulnerable to numerical errors in the case of very long extrapolation intervals. To mitigate this issue and to further increase the accuracy of the predictions, a viable alternative is to use the concept of extrapolation with quadratic interpolation, which will be introduced in the next section.

7.5.4 *Quadratic Interpolation*

The idea underlying quadratic interpolation is to predict the values of the output variable at two regularly spaced time steps far ahead in the future. Between those output values and the one at the current time step, the output trajectory can be approximated by a quadratic function. Specifically, this is done for the values at time steps ℓ_0, $\ell_0 + n_s$, and $\ell_0 + 2n_s$, where the design parameter $n_s \in \mathbb{N}$ determines the interval between those time steps. It is typically in the range of 5–25 steps. The intersection points with the bounds are found through quadratic interpolation.

To predict the output values at $\ell_0 + n_s$ and $\ell_0 + 2n_s$, we discretize the continuous-time controller model (7.14) with the sampling interval $n_s T_s$. As a result, two (instead of $2n_s$) prediction steps are required to compute $\boldsymbol{y}(\ell_0 + n_s)$ and $\boldsymbol{y}(\ell_0 + 2n_s)$.

As for quadratic extrapolation, the trajectory of the output variable y_j is given by

$$y_j(\ell_0 + n_j) = a_j n_j^2 + b_j n_j + c_j \,. \tag{7.55}$$

The coefficients a_j, b_j, and c_j are now obtained from the equations

$$y_j(\ell_0) = c_j \tag{7.56a}$$

$$y_j(\ell_0 + n_s) = a_j n_s^2 + b_j n_s + c_j \tag{7.56b}$$

$$y_j(\ell_0 + 2n_s) = 4a_j n_s^2 + 2b_j n_s + c_j \,. \tag{7.56c}$$

Solving the equation system (7.56) leads to the coefficients

$$a_j = \frac{1}{2n_s^2}(y_j(\ell_0) - 2y_j(\ell_0 + n_s) + y_j(\ell_0 + 2n_s)) \tag{7.57a}$$

$$b_j = \frac{1}{2n_s}(-3y_j(\ell_0) + 4y_j(\ell_0 + n_s) - y_j(\ell_0 + 2n_s)) \tag{7.57b}$$

$$c_j = y_j(\ell_0). \tag{7.57c}$$

The mechanism for determining the desired intersection point of the predicted output trajectory with its bounds is the same as for quadratic extrapolation (see Sect. 7.5.3).

Example 7.7 *To illustrate the concept of quadratic interpolation, consider again the extension problem of Example 7.5. We set n_s to 12. As shown in Fig. 7.10, the output values of y_j are predicted at time steps $\ell_2 = \ell_0 + n_s$ and $\ell_3 = \ell_0 + 2n_s$ using the discrete-time controller model with the sampling interval n_sT_s. Quadratic interpolation leads to the predicted output trajectory, which is shown as the dash-dotted line in Fig. 7.10(a). Quadratic interpolation provides a very accurate approximation of the nonlinear output trajectory, which is shown as the solid line, making both lines almost indistinguishable. The quadratically interpolated output trajectory is predicted to remain within its bounds for $n_j = 25$ steps, that is, until time step $\ell_1 = k + 26$.*

Quadratic interpolation provides very accurate predictions within the interval $[\ell_0, \ell_0 + 2n_s]$ over which the interpolation is performed. Indeed, its approximation error is typically below the length of the sampling interval T_s, and it is thus often negligible. Nevertheless, in the case of very long extension legs, the predicted intersection point might occur at a time step that significantly exceeds $\ell_0 + 2n_s$. This might result in a noticeable approximation error.

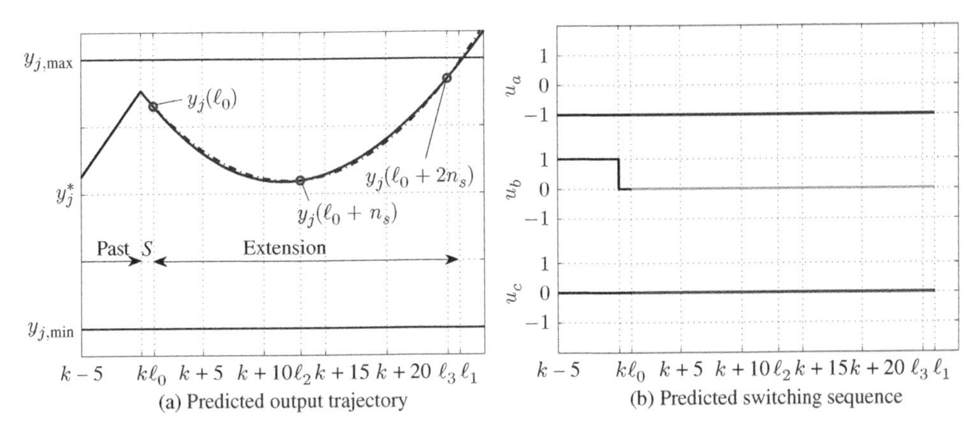

Figure 7.10 Quadratic interpolation: Predicted trajectory (dash-dotted line) of the jth output variable y_j starting at time step ℓ_0. The interpolation is based on the output values at time steps ℓ_0, $\ell_2 = \ell_0 + n_s$, and $\ell_3 = \ell_0 + 2n_s$, where $\ell_0 = k + 1$, $\ell_2 = k + 13$, and $\ell_3 = k + 25$. The output variable is predicted to violate its upper bound between time steps $\ell_1 = \ell_0 + n_j$ and $\ell_1 + 1$, where $\ell_1 = k + 26$. The nonlinear trajectory of y_j is indicated by a solid line

To mitigate this issue and to avoid the adoption of time-varying n_s, the interpolation procedure can be repeated over a shifted time interval. More specifically, if the intersection point is predicted to occur at the time step $\ell_1 \gg \ell_0 + 2n_s$, the value of the output variable at $\ell_0 + 3n_s$ can be predicted. With this, quadratic interpolation can be performed using the output values at time steps $\ell_0 + n_s$, $\ell_0 + 2n_s$, and $\ell_0 + 3n_s$, thereby achieving a more accurate prediction of the intersection point. This procedure can be repeated as required. For more details on *iterative* quadratic interpolation, the reader is referred to [17] and [18].

If an extension step is followed by a switching transition, the state vector $\boldsymbol{x}$ is required at the end of the extension step at time step $\ell_0 + n$. To determine $\boldsymbol{x}(\ell_0 + n)$, the extension methods described previously can be directly applied.

7.6 Summary and Discussion

The MPDTC control problem of maintaining the output variables within their bounds while minimizing the switching effort was introduced in this chapter. Based on first principles, a discrete-time nonlinear model of the drive system was derived. Similarly, the switching losses in the inverter could be predicted using first principles. The control objectives were mapped into a cost function, which was minimized subject to the evolution of the dynamic drive model and constraints imposed on the output variables and the switch positions. This led to the naive optimization problem (7.26), which cannot be solved in real time but which serves as a starting point to devise an algorithm that derives a suboptimal solution to the optimization problem.

As the objective is to maintain the output variables within their bounds while minimizing the switching effort, a promising heuristic technique is to consider switching transitions only when bound violations are about to occur and to freeze the switch positions for the remainder of the prediction horizon. This policy can be described by the switching horizon with the elements *S* and *E*, which represent switching steps and extension legs, respectively. The extension legs can be addressed in a computationally efficient manner through the use of linear and quadratic extrapolation and interpolation techniques that predict future bound violations.

The imposition of bounds on the output variables and the notion of extension legs allow us to achieve very long prediction horizons at a modest computational burden. The use of long prediction horizons enables MPDTC to make better educated decisions. For example, a certain switching sequence might appear to be prohibitively expensive in terms of the switching effort when considering a short prediction horizon. Over a longer time period, however, this switching sequence might turn out to be very cheap. To provide the MPDTC algorithm with the capability to choose this switching sequence, long prediction horizons are required. This characteristic and the benefit of long prediction horizons are illustrated in the following example.

Example 7.8 *Consider the switching horizon* SSE *and the three switching sequences* $\boldsymbol{U}_i(k)$, $i \in \mathcal{I} = \{1, 2, 3\}$. *The latter are shown in Fig. 7.11 along with the output trajectories they induce. In this example, to simplify the exposition, we neglect the neutral point potential, which is treated in exactly the same way as the torque and the stator flux magnitude.*

$\boldsymbol{U}_1(k)$ *is not a candidate switching sequence, because it violates the lower torque bound after the first switching transition. In contrast to this,* $\boldsymbol{U}_2(k)$ *and* $\boldsymbol{U}_3(k)$ *are candidate sequences. Their torque and stator flux trajectories are extended using linear extrapolation.*

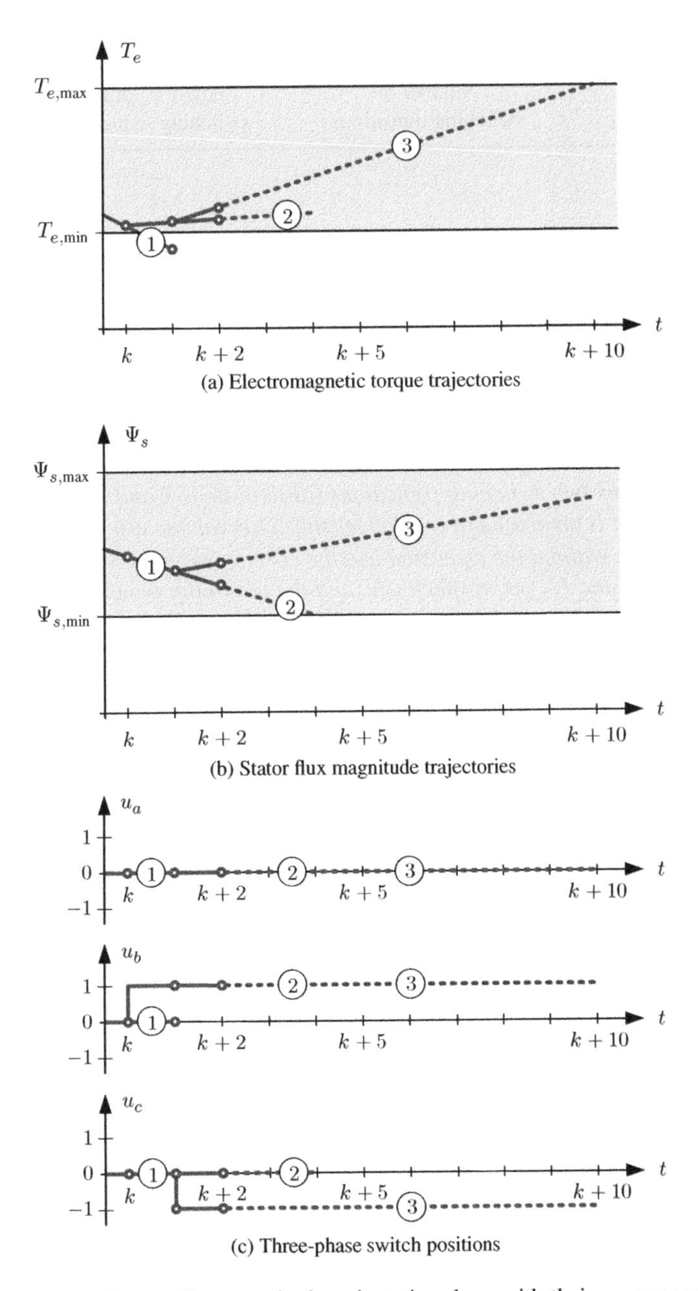

Figure 7.11 Torque and stator flux magnitude trajectories along with their corresponding switching sequences with the switching horizon *SSE* in Example 7.8. The three switching sequences $\boldsymbol{U}_1(k)$, $\boldsymbol{U}_2(k)$, and $\boldsymbol{U}_3(k)$ are indicated by their numbers. The switching steps are indicated by solid lines, while the extension legs are depicted by dashed lines. The regions between the upper and lower (hysteresis) bounds are shaded

Table 7.2 Characteristics of the three switching sequences in Example 7.8

Index i of the switching sequence	Number of switching transitions	Length N_i of the switching sequence	Cost J_i
1	—	—	—
2	1	4	0.25
3	2	10	0.2

It can be seen that the total lengths of the trajectories amount to 4 and 10 steps, respectively, while the underlying switching sequences require 1 and 2 switching transitions. These numerical results are summarized in Table 7.2.

We choose the cost function $J = J_f$, where J_f captures the short-term switching frequency (see (7.31)). The minimization of J results in the sequence $\boldsymbol{U}_3(k)$ being the optimal one. Even though $\boldsymbol{U}_3(k)$ requires two switching transitions (at time steps k and $k+1$), this investment in the switching effort is amortized over a longer time interval because of the longer length of the output trajectory. Without the extension leg, the corresponding cost expressions would be 0.5 and 1 for $\boldsymbol{U}_2(k)$ and $\boldsymbol{U}_3(k)$, respectively, and the controller would have selected $\boldsymbol{U}_2(k)$ as the optimal sequence. In the long run, however, $\boldsymbol{U}_2(k)$ is clearly inferior to $\boldsymbol{U}_3(k)$. This motivates the notion of the extension leg and the use of long prediction horizons.

The implementation of the MPDTC algorithm on a control platform gives rise to a number of delays, with the most prominent one relating to the controller computation time. The uplink communication delays between the (current and voltage) sensors and the controller computation unit are also often significant. Similar delays are imposed by the downlink communication from the controller computation unit to the gate drivers. Provided that these delays are constant and known, they can be compensated for by an initial state prediction stage. This delay compensation scheme is described in detail in [9] and in Sect. 4.2.8 along with the common causes of delays.

Appendix 7.A: Controller Model of the NPC Inverter Drive System

For the NPC drive model (7.14) and (7.16), the continuous-time system and input matrices of the machine model are

$$\boldsymbol{F}_m = \begin{bmatrix} -R_s\frac{X_r}{D} & 0 & R_s\frac{X_m}{D} & 0 \\ 0 & -R_s\frac{X_r}{D} & 0 & R_s\frac{X_m}{D} \\ R_r\frac{X_m}{D} & 0 & -R_r\frac{X_s}{D} & -\omega_r \\ 0 & R_r\frac{X_m}{D} & \omega_r & -R_r\frac{X_s}{D} \end{bmatrix} \tag{7.A.1a}$$

$$\boldsymbol{G}_m = \frac{v_{\mathrm{dc}}}{6}\begin{bmatrix} 2 & -1 & -1 \\ 0 & \sqrt{3} & -\sqrt{3} \\ 0 & 0 & 0 \\ 0 & 0 & 0 \end{bmatrix}, \tag{7.A.1b}$$

respectively, and the continuous-time system vector of the inverter model is

$$\boldsymbol{f}_i(\boldsymbol{u}) = \frac{1}{2X_c D}|\boldsymbol{u}|^T \,\tilde{\boldsymbol{K}}^{-1}\left[X_r\boldsymbol{I}_2 \;\; -X_m\boldsymbol{I}_2\right] \tag{7.A.2a}$$

$$= \frac{1}{4X_c D}|\boldsymbol{u}|^T \begin{bmatrix} 2X_r & 0 & -2X_m & 0 \\ -X_r & \sqrt{3}X_r & X_m & -\sqrt{3}X_m \\ -X_r & -\sqrt{3}X_r & X_m & \sqrt{3}X_m \end{bmatrix} \tag{7.A.2b}$$

$$= \frac{1}{4X_c D}\left[X_r(2|u_a| - |u_b| - |u_c|) \;\; \sqrt{3}X_r(|u_b| - |u_c|) \;\; \dots\right]. \tag{7.A.2c}$$

Note that $|\boldsymbol{u}|$ denotes the component-wise absolute value of the inverter switch position, that is, $|\boldsymbol{u}| = [|u_a| \; |u_b| \; |u_c|]^T$. The vector-valued output function of the drive model is given by

$$\boldsymbol{h}(\boldsymbol{x}) = \begin{bmatrix} \frac{1}{\text{pf}}\frac{X_m}{D}(x_2x_3 - x_1x_4) \\ \sqrt{x_1^2 + x_2^2} \\ x_5 \end{bmatrix}, \tag{7.A.3}$$

where x_j refers to the jth component of the vector $\boldsymbol{x}$.

References

[1] G. Papafotiou, T. Geyer, and M. Morari, "Optimal direct torque control of three-phase symmetric induction motors," Tech. Rep. AUT03-07, Automatic Control Laboratory, ETH Zurich, http://control.ee.ethz.ch, 2003.

[2] G. Papafotiou, T. Geyer, and M. Morari, "Optimal direct torque control of three-phase symmetric induction motors," in *Proceedings of IEEE Conference on Decision and Control* (Atlantis, Bahamas), Dec. 2004.

[3] G. Papafotiou, T. Geyer, and M. Morari, "A hybrid model predictive control approach to the direct torque control problem of induction motors," *Int. J. Robust Nonlinear Control*, vol. 17, pp. 1572–1589, Nov. 2007.

[4] T. Geyer and G. Papafotiou, "Direct torque control for induction motor drives: A model predictive control approach based on feasibility," in *Hybrid systems: Computation and control* (M. Morari and L. Thiele, eds.), vol. 3414 of *LNCS*, pp. 274–290, Heidelberg: Springer, Mar. 2005.

[5] T. Geyer, G. Papafotiou, and M. Morari, "Method for operating a rotating electrical machine." EP patent 1 670 135, US patent 7 256 561 and JP patent 4 732 883, 2004.

[6] T. Geyer, *Low complexity model predictive control in power electronics and power systems*. PhD thesis, Automatic Control Laboratory, ETH Zurich, 2005.

[7] T. Geyer, G. Papafotiou, and M. Morari, "Model predictive direct torque control—Part I: Concept, algorithm and analysis," *IEEE Trans. Ind. Electron.*, vol. 56, pp. 1894–1905, Jun. 2009.

[8] J. Kley, G. Papafotiou, K. Papadopoulos, P. Bohren, and M. Morari, "Performance evaluation of model predictive direct torque control," in *Proceedings of Power Electronics Specialists Conference* (Rhodes, Greece), pp. 4737–4744, Jun. 2008.

[9] G. Papafotiou, J. Kley, K. G. Papadopoulos, P. Bohren, and M. Morari, "Model predictive direct torque control—Part II: Implementation and experimental evaluation," *IEEE Trans. Ind. Electron.*, vol. 56, pp. 1906–1915, Jun. 2009.

[10] T. Geyer, "Generalized model predictive direct torque control: Long prediction horizons and minimization of switching losses," in *Proceedings of the IEEE Conference on Decision Control* (Shanghai, China), pp. 6799–6804, Dec. 2009.

[11] S. Mastellone, G. Papafotiou, and E. Liakos, "Model predictive direct torque control for MV drives with *LC* filters," in *Proceedings of European Power Electronic Conference* (Barcelona, Spain), pp. 1–10, Sep. 2009.

[12] J. Holtz and S. Stadtfeld, "A predictive controller for the stator current vector of AC machines fed from a switched voltage source," in *Proceedings of IEEE International Power Electronics Conference* (Tokyo, Japan), pp. 1665–1675, Apr. 1983.

[13] J. Holtz and S. Stadtfeld, "Field-oriented control by forced motor currents in a voltage fed inverter drive," in *Proceedings of IFAC Symposium* (Lausanne, Switzerland), pp. 103–110, Sep. 1983.

[14] A. Khambadkone and J. Holtz, "Low switching frequency and high dynamic pulse width modulation based on field-orientation for high-power inverter drive," *IEEE Trans. Power Electron.*, vol. 7, pp. 627–632, Oct. 1992.

[15] J. Scoltock, T. Geyer, and U. K. Madawala, "A comparison of model predictive control schemes for MV induction motor drives," *IEEE Trans. Ind. Inf.*, vol. 9, pp. 909–919, May 2013.

[16] T. Geyer, R. P. Aguilera, and D. E. Quevedo, "On the stability and robustness of model predictive direct current control," in *Proceedings of IEEE International Conference Industrial Technology* (Cape Town, South Africa), Feb. 2013.

[17] Y. Zeinaly, T. Geyer, and B. Egardt, "Trajectory extension methods for model predictive direct torque control," in *Proceedings of Applied Power Electronics Conference and Exposition* (Fort Worth, TX, USA), pp. 1667–1674, Mar. 2011.

[18] T. Geyer, "Computationally efficient MPDTC." EP patent 2 348 631, EP patent 2 528 225 and US patent 13 010 809.

8

Performance Evaluation of Model Predictive Direct Torque Control

The performance of model predictive direct torque control (MPDTC) is examined in this chapter, using three-level and five-level inverters connected to medium-voltage (MV) induction machines as case studies. Two different cost functions are examined. The first one minimizes the switching frequency, while the second variety focuses on the switching losses. By using MPDTC instead of direct torque control (DTC), the steady-state performance in terms of harmonic distortions and switching losses can be significantly improved, particularly when adopting long prediction horizons. The performance during torque steps is similar for both control methods.

8.1 Performance Evaluation for the NPC Inverter Drive System

The performance of MPDTC during steady-state operation and torque transients is compared with that of DTC. As a case study, consider a three-level, neutral-point-clamped (NPC) inverter with a floating neutral point potential and the nominal dc-link voltage $V_{dc} = 4840$ V. The inverter uses only two *di/dt* snubbers—one in the upper and the other in the lower half of the inverter. This restriction imposes additional switching constraints. To minimize the switching losses of the gate-commutated thyristors (GCTs), the switching frequency is limited to a few hundred hertz.

The inverter drives an MV squirrel-cage induction machine, which is rated at 3.3 kV, 50 Hz, and 2 MVA. This case study is described in Sect. 2.5.2. The parameters of the inverter, machine, switching devices, and their losses along with the base quantities of the per unit (pu) system are provided in Sect. 2.5.1. The corresponding continuous-time models are summarized in Sect. 7.2.1. The discrete-time controller model and the model of the switching effort are derived in Sects. 7.2.3 and 7.2.4.

The DTC control objectives are to keep the three output variables, namely the electromagnetic torque, the magnitude of the stator flux, and the neutral point potential, within given (hysteresis) bounds. In MPDTC, these objectives are inherited from DTC. The torque and flux bounds indirectly determine the stator current ripple and thus the current distortions.

Model Predictive Control of High Power Converters and Industrial Drives, First Edition. Tobias Geyer.

Companion Website: www.wiley.com/go/geyermodelpredictivecontrol

In addition, we aim to minimize the inverter losses. An indirect way of doing this is to minimize the (short-term) average switching frequency. We will see that by directly targeting the switching losses in the MPDTC cost function, the losses can be reduced more effectively than when minimizing the switching frequency.

8.1.1 Simulation Setup

A detailed MATLAB/Simulink model of the drive system was used for the performance evaluation. The block diagram of the DTC scheme is shown in Fig. 3.29, and the DTC hysteresis controllers and look-up table are described in Sect. 3.6.3. For MPDTC, the look-up table with the DTC strategy was replaced by a function that runs the MPDTC algorithm at each sampling instant. The MPDTC algorithm was explained in the previous chapter. As shown in Fig. 7.1, and similar to DTC, outer control loops adjust the references for the torque and stator flux magnitude and the widths of the bounds.

MPDTC is based on the cost function $J = J_{\text{sw}} + J_{\text{bnd}} + J_t$, which consists of three terms. The first term minimizes either the switching frequency or the switching losses. Both options will be considered in this section. The second term $J_{\text{bnd}} = \boldsymbol{q}^T \boldsymbol{\epsilon}$ penalizes root mean square (rms) bound violations of the controlled variable $\boldsymbol{y}$. We set the corresponding penalty $\boldsymbol{q}$ to zero during steady-state operation and to $\boldsymbol{q} = [1000\ 0\ 0]^T$ during transients, thus penalizing bound violations of the torque. The third term J_t, which can be used to reduce the likelihood of deadlocks, is not used. For an in-depth treatment of deadlocks and techniques to avoid them, the reader is referred to Sects. 9.3 and 9.5.

The following performance evaluation compares the switching frequency and the switching losses between DTC and MPDTC while operating at a similar torque and flux magnitude ripple. To achieve similar ripples and to partly account for DTC's tendency to violate the torque and stator flux magnitude bounds, the bounds were relaxed for MPDTC. Specifically, the torque bounds were widened by ± 0.03 pu and the bounds of the stator flux magnitude were relaxed by ± 0.004 pu. The bounds of the neutral point potential were set to ± 0.05 pu for both control schemes.

8.1.2 Steady-State Operation

8.1.2.1 Operation at 70% Speed

We start by investigating the steady-state performance of the drive when operating at 70% speed and rated torque. For the switching horizon *eSESESE* and the cost function minimizing the switching losses, we compare DTC and MPDTC with each other. Figures 8.1–8.7 show selected waveforms over one fundamental period. Figures 8.1 and 8.2 show the torque and the stator flux magnitude together with their upper and lower bounds. These figures underline the observation that DTC switches only after a torque or flux bound has been violated, while MPDTC predicts future bound violations and switches proactively before these are violated. Because of this, it was possible to widen the torque and flux bounds for MPDTC. Despite this relaxation, the resulting ripples are still slightly smaller for MPDTC, which achieves a small reduction in the current and torque distortions when compared to DTC. This will be shown later in Table 8.1.

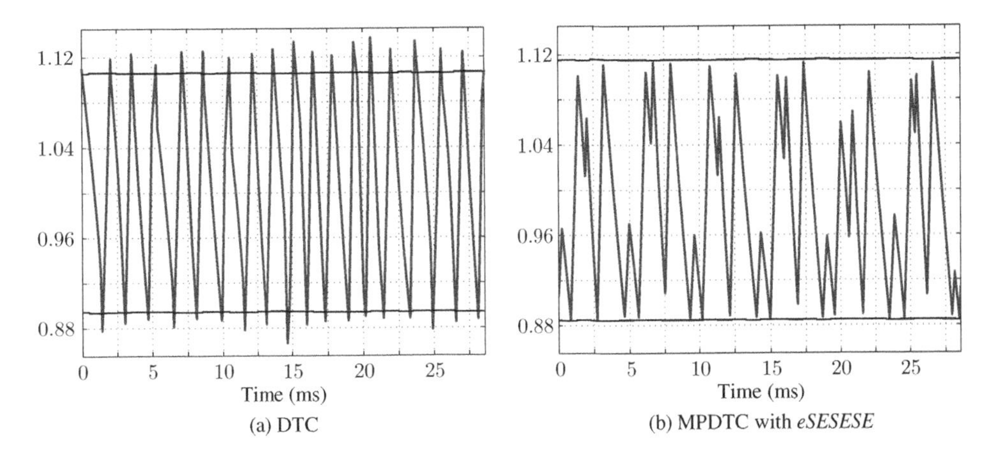

(a) DTC (b) MPDTC with *eSESESE*

Figure 8.1 Electromagnetic torque for DTC and MPDTC minimizing the switching losses, when operating at 70% speed and rated torque

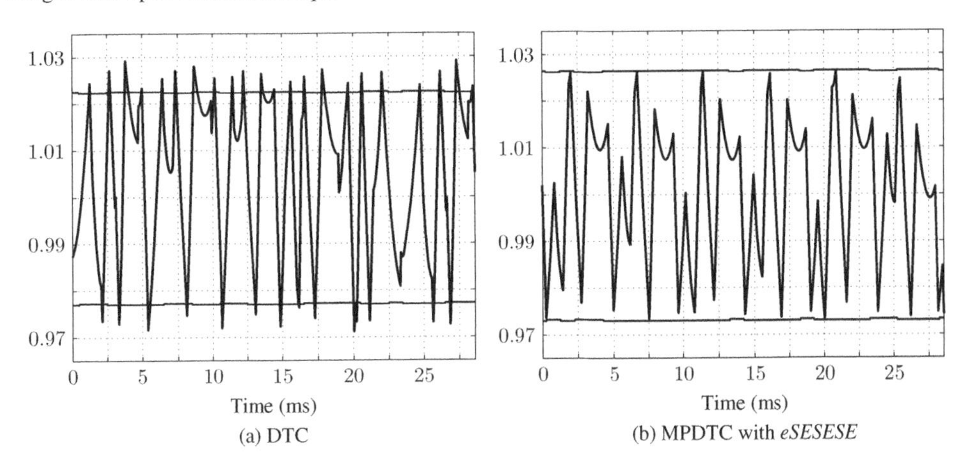

(a) DTC (b) MPDTC with *eSESESE*

Figure 8.2 Stator flux magnitude of the simulation in Fig. 8.1

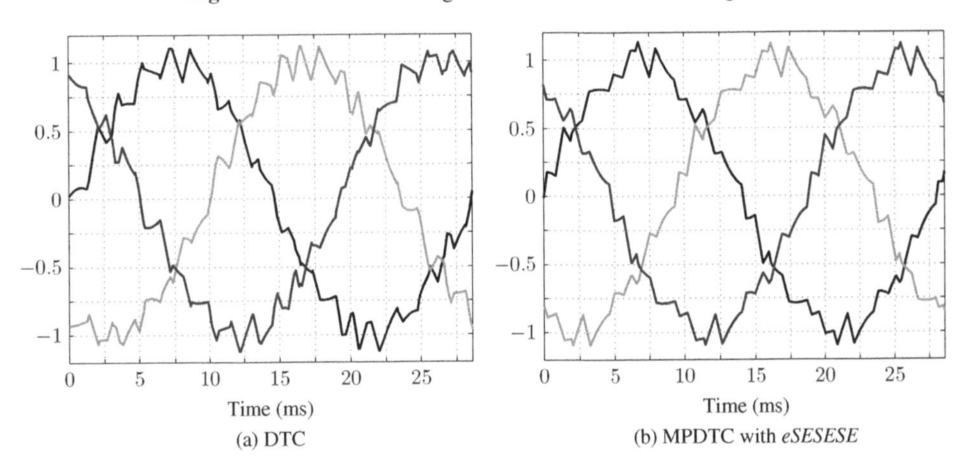

(a) DTC (b) MPDTC with *eSESESE*

Figure 8.3 Three-phase stator currents of the simulation in Fig. 8.1

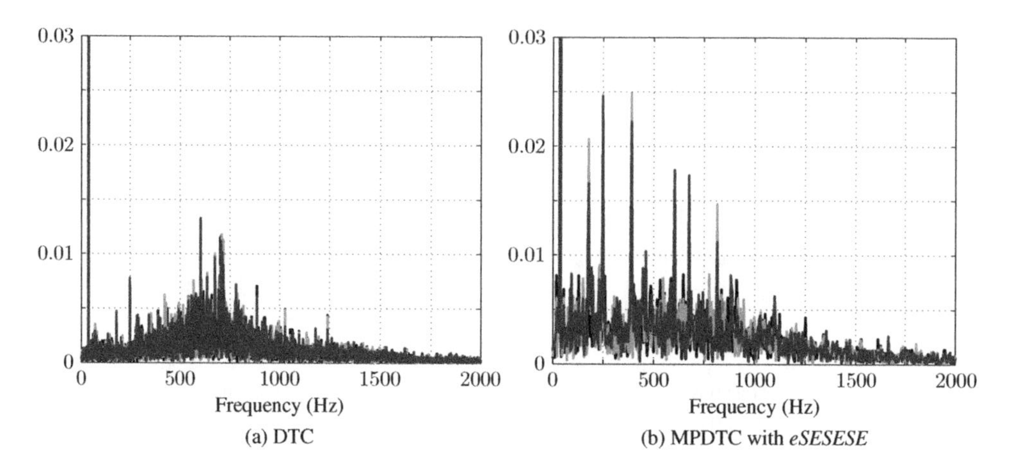

Figure 8.4 Spectra of the three-phase stator currents shown in Fig. 8.3

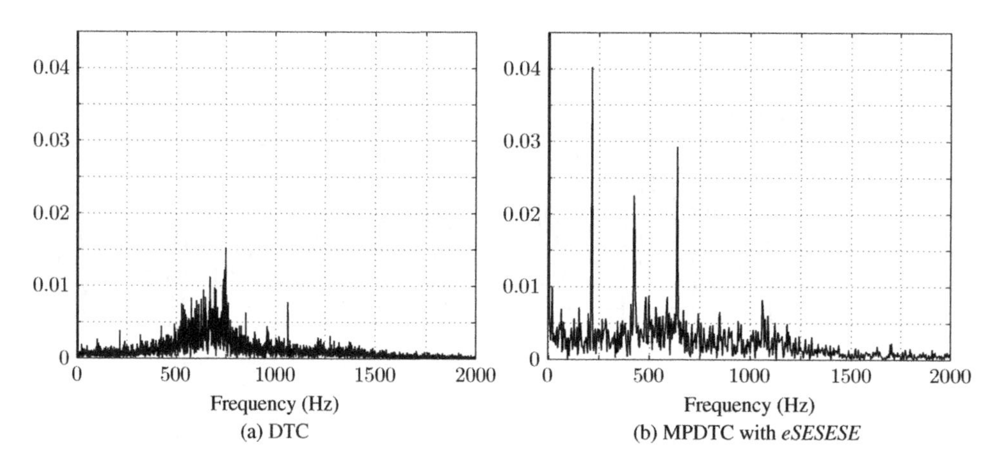

Figure 8.5 Spectra of the electromagnetic torque shown in Fig. 8.1

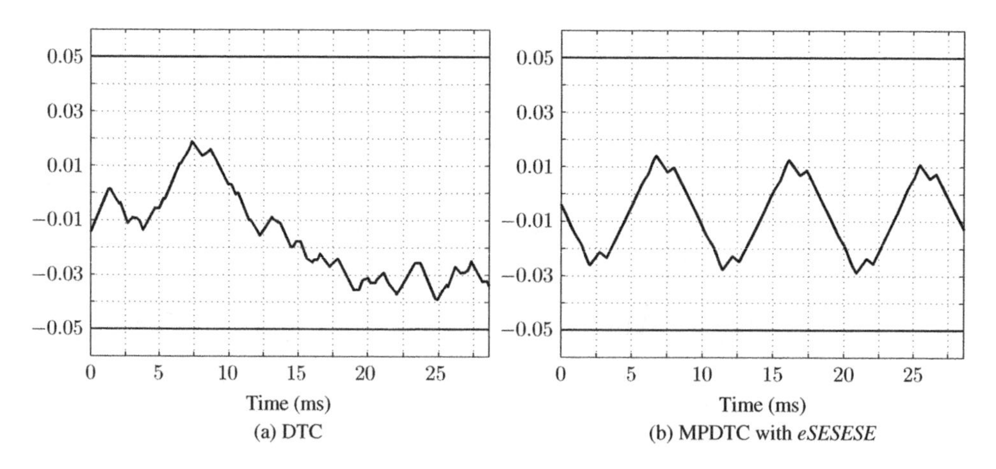

Figure 8.6 Neutral point potential of the simulation in Fig. 8.1

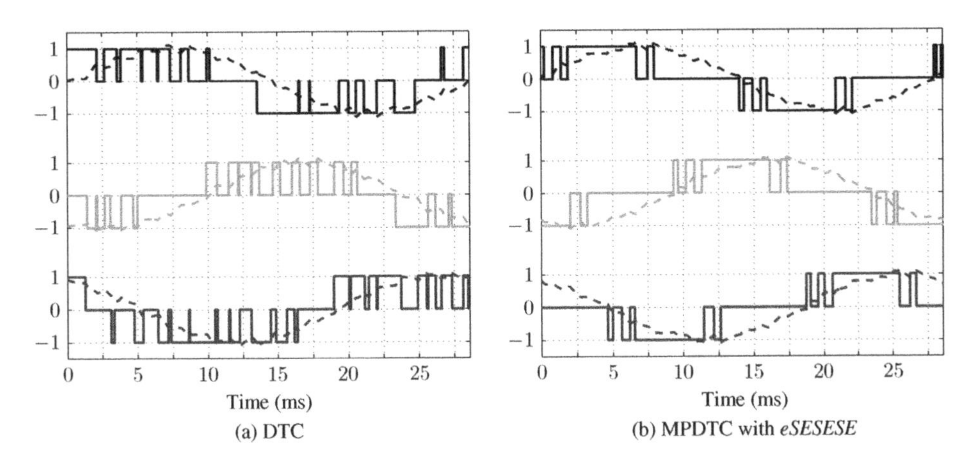

Figure 8.7 Three-phase switch positions and the three-phase stator currents (shown as dashed lines) of the simulation in Fig. 8.1

Table 8.1 Comparison of DTC with MPDTC using different switching horizons and control objectives

Control scheme	Switching horizon	Control objective	Prediction horizon		Performance (%)			
			average	maximum	P_{sw}	f_{sw}	I_{TDD}	T_{TDD}
DTC	—	—	—	—	100	100	100	100
MPDTC	*eSSE*	f_{sw}	25.1	95	64.6	65.0	95.2	86.5
MPDTC	*eSSESE*	f_{sw}	53.4	114	49.7	54.9	95.3	87.1
MPDTC	*eSESESE*	f_{sw}	73.6	112	47.9	52.9	93.9	87.3
MPDTC	*eSSE*	P_{sw}	23.5	87	58.8	77.8	95.9	93.3
MPDTC	*eSSESE*	P_{sw}	50.8	108	49.0	63.6	95.5	87.4
MPDTC	*eSESESE*	P_{sw}	72.2	123	41.9	56.9	92.6	88.9

The fourth and fifth columns indicate the average and maximum lengths of the achieved prediction horizon. The last four columns relate to the switching losses P_{sw}, switching frequency f_{sw}, current TDD I_{TDD}, and torque TDD T_{TDD}, using DTC as a baseline.

The three-phase stator currents are shown in Fig. 8.3. The fundamental current component is 1 pu, which is in line with operation at rated torque. The harmonic spectrum was computed for each phase of the stator current using a Fourier transformation of the current waveform. The spectrum of each phase is shown separately in Fig. 8.4.

As DTC lacks periodicity in the switching signal, it produces a current spectrum that is mostly flat. The largest amplitudes of the current harmonics are found at frequencies around twice the switching frequency. Distinctive current harmonics at odd and non-triplet multiples of the fundamental frequency of 35 Hz nevertheless exist, such as the 5th, 7th, 17th, and 19th harmonics. Notwithstanding this, the amplitudes of the current harmonics are below 1.5% of the fundamental current component amplitude (of 1 pu).

When adopting a long prediction horizon and minimizing the switching losses, MPDTC concentrates the harmonic energy into discrete current harmonics. All odd and non-triplen harmonics up to the 23rd harmonic are pronounced and can be easily identified in Fig. 8.4(b). This implies that MPDTC commands switching transitions that exhibit a certain degree of periodicity during steady-state operation.

Similarly, the torque spectrum of DTC is mostly flat, while that of MPDTC features large torque harmonics, as can be seen in Fig. 8.5. The 5th and the 7th current harmonics form a pair that creates the 6th harmonic in the torque. The 12th and 18th torque harmonics are formed accordingly. These torque harmonics at integer multiples of $6f_1$, where f_1 denotes the fundamental frequency, are pronounced in MPDTC. Overall, the harmonic torque energy is lower in MPDTC than in DTC, which is evidenced by the fact that MPDTC lowers the torque's total demand distortion (TDD) with respect to DTC (see Table 8.1).

Periodicity can also be observed in the evolution of the neutral point potential. As can be seen in Fig. 8.6, MPDTC produces a distinctive third harmonic in the neutral point potential. This stands in contrast to DTC. Nevertheless, both methods maintain the neutral point potential well within its upper and lower bounds.

The last figure, Fig. 8.7, compares the three-phase switch positions that are issued by the two control methods. The switch positions are plotted with their respective phase currents. As can be seen in Fig. 8.7(a), DTC tends to switch continuously over time regardless of the phase current. The torque, flux, and neutral point potential are considered independently of each other.

In contrast, MPDTC considers all three output variables simultaneously in a multiple-input multiple-output (MIMO) control approach and optimizes its switching decisions such that the switching losses are minimized. As can be observed in Figs. 8.1 and 8.2, this enables MPDTC to keep the torque and flux within their bounds for longer before an imminent bound violation triggers a switching transition. MPDTC is particularly good at exploiting the convex curvature of the flux trajectory. As a result, MPDTC requires significantly fewer switching transitions for the same torque and flux ripple, as can be observed in Fig. 8.7(b).

When minimizing the switching losses in MPDTC, about half of the switching transitions are centered at the phase currents' zero crossings, where they incur virtually no switching losses, while the other half are issued close to the peak currents. This can be observed in Fig. 8.7(b). Each of the two groups of switching transitions covers approximately 30° of the fundamental period in each phase. In the remaining 60° between these clusters of switching transitions, because of the 120° phase shift between the phases, the other two phases provide switching transitions. As a result, switching transitions occur regularly in the three-phase system, and the torque, stator flux magnitude, and neutral point potential are kept permanently under closed-loop control. At the same time, the switching losses are reduced to a minimum; MPDTC more than halves the switching losses at this operating point.

In a way, the penalty on the switching losses imposes a time-varying penalty on the switching transitions that is synchronized through the phase currents with the fundamental period. This characteristic enhances the periodicity of the switching pattern, which is reflected in the creation of distinct harmonics in the current and torque spectra.

Figure 8.8 compares the switching patterns that result from MPDTC minimizing either the switching losses or the switching frequency. When the latter policy is adopted, a significant proportion of the switching transitions occur when the phase currents are high. The periodicity in the switching pattern is less pronounced, resulting in harmonic current and torque

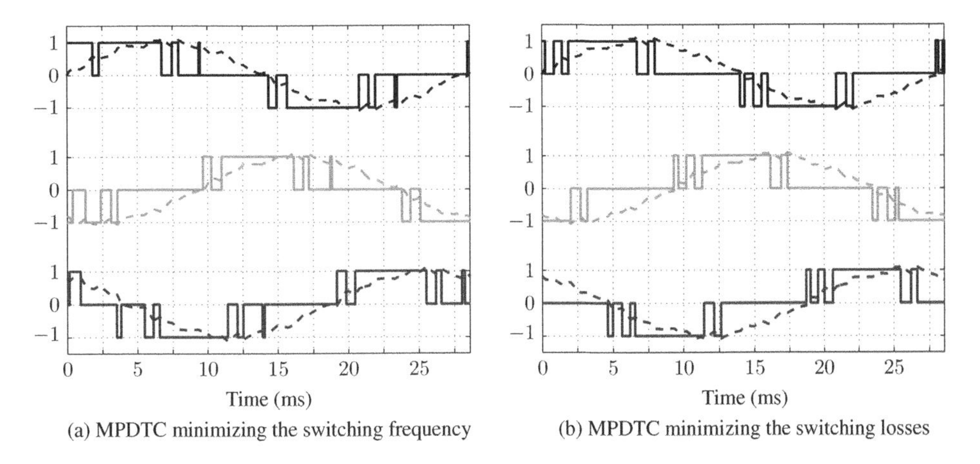

(a) MPDTC minimizing the switching frequency (b) MPDTC minimizing the switching losses

Figure 8.8 Three-phase switch positions and the three-phase stator currents (shown as dashed lines) for MPDTC with the switching horizon *eSESESE* when minimizing either the switching frequency or the switching losses. Note that Fig. 8.8(b) is a repetition of Fig. 8.7(b) to allow a side-by-side comparison

spectra with less distinctive discrete harmonics. Nevertheless, MPDTC minimizing the switching frequency is effective at this operation point in that it almost halves the switching frequency and thus the switching losses, when compared to DTC.

Table 8.1 summarizes the performance of MPDTC for various switching horizons and for both minimization criteria (either the switching frequency or the switching losses). MPDTC is compared with DTC in terms of the switching losses, switching frequency, and the current and torque distortions. DTC is used as a baseline and the MPDTC performance indices are normalized with respect to this baseline. The simulations refer to steady-state operation at 70% speed and rated torque.

When adopting the short switching horizon *eSSE* and minimizing the switching frequency, MPDTC is capable of reducing both the switching frequency and the switching losses by about one-third. Increasing the switching horizon to *eSSESE* reduces these quantities by another 20%. The average length of the prediction horizon is doubled from 25 steps to more than 50. Further extending the switching horizon to *eSESESE* increases the average length of the prediction horizon by another 50%, but the improvement in the switching frequency and the switching losses is minor. In general, such very long prediction horizons often achieve diminishing returns in terms of the performance gain.

Penalizing the switching losses (instead of the switching frequency) further reduces the switching losses. For the switching horizon *eSESESE*, for example, the switching losses can be reduced by another 13%. The switching frequency, however, tends to be slightly higher. In all cases, the current and torque TDD is below that of DTC, particularly when adopting long switching horizons.

8.1.2.2 Operation Over a Range of Speed Operating Points

To provide a more comprehensive analysis of the performance benefits of MPDTC, we compare it with DTC over a range of operating points. The speed is varied between 0.55 and 1 pu in steps of 0.01 pu while operating at rated torque. The torque and stator flux bounds are relaxed again for MPDTC by ± 0.03 and ± 0.004 pu, respectively.

In order to facilitate the comparison, we introduce the two performance metrics

$$c_f = I_{\text{TDD}} \cdot f_{\text{sw}} \tag{8.1a}$$

$$c_P = I_{\text{TDD}} \cdot P_{\text{sw}}, \tag{8.1b}$$

which are the product of the current TDD on the one side and the switching frequency or the switching losses on the other side. Similar performance metrics were introduced in Sect. 3.5 for PWM.

We consider short, medium, and long horizons for MPDTC. We define the switching horizon *eSE* as the short horizon and the switching horizons *eSSE* and *eSESE* to be medium horizons. For each speed operating point, we take the horizon that minimizes the corresponding metric (8.1). More specifically, when using MPDTC that minimizes the switching frequency, we employ c_f as a metric. Conversely, when minimizing the switching losses, we use c_P. The switching horizons *eSSESE*, *eSESESE*, and *eSSESESE* constitute long horizons. As for medium horizons, we take the one that minimizes the respective metric.

MPDTC minimizing the switching frequency is compared to DTC in Fig. 8.9. The two metrics are normalized with respect to DTC. By minimizing the switching frequency, the switching losses are also reduced. In general, the reduction in terms of the switching frequency and switching losses is the same; this can be seen when comparing Fig. 8.9(a) with Fig. 8.9(b). Even with the very short switching horizon *eSE*, MPDTC is capable of significantly improving on DTC by reducing the switching frequency and the switching losses by about one-third. Nevertheless, the performance for the switching horizon *eSE* seems to be dominated by the adverse impact of deadlocks, which are relatively frequent for such short horizons. Specifically, deadlocks abound between 0.65 and 0.7 pu and around 0.8 pu speed. The issue of deadlocks will be explained in Sect. 9.3 and techniques to avoid them will be proposed in Sect. 9.5. Such techniques are not employed here.

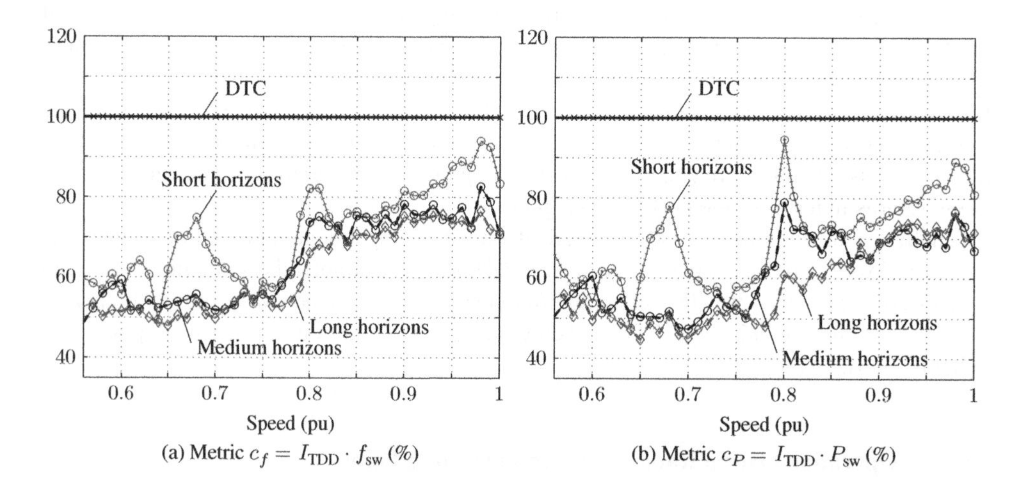

Figure 8.9 Steady-state performance of MPDTC minimizing the switching *frequency*

Except for the speed operating point around 0.8 pu, medium-length horizons largely avoid deadlocks, thus improving the performance of MPDTC. Long horizons almost completely avoid deadlocks, thus avoiding the performance deterioration around 80% speed. Disregarding deadlocks, we conclude that very long prediction horizons are not mandatory to achieve an excellent performance when minimizing the switching frequency in MPDTC. Medium horizons, such as *eSSE* and *eSESE*, are in many cases sufficient. This observation will also be made for the five-level inverter case study in Sect. 8.2.4.

We also observe that the two performance metrics can be halved with MPDTC in the speed range between 0.55 and 0.8 pu. As we have seen, this implies that the switching frequency and hence the switching losses can be halved, while keeping the current distortions at the same level. The reverse also holds true in many cases, allowing one to drastically reduce the current distortions while leaving the switching frequency unchanged. The performance improvement is smaller at higher speeds, amounting to a reduction of the metrics between a quarter and a third.

Penalizing the switching losses in the MPDTC cost function alters the performance results significantly. As shown in Fig. 8.10, medium horizons fully resolve the issue of deadlocks and provide a smooth performance metric with the speed as the argument. Long horizons significantly improve on medium horizons by further reducing the switching losses (see Fig. 8.10(b)). In many cases, however, MPDTC requires a slight increase in the switching frequency to achieve this reduction in the switching losses. This characteristic can be observed in Fig. 8.10(a). Nonetheless, the switching losses (and not the switching frequency) relate to the efficiency of the inverter and its cooling capability, and thus constitute the main quantity to be reduced.

The losses metric c_P can be reduced by about 60% with respect to DTC when operating in the medium-speed region between 0.6 and 0.75 pu. In the high-speed region above 0.8 pu, the reduction amounts to at least one-third. We conclude that, when minimizing the switching losses, very long switching horizons carry a performance benefit over horizons of medium lengths. The successful minimization of the switching losses requires the placement of the

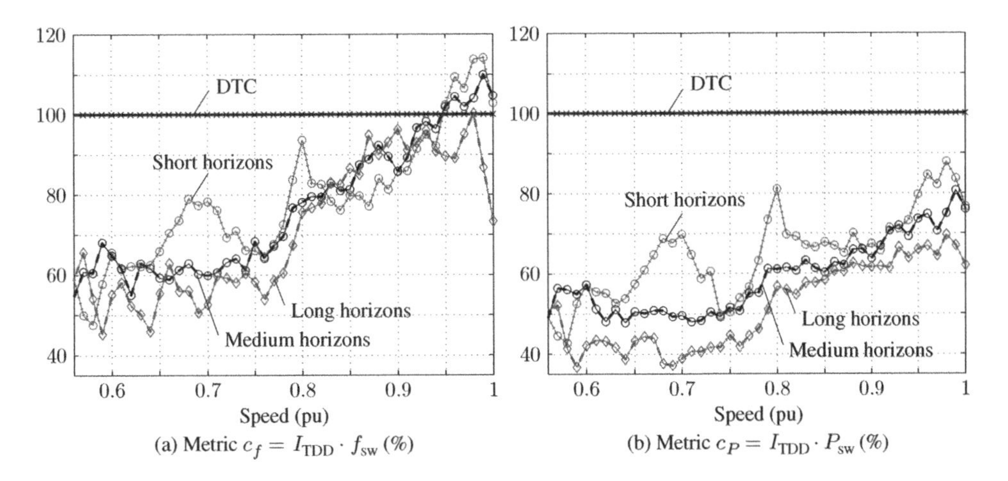

(a) Metric $c_f = I_{\text{TDD}} \cdot f_{\text{sw}}$ (%) (b) Metric $c_P = I_{\text{TDD}} \cdot P_{\text{sw}}$ (%)

Figure 8.10 Steady-state performance of MPDTC minimizing the switching *losses*

switching instants at positions where the phase current is small while keeping the three output variables within their bounds. This is a complex optimization problem that benefits from long horizons.

8.1.3 Operation during Transients

The dynamic performance of MPDTC during torque reference steps is investigated and compared with that of DTC. For MPDTC, we choose the switching horizon *eSESE*, the penalty on rms bound violations $\boldsymbol{q} = [1000\ 0\ 0]^T$, and slightly widen the torque and flux bounds as previously. While operating at 70% speed, we reverse the torque by applying torque steps from 1 to -1 pu and back to 1 pu. Both control methods use the same initial conditions for the machine and the neutral point potential.

The torque response is shown in Fig. 8.11. Both control methods are effectively equally fast. During the negative torque step, the torque settling times of DTC and MPDTC amount to 0.8 and 1.1 ms, respectively, while during the positive torque step the respective time intervals are 4.2 and 3.6 ms. The negative torque transient is much shorter than the positive one because ample voltage margin is available at close to nominal speed when reversing the torque. Specifically, both control methods temporarily invert the stator voltage during the negative step, as can be observed in Fig. 8.12. This figure also indicates that MPDTC switches significantly less often than DTC. Over the short time window of 30 ms, MPDTC reduces the switching frequency and the switching losses by 40% compared to DTC.

As shown in Fig. 8.13, both control methods keep the stator flux magnitude well within its bounds. The same applies to the neutral point potential, which is not shown here. The rapid torque response is also visible when examining the three-phase stator currents, which are shown in Fig. 8.14. Temporary overcurrent conditions are avoided in both cases.

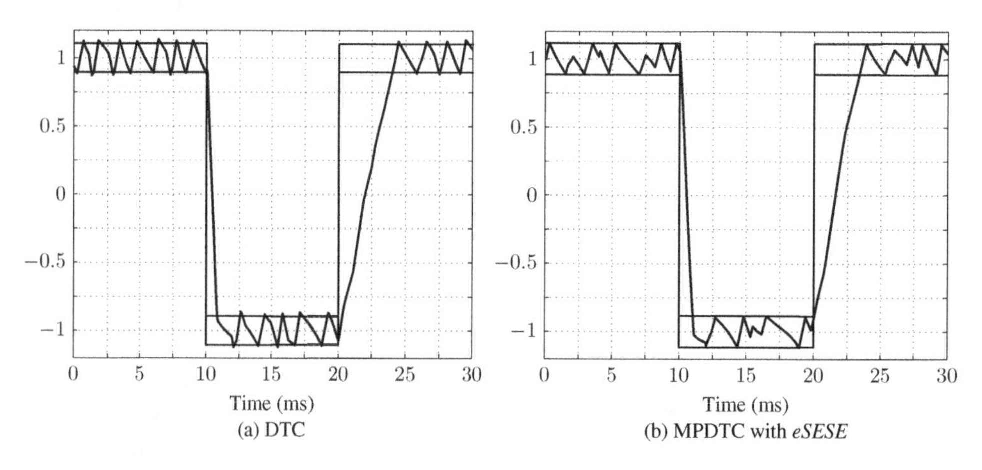

Figure 8.11 Electromagnetic torque for DTC and MPDTC minimizing the switching frequency, when operating at 70% speed. Torque steps of magnitude 2 pu are applied at time instants 10 and 20 ms

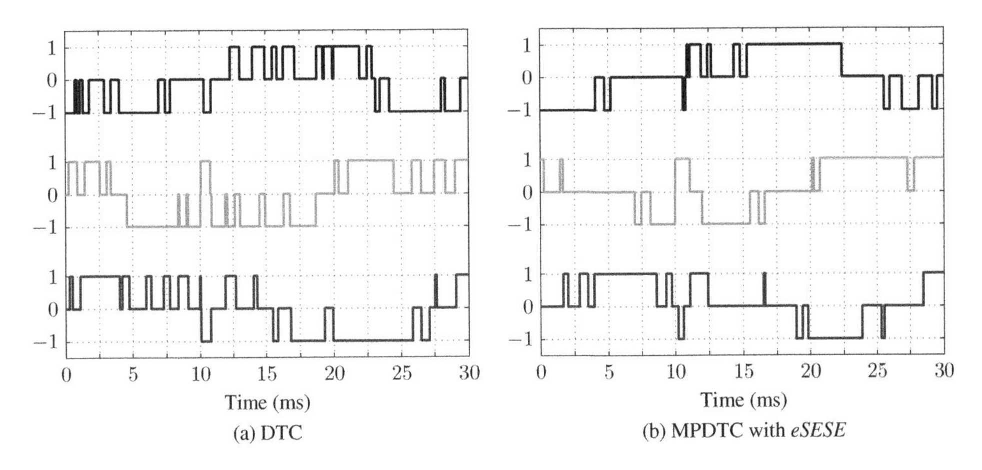

Figure 8.12 Three-phase switch positions of the torque step simulation in Fig. 8.11

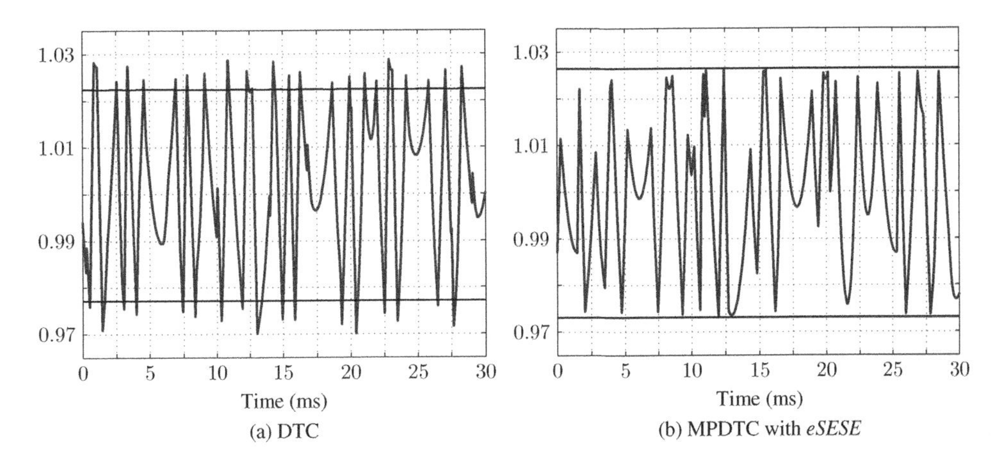

Figure 8.13 Stator flux magnitude of the torque step simulation in Fig. 8.11

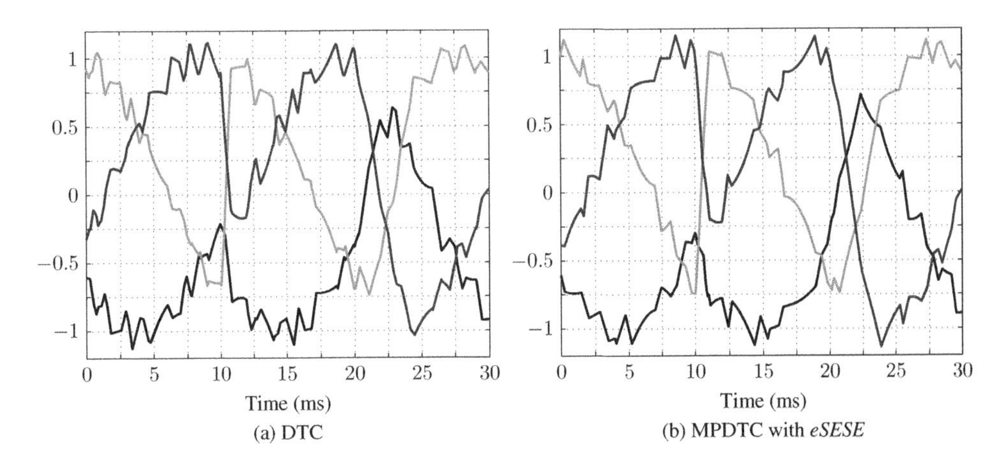

Figure 8.14 Three-phase stator currents of the torque step simulation in Fig. 8.11

8.2 Performance Evaluation for the ANPC Inverter Drive System

The five-level active neutral-point-clamped (ANPC) inverter is a recent topology [1, 2] that extends the classic three-level NPC converter [3]. The NPC diodes are replaced by active switches as in [4] and floating capacitors are added to each phase, similar to a flying capacitor converter [5]. The five-level ANPC inverter topology is reviewed in detail in Sect. 2.4.2.

An MV induction machine rated at 6 kV and 1 MVA is connected to this inverter, forming a variable-speed drive (VSD) system, which is shown in Fig. 8.15. The control problem of this five-level ANPC drive extends that of the three-level NPC drive summarized in Sect. 7.2.2. In addition to the electromagnetic torque, stator flux magnitude, and neutral point potential, the three voltages of the phase capacitors need to be balanced around their references. The control problem thus involves two machine quantities and four inverter voltages.

A number of control and modulation strategies have been proposed for the five-level ANPC topology. All of these approaches divide the control and modulation problem into two hierarchical layers. The *upper* layer controls the machine currents by manipulating the three-phase inverter voltages. To achieve this, control and modulation schemes, which were originally developed for two- and three-level converters, were extended to five levels. This includes DTC [1], selective harmonic elimination, and optimized pulse patterns [6]. The *lower* layer maps the upper layer's differential-mode voltage command into inverter gating signals. By exploiting the redundancy in the phase voltages, the four internal inverter voltages can be balanced around their respective references, as was shown, for example, in [1]. A hierarchical control architecture based on DTC is shown in Fig. 8.16.

Owing to its ability to handle complex multiobjective drive control problems, its very fast torque response, and its ability to achieve low switching frequencies and losses, MPDTC appears to be an ideal candidate to address the control and modulation problem of the five-level ANPC inverter. In particular, MPDTC allows one to formulate and solve the control and modulation problem in one computational stage, thus addressing the torque and flux control problem as well as the balancing of the internal inverter voltages in a combined manner. As a result,

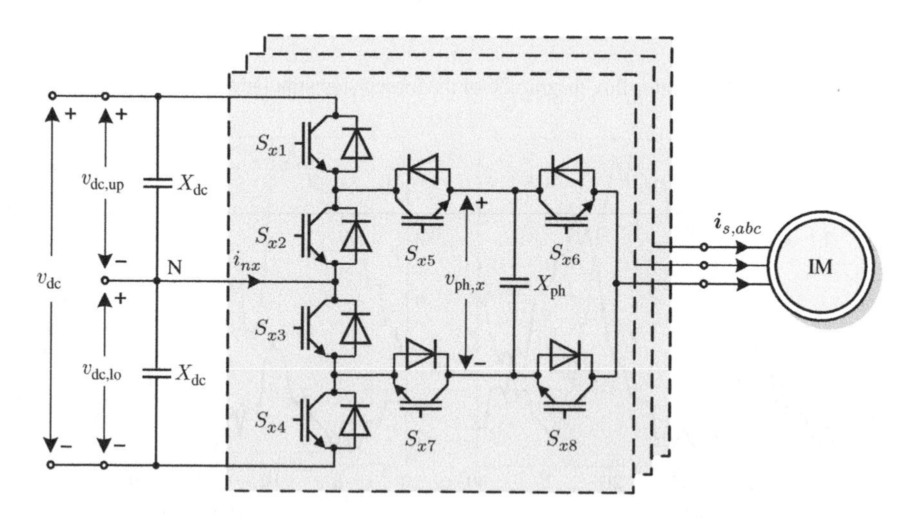

Figure 8.15 Five-level ANPC inverter drive system with an induction machine (IM)

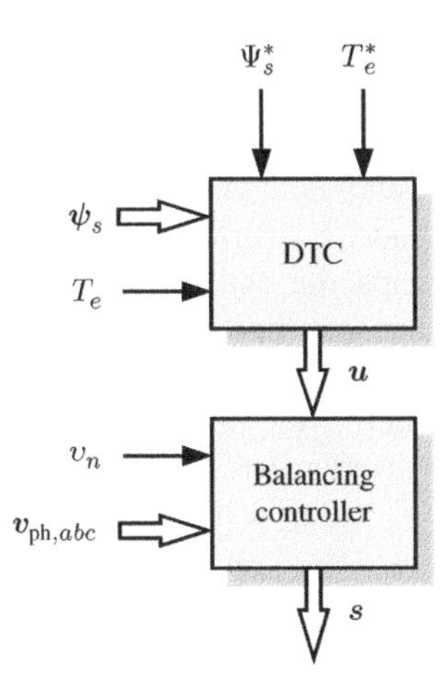

Figure 8.16 Hierarchical control architecture used for the five-level ANPC inverter drive system. DTC serves as the control and modulation method used in the upper control layer

the limitation inherently imposed by separating the control and modulation problem into two layers, in which the set of available control actions is inevitably reduced, is overcome. This results in a performance advantage.

In this section, we summarize the controller model and adapt and extend MPDTC to the five-level topology. Using a 1 MVA drive system with an induction machine as a case study, we compare the performances of MPDTC and DTC at steady-state operation and during transients. The comparison indicates that MPDTC achieves a reduction in the current TDD by 50% and more for the same switching frequency while preserving the very fast torque response of DTC.

8.2.1 Controller Model

The internal prediction model on which MPC relies to predict the future drive trajectories consists of three parts—the machine model, the inverter model, and the inverter's switching restrictions. We use the same machine model as in Sects. 7.2.1 and 7.2.3. Specifically, we define the machine state vector

$$\boldsymbol{x}_m = [\psi_{s\alpha}\ \psi_{s\beta}\ \psi_{r\alpha}\ \psi_{r\beta}]^T, \tag{8.2}$$

which comprises the stator and the rotor flux linkages in orthogonal coordinates. For the inverter, we define the state vector

$$\boldsymbol{x}_i = [v_{\text{ph},a}\ v_{\text{ph},b}\ v_{\text{ph},c}\ v_n]^T, \tag{8.3}$$

with the phase capacitor voltages in the phases a, b, and c, and the neutral point potential. As input vector, we choose the three-phase switch position

$$\boldsymbol{s} = \boldsymbol{s}_{abc} = [s_a\ s_b\ s_c]^T \in \{0, 1, \dots, 7\}^3 \tag{8.4}$$

rather than the phase-level vector $\boldsymbol{u} = \boldsymbol{u}_{abc} \in \{-2, -1, 0, 1, 2\}^3$. The former encompasses the single-phase redundancy that is inherent to the five-level topology, while the latter masks this type of redundancy and exhibits only three-phase redundancy. The choice of $\boldsymbol{s}$ as the input vector enables MPDTC to exploit both kinds of redundancies. For the definitions of the three-phase vectors $\boldsymbol{s}$ and $\boldsymbol{u}$, the reader is referred to Sect. 2.4.2.

The fluctuations of the phase capacitor and dc-link voltages are often significant in the five-level ANPC inverter. To ensure that the controller model provides predictions that are sufficiently accurate, we consider these fluctuations in the predicted stator voltage in (7.1). Specifically, using Table 2.7, we derive the three-phase stator voltage as a function of the three-phase switch position $\boldsymbol{s}$, the inverter states $\boldsymbol{x}_i$, and the total dc-link voltage v_{dc}. To simplify the notation, we usually drop v_{dc} from the list of arguments and simply write $\boldsymbol{v}_{s,abc}(\boldsymbol{s}, \boldsymbol{x}_i)$ for the three-phase stator voltage.

This allows us to write the machine's differential equations (7.4a) and (7.4b) in the state-space form

$$\frac{\mathrm{d}\boldsymbol{x}_m(t)}{\mathrm{d}t} = \boldsymbol{F}_m \boldsymbol{x}_m(t) + \boldsymbol{G}_m \boldsymbol{v}_{s,abc}(\boldsymbol{s}(t), \boldsymbol{x}_i(t)). \tag{8.5}$$

The system and input matrices $\boldsymbol{F}_m$ and $\boldsymbol{G}_m$ are provided in Appendix 8.A.

The dynamic model of the inverter captures the evolution of the three phase capacitor voltages and of the neutral point potential. These four inverter voltages depend on the switch position and the stator currents. As the latter can be expressed as a linear combination of the machine vector (see (2.53)), we can state the derivative of the inverter state vector

$$\frac{\mathrm{d}\boldsymbol{x}_i(t)}{\mathrm{d}t} = \boldsymbol{F}_i(\boldsymbol{s}(t))\, \boldsymbol{x}_m(t) \tag{8.6}$$

as a function of the machine state vector and the switch position. The system matrix $\boldsymbol{F}_i(\boldsymbol{s})$ is derived in Appendix 8.A. It is a function of the three-phase switch position $\boldsymbol{s}$.

In order to derive the continuous-time state-space representation of the ANPC drive system, we combine the state-space model of the machine (8.5) with that of the inverter (8.6):

$$\frac{\mathrm{d}\boldsymbol{x}(t)}{\mathrm{d}t} = \boldsymbol{F}(\boldsymbol{s}(t))\, \boldsymbol{x}(t) + \boldsymbol{G}\, \boldsymbol{v}_{s,abc}(\boldsymbol{s}(t), \boldsymbol{x}(t)) \tag{8.7a}$$

$$\boldsymbol{y}(t) = \boldsymbol{h}(\boldsymbol{x}(t)). \tag{8.7b}$$

The state vector of the drive system

$$\boldsymbol{x} = [\boldsymbol{x}_m^T\ \boldsymbol{x}_i^T]^T \tag{8.8}$$

is defined as the concatenation of the machine and inverter state vectors. The system and input matrices are

$$\boldsymbol{F}(\boldsymbol{s}) = \begin{bmatrix} \boldsymbol{F}_m & \boldsymbol{0}_{4\times4} \\ \boldsymbol{F}_i(\boldsymbol{s}) & \boldsymbol{0}_{4\times4} \end{bmatrix} \text{ and } \boldsymbol{G} = \begin{bmatrix} \boldsymbol{G}_m \\ \boldsymbol{0}_{4\times3} \end{bmatrix}. \tag{8.9}$$

The output vector

$$\boldsymbol{y} = [T_e\ \Psi_s\ v_{\text{ph},a}\ v_{\text{ph},b}\ v_{\text{ph},c}\ v_n]^T \tag{8.10}$$

comprises the electromagnetic torque, the stator flux magnitude, the three phase capacitor voltages, and the neutral point potential. The output function $\boldsymbol{h}(\boldsymbol{x})$ is provided in Appendix 8.A.

The zeros in the input matrix $\boldsymbol{G}$ conceal the fact that the evolution of the four inverter state variables strongly depends on the switch position $\boldsymbol{s}$. Indeed, this dependency is manifested by the inverter system matrix $\boldsymbol{F}_i(\boldsymbol{s})$ being a function of $\boldsymbol{s}$.

In a last step, we derive the discrete-time representation of the drive model by integrating (8.7) from $t = kT_s$ to $t = (k+1)T_s$ using the forward Euler discretization method. This leads to

$$\boldsymbol{x}(k+1) = \boldsymbol{A}(\boldsymbol{s}(k))\, \boldsymbol{x}(k) + \boldsymbol{B}\, \boldsymbol{v}_{s,abc}(\boldsymbol{s}(k), \boldsymbol{x}(k)) \tag{8.11a}$$

$$\boldsymbol{y}(k) = \boldsymbol{h}(\boldsymbol{x}(k)) \tag{8.11b}$$

with the discrete-time matrices

$$\boldsymbol{A}(\boldsymbol{s}(k)) = \boldsymbol{I}_8 + \boldsymbol{F}(\boldsymbol{s}(k))\, T_s \ \text{ and } \ \boldsymbol{B} = \boldsymbol{G}\, T_s. \tag{8.12}$$

The switching restrictions are stated in Sect. 2.4.2. Specifically, single-phase switching restrictions are imposed (see Fig. 2.24), as well as restrictions on the three-phase switching transitions (see Table 2.8). Both types of restrictions depend on the switch positions applied during the past 50 μs and on the sign of the stator current. Assuming a sampling interval of 25 μs, the switching restriction at time step k can be formally stated as

$$\boldsymbol{s}(k) \in \boldsymbol{\mathcal{S}}(\boldsymbol{i}_{s,abc}(k), \boldsymbol{s}(k-1), \boldsymbol{s}(k-2)). \tag{8.13}$$

In equation (8.13), $\boldsymbol{\mathcal{S}}$ denotes the set of allowed three-phase switch positions the inverter may transition to from $\boldsymbol{s}(k-1)$.

In summary, the controller model includes the standard dynamic model of an induction machine with four state variables, the inverter dynamics, and restrictions on the switching transitions. If required, the model can be extended to include the speed dynamic (7.4c).

8.2.2 *Modified MPDTC Algorithm*

We distinguish between two groups of switches in phase x, with $x \in \{a, b, c\}$. The dc-link and the switches S_{x1}–S_{x4} form the ANPC part, while the phase capacitor and the switches S_{x5}–S_{x8} constitute the flying capacitor (FC) part. Each one of the S_{x1}–S_{x4} switches consists of two series-connected insulated-gate bipolar transistors (IGBTs), while the switches S_{x5}–S_{x8} correspond to single IGBTs. Slightly abusing the notation, we introduce the variables

$$\Delta s_{\text{ANPC},x}(k) = f_{\text{ANPC}}(i_{sx}(k), s_x(k), s_x(k-1)) \tag{8.14a}$$

$$\Delta s_{\text{FC},x}(k) = f_{\text{FC}}(i_{sx}(k), s_x(k), s_x(k-1)), \tag{8.14b}$$

which capture at the discrete time step k the per-phase *on* (or *off*) transitions in the ANPC and in the FC part, respectively. The functions f_{ANPC} and f_{FC} are implicitly defined in Fig. 2.24. Their outputs are 0, 1, or 2. We also define the vectors

$$\Delta \boldsymbol{s}_{\text{ANPC}} = \begin{bmatrix} \Delta s_{\text{ANPC},a} \\ \Delta s_{\text{ANPC},b} \\ \Delta s_{\text{ANPC},c} \end{bmatrix} \ \text{ and } \ \Delta \boldsymbol{s}_{\text{FC}} = \begin{bmatrix} \Delta s_{\text{FC},a} \\ \Delta s_{\text{FC},b} \\ \Delta s_{\text{FC},c} \end{bmatrix}. \tag{8.15}$$

The IGBTs of the FC part bear the majority of the switching burden. It is thus particularly important to minimize the switching frequency of these switches. This motivates the cost function

$$J = J_{\text{sw}} + J_{\text{bnd}} + \lambda_n (\upsilon_n(k + N_i))^2 \tag{8.16}$$

with

$$J_{\text{sw}} = \frac{1}{N_i} \sum_{\ell=k}^{k+N_i-1} \lambda_s \|\Delta \boldsymbol{s}_{\text{ANPC}}(\ell)\|_1 + \|\Delta \boldsymbol{s}_{\text{FC}}(\ell)\|_1. \tag{8.17}$$

The first term in (8.16) represents the (short-term) switching frequency over the length N_i of the ith candidate switching sequence. We use the tuning parameter $0 \leq \lambda_s < 1$ to discount switching transitions in the ANPC part. The second term $J_{\text{bnd}} = \boldsymbol{q}^T \boldsymbol{\epsilon}$ penalizes bound violations of the output variables. Similar to (7.33), we define the normalized rms bound violations as

$$\boldsymbol{\epsilon}(k) = [\epsilon_T(k) \ \epsilon_\Psi(k) \ \epsilon_a(k) \ \epsilon_b(k) \ \epsilon_c(k) \ \epsilon_\upsilon(k)]^T. \tag{8.18}$$

The third term in the cost function adds a terminal weight on the neutral point potential, by penalizing the potential's deviation from zero at the end of the switching sequence. The penalty is adjusted using the weight $\lambda_n \geq 0$, which is significantly smaller than 1.

The purpose of this third term is to reduce the likelihood of infeasibilities or deadlocks, that is, situations in which the set of candidate switching sequences is empty. Such scenarios tend to occur when two or more output variables are close to their bounds. In most cases, the neutral point potential and one of the phase capacitor voltages act as antagonists. The likelihood of such events can be largely reduced by adding a terminal weight on the neutral point potential to the cost function. This penalty provides an incentive for MPDTC to drive the neutral point potential closer to zero whenever the predicted increase in the switching frequency is minor. For more details on the phenomenon of deadlocks and strategies to avoid them, the reader is referred to Sects. 9.3 and 9.5. In the case of a deadlock, the deadlock resolution strategy described in Sect. 9.4 is employed.

The MPDTC algorithm in Sect. 7.4.5 is adapted to the five-level ANPC inverter drive system by performing the following three modifications. First, the controller manipulates the three-phase switch position $\boldsymbol{s}$, with $\boldsymbol{s} \in \{0, 1, \ldots, 7\}^3$. This implies that the switching sequence is

$$\boldsymbol{S}(k) = [\boldsymbol{s}^T(k) \quad \boldsymbol{s}^T(k+1) \ \cdots \ \boldsymbol{s}^T(k+N-1)]^T, \tag{8.19}$$

which replaces (7.36). Second, when switching at time step ℓ, only the admissible switch positions $\boldsymbol{s}(\ell)$ in the set $\mathcal{S}(\boldsymbol{i}_{s,abc}(\ell), \boldsymbol{s}(\ell-1), \boldsymbol{s}(\ell-2))$ are considered, generalizing the switching constraint (7.37). Third, the controller model in (7.38) and (7.40b)–(7.40c) is replaced by (8.11).

8.2.3 Simulation Setup

In the remainder of this section, the performance of MPDTC is evaluated and benchmarked with DTC for an MV five-level ANPC inverter drive system. More specifically, a 6 kV, 50 Hz squirrel-cage induction machine rated at 1 MVA is used. The inverter's nominal dc-link voltage is 9.8 kV. The drive system case study is summarized in Sect. 2.5.3. The parameters of the machine and inverter are summarized in Tables 2.12 and 2.13.

A very accurate and detailed MATLAB/Simulink model of the drive system was used for this comparison to ensure a simulation setup as realistic as possible. This model includes an observer for the motor flux linkages and outer control loops that adjust the (time-varying) bounds on the torque and stator flux magnitude. The optional speed encoder is not used. The induction machine model includes the saturation of the machine's magnetic material and the changes of the rotor resistance because of the skin effect. Measurement delays and the controller's computational delay are explicitly modeled. The Simulink model includes an active front end (AFE) with a transformer and a grid model. The AFE regulates the total dc-link voltage and the reactive power injected into the grid. The AFE does not control the neutral point potential, but it nevertheless has a significant influence on it.

As a benchmark control scheme, the DTC scheme described in Sect. 3.6.3 is extended to the five-level inverter. The torque and flux hysteresis controllers are augmented with additional hysteresis bands, and the DTC look-up table is refined accordingly. The hierarchical control structure shown in Fig. 8.16 is adopted. DTC controls the machine's torque and degree of magnetization through the manipulation of the phase levels $\boldsymbol{u}$. An underlying balancing controller maintains the three phase capacitor voltages and the neutral point potential within their bounds. The balancing controller exploits the single-phase and three-phase redundancies inherent to the five-level ANPC topology and issues the three-phase switch position $\boldsymbol{s}$. This two-tiered control scheme is described in [1].

In the case of MPDTC, the Simulink block with the DTC and balancing scheme was replaced by a function that runs the MPDTC algorithm at each sampling instant. The weights in the cost function (8.16) are set to $\lambda_s = 0.1$ and $\lambda_n = 0.1$. The penalty on the bound violation is set to $\boldsymbol{q} = \boldsymbol{0}_{6\times1}$ during steady-state operation. During torque transients, however, it is set to $\boldsymbol{q} = [1000\,0\ \cdots\ 0]^T$, strongly penalizing bound violations of the torque.

8.2.4 Steady-State Operation

The performance comparison between MPDTC and DTC focuses on the current and torque TDDs, I_{TDD} and T_{TDD}, respectively, and the following three switching frequencies: the average of all 36 device switching frequencies, $f_{\text{sw,avg}}$; the switching frequency of the IGBT pairs S_{x1}–S_{x4} (ANPC part of the inverter), $f_{\text{sw,ANPC}}$; and the switching frequency of the IGBTs S_{x5}–S_{x8} (FC part), $f_{\text{sw,FC}}$.

DTC imposes the symmetrical bounds $0.25v_{\text{dc}} \pm 0.055$ pu on the phase capacitor voltages. These bounds are centered at a quarter of the dc-link voltage. The neutral point potential is kept within ± 0.09 pu. Both sets of bounds are inherited by MPDTC. The bounds on the electromagnetic torque and the stator flux magnitude are set by the outer DTC loops. In order to reduce the current and torque distortions, the widths of the torque and flux bounds were multiplied by 0.774 and 0.427, respectively. This resulted in the effective torque and flux bounds of about 1 ± 0.04 and 1 ± 0.005 pu, respectively, when operating at rated torque.

8.2.4.1 Operation at Nominal Speed

Consider steady-state operation at nominal speed and rated torque. Table 8.2 shows selected simulation results that compare the closed-loop performances of DTC and MPDTC. Owing to the tightening of the bounds and because MPDTC adheres more closely to the bounds, the

Table 8.2 Comparison between DTC and MPDTC with various switching horizons. The comparison is made at nominal speed and rated torque in terms of the current and torque TDDs, I_{TDD}, and T_{TDD}, respectively, which are given in percent, using DTC as a baseline

Control scheme	Switching horizon	Average prediction horizon N_p	I_{TDD} (%)	T_{TDD} (%)	$f_{\text{sw,avg}}$ (Hz)	$f_{\text{sw,ANPC}}$ (Hz)	$f_{\text{sw,FC}}$ (Hz)
DTC	—	—	100	100	421	315	634
MPDTC	*eSE*	8.4	48.6	49.9	383	272	605
MPDTC	*eSSE*	13.7	47.7	50.5	350	248	555
MPDTC	*eSESE*	19.7	47.1	49.9	337	238	534
MPDTC	*eSESESE*	30.4	45.8	48.7	326	229	519

The average switching frequency $f_{\text{sw,avg}}$, the switching frequency of the ANPC part $f_{\text{sw,ANPC}}$, and the switching frequency of the FC part $f_{\text{sw,FC}}$ are also shown.

current and torque TDDs are more than halved for MPDTC, while the switching frequency is also reduced.

As the switching horizon is extended, the resulting prediction horizon grows accordingly, enabling MPDTC to look further ahead and to achieve a significant reduction in the switching frequencies. For MPDTC with the switching horizon *eSE*, for example, the average switching frequency and thus the switching losses can be reduced by almost 10% with respect to DTC. In contrast, when adopting the long switching horizon *eSESESE*, the average switching frequency is lowered by more than 20%.

The IGBTs in the FC part carry the majority of the switching burden and constitute the limiting factor. Their switching frequencies can be reduced by 5% by MPDTC with *eSE* and by almost 20% when using *eSESESE*. This is a noteworthy result, because the FC switches are predominantly used to balance the phase capacitor voltages and the controller has only few degrees of freedom to improve the balancing. As approximately half of the switching transitions in the FC part are triggered by internal inverter voltages approaching their bounds, a 10% reduction may enable one to tighten the bounds on the torque and stator flux by another 20% and to reduce the corresponding distortions accordingly. Interestingly, as the switching horizon is extended, the current TDD also drops slightly. This indicates that the torque and stator flux magnitude are kept more tightly within their bounds when using long prediction horizons.

Figures 8.17(a) and 8.18(a) show waveforms for DTC operating at nominal speed and rated torque. Significant violations of the torque and stator flux bounds occur because of the fact that DTC switches only *after* a bound has been violated. The phase currents in Fig. 8.19(a) exhibit a noticeable current ripple. The torque and current spectra in Figs. 8.20(a) and 8.21(a) were computed using a Fourier transformation. The amplitudes of the harmonics are small and below 2% of the 1 pu amplitudes of the rated torque and phase currents.

The balancing controller underlying DTC keeps the neutral point potential well within its bounds, except for rare violations. The balancing controller imposes inner and outer bounds on the phase capacitor voltages to ensure that the voltages are always kept within their outer bounds. However, the use of additional inner bounds implies that the outer bounds are not always fully utilized. This can be seen in Fig. 8.23(a), which only shows the outer bounds. The phase levels and switch positions are shown in Figs. 8.24(a) and 8.25(a). Three different switching frequencies that are obtained with DTC are summarized in the first line in Table 8.2.

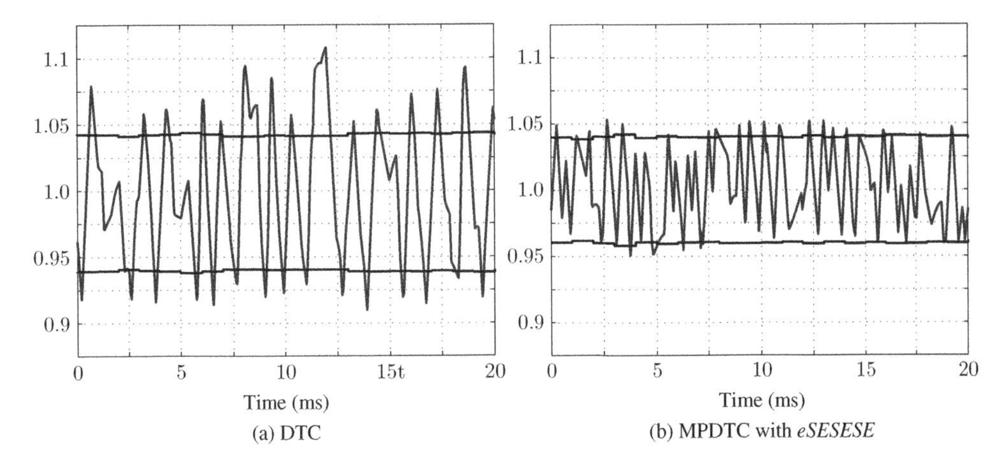

Figure 8.17 Electromagnetic torque for DTC and MPDTC minimizing the switching frequency, when operating at rated speed and rated torque

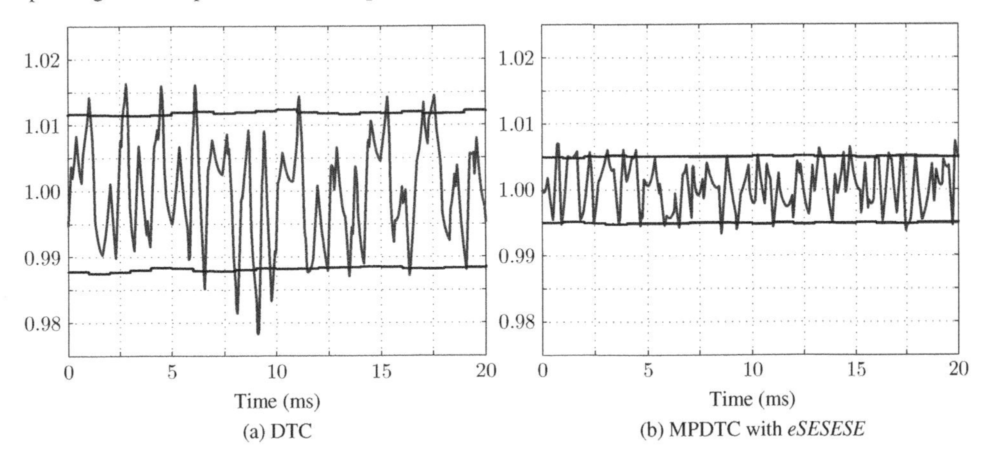

Figure 8.18 Stator flux magnitude of the simulation in Fig. 8.17

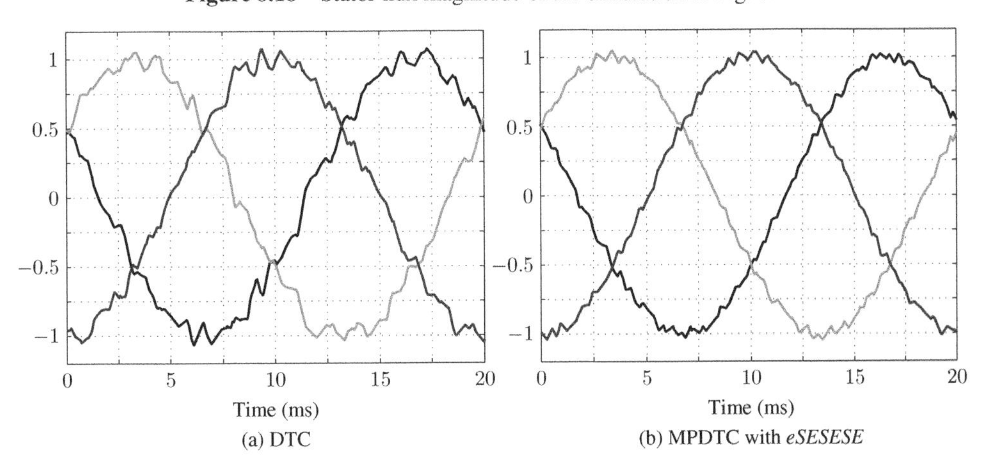

Figure 8.19 Three-phase stator currents of the simulation in Fig. 8.17

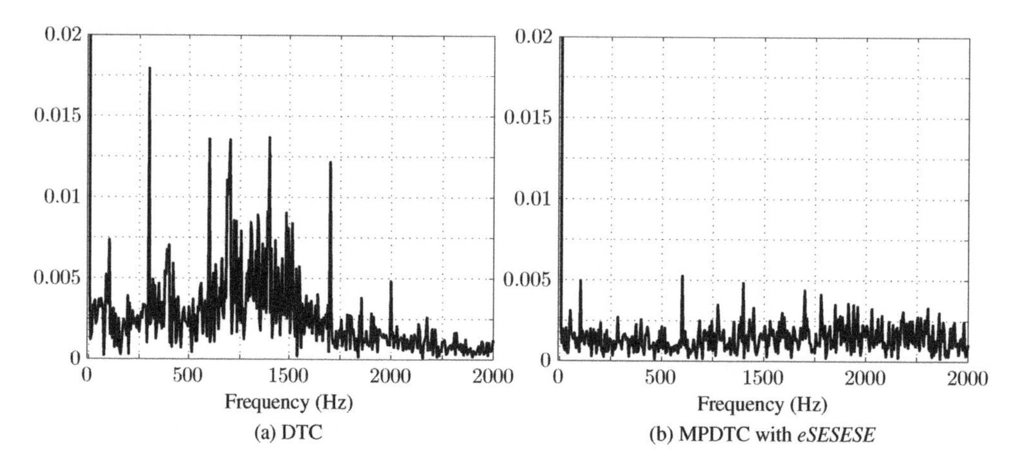

(a) DTC (b) MPDTC with *eSESESE*

Figure 8.20 Spectra of the electromagnetic torque shown in Fig. 8.17

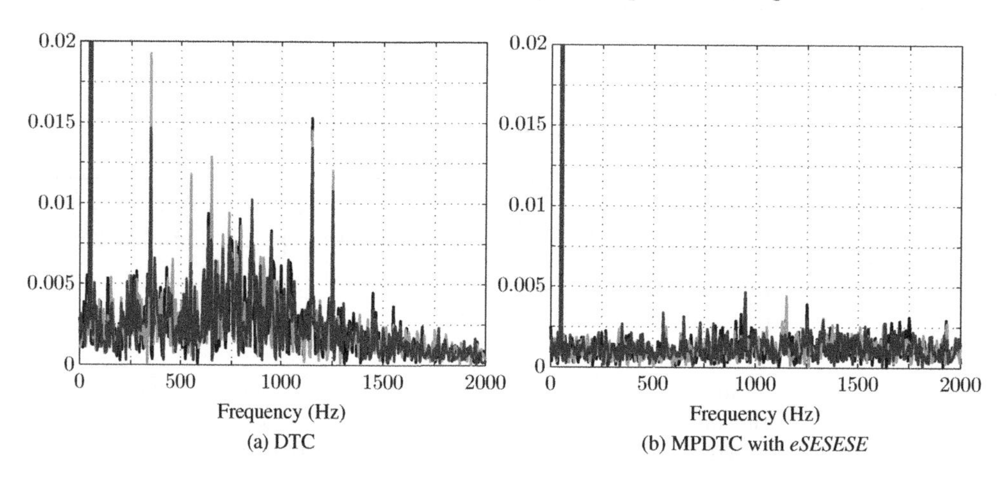

(a) DTC (b) MPDTC with *eSESESE*

Figure 8.21 Spectra of the three-phase stator currents shown in Fig. 8.19

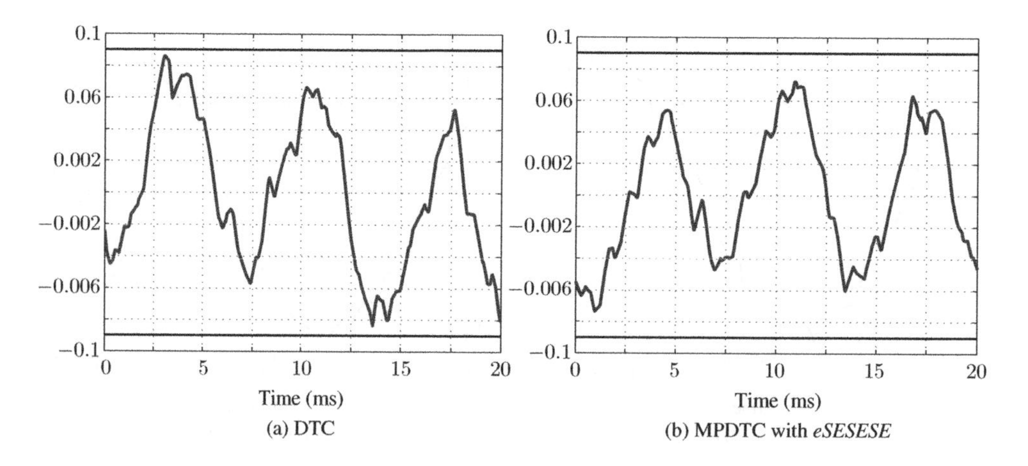

(a) DTC (b) MPDTC with *eSESESE*

Figure 8.22 Neutral point potential of the simulation in Fig. 8.17

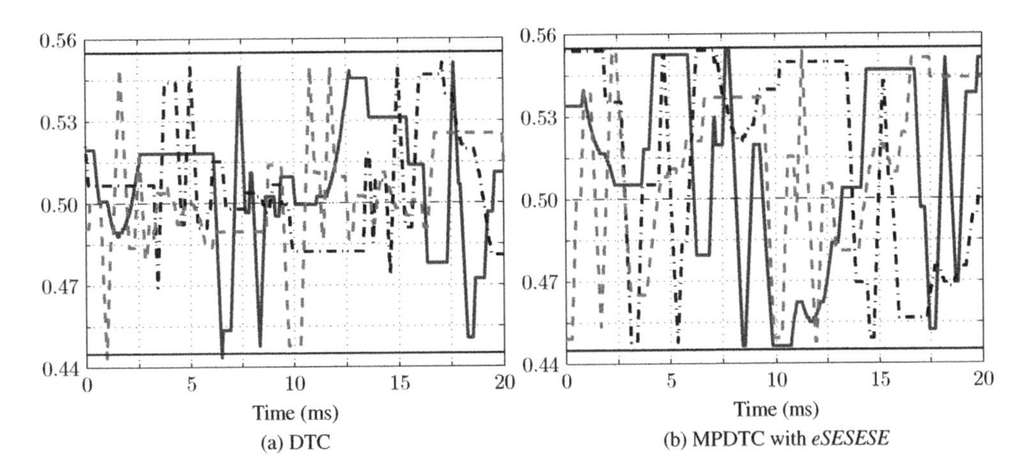

Figure 8.23 Phase capacitor voltages of the simulation in Fig. 8.17

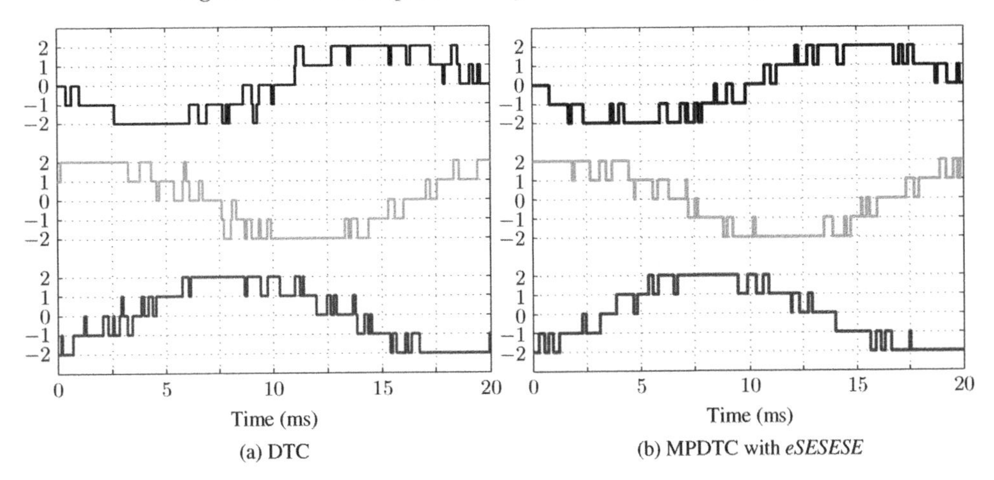

Figure 8.24 Phase levels $\boldsymbol{u}$ of the simulation in Fig. 8.17

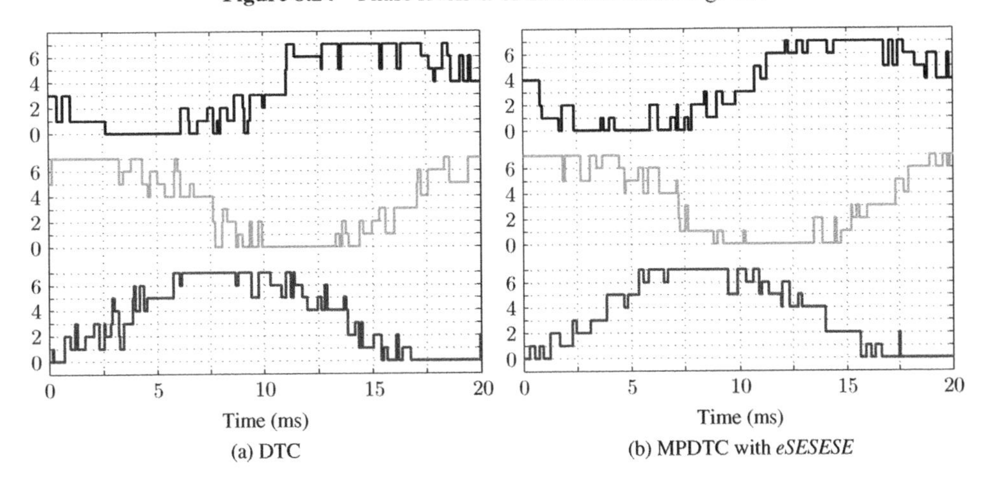

Figure 8.25 Switch positions $\boldsymbol{s}$ of the simulation in Fig. 8.17

Figures 8.17(b)–8.25(b) show the corresponding waveforms for MPDTC with the switching horizon *eSESESE*. The long prediction horizon and the internal controller model enable MPDTC to make educated switching decisions. The small bound violations are due to uncompensated measurement and computation delays. When compared with DTC, it was possible to tighten the bounds on the torque and stator flux magnitude while maintaining (or even slightly reducing) the switching frequency. The tightening of the bounds significantly reduces the current ripple (see Fig. 8.19(b)) and effectively halves the current and torque TDDs, as shown in Table 8.2.

The torque and current spectra are considerably flatter and below 0.5% of the rated torque and phase currents, respectively, as can be seen in Figs. 8.20(b) and 8.21(b). The neutral point potential is kept well within its bounds (see Fig. 8.22(b)), despite the unmodeled interference from the AFE. The bound width on the phase capacitor voltages is fully utilized, but not violated (see Fig. 8.23(b)), because potential violations are predicted and can thus be avoided by MPDTC. The phase levels and switch positions are shown in Figs. 8.24(b) and 8.25(b). Compared to those of DTC, these waveforms exhibit a smoother shape and approximate sinusoidal waveforms more accurately.

8.2.4.2 Operation Over a Range of Speed Operating Points

Next, the performance of the two control and modulation schemes is compared over a range of speed operating points. The speed is varied between 0.5 and 1.1 pu in steps of 0.05 pu while operating at rated torque. For MPDTC, the torque and stator flux bounds are multiplied by the same factors (0.774 for the torque and 0.427 for the flux magnitude) as during operation at nominal speed. The bounds on the four internal inverter voltages remain unchanged. The short switching horizon *eSE* and the medium switching horizon *eSESE* are considered for MPDTC.

Figure 8.26(a) shows the current TDD achieved by MPDTC in percent using DTC as a baseline. As can be seen, for this set of bounds, the current TDD is halved throughout the considered

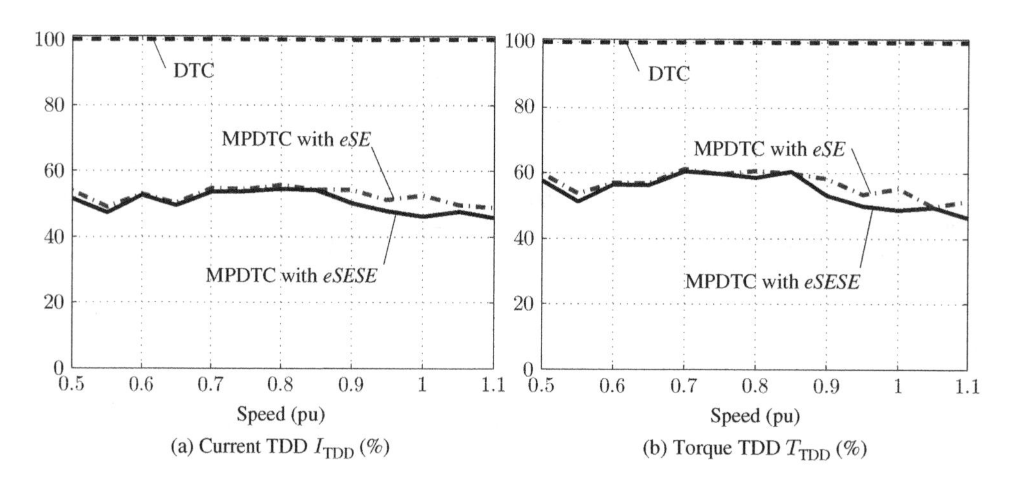

Figure 8.26 Steady-state performance comparison between MPDTC and DTC when varying the speed setpoint between 0.5 and 1.1 pu and operating at rated torque. The current and torque TDDs are given in percent, using DTC as a baseline

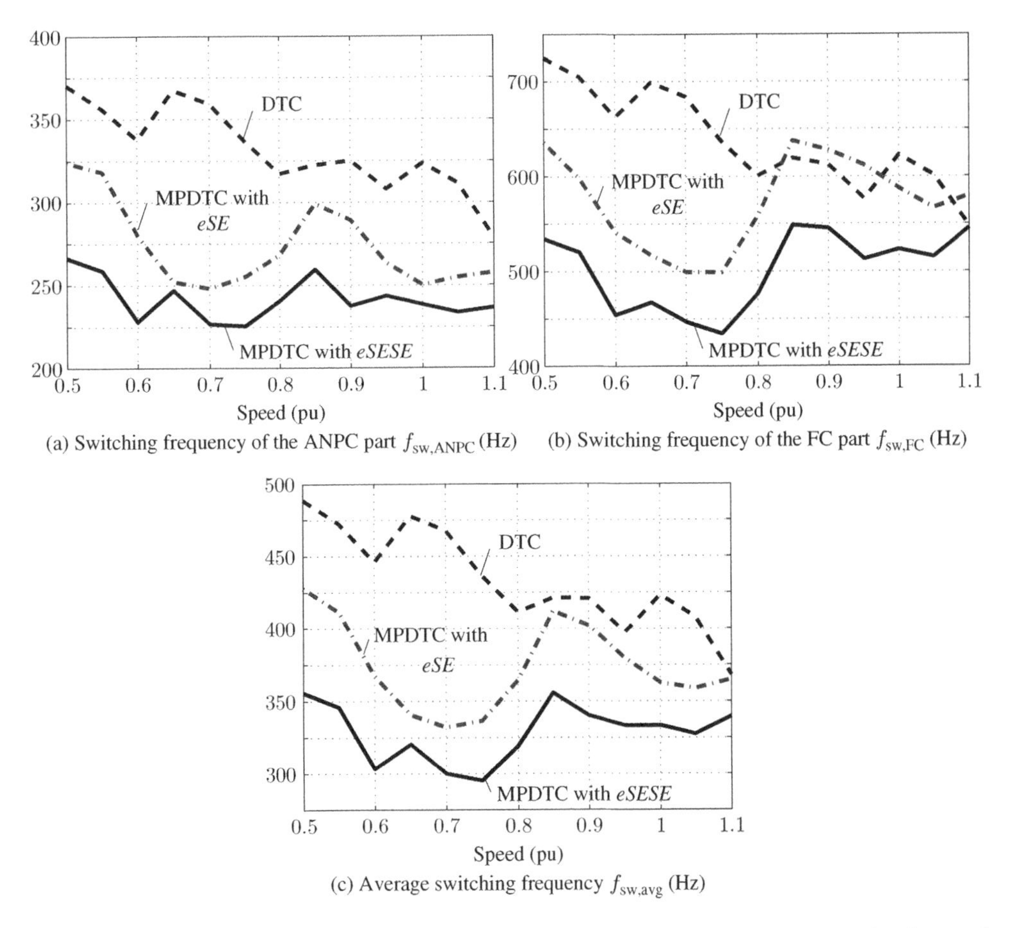

(a) Switching frequency of the ANPC part $f_{sw,ANPC}$ (Hz) (b) Switching frequency of the FC part $f_{sw,FC}$ (Hz)

(c) Average switching frequency $f_{sw,avg}$ (Hz)

Figure 8.27 Steady-state performance comparison between MPDTC and DTC when varying the speed setpoint between 0.5 and 1.1 pu and operating at rated torque. The switching frequencies are given in hertz

range of operating points. Similarly, the torque TDD shown in Fig. 8.26(b) is reduced by 40–50%. The improvement in the current and torque TDDs is independent of the speed in this experiment. This is due to our choice of using the same factor, by which the MPDTC bounds are tightened with respect to DTC, at all operating points.

The three key switching frequencies are shown in Fig. 8.27, namely the device switching frequencies of the ANPC and FC parts, and the average device switching frequency. For speeds below 0.8 pu, MPDTC with the switching horizon *eSE* greatly reduces the switching frequencies. Specifically, the pivotal switching frequency of the IGBTs in the FC part is decreased by up to 180 Hz (see Fig. 8.27(b)), whereas the average switching frequency is reduced by up to 135 Hz (see Fig. 8.27(c)). Even though the switching frequency in the ANPC part is only weakly penalized with a discount factor of 0.1, the switching frequency reduction of up to 110 Hz in the ANPC part is also significant, as can be seen in Fig. 8.27(a).

Table 8.3 Steady-state comparison between DTC and MPDTC operating at rated torque

Control scheme	Switching horizon	I_{TDD} (%)	T_{TDD} (%)	$f_{\text{sw,avg}}$ (Hz)	$f_{\text{sw,ANPC}}$ (Hz)	$f_{\text{sw,FC}}$ (Hz)
DTC	—	100	100	434	332	638
MPDTC	*eSE*	52.4	56.6	374	274	574
MPDTC	*eSESE*	50.4	54.5	328	242	502

The performance metrics are averaged over 13 speed operating points between 0.5 and 1.1 pu.

It appears that MPDTC provides the largest performance benefit at around 0.7 pu speed. A similar observation has been made when applying MPDTC to an NPC inverter drive system (see Sect. 8.1). The reductions in the switching frequencies are less pronounced for speeds above 0.8 pu. Nevertheless, in addition to the switching frequency reductions, the current and torque TDDs are reduced by 40–50% throughout the range of investigated speed operating points, as discussed before.

Adopting longer switching horizons further reduces the switching frequencies. Using *eSESE* instead of *eSE*, for example, reduces on average the switching frequency of the FC devices by another 70 Hz. The switching frequency of the ANPC part is further reduced by 30 Hz, and the average switching frequency is reduced by another 45 Hz. As can be seen in Fig. 8.27, this switching frequency difference is effectively independent of the speed operating point.

Table 8.3 summarizes the *average* performance improvement that MPDTC with the switching horizons *eSE* and *eSESE* achieves over DTC at steady-state operation. For this, the results shown in Figs. 8.26 and 8.27 were averaged over the speed range 0.5–1.1 pu. We conclude that MPDTC with the medium switching horizon *eSESE* achieves an average reduction of the current and torque TDDs by about 45%. At the same time, the average device switching frequency of the FC part is reduced by 136 Hz and the average switching frequency of all semiconductor switches is lowered by 106 Hz.

8.2.5 *Operation during Transients*

Figures 8.28–8.30 compare the performances of DTC and MPDTC during torque transients. Steps of magnitude 1 pu are applied to the torque reference when operating at nominal speed. Both control schemes fully exploit the available dc-link voltage by temporarily inverting the voltage applied to the machine. As a result, DTC and MPDTC provide similar fast torque responses. The torque settling time for negative torque steps is around 0.4 ms, while it is about 1.5 ms for the positive torque step. Overshoots in the torque occur in both schemes, which appear to be a result of the switching restrictions and delays. The other output variables are kept well within their bounds, and unnecessary switching is avoided.

The MPDTC results are based on the switching horizon *eSSE*. Shorter horizons, such as *eSE*, tend to result in slower torque responses, because very short switching horizons in connection with the switching restrictions limit the set of voltage vectors that can be considered by MPDTC.

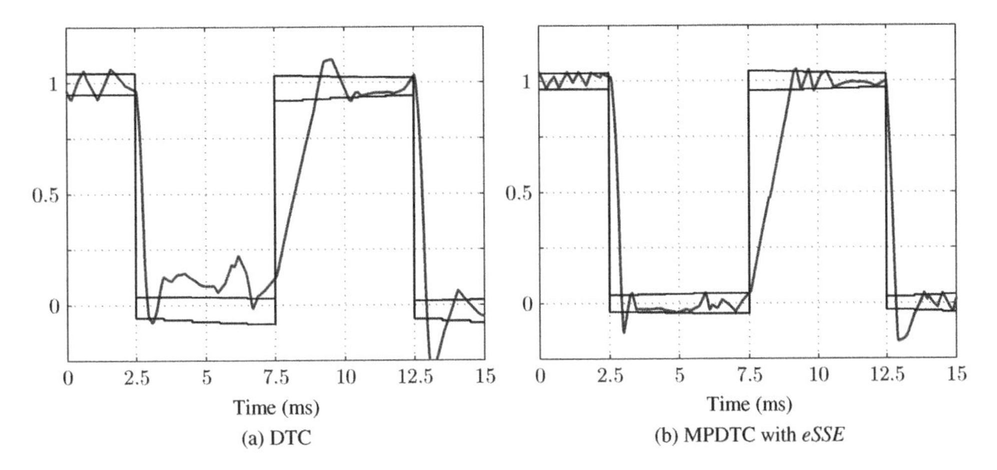

Figure 8.28 Electromagnetic torque for DTC and MPDTC minimizing the switching frequency, when operating at nominal speed. Torque steps of magnitude 1 pu are applied every 5 ms

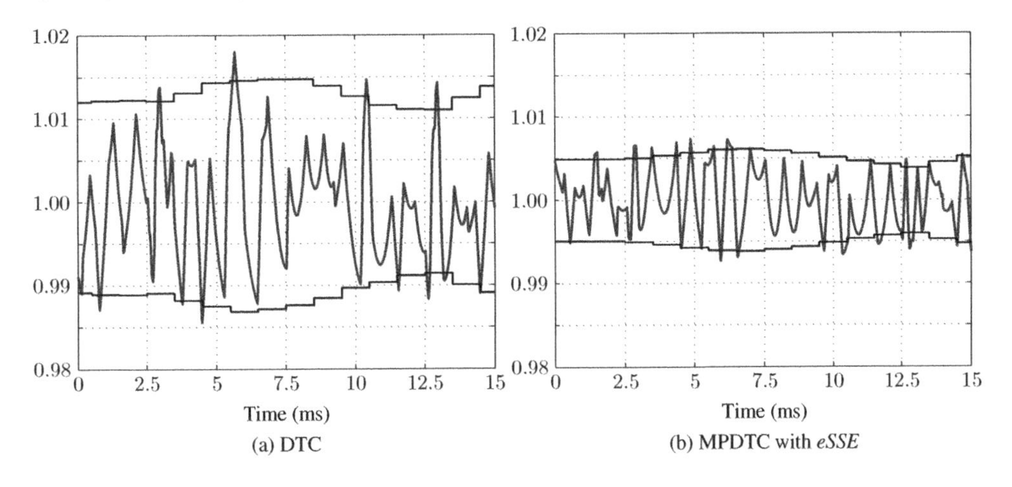

Figure 8.29 Stator flux magnitude of the simulation in Fig. 8.28

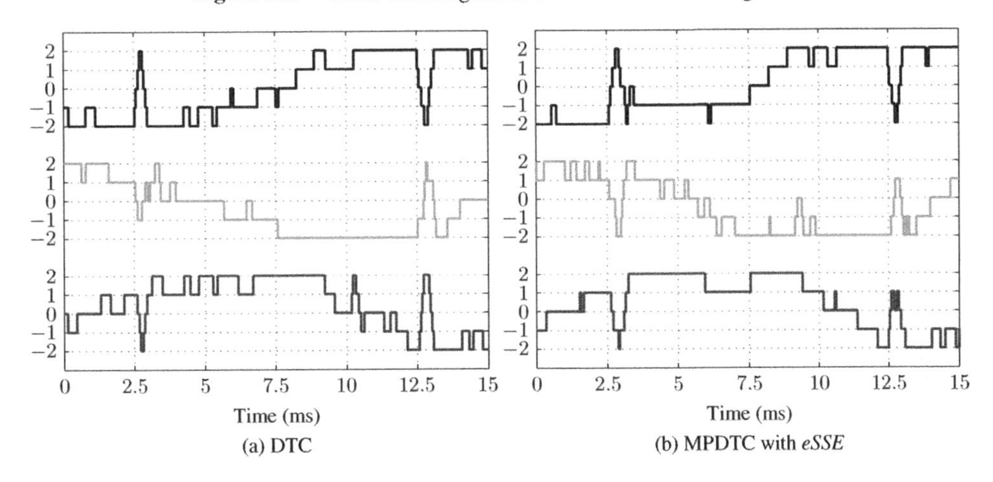

Figure 8.30 Phase levels $\boldsymbol{u}$ of the simulation in Fig. 8.28

8.3 Summary and Discussion

The performance of MPDTC was evaluated in this chapter for MV induction motor drives based on the classic three-level NPC inverter and the recent five-level ANPC inverter. During steady-state operation, the main performance criteria are the switching frequency, the switching losses in the semiconductors, and the current and torque distortions. For the NPC inverter, the focus was on reducing the switching frequency and the losses. Compared to DTC, these can be reduced by at least 25%. In the medium-speed regime, reductions of up to 60% are possible. To minimize the switching losses as much as possible, the losses should be penalized in the MPDTC cost function, and long switching horizons are required. Conversely, the harmonic distortions can be reduced while maintaining the switching frequency or losses at the level of DTC. Improvements of 30% and more are often achieved. To achieve low current distortions, the torque and flux bounds have to be tightened. Very tight bounds might necessitate an increase in the sampling frequency.

It is relatively straightforward to adapt MPDTC to the five-level ANPC inverter drive system. The focus here was on reducing the harmonic distortions. Compared to standard DTC, the current and torque distortions can be halved while maintaining—and in many cases reducing—the switching frequency. The opposite, that is, halving the switching frequency for the same harmonic distortions, is not possible with this inverter, because a significant proportion of the switching effort is required to maintain the phase capacitor voltages within their upper and lower bounds. We had seen for the NPC inverter that, when minimizing the switching frequency in the cost function, medium prediction horizons are often sufficient. This observation can also be made for the five-level inverter.

During transients such as step changes in the torque reference, MPDTC provides as fast a dynamic torque response as DTC. The torque settling times of these two control schemes are very similar. This was shown for the three-level as well as the five-level inverter drive system.

MPDTC is attractive also in light of its simple outer control loops. A field-oriented reference frame is not required, and the torque is directly set by the speed controller. The outer control loops thus require almost no tuning and no drive parameters. Tuning is, however, required to choose appropriate bounds on the torque and flux magnitude. In this chapter, we have bypassed this issue and simply adopted the—somewhat modified—DTC bounds. If the current distortions per switching frequency (or losses) are the main concern, a revised set of bounds often leads to better results. A derivative of MPDTC will be introduced in Sect. 11.1, which directly controls the ripple current and thus the current TDD. Requiring only one tuning parameter, this MPDTC derivative permits a simple and straightforward tuning procedure.

Even though only simulation results have been provided in this chapter, a preliminary version of MPDTC has been successfully implemented and tested on an MV drive system with an NPC inverter. This version is akin to MPDTC with the switching horizon *SE*, but enumerates all 27 switching transitions in the switching step *S*. For switch positions that are unattainable within one switching transition, a feasible switching sequence is established that involves multiple control cycles and links the current switch position with the desired one. This preliminary MPDTC version is explained in detail in [7] and [8].

The experimental results were obtained for a 3.3 kV induction motor, whose parameters are summarized in Sect. 2.5.1. In fact, the parameters of this electrical machine form the basis of the NPC inverter drive case study, which is used throughout this book. During the experiments, power levels of up to 1 MVA were achieved, as described in [9]. For the five-level ANPC inverter, an implementation of MPDTC on a small field-programmable gate array (FPGA) is a feasible task [10] when using the short switching horizon *eSE*. Experimental results have, however, not yet been reported.

MPDTC can be adapted to other power electronic topologies and machines. An *LC* filter is placed between the NPC inverter and the induction machine in [11]. MPDTC can be adapted to this drive setup by controlling the virtual inverter torque and the virtual inverter flux instead of the corresponding machine quantities. The extension of MPDTC to a five-level inverter driving a high-frequency induction machine was shown in [12]. Each phase leg of the inverter comprises a three-level H-bridge module. Alternatively, MPDTC can be used for permanent magnet synchronous machines (PMSMs). Performance improvements over DTC that are similar to those discussed in Sect. 8.1 can be achieved [13].

Irrespective of the successful implementation of MPDTC with short switching horizons, MPDTC with long horizons and full enumeration is computationally demanding and requires a powerful control platform. To facilitate the implementation of MPDTC on a control platform with modest computational power, techniques from mathematical programming such as branch-and-bound can be used. As exemplified in Chap. 10, such techniques reduce the computation time by an order of magnitude while causing only a negligible impact on the closed-loop performance.

Appendix 8.A: Controller Model of the ANPC Inverter Drive System

The continuous-time state-space matrices of the five-level ANPC drive model (8.7) and (8.9) are derived in this appendix. The system matrix of the machine $\boldsymbol{F}_m$ is the same as in (7.A.1a). The three-phase stator voltage $\boldsymbol{v}_{s,abc}(\boldsymbol{s}, \boldsymbol{x}_i)$ is constructed from the phase capacitor voltages $v_{\text{ph},x}$ and the neutral point potential v_n, which are part of the inverter state vector $\boldsymbol{x}_i$. Based on v_n and the total dc-link voltage v_{dc}, the upper and lower dc-link voltages $v_{\text{dc,up}} = 0.5v_{\text{dc}} - v_n$ and $v_{\text{dc,lo}} = 0.5v_{\text{dc}} + v_n$ are derived. Table 2.7 provides then $\boldsymbol{v}_{s,abc}(\boldsymbol{s}, \boldsymbol{x}_i)$.

The input matrix transforms the three-phase stator voltage into orthogonal coordinates and applies the transformed stator voltage to the stator flux dynamics. This leads to

$$\boldsymbol{G}_m = \frac{1}{3}\begin{bmatrix} 2 & -1 & -1 \\ 0 & \sqrt{3} & -\sqrt{3} \\ 0 & 0 & 0 \\ 0 & 0 & 0 \end{bmatrix}. \tag{8.A.1}$$

To derive the inverter model, we introduce for the phase capacitor dynamics the auxiliary logic variables

$$\delta_{\text{ph},x} = \begin{cases} 1, & \text{if } s_x \in \{2,6\} \\ -1, & \text{if } s_x \in \{1,5\} \\ 0, & \text{else,} \end{cases} \tag{8.A.2}$$

with $x \in \{a,b,c\}$. This allows us to write the differential equations of the capacitor voltages (2.91) in the compact form

$$\frac{dv_{\text{ph},x}}{dt} = \frac{1}{X_{\text{ph}}}\delta_{\text{ph},x} i_{sx}. \tag{8.A.3}$$

Recall that X_{ph} denotes the pu equivalence of the phase capacitance, and i_{sx} is the stator current in phase x.

Similarly, we introduce the three logic variables

$$\delta_{nx} = \begin{cases} 1, & \text{if } s_x \in \{2,3,4,5\} \\ 0, & \text{else} \end{cases} \tag{8.A.4}$$

for the neutral point potential and rewrite the differential equation (2.93) as

$$\frac{dv_n}{dt} = -\frac{1}{2X_{\text{dc}}}(\delta_{na} i_{sa} + \delta_{nb} i_{sb} + \delta_{nc} i_{sc}). \tag{8.A.5}$$

The three-phase stator current can be expressed in terms of the stator and rotor flux vectors in orthogonal coordinates. These flux vectors form the state vector of the machine $\boldsymbol{x}_m$. Combining (2.53) with the pseudo-inverse of the Clarke transformation $\tilde{\boldsymbol{K}}^{-1}$ allows us to write

$$\boldsymbol{i}_{s,abc} = \tilde{\boldsymbol{K}}^{-1} \frac{1}{D} \left[X_r \boldsymbol{I}_2 \;\; -X_m \boldsymbol{I}_2 \right] \boldsymbol{x}_m, \tag{8.A.6}$$

where $\boldsymbol{I}_2$ denotes the 2×2 identity matrix.

The continuous-time system matrix of the inverter

$$\boldsymbol{F}_i(\boldsymbol{s}) = \begin{bmatrix} \frac{\delta_{\text{ph},a}}{X_{\text{ph}}} & 0 & 0 \\ 0 & \frac{\delta_{\text{ph},b}}{X_{\text{ph}}} & 0 \\ 0 & 0 & \frac{\delta_{\text{ph},c}}{X_{\text{ph}}} \\ -\frac{\delta_{na}}{2X_{\text{dc}}} & -\frac{\delta_{nb}}{2X_{\text{dc}}} & -\frac{\delta_{nc}}{2X_{\text{dc}}} \end{bmatrix} \tilde{\boldsymbol{K}}^{-1} \frac{1}{D} \left[X_r \boldsymbol{I}_2 \;\; -X_m \boldsymbol{I}_2 \right] \tag{8.A.7}$$

is obtained by inserting (8.A.6) into (8.A.3) and (8.A.5). This matrix is dependent on the auxiliary logic variables, which in turn depend on the three-phase switch position $\boldsymbol{s}$.

The vector-valued output function of the complete drive model is given by

$$\boldsymbol{h}(\boldsymbol{x}) = \left[\frac{1}{\text{pf}} \frac{X_m}{D}(x_2 x_3 - x_1 x_4) \;\; \sqrt{x_1^2 + x_2^2} \;\; x_5 \;\; x_6 \;\; x_7 \;\; x_8 \right]^T, \tag{8.A.8}$$

where x_j refers to the jth component of the state vector $\boldsymbol{x} = [\boldsymbol{x}_m^T \; \boldsymbol{x}_i^T]^T$.

References

[1] F. Kieferndorf, M. Basler, L. Serpa, J.-H. Fabian, A. Coccia, and G. Scheuer, "A new medium voltage drive system based on ANPC-5L technology," in *Proceedings of IEEE International Conference on Industrial Technology* (Viña del Mar, Chile), pp. 605–611, Mar. 2010.

[2] P. Barbosa, P. Steimer, J. Steinke, L. Meysenc, M. Winkelnkemper, and N. Celanovic, "Active neutral-point-clamped multilevel converters," in *Proceedings of IEEE Power Electronics Specialists Conference* (Recife, Brasil), pp. 2296–2301, Jun. 2005.

[3] A. Nabae, I. Takahashi, and H. Akagi, "A new neutral-point-clamped PWM inverter," *IEEE Trans. Ind. Appl.*, vol. IA-17, pp. 518–523, Sep./Oct. 1981.

[4] T. Brückner, S. Bernet, and H. Guldner, "The active NPC converter and its loss-balancing control," *IEEE Trans. Ind. Electron.*, vol. 52, pp. 855–868, Jun. 2005.

[5] T. Meynard and H. Foch, "Multilevel conversion: High voltage choppers and voltage source inverters," in *Proceedings of IEEE Power Electronics Specialists Conference*, pp. 397–403, Jun. 1992.

[6] J. Meili, S. Ponnaluri, L. Serpa, P. K. Steimer, and J. W. Kolar, "Optimized pulse patterns for the 5-level ANPC converter for high speed high power applications," in *Proceedings of IEEE Industrial Electronics Society Annual Conference*, pp. 2587–2592, 2006.
[7] T. Geyer, *Low complexity model predictive control in power electronics and power systems*. PhD thesis, Automatic Control Laboratory ETH Zurich, 2005.
[8] T. Geyer, G. Papafotiou, and M. Morari, "Model predictive direct torque control—Part I: Concept, algorithm and analysis," *IEEE Trans. Ind. Electron.*, vol. 56, pp. 1894–1905, Jun. 2009.
[9] G. Papafotiou, J. Kley, K. G. Papadopoulos, P. Bohren, and M. Morari, "Model predictive direct torque control—Part II: Implementation and experimental evaluation," *IEEE Trans. Ind. Electron.*, vol. 56, pp. 1906–1915, Jun. 2009.
[10] J. Vallone, T. Geyer, and E. Rohr, "FPGA-based model predictive control," EP patent application 15 172 241 A1, 2015.
[11] S. Mastellone, G. Papafotiou, and E. Liakos, "Model predictive direct torque control for MV drives with *LC* filters," in *Proceedings of European on Power Electronics Conference* (Barcelona, Spain), pp. 1–10, Sep. 2009.
[12] T. Geyer and G. Papafotiou, "Model predictive direct torque control of a variable speed drive with a five-level inverter," in *Proceedings of IEEE Industrial Electronics Society Annual Conference* (Porto, Portugal), pp. 1203–1208, Nov. 2009.
[13] T. Geyer, G. A. Beccuti, G. Papafotiou, and M. Morari, "Model predictive direct torque control of permanent magnet synchronous motors," in *Proceedings of IEEE Energy Conversion Congress and Exposition* (Atlanta, GA, USA), pp. 199–206, Sep. 2010.

9

Analysis and Feasibility of Model Predictive Direct Torque Control

Direct torque control (DTC) and model predictive direct torque control (MPDTC) impose upper and lower bounds on the torque and stator flux magnitude. As shown in Sect. 9.1, these bounds can be translated into equivalent bounds on the stator flux components in orthogonal coordinates. This gives rise to the target set within which DTC and MPDTC maintain the stator flux vector. Section 9.2 derives and visualizes the state-feedback control law of MPDTC around this target set. The control law is computed in an offline procedure. The availability of the control law allows one to analyze the controller, and to illustrate and better understand its behavior and decision-making process.

Similar to DTC, the MPDTC algorithm occasionally runs into situations in which the control problem does not permit a solution and is thus infeasible. These so-called *deadlocks* refer to instances in which no switching sequence exists that keeps the controlled variables within their bounds or, when a bound has been violated, reduces the bound violation. As with DTC, these deadlocks are induced by the imposition of bounds on the controlled variables and the fact that the number of available voltage vectors is finite. The root cause of deadlocks is analyzed in Sect. 9.3.

In the case of a deadlock, one option is to relax the bounds and to minimize the predicted bound violation instead of the switching effort. This so-called *deadlock resolution* strategy, which is proposed in Sect. 9.4, is executed until the deadlock has been resolved. However, the execution of this resolution strategy often leads to a spike in the instantaneous switching frequency, which we refer to as a *switching burst*. In the worst case scenario, such a switching burst could lead to a trip of the drive. Section 9.5 proposes methods that aim to avoid deadlocks and their associated switching bursts. The proposed methods are based on the notions of terminal weights and terminal constraints. These measures drastically reduce the likelihood of deadlocks and—in many cases—avoid them altogether.

Throughout this chapter, we will consider a three-level, neutral-point-clamped (NPC) inverter with a medium-voltage (MV) induction machine as the basis for our analysis. The MPDTC control problem of this drive system is to maintain the electromagnetic torque, the stator flux magnitude, and the floating neutral point potential within given bounds.

Model Predictive Control of High Power Converters and Industrial Drives, First Edition. Tobias Geyer.

Companion Website: www.wiley.com/go/geyermodelpredictivecontrol

9.1 Target Set

We start by introducing the concept of the target set for MPDTC. The target set will be instrumental when deriving and analyzing the state-feedback control law, and when analyzing and avoiding deadlocks.

For a given rotor flux vector $\boldsymbol{\psi}_r$ with the magnitude $\Psi_r = \|\boldsymbol{\psi}_r\|$, the references on the electromagnetic torque T_e^* and stator flux magnitude Ψ_s^* can be translated into an equivalent reference for the stator flux vector $\boldsymbol{\psi}_s^*$. To accomplish this, it is convenient to work in a dq reference frame, whose d-axis is aligned with the rotor flux vector, as shown in Fig. 9.1. Using the expressions (7.6) and (7.7) for the torque and stator flux magnitude, it is straightforward to derive the d- and q-components of the stator flux reference vector:

$$\psi_{sd}^* = \sqrt{(\Psi_s^*)^2 - \left(\text{pf}\,\frac{D}{X_m \Psi_r} T_e^*\right)^2} \tag{9.1a}$$

$$\psi_{sq}^* = \frac{\text{pf}\,D}{X_m \Psi_r} T_e^*. \tag{9.1b}$$

In Sect. 7.3.2, we defined $T_{e,\min}$ and $T_{e,\max}$ as the lower and upper bounds on the electromagnetic torque. The bounds on the magnitude of the stator flux vector were defined accordingly as $\Psi_{s,\min}$ and $\Psi_{s,\max}$. The bounds on the neutral point potential are given by $v_{n,\min}$ and $v_{n,\max}$.

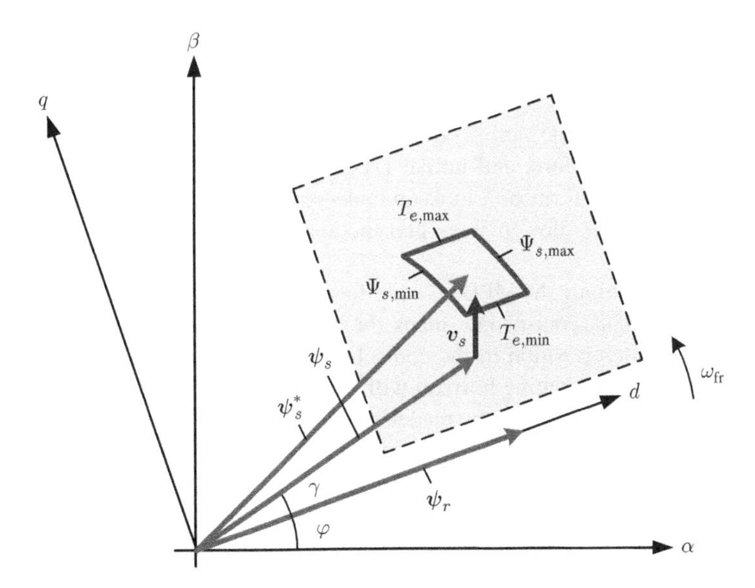

Figure 9.1 Stator and rotor flux vectors $\boldsymbol{\psi}_s$ and $\boldsymbol{\psi}_r$ in the dq reference frame, which rotates with the angular velocity ω_{fr}. The target set around the stator flux reference $\boldsymbol{\psi}_s^*$ is indicated by solid lines, which correspond to the upper and lower bounds on the torque and stator flux magnitude, respectively. The stator flux vector is driven into the target set by the voltage vector $\boldsymbol{v}_s$. The dashed lines indicate the rectangular set for which the state-feedback control law will be derived

The neutral point potential at time t is obtained by integrating (7.3), which leads to

$$v_n(t) = v_n(t_0) + \frac{1}{2X_c}\int_{t_0}^{t} |\boldsymbol{u}(\tau)|^T \boldsymbol{i}_{s,abc}(\tau)\mathrm{d}\tau. \tag{9.2}$$

As the neutral point potential is the integral of the component-wise absolute value of the three-phase switch position $\boldsymbol{u} = [u_a\ u_b\ u_c]^T$ multiplied with the three-phase stator currents $\boldsymbol{i}_{s,abc}$, it cannot be easily represented in the stator flux plane. We thus restrict the discussion in this and the following section to the electromagnetic torque and the stator flux magnitude.

The upper and lower torque and flux magnitude bounds can be translated into the stator flux plane with the help of (9.1). Because of the linear relationship between the torque and the q-component of the stator flux vector (see (9.1b)), the torque bounds form lines that run parallel to the rotor flux vector. The upper and lower bounds of the stator flux magnitude form concentric circles around the origin. As depicted in Fig. 9.1, the four bounds form a *target set* for the stator flux vector. Keeping the latter within this set is equivalent to maintaining the electromagnetic torque and the stator flux magnitude within their respective upper and lower bounds. When this is achieved, the desired torque will be generated and the machine will be appropriately magnetized.

At steady-state operating conditions, the target set rotates in synchronism with the rotor flux vector, while it remains stationary within the dq reference frame. During transients, such as torque steps, the target set is moved along the q-axis. In the presence of large steps, this movement will result in the stator flux vector violating the target set. To ensure a minimum torque settling time and to avoid too high or too low a stator flux magnitude, the stator flux vector needs to be driven back into the target set as quickly as possible by an appropriate voltage vector $\boldsymbol{v}_s$.

9.2 The State-Feedback Control Law

MPDTC is based on an online optimization stage that computes a suitable inverter switch position in real time. Therefore, and unlike DTC, the MPDTC control law is not directly available—for example, in form of a look-up table—and thus cannot be analyzed and illustrated. This complicates the design stage and the analysis of the MPDTC decision-making process.

To rectify this shortcoming, the MPDTC state-feedback control law will be computed, illustrated, and analyzed. This control law maps the state vector over the state space and the previously applied switch position to the control input (the switch position). The impact of varying the length of the switching horizon will also be shown. The information and insight obtained is not only meant to further the reader's understanding of MPDTC, but it is also envisioned that this will facilitate future revisions and improvements of the MPDTC algorithm, for example, by further reducing its computational burden.

When formulating model predictive control (MPC) problems for linear and piecewise affine (linear plus offset) systems with piecewise affine constraints, the state-feedback control law can be computed in a mathematically elegant way. The resulting control input is piecewise affine in the state vector. Specifically, the state space is divided into regions,[1] and the control input of each region is an affine function of the state vector. For more details on the

[1] If the 1- or ∞-norm is used in the cost function, the regions are polyhedra, whereas for the 2-norm, non-convex ellipsoidal regions may emerge.

state-feedback control law of MPC for piecewise affine systems, the reader is referred to [1] and the references therein. The multiparametric MATLAB toolbox [2] provides a powerful set of tools to compute and analyze such solutions.

For power electronics systems, state-feedback control laws can be derived by approximating the nonlinearities by piecewise affine functions, modeling the system in the mixed logical dynamical (MLD) framework [3], formulating the MPC control problem as a closed-form optimization problem, and using a modified version of the multiparametric toolbox to derive the state-feedback control law. Owing to its daunting computational complexity, this approach is feasible only for two-level inverters [4, 5] and dc–dc converters [6]. A problem-specific computational scheme, which exploits the structure of the drive control problem, is proposed in [7]. The derivation of a state-feedback control law for a simplified field-oriented control (FOC) problem is described in [8].

These standard techniques are not applicable for computing the state-feedback control law of MPDTC, because the MPDTC optimization problem (7.26) is solved in an approximate manner by adopting the notion of trajectory extension and considering switching transitions only when the output variables are close to their bounds. For details on the MPDTC algorithm, the reader is referred to Sect. 7.4.5.

9.2.1 *Preliminaries*

As a case study, consider an MV drive system consisting of a three-level NPC inverter with an induction machine. This case study is described in detail in Sect. 2.5.1 along with the corresponding parameters and the base quantities of the per unit (pu) system. To simplify the case study, the inverter is fed by two dc voltage sources, removing the need to actively balance the neutral point potential around zero. This allows us to remove the neutral point potential from the state and output vectors, reducing their dimensions by one.

Consider the cost function

$$J = J_{\text{sw}} + J_{\text{bnd}} + J_t \tag{9.3}$$

as defined in (7.30). We target the switching losses and set $J_{\text{sw}} = J_P$, which represents the switching power losses over the length of the prediction horizon (see (7.32)). The term $J_{\text{bnd}} = \boldsymbol{q}^T \boldsymbol{\epsilon}(k)$ penalizes the predicted root mean square (rms) bound violation of the output variables, which we define as

$$\boldsymbol{\epsilon}(k) = [\epsilon_T(k)\ \epsilon_\Psi(k)]^T, \tag{9.4}$$

similar to (7.3.3). We set the corresponding penalty to $\boldsymbol{q} = [2\ 2]^T$. The third term, J_t, is not used and set to zero.

The MPDTC algorithm minimizes the cost function (9.3) subject to the evolution of the controller model (7.26b)–(7.26c) and two additional sets of constraints: The switching constraints (7.26e)

$$\boldsymbol{u}(\ell) \in \boldsymbol{\mathcal{U}} \text{ and } \|\Delta \boldsymbol{u}(\ell)\|_\infty \leq 1, \tag{9.5}$$

with $\boldsymbol{\mathcal{U}} = \{-1, 0, 1\}^3$ and $\Delta \boldsymbol{u}(\ell) = \boldsymbol{u}(\ell) - \boldsymbol{u}(\ell - 1)$, and the constraints on the output variables (7.26d)

$$\begin{cases} \varepsilon_j(\ell+1) = 0, & \text{if } \varepsilon_j(\ell) = 0 \\ \varepsilon_j(\ell+1) < \varepsilon_j(\ell), & \text{if } \varepsilon_j(\ell) > 0. \end{cases} \tag{9.6}$$

Here, the nonnegative ε_j denotes the degree by which the jth output variable violates one of its bounds. In (7.28), for example, we had defined the degree of the torque bound violation as

$$\varepsilon_T(\ell) = \frac{1}{T_{e,\max} - T_{e,\min}} \begin{cases} T_e(\ell) - T_{e,\max} & \text{if } T_e(\ell) > T_{e,\max} \\ T_{e,\min} - T_e(\ell) & \text{if } T_e(\ell) < T_{e,\min} \\ 0 & \text{else}\,, \end{cases} \tag{9.7}$$

where ε_T is normalized by the bound width. The degree of the bound violation for the stator flux magnitude is defined accordingly and is denoted by ε_Ψ. The output constraint (9.6) ensures that the output variables at time step $\ell + 1$ are either kept within their bounds or, if a bound has been violated, are moved closer to the violated bound at every time step within the prediction horizon.

We restrict the analysis to operation at steady state. The dimension of the state space for which the control law is derived can be reduced thanks to the following observations. The machine operates at a constant rotor flux magnitude and at a constant speed. By treating these two quantities as parameters, the machine state can be fully described by the stator flux vector and the angular position of the rotor flux. This is conveniently done in the rotating dq reference frame, which is aligned with the rotor flux vector and rotates synchronously with it. The (redefined) state vector $\boldsymbol{x}(k)$ of the machine is given by the two components of the stator flux vector in dq, $\boldsymbol{\psi}_{s,dq}(k)$, and the rotor flux angle $\varphi(k)$. As the neutral point potential is fixed to zero, the inverter state is fully described by the switch position $\boldsymbol{u}(k-1)$, which was selected in the previous control cycle.

Unless otherwise stated, the operating point is at nominal speed $\omega_r = 1$ pu and rated torque, the rotor flux angle is zero, and the applied switch position is $\boldsymbol{u}(k-1) = [-1\ 0\ -1]^T$. The magnitude of the rotor flux vector is $\Psi_r = 0.92$ pu and the stator flux reference vector in the rotating dq reference frame is $\boldsymbol{\psi}^*_{s,dq} = [0.972\ 0.235]^T$ pu for the considered machine. The bounds on the electromagnetic torque are chosen as $T_{e,\min} = 0.85$ pu and $T_{e,\max} = 1.15$ pu, whereas the bounds on the stator flux magnitude are $\Psi_{s,\min} = 0.97$ pu and $\Psi_{s,\max} = 1.03$ pu. These values define the target set around $\boldsymbol{\psi}^*_{s,dq}$, which is shown in Fig. 9.1. The sampling interval is set to $T_s = 25$ µs.

In a first step, the control law is derived for a given rotor flux angle and for a subset of the state space. The boundary of this set is indicated by dashed lines in Fig. 9.1. The set is centered on the stator flux reference vector, with its edges parallel to the d- and q-axes. The lengths of its edges are (arbitrarily) chosen as 0.16 pu.

9.2.2 *Control Law for a Given Rotor Flux Vector*

The state-feedback control law constitutes the mapping from the stator flux vector in dq, the rotor flux angle, and the previously chosen switch position to the optimal three-phase switch position $\boldsymbol{u}_{\text{opt}}$. This can be written at time step k as

$$\boldsymbol{u}_{\text{opt}}(k) = \boldsymbol{f}_{\text{MPC}}(\boldsymbol{\psi}_{s,dq}(k), \varphi(k), \boldsymbol{u}(k-1)). \tag{9.8}$$

Recall that we assume steady-state operation and treat the magnitude of the rotor flux vector and the speed as parameters. The function $\boldsymbol{f}_{\text{MPC}}$ can be evaluated by executing the MPDTC algorithm, which is stated in Sect. 7.4.5.

In order to compute the control law in an offline procedure, we consider stator flux vectors within the rectangular set, which is indicated by the dashed lines in Fig. 9.1. We resort to the technique of gridding and generate a fine set of grid points along the d- and q-axes. These points correspond to stator flux components within the rectangular set. The grid points, along with the rotor flux angle φ, define the machine's state. For a given switch position $\boldsymbol{u}(k-1)$, the optimal control input $\boldsymbol{u}_{\text{opt}}(k)$ can be computed for each grid point, which yields the state-feedback control law. The latter can be stored in a look-up table.

9.2.2.1 State-Feedback Control Law for a Short Switching Horizon

Several control laws that resulted from this procedure are shown in Figs. 9.2 and 9.3 for MPDTC with the short switching horizon *SE*. The optimal three-phase switch positions $\boldsymbol{u}_{\text{opt}}(k)$ are plotted in the two-dimensional space that is spanned by the d- and q-components of the stator flux vector. The different shades of gray refer to different switch positions. As can be seen, neighboring stator flux vectors relate to the same switch position forming distinctive regions in the state space that share the same control input. The switch positions of these regions are indicated by the symbols +, 0, and −. For example, 00− refers to $\boldsymbol{u}_{\text{opt}}(k) = [0\ 0\ -1]^T$.

The target set is shown as the slightly curved parallelogram with solid lines. The arrows correspond to the voltage vectors in the dq frame. These arrows highlight the different velocities and directions in which the various voltage vectors drive the stator flux vector relative to the rotating dq reference frame. Specifically, the length of the arrows indicates the amount by which the stator flux vector is moved within 100 μs.

Moreover, selected predicted stator flux trajectories are shown for several regions. Every second sampling instant (i.e., every 50 μs) along the trajectories is indicated by a dot. These trajectories start at selected stator flux vectors and terminate when a bound is about to be violated, thus predicting that switching will be required at this point in the future. In Fig. 9.2(a), for example, the stator flux trajectory starting in the lower right region with $\boldsymbol{u}_{\text{opt}}(k) = [-1\ 1\ -1]^T$ is 53 steps or 1.325 ms long. Also note that in the dq reference frame, in general, voltage vectors move the stator flux along curved rather than straight trajectories.

9.2.2.2 Analysis and Observations

In the following, details about the individual control laws in Figs. 9.2 and 9.3 are provided. The control laws are based on the assumptions and settings stated in Sect. 9.2.1.

The resulting regions have clearly defined borders, within which the same control input (switch position) is used. This forms distinct areas in the state space. When the stator flux vector at time step k is within the target set, switching is not required and thus avoided. In Fig. 9.2(a), for example, $\boldsymbol{u}_{\text{opt}}(k) = [-1\ 0\ -1]^T$ is chosen, which results in the almost vertical stator flux trajectory. As a result, within the target set, the control law heavily depends on $\boldsymbol{u}(k-1)$, because this largely determines the switching losses and thus the overall cost. This characteristic of MPDTC will be explained in more detail later in this section.

The controller predicts when the target set will be violated and aims to switch such that any violation is avoided. As an example of this, consider the lower edge of the target set in Fig. 9.2(a), which refers to the lower torque bound. Instead of choosing

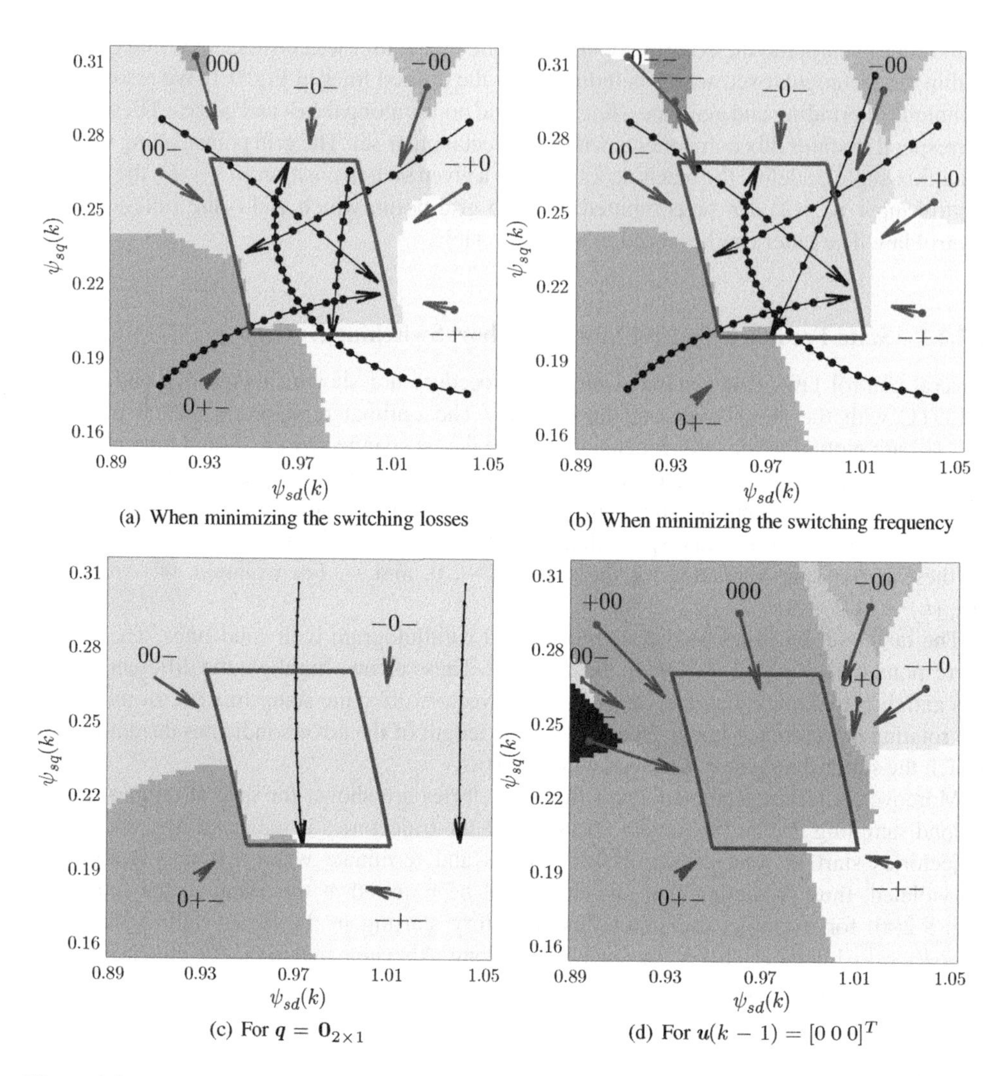

Figure 9.2 State-feedback control laws, that is, inverter switch position $\boldsymbol{u}_{\text{opt}}(k)$, as a function of the stator flux vector $\boldsymbol{\psi}_{s,dq}(k)$, the rotor flux angle $\varphi(k)$, and the inverter switch position $\boldsymbol{u}(k-1)$. The rotor flux angle is $\varphi(k) = 0°$, the speed operating point is $\omega_r = 1$ pu, and the applied switch position is $\boldsymbol{u}(k-1) = [-1\ 0\ -1]^T$, unless otherwise stated. Predicted stator flux trajectories are shown as curved lines with dots, while the target set is indicated by the parallelogram. The arrows indicate the voltage vectors

$\boldsymbol{u}_{\text{opt}}(k) = [-1\ 0\ -1]^T$ throughout the target set, switching is performed preemptively when the stator flux is one sampling interval away from the lower torque bound. This time interval translates to different distances in the state space, depending on the velocity of the voltage vector relative to the dq frame. This can be observed when comparing Figs. 9.2(a) and (d). The default voltage vector $\boldsymbol{u}_{\text{opt}}(k) = [0\ 0\ 0]^T$ in Fig. 9.2(d) points roughly in the same direction as $\boldsymbol{u}_{\text{opt}}(k) = [-1\ 0\ -1]^T$, but its velocity is (relative to the reference frame)

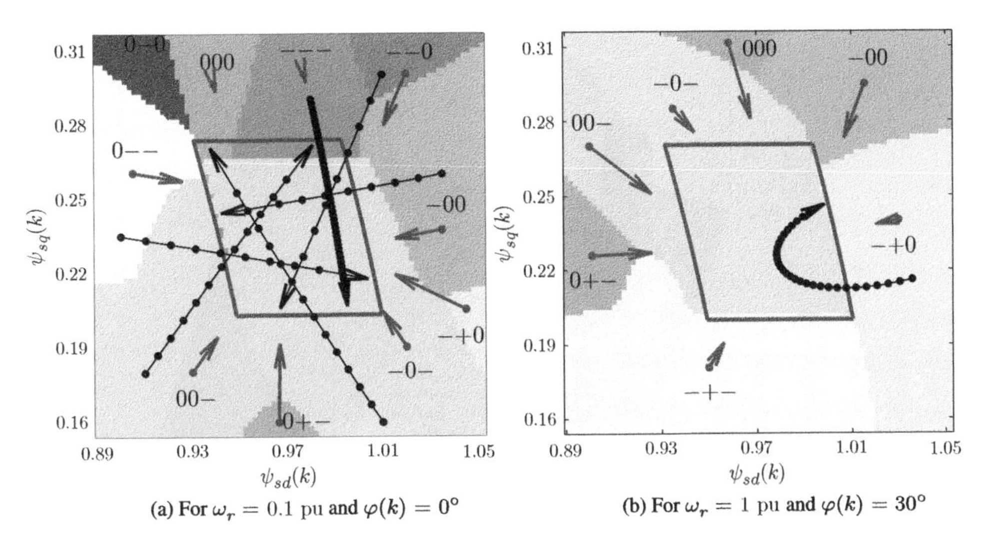

(a) For $\omega_r = 0.1$ pu and $\varphi(k) = 0°$

(b) For $\omega_r = 1$ pu and $\varphi(k) = 30°$

Figure 9.3 State-feedback control laws, that is, inverter switch position $\boldsymbol{u}_{\text{opt}}(k)$, as a function of the stator flux vector $\boldsymbol{\psi}_{s,dq}(k)$, the rotor flux angle $\varphi(k)$, and the inverter switch position $\boldsymbol{u}(k-1)$. Predicted stator flux trajectories are shown as curved lines with dots, while the target set is indicated by the parallelogram. The arrows indicate the voltage vectors

significantly higher. As a result, the band around the lower torque bound, in which switching is performed, is accordingly larger in Fig. 9.2(d).

When the stator flux vector significantly violates the target set, however, the differential mode of the control laws tend to become similar, irrespective of $\boldsymbol{u}(k-1)$. This can be seen when comparing Figs. 9.2(a) and (d), which only differ with respect to $\boldsymbol{u}(k-1)$. As an example for this, consider the region with $\boldsymbol{u}_{\text{opt}}(k) = [0\ 1\ 0]^T$ in Fig. 9.2(d), which corresponds to the region with $\boldsymbol{u}_{\text{opt}}(k) = [-1\ 0\ \ -1]^T$ in Fig. 9.2(a). The voltage vectors have the same differential-mode voltage but a different common mode.

The reason for this characteristic is that the bound violation term J_{bnd} dominates over the switching effort in the cost function when the stator flux vector is well outside its bounds. As J_{bnd} is independent of $\boldsymbol{u}(k-1)$, switching is performed almost regardless of the applied switch position. Moreover, the stator flux vector is manipulated by the differential-mode component of the voltage vector, not its common-mode content.

When minimizing the switching frequency instead of the switching losses, only minor alterations in the resulting control law result, as shown in Fig. 9.2(b). Differences arise mostly with regard to the common mode of the voltage vectors, as can be seen in the upper left corner of the figure. When a switching transition from $\boldsymbol{u}(k-1) = [-1\ 0\ \ -1]^T$ to a zero vector is required, two options exist, namely $\boldsymbol{u}(k) = [-1\ \ -1\ \ -1]^T$ and $\boldsymbol{u}(k) = [0\ 0\ 0]^T$. The first option involves only one switching transition, which is preferable when minimizing the switching frequency. The second option involves two switching transitions with—in this particular case—very small currents in the corresponding phases. Therefore, when minimizing the switching losses, it is here advantageous to switch twice.

These differences are also reflected in Fig. 9.4, which shows the predicted switching efforts for the two control laws discussed earlier. The predicted switching losses in kilowatts are

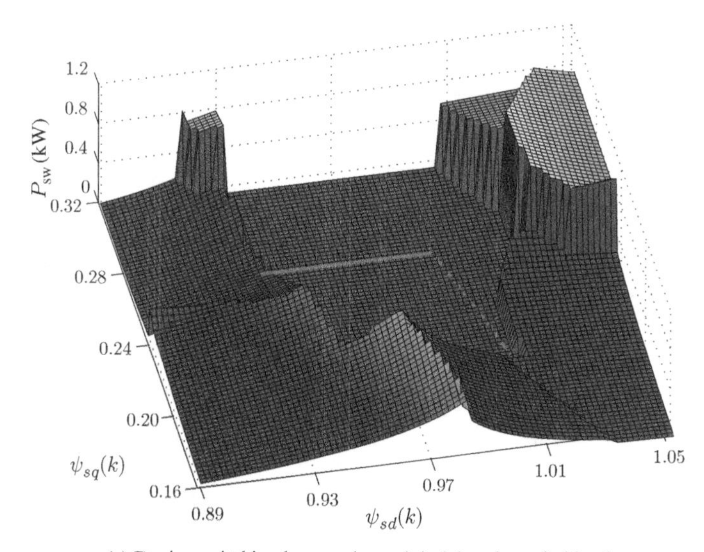

(a) Device switching losses when minimizing the switching losses

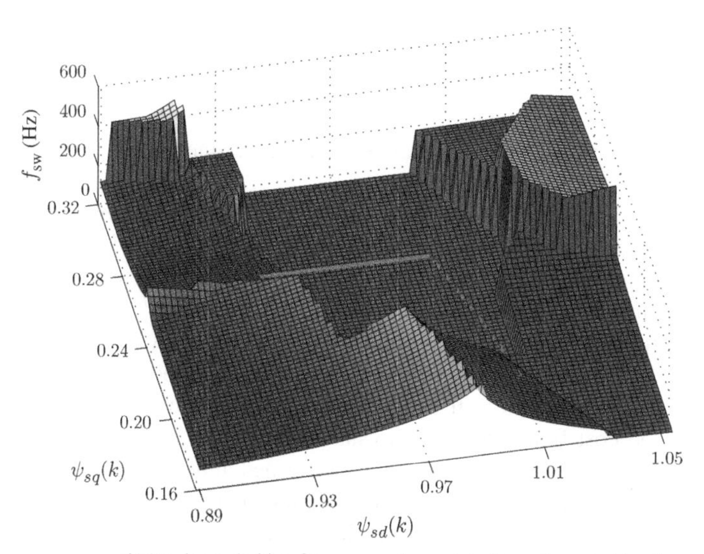

(b) Device switching frequency when minimizing the switching frequency

Figure 9.4 Predicted switching effort, discounted over the prediction horizon, as a function of the stator flux vector $\boldsymbol{\psi}_{s,dq}(k)$, the rotor flux angle $\varphi(k) = 0°$, and the switch position $\boldsymbol{u}(k-1) = [-1\ 0\ -1]^T$. The target set is indicated by the parallelogram. The two figures relate to Fig. 9.2(a) and (b), respectively

obtained by dividing J_{sw} by 1000 T_s. A subsequent division by 12 yields the average switching losses per semiconductor device,[2] which are depicted in Fig. 9.4(a). The device switching frequency is obtained accordingly.

[2] Recall that an NPC inverter is used with 12 integrated-gate-commutated thyristors (IGCTs).

It can be seen that the surfaces of the switching efforts are smooth within each region. When moving from one region to a neighboring one, the transition is smooth, provided that both control laws meet the constraint (9.6) at the intersection. As an example, consider the regions with the control inputs $\boldsymbol{u}_{\text{opt}}(k) = [0\ 1\ -1]^T$ and $\boldsymbol{u}_{\text{opt}}(k) = [-1\ 1\ -1]^T$. If, however, one of the control inputs ceases to meet the constraint (9.6), then, when moving from one region to a neighboring one, the switching effort at the transition changes in a stepwise manner. This can be seen at the boundary between the regions with $\boldsymbol{u}_{\text{opt}}(k) = [-1\ 0\ -1]^T$ and $\boldsymbol{u}_{\text{opt}}(k) = [0\ 1\ -1]^T$. When moving from the first region to the second one, the control input ceases to meet this constraint, triggering a switching transition and a stepwise change in the switching effort.

Next, consider the control law depicted in Fig. 9.2(c), which is obtained by setting the weight q to zero. As a result, only the switching losses are penalized, but no incentive is provided to move the stator flux vector quickly back into the target set. This greatly enlarges the region in which the previously applied control input is maintained, that is, $\boldsymbol{u}_{\text{opt}}(k) = \boldsymbol{u}(k-1)$. In this region, as exemplified for the two predicted stator flux trajectories shown in Fig. 9.2(c), the degree of the bound violation decreases at every time step. The second constraint in (9.6) is thus met, but the convergence rate is small for the right trajectory. Note that this trajectory terminates when the lower torque bound—and hence the constraint (9.6)—is about to be violated.

Figure 9.3(a) shows the control law when lowering the speed to $\omega_r = 0.1$ pu. The stator flux trajectories are now effectively straight lines, and the zero voltage vector leads to a very slow stator flux movement relative to the dq reference frame.

So far, we have investigated control laws only for the case where the rotor flux angle is $\varphi(k) = 0°$. Figure 9.3(b) shows the control law for $\varphi(k) = 30°$ at nominal speed. When compared to the control law for $\varphi(k) = 0°$ in Fig. 9.2(a), the nonzero voltage vectors are rotated and the regions are deformed accordingly.

9.2.2.3 Illustration of the Control Law Derivation

We now provide additional insight into the derivation of the state-feedback control law. For this, consider in Fig. 9.2(a) the control law along the (not shown) line given by $\psi_{sd} \in [0.89, 1.05]$ pu and $\psi_{sq} = 0.235$ pu, which corresponds to the torque reference. This line is equivalent to a one-dimensional slice through the stator flux plane. As mentioned previously, the applied switch position is $\boldsymbol{u}(k-1) = [-1\ 0\ -1]^T$, from which transitions to 11 different switch positions are possible, in accordance with the constraint (9.5). We consider only four options in Fig. 9.5. We either keep the switch position that is currently applied, or we switch to one of the three new switch positions by considering

$$\boldsymbol{u}(k) \in \{\boldsymbol{u}_1, \boldsymbol{u}_2, \boldsymbol{u}_3, \boldsymbol{u}_4\} \tag{9.9}$$

with $\boldsymbol{u}_1 = [-1\ 0\ -1]^T$, $\boldsymbol{u}_2 = [0\ 0\ -1]^T$, $\boldsymbol{u}_3 = [-1\ 1\ -1]^T$ and $\boldsymbol{u}_4 = [0\ 1\ -1]^T$.

For certain stator flux vectors, some of these switch positions lead to a violation of the constraint (9.6) and thus cannot be applied. Keeping the applied switch position for $\psi_{sd} < 0.94$ pu, for example, would violate the constraint (9.6).

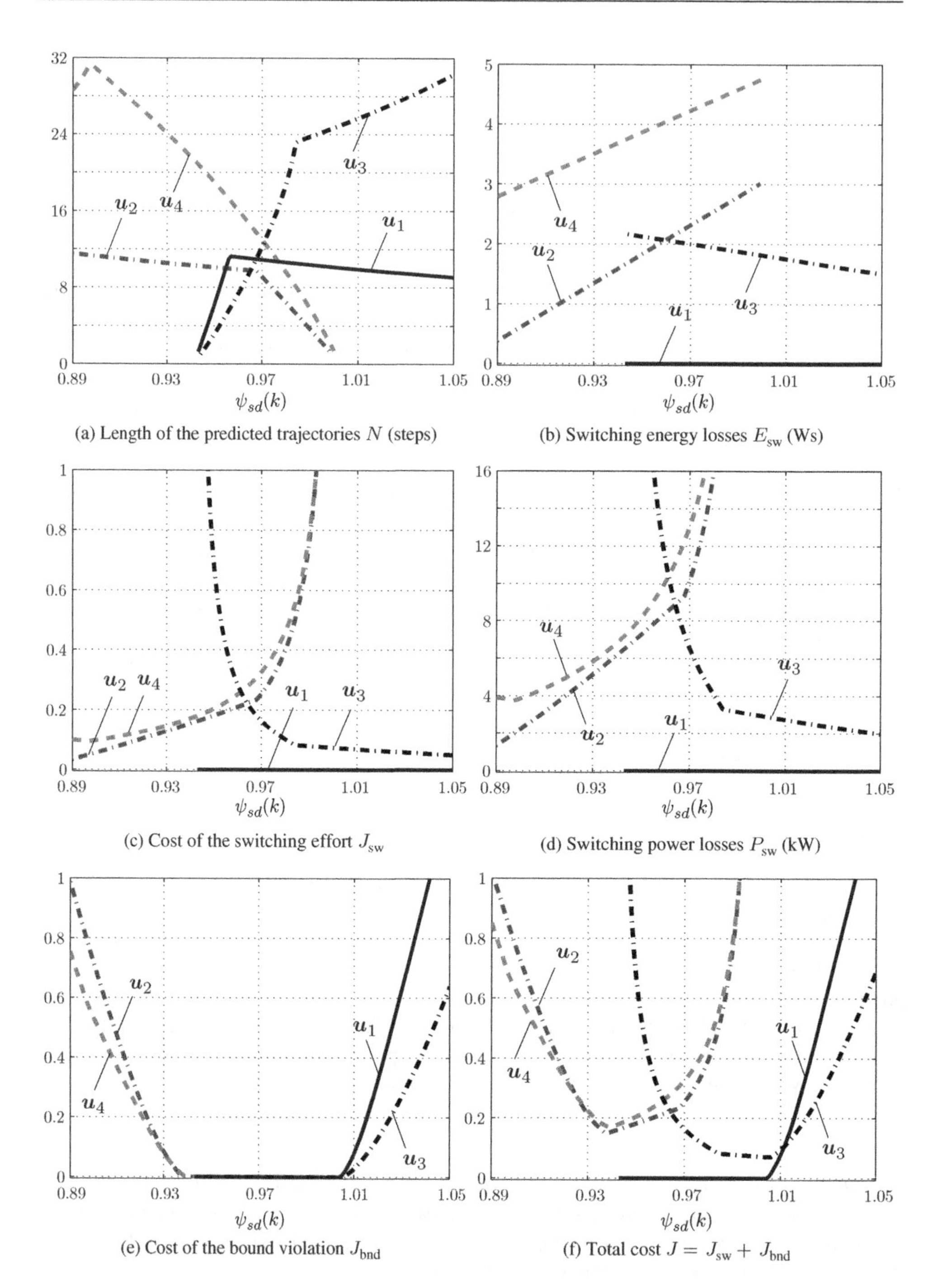

Figure 9.5 Visualization of the control law derivation along the line $\psi_{sd} \in [0.89, 1.05]$ pu and $\psi_{sq} = 0.235$ pu in Fig. 9.2(a), with MPDTC minimizing the switching losses and the switching horizon *SE*. Starting with the switch position $\boldsymbol{u}(k-1) = [-1\ 0\ -1]^T$, 4 out of the 12 possible switch positions are considered at time step k: $\boldsymbol{u}_1$ (no switching), $\boldsymbol{u}_2$ (switching in phase a), $\boldsymbol{u}_3$ (switching in phase b), and $\boldsymbol{u}_4$ (switching in phases a and b)

Figure 9.5(a) shows the lengths of the predicted stator flux trajectories. These lines are slightly curved because of the rotation of the reference frame. The distinctive changes in the slopes are caused by changes in the bound at which the trajectory terminates. Below $\psi_{sd} = 0.955\,\text{pu}$, for example, the stator flux trajectory corresponding to $\boldsymbol{u}_1$ terminates at the lower flux bound, while above this threshold it terminates at the lower torque bound (see also Fig. 9.2(a)).

The switching energy losses (in watt-second) depend on the commutated stator current, which is a linear combination of the stator and rotor flux vectors. The switching energy losses thus depend linearly on the stator flux components. This is confirmed by the distinctively straight lines in Fig. 9.5(b).

The cost of the switching effort J_{sw} in Fig. 9.5(c) is obtained by dividing the switching energy losses by the trajectory lengths, as explained earlier. As a result, these costs are—similar to the trajectory lengths—slightly curved lines with discontinuities. The switching power losses in Fig. 9.5(d) are obtained by scaling Fig. 9.5(c), as described in the previous section.

The cost of the bound violation J_{bnd} is zero as long as the stator flux trajectory remains within the target set. This is the case when the initial stator flux vector is within the set, as shown in Fig. 9.5(e). As the starting point of the stator flux trajectory moves away from the target set, the cost of the bound violation quickly increases as a result of the rms formulation used in J_{bnd} (see (7.33)). The slopes differ between the various switch positions, according to the predicted rms violation of the bounds. For $\psi_{sd} > 1.005\,\text{pu}$, for example, the switch position $\boldsymbol{u}(k) = [-1\;1\;\;-1]^T$ brings the stator flux vector back into the target set significantly faster than $\boldsymbol{u}(k) = [-1\;0\;\;-1]^T$ does. This is clearly visible in Fig. 9.2(a) and is illustrated in Fig. 9.5(e), in that the former switch position entails a lower penalty on the bound violation.

The total cost J in Fig. 9.5(f) is the sum of the costs of the switching effort and of the bound violation. By minimizing the total cost, the optimal control input $\boldsymbol{u}_{\text{opt}}(k)$ is derived. For $\psi_{sd} < 0.94\,\text{pu}$, $\boldsymbol{u}(k) = [0\;0\;\;-1]^T$ and $\boldsymbol{u}(k) = [0\;1\;\;-1]^T$ yield similar costs. The first switch position incurs a lower switching effort but is slower in bringing the stator flux vector back into the target set. Therefore, in the interval $0.925\,\text{pu} \leq \psi_{sd} < 0.94\,\text{pu}$, the former is chosen as the optimal control input $\boldsymbol{u}_{\text{opt}}(k)$, while the latter is optimal for $\psi_{sd} < 0.925\,\text{pu}$. Within the target set and when slightly violating the upper flux bound, that is, for $0.94\,\text{pu} \leq \psi_{sd} \leq 1.005\,\text{pu}$, it is optimal not to switch, that is, to use $\boldsymbol{u}_{\text{opt}}(k) = \boldsymbol{u}(k-1)$. For significant violations of the upper flux bound, that is, for $\psi_{sd} > 1.005\,\text{pu}$, $\boldsymbol{u}_{\text{opt}}(k) = [-1\;1\;\;-1]^T$ is optimal.

9.2.2.4 Analysis for Longer Switching Horizons

The analysis has so far focused on the switching horizon *SE*. Longer switching horizons are considered in the following, using *SESE* as an illustrative example. The same assumptions as previously mentioned are used, namely the switching losses are minimized, the rotor's angular position is $\varphi(k) = 0°$, and the previously applied switch position is $\boldsymbol{u}(k-1) = [-1\;0\;\;-1]^T$.

Example state-feedback control laws are shown in Fig. 9.6. Several predicted stator flux trajectories are shown, of which every second sampling instant is indicated by a small circle. Three features distinguish the control law with the switching horizon *SESE* from that with *SE*.

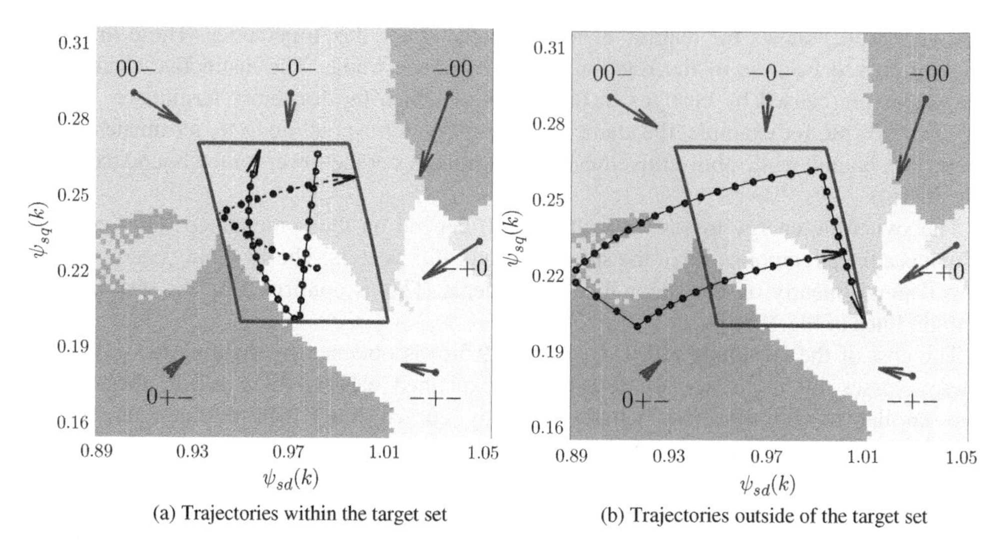

(a) Trajectories within the target set

(b) Trajectories outside of the target set

Figure 9.6 State-feedback control law for the switching horizon *SESE* and the inverter switch position $\boldsymbol{u}(k-1) = [-1\ 0\ -1]^T$. This figure corresponds to Fig. 9.2(a) with the switching horizon *SE*

First, switching is scheduled to be performed twice within the prediction horizon, namely at the current time step k and again when a bound is predicted to be hit. As a result, two different switch positions are used within the prediction horizon, leading to distinctive vertices in the predicted stator flux trajectories. The control law refers to the first switch position, that is, to the optimal switch position at time step k, $\boldsymbol{u}_{\text{opt}}(k)$. Assuming that the second switching transition occurs at time step $\ell > k$, the second predicted switch position $\boldsymbol{u}(\ell)$ cannot be directly observed from the control map in Fig. 9.6. It can be reconstructed, though, from the direction and velocity of the predicted stator flux trajectory. In general, $\boldsymbol{u}(\ell)$ does not coincide with the switch position $\boldsymbol{u}_{\text{opt}}(k)$ of the region in which the second switching is predicted to occur. As an example, consider the dotted predicted trajectory in Fig. 9.6(a) and its switching transition at the lower flux bound. The control law associated with the region in which this transition is predicted to occur is $\boldsymbol{u}_{\text{opt}}(k) = [0\ 0\ -1]^T$, while the second switch position is $\boldsymbol{u}(\ell) = [0\ 1\ -1]^T$.

Second, switching is also performed well within the target set, as can be seen in Fig. 9.6(a). Consider the predicted trajectory with the straight downward-pointing line, for which switching is postponed until the lower torque bound is about to be hit. When moving toward this bound, the number of steps over which the switching effort can be depreciated becomes smaller and smaller. At a certain point, switching preemptively becomes cheaper than further delaying the switching transition, because the switching energy losses associated with the new trajectory are likely to be low and can be depreciated over a long trajectory. As a result, the region with the control input $\boldsymbol{u}_{\text{opt}}(k) = [-1\ 1\ -1]^T$ is extended well into the target set. This is exemplified by the dotted trajectory. Therefore, when optimizing over multiple switching transitions, MPDTC may choose to switch preemptively. This is an important characteristic of MPDTC, which is due to the imposition of bounds, the discrete nature of the switch positions, and the variable-length prediction horizons.

Third, some regions may not have well-defined boundaries, as can be observed in Fig. 9.6(b) for the regions with $\boldsymbol{u}_{\text{opt}}(k) = [0\ 0\ -1]^T$ and $\boldsymbol{u}_{\text{opt}}(k) = [0\ 1\ -1]^T$. Two example trajectories are shown, which start from very similar stator flux positions and provide—despite their different switching sequences—very similar overall costs. By perturbing $\boldsymbol{\psi}_{s,dq}(k)$ slightly, either one or the other switching sequence is selected. This phenomenon results from the fact that MPDTC operates in the discrete-time domain and that the trajectory length is a natural rather than a real number. It is obvious that the length of the upper trajectory is very sensitive to small perturbations in $\boldsymbol{\psi}_{s,dq}(k)$—shifting $\boldsymbol{\psi}_{s,dq}(k)$ slightly along the d-axis, for example, has a major influence on the length of the downward-pointing second part of the trajectory. Reducing the length of the sampling interval mitigates this issue.

Both trajectories have effectively the same cost and thus provide the same performance, making the choice between the two irrelevant. It is, however, desirable that MPDTC adheres to a strategy once selected and avoids switching repeatedly between one strategy and another. This can be easily enforced by re-evaluating the control input only once a bound of the target set is about to be violated. This is achieved through the optional Step 1b in the MPDTC algorithm in Sect. 7.4.5. This policy also prevents MPDTC from preemptively switching when the stator flux vector is located well within the target window, as discussed in the previous paragraph.

9.2.3 *Control Law along an Edge of the Target Set*

We have seen in Sect. 9.2.2 that during steady-state operation, when the stator flux vector is kept within the target set and the switching horizon *SE* is used, switching is essentially performed only along the edges of the target set.

To gain insight into the dependence of the control law when varying the rotor flux angle, one can compute the control law for different angular positions of the rotor flux vector $\varphi(k)$, as exemplified in Fig. 9.3(b). An alternative approach is to compute the control law over a two-dimensional space that is spanned by the rotor flux angle and the position along one of the edges of the target set. This is done separately for each one of the four edges. The lower flux bound, for example, can be parameterized in polar coordinates using the amplitude $\Psi_s = \Psi_{s,\min}$ and the load angle $\gamma(k)$, which has been previously defined as the angle between the stator and rotor flux vectors. Therefore, for the lower flux bound, the control law can be derived as a function of the rotor flux angle $\varphi(k)$ and the load angle $\gamma(k)$.

The resulting control law is shown in Fig. 9.7. As expected, the control law for $\varphi(k) = 0°$ in Fig. 9.7 is identical to that in Fig. 9.2(a) along the lower flux bound (left edge) of the target set. The same holds true for $\varphi(k) = 30°$ and Fig. 9.3(b). Owing to the properties of symmetry, it suffices to compute the control law over an angle span of 60° for $\varphi(k)$ to fully characterize the controller.

9.3 Analysis of the Deadlock Phenomena

Despite the performance benefits of MPDTC, the algorithm occasionally runs into the so-called infeasible states or deadlocks. An infeasible state at time step k is the combination of a state vector $\boldsymbol{x}(k)$ and a switch position $\boldsymbol{u}(k-1)$ for which no switching sequence exists that meets

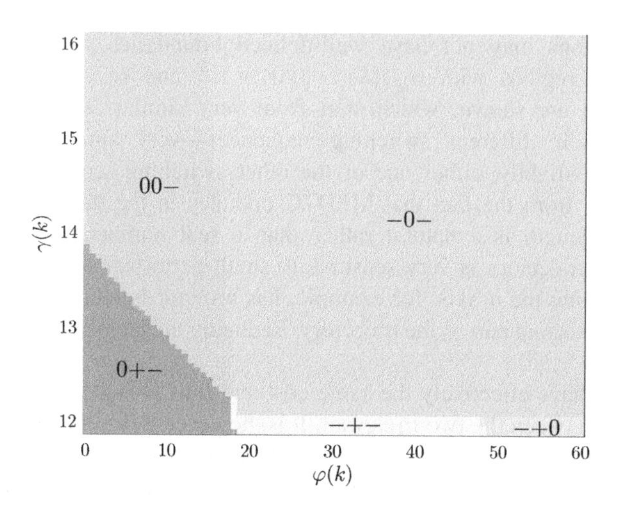

Figure 9.7 State-feedback control law along the lower flux bound of the target set for $\boldsymbol{u}(k-1) = [-1\ 0\ -1]^T$, where $\varphi(k)$ represents the angular position of the rotor flux vector and $\gamma(k)$ is the (load) angle between the stator and rotor flux vectors. Both angles are given in degrees. The switching horizon *SE* is used

the constraints (9.6) on the output variables. This implies that the control problem (7.26) is infeasible, not permitting a solution.

9.3.1 Root Cause Analysis of Deadlocks

In this section, we provide insight into the nature of the deadlocks along with a root cause analysis. We will see that deadlocks are caused by the combination of the output variables being constrained between upper and lower bounds and the fact that the switch positions are restricted to a finite and discrete-valued set. Constraints on the allowed switching transitions, which limit the set of voltage vectors that can be reached within one time step, further aggravate the problem, while long switching horizons alleviate it.

We focus again on MPDTC for a three-level inverter, but consider now the case study in Sect. 2.5.2, which is based on an NPC inverter comprising only one di/dt snubber per inverter half. Unlike stated in Sect. 2.5.2, we set the nominal dc-link voltage to 5.2 kV, which is equivalent to 1.93 pu. The parameters of the MV induction machine are given in Table 2.10. In the first step, we ignore the neutral point potential and only focus on the controlled machine quantities, that is, the electromagnetic torque and the stator flux magnitude.

We start by determining the set of voltage vectors that achieve a constant torque. To facilitate this, we neglect the discrete nature of the voltage vectors, that is, we assume that the inverter can produce at each phase terminal real-valued voltages between $-0.5v_{\text{dc}}$ and $0.5v_{\text{dc}}$. As a result, the discrete voltage vectors shown in Fig. 9.8 are *relaxed* to the set of real-valued vectors, which is enclosed by the dashed hexagon. The corner points of the hexagon have the Euclidean distance $2/3v_{\text{dc}}$ to the origin. Consider the orthogonal and rotating dq reference frame, which is aligned with the rotor flux vector $\boldsymbol{\psi}_r$ and rotates at the electrical angular speed of the rotor ω_r. In this reference frame, we denote the relaxed voltage vectors by $\tilde{\boldsymbol{v}}_s = [\tilde{v}_{sd}\ \tilde{v}_{sq}]^T$.

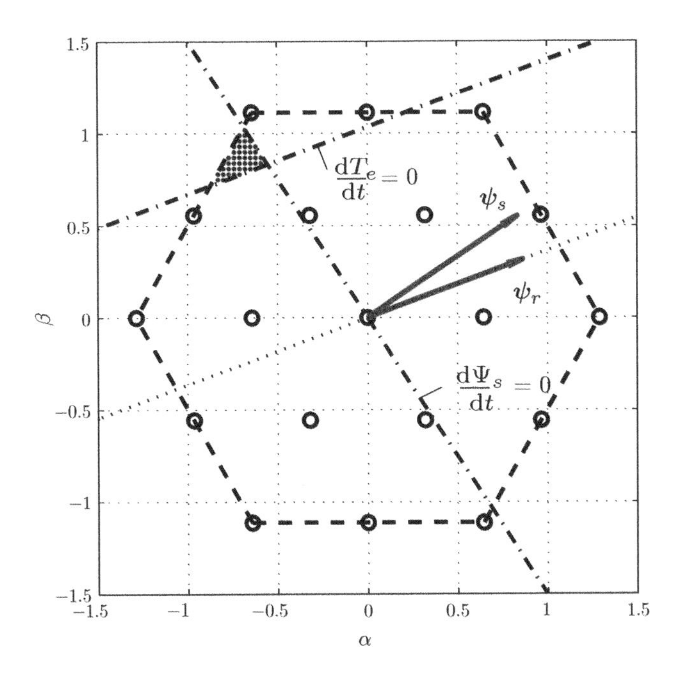

Figure 9.8 Geometric analysis of the root cause of deadlocks related to the torque and the stator flux magnitude. The (discrete) voltage vectors are shown as small circles, and the set of relaxed voltage vectors is enclosed by the dashed hexagon. Assuming nominal speed and rated torque, the constant torque and stator flux magnitude lines are shown as dash-dotted lines. The constant torque line is parallel to the rotor flux vector. The set of relaxed voltage vectors that increase the torque and reduce the stator flux magnitude is indicated by the dotted region

Recall the definition of the electromagnetic torque in (2.56)

$$T_e = \frac{1}{\text{pf}} \frac{X_m}{D} \boldsymbol{\psi}_r \times \boldsymbol{\psi}_s = \frac{1}{\text{pf}} \frac{X_m}{D} (\psi_{rd}\psi_{sq} - \psi_{rq}\psi_{sd}), \tag{9.10}$$

which is the cross product of the rotor and stator flux linkage vectors $\boldsymbol{\psi}_s = [\psi_{sd}\ \psi_{sq}]^T$ and $\boldsymbol{\psi}_r = [\psi_{rd}\ \psi_{rq}]^T$. The derivative of the torque with respect to time is given by

$$\frac{\mathrm{d}T_e}{\mathrm{d}t} = \frac{1}{\text{pf}} \frac{X_m}{D} \left(\psi_{rd} \frac{\mathrm{d}\psi_{sq}}{\mathrm{d}t} - \psi_{rq} \frac{\mathrm{d}\psi_{sd}}{\mathrm{d}t} \right), \tag{9.11}$$

where we have taken advantage of the fact that the derivatives of the rotor flux vector in the synchronously rotating reference frame are zero. We insert the d- and q-components of the stator flux derivative (2.55a) into (9.11). The stator and rotor resistances are typically very small in an MV setting and can thus be neglected for the purpose of this investigation. This leads to the torque derivative

$$\frac{\mathrm{d}T_e}{\mathrm{d}t} = \frac{1}{\text{pf}} \frac{X_m}{D} (\tilde{v}_{sq}\psi_{rd} - \tilde{v}_{sd}\psi_{rq} - \omega_r(\psi_{sd}\psi_{rd} + \psi_{sq}\psi_{rq})) \tag{9.12a}$$

$$= \frac{1}{\text{pf}} \frac{X_m}{D} (\boldsymbol{\psi}_r \times \tilde{\boldsymbol{v}}_s - \omega_r \boldsymbol{\psi}_s^T \boldsymbol{\psi}_r). \tag{9.12b}$$

Setting the torque derivative to zero simplifies (9.12) to

$$\boldsymbol{\psi}_r \times \tilde{\boldsymbol{v}}_s = \omega_r \boldsymbol{\psi}_s^T \boldsymbol{\psi}_r. \tag{9.13}$$

Considering the d- and q-components of the relaxed voltage vector as free variables, (9.13) describes a line in the dq-plane that is parallel to the rotor flux vector. The line's distance to the origin is given by $|\omega_r \boldsymbol{\psi}_s^T \boldsymbol{\psi}_r| / \|\boldsymbol{\psi}_r\|$. We refer to this line as the constant *torque* line. Voltage vectors that lie below the constant torque line (including the origin) decrease the torque, while voltage vectors beyond the line increase it.

At steady-state operating conditions, the torque and the stator flux magnitude are tightly kept within bounds around their references. This ensures that the inner product of the two fluxes in (9.13) is effectively constant. As a result, the distance of the constant torque line from the origin is proportional to the speed ω_r and therefore to the modulation index. Increasing the speed moves the constant torque line away from the origin, reducing the number of discrete-valued voltage vectors that are available to increase the torque if required. Thus the higher the angular speed, the more difficult the control problem becomes to solve.

The magnitude of the stator flux vector is

$$\Psi_s = \|\boldsymbol{\psi}_s\| = \sqrt{(\psi_{sd})^2 + (\psi_{sq})^2}, \tag{9.14}$$

and its derivative with respect to time can be computed as

$$\frac{\mathrm{d}\Psi_s}{\mathrm{d}t} = \frac{1}{\Psi_s}\left(\psi_{sd}\frac{\mathrm{d}\psi_{sd}}{\mathrm{d}t} + \psi_{sq}\frac{\mathrm{d}\psi_{sq}}{\mathrm{d}t}\right). \tag{9.15}$$

We use again the derivative of the stator flux vector (2.55a) and set the stator and rotor resistances to zero. This simplifies (9.15) to

$$\frac{\mathrm{d}\Psi_s}{\mathrm{d}t} = \frac{\psi_{sd}\tilde{v}_{sd} + \psi_{sq}\tilde{v}_{sq}}{\Psi_s} = \frac{\boldsymbol{\psi}_s^T \tilde{\boldsymbol{v}}_s}{\Psi_s}. \tag{9.16}$$

Setting the derivative of the stator flux magnitude to zero yields the compact expression

$$\boldsymbol{\psi}_s^T \tilde{\boldsymbol{v}}_s = 0. \tag{9.17}$$

The relaxed voltage vectors that fulfill (9.17) form a line in the dq-plane, which is perpendicular to the stator flux vector and passes through the origin. This line is referred to as the constant *stator flux magnitude* line. Voltage vectors that lie on the same side as the stator flux vector increase the stator flux magnitude, while voltage vectors that lie on the opposite side decrease it.

Example 9.1 *Consider the NPC inverter drive system operating at nominal speed and rated torque. Assume that the torque has hit its lower bound and is commanded to increase, while the stator flux magnitude has hit its upper bound, necessitating a decrease in the flux magnitude. Consider the stator and rotor flux vectors shown in Fig. 9.8, which relate to a fully magnetized machine and to the rated torque. The constant torque and stator flux magnitude lines are shown as dash-dotted lines.*

The set of (relaxed) voltage vectors that increase the torque and reduce the stator flux magnitude is indicated by the dotted area. As can be seen, in this example, the dotted set contains no discrete voltage vector. This implies that the MPDTC control problem permits no solution that fulfills the requirements imposed on the torque and stator flux magnitude. The control problem for the given flux vectors cannot be solved and is thus infeasible, giving rise to a deadlock.

The previous analysis focused on the torque and stator flux magnitude. When the neutral point potential is also taken into account and when it is close to one of its bounds, further restrictions are imposed on the voltage vectors. Depending on the sign of the phase current, each switch position that corresponds to at least one phase connected to the neutral point has a specific influence on the neutral point potential. Even if a voltage vector is available that satisfies the requirements for the torque and the stator flux magnitude, the corresponding switch position might lead to a violation of the neutral point potential's bounds. The addition of switching restrictions imposed by the use of two di/dt snubbers and the ruling out of switching transitions between the upper and the lower dc-link rails further reduce the set of available voltage vectors.

9.3.2 Location of Deadlocks

To determine the location of deadlocks, MPDTC with the switching horizon *SSE* minimizing the switching frequency was run for the NPC drive system at steady-state operating conditions. The penalties J_{bnd} and J_t in the cost function (9.3) were set to zero. The switching transitions were limited to the set of transitions shown in Fig. 2.21. One thousand fundamental periods were simulated at nominal speed and at rated torque. The torque was kept within the symmetric bounds $T_{e,\min} = 0.88$ pu and $T_{e,\max} = 1.12$ pu, whereas asymmetric bounds at $\Psi_{s,\min} = 0.97$ pu and $\Psi_{s,\max} = 1.015$ pu were used for the stator flux magnitude. The bounds $v_{n,\max} = -v_{n,\min} = 0.04$ pu were imposed on the neutral point potential. We also define the set

$$\mathcal{Y} = [T_{e,\min}, T_{e,\max}] \times [\Psi_{s,\min}, \Psi_{s,\max}] \times [v_{n,\min}, v_{n,\max}], \tag{9.18}$$

which is formed by the upper and lower bounds on the torque, stator flux magnitude, and neutral point potential.

Figure 9.9 depicts the resulting deadlocks within the torque and stator flux magnitude bounds $[T_{e,\min}, T_{e,\max}]$ and $[\Psi_{s,\min}, \Psi_{s,\max}]$, respectively. This two-dimensional set is the projection of the three-dimensional set $\mathcal{Y}$ onto the torque and stator flux magnitude subspace. It is indicated by the large rectangle. To retrieve some of the information lost because of this projection, we categorize the deadlocks in terms of the neutral point potential at the time the deadlock occurs. To this end, we divide the deadlocks into the two types M and N.

$$\begin{aligned}
&\text{Type } M\ (\bullet) && \text{if } v_n \in [v_{n,\min} + \Delta v_n, v_{n,\max} - \Delta v_n], \\
&\text{Type } N_a\ (\blacktriangle) && \text{if } v_n \geq v_{n,\max} - \Delta v_n, \\
&\text{Type } N_b\ (\blacktriangledown) && \text{if } v_n \leq v_{n,\min} + \Delta v_n.
\end{aligned} \tag{9.19}$$

The parameter

$$\Delta v_n = 0.0125(v_{n,\max} - v_{n,\min}) = 0.001 \text{ pu} \tag{9.20}$$

defines a thin region around the upper and lower bounds of the neutral point potential.

The first type of deadlock is characterized by the neutral point potential being well within its bounds. Therefore, only the two output variables of the machine give rise to the deadlock. We refer to these deadlocks as Type *M* deadlocks, where *M* refers to the machine.

The second type of deadlocks corresponds to situations in which the neutral point potential is close to, or violates its upper or lower bound. Owing to the fact that the neutral point potential

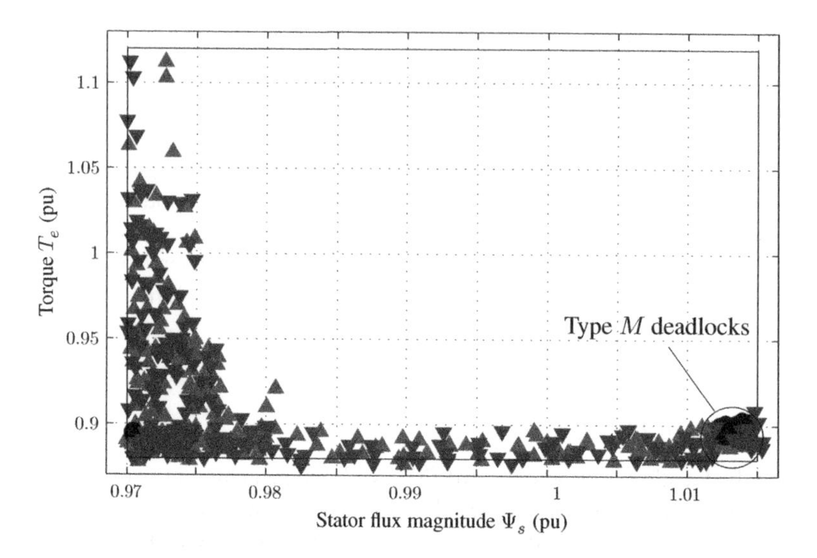

Figure 9.9 Deadlocks within the torque and stator flux magnitude bounds when operating at nominal speed and rated torque. The Type M (indicated by circles) and Type N deadlocks (indicated by upward and downward-pointing triangles) form clearly visible clusters

is involved in the deadlock, we refer to this as Type N deadlocks. Depending on whether the neutral point potential is at its upper or lower bound, we distinguish between the two subtypes N_a and N_b, respectively.

Figure 9.9 reveals that Type M deadlocks are concentrated in the lower right corner. In this region, the electromagnetic torque is close to its lower bound and a voltage vector is required that increases the torque. On the other hand, the stator flux magnitude is at its upper bound and must be reduced. This situation was discussed in Example 9.1 in the previous section.

Type N deadlocks arise close to the lower bounds of the torque and flux magnitude, particularly in the lower left corner, where both machine variables are close to their lower bounds (see Fig. 9.9). As the neutral point potential is close to one of its bounds, this restricts the choice of admissible voltage vectors. The constraints on the switching transitions further restrict the available set.

Type N deadlocks are significantly more frequent than Type M deadlocks, which amount only to about 20% of the total number of deadlocks. We also observe that Type M and N deadlocks occur at different locations within the set $\mathcal{Y}$, suggesting that they ought to be handled separately, as will be discussed in Sect. 9.5.

An alternative representation of the deadlocks is provided in Fig. 9.10, which shows the deadlocks as a function of the stator flux magnitude $\Psi_s \in [\Psi_{s,\min}, \Psi_{s,\max}]$ and the rotor flux angle $\varphi \in [-180°, 180°]$ with respect to the α-axis of the orthogonal coordinate system. The set shown in this figure can be interpreted as the projection of $\mathcal{Y}$ onto the one-dimensional stator flux magnitude space, while showing this projecting as a function of the rotor flux angle. It can be seen that Type M deadlocks appear every 60° as a result of the 60°-symmetry that is inherent to the voltage vectors. Type N deadlocks also appear every 60°, but they alternate between the two subtypes. Specifically, Type N_a and N_b deadlocks occur every 120° because of the significant third-harmonic component that is present in the neutral point potential.

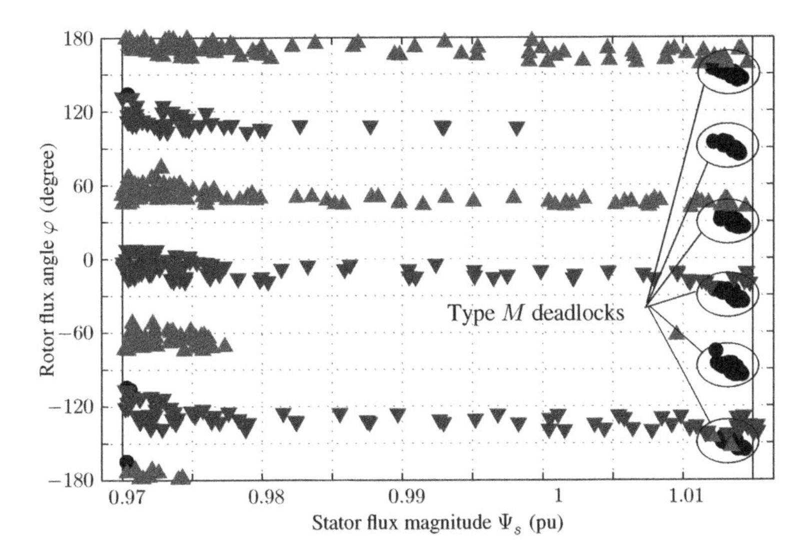

Figure 9.10 Deadlocks within the stator flux magnitude bounds when varying the rotor flux angle φ. Operation is at nominal speed and rated torque. The Type M (indicated by circles) and Type N deadlocks (indicated by upward and downward-pointing triangles) form clearly visible clusters

So far we have considered operation at nominal speed only. Figure 9.11 shows the influence speed variations have on the frequency of deadlocks f_{DL}, which is the (average) number of deadlocks per second. Lowering the speed does not necessarily reduce the frequency of deadlocks, because Type N deadlocks dominate. When varying the speed, the distribution of the deadlocks in the torque and stator flux magnitude plane is qualitatively similar to those shown in Figs. 9.9 and 9.10.

9.4 Deadlock Resolution

The previous section has shown that situations exist in which no switch position is available that keeps the three output variables within their bounds or, when one or more of them violate a bound, reduces the bound violations at each time step. In case no candidate switching sequence exists that meets the switching and output constraints (9.5) and (9.6), the optimization problem (7.26) does not permit a solution and is infeasible.

To resolve this issue, several options exist. One is to widen (some of) the bounds on the output variables. However, the torque and flux bounds implicitly define bounds on the stator currents and thus constrain the peak current. Widening one of those bounds carries the risk of causing too high a phase current in the inverter. The bounds on the neutral point potential limit the peak blocking voltage of the gate-commutated thyristors (GCTs). Temporarily widening any of the bounds on the output variables has to be done with great care, to avoid causing an overcurrent or overvoltage trip.

An alternative, and preferred, option is to change the optimality criterion in the optimization problem and to temporarily refrain from minimizing the switching effort. During a deadlock

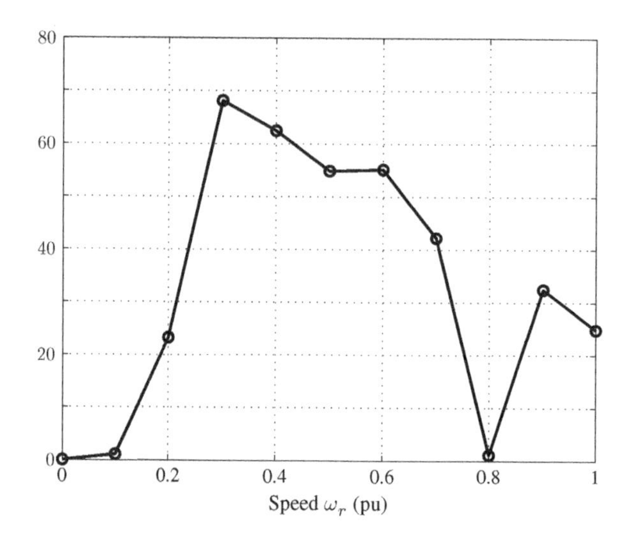

Figure 9.11 Frequency of deadlocks f_{DL} (Hz) as a function of the speed ω_r, when using the original version of MPDTC and operating at rated torque

situation, the quick resolution of the deadlock is of the highest importance, because significant bound violations might lead to a trip of the drive. We thus choose to minimize the predicted bound violation until all output variables are moved back within their bounds and the deadlock has been resolved. To this end, we replace the hard constraint (9.6) by a soft constraint that penalizes the worst predicted violation. This is achieved by the new cost function

$$J_{\mathrm{DL}} = \| \sum_{\ell=k+1}^{k+N_p} \boldsymbol{\varepsilon}(\ell) \|_{\infty}, \tag{9.21}$$

where the vector $\boldsymbol{\varepsilon} = [\varepsilon_T \ \varepsilon_\Psi \ \varepsilon_v]^T$ denotes the degrees of the bound violations for the three output variables. For the definition of ε_T the reader is referred to (7.28). The variables ε_Ψ and ε_v are defined accordingly. Recall that the bound violations are normalized with respect to the bound width, that is, the distance between the upper and the lower bounds.

Over the prediction horizon N_p, the predicted bound violations are summed up for each output variable, yielding a vector of dimension 3×1. The ∞-norm provides the maximum value of this vector, which is a scalar. We therefore aim at minimizing the worst violation. Note that the bound violations are, by definition, nonnegative, and the elements of the sum in (9.21) are thus also nonnegative.

We also redefine the switching horizon and use only the switching elements *S*. Typically, we set the switching horizon to *S* or *SS* during a deadlock. The switching horizon *eSSESE*, for example, turns into *SS* and the prediction horizon is set to $N_p = 2$. A lower torque bound violation is shown in Fig. 9.12 for $N_p = 2$. The degrees of the normalized torque violations are predicted at time steps $k + 1$ and $k + 2$, and their sum is penalized in the cost function J_{DL}. The same is performed for the stator flux magnitude and the neutral point potential. A switching sequence of two steps is derived that meets the switching constraint (9.5) and minimizes the cost function (9.21).

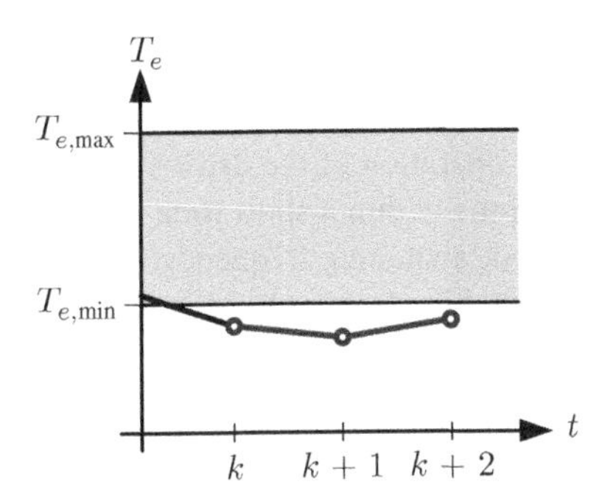

Figure 9.12 Torque violating the lower torque bound and a predicted torque trajectory that minimizes future violations when considering the switching horizon *SS*

The MPDTC deadlock resolution algorithm is a derivative of the standard MPDTC algorithm of Sect. 7.4.5. Note that the sequence of actions A holds only *S* elements.

1. The root node is initialized and pushed onto the stack.

2a. The top node i with a nonempty sequence of actions, $A_i \neq \emptyset$, is taken from the stack.

2b. The first element is read out from A_i and removed. All admissible switching transitions are enumerated according to (7.37). The state and output vectors at the next time step are predicted for each admissible switching transition using (7.38). The new node j is created for each switching transition. The node i is removed.

2c. The newly created nodes are pushed onto the stack.

2d. If at least one node with a nonempty set of actions A remains, the algorithm proceeds with Step 2a, or else it proceeds with Step 3a.

   The result of Step 2 are the leaf nodes $i \in \mathcal{I}$, where $\mathcal{I}$ is an index set. These nodes correspond to the switching sequences $\boldsymbol{U}_i(k)$.

3a. For each leaf node $i \in \mathcal{I}$, the associated cost $J_{\mathrm{DL},i}$ is computed.

3b. The leaf node with the index

$$i = \arg \min_{i \in \mathcal{I}} J_{\mathrm{DL},i} \tag{9.22}$$

   is chosen that has the minimal cost value. The associated switch position at time step k is read out and set as the optimal one, $\boldsymbol{u}_{\mathrm{opt}}(k) = \boldsymbol{u}_i(k)$.

4. The switch position $\boldsymbol{u}_{\mathrm{opt}}(k)$ is applied to the inverter. The standard MPDTC algorithm is executed at the next time step $k+1$. If it yields an empty set of candidate switching sequences, the deadlock resolution algorithm is called upon.

This deadlock resolution algorithm is formulated in a fairly general way. The algorithm can be significantly simplified when choosing the switching horizon *S* during deadlocks. In this case, all switching transitions at time step k have already been enumerated and the output vectors at time step $k+1$ have been predicted by the MPDTC algorithm. When memorizing these, in the case of a deadlock, one only needs to apply the revised cost function J_{DL} and minimize it according to (9.22). Additional enumerations and predictions are not required, thus keeping the additional computational burden required to resolve the deadlock at a minimum. This special case of the deadlock resolution algorithm was initially proposed in [9].

9.5 Deadlock Avoidance

The deadlock resolution strategy proposed in the previous section reliably resolves all deadlocks and ensures the continuous operation of the drive system. However, this strategy often requires several switching transitions within a short time interval to resolve a deadlock. This leads to a spike in the instantaneous switching frequency.

We define the *instantaneous* switching frequency as the average number of switching events (of all switching devices) over a time window of 1 ms. Figure 9.13 shows the instantaneous switching frequency of MPDTC minimizing the switching frequency with the switching horizon *SSE*. Characteristic spikes in the instantaneous switching frequency can be observed, which we refer to as *switching bursts*. These bursts correlate with the occurrence of deadlocks, which are denoted by squares. This indicates that deadlocks cause switching bursts; avoiding deadlocks avoids these bursts, as will be shown in Sect. 9.5.2.

For the safe operation of the inverter, it is mandatory to avoid bursts in the instantaneous switching frequency. Switching bursts can lead to the overheating of semiconductor switches and might prevent the gate drivers of the GCTs from fully recharging before the next switching transition. Therefore, the number of switching events per millisecond is monitored by a protection mechanism. In the worst case, switching bursts trigger this protection mechanism and lead to the tripping of the drive system.

9.5.1 Deadlock Avoidance Strategies

In the following, three families of deadlock avoidance strategies are introduced. These approaches are based on terminal soft constraints (Approach A), terminal weights (Approach B), and exact deadlock prediction (Approach C), respectively. We will see that Approaches A and C can be applied to Type *M* and *N* deadlocks, while Approach B is

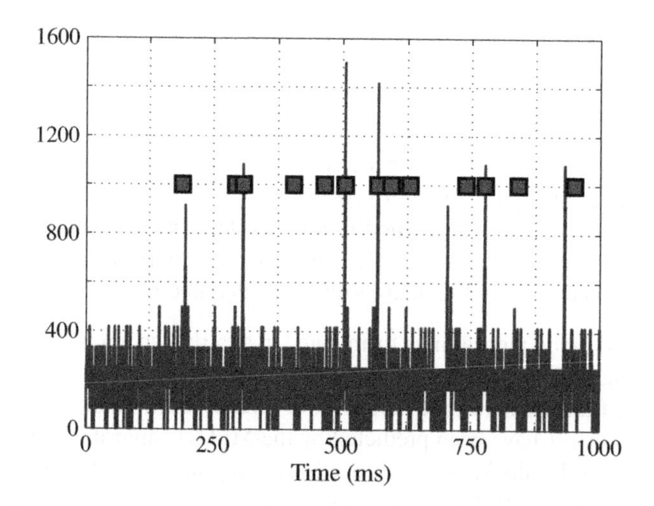

Figure 9.13 Instantaneous switching frequency (Hz) when operating the NPC inverter drive system at nominal speed and rated torque for 1 s. The squares indicate deadlocks

restricted to Type *N* deadlocks. To simplify the exposition, we set the bound violation term J_{bnd} in the cost function (9.3) to zero.

9.5.1.1 Approach A_1: Terminal Soft Constraint on the Torque and Stator Flux Magnitude

The terminal *soft constraint*

$$J_t = \begin{cases} \lambda_m & \text{if } \boldsymbol{y}(k+N_i) \in \mathcal{Y}_c \\ 0 & \text{else} \end{cases} \tag{9.23}$$

is added to the cost function (9.3). This soft constraint adds the large penalty $\lambda_m \gg 0$ to the cost function if the output vector at the end of the switching sequence (recall that the length of the *i*th switching sequence is N_i) is within the set $\mathcal{Y}_c$. We refer to this set as the *critical region* for the torque and stator flux magnitude

$$\mathcal{Y}_c = \{\boldsymbol{y} \mid T_e \leq T_{e,\min} + \Delta T_e\} \cap \{\boldsymbol{y} \mid \Psi_s \geq \Psi_{s,\max} - \Delta\Psi_s\} \cap \mathcal{Y}. \tag{9.24}$$

The set $\mathcal{Y}_c$ is a subset of $\mathcal{Y}$ and comprises the bottom right corner of Fig. 9.9. The positive parameters ΔT_e and $\Delta\Psi_s$ are chosen such that all Type *M* deadlocks are covered, while $\mathcal{Y}_c$ is as small as possible.

The terminal soft constraint (9.23) has the following effect on the selection process of the optimal switching sequence. If at least one sequence exists such that $J_t = 0$ holds, which implies that its output trajectory does not terminate in the critical region, then only switching sequences are considered that meet the constraint

$$\boldsymbol{y}(k+N_i) \in \mathcal{Y} \setminus \mathcal{Y}_c. \tag{9.25}$$

In general, this reduces the set of candidate switching sequences. Out of this set, the sequence with the lowest switching effort J_{sw} is chosen as the optimal sequence. All switching sequences with $J_t = \lambda_m$ are by definition suboptimal.

If no sequence with $J_t = 0$ exists, that is, all sequences drive the outputs into the critical region, then (9.25) is implicitly relaxed and all candidate switching sequences are considered. The penalty λ_m then adds only an offset to the cost of all candidate switching sequences. The proposed method performs well only if the region corresponding to Type *M* deadlocks is well defined and small compared to $\mathcal{Y}$, because MPDTC forgoes a certain degree of freedom, potentially impacting on its performance. In our case, these two conditions are met.

9.5.1.2 Approach A_2: Terminal Soft Constraint on the Torque, Stator Flux Magnitude and Neutral Point Potential

As Type *N* deadlocks dominate, it is expedient to avoid the neutral point potential from hitting its bounds. This can be achieved by augmenting the terminal soft constraint (9.23) to also address the neutral point potential and therefore Type *N* deadlocks. The critical region is modified to $\mathcal{Y}'_c = \mathcal{Y}_c \cap \mathcal{Y}_n$ with

$$\mathcal{Y}_n = \{\boldsymbol{y} \mid \upsilon_n \leq \upsilon_{n,\min} + \Delta\upsilon_n\} \cup \{\boldsymbol{y} \mid \upsilon_n \geq \upsilon_{n,\max} - \Delta\upsilon_n\}, \tag{9.26}$$

with $\Delta\upsilon_n$ as defined in (9.20).

9.5.1.3 Approach B: Terminal Weight on the Neutral Point Potential

Imposing a terminal constraint on the neutral point potential might merely, in effect, tighten the bounds imposed on the neutral point potential, rather than avoid the drive system from running into deadlocks. A quadratic penalty can be a gentler and more suitable means to keep the neutral point potential close to its reference and away from its bounds. Focusing on the neutral point potential's trajectory, we add the quadratic terminal weight (or penalty)

$$J_t = \lambda_n (v_n(k + N_i))^2 \tag{9.27}$$

to the cost function (9.3), with $\lambda_n \geq 0$ being a tuning parameter. As a result, trajectories of the neutral point potential ending close to the reference at zero are penalized only a little, whereas trajectories with significant deviations are penalized severely.

9.5.1.4 Approach A_1B: Combination of Approaches A_1 and B

The terminal soft constraint on the torque and stator flux trajectories (Approach A_1) can be combined with the terminal weight on the neutral point potential (Approach B) by adding both (9.23) and (9.27) to the cost function (9.3).

9.5.1.5 Approach C_1: Deadlock Prediction at Time Step $k + N_i$

Approach C_1 adds a post-processing step to the MPDTC algorithm. Once the candidate switching sequences have been enumerated and the tentative optimal switching sequence $\boldsymbol{U}_{\text{opt}}(k)$ has been determined, a deadlock prediction procedure is executed that attempts to ensure that the chosen switching sequence will not lead to a deadlock. Specifically, we replace Step 3b of the MPDTC algorithm in Sect. 7.4.5 by the following procedure:

3b. The leaf node with the index

$$i = \arg \min_{i \in \mathcal{I}} J_i$$

is chosen, which has the lowest cost value. The associated terminal state vector $\boldsymbol{x}_i(k + N_i)$ and the switching sequence $\boldsymbol{U}_i(k)$ starting at time step k are read out.

3c. Using $\boldsymbol{x}_i(k + N_i)$ as the new initial state, the algorithm considers switching sequences $\boldsymbol{U}(k + N_i)$ that start at time step $k + N_i$. A short switching horizon such as *eSE* suffices. The algorithm determines whether at least one candidate switching sequence $\boldsymbol{U}(k + N_i)$ exists that meets the switching and output constraints (9.5) and (9.6).

3d. The existence of such a $\boldsymbol{U}(k + N_i)$ serves as a proof that the switching sequence $\boldsymbol{U}_i(k)$ will *not* lead into a deadlock.

- If this is the case, the algorithm proceeds with $\boldsymbol{U}_i(k)$, reads out the first element from it, sets it as the optimal one, $\boldsymbol{u}_{\text{opt}}(k) = \boldsymbol{u}_i(k)$, and continues with Step 4 of the MPDTC algorithm in Sect. 7.4.5.
- If such a $\boldsymbol{U}(k + N_i)$ was not found, $\boldsymbol{U}_i(k)$ is discarded, the corresponding index i is removed from the index set $\mathcal{I}$, and the algorithm proceeds with Step 3b by considering the switching sequence with the second lowest cost.

In this way, candidate switching sequences are analyzed in an ascending order of their cost, starting with the sequence with the lowest cost, until one is found that provides a certificate that it will not lead into a deadlock. If no such sequence exists, the one with the lowest cost is selected, similar to Approach A_1. Note that this certificate is only valid at steady-state operating conditions and, strictly speaking, only under nominal conditions, which imply negligible model mismatches and no measurement or flux observer errors.

9.5.1.6 Approach C_2: Deadlock Prediction at Time Step $k+1$

A modified version of Approach C_1 uses $\boldsymbol{x}_i(k+1)$ as the new initial state, rather than $\boldsymbol{x}_i(k+N_i)$. Looking only one step ahead renders the deadlock prediction strategy highly robust, because $\boldsymbol{x}_i(k+1)$ can be predicted very accurately.

9.5.2 *Performance Evaluation*

The closed-loop performance of the proposed deadlock avoidance strategies is evaluated now using simulations. Specifically, the influence of the deadlock avoidance schemes on the frequency of deadlocks, the occurrence of switching bursts, the device switching frequency, and the current and torque TDDs are investigated. The NPC inverter drive system with one di/dt snubber per dc-link half is used as a case study, as explained in Sect. 9.3.1. All simulations were run at the rated torque, using the switching horizon *SSE* and the sampling interval $T_s = 25$ μs.

9.5.2.1 Effect on the Frequency of Deadlocks

Figure 9.14 depicts the frequency of deadlocks as a function of the speed ω_r. The solid lines refer to the original MPDTC algorithm, which serves as a benchmark. Figure 9.14(a) presents the results for Approaches A_1, A_2, and B, whereas Fig. 9.14(b) focuses on Approaches A_1B, C_1, and C_2. Approach A_1B avoids all deadlocks except at $\omega_r = 0.6$ pu. Considering the results of Approaches A_1 and B separately, one can see that Approach A_1B is the synergy of the two strategies, combining their benefits. A detailed analysis shows that Approach A_1 significantly reduces Type *M* deadlocks, while increasing Type *N* deadlocks, such that the overall result does not exhibit a significant improvement. If Approach A_1 is, however, combined with Approach B, which reliably resolves all Type *N* deadlocks, this negative effect is compensated for, resulting in a very good performance. Approaches A_2 and C_1 work nearly as well as A_1B in terms of deadlock reduction, while Approach C_2 is less successful.

9.5.2.2 Effect on Switching Bursts

Repeating Fig. 9.13 to allow a side-by-side comparison, Fig. 9.15(a) shows the instantaneous switching frequency for the original MPDTC algorithm. Switching bursts are clearly identifiable and the deadlocks are marked by squares. Approach A_1B, however, successfully avoids all deadlocks and all switching bursts, as can be seen in Fig. 9.15(b). This confirms the hypothesis that, by preventing the MPDTC algorithm from running into deadlocks, the switching bursts are also avoided.

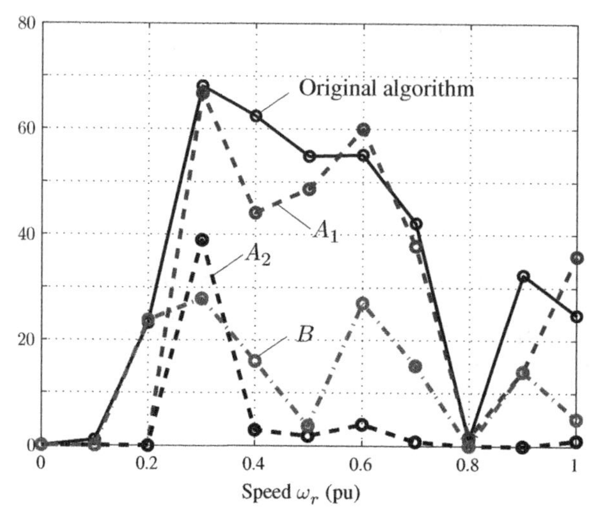

(a) Approaches A_1, A_2, and B compared with the original MPDTC algorithm

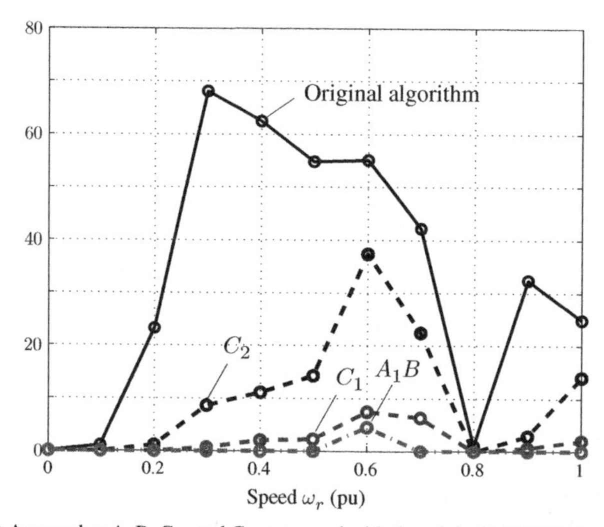

(b) Approaches A_1B, C_1, and C_2 compared with the original MPDTC algorithm

Figure 9.14 Frequency of deadlocks f_{DL} (Hz) as a function of the speed ω_r for Approaches A_1, A_2, B, A_1B, C_1, and C_2. The frequency of deadlocks resulting from the original MPDTC algorithm is also shown

For the other promising Approaches A_2, C_1, and C_2, however, switching bursts are not fully avoided. It appears that, for these approaches, the switching effort that is required to avoid deadlocks is similar to that needed to resolve deadlocks using the deadlock resolution strategy. In other words, for Approaches A_2, C_1, and C_2, the switching bursts are only shifted in time, but not prevented.

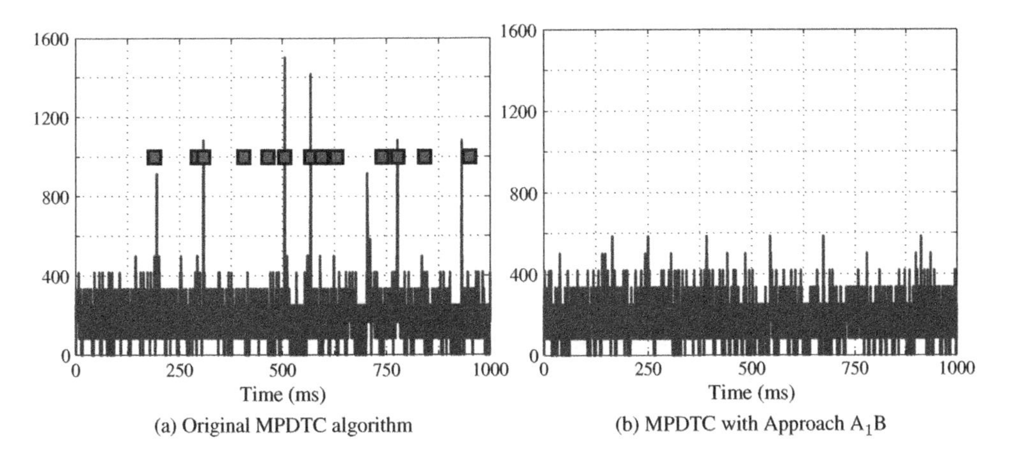

Figure 9.15 Instantaneous switching frequency (Hz) when operating the NPC inverter drive system at nominal speed and rated torque for 1 s. The squares indicate deadlocks

Additional investigations revealed that Type *N* deadlocks are the dominant cause of the switching bursts and that these deadlocks are, in general, more difficult to resolve than Type *M* deadlocks, which are caused by the torque and stator flux magnitude. It is therefore of paramount importance to efficiently avoid Type *N* deadlocks. Only Approach B, which is based on a terminal weight on the neutral point potential, achieves this consistently. The terminal weight provides MPDTC with an incentive to move the neutral point potential closer to its reference whenever the additional switching effort required to do so is minor. Provided that the weight λ_n is not overly small, this mechanism prevents the neutral point potential from hitting its bounds.

9.5.2.3 Effect of Approach A_1B on the Performance

As only Approach A_1B avoids deadlocks as well as switching bursts, the subsequent investigation is restricted to this deadlock avoidance scheme. In the following, the effect of A_1B on the frequency of deadlocks f_{DL}, the switching frequency f_{sw}, the stator current TDD I_{TDD}, and the electromagnetic torque TDD T_{TDD} is investigated and discussed. Using the original MPDTC algorithm as a baseline, the latter three performance values are normalized and their percentage-wise deviation from the original algorithm is considered. Specifically, we define the *relative* switching frequency

$$f_{\text{sw}}^{\text{rel}} = \frac{f_{\text{sw}}^{\text{A}_1\text{B}} - f_{\text{sw}}^{\text{org}}}{f_{\text{sw}}^{\text{org}}}. \tag{9.28}$$

The relative current and torque TDDs are defined accordingly.

Figure 9.16 shows the frequency of deadlocks and the relative switching frequency as a function of the terminal weight λ_n that is imposed on the neutral point potential at the end of its predicted trajectory. The three different speeds $\omega_r \in \{0.3, 0.6, 1\}$ pu are investigated. As shown in Fig. 9.16(a), as λ_n is increased, the neutral point potential is kept more tightly around its reference, and the number of deadlocks is reduced accordingly. Reducing the deadlocks

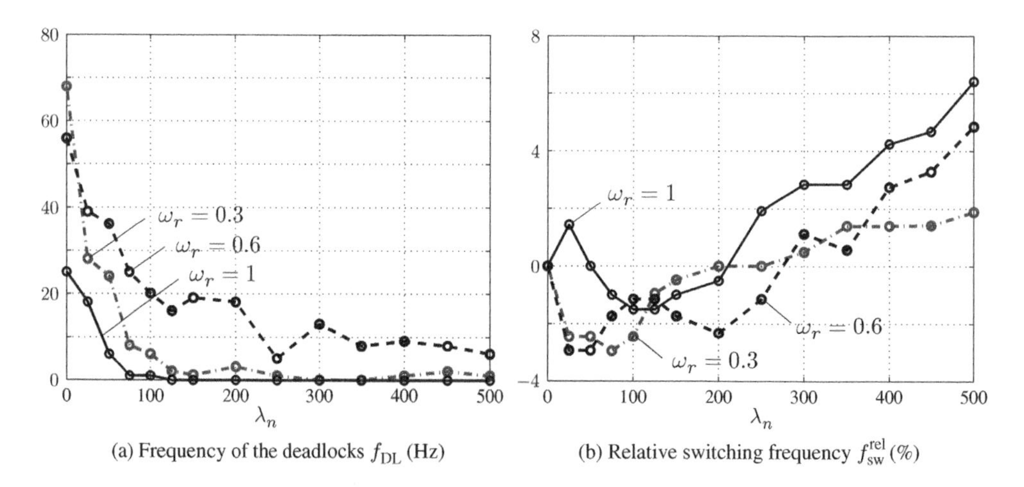

(a) Frequency of the deadlocks f_{DL} (Hz)

(b) Relative switching frequency $f_{\text{sw}}^{\text{rel}}$ (%)

Figure 9.16 Approach A_1B: Frequency of the deadlocks and relative switching frequency as a function of the terminal weight λ_n on the neutral point potential for the three different speed setpoints $\omega_r \in \{0.3, 0.6, 1\}$ pu

Table 9.1 Approach A_1B: Performance metrics as a function of the angular speed ω_r, using the original MPDTC algorithm as a baseline

ω_r (pu)	$f_{\text{sw}}^{\text{rel}}$ (%)	$I_{\text{TDD}}^{\text{rel}}$ (%)	$T_{\text{TDD}}^{\text{rel}}$ (%)	f_{DL} (Hz)	λ_n
0.1	0	−0.5	0	0	25
0.2	0	−1.2	−0.3	0	75
0.3	0.5	−0.9	1.4	0	300
0.4	−0.9	−2	0.8	0	150
0.5	−2.5	−1	0.3	0	125
0.6	−1.1	−0.4	−1.6	4.5	250
0.7	0	−0.6	0.2	0	50
0.8	0	0	0	0	0
0.9	−1.3	−0.25	−0.2	0	75
1	−1.5	−0.9	−0.6	0	125

tends to also reduce the switching frequency, as evidenced by Fig. 9.16(b), provided that λ_n is relatively small. Large λ_n entail a higher control effort, which results in an increase in the switching frequency.

Figure 9.16 is complemented by Table 9.1, which lists the relative performance of Approach A_1B (with respect to the original MPDTC algorithm) for the whole range of speed operating points, while operating at rated torque. The table reveals that the benefit of the terminal weight λ_n is similar for all speed setpoints. To obtain excellent results, however, λ_n has to be adjusted depending on the speed, as summarized in Table 9.1. Figure 9.16 indicates that this tuning of λ_n is relatively straightforward. As the weight λ_n is increased, the frequency of deadlocks first drops steeply, and then remains mostly flat. On the other hand, the impact of too high a λ_n on the switching frequency is modest.

Table 9.1 also indicates that the influence of the terminal weight λ_n on the current distortions is small, because the latter primarily depends on the width of the bounds that are imposed on the stator flux magnitude. Nevertheless, smaller fluctuations of the neutral point potential tend to reduce the distortions in the applied stator voltages and avoiding deadlocks prevents potential violations of the bounds. These two aspects have a positive influence on the current TDD.

9.6 Summary and Discussion

This chapter focused on two major topics for MPDTC—analysis of the control law and feasibility of the control problem. Both aspects will be summarized and discussed in the following.

9.6.1 Derivation and Analysis of the State-Feedback Control Law

Unlike FOC and DTC, the control law is not directly available in MPDTC. Section 9.2 showed a straightforward method to compute the state-feedback control law and—by analyzing and interpreting it—has provided additional insight into MPDTC.

The derivation and visual representation of the control law is paramount during the design phase of the controller, because it enables one to analyze and understand the controller's choices, to assess the impact that different cost functions have on the closed-loop behavior, to understand the impact switching constraints have, and to evaluate the influence of phenomena such as model uncertainties, observer noise, and unaccounted for dc-link voltage fluctuations.

Along with the plotting and analysis of the predicted output trajectories and switching sequences, the availability of this method constitutes one of the main advantages of MPC in general, and MPDTC in particular, over classic control methods. For the latter, the design and tuning process is usually restricted to running closed-loop simulations, and to iterate on a trial and error basis. Furthermore, with this tool at one's disposal, the following tasks are envisioned to be achievable in the future: a further reduction of the computational effort, the derivation of switching heuristics, a further improvement of the closed-loop performance, and a detailed feasibility analysis.

The techniques proposed in this section are directly applicable to other predictive control methods, including predictive control with reference tracking, which is discussed in Chaps. 4–6, and derivatives of MPDTC. Two of them, model predictive direct current control (MPDCC) and model predictive direct power control (MPDPC), are described in Chap. 11. The third one, model predictive direct balancing control (MPDBC), has been proposed in [10]. It is also straightforward to address other multilevel inverter topologies and to include the neutral point potential in the considerations. Since its inception, the derivation of the state-feedback control as described in this section has proven to be instrumental in analyzing and improving MPDTC.

9.6.2 Deadlock Analysis, Resolution, and Avoidance

For an NPC inverter drive system, Sect. 9.3 has shown that the deadlocks encountered during the execution of the MPDTC algorithm can be classified into two groups. The first group comprises the deadlocks that are exclusively caused by the interplay between the torque and the stator flux magnitude of the machine. The neutral point potential is involved in the second group of deadlocks. When the neutral point potential is close to one of its bounds or even violates one of them, it acts as an antagonist to the torque or stator flux magnitude.

A computationally simple yet effective deadlock resolution mechanism was discussed in Sect. 9.4. All simulation results shown in this book that are based on the MPDTC family use this mechanism.

Even though deadlocks can be resolved, they often trigger bursts in the instantaneous switching frequency. To prevent the MPDTC algorithm from running into deadlocks, a combination of terminal *constraints* on the electromagnetic torque and stator flux magnitude together with a terminal *weight* on the neutral point potential has been proposed as Approach A_1B in Sect. 9.5.1. This minor modification to the original MPDTC algorithm successfully avoids virtually all deadlocks over the whole speed range. As a result, the switching bursts are also avoided. Moreover, a minute reduction in the switching frequency and the current distortions is, in many cases, observable.

This deadlock avoidance method is also applicable to drives with more complicated inverter topologies such as the five-level ANPC inverter drive system (see the case study described in Sect. 2.5.3). Unlike for the NPC inverter, for which only the neutral point potential must be balanced, in the case of the five-level topology also its three phase capacitor voltages must be controlled. This greatly complicates the task of the controller. When using MPDTC for the drive, the introduction of a terminal *weight* on the neutral point potential maintains the latter closer to its reference and effectively avoids almost all deadlocks. Note that for the results depicted in Sect. 8.2, such a terminal weight has been used.

Alternatively, the control problem of the five-level ANPC drive can be divided into an upper-level machine controller and a lower-level inverter balancing task, which is performed by MPDBC [10]. For the latter, a terminal *weight* on the neutral point potential greatly reduces the balancing effort that is required for the neutral point potential. At low speed, a penalty on the common-mode voltage is imposed. As a result, the deadlocks and switching bursts are not only avoided, but a significant improvement in the overall performance is also observable. At nominal speed, with respect to the baseline MPDBC method, the average switching frequency and the current TDD are reduced by 20% and 13%, respectively, as indicated in [11].

The proposed deadlock avoidance strategies can also be directly applied to the other members of the MPDTC family, notably to MPDCC (see Sect. 11.1) and to MPDPC. The latter is the adaptation of MPDTC for grid-connected converters (see Sect. 11.2).

References

[1] F. Borrelli, M. Baotić, A. Bemporad, and M. Morari, "Dynamic programming for constrained optimal control of discrete-time linear hybrid systems," *Automatica*, vol. 41, pp. 1709–1721, Oct. 2005.

[2] M. Kvasnica, P. Grieder, M. Baotić, and M. Morari, "Multi parametric toolbox (MPT)," in *Hybrid systems: Computation and control* (R. Alur and G. Pappas, eds.), vol. 2993 of LNCS, pp. 448–462, Heidelberg: Springer, 2004. http://control.ee.ethz.ch/mpt.

[3] A. Bemporad and M. Morari, "Control of systems integrating logic, dynamics and constraints," *Automatica*, vol. 35, pp. 407–427, March 1999.

[4] T. Geyer, *Low complexity model predictive control in power electronics and power systems*. PhD thesis, Automatic Control Laboratory ETH Zurich, 2005.

[5] G. Papafotiou, T. Geyer, and M. Morari, "A hybrid model predictive control approach to the direct torque control problem of induction motors," *Int. J. Robust Nonlinear Control*, vol. 17, pp. 1572–1589, Nov. 2007.

[6] T. Geyer, G. Papafotiou, R. Frasca, and M. Morari, "Constrained optimal control of the step-down DC-DC converter," *IEEE Trans. Power Electron.*, vol. 23, pp. 2454–2464, Sep. 2008.

[7] T. Geyer and G. Papafotiou, "Direct torque control for induction motor drives: A model predictive control approach based on feasibility," in *Hybrid systems: Computation and control* (M. Morari and L. Thiele, eds.), vol. 3414 of LNCS, pp. 274–290, Heidelberg: Springer, Mar. 2005.

[8] A. Linder and R. Kennel, "Model predictive control for electrical drives," in *Proceedings of IEEE Power Electronics Specialists Conference* (Recife, Brasil), pp. 1793–1799, 2005.

[9] G. Papafotiou, J. Kley, K. G. Papadopoulos, P. Bohren, and M. Morari, "Model predictive direct torque control—Part II: Implementation and experimental evaluation," *IEEE Trans. Ind. Electron.*, vol. 56, pp. 1906–1915, Jun. 2009.

[10] F. Kieferndorf, P. Karamanakos, P. Bader, N. Oikonomou, and T. Geyer, "Model predictive control of the internal voltages of a five-level active neutral point clamped converter," in *Proceedings of IEEE Energy Conversion Congress and Exposition* (Raleigh, NC, USA), Sep. 2012.

[11] T. Burtscher and T. Geyer, "Deadlock avoidance in model predictive direct torque control," *IEEE Trans. Ind. Appl.*, vol. 49, pp. 2126–2135, Sep./Oct. 2013.

10

Computationally Efficient Model Predictive Direct Torque Control

The computational complexity of model predictive direct torque control (MPDTC) is proportional to the number of admissible switching transitions to the power of the number of switching events considered within the prediction horizon. The former, the number of switching transitions per time step, is determined by the inverter topology—most prominently by the number of available voltage levels. The latter, the number of switching events, is set by the switching horizon. On one hand, long switching horizons greatly boost the performance of MPDTC, in the sense that the switching losses, the current distortions, or the torque distortions can be significantly reduced, as shown in Chap. 8. On the other hand, long switching horizons lead to a combinatorial explosion of the number of admissible switching sequences to be explored.

Until now, finding the optimal switching sequence required the investigation of all admissible switching sequences by the MPDTC algorithm (see Sect. 7.4.5). This brute-force concept of full enumeration becomes computationally very expensive, and thus often prohibitive, for long switching horizons. Specifically, when running MPDTC with very long switching horizons, the combinatorial explosion slows down the simulations significantly. For a real-time implementation of MPDTC on control hardware, the achievable switching horizons are often limited to relatively short horizons, such as *SE*, *SSE*, or *SESE*.

This shortcoming motivates the techniques presented in this chapter, which drastically reduce the number of switching sequences to be explored and thus lessen the computational burden of MPDTC. The first technique, branch-and-bound, uses upper and lower bounds on the cost function to discard large parts of the search tree. A simple branching heuristic is used to select promising parts of the search tree. As a result, the *optimal* solution is found more quickly and the *average* number of computations is reduced. To limit the *maximum* number of computations, the optimization procedure can be stopped if the number of computational steps exceeds a certain threshold. Despite the possibility of suboptimal results, the performance deterioration is small, provided that the threshold is chosen carefully. Alternatively, one can choose to stop if the incumbent best solution is sufficiently close to the optimum.

Model Predictive Control of High Power Converters and Industrial Drives, First Edition. Tobias Geyer.

Companion Website: www.wiley.com/go/geyermodelpredictivecontrol

Performance evaluations indicate that these techniques reduce the computation time by an order of magnitude when compared to full enumeration. MPDTC with long switching horizons is thus expected to become implementable on the control hardware available today, allowing one to take full advantage of its performance benefits.

In order to reduce the computational burden for direct model predictive control (MPC) schemes, a commonly followed approach is to restrict the search space *a priori*. For one-step predictive control with reference tracking, for example, the authors of [1] propose to restrict the set of voltage vectors that are explored to the vectors neighboring the currently applied vector. Instead of ruling out specific voltage vectors *a priori*, the branch-and-bound technique presented in this chapter removes voltage vectors dynamically *during* the optimization stage, basing the line of reasoning on the cost function rather than on specific voltage vectors. The proposed method thus appears to be more elegant, less restrictive, and more flexible than previously reported approaches.

After revisiting the drive control problem and defining terminology in Sect. 10.1, Sect. 10.2 proposes a computationally efficient version of MPDTC based on branch-and-bound with an upper bound on the number of computations. Computational results are presented in Sect. 10.3, and the implications of the revised MPDTC algorithm are discussed and conclusions are drawn in Sect. 10.4. For an introduction to the concept of branch-and-bound, the reader is referred to Sect. 3.8.

10.1 Preliminaries

Recall the MPDTC control problem of maintaining the so-called output variables, namely the electromagnetic torque, the length (or magnitude) of the stator flux vector, and the neutral point potential within given upper and lower bounds. Moreover, the switching losses of the inverter are to be minimized. To achieve this, we adopt in this chapter the cost function

$$J = \frac{1}{N_p} \sum_{\ell=k}^{k+N_p-1} e_{\text{sw}}(\boldsymbol{x}(\ell), \boldsymbol{u}(\ell), \boldsymbol{u}(\ell-1)), \tag{10.1}$$

as defined in (7.32). Cost function terms on bound violations and terminal weights as in (7.30) are not considered. Note that we omit, as in Chap. 7, the scaling of the cost J by the sampling interval T_s to reduce the computational burden. When plotting J, however, we typically divide it by the sampling interval T_s. The cost can then be stated in terms of the unit watt.

Recall the notions of the switching horizon and of the search tree, which were introduced in Sects. 7.4.3 and 7.4.4, respectively. The MPDTC algorithm with full enumeration in Sect. 7.4.5 serves as a baseline. Its computational burden is proportional to the total number of nodes in the search tree.

Before proceeding, we introduce the terminology that will be required in this chapter. In doing so, we follow [2]. We distinguish between incomplete and complete candidate switching sequences. The latter have been fully computed for the whole switching horizon, while for incomplete candidate switching sequences some actions, in the form of switching transitions or extension legs, are left. Complete candidate switching sequences correspond to leaf nodes, while incomplete ones are bud nodes. As explained in Sect. 7.4.4, each node corresponds to an (incomplete) candidate switching sequence. This allows us to use the index i to denote the ith switching sequence as well as the ith node. We define the following terms:

- $J_i = E_{\text{sw},i}/N_i$ is the cost associated with a (complete) candidate switching sequence, where $E_{\text{sw},i}$ is the sum of the switching energy losses and N_i is the length of the switching sequence.
- $N_{\max}$ is an upper bound on the (maximum) length of the prediction horizon, that is, it is assumed that $N_i \leq N_{\max}$ holds for all i.
- $J_{i,\min} = E_{\text{sw},i}/N_{\max}$ is a lower bound on the cost of the ith incomplete switching sequence, where $E_{\text{sw},i}$ is the sum of the switching energy losses incurred so far for this sequence. As $E_{\text{sw},i}$ increases monotonically as the ith switching sequence is extended and $N_i \leq N_{\max}$ holds by definition, it also holds that $J_{i,\min} \leq J_i$, that is, $J_{i,\min}$ always serves as a lower bound on the cost J_i.
- J_{opt} is the optimal (minimum) cost of *all* complete candidate switching sequences. This cost is available only when all candidate switching sequences have either been fully explored or certificates have been obtained that the unexplored ones are suboptimal.
- $J_{\max}$ denotes the incumbent minimal cost, that is, the smallest cost found so far for all complete candidate switching sequences. This cost constitutes an upper bound on the optimal cost to be found, that is, $J_{\max} \geq J_{\text{opt}}$.
- $J_{\min}$ refers to the minimum of all lower bounds $J_{i,\min}$. It holds that $J_{\min} \leq J_{\text{opt}}$.

In summary, we will use one static upper bound, $N_{\max}$, and the two dynamic bounds, $J_{\min}$ and $J_{\max}$, which bound the optimal cost J_{opt}, which is to be found. By definition, the latter is bounded by

$$J_{\min} \leq J_{\text{opt}} \leq J_{\max}. \tag{10.2}$$

10.2 MPDTC with Branch-and-Bound

The general concept of branch-and-bound has been summarized in Sect. 3.8. In the following, the branch-and-bound concept will be tailored to the peculiarities of MPDTC. The modified concept will be introduced in an intuitively accessible way, through the provision of several examples.

10.2.1 Principle and Concept

The cost in (10.1) directly relates to the switching losses. When constructing candidate switching sequences along a time axis, which starts at time step k and extends into the future, the evolution of the cost over time is, unfortunately, neither smooth nor monotonic. This is illustrated in the following example.

Example 10.1 *Consider the switching horizon* SESE *and the evolution of the cost over time. Switching Sequence 1a in Fig. 10.1 switches at time step k with the switching energy losses $e_{sw,1}$. The switching sequence is extended from time step $k+1$ onward, which reduces the cost in terms of the switching power losses by distributing the switching energy losses over a longer time interval. The extension leg terminates when a bound is hit between time steps $k+7$ and $k+8$, triggering another switching transition.*

Assume that at time step $k+7$ two switching transitions are admissible. These carry the switching energy losses $e_{sw,2}$ and $e_{sw,3}$, respectively, and create two possibilities on how *to extend Sequence 1. This leads to the Sequences 1 and 2, which are complete candidate switching sequences with no actions remaining. Their first parts coincide with Sequence 1a, which is an incomplete candidate switching sequence.*

We conclude that the cost increases in a step-like manner when switching transitions occur. When the sequence is extended using extrapolation, the cost decreases smoothly, because the switching losses are depreciated over a longer time interval. This non-monotonic characteristic of the cost over time necessitates the introduction of $N_{\max}$, which provides an upper bound on the length of the switching sequences. Consider again the previous example.

Example 10.2 *In Fig. 10.1, the cost associated with the complete candidate Sequence 1 is* $J_1 = (e_{sw,1} + e_{sw,2})/N_1$, *with* $N_1 = 12$ *time steps. The incumbent minimal cost (and the upper bound) is* $J_{max} = J_1$. *Having computed the second switching transition at* $k+7$ *with the energy losses* $e_{sw,3}$, *one can try to find a proof* before *extending Sequence 2 that this sequence, when completed, will only lead to a suboptimal solution that is inferior to the incumbent optimum. This proof can be found by computing the lower bound on the cost for Sequence 2, which is given by* $J_{2,min} = (e_{sw,1} + e_{sw,3})/N_{max}$. *As* $J_{2,min}$ *is equal to or exceeds* J_{max}, *the remainder of this sequence can be discarded and removed from the search tree. If this were not the case, however, the sequence would have to be further considered.*

The same reasoning applies to the dash-dotted Sequence 3a and its child Sequences 3 and 4. Having computed the first switching transition with the losses $e_{sw,4}$, *the whole subtree, starting at this node, can be discarded, because* $J_{4,min} = e_{sw,4}/N_{max}$ *exceeds* J_{max}.

These two examples provide an indication of how bounding can be accomplished in MPDTC. More specifically, the branch-and-bound algorithm tailored to the MPDTC problem

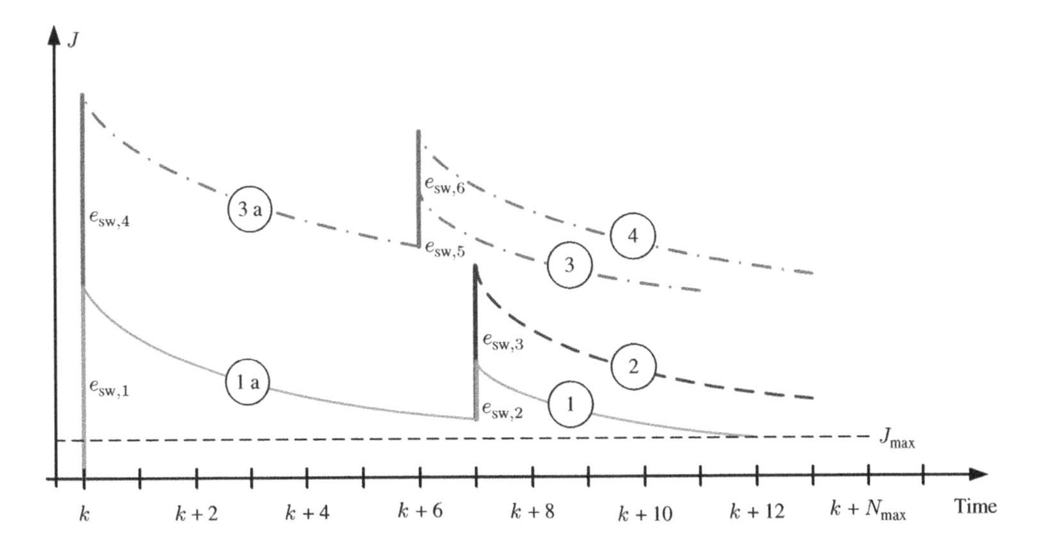

Figure 10.1 Cost J (W) when extending switching sequences over future time steps, where $e_{sw,j}$ denotes the jth switching energy loss (Ws). The incumbent minimal cost $J_{\max} = (e_{sw,1} + e_{sw,2})/12$ refers to Sequence 1. $N_{\max} = 14$ denotes the upper bound on the length of the switching sequences

setup can be described as follows. Compute the switching sequences, the associated output trajectories, and the cost step by step as the search tree is explored from its root node to the terminal nodes (leaves). Consider the bud node i, which corresponds to an incomplete candidate switching sequence. If the lower bound on its final cost $J_{i,\min}$ exceeds that of the lowest cost found so far, $J_{\max}$, a certificate has been found that it is suboptimal. This allows us to discard this bud node and to prune the attached unexplored part of the search tree. If a candidate switching sequence is completed, compute its cost J_i and update the incumbent minimal cost (if required) by setting $J_{\max} = \min(J_{\max}, J_i)$. The algorithm summarized in Sect. 7.4.5 can easily be augmented by this branch-and-bound methodology, as will be shown in Sect. 10.2.4.

10.2.2 Properties of Branch-and-Bound

The concept of branch-and-bound is further explained and illustrated in the following example.

Example 10.3 *Consider a three-level inverter with an induction machine. This case study is described and defined in Sect. 2.5.2. The machine is operated at 60% speed and rated torque, using MPDTC with the switching horizon* eSSESE. *For a specific instance of the optimization problem at time step k, the induced search tree contains 730 nodes. Using full enumeration, all 730 nodes are explored. As shown in Fig. 10.2(a), the incumbent minimal cost J_{max} drops fairly quickly, but the minimal cost $J_{opt} = 2.25\,kW$ is only found after having almost fully explored the search tree. The optimal switch position $\boldsymbol{u}_{opt}$, which is the first element in the optimal switching sequence $\boldsymbol{U}_{opt}$, will already have been found after having explored 221 nodes. Nevertheless, to obtain a certificate that this is indeed the optimal switch position, the search tree has to be fully explored.*

In contrast to this, with branch-and-bound, promising nodes are explored first and clearly suboptimal parts of the search tree are pruned. As a result, the optimal cost J_{opt} and the optimal

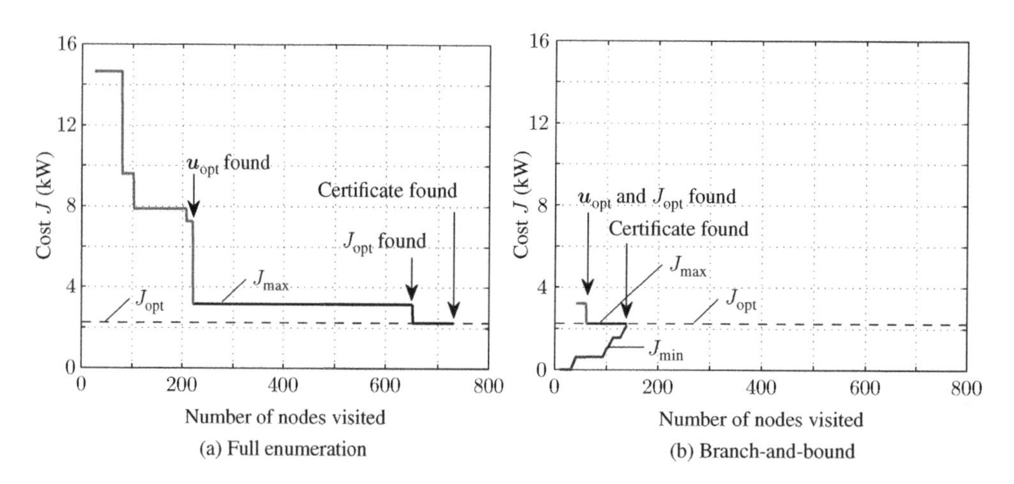

Figure 10.2 Evolution of the optimal cost (in kilowatt) when solving one instance of the MPDTC optimization problem, using full enumeration (a) and branch-and-bound (b). The incumbent minimal cost $J_{\max}$ is shown versus the number of nodes visited

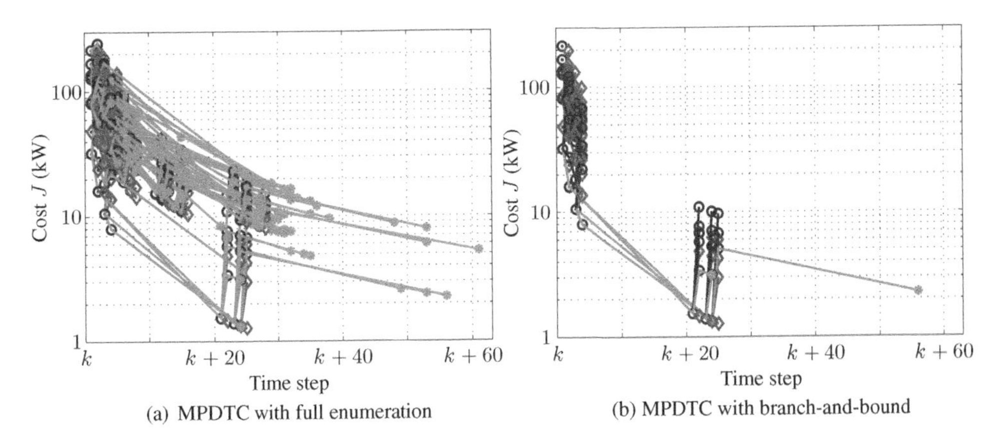

Figure 10.3 Costs of all switching sequences when solving one instance of the MPDTC optimization problem, using full enumeration (a) and branch-and-bound (b). The costs are shown as a function of the time step within the prediction horizon. Complete candidate switching sequences terminate with a large dot, incomplete (i.e., pruned) candidate sequences end with a circle, and non-candidate sequences end with a diamond. Note the logarithmic scaling of the cost

switch position $\boldsymbol{u}_{opt}$ are found significantly earlier—in this example after only 61 nodes. Some additional nodes need to be explored to prove that this is indeed the optimum. This certificate is obtained after visiting a total of 140 nodes.

Even when using branch-and-bound, all of the nodes induced by the first part of the switching horizon, that is, eS, *and most of the nodes corresponding to the second switching transition* S *have to be explored, because bounding is not effective for the first part of the switching horizon. This can be clearly seen in Fig. 10.3 for the prediction interval from k to $k + 5$. For the second half of the switching horizon, that is,* ESE, *which corresponds in this case to the prediction interval roughly from $k + 6$ to $k + 62$, bounding is very effective. This significantly reduces the number of nodes that are explored and prevents the algorithm from exploring suboptimal nodes with costs that are higher than the incumbent minimum. As a result, when using branch-and-bound, in this example less than 20% of the search tree needs to be explored.*

A few remarks concerning branch-and-bound are required. This algorithm does *not* affect the optimality of the solution, that is, the same optimal switching sequence will be found as with full enumeration. In general, branch-and-bound methods drastically reduce the *average* computation time when compared to full enumeration. Yet, in the worst case, despite branch-and-bound techniques, a full enumeration of the search tree might be required to find not only the optimum but also a proof (certificate) that the optimum has been found. Such a certificate is provided when no more bud nodes exist with $J_{i,\min} < J_{\max}$. The optimal switching sequence is usually found relatively early during the search process. Moreover, because we require only the first element of this sequence, that is, the optimal switch position $\boldsymbol{u}_{\text{opt}}$, the solution is actually found even faster, as indicated in Fig. 10.2. We will exploit this characteristic in the next section.

At each step during the optimization procedure, the optimal cost J_{opt} is lower and upper bounded according to (10.2). As the optimization proceeds, these bounds are tightened. The bounds provide information on how close the incumbent minimal cost is to the optimum. This can be seen in Fig. 10.2(b), where the upper line refers to $J_{\max}$ and the lower one corresponds to $J_{\min}$. Both lines converge to the optimal cost J_{opt}, which is given by the dashed line.

Branch-and-bound works best if the upper and lower bounds are tight. A tight upper bound $J_{\max}$ is achieved by finding a close-to-optimal leaf node with a low cost at an early stage of the optimization. To achieve this, depth-first search techniques can be employed and the optimal switching sequence from the previous time step $k-1$ can be used to warm-start the optimization procedure. A tight lower bound $J_{\min}$ is the result of a tight upper bound on the maximum length of the prediction horizon $N_{\max}$. During the optimization process, branching heuristics can help to identify the most promising nodes and to explore them first. One simple heuristic we will use is to first consider the bud node i with the lowest bound on its cost $J_{i,\min}$.

10.2.3 Limiting the Maximum Number of Computations

In a practical controller implementation, only a limited number of computations can be performed within the time interval that is available to compute the control input. Therefore, it might be necessary to limit the maximum number of computational steps or to impose an upper bound on the computation time. Aborting the branch-and-bound optimization before a certificate of optimality has been obtained might lead to suboptimal results, that is, switching sequences that yield a higher cost than the optimal sequence. Therefore, a conservative implementation adopts a fairly short switching horizon, so as to ensure that the search tree can always be fully enumerated in the time available and that the optimal solution is thus found under all circumstances.

On the other hand, by lowering the switching effort for the same distortion levels, or vice versa, long horizons offer a significant performance gain, as illustrated in Sect. 8.1. Therefore, it might be beneficial to adopt very long horizons, to impose an upper bound on the number of computations, and to accept that the result lacks—in some cases—a certificate of optimality or even optimality. Yet, as explained in the previous section, in most cases the optimum will have already been found, despite the missing certificate.

To this end, stopping criteria can be added to the MPDTC algorithm. For example, an upper bound can be imposed on the number of nodes that are explored or on the elapsed computation time. If this number or time is exceeded, the optimization procedure is stopped, and the switching sequence with the incumbent minimum cost is accepted as the solution. Alternatively, one may run the optimization procedure for as long as possible, for example, until an interrupt is received to stop it. This allows one to reduce idle processor time and to invest this time on improving the incumbent optimal solution.

When using such stopping criteria, it is of paramount importance to ensure that in all circumstances a switching sequence is found that ensures feasibility (the first switch position is admissible, i.e., meets all switching constraints), stability (the output variables either stay within their bounds or they are brought closer to the bounds if they have been violated), and good performance (the predicted cost is small). One way to achieve this is to solve the MPDTC control problem in two stages. In the first stage, a short switching horizon is adopted, which results in an optimization problem that is guaranteed to be solvable within the available time. One could use MPDTC with full enumeration and a short switching horizon, such as *eSSE*. The solution of this first stage is an optimal switching sequence, which—by definition—ensures feasibility, stability, and good performance.

The second stage uses a long switching horizon such as *eSSESESE* and computes in the remaining time a switching sequence that improves on the short horizon solution, by further reducing the cost of the switching sequence obtained in the first stage. If the switching horizon of the first stage is equal to the first part of the second stage's switching horizon, the computational results obtained in the first stage can be used to warm-start the second stage. In particular, the leaf nodes of the first stage turn into bud nodes for the second stage, to which only the second part of the switching horizon (in this example *SESE*) needs to be applied.

The computational complexity of a two-staged approach tends to be only slightly higher than that of a single long-horizon stage with branch-and-bound. This is due to fact that the number of nodes in the search tree of the first stage is one or two orders of magnitude smaller than the number of nodes in the second stage. Moreover, bounding tends to be less effective in the first part of a long switching horizon. This is due to the lower bound $J_{i,\min}$ being not tight, because the sum of the switching energy losses $E_{\mathrm{sw},i}$ is depreciated over the upper bound on the switching sequence length $N_{\max}$.

Another alternative is to stop once a guarantee of closeness to optimality has been obtained. Closeness to optimality can be defined via the cost. For example, an acceptable deviation from optimality could be 5%, and the optimization procedure could be stopped once $J_{\min} \geq 0.95 J_{\max}$ is met, that is, when the lower bound on the cost of uncompleted switching sequences is within 5% of the cost of the best completed switching sequence obtained so far. However, such a stopping criterion has to be chosen carefully to ensure that the optimization algorithm always concludes within the available time.

10.2.4 Computationally Efficient MPDTC Algorithm

An MPDTC algorithm based on full enumeration has been proposed in Sect. 7.4.5. In this section, a computationally efficient version of this algorithm is proposed, which is based on a tailored branch-and-bound technique and reduces the average computational burden. By imposing the upper bound $\kappa \leq \kappa_{\max}$ on the number of nodes κ explored, the maximum computational burden can be also limited.

1a. The root node is initialized and pushed onto the stack. The incumbent minimal cost and the node counter are set to $J_{\max} = \infty$ and $\kappa = 0$.
1b. Optional step: The output trajectories are extended using (7.40). If the length of the extended trajectory exceeds a given threshold, $\boldsymbol{u}_{\mathrm{opt}} = \boldsymbol{u}(k-1)$ is set and the algorithm proceeds with Step 4. Otherwise, the extension leg is discarded and the algorithm proceeds with Step 2a.
2a. The top node i with the lower bound on the cost $J_{i,\min}$ and with a nonempty sequence of actions, $A_i \neq \emptyset$, is taken from the stack.
2b. The first element is read out from A_i and removed. Recall that A_i is a string with the elements S, E, and e.
   - For S, all admissible switching transitions are enumerated according to (7.37). The state and output vectors at the next time step are predicted for each admissible switching transition using (7.38). If the output constraint (7.39) is met, the new node j is created. Assuming that the switching transition occurred at time step ℓ,

the switching losses of the semiconductor switches are predicted with the help of (7.24) and Table 2.5. This yields the losses $e_{sw}(\ell)$, which are added to the sum of the switching losses $E_{sw,i}$ incurred so far for this switching sequence by setting $E_{sw,j} = E_{sw,i} + e_{sw}(\ell)$. The node i is removed, but several child nodes have been created. The node counter κ is increased accordingly.

- For E, the output trajectories are extended using (7.40) and the node i is updated. A new node is not created.
- For e, the node i is kept and the optional extension leg is ignored. The new node j is created as a copy of the node i, and its trajectories are extended using (7.40). The node counter is updated according to $\kappa = \kappa + 1$.

2c. The cost expressions are updated and pruning is performed.

- For leaf nodes: The incumbent minimal cost is updated according to $J_{max} = \min(J_{max}, J_i)$, where $J_i = E_{sw,i}/N_i$.
- For bud nodes: The lower bounds $J_{i,min} = E_{sw,i}/N_{max}$ are computed. Bud nodes with $J_{i,min} \geq J_{max}$ are removed (i.e., pruned).

2d. The newly created and updated nodes are pushed onto the stack. By definition, these nodes relate to candidate switching sequences.

2e. If at least one node i with a nonempty set of actions A_i remains, and if $\kappa < \kappa_{max}$, the algorithm proceeds with Step 2a, or else it proceeds with Step 3a.

The result of Step 2 are the leaf nodes $i \in \mathcal{I}$, where $\mathcal{I}$ is an index set. These nodes correspond to the candidate switching sequences $\boldsymbol{U}_i(k)$.

3a. For each leaf node $i \in \mathcal{I}$, the associated cost $J_i = E_{sw,i}/N_i + J_{bnd,i} + J_{ti}$ is computed, as defined in (7.30) and (7.32). Note that N_i is the length of the switching sequence $\boldsymbol{U}_i$.

3b. The leaf node i with $J_i = J_{max}$ is chosen. The associated switch position at time step k is read out and set as the optimal one, $\boldsymbol{u}_{opt} = \boldsymbol{u}_i(k)$.

4. The switch position $\boldsymbol{u}_{opt}$ is applied to the inverter, and this procedure is executed again at the next time step $k + 1$.

This algorithm uses a branching heuristic in Step 2a, performs bounding in Step 2c, and limits the computational burden in Step 2e. As for the original MPDTC algorithm, the switching frequency can be minimized instead of the switching losses. For this, replace $E_{sw,i}$ by the number of commutations $S_{sw,i}$, add the incremental number of commutations $S_{sw,i} = S_{sw,i} + \|\Delta \boldsymbol{u}(\ell)\|_1$ in Step 2b, and use in Step 3 the cost $J_i = S_{sw,i}/N_i + J_{bnd,i} + J_{ti}$, which approximates the average switching frequency over the length of the switching sequence.

The MPDTC algorithm is based on the notion of nodes in the search tree. For the computationally efficient MPDTC algorithm, these nodes are the same as for the original MPDTC algorithm based on full enumeration, except for the addition of the lower bound on the ith node, $J_{i,min}$. Specifically, at time step ℓ_0, $\ell_0 = k, k+1, \ldots$, such a node is specified by the 10-tuple[1] $(\boldsymbol{u}(\ell_0 - 1), \boldsymbol{x}(\ell_0), \boldsymbol{y}(\ell_0 - 1), \boldsymbol{y}(\ell_0), E_{sw}, S_{sw}, \ell_0, A, J_{min}, \boldsymbol{u}(k))$, as defined in Sects. 7.4.4 and 10.1. Two global variables are required, namely the node counter κ and the incumbent minimal cost J_{max}.

[1] Note that in the node either the switching losses E_{sw} or the number of commutations S_{sw} is required, allowing one to reduce the node to a 9-tuple.

We assumed in Step 3 that at least one leaf node exists, that is, that the set $\mathcal{I}$ is nonempty. In case of overly tight $\kappa_{\max}$ or deadlocks, however, $\mathcal{I}$ might be empty. In this case, the deadlock resolution algorithm stated in Sect. 9.4 is executed in lieu of Step 3.

10.3 Performance Evaluation

10.3.1 Case Study

As a case study, consider again the drive system summarized in Sect. 2.5.2, which encompasses a three-level voltage source inverter with two *di/dt* snubbers, a medium-voltage (MV) induction machine, and a constant mechanical load. The detailed parameters of the drive can be found in Table 2.10. A diode front end is used, which has the nominal dc-link voltage $V_{\text{dc}} = 4294$ V. At 60% speed with a 100% torque setpoint, the steady-state performance of direct torque control (DTC) was compared with that of computationally efficient MPDTC for short and long switching horizons.

A detailed MATLAB/Simulink model of the drive system was used for the performance evaluation, which is similar to the DTC block diagram shown in Fig. 3.29. Outer control loops adjust the references for the torque and stator flux magnitude and the widths of the bounds. The DTC hysteresis controllers and look-up table are described in Sect. 3.6.3. For MPDTC, the look-up table with the DTC strategy was replaced by a function that runs the computationally efficient version of the MPDTC algorithm at each sampling instant.

The general form of the MPDTC cost function is $J = J_{\text{sw}} + J_{\text{bnd}} + J_t$ (see (7.30)). The first term penalizes the switching losses. The second and third terms are not used here and are set to zero. The torque and flux bounds were widened for MPDTC by 0.015 and 0.005 pu, respectively, to account for DTC's imminent violations of the bounds. As a result, DTC and MPDTC yield similar total demand distortions (TDDs) of the stator current, while for MPDTC with long horizons, the torque TDD is slightly lower than for DTC. Note that the torque and especially the flux bounds are asymmetric. The bounds on the neutral point potential were set to ± 0.05 pu for both control schemes.

10.3.2 Performance Metrics during Steady-State Operation

The steady-state performance of DTC can be assessed from Fig. 10.4, which over a fundamental period shows the waveforms of the electromagnetic torque, the magnitude of the stator flux, the neutral point potential, and the three-phase switch positions along with the stator currents. The upper and lower bounds on the torque, flux magnitude, and neutral point potential are also shown. It can be seen that DTC reacts only once an output variable has violated its bound and that the neutral point potential does not fully exploit the width of the bounds imposed upon it. The switching frequency is relatively high, and the switching transitions are almost equally spaced, regardless of the phase current's magnitude. As a result, the switching losses are fairly high.

Figure 10.5 shows the steady-state waveforms for the computationally efficient MPDTC scheme with the short switching horizon *eSSE*. As can be seen, the bounds are well respected and the switching frequency is reduced significantly. More specifically, with respect to DTC,

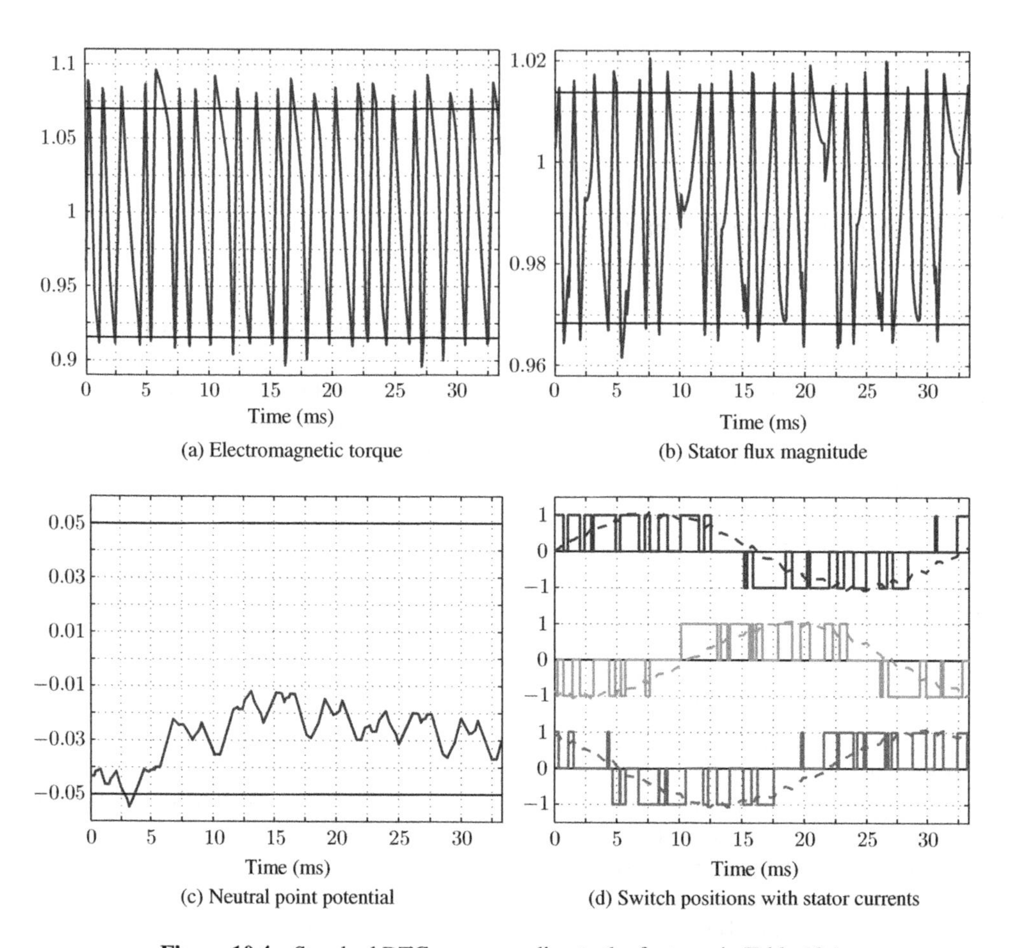

Figure 10.4 Standard DTC, corresponding to the first row in Table 10.1

the switching frequency is lowered by at least 25% and the switching losses are reduced by 40%, as summarized in Table 10.1.

The standard MPDTC algorithm based on the full enumeration of the search tree requires the exploration of up to 277 nodes to achieve this result, as shown in the first row in Table 10.2. By adopting branch-and-bound techniques, the switching heuristic, and the conservative bound $N_{\mathrm{max}} = 100$, the average number of nodes that is visited is reduced by a quarter while always providing the optimal solution. Yet, the maximum number of nodes explored remains effectively the same. To reduce the latter significantly, a tight bound on the length of the switching sequences, such as $N_{\mathrm{max}} = 50$, has to be chosen along with an upper bound on the number of nodes explored, such as $\kappa_{\mathrm{max}} = 50$. This leads to suboptimal results—in almost 8% of the cases a suboptimal $\boldsymbol{u}(k)$ is computed (see Table 10.1)—but this appears, at least in this particular case, to barely affect the performance. As a result, the maximum computational burden is reduced by 82%, from 277 down to 50 nodes explored.

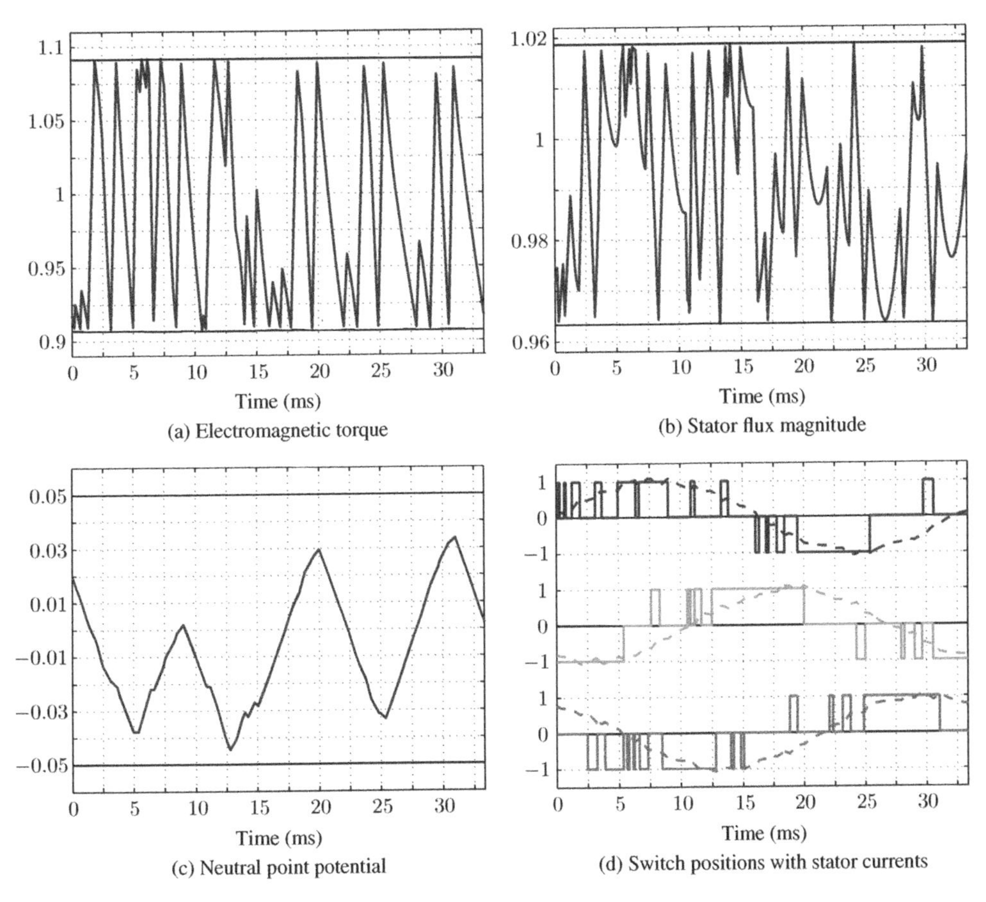

Figure 10.5 Computationally efficient MPDTC with the switching horizon *eSSE*, corresponding to the fifth row in Table 10.1 and the fourth row in Table 10.2

For the long switching horizon *eSSESESE*, the switching losses are reduced by another 35% with respect to MPDTC with *eSSE* and by 60% when compared to DTC. This is achieved, as can be observed from Fig. 10.6(d), by reducing the switching frequency by another 30% and by carefully redistributing the remaining switching transitions along the time axis. As a result, about half of the transitions occur when the respective phase current and, hence, the incurred switching losses are small. Note that the switching losses are not reduced at the expense of higher current and torque TDDs. Interestingly, as the switching horizon is extended, the current and torque distortions tend to get smaller, as can be seen in Table 10.1. At the same time, the switching pattern tends to become more repetitive, resulting in a current spectrum with distinctive harmonics. These include pronounced 7th, 11th, 17th, and 19th harmonics (see Fig. 10.7(c)). The repetitiveness in the switching pattern is also reflected in the evolution of the neutral point potential (see Fig. 10.6(c)).

As summarized in Table 10.2, the computational burden of MPDTC with full enumeration and the long switching horizon *eSSESESE* is exorbitant—its search trees encompass

Table 10.1 Comparison of the performance of DTC, full-enumeration MPDTC, and computationally efficient MPDTC with the upper bounds $N_{\max}$ and $\kappa_{\max}$ on the horizon length and the number of nodes explored, respectively

Control scheme	Switching horizon	$N_{\max}$	$\kappa_{\max}$	$\boldsymbol{u}_{\text{opt}}$ found (%)	Performance (%)			
					P_{sw}	f_{sw}	I_{TDD}	T_{TDD}
DTC	—	—	—	—	100	100	100	100
MPDTC	*eSSE*	—	—	100	57.3	71.2	103	98.4
MPDTC	*eSSE*	100	—	100	57.3	71.2	103	98.4
MPDTC	*eSSE*	50	—	97.4	57.7	73.4	103	102
MPDTC	*eSSE*	50	50	92.2	58.3	74.1	104	103
MPDTC	*eSSESESE*	—	—	100	37.9	48.9	97.0	92.0
MPDTC	*eSSESESE*	150	—	100	37.9	48.9	97.0	92.0
MPDTC	*eSSESESE*	110	—	96.7	40.9	50.0	99.5	92.2
MPDTC	*eSSESESE*	110	600	92.1	38.6	51.4	97.3	94.0

The fifth column states the probability that the optimal $\boldsymbol{u}(k)$ is found at each control cycle. The remaining four columns relate to the switching losses P_{sw}, switching frequency f_{sw}, current TDD I_{TDD}, and torque TDD T_{TDD}, using DTC as a baseline.

Table 10.2 Comparison of the computational burden of full-enumeration MPDTC and computationally efficient MPDTC with the upper bounds $N_{\max}$ and $\kappa_{\max}$ on the horizon length and the number of nodes explored, respectively

Control scheme	Switching horizon	$N_{\max}$	$\kappa_{\max}$	Prediction horizon		Nodes explored	
				Average	Maximum	Average	Maximum
MPDTC	*eSSE*	—	—	26.6	96	112	277
MPDTC	*eSSE*	100	—	25.7	92	86.9	275
MPDTC	*eSSE*	50	—	25.6	96	64.3	249
MPDTC	*eSSE*	50	50	22.0	97	43.6	50
MPDTC	*eSSESESE*	—	—	98.2	150	3246	7693
MPDTC	*eSSESESE*	150	—	98.2	150	1884	6806
MPDTC	*eSSESESE*	110	—	101	157	1102	4756
MPDTC	*eSSESESE*	110	600	88.0	152	483	600

The fifth and sixth columns indicate the average and maximum lengths of the achieved prediction horizon. The last two columns state the number of explored nodes in the search tree, which is proportional to the computational burden. The rows in this table correspond to the rows in Table 10.1.

almost 8000 nodes. Branch-and-bound with the upper bound $N_{\max} = 150$ on the length of the prediction horizon cuts down the average number of nodes explored and thus the computation time by 40%, but the maximum number of nodes is barely affected. The tight upper bound $N_{\max} = 110$ achieves a more significant reduction of the average and maximum computations by 65% and 40%, respectively. This, however, is achieved at the expense of not always obtaining the optimal switch position (see Table 10.1). The impact on the closed-loop

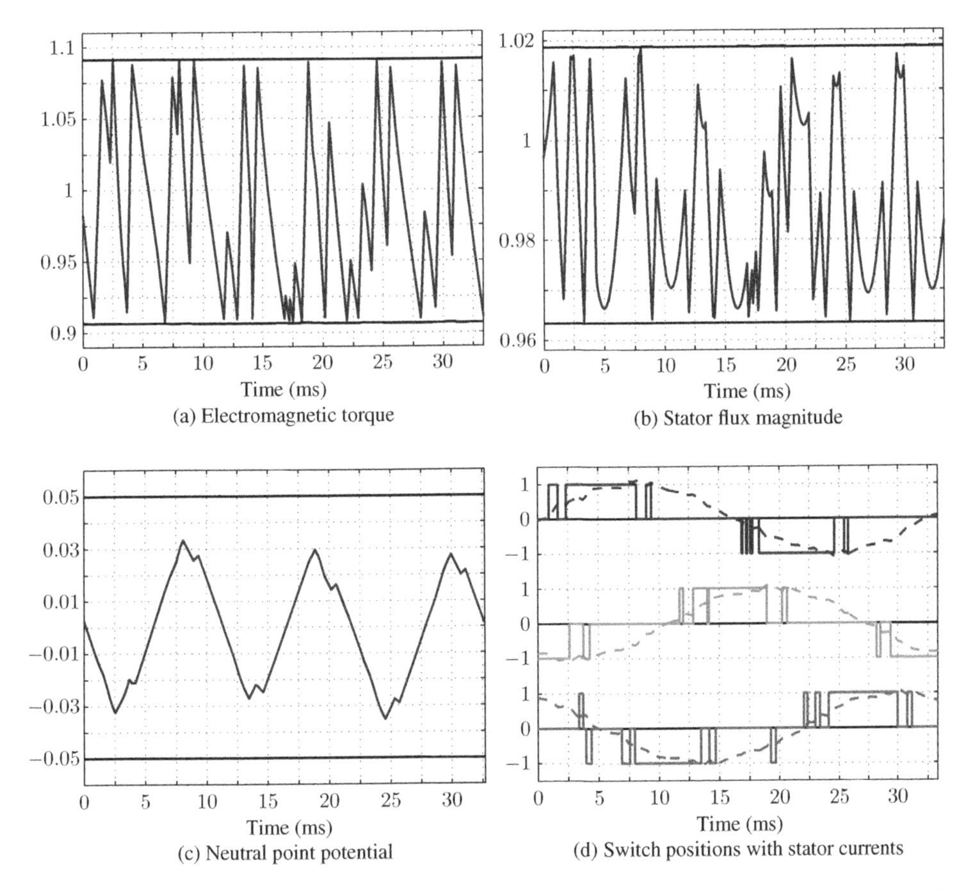

Figure 10.6 Computationally efficient MPDTC with the switching horizon *eSSESESE*, corresponding to the last rows in Tables 10.1 and 10.2

performance, such as the switching losses, switching frequency, current TDD, and torque TDD, is nevertheless small, as depicted in Table 10.1.

Despite the use of branch-and-bound, the maximum computational burden remains high. To drastically reduce the latter by more than 90%, a fairly low upper bound on the number of explored nodes, $\kappa_{\max} = 600$, is required. As for the *eSSE* case, the impact on the performance appears to be small, with the switching effort and the TDDs of the stator currents and the torque deteriorating by at most 2% and thus being only mildly affected (see Table 10.1). This minor deterioration contrasts with the observation that in 8% of the cases a suboptimal switch position is applied to the drive system.

10.3.3 Computational Metrics during Steady-State Operation

We have seen in the previous section that the addition of a tight upper bound $\kappa_{\max}$ on the number of nodes explored in MPDTC with branch-and-bound has only a relatively minor impact on

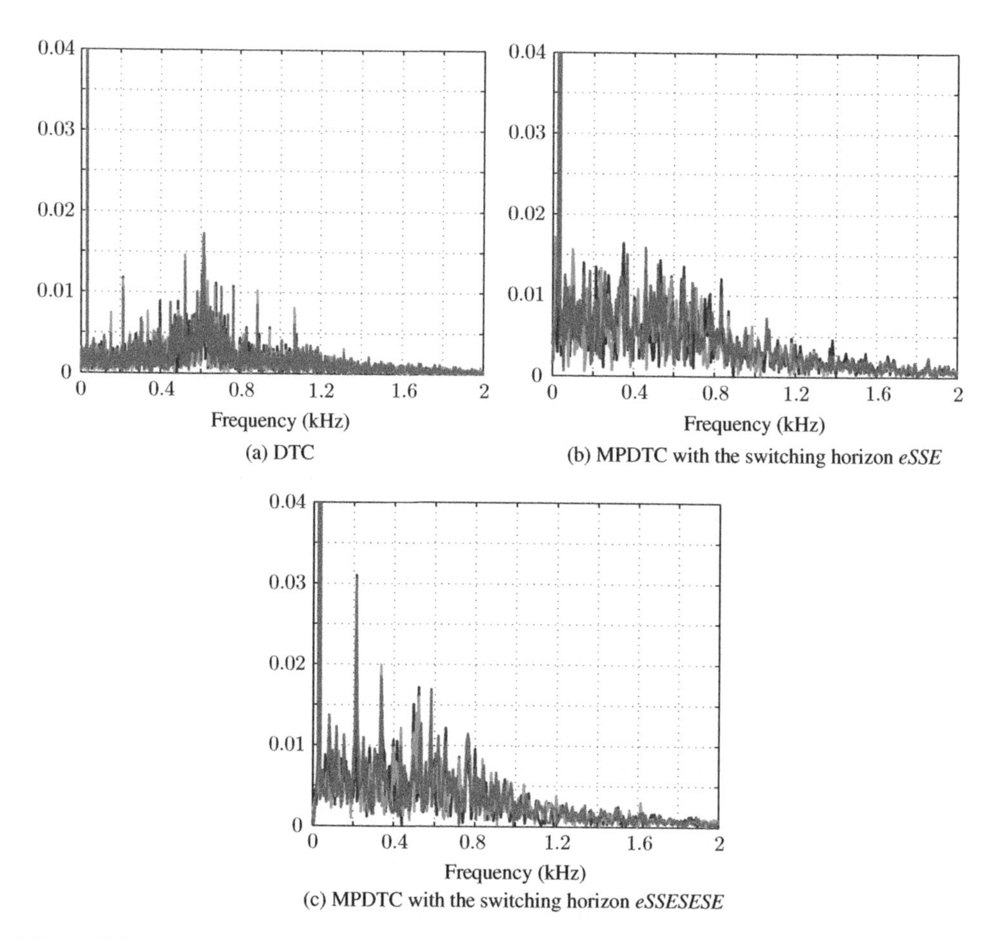

Figure 10.7 Spectra of the three-phase stator currents for DTC and computationally efficient MPDTC with the switching horizons *eSSE* and *eSSESESE*, respectively. Note that the current TDDs are effectively the same

the closed-loop performance of the drive system in terms of switching losses and distortions. This characteristic can be understood by investigating the probability that the optimal cost will be found when exploring a certain number of nodes. To determine this, MPDTC with the switching horizon *eSSESESE* was simulated again. At each time step, the number of nodes that had been explored when finding the optimal cost for the first time was recorded. These numbers were grouped into bins of width 100, and their sum was normalized to 100%. The resulting histograms are shown in Fig. 10.8.

These histograms depict the probability that the optimal cost J_{opt}—and thus the optimal switching sequence $\boldsymbol{U}_{\text{opt}}$—will be found when exploring a certain number of nodes. The vertical lines denote the 50th, 95th, and 99th percentiles. The 95th percentile, for example, indicates the number of nodes such that in 95% of the cases, at most this number of nodes is required to obtain the optimal cost. The histogram is relatively flat for standard full-enumeration MPDTC, as shown in Fig. 10.8(a), with optimality being achieved in

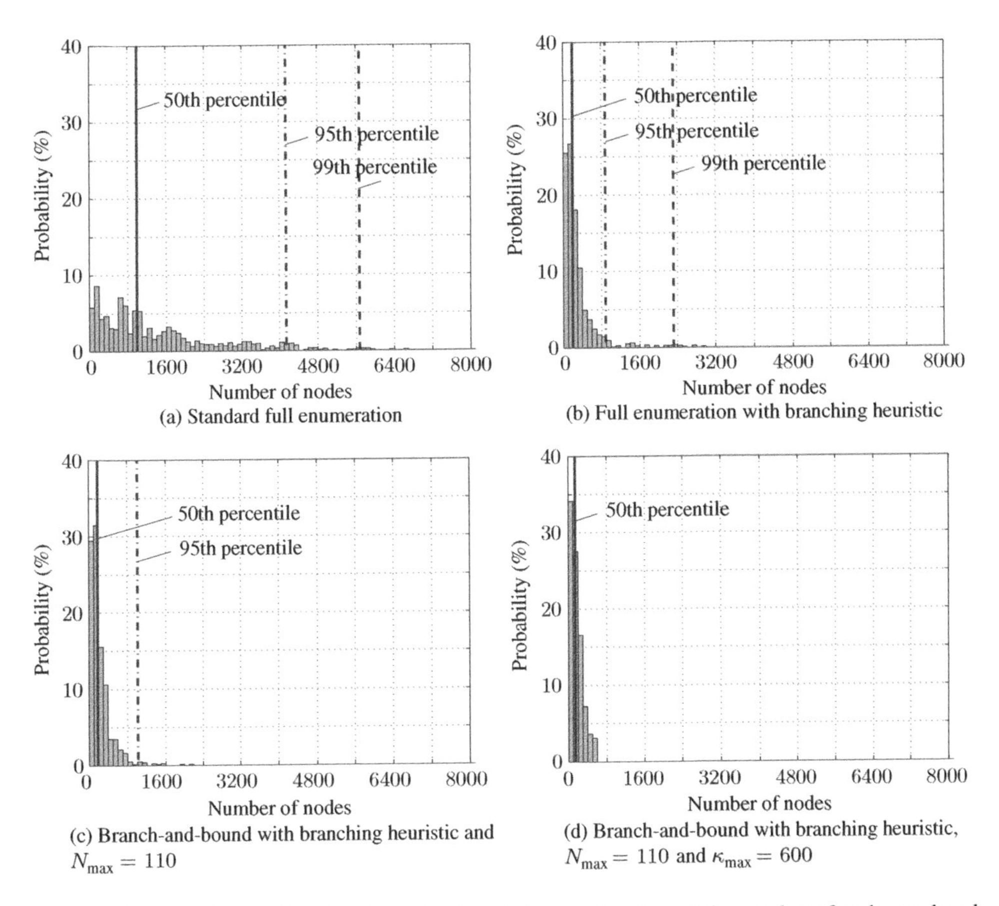

Figure 10.8 Probability of finding the optimal cost J_{opt} as a function of the number of nodes explored for MPDTC with the switching horizon *eSSESESE*

95% of the cases after the exploration of 4150 nodes. Adding the branching heuristic to full-enumeration MPDTC significantly increases the probability of finding the optimal cost during the early stages of the optimization process, as indicated by Fig. 10.8(b). The 95th percentile corresponds to 885 nodes.

Further pruning suboptimal parts of the search tree with the help of the tight upper bound $N_{max} = 110$ shifts the 95th percentile to 1017 nodes (see Fig. 10.8(c)). Note that the 99th percentile does not exist, because the optimal cost is found in only 96.7% of the cases, as seen from Table 10.1. When imposing the upper bound $\kappa_{max} = 600$ on the number of explored nodes, the optimal cost is found in only 92.1% of the cases and the 95th percentile does not exist. A comparison between Fig. 10.8(c) and (d) shows that the application of switching commands that correspond to suboptimal costs slightly modifies the histogram. Applying suboptimal switching commands in closed-loop operation modifies the state and output trajectories of the drive system. This also implies that the search trees of the MPDTC optimization problems differ from the cases in which the optimal cost has been found.

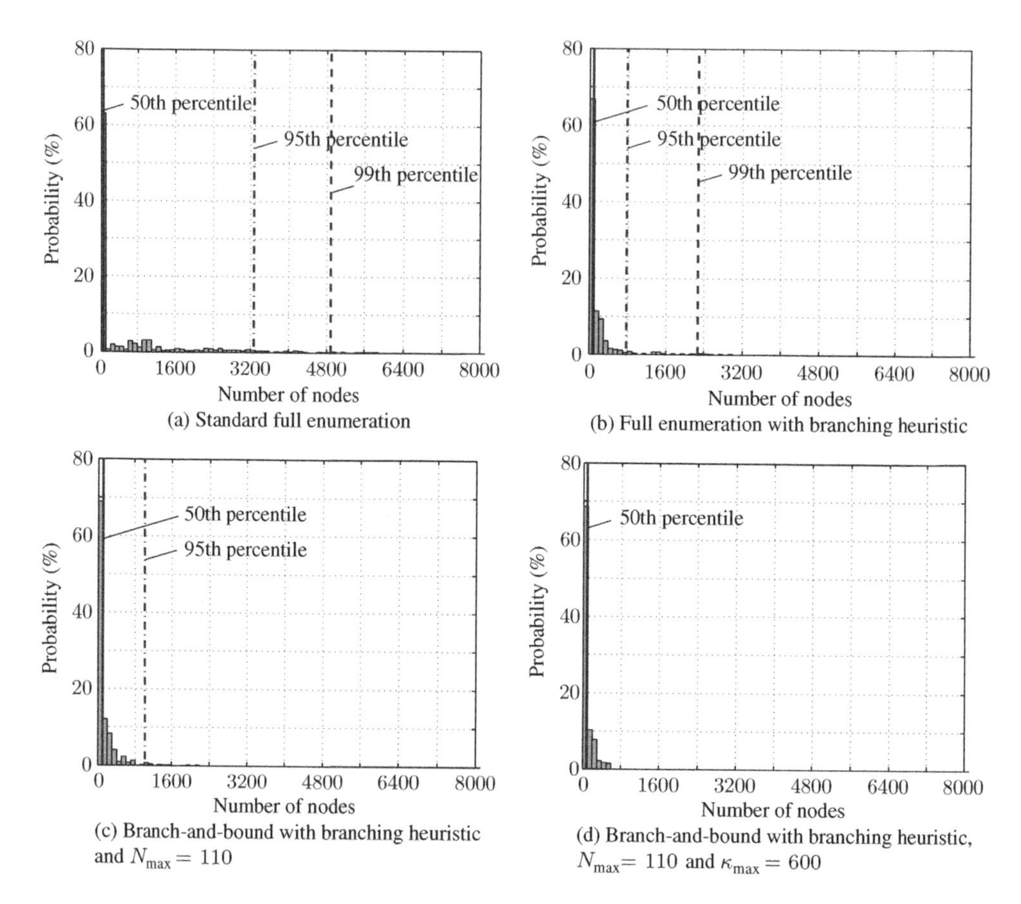

Figure 10.9 Probability of finding the optimal switch position $\boldsymbol{u}_{opt}$ as a function of the number of nodes explored for MPDTC with the switching horizon *eSSESESE*

So far, we have investigated the probability that the optimal cost and thus the optimal switching sequence $\boldsymbol{U}_{opt}$ is found. In the end, however, only the first element of this switching sequence, $\boldsymbol{u}_{opt}$, is applied to the inverter. The corresponding histograms, which indicate after how many steps the optimal switch position $\boldsymbol{u}_{opt}$ is found, are shown in Fig. 10.9. On average, MPDTC with full enumeration finds the optimal switch position quickly, but because of the long tail the 95th percentile is at 3270 nodes (see Fig. 10.9(a)). This indicates that the imposition of an upper bound on the nodes explored is either impractical or would lead to a significant deterioration of the closed-loop performance.

The application of branching heuristics and the pruning of suboptimal parts of the search tree largely removes the long tail and increases the probability of finding the optimal switch position during an early stage of the optimization process. As shown in Fig. 10.9(c), the 95th percentile is shifted to 970 nodes. This served as a motivation to place the upper bound at $\kappa_{max} = 600$.

Even though full-enumeration MPDTC often finds the optimal switching sequence early on, the whole search tree needs to be explored in order to provide a certificate of optimality. As

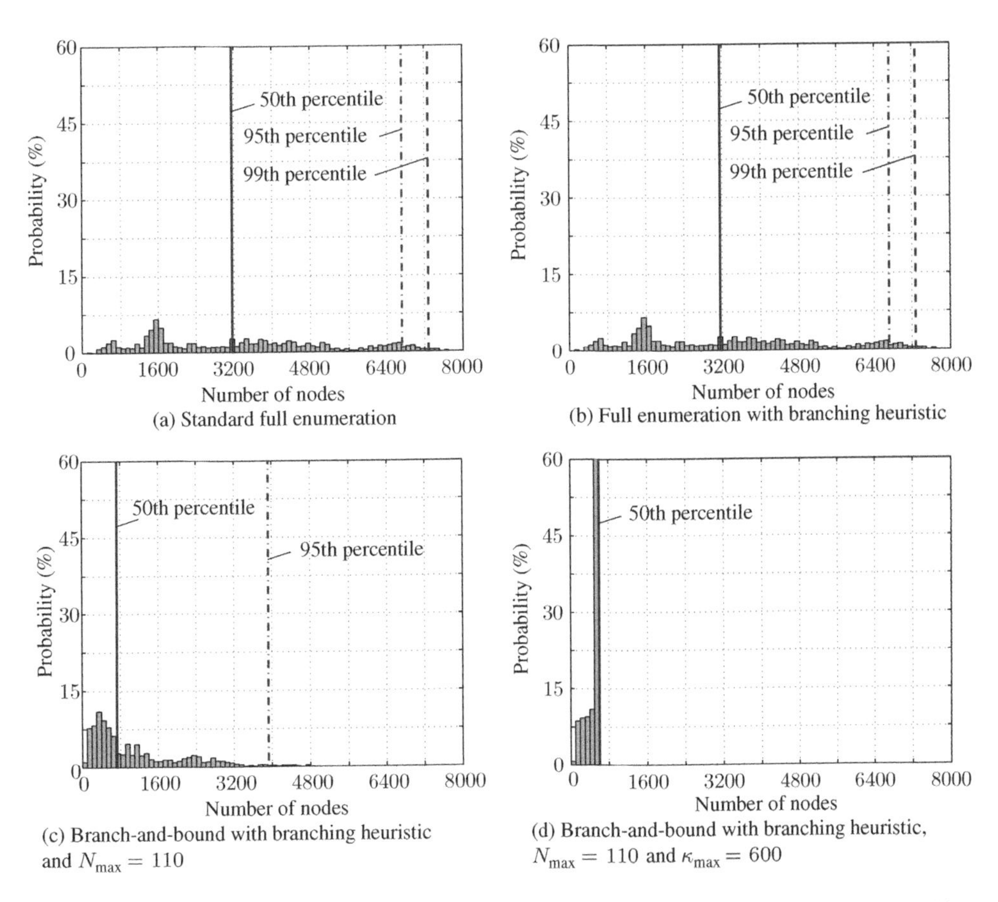

Figure 10.10 Probability of the number of nodes that need to be explored to solve the MPDTC problem with the switching horizon *eSSESESE*

shown in Fig. 10.10(a), up to 7700 nodes must be explored to achieve this. The same applies to the full-enumeration MPDTC with the branching heuristic (see Fig. 10.10(b)). For MPDTC with branch-and-bound, the branching heuristic and $N_{max} = 110$, the maximum number of nodes to be explored until a certificate of optimality is found remains high, see Fig. 10.10(c), but the median (i.e., the 50th percentile) is greatly reduced from 3200 to 726 nodes. Note that the average number of nodes explored is 1102, as stated in Table 10.2. Therefore, to achieve a hard limit on the computational burden, the upper bound $\kappa_{max} = 600$ is required. As discussed previously, the optimal switch position $\boldsymbol{u}_{opt}$ is found in most cases before $\kappa_{max} = 600$ is reached, which explains why the impact on the closed-loop performance is minor.

10.4 Summary and Discussion

This chapter proposed a modified version of the MPDTC algorithm based on branch-and-bound techniques, which have been adopted from mathematical programming. The proposed branch-and-bound algorithm is based on three techniques. First, for each (incomplete) candidate switching sequence, a lower bound on its final cost is computed. Out of the pool of (incomplete) candidate switching sequences, the one with the lowest bound on its cost is chosen to be extended (by applying switching transitions or extension legs to it).

Second, the lowest cost of fully computed candidate switching sequences is monitored. This serves as an upper bound on the final, optimal cost. If the lower cost bound of an incomplete switching sequence exceeds the upper bound on the optimal cost, this switching sequence is suboptimal and it can be removed from further consideration. In doing so, optimality of the final solution is not impacted on, provided that the upper bound on the length of the switching sequences has been correctly chosen. Third, an upper bound on the number of nodes to be explored is imposed. Recall that the number of visited nodes directly corresponds to the computational burden of the MPDTC algorithm. By imposing such an upper bound, the real-time computational burden of the algorithm is limited. This mechanism can be used to ensure that the algorithm always terminates within the allocated time frame.

The first two techniques increase the probability that the optimal switch position is found during an early stage of the optimization process. This enables the use of the third technique, namely the addition of an upper bound on the nodes to be explored, without significantly affecting the closed-loop performance.

Simulation results suggest that the proposed branch-and-bound techniques achieve a reduction of the worst case computational effort by an order of magnitude. In general, the longer the prediction horizon, the more significant is the percentage-wise reduction of the computational burden. This observation makes these techniques particularly attractive for MPDTC with very long prediction horizons. In particular, branch-and-bound methods can make long-horizon MPDTC computationally feasible, enabling one to take full advantage of its performance benefits.

The techniques presented here are equally applicable to the adaptation of MPDTC to the current control problem—model predictive direct current control (MPDCC) (see Sect. 11.1). When considering inverter topologies with a higher number of switching levels, such as five-level topologies, the benefit in terms of a reduction of the computational effort is expected to become even more prominent.

References

[1] M. J. Duran, J. Prieto, F. Barrero, and S. Toral, "Predictive current control of dual three-phase drives using restrained search techniques," *IEEE Trans. Ind. Electron.*, vol. 58, pp. 3253–3263, Aug. 2011.

[2] I. Nowak, *Relaxation and decomposition methods for mixed integer nonlinear programming*. Cambridge, MA: Birkhäuser Verlag, 2000.

11

Derivatives of Model Predictive Direct Torque Control

Model predictive direct torque control (MPDTC), as introduced in Chap. 7, can be considered a refinement of direct torque control (DTC). MPDTC inherits DTC's principle to keep the electromagnetic torque, the stator flux magnitude, and the neutral point potential within given bounds, but MPDTC replaces the DTC hysteresis controller and the look-up table by an optimization problem. The latter consists of three parts: a cost function that captures the predicted switching effort, a controller model of the drive system that endows the controller with the ability to predict the system response to possible switching sequences, and a set of constraints that is imposed on the controlled output variables and the allowed switching transitions.

The switch position is determined by solving this optimization problem online without requiring a modulator. MPDTC considers switching transitions only when one of the output variables is close to one of its bounds; otherwise, the switch position is kept constant. In doing so, very long prediction horizons can be achieved by considering only a few switching transitions. The former determines the closed-loop performance, while the latter directly relates to the computational burden of solving the optimization problem.

The fundamental principle of MPDTC can be adapted to other control problems. Specifically, instead of controlling the torque and the stator flux magnitude, the stator currents of an electrical machine can be kept within upper and lower bounds. The currents of grid-connected converters can be controlled in a similar manner. We refer to this as *model predictive direct current control* (MPDCC). Alternatively, the real and reactive power can be controlled in a converter, by imposing upper and lower bounds on these quantities. This concept is known as *model predictive direct power control* (MPDPC), which generalizes direct power control (DPC). Both MPDCC and MPDPC are introduced in this chapter along with detailed performance evaluations. The chapter concludes with a comparison of the shape of the bounds of MPDTC, MPDCC, and MPDPC.

Model Predictive Control of High Power Converters and Industrial Drives, First Edition. Tobias Geyer.

Companion Website: www.wiley.com/go/geyermodelpredictivecontrol

11.1 Model Predictive Direct Current Control

This section proposes a model predictive current controller with very long prediction horizons in the range of 100 time steps. The proposed MPDCC scheme keeps the stator currents within hexagonal bounds around their references in the stationary orthogonal reference frame. Similarly, the neutral point potential is balanced around zero by imposing upper and lower bounds on it. The switch positions are chosen such that either the switching losses or the switching frequency is minimized.

By addressing the current control and the modulation problems in one computational stage, the harmonic current distortion and the switching losses can be reduced simultaneously, when compared to classical modulation schemes such as carrier-based pulse width modulation (CB-PWM) or space vector modulation (SVM). Indeed, at low pulse numbers, the ratio between harmonic distortions and switching losses is similar to that obtained with optimized pulse patterns (OPPs). During transients, however, very fast current responses are achieved.

This MPDCC scheme can be considered as an adaptation of MPDTC to the current control problem by modifying the control objectives. Instead of controlling the torque and stator flux magnitude, the stator currents are controlled. A preliminary MPDCC scheme based on the initial MPDTC [1] algorithm minimizing the inverter switching frequency and using a relatively short prediction horizon was presented in [2] for a two-level inverter.

Nevertheless, the idea of keeping the stator currents of a machine within bounds, predicting future current trajectories using a model, and minimizing a cost function had already been proposed in the 1980s [3–5], albeit with short prediction horizons and only for two-level converters. Moreover, either circular or rectangular (rather than hexagonal) bounds were imposed on the stator currents in the stationary orthogonal reference frame. These types of bounds lead to a simple control problem formulation, but they are suboptimal in terms of the harmonic current distortions. For an in-depth comparison of MPDCC with these early predictive current control methods, the reader is referred to [6].

11.1.1 Case Study

Consider as a case study the medium-voltage (MV) inverter drive system shown in Fig. 11.1. The MV induction machine is connected to a three-level neutral-point-clamped (NPC) inverter with a floating neutral point potential. The voltages over the upper and lower dc-link capacitors are given by $v_{\mathrm{dc,up}}$ and $v_{\mathrm{dc,lo}}$, respectively. Their sum is the total (instantaneous) dc-link voltage $v_{\mathrm{dc}} = v_{\mathrm{dc,up}} + v_{\mathrm{dc,lo}}$. We use the symbol V_{dc} to refer to the nominal dc-link voltage.

The per-phase switch positions u_a, u_b, and u_c are restricted to the set $\{-1, 0, 1\}$, and the three-phase switch position of the inverter is given by $\boldsymbol{u}_{abc} = [u_a \; u_b \; u_c]^T$. The stator voltage in the stationary orthogonal coordinate systems is

$$\boldsymbol{v}_s = \frac{1}{2} v_{\mathrm{dc}} \, \tilde{\boldsymbol{K}} \, \boldsymbol{u}_{abc}, \tag{11.1}$$

where $\boldsymbol{v}_s = [v_{s\alpha} \; v_{s\beta}]^T$, and $\tilde{\boldsymbol{K}}$ is the transformation matrix (2.13) of the reduced Clarke transformation (2.12).

The evolution of the neutral point potential of the inverter

$$v_n = \frac{1}{2}(v_{\mathrm{dc,lo}} - v_{\mathrm{dc,up}}) \tag{11.2}$$

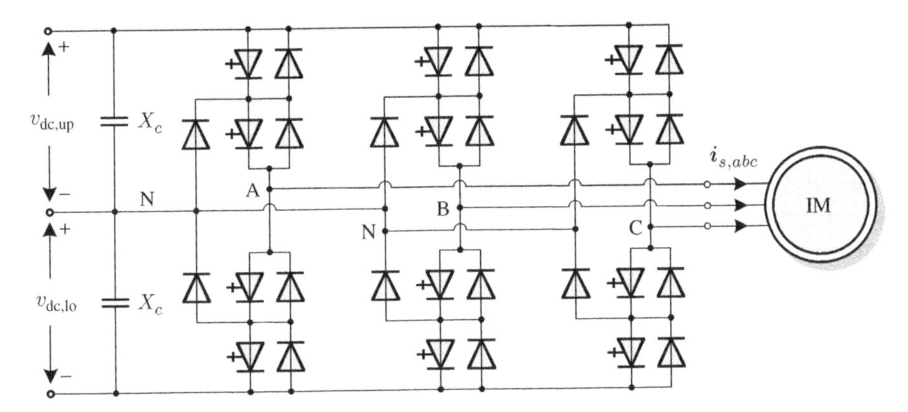

Figure 11.1 Three-level neutral-point-clamped (NPC) voltage source inverter driving an induction machine (IM)

is governed by the differential equation

$$\frac{dv_n}{dt} = \frac{1}{2X_c}|\boldsymbol{u}_{abc}|^T \boldsymbol{i}_{s,abc}. \tag{11.3}$$

Adopting the per unit (pu) system, X_c denotes the pu equivalence of one of the two dc-link capacitors. The component-wise absolute value of the inverter switch position is defined as $|\boldsymbol{u}_{abc}| = [|u_a| \ |u_b| \ |u_c|]^T$, and the three-phase stator current is given by $\boldsymbol{i}_{s,abc} = [i_{sa} \ i_{sb} \ i_{sc}]^T$. For more details on the NPC inverter, see Sect. 2.4.1. The details of the drive system and its parameters are provided in Sect. 2.5.1.

Switching losses arise in the inverter when turning the semiconductors on or off and commutating the phase current with a nonzero blocking voltage. These losses depend on the applied voltage, the commutated current, and the semiconductor's characteristics. Considering integrated-gate-commutated thyristors (IGCTs), with the gate-commutated thyristor (GCT) being the semiconductor switch, the turn-on and turn-off losses can be well approximated to be linear in the dc-link voltage and the phase current. The reverse recovery losses of the diodes are nonlinear in the commutated current. For more details on semiconductors and their associated switching losses, the reader is referred to Sect. 2.3. The specific switching losses of the NPC inverter are explained in Sects. 2.4.1 and 2.5.1.

State-space models of induction machines were derived in Sect. 2.2. For the current control problem at hand, it is convenient to adopt the stationary $\alpha\beta$ reference frame and to choose the stator current $\boldsymbol{i}_s = [i_{s\alpha} \ i_{s\beta}]$ and the rotor flux $\boldsymbol{\psi}_r = [\psi_{r\alpha} \ \psi_{r\beta}]^T$ as state variables. The continuous-time machine model of a squirrel-cage induction machine

$$\frac{d\boldsymbol{i}_s}{dt} = -\frac{1}{\tau_s}\boldsymbol{i}_s + \left(\frac{1}{\tau_r}\boldsymbol{I}_2 - \omega_r \begin{bmatrix} 0 & -1 \\ 1 & 0 \end{bmatrix}\right) \frac{X_m}{D}\boldsymbol{\psi}_r + \frac{X_r}{D}\boldsymbol{v}_s \tag{11.4a}$$

$$\frac{d\boldsymbol{\psi}_r}{dt} = \frac{X_m}{\tau_r}\boldsymbol{i}_s - \frac{1}{\tau_r}\boldsymbol{\psi}_r + \omega_r \begin{bmatrix} 0 & -1 \\ 1 & 0 \end{bmatrix} \boldsymbol{\psi}_r \tag{11.4b}$$

$$\frac{d\omega_r}{dt} = \frac{1}{M}(T_e - T_\ell) \tag{11.4c}$$

is obtained from (2.59) by setting the rotor voltage $\boldsymbol{v}_r$ and the angular speed of the reference frame ω_{fr} to zero.

All parameters and variables are normalized using the pu system, including the time axis. The parameters of the machine model include the stator and rotor resistances R_s and R_r, the stator and rotor leakage reactances X_{ls} and X_{lr}, and the main reactance X_m. The stator and rotor self-reactances are defined as $X_s = X_{ls} + X_m$ and $X_r = X_{lr} + X_m$, respectively. The determinant is $D = X_s X_r - X_m^2$. We have previously introduced the transient stator time constant and the rotor time constant in (2.60) as

$$\tau_s = \frac{X_r D}{R_s X_r^2 + R_r X_m^2} \text{ and } \tau_r = \frac{X_r}{R_r}. \tag{11.5}$$

Note that all rotor quantities are referred to the stator circuit, and $\boldsymbol{I}_2$ denotes the two-dimensional identity matrix.

Furthermore, ω_r denotes the electrical angular rotor speed, M is the moment of inertia, and T_ℓ is the load torque. The electromagnetic torque in terms of the stator current and the rotor flux vector is

$$T_e = \frac{1}{\text{pf}}\frac{X_m}{X_r}\boldsymbol{\psi}_r \times \boldsymbol{i}_s = \frac{1}{\text{pf}}\frac{X_m}{X_r}(\psi_{r\alpha} i_{s\beta} - \psi_{r\beta} i_{s\alpha}) \tag{11.6}$$

according to (2.61).

Equations (11.4)–(11.6) represent the standard dynamic model of a squirrel-cage induction machine, where the saturation of the machine's magnetic material, the changes of the rotor resistance due to the skin effect, and the changes of the stator resistance due to temperature are neglected.

11.1.2 Control Problem

The control problem considered in this section has three aspects. First, the stator currents $\boldsymbol{i}_s$ of the machine must be regulated along their time-varying references $\boldsymbol{i}_s^*$. At steady-state operating conditions, the main performance metric is the harmonic distortion of the current, that is, the total demand distortion (TDD) of the current (3.2). By minimizing the current TDD, the copper losses and, therefore, the thermal losses in the machine windings are reduced. During transients, a high dynamic performance must be ensured with short settling times in the range of a few milliseconds. To achieve this, the fast regulation of the stator currents to their new references is required.

Second, the switching losses in the semiconductors are to be minimized. An indirect way of achieving this is to reduce the device switching frequency, while a direct way is to minimize the predicted switching losses. Third, the neutral point potential of the inverter has to be balanced around zero.

This control problem is described in detail in Sect. 4.2.2. In particular, as shown in Fig. 4.11, the inner stator current control loop is augmented in a cascaded controller fashion by outer control loops that operate in the rotating dq reference frame. These outer loops comprise rotor flux, torque, and speed controllers, which are typically based on proportional–integral (PI) controllers with appropriate feedforward terms.

We have seen in Sect. 3.5 that a fundamental trade-off exists between the current TDD and the switching effort. Lower current TDDs imply higher switching frequencies and losses, and vice versa. We will show in Sect. 11.1.8 that the current TDD is proportional to the current ripple, which is the difference between the stator current and its reference. This allows us to consider the current ripple in the MPDCC problem formulation instead of the current TDD. To simplify the tuning procedure, we fix the current ripple (and thus the current TDD) by imposing upper and lower bounds on the stator currents. The second trade-off quantity—the switching effort—can then be minimized.

11.1.3 Formulation of the Stator Current Bounds

The bounds on the stator currents can be imposed in various ways. Possible bound shapes include circles, rectangles, and hexagons, which can be imposed either in the stationary or rotating coordinate system. The most prominent bound shapes will be discussed in this section.

We consider the use of symmetric bounds around the stator current reference. Let

$$i_{\mathrm{rip},a} = i_{sa}^* - i_{sa} \tag{11.7}$$

denote the ripple current in phase a. The ripple currents in phases b and c are defined accordingly. We introduce the positive parameter δ_i to describe the difference between the upper (or lower) bound and the current reference.

The natural choice [7] is to impose upper and lower bounds on the abc ripple currents in the form

$$|i_{\mathrm{rip},a}| \leq \delta_i \tag{11.8a}$$

$$|i_{\mathrm{rip},b}| \leq \delta_i \tag{11.8b}$$

$$|i_{\mathrm{rip},c}| \leq \delta_i\,. \tag{11.8c}$$

Figure 11.2(a) shows the a-, b-, and c-axes, which are displaced by $2\pi/3$ with respect to each other. The upper and lower bounds form lines that are perpendicular to these axes and displaced by δ_i from the origin. The set of ripple currents that meets the constraints (11.8) lies between these lines. The latter are edges (or facets) that form a polygon, which is shown in Fig. 11.2(a) as the gray hexagon.

Using the reduced Clarke transformation (2.12), the abc ripple currents in (11.8) can be expressed in terms of the $\alpha\beta$ ripple currents. This translates the bounds from the abc coordinate system into the stationary $\alpha\beta$ reference frame, in which we have

$$|i_{\mathrm{rip},\alpha}| \leq \delta_i \tag{11.9a}$$

$$|i_{\mathrm{rip},\alpha} - \sqrt{3} i_{\mathrm{rip},\beta}| \leq 2\delta_i \tag{11.9b}$$

$$|i_{\mathrm{rip},\alpha} + \sqrt{3} i_{\mathrm{rip},\beta}| \leq 2\delta_i\,. \tag{11.9c}$$

It turns out that the bounds (11.9b) and (11.9c) can also be imposed by the compact statement $|i_{\mathrm{rip},\alpha}| + \sqrt{3}|i_{\mathrm{rip},\beta}| \leq 2\delta_i$.

The set of ripple currents in $\alpha\beta$ that meets (11.9) is depicted in Fig. 11.2(b) as the gray polygon. It is clear that symmetric upper and lower bounds imposed on the three-phase currents in the abc coordinate system are equivalent to hexagonal bounds in $\alpha\beta$ coordinates. Moreover,

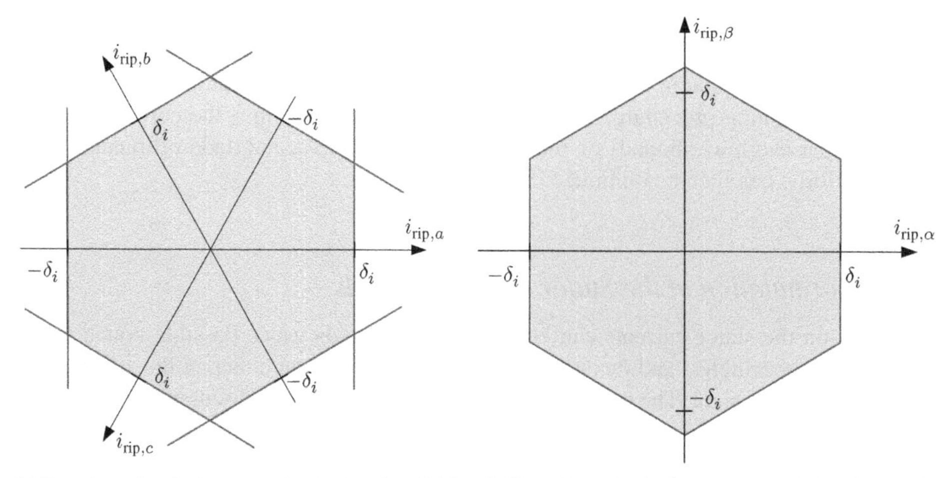

(a) Bounds on the ripple current in abc based on (11.8) (b) Bounds on the ripple current in $\alpha\beta$ based on (11.9)

Figure 11.2 Bounds on the ripple current. The upper and lower bounds on the phase currents are translated into hexagonal bounds in the stationary orthogonal reference frame

because the star point of the machine is not connected, the sum of the three-phase currents is zero at any instant in time, that is, it is common-mode-free. This statement also holds true for the ripple currents.

Alternatively, the set defined by the bounds (11.8) can be represented in an abc system, in which the three axes are orthogonal to each other. As the common-mode component of the ripple current is zero, this set is of dimension two. Figure 11.3(a) shows the projection of this set onto a plane spanned by orthogonal a- and b-axes, while Fig. 11.3(b) shows the projection onto the bc-plane. The projection onto the ac-plane is the same as in Fig. 11.3(a).

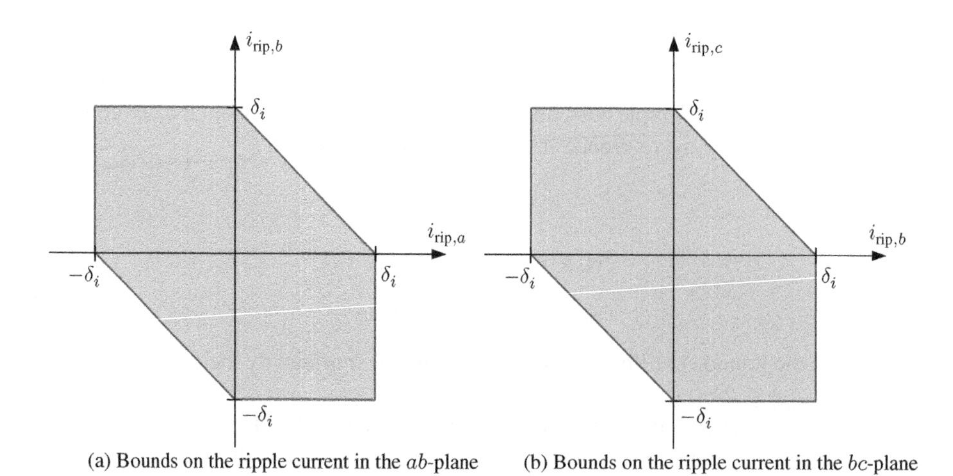

(a) Bounds on the ripple current in the ab-plane (b) Bounds on the ripple current in the bc-plane

Figure 11.3 Projection of the ripple current bounds (11.8) onto the ab- and bc-planes

The fact that the ripple current is common-mode-free adds a coupling constraint between the phases, which brings about non-square sets in the ab-, bc-, and ac-planes. The ripple current in phase c, for example, is given by $i_{\text{rip},c} = -i_{\text{rip},a} - i_{\text{rip},b}$. The upper constraint in (11.8c), $i_{\text{rip},c} \leq \delta_i$, is equivalent to the constraint

$$i_{\text{rip},b} \geq -i_{\text{rip},a} - \delta_i \tag{11.10}$$

in the ab-plane, which removes the bottom left part from the ripple current set in Fig. 11.3(a). Similarly, the lower constraint in (11.8c), $i_{\text{rip},c} \geq -\delta_i$, removes the top right part from the ripple current set.

Conversely, one might impose upper and lower bounds on the currents in the $\alpha\beta$ reference frame, as proposed, for example, in [2]:

$$|i_{\text{rip},\alpha}| \leq \delta_i \tag{11.11a}$$

$$|i_{\text{rip},\beta}| \leq \delta_i. \tag{11.11b}$$

These constraints and the square set they form are visualized in Fig. 11.4(a). The hexagonal set that corresponds to (11.9) is indicated by dotted lines. Square bounds in the $\alpha\beta$ frame relate to non-constant bounds in the abc system.

It is obvious that the two constraint formulations (11.8) and (11.11) lead to different sets. By definition, the harmonic current distortion relates to the ripple current in the abc system. Thus, from a TDD perspective, it is advantageous to impose the constraint (11.8) rather than (11.11). This is confirmed by simulation results, even though the differences are relatively small, amounting to only a few percent.

If an approximation of the hexagon is desired, circular bounds of the form

$$i_{\text{rip},\alpha}^2 + i_{\text{rip},\beta}^2 \leq \delta_i^2 \tag{11.12}$$

provide a viable alternative. The circular set they define is shown in Fig. 11.4(b). Circular bounds were initially proposed in [3].

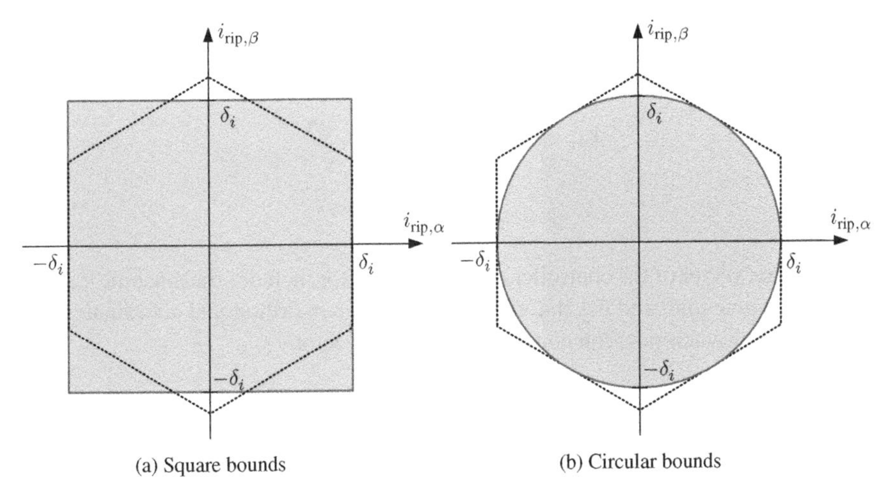

Figure 11.4 Alternative definitions of bounds on the ripple current in the stationary orthogonal reference frame and comparison with the hexagonal bounds (dotted lines)

Bounds can also be imposed on the ripple current in an orthogonal and rotating dq reference frame. When aligning the reference frame with the rotor flux vector, as proposed in [4], the torque can be controlled by imposing upper and lower bounds on the q-component of the stator current. Similarly, upper and lower bounds can be imposed on the d-component of the stator current to control the magnetization of the machine. Strictly speaking, however, radial bounds should be adopted, as discussed in Sect. 9.1. It should be clear from the previous discussion that rectangular bounds tend to yield inferior current TDDs to hexagonal and circular bounds.

As the stator currents of the machine model will be formulated in the three-phase abc system, it is convenient to also formulate the current constraints in this system. Therefore, we adopt the constraint formulation (11.8) in MPDCC. In order to impose upper and lower bounds on the ripple currents, the trajectory of the stator current reference must be predicted over the time interval of the prediction horizon. To this end, we assume that the current references are sinusoidal waveforms with the angular stator frequency ω_s. The evolution of the stator current reference in orthogonal coordinates is then described by the differential equation

$$\frac{\mathrm{d}\boldsymbol{i}_s^*(t)}{\mathrm{d}t} = \boldsymbol{F}_r \boldsymbol{i}_s^*(t) \text{ with } \boldsymbol{F}_r = \omega_s \begin{bmatrix} 0 & -1 \\ 1 & 0 \end{bmatrix}. \tag{11.13}$$

11.1.4 Controller Model

MPDCC relies on a model of the physical drive system to predict future trajectories of the stator current and neutral point potential. As the prediction horizon is in the range of a few milliseconds, we may assume that the rotor speed is effectively constant within the prediction horizon. This allows us to treat the speed as a time-varying parameter and to reduce the dimension of the state-space model. Nevertheless, for highly dynamic drives or for drives with little inertia, it might be necessary to include the speed as an additional state in the model.

We choose the stator current and the rotor flux as state variables for the electrical machine. To simplify the imposition of upper and lower bounds on the three-phase stator currents, we model the stator currents in the three-phase abc system. This choice also tends to simplify the extension mechanism in MPDCC (see the corresponding description in Sect. 7.5 for MPDTC). The rotor flux vector is modeled in stationary orthogonal coordinates. To this end, we define the machine state vector

$$\boldsymbol{x}_m = [i_{sa}\ i_{sb}\ i_{sc}\ \psi_{r\alpha}\ \psi_{r\beta}]^T \tag{11.14}$$

and choose

$$\boldsymbol{x} = [\boldsymbol{x}_m^T\ v_n\ (\boldsymbol{i}_s^*)^T]^T \tag{11.15}$$

as the overall state vector of the controller model. This vector includes the machine state vector, the neutral point potential, and the stator current reference in orthogonal coordinates.

The three-phase switch position constitutes the input vector

$$\boldsymbol{u} = \boldsymbol{u}_{abc} = [u_a\ u_b\ u_c]^T \in \{-1, 0, 1\}^3. \tag{11.16}$$

The three-phase ripple current of the stator along with the neutral point potential forms the output vector

$$\boldsymbol{y} = [i_{\text{rip},a}\ i_{\text{rip},b}\ i_{\text{rip},c}\ v_n]^T. \tag{11.17}$$

The continuous-time state-space representation of the drive system is given by

$$\frac{\mathrm{d}\boldsymbol{x}(t)}{\mathrm{d}t} = \boldsymbol{F}(\boldsymbol{u}(t))\,\boldsymbol{x}(t) + \boldsymbol{G}\,\boldsymbol{u}(t) \tag{11.18a}$$

$$\boldsymbol{y}(t) = \boldsymbol{C}\boldsymbol{x}(t) \tag{11.18b}$$

with the system, input, and output matrices

$$\boldsymbol{F}(\boldsymbol{u}) = \begin{bmatrix} \boldsymbol{F}_m & \boldsymbol{0}_{5\times 1} & \boldsymbol{0}_{5\times 2} \\ \boldsymbol{f}_i(\boldsymbol{u}) & 0 & \boldsymbol{0}_{1\times 2} \\ \boldsymbol{0}_{2\times 5} & \boldsymbol{0}_{2\times 1} & \boldsymbol{F}_r \end{bmatrix}, \quad \boldsymbol{G} = \begin{bmatrix} \boldsymbol{G}_m \\ \boldsymbol{0}_{1\times 3} \\ \boldsymbol{0}_{2\times 3} \end{bmatrix}, \tag{11.19a}$$

$$\boldsymbol{C} = \begin{bmatrix} -1 & 0 & 0 & 0\;0\;0 & 1 & 0 \\ 0 & -1 & 0 & 0\;0\;0 & -1/2 & \sqrt{3}/2 \\ 0 & 0 & -1 & 0\;0\;0 & -1/2 & -\sqrt{3}/2 \\ 0 & 0 & 0 & 0\;0\;1 & 0 & 0 \end{bmatrix}. \tag{11.19b}$$

The machine's system and input matrices $\boldsymbol{F}_m$ and $\boldsymbol{G}_m$, and the system vector of the inverter $\boldsymbol{f}_i(\boldsymbol{u})$ are derived and provided in Appendix 11.A. The submatrix $\boldsymbol{F}_r$ can be found in (11.13). The output matrix $\boldsymbol{C}$ directly follows from the definition of the three-phase ripple current

$$\boldsymbol{i}_{\text{rip},abc} = \tilde{\boldsymbol{K}}^{-1}\boldsymbol{i}_s^* - \boldsymbol{i}_{s,abc}, \tag{11.20}$$

see also (11.7). The pseudo-inverse of the reduced Clarke transformation $\tilde{\boldsymbol{K}}^{-1}$ (see (2.13)) translates the stator current reference from orthogonal $\alpha\beta$ coordinates to the three-phase abc system.

MPDCC requires the discrete-time representation of the state-space model (11.18), which is valid at the discrete-time steps $t = kT_s$, where $k \in \mathbb{N}$ denotes the time step and T_s is the sampling interval. To time-discretize (11.18), we follow the procedure that was proposed in Sect. 7.2.3 for MPDTC. Specifically, we use the exact Euler discretization method for the machine model, forward Euler discretization for the neutral point potential, and exact Euler discretization for the prediction of the stator current reference. This leads to the discrete-time model

$$\boldsymbol{x}(k+1) = \boldsymbol{A}(\boldsymbol{u}(k))\,\boldsymbol{x}(k) + \boldsymbol{B}\,\boldsymbol{u}(k) \tag{11.21a}$$

$$\boldsymbol{y}(k) = \boldsymbol{C}\boldsymbol{x}(k) \tag{11.21b}$$

with the matrices

$$\boldsymbol{A}(\boldsymbol{u}) = \begin{bmatrix} \boldsymbol{A}_m & \boldsymbol{0}_{5\times 1} & \boldsymbol{0}_{5\times 2} \\ \boldsymbol{f}_i(\boldsymbol{u})T_s & 1 & \boldsymbol{0}_{1\times 2} \\ \boldsymbol{0}_{2\times 5} & \boldsymbol{0}_{2\times 1} & \boldsymbol{A}_r \end{bmatrix} \text{ and } \boldsymbol{B} = \begin{bmatrix} \boldsymbol{B}_m \\ \boldsymbol{0}_{1\times 3} \\ \boldsymbol{0}_{2\times 3} \end{bmatrix}. \tag{11.22}$$

The submatrices are given by

$$\boldsymbol{A}_m = \boldsymbol{e}^{\boldsymbol{F}_m T_s}, \boldsymbol{A}_r = \boldsymbol{e}^{\boldsymbol{F}_r T_s} \text{ and } \boldsymbol{B}_m = -\boldsymbol{F}_m^{-1}\,(\boldsymbol{I}_5 - \boldsymbol{A}_m)\,\boldsymbol{G}_m. \tag{11.23}$$

11.1.5 Control Problem Formulation

As stated in Sect. 11.1.2, the control objectives are to keep the instantaneous stator current components within given bounds around their respective references and to balance the neutral point potential around zero, while minimizing the switching losses. These control objectives are mapped into a cost function that yields a scalar cost (here the short-term switching losses) that is minimized subject to the dynamic evolution of the internal controller model of the drive system and subject to constraints. This leads to the optimization problem

$$\boldsymbol{U}_{\text{opt}}(k) = \arg \underset{\boldsymbol{U}(k)}{\text{minimize}}\ J \tag{11.24a}$$

$$\text{subject to } \boldsymbol{x}(\ell+1) = \boldsymbol{A}(\boldsymbol{u}(\ell))\boldsymbol{x}(\ell) + \boldsymbol{B}\boldsymbol{u}(\ell) \tag{11.24b}$$

$$\boldsymbol{y}(\ell+1) = \boldsymbol{C}\boldsymbol{x}(\ell+1) \tag{11.24c}$$

$$\begin{cases} \varepsilon_j(\ell+1) = 0, & \text{if } \varepsilon_j(\ell) = 0 \\ \varepsilon_j(\ell+1) < \varepsilon_j(\ell), & \text{if } \varepsilon_j(\ell) > 0 \end{cases} \tag{11.24d}$$

$$\boldsymbol{u}(\ell) \in \boldsymbol{\mathcal{U}}\,, \ \ ||\Delta \boldsymbol{u}(\ell)||_\infty \le 1 \tag{11.24e}$$

$$\forall \ell = k, \dots, k+N_p-1, \forall j = 1,2,3,4. \tag{11.24f}$$

The cost function J is minimized subject to three sets of constraints. The equality constraints (11.24b) and (11.24c) represent the controller model and thus describe the dynamic evolution of the drive system (and its current references) over the prediction horizon N_p when applying the sequence of control inputs (or switching sequence)

$$\boldsymbol{U}(k) = [\boldsymbol{u}^T(k)\ \boldsymbol{u}^T(k+1)\ \dots\ \boldsymbol{u}^T(k+N_p-1)]^T. \tag{11.25}$$

The second set of constraints (11.24d) imposes bounds on the output variables. As shown in Fig. 11.2(a), we impose symmetric upper and lower bounds on the three-phase stator ripple current. For the ripple current in phase a (see (11.7)), we define the nonnegative variable

$$\varepsilon_a(\ell) = \frac{1}{2\delta_i}\begin{cases} |i_{\text{rip},a}(\ell)| - \delta_i & \text{if } |i_{\text{rip},a}(\ell)| > \delta_i \\ 0 & \text{else}, \end{cases} \tag{11.26}$$

which is normalized by the bound width. The variables ε_b and ε_c are defined accordingly. Similarly, symmetric upper and lower bounds are imposed on the neutral point potential v_n, using the parameter δ_v and the degree of the bound violation ε_v.

We aggregate these bound violations to the vector of bound violations

$$\boldsymbol{\varepsilon} = [\varepsilon_a\ \varepsilon_b\ \varepsilon_c\ \varepsilon_v]^T. \tag{11.27}$$

Its first three components relate to the three-phase stator current, while the fourth component denotes the degree of the bound violation of the neutral point potential. We use the index j to denote the jth component of the vector $\boldsymbol{\varepsilon}$.

The third set of constraints (11.24e) limits the manipulated variable $\boldsymbol{u}$ to the integer values $\{-1, 0, 1\}^3$, which are available for a three-level inverter. Switching between the upper and the lower rail is inhibited by the second constraint in (11.24e), where we have introduced $\Delta \boldsymbol{u}(\ell) = \boldsymbol{u}(\ell) - \boldsymbol{u}(\ell - 1)$.

The cost function J is the same as for MPDTC. It consists of the three terms

$$J = J_{\text{sw}} + J_{\text{bnd}} + J_t \,. \tag{11.28}$$

The switching effort J_{sw} represents either the short-term switching frequency (7.31) or the switching power losses (7.32), as described in detail in Sect. 7.3.3.

The second term in the cost function $J_{\text{bnd}} = \boldsymbol{q}^T \boldsymbol{\epsilon}$ penalizes the root-mean-square (rms) bound violation of the output vector, which is defined as

$$\boldsymbol{\epsilon} = \left[\epsilon_a \; \epsilon_b \; \epsilon_c \; \epsilon_v\right]^T . \tag{11.29}$$

Its first three components relate to the three-phase ripple current, while the fourth one is the rms bound violation of the neutral point potential. The four components are defined in accordance with (7.35).

The last term in the cost function J_t is an optional penalty on the predicted output quantities that is imposed on them at the end of the prediction horizon. This term can be used to reduce the likelihood of the so-called deadlocks. For more details on the formulation of the cost function, deadlocks, and schemes to avoid them, the reader is referred to Sects. 7.3.3, 9.3, and 9.5.

11.1.6 MPDCC Algorithm

The block diagram of MPDCC including the outer flux and speed control loops is provided in Fig. 4.11. The MPDCC algorithm is similar to the MPDTC algorithm described in Sect. 7.4.5. In particular, the MPDCC algorithm adopts the concepts of the switching and prediction horizons, the full enumeration of candidate switching sequences in a search tree, the extension of output trajectories until a bound violation is predicted to occur, and the target set.

The main difference between the MPDCC and the MPDTC algorithms concerns the characteristic and handling of the bounds on the output variables. In MPDTC, the references on the output variables are constant within the prediction horizon, which implies that the upper and lower bounds are also constant. This greatly simplifies the handling of the bounds.

In MPDCC, on the other hand, the bounds on the stator currents are time-varying. The upper bound in phase a, for example, is equal to the phase a current reference plus δ_i. This is exemplified in Fig. 11.5(a). Time-varying bounds complicate the extension step. To mitigate this issue, we choose as output variables the stator *ripple* currents rather than the stator currents. This renders the upper and lower bounds time-invariant and constant throughout the prediction horizon, as shown in Fig. 11.5(b). As a result, the MPDTC extension mechanism as described in Sect. 7.5 is also applicable to MPDCC.

Nevertheless, the future stator current references need to be predicted. This adds two additional state variables to the controller model and makes the control problem formulation (11.24) more complex to solve. To reduce the computational burden, the controller model can be split into two. One model captures the evolution of the stator current reference and rotor flux vector. The weak coupling from the stator to the rotor flux can either be neglected, or a nominal stator flux evolution can be assumed, which relates the stator flux to the stator current reference. The response of this model is independent of the switching sequence and needs to be computed only once at time step k.

The second model predicts the trajectories of the ripple current and the neutral point potential, taking into account the predicted trajectories of the rotor flux and stator current reference.

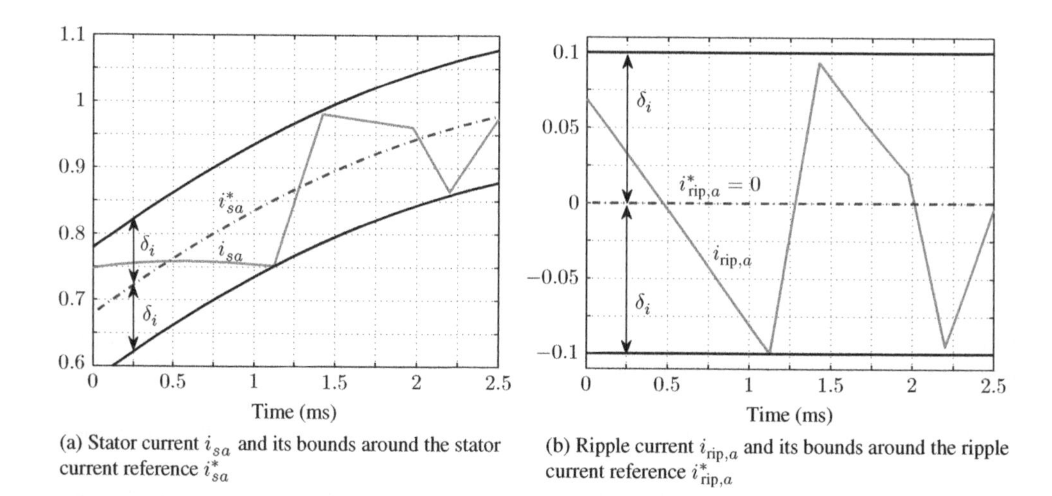

(a) Stator current i_{sa} and its bounds around the stator current reference i^*_{sa}

(b) Ripple current $i_{\text{rip},a}$ and its bounds around the ripple current reference $i^*_{\text{rip},a}$

Figure 11.5 Imposition of upper and lower bounds on the stator current and its corresponding ripple current. Phase a is used as an example

This model uses the three-phase stator current and the neutral point potential as state variables, the three-phase switch position as input variable, and the rotor flux and stator current reference in orthogonal coordinates as parameters.

Alternatively, MPDCC can be formulated in a dq reference frame, which rotates synchronously with the rotor flux vector. The current references are constant in this reference frame, and so are the upper and lower bounds. The hexagonal bounds, however, rotate. A possible simplification is to approximate the hexagon by a circle, as shown in Fig. 11.4(b) and originally proposed in [3].

Moreover, when formulating the controller model in the dq reference frame, the three-phase switch position needs to be transformed into the rotating frame. This is achieved with the Park transformation, which requires the angular position of the reference frame. Furthermore, to predict the evolution of the neutral point potential, the stator currents are required either in the three-phase abc system or in the stationary $\alpha\beta$ reference frame. This necessitates the use of the inverse Park transformation, further complicating the prediction of the state and output vectors.

We conclude that, when choosing the reference frame, a trade-off emerges between the necessity to predict stator current reference trajectories and the frequent use of the (inverse) Park transformation. In this section, we opt for the simplest approach, using one controller model with an eight-dimensional state vector. The latter includes the three-phase stator current, rotor flux in orthogonal coordinates, neutral point potential, and stator current reference in orthogonal coordinates (see (11.15)).

11.1.7 Performance Evaluation

Consider the three-level NPC voltage source inverter with a floating neutral point potential shown in Fig. 11.1. The total dc-link voltage is equal to its nominal value $V_{\text{dc}} = 5.2\,\text{kV}$. The inverter is connected to an MV induction machine with a constant mechanical load. The drive system case study is summarized in Sect. 2.4.1, and the detailed parameters of the machine and inverter are provided in Table 2.10. The MPDCC cost function minimizes the predicted switching power losses. Bound violations and output quantities at the end of the prediction horizon are not penalized. Therefore, the cost function is simply given by $J = J_P$, with J_P as in (7.32).

The stator current reference is provided by outer control loops, which are shown in Fig. 4.11. Symmetric bounds around the three-phase stator current references are imposed, and the ripple current in each phase is kept within $[-\delta_i,\ \delta_i]$. The width $2\delta_i$ of the current bounds determines the trade-off between the switching losses and the current distortions. This will be shown in Sect. 11.1.8, making δ_i the main tuning parameter of MPDCC. A second parameter defines the bounds on the neutral point potential, which is set to $\delta_v = 0.04$ pu. This implies that the neutral point potential is kept within the range $[-0.04,\ 0.04]$ pu. Operation at steady-state conditions is ensured by starting the main simulation only after initial transients have settled down. The sampling interval is set to $T_s = 25$ μs.

11.1.7.1 Steady-State Operation

Closed-loop simulations at 60% speed and rated torque are shown in the following to assess the steady-state performance of MPDCC. The key performance criteria are the switching losses in the inverter, the average switching frequency per semiconductor device, and the harmonic current and torque distortions. The performance evaluation is done for switching horizons of varying length and for various current bound widths. MPDCC is benchmarked with two well-established control and modulation methods: field-oriented control (FOC) with SVM and scalar control with OPPs.

The SVM switching signals are generated by a three-level, asymmetric, regularly sampled CB-PWM. The two triangular carriers are in phase and have the carrier frequency f_c. The three-phase reference voltage is augmented by the common-mode voltage term (3.16), which is of the min/max type plus a modulus operation. The resulting gating signals are the same as for SVM, as explained in Sect. 3.3.2. We will therefore refer to this modulation method as SVM.

Alternatively, OPPs can be calculated in an offline procedure by computing the optimal switching angles over a quarter of the fundamental period for all possible operating points. This minimizes the current distortions for a given pulse number d (number of switching transitions per phase within a quarter of the fundamental period). The concept of OPPs is summarized in detail in Sect. 3.4.

Table 11.1 summarizes selected closed-loop simulations by stating the controller settings and the resulting performance metrics. The latter are provided as absolute values. Three sets of simulations were run. The first set uses a switching frequency of 60 Hz, while the second and third sets result in switching losses of around 3.1 and 7.8 kW, respectively. Table 11.2 summarizes the same simulation results as in Table 11.1, but represents the performance metrics relative to the SVM baseline.

Selected waveforms over one fundamental period of 33.3 ms are shown in Figs. 11.6(a)–11.11(a) for SVM based on the carrier frequency $f_c = 270$ Hz. The repetitive nature of the switching pattern implies that the corresponding harmonic spectra are discrete. Figure 11.11(a) shows the switching pattern along with the phase currents. Owing to the fixed modulation cycle, the switching transitions are evenly distributed over the fundamental period, which

Table 11.1 Comparison of MPDCC with SVM and OPPs using absolute values for the performance metrics

Control scheme	Control setting	Switching horizon	Average prediction horizon (steps)	P_{sw} (kW)	f_{sw} (Hz)	I_{TDD} (%)	T_{TDD} (%)
SVM	$f_c = 90$ Hz	—	—	1.25	60.0	17.5	5.77
MPDCC	$\delta_i = 0.21$ pu	*eSE*	63.2	1.00	61.0	10.7	5.09
OPP	$d = 2$	—	—	1.42	60.0	10.4	5.14
SVM	$f_c = 270$ Hz	—	—	3.04	150	8.63	3.28
MPDCC	$\delta_i = 0.144$ pu	*eSE*	27.6	3.06	170	10.4	8.29
MPDCC	$\delta_i = 0.099$ pu	*eSESE*	55.4	3.08	162	6.42	4.17
MPDCC	$\delta_i = 0.095$ pu	*eSESESE*	82.4	3.04	166	6.09	4.01
OPP	$d = 5$	—	—	3.11	150	5.51	2.80
SVM	$f_c = 720$ Hz	—	—	7.98	375	3.13	1.33
MPDCC	$\delta_i = 0.054$ pu	*eSE*	11.6	7.76	406	3.67	2.85
MPDCC	$\delta_i = 0.042$ pu	*eSESE*	21.8	7.66	420	2.68	1.74
MPDCC	$\delta_i = 0.040$ pu	*eSESESE*	31.1	7.95	470	2.54	1.67
OPP	$d = 12$	—	—	7.70	360	2.37	1.48

These metrics include the switching losses P_{sw}, the switching frequency f_{sw}, the current TDD I_{TDD}, and the torque TDD T_{TDD}. The three sets of comparisons refer to a switching frequency of 60 Hz, and to switching losses of around 3.1 and 7.8 kW, respectively. The operating point is at 60% speed and rated torque.

Table 11.2 Comparison of MPDCC with SVM and OPPs using relative values for the performance metrics

Control scheme	Control setting	Switching horizon	Average prediction horizon (steps)	P_{sw} (%)	f_{sw} (%)	I_{TDD} (%)	T_{TDD} (%)
SVM	$f_c = 90$ Hz	—	—	100	100	100	100
MPDCC	$\delta_i = 0.21$ pu	*eSE*	63.2	80.4	102	60.8	88.2
OPP	$d = 2$	—	—	114	100	59.5	89.1
SVM	$f_c = 270$ Hz	—	—	100	100	100	100
MPDCC	$\delta_i = 0.144$ pu	*eSE*	27.6	101	113	120	253
MPDCC	$\delta_i = 0.099$ pu	*eSESE*	55.4	102	108	74.4	127
MPDCC	$\delta_i = 0.095$ pu	*eSESESE*	82.4	100	111	70.6	122
OPP	$d = 5$	—	—	103	100	63.8	85.4
SVM	$f_c = 720$ Hz	—	—	100	100	100	100
MPDCC	$\delta_i = 0.054$ pu	*eSE*	11.6	97.2	108	117	215
MPDCC	$\delta_i = 0.042$ pu	*eSESE*	21.8	95.9	112	85.5	131
MPDCC	$\delta_i = 0.040$ pu	*eSESESE*	31.1	99.6	125	81.2	126
OPP	$d = 12$	—	—	96.5	96.0	75.7	111

The same results as in Table 11.1 are shown, using SVM as a baseline.

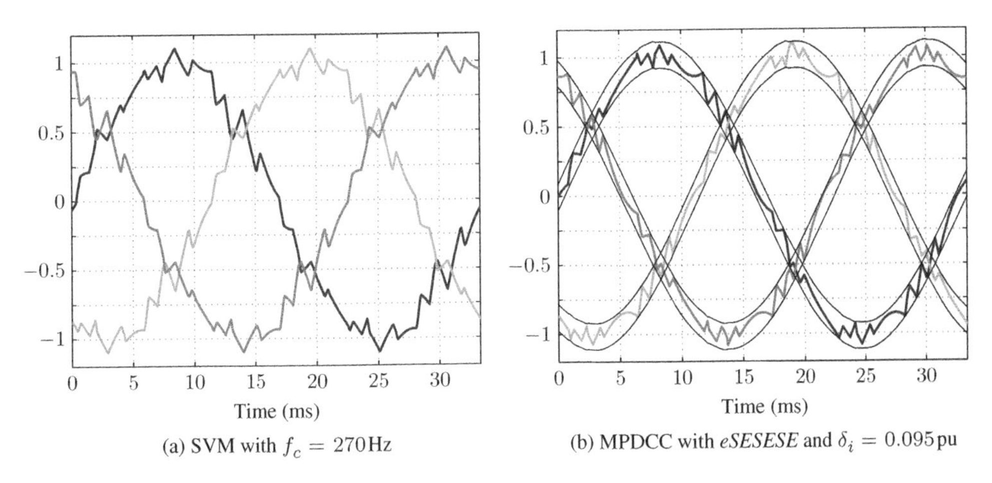

(a) SVM with $f_c = 270\,\text{Hz}$ (b) MPDCC with *eSESESE* and $\delta_i = 0.095\,\text{pu}$

Figure 11.6 Three-phase stator currents for SVM and MPDCC, when operating at 60% speed and rated torque. Both schemes yield the same switching losses

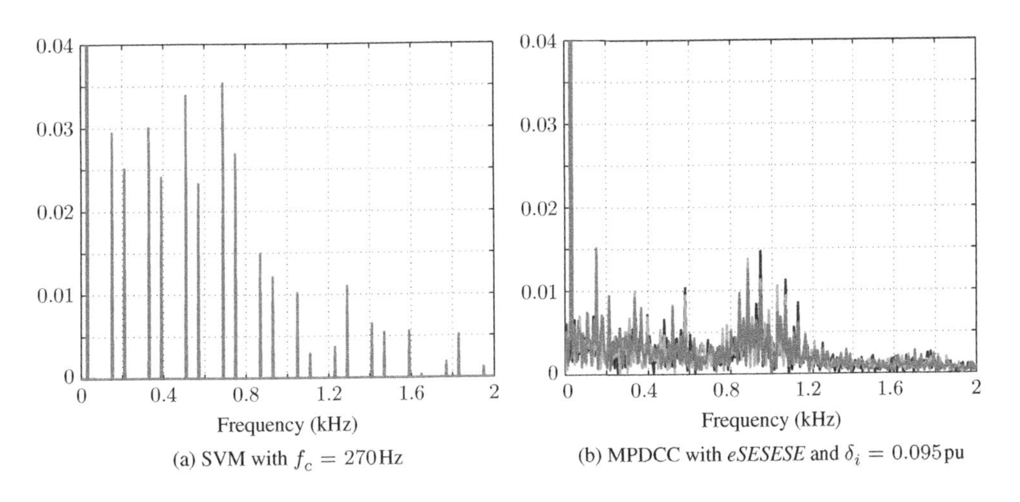

(a) SVM with $f_c = 270\,\text{Hz}$ (b) MPDCC with *eSESESE* and $\delta_i = 0.095\,\text{pu}$

Figure 11.7 Spectra of the three-phase stator currents that are shown in Fig. 11.6

implies that several transitions occur at high phase currents. The resulting switching losses are 3.04 kW, and the current TDD is 8.63%, as summarized in Table 11.1.

The MPDCC current ripple can be adjusted by tuning the stator current bounds such that similar switching losses are obtained. As the switching horizon is increased from *eSE* to *eSESESE*, the average prediction horizon increases, allowing MPDCC to make better informed decisions by looking further into the future. As a result, the bounds can be tightened, thereby reducing the harmonic current and torque distortions, while keeping the switching losses constant. Figures 11.6(b)–11.11(b) show selected waveforms for MPDCC with the long switching horizon *eSESESE* and the bound parameter $\delta_i = 0.095\,\text{pu}$.

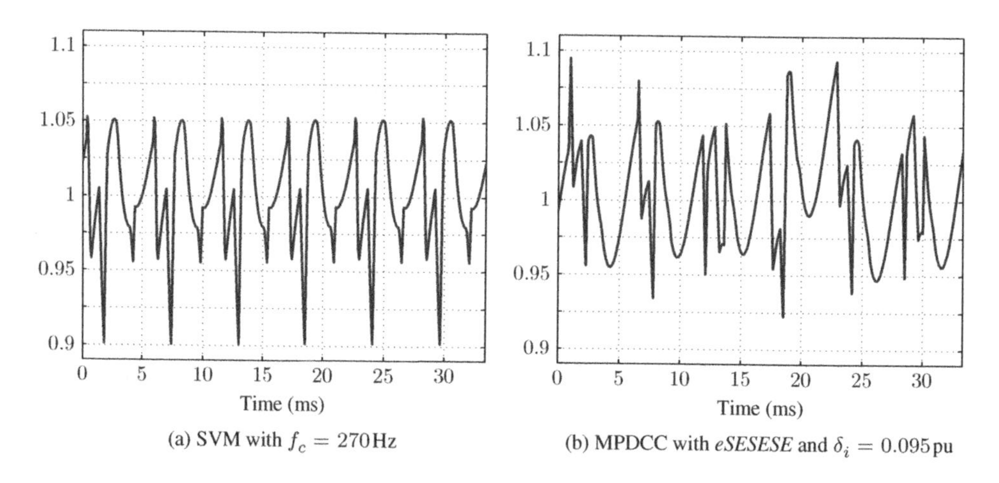

(a) SVM with $f_c = 270\,\text{Hz}$ (b) MPDCC with *eSESESE* and $\delta_i = 0.095\,\text{pu}$

Figure 11.8 Electromagnetic torque that corresponds to the simulations in Fig. 11.6

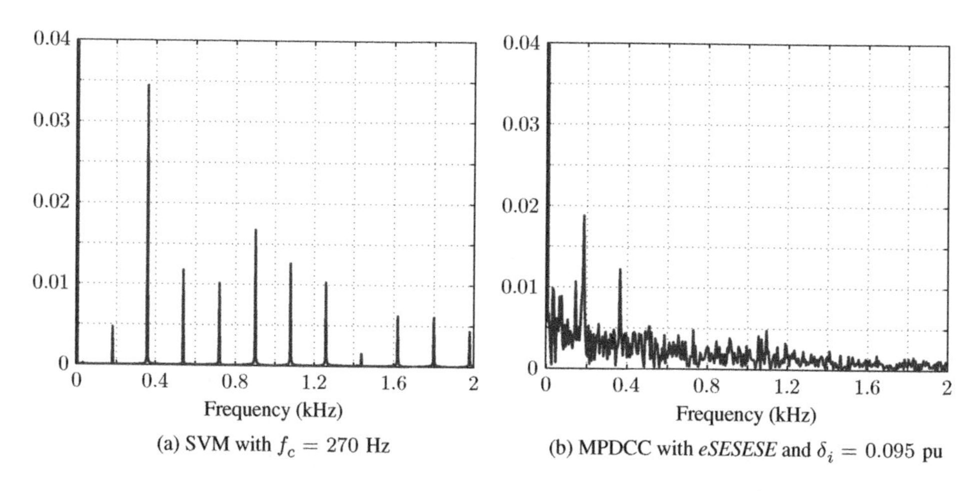

(a) SVM with $f_c = 270$ Hz (b) MPDCC with *eSESESE* and $\delta_i = 0.095$ pu

Figure 11.9 Spectra of the electromagnetic torque that is shown in Fig. 11.8

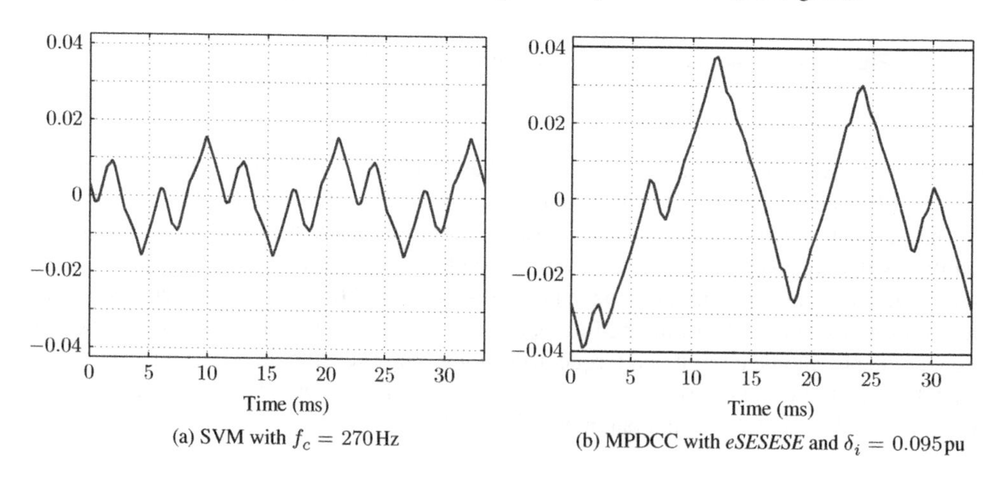

(a) SVM with $f_c = 270\,\text{Hz}$ (b) MPDCC with *eSESESE* and $\delta_i = 0.095\,\text{pu}$

Figure 11.10 Neutral point potential, which corresponds to the simulations in Fig. 11.6

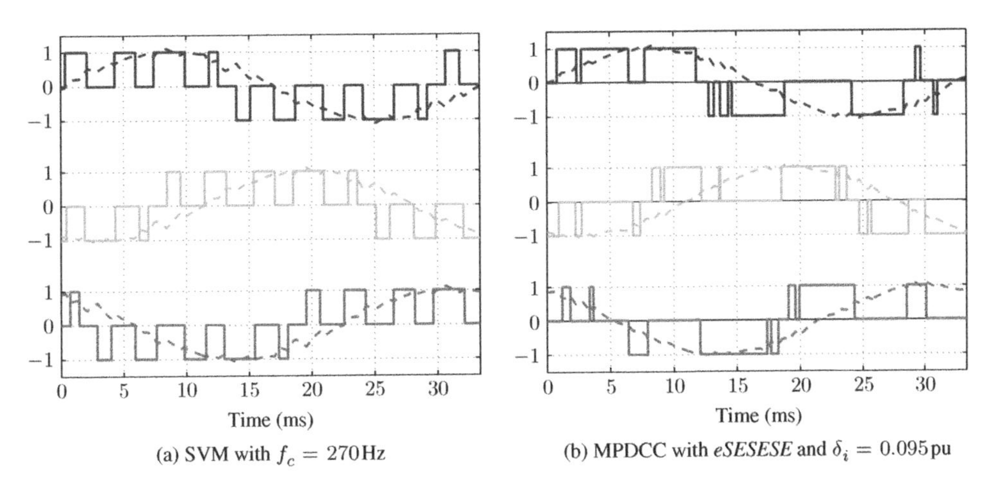

(a) SVM with $f_c = 270\,\text{Hz}$ (b) MPDCC with *eSESESE* and $\delta_i = 0.095\,\text{pu}$

Figure 11.11 Three-phase switch positions and phase currents (dashed lines), which correspond to the simulations in Fig. 11.6

For the same switching losses, the current distortions are reduced by 30%. The torque distortions are, however, 22% worse than for SVM. The switching frequency also tends to be higher than for SVM, because it is not directly penalized in the cost function. By arranging the switching pattern such that a significant proportion of the switching transitions occurs when the phase currents and thus the switching losses are small, the switching losses are kept at the same level as for SVM despite the higher switching frequency. Interestingly enough, MPDCC with long horizons achieves almost the same switching losses and current distortions as the OPP with pulse number $d = 5$ (3.04 kW vs 3.11 kW and 6.09% vs 5.51%, respectively). The torque distortions are, however, significantly worse (4.01% vs 2.80%), and the switching frequency is slightly higher (166 Hz vs 150 Hz), as stated in Table 11.1.

Alternatively, one may wish to minimize the switching losses with regard to SVM, possibly at the expense of pronounced current distortions. As an example, consider again SVM based on $f_c = 270\,\text{Hz}$. MPDCC with the switching horizon *eSE* and the bound parameter $\delta_i = 0.21\,\text{pu}$ leads to 24% higher current distortions (10.7% instead of 8.63%), but the switching losses are reduced from 3.04 to 1.0 kW, that is, by 67%! In this case, MPDCC actually outperforms the OPP with pulse number $d = 2$. For similar current and torque distortions, MPDCC reduces the switching losses from 1.42 to 1.0 kW, that is, by 30%. This might appear to be counterintuitive, as it is often assumed that OPPs provide the upper bound on the achievable steady-state performance of a modulator.

Recall that the OPPs were computed by minimizing the current distortions for a given pulse number (or switching frequency), disregarding the switching losses. By taking the switching losses into account and by rearranging the switching transitions accordingly, MPDCC is able to achieve similarly low distortions, while further reducing the switching losses. This beneficial characteristic is exemplified qualitatively in Fig. 11.12, which compares the switching patterns of the OPP and MPDCC when operating at a switching frequency of 60 Hz. Table 11.3 provides a quantitative comparison of the switching losses in phase a for half a fundamental period. Owing to the observed half-wave and three-phase symmetry, the ratio between the switching losses is the same for the three-phase system.

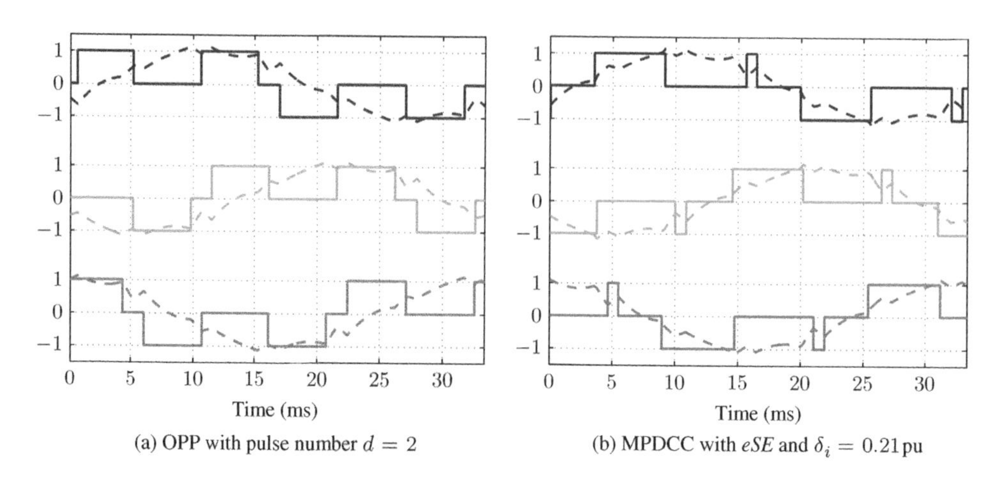

(a) OPP with pulse number $d = 2$

(b) MPDCC with *eSE* and $\delta_i = 0.21\,\text{pu}$

Figure 11.12 Three-phase switch positions and phase currents (dashed lines) for an OPP and MPDCC. Both schemes yield the same switching frequency of 60 Hz and similar distortion levels, but MPDCC reduces the switching losses by 30%

Table 11.3 Switching losses resulting from the first four switching transitions of the phase a switching pattern shown in Fig. 11.12

Quantity	OPP switching transitions				MPDCC switching transitions			
i_{sa} (pu)	−0.65	0.55	0.93	0.93	0.14	1.15	0.33	0.60
e_{on} (J)			0.16		0.02		0.06	
e_{off} (J)	1.48	1.25		2.12		2.62		1.37
e_{rr} (J)			2.78		0.41		0.99	
$\sum e$ (J)		7.79				5.47		

The variables e_{on}, e_{off}, and e_{rr} denote the GCT turn-on, GCT turn-off, and the diode reverse recovery losses, respectively. The switching losses of MPDCC are 30% lower than those of the OPP.

The benefit of MPDCC is particularly pronounced when operating at low pulse numbers. For a switching frequency of 60 Hz, MPDCC reduces the switching losses by 20% and the current TDD by about 40%, when compared to SVM that is based on a carrier frequency of 90 Hz. At higher switching frequencies, however, the performance gain achieved with MPDCC is less significant. This can be seen in Table 11.2 when benchmarking MPDCC with respect to SVM based on the carrier frequency $f_c = 720\,\text{Hz}$. MPDCC with the switching horizon *eSESESE* reduces the current distortions by 19% for the same switching losses. The switching frequency and the torque distortions are, however, both increased by about 25%.

OPPs maintain a small edge over long-horizon MPDCC at high switching frequencies. For similar switching losses, the OPP reduces the current and torque distortions by 7% and 11%, respectively, and the switching frequency by 23%, when compared to MPDCC with the switching horizon *eSESESE*.

11.1.7.2 Operation during Transients

The dynamic response of MPDCC to step changes in the torque reference is explored in the following. The switching horizon *eSSE* and the current bound parameter $\delta_i = 0.08$ pu are chosen. The bounds on the neutral point potential remain at $\delta_v = 0.04$ pu. During steady-state operation, these controller settings result in the switching losses $P_{\text{sw}} = 4.93$ kW, the switching frequency $f_{\text{sw}} = 240$ Hz, and the current TDD $I_{\text{TDD}} = 5.48\%$, which are typical for an IGCT-based MV drive. The second term in the cost function $J_{\text{bnd}} = \boldsymbol{q}^T \boldsymbol{\epsilon}$, which penalizes the rms bound violation of the outputs, is utilized to shorten the torque transients. By setting the penalty to $\boldsymbol{q} = 1000 \cdot [1\,1\,1\,0]^T$, we heavily penalize violations of the bounds on the three-phase ripple current.

At 60% speed, torque reference steps of magnitude 1 pu are imposed. As shown in Fig. 11.13, a very fast current and thus torque response is achieved. During the negative torque steps, sufficient voltage margin is available and MPDCC temporarily inverts the

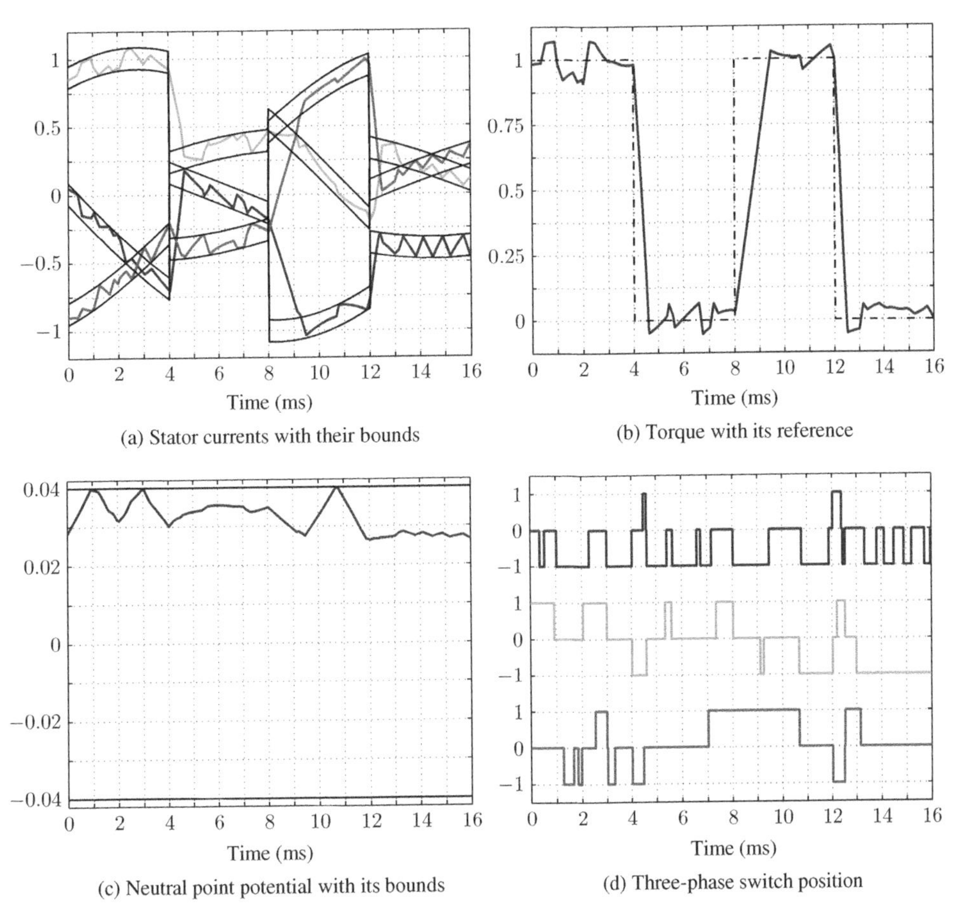

Figure 11.13 Dynamic response of MPDCC to torque steps of magnitude 1 pu when operating at 60% speed

applied stator voltage. As a result, the torque transients are below 0.6 ms. For $\boldsymbol{q} = \mathbf{0}_{4\times 1}$, a slightly more sluggish response would have resulted, with the torque transients lasting for 1.3 ms. Little voltage margin is available during the positive torque step. The transient lasts for 1.4 ms and is thus almost three times longer than during the negative torque steps. Setting $\boldsymbol{q}$ to zero has no influence on the control actions during the positive torque step.

These results indicate that MPDCC is as fast as deadbeat and hysteresis control schemes. Excessive switching during the transients is avoided, as can be seen from Fig. 11.13(d). Note that in this example, the neutral point potential is close to its upper bound, necessitating switching transitions that modify the common-mode voltage at time instants 1, 3, and 10.7 ms to avoid the neutral point potential from violating its upper bound.

11.1.8 Tuning

The role of the bound width on the stator ripple current is investigated now. At 60% speed and rated torque, closed-loop simulations for MPDCC with the switching horizon *eSESE* were run while varying the current bound parameter δ_i between 0.02 and 0.23 pu. The resulting current and torque TDDs along with the associated switching frequency and losses are shown in Fig. 11.14. These four quantities are normalized with respect to their maximum values and are provided in percent.

The bound width directly determines the amplitude of the stator ripple current. Because of the relationship between the rms value of the ripple and the TDD, one would expect that δ_i also determines the harmonic current distortions. This is indeed the case, as can be seen in the figure. In particular, the current distortions are effectively linear in the current bound

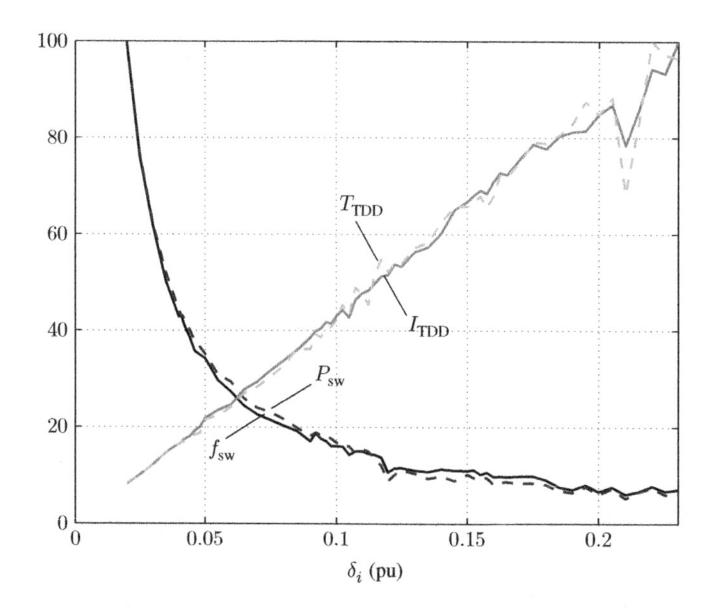

Figure 11.14 Tuning of MPDCC using the bound parameter δ_i. The current and torque TDD, the switching losses, and the switching frequency are given in percent and are normalized with respect to their maximum value in the interval $\delta_i \in [0.02, 0.23]$ pu

parameter up to 0.2 pu. Because the torque is the cross product between the stator current and the (sinusoidal) rotor flux components in orthogonal coordinates, the torque distortions are also proportional to δ_i. For very wide bounds, and when approaching six-step operation, however, the relationship between the bound width and the harmonic current and torque distortions becomes nonlinear.

We have seen in Sect. 3.5 for CB-PWM that the products of the current distortions on one hand and the switching frequency and losses on the other, $I_{\text{TDD}} \cdot f_{\text{sw}}$ and $I_{\text{TDD}} \cdot P_{\text{sw}}$, are invariant under some minor assumptions. The same applies to the torque TDD. Indeed, one would expect that similar statements also hold true for other control and modulation methods, including MPDCC. This is indeed the case, as an analysis of the data shown in Fig. 11.14 reveals. Specifically, the switching losses and switching frequency are hyperbolic functions of the bound parameter δ_i; that is, they are inversely proportional to δ_i.

We conclude that in MPDCC the width of the ripple current bounds is a tuning parameter that adjusts the trade-off between the harmonic distortions and the switching effort (which is either the switching frequency or the switching losses). This tuning parameter is equivalent to the penalty λ_u on switching transitions, which is used in the cost function of direct MPC with current reference tracking (see the introductory Sects. 4.1 and 4.2 and Chap. 5). In CB-PWM, this trade-off is adjusted by the carrier frequency and in OPPs by the pulse number d.

Three additional parameters are used in MPDCC. The bounds on the neutral point potential at $\pm\delta_v$ are chosen such that, for a given dc-link voltage, the blocking voltages over the GCTs are limited to values that are within their safe operating area. The sampling interval T_s should be set to as small a value as possible in order to conceal the adverse effect of restricting the switching transitions to discrete-time instants. For switching frequencies of a few hundred hertz, a sampling interval of 25 μs is appropriate. The switching horizon should be as long as possible. To limit the considerable computational burden long switching horizons entail, MPDCC can be augmented with branch-and-bound methods. Such techniques are described in Chap. 10 for MPDTC, but they are also applicable to MPDCC.

11.2 Model Predictive Direct Power Control

So far, we have almost exclusively focused on machine-side inverters in the setting of variable-speed drives (VSDs). Their counterpart on the grid side, however, is of equal importance. More specifically, MV grid-connected converters are used as active front ends in VSDs [8, 9] to connect the dc-link stage to the grid. A VSD with an active front end is shown in Fig. 1.1. Another important application of grid-connected converters is the integration of renewable energy sources in the grid [10]. Grid-connected converters are also used for energy storage systems [10] and uninterruptible power supplies, and to enhance the power quality [11] at the point of common coupling (PCC).

The control objective of grid-connected converters is to maintain the dc-link voltage close to its reference by manipulating the flow of real power through the converter [12]. The reactive power reference is usually set to zero, in order to achieve a power factor of 1. The most widely used control technique is voltage-oriented control (VOC) [13, 14]. An orthogonal dq reference frame is established that rotates in synchronism with the voltage of the PCC. The real and reactive power components are decoupled in this reference frame, and the power set points are translated into equivalent set points for the dq currents.

To control the real and reactive power, two current control loops are devised in the rotating reference frame. The current controllers manipulate voltage references that are fed into a modulator, which in turn generates the switching signals. VOC can be interpreted as the grid-side equivalent of FOC. As in FOC, the two controlled quantities—the electromagnetic torque and flux magnitude—are controlled *indirectly* via their corresponding current components. For a summary of FOC, the reader is referred to Sect. 3.6.2.

An alternative method to control grid-connected converters is DPC, which is the grid-side equivalent of DTC. DPC *directly* controls the instantaneous real and reactive power by imposing hysteresis bounds on these components. A look-up table is utilized to select an appropriate converter switch position. DPC was proposed in [15] for two-level converters. Rather than measuring the PCC voltage, the latter is reconstructed using the equivalent grid inductance, the derivative of the converter current, and the switched converter voltage.

The concept of DPC can be augmented with the notion of virtual flux vectors [16]. As current derivatives are no longer required to estimate the real and reactive power components, the level of noise on the estimated power can be reduced. DPC based on virtual flux vectors, which we refer to as VF-DPC, was proposed in [17]. Note that a related control method had been proposed almost a decade earlier in [18]. Instead of directly controlling the real and reactive power, these quantities are controlled *indirectly* in [18] through the angular position and the magnitude of the virtual converter flux vector with respect to the virtual grid flux vector.

By imposing hysteresis bounds on the neutral point potential and by exploiting the redundancy in the voltage vectors with regard to their common-mode component, VF-DPC can be extended to three-level NPC converters [19]. Similarly, for five-level active NPC converters, the voltages of the phase capacitors can be maintained close to their nominal values by imposing hysteresis bounds on them and by extending the VF-DPC scheme [20]. To address converters that are connected via an intermediate LC filter to the grid, the VF-DPC scheme needs to be augmented by an active damping mechanism, as described in [21].

Similar to DTC, DPC achieves a superior performance during transients and dynamic operation. The switching frequency is, however, variable, and the harmonic spectrum is nondeterministic. To achieve a constant switching frequency and a deterministic harmonic spectrum, a modulator can be added. To this end, [22] replaces the hysteresis controllers by PI controllers and the look-up table by a space vector modulator (SVM). In a further step, the PI controllers can be replaced by deadbeat controllers [23].

Recently, predictive control methods have been proposed that extend the DPC concept. In [24], the switching instants of a given sequence of voltage vectors are manipulated with the objective of minimizing the predicted rms ripple on the real and reactive power. A constant switching frequency is achieved without explicitly using a modulator. A one-step predictive controller was proposed in [25]. The predicted deviations of the real and reactive power components from their references are penalized in a cost function, and switch positions are chosen that minimize this cost function.

In the same way that the control concept of DTC can be translated to DPC, MPDTC can be adapted to grid-connected converters, giving rise to MPDPC. MPDPC can be interpreted as an extension of DPC, in which the hysteresis controllers and the look-up table are replaced by an online optimization stage, while the symmetrical bounds on the real and reactive power components are inherited from DPC. Unlike other predictive DPC methods, MPDPC is well suited to achieve very long prediction horizons, thanks to the notion of the switching horizon and the extension mechanism (see Sects. 7.4.3 and 7.5). This facilitates the reduction of either the harmonic current distortions or the switching losses of the semiconductor devices while preserving the superior dynamic performance of DPC.

11.2.1 Case Study

As a case study, we consider in the second part of this chapter a 9 MVA NPC converter with a line-to-line rms voltage of 3.3 kV. As shown in Fig. 2.28, the grid-connected converter is connected via a transformer to the PCC. The resistances, reactances, and voltage sources to the right of the PCC model the distribution and transmission system of the grid. We refer these quantities to the secondary side of the transformer and adopt a pu system based on the rated values in Table 2.14. The transformer, PCC, and grid can then be represented by the equivalent resistance $R = 0.015$ pu, reactance $X = 0.15$ pu, and three-phase grid voltage $\boldsymbol{v}_{g,abc} = [v_{ga}\ v_{gb}\ v_{gc}]^T$ with the amplitude 1 pu. This equivalent representation of the converter system is shown in Fig. 11.15, thereby repeating Fig. 2.29 for the reader's convenience.

We define the three-phase converter current as $\boldsymbol{i}_{c,abc} = [i_{ca}\ i_{cb}\ i_{cc}]^T$. Recall that $v_{\text{dc,up}}$ and $v_{\text{dc,lo}}$ denote the voltages over the upper and lower dc-link capacitors. The instantaneous dc-link voltage is given by $v_{\text{dc}} = v_{\text{dc,up}} + v_{\text{dc,lo}}$, while V_{dc} denotes the nominal dc-link voltage.

The three-phase switch position $\boldsymbol{u}_{abc} = [u_a\ u_b\ u_c]^T$ is restricted to the set $\{-1, 0, 1\}^3$. Analogous to (11.1), the converter voltage in stationary orthogonal coordinates is

$$\boldsymbol{v}_c = \frac{1}{2} v_{\text{dc}}\, \tilde{\boldsymbol{K}} \boldsymbol{u}_{abc} \tag{11.30}$$

with $\boldsymbol{v}_c = [v_{c\alpha}\ v_{c\beta}]^T$. $\tilde{\boldsymbol{K}}$ is the reduced Clarke transformation matrix (see (2.13)). The evolution of the neutral point potential $v_n = \frac{1}{2}(v_{\text{dc,lo}} - v_{\text{dc,up}})$ is described by the differential equation

$$\frac{\mathrm{d}v_n}{\mathrm{d}t} = \frac{1}{2X_c} |\boldsymbol{u}_{abc}|^T\, \boldsymbol{i}_{c,abc}, \tag{11.31}$$

where $|\boldsymbol{u}_{abc}|$ denotes the component-wise absolute value of the three-phase switch position. For the definition of the remaining NPC converter-related variables and parameters, the reader is referred to Sects. 11.1.1 and 2.4.1. The parameters of the case study are summarized in Table 2.15.

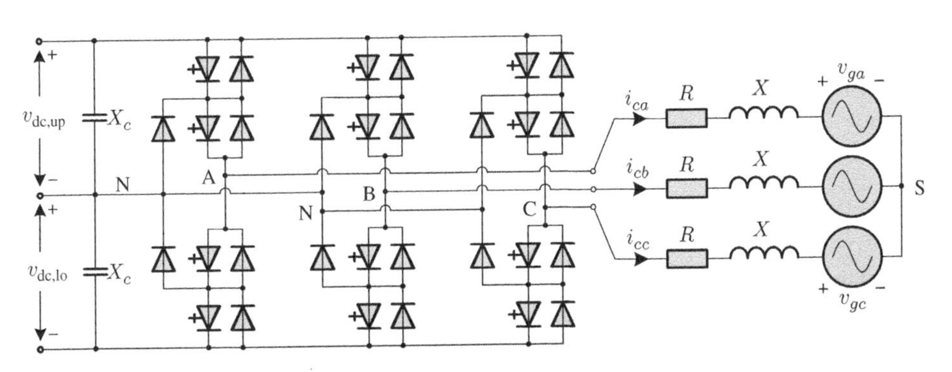

Figure 11.15 Grid-connected NPC converter system with a simplified grid representation

The phase a converter voltage v_{ca} is the voltage between the phase terminal A and the neutral point potential N, whereas v_0 is the voltage between the star point S of the grid voltage and N. In the pu system, the continuous-time differential equation of the phase a converter current is

$$v_{ca} = Ri_{ca} + X\frac{di_{ca}}{dt} + v_{ga} + v_0. \tag{11.32}$$

The differential equations for phases b and c are defined accordingly. By defining $\boldsymbol{v}_{c,abc} = [v_{ca}\ v_{cb}\ v_{cc}]^T$ and $\boldsymbol{v}_0 = [v_0\ v_0\ v_0]^T$, we can write the three differential equations in matrix notation

$$\frac{d\boldsymbol{i}_{c,abc}}{dt} = -\frac{R}{X}\boldsymbol{i}_{c,abc} - \frac{1}{X}\boldsymbol{v}_{g,abc} + \frac{1}{X}\boldsymbol{v}_{c,abc} - \frac{1}{X}\boldsymbol{v}_0. \tag{11.33}$$

By left-multiplying (11.33) with the reduced Clarke transformation matrix $\tilde{\boldsymbol{K}}$ (see (2.11)), we can express the differential equation in the stationary reference frame as

$$\frac{d\boldsymbol{i}_c}{dt} = -\frac{R}{X}\boldsymbol{i}_c - \frac{1}{X}\boldsymbol{v}_g + \frac{1}{X}\boldsymbol{v}_c. \tag{11.34}$$

To this end, we have introduced $\boldsymbol{i}_c = [i_{c\alpha}\ i_{c\beta}]^T$ and $\boldsymbol{v}_g = [v_{g\alpha}\ v_{g\beta}]^T$. The Clarke transformation also removed the star point voltage $\boldsymbol{v}_0$ from the right-hand side of (11.34), because $\tilde{\boldsymbol{K}}\boldsymbol{v}_0 = [0\ 0]^T$. In the absence of a fault, $i_{ca} + i_{cb} + i_{cc} = 0$ holds for the converter current. With the zero-sequence current being zero, (11.34) fully describes the dynamics of the converter currents.

We assume that the grid voltage amplitude in each phase is 1 pu and that the phase shift between the phases is 120°. The evolution of the grid voltage can then be described in the stationary reference frame by the differential equation

$$\frac{d\boldsymbol{v}_g}{dt} = \boldsymbol{F}_r\boldsymbol{v}_g \text{ with } \boldsymbol{F}_r = \omega_g \begin{bmatrix} 0 & -1 \\ 1 & 0 \end{bmatrix}, \tag{11.35}$$

where $\omega_g = 2\pi f_g$ denotes the angular grid frequency.

11.2.2 Control Problem

In order to formulate the control problem, we require definitions for the *instantaneous* real and reactive power (rather than the average power). Starting from the three-phase grid voltage and converter current, the instantaneous real and reactive power are derived in Appendix 11.B. These power components are defined in the stationary orthogonal reference frame and the pu system as

$$P = v_{g\alpha}i_{c\alpha} + v_{g\beta}i_{c\beta} \tag{11.36a}$$

$$Q = v_{g\alpha}i_{c\beta} - v_{g\beta}i_{c\alpha}, \tag{11.36b}$$

in accordance with [26]. Active power corresponds to the current component that is in phase with the voltage. Reactive power relates to the current component that is orthogonal to (or 90° out of phase with) the voltage. Different sign conventions exist for the reactive power. According to the definition in (11.36b), positive reactive power relates to a capacitive load, in which the current leads the voltage. Negative reactive power, on the other hand, corresponds to an inductive load and a lagging current. For a comprehensive review of the notion of instantaneous real and reactive power, see [27, Appendix B] and the literature cited therein.

Strictly speaking, the real and reactive power should be controlled at the PCC. In the simplified representation in Fig. 11.15, the transformer and the grid have been replaced by the equivalent reactance X and resistance R, thereby removing the PCC. We thus define and control the power at the grid voltage sources.

The primary control objective of MPDPC is inherited from DPC, which is to maintain the real and reactive power within symmetrical bounds. These bounds are defined around their respective references P^* and Q^*. The real power reference is set by an external control loop, which regulates the dc-link voltage around its nominal value by manipulating this reference. The reactive power reference is usually set to zero to achieve unity power factor operation. Furthermore, symmetrical bounds around zero are imposed on the neutral point potential.

MPDPC selects three-phase switch positions with the aim of keeping these three output variables, namely the real power, the reactive power, and the neutral point potential, within their respective bounds. MPDPC minimizes the switching effort at the same time, either in the form of the switching frequency or the switching losses.

11.2.3 Controller Model

The controller model captures the current dynamic, the grid voltage evolution, and the dynamic of the neutral point potential. We therefore choose the state vector

$$\boldsymbol{x} = [i_{c\alpha}\ i_{c\beta}\ v_{g\alpha}\ v_{g\beta}\ v_n]^T \tag{11.37}$$

with the converter current and grid voltage in orthogonal coordinates as well as the neutral point potential. The input vector is the three-phase switch position

$$\boldsymbol{u} = \boldsymbol{u}_{abc} = [u_a\ u_b\ u_c]^T \in \{-1, 0, 1\}^3. \tag{11.38}$$

The output vector

$$\boldsymbol{y} = [P\ Q\ v_n]^T \tag{11.39}$$

comprises the (instantaneous) real and reactive power and the neutral point potential.

The continuous-time state-space representation of the grid-connected converter system is

$$\frac{\mathrm{d}\boldsymbol{x}(t)}{\mathrm{d}t} = \boldsymbol{F}(\boldsymbol{u}(t))\ \boldsymbol{x}(t) + \boldsymbol{G}\ \boldsymbol{u}(t) \tag{11.40a}$$

$$\boldsymbol{y}(t) = \boldsymbol{h}(\boldsymbol{x}(t)) \tag{11.40b}$$

with the system and input matrices

$$\boldsymbol{F}(\boldsymbol{u}) = \begin{bmatrix} \boldsymbol{F}_g & \boldsymbol{0}_{4\times 1} \\ \boldsymbol{f}_c(\boldsymbol{u}) & 0 \end{bmatrix} \text{ and } \boldsymbol{G} = \begin{bmatrix} \boldsymbol{G}_g \\ \boldsymbol{0}_{1\times 3} \end{bmatrix}. \tag{11.41}$$

The matrices $\boldsymbol{F}_g$ and $\boldsymbol{G}_g$ are derived and provided in Appendix 11.C along with the input-dependent row vector $\boldsymbol{f}_c(\boldsymbol{u})$ and the state-dependent output function $\boldsymbol{h}(\boldsymbol{x})$.

The discrete-time model can be derived in the same way as for MPDTC in Sect. 7.2.3 and MPDCC in Sect. 11.1.4. The exact Euler discretization method is applied to the dynamics of the converter current and grid voltage, whereas forward Euler is used for the neutral point potential. The resulting discrete-time representation of (11.40) is

$$\boldsymbol{x}(k+1) = \boldsymbol{A}(\boldsymbol{u}(k))\,\boldsymbol{x}(k) + \boldsymbol{B}\,\boldsymbol{u}(k) \tag{11.42a}$$

$$\boldsymbol{y}(k) = \boldsymbol{h}(\boldsymbol{x}(k)) \tag{11.42b}$$

with the matrices

$$\boldsymbol{A}(\boldsymbol{u}) = \begin{bmatrix} \boldsymbol{A}_g & \boldsymbol{0}_{4\times 1} \\ \boldsymbol{f}_c(\boldsymbol{u})T_s & 1 \end{bmatrix} \text{ and } \boldsymbol{B} = \begin{bmatrix} \boldsymbol{B}_g \\ \boldsymbol{0}_{1\times 3} \end{bmatrix} \tag{11.43}$$

and the sampling interval T_s. The submatrices are given by

$$\boldsymbol{A}_g = \boldsymbol{e}^{\boldsymbol{F}_g T_s} \text{ and } \boldsymbol{B}_g = -\boldsymbol{F}_g^{-1}\,(\boldsymbol{I}_4 - \boldsymbol{A}_g)\,\boldsymbol{G}_g\,. \tag{11.44}$$

11.2.4 Control Problem Formulation

A cost function capturing the switching effort is minimized subject to the evolution of the controller model and constraints. The resulting optimization problem underlying MPDPC is effectively the same as for MPDTC in (7.26). Nevertheless, for the sake of completeness, the MPDPC optimization problem is stated next.

$$\boldsymbol{U}_{\text{opt}}(k) = \arg \underset{\boldsymbol{U}(k)}{\text{minimize}}\; J \tag{11.45a}$$

$$\text{subject to } \boldsymbol{x}(\ell+1) = \boldsymbol{A}(\boldsymbol{u}(\ell))\boldsymbol{x}(\ell) + \boldsymbol{B}\boldsymbol{u}(\ell) \tag{11.45b}$$

$$\boldsymbol{y}(\ell+1) = \boldsymbol{h}(\boldsymbol{x}(\ell+1)) \tag{11.45c}$$

$$\begin{cases} \varepsilon_j(\ell+1) = 0, & \text{if } \varepsilon_j(\ell) = 0 \\ \varepsilon_j(\ell+1) < \varepsilon_j(\ell), & \text{if } \varepsilon_j(\ell) > 0 \end{cases} \tag{11.45d}$$

$$\boldsymbol{u}(\ell) \in \boldsymbol{\mathcal{U}}\,,\;\; \|\Delta \boldsymbol{u}(\ell)\|_\infty \leq 1 \tag{11.45e}$$

$$\forall \ell = k, \ldots, k+N_p-1, \forall j = 1,2,3. \tag{11.45f}$$

The discrete-time controller model is given in (11.42). The constraints (11.45d) impose upper and lower bounds on the three output variables. These bounds are symmetric around their references. For MPDPC, the references can be assumed to be constant within the prediction horizon. The vector of bound violations is defined as

$$\boldsymbol{\varepsilon} = [\varepsilon_P\; \varepsilon_Q\; \varepsilon_v]^T\,. \tag{11.46}$$

We use the index $j \in \{1,2,3\}$ to denote the jth component of ε. These components are non-negative and normalized by two times the bound widths. The degree of the bound violation for the real power, for example, is defined as

$$\varepsilon_P(\ell) = \frac{1}{2\delta_P} \begin{cases} |P^*(k) - P(\ell)| - \delta_P & \text{if } |P^*(k) - P(\ell)| > \delta_P \\ 0 & \text{else}, \end{cases} \tag{11.47}$$

where $\delta_P > 0$ denotes the difference between the upper (or lower) bound and the reference. The degrees of the bound violations for the reactive power and the neutral point potential are defined accordingly, using the bound width parameters $\delta_Q > 0$ and $\delta_\upsilon > 0$, respectively.

As for MPDTC and MPDCC, the cost function

$$J = J_{\text{sw}} + J_{\text{bnd}} + J_t \tag{11.48}$$

of MPDPC has three terms. J_{sw} captures the switching effort in terms of either the short-term switching frequency or the switching losses. The second term is defined as $J_{\text{bnd}} = \boldsymbol{q}^T \boldsymbol{\epsilon}$ with

$$\boldsymbol{\epsilon} = [\epsilon_P \; \epsilon_Q \; \epsilon_\upsilon]^T . \tag{11.49}$$

Its three components denote the rms bound violations of the real and reactive power and the neutral point potential. They are defined similar to (7.35). The third term J_t serves as a means to reduce the likelihood of deadlocks (see Sect. 9.4).

This section has introduced the MPDPC optimization problem by defining the variables and parameters that are unique to MPDPC. For a more comprehensive description of the optimization problem and its underlying rationale, the reader is referred to the control problem formulation of MPDTC in Sect. 7.3.

To solve this optimization problem, the MPDTC algorithm can be adapted to the grid side in a straightforward manner. The MPDTC and MPDPC algorithms are formulated in the stationary orthogonal reference frame, and the references and bounds on the output variables are constant within the prediction horizon. This simplifies the algorithm compared to MPDCC, in which the current bounds are time-varying. A detailed description of the MPDTC algorithm is provided in Sect. 7.4, including the nomenclature and the concepts of the switching horizon, search tree, and full enumeration.

11.2.5 Performance Evaluation

A brief performance evaluation of MPDPC is provided here for the grid-connected NPC converter system shown in Fig. 11.15. The 9 MVA converter with a floating neutral point potential and the nominal dc-link voltage $V_{\text{dc}} = 5.2$ kV is connected to the secondary side of a step-down transformer with a line-to-line rms voltage of 3.3 kV. The transformer and grid reactances are referred to the secondary side of the transformer, with their sum being equal to $X = 0.15$ pu. The grid frequency is $f_g = 50$ Hz. The pu system and the parameters of this case study are provided in Tables 2.14 and 2.15, respectively.

Regarding the controller settings for MPDPC, we choose to minimize the switching losses in the cost function. The penalty on rms bound violations is set to $\boldsymbol{q} = 1000 \cdot [1\;1\;0]^T$ during transients. When operating at steady state, $\boldsymbol{q}$ is set to zero. As will be explained at the end of this chapter, we choose equal bound widths for the real and reactive power, setting $\delta_P = \delta_Q$. The bound width parameter for the neutral point potential is set to $\delta_\upsilon = 0.03$ pu, and the sampling interval is $T_s = 25$ μs.

11.2.5.1 Steady-State Operation

While operating at steady state, we compare the switching losses, the switching frequency, and the current distortions of MPDPC with those of SVM. For SVM, we consider the equivalent

Table 11.4 Comparison of MPDPC with SVM using absolute values for the switching losses P_{sw}, the switching frequency f_{sw}, and the current TDD I_{TDD}

Control scheme	Control setting	Switching horizon	Average prediction horizon (steps)	P_{sw} (kW)	f_{sw} (Hz)	I_{TDD} (%)
SVM	$f_c = 750\,\text{Hz}$	—	—	27.0	400	7.79
MPDPC	$\delta_P = \delta_Q = 0.10\,\text{pu}$	*eSE*	11.2	24.2	422	7.81
MPDPC	$\delta_P = \delta_Q = 0.10\,\text{pu}$	*eSESE*	25.4	22.9	368	7.46
MPDPC	$\delta_P = \delta_Q = 0.083\,\text{pu}$	*eSESE*	21.3	27.3	447	6.27

The operating point is at nominal real power and zero reactive power.

Table 11.5 Comparison of MPDPC with SVM using relative values for the performance metrics

Control scheme	Control setting	Switching horizon	Average prediction horizon (steps)	P_{sw} (%)	f_{sw} (%)	I_{TDD} (%)
SVM	$f_c = 750\,\text{Hz}$	—	—	100	100	100
MPDPC	$\delta_P = \delta_Q = 0.10\,\text{pu}$	*eSE*	11.2	89.6	106	100
MPDPC	$\delta_P = \delta_Q = 0.10\,\text{pu}$	*eSESE*	25.4	84.8	92.0	95.8
MPDPC	$\delta_P = \delta_Q = 0.083\,\text{pu}$	*eSESE*	21.3	101	112	80.5

The same results as in Table 11.4 are shown, using SVM as a baseline.

carrier frequency of $f_c = 750\,\text{Hz}$, which results in a switching frequency for each semiconductor device of 400 Hz. As summarized in Table 11.4, the converter switching losses amount to 27 kW, and the current TDD is 7.79% for SVM. With MPDPC, with the short switching horizon *eSE* and $\delta_P = \delta_Q = 0.1\,\text{pu}$, similar current distortions are achieved while the switching losses can be reduced to 24.2 kW. As stated in Table 11.5, this is a reduction of 10%. The switching frequency is, however, slightly higher.

Longer switching horizons, such as *eSESE*, further improve the steady-state performance of MPDPC. Compared to SVM, the switching losses and the switching frequency are reduced by 15% and 8%, respectively, while the current distortions are reduced by 4% (see Table 11.5). By tightening the bounds, the switching loss reduction achieved by MPDPC can be translated into a further reduction of the current distortions. For the bounds $\delta_P = \delta_Q = 0.083\,\text{pu}$, for example, the same switching losses as in SVM result, but the current TDD is reduced by almost 20%. The device switching frequency is, however, 12% higher than for SVM.

Figures 11.16–11.20 compare the steady-state waveforms of SVM and MPDPC over one fundamental period. The settings for SVM and MPDPC are stated in Table 11.4 in the first and the last row, respectively. Both control and modulation schemes operate at the same converter switching losses of 27 kW. The instantaneous real and reactive power waveforms are shown in Fig. 11.16. Compared to SVM, MPDPC reduces the ripple on the real and reactive power by 11% and 41%, respectively. As can be seen in Fig. 11.17, this reduction in the power ripple translates into a slight reduction of the current ripple. The current spectra of the two methods are very different though (see Fig. 11.18). SVM produces the well-known discrete spectrum with the harmonic content limited to integer multiples of the fundamental and

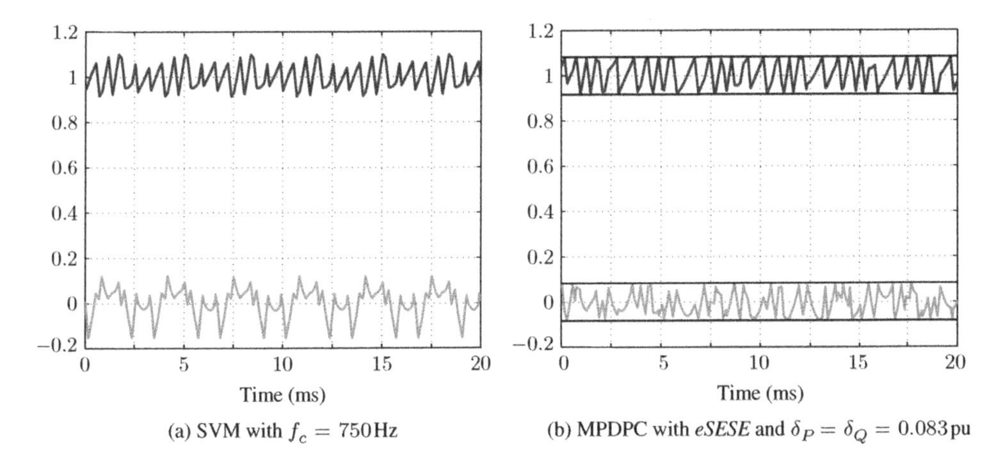

Figure 11.16 Instantaneous real and reactive power for SVM and MPDPC, when operating at nominal real power and zero reactive power. Both schemes yield the same switching losses

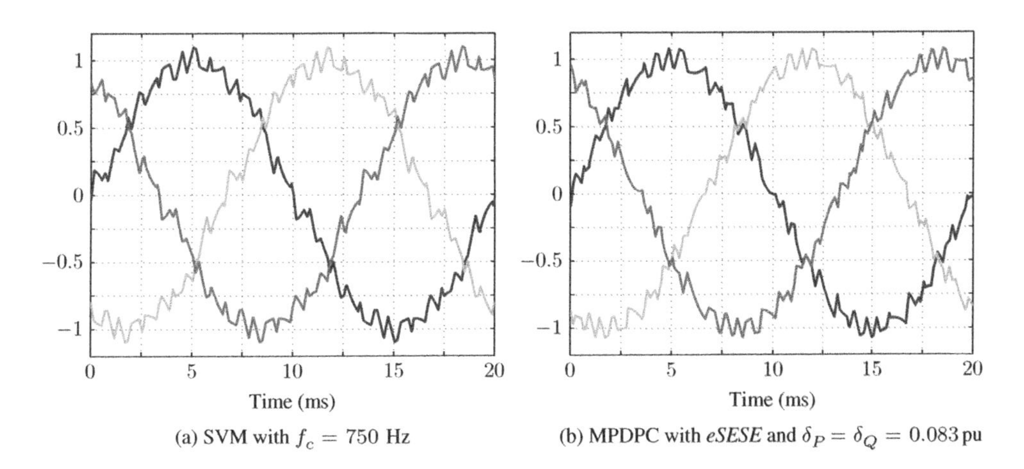

Figure 11.17 Three-phase converter currents, which correspond to the simulations in Fig. 11.16

carrier frequencies. In contrast, MPDPC—like MPDTC and MPDCC—produces an almost flat harmonic spectrum with few distinct harmonics. The neutral point potential in Fig. 11.19 is balanced around zero for both methods. As relatively wide bounds were chosen, the ripple is higher for MPDPC than for SVM.

Figure 11.20 shows the three-phase switch positions as solid lines and the respective phase currents as dashed lines. Owing to the fixed-length modulation cycle for SVM, its switching transitions are evenly spread along the fundamental period, irrespective of the switching losses. In MPDPC, however, the switching losses are considered and minimized. As a result, as can be seen in Fig. 11.20(b), MPDPC shifts the majority of the switching transitions from time instants with high phase currents to time instants when the currents are low. Despite the switching

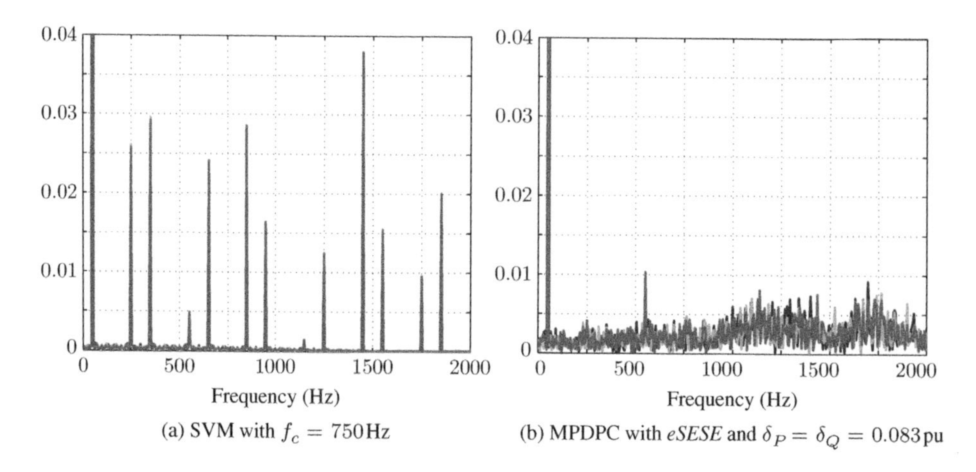

Figure 11.18 Spectra of the three-phase converter currents, which are shown in Fig. 11.17

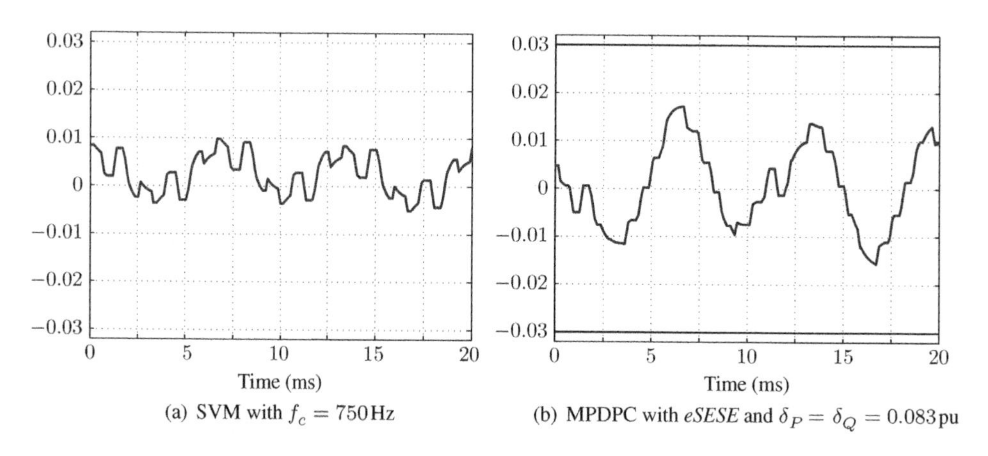

Figure 11.19 Neutral point potential, which corresponds to the simulations in Fig. 11.16

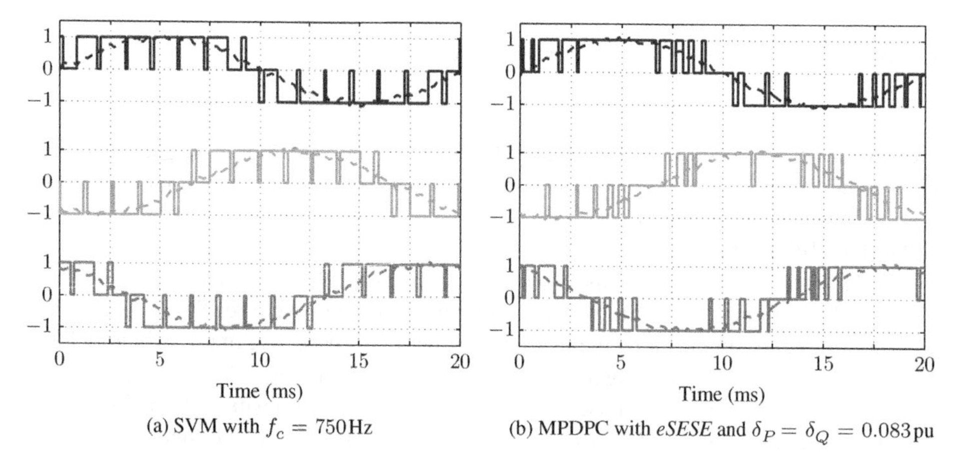

Figure 11.20 Three-phase switch positions (solid lines) and converter currents (dashed lines), which correspond to the simulations in Fig. 11.16

frequency being 10% higher for MPDPC than for SVM, the switching losses are the same. The additional switching transitions are used to reduce the ripple on the real and reactive power, thus reducing the current TDD.

11.2.5.2 Grid Standards

Tables 11.4 and 11.5 suggest that MPDPC might have an edge over SVM, because it either reduces the switching losses or lowers the current distortions. The latter is, however, of secondary importance for grid-connected converters. Instead, as summarized in Sect. 3.1.2, such converters must meet certain grid standards at the PCC. These grid standards impose upper bounds on individual current and voltage harmonics. Commonly imposed grid standards are the IEEE 519 standard [28] for current harmonics and the IEC 61000-2-4 standard [29] for voltage harmonics.

The IEEE 519 standard specifies limits on the current harmonics as a function of the short-circuit ratio k_{sc}. The latter was defined in (2.98). In this case study, it is equal to 20. The corresponding limits on the current harmonics are shown in Fig. 3.1 for harmonics of integer order. Recall that the order of a harmonic is the frequency ratio between the harmonic and the fundamental component. Harmonics of noninteger order are lumped to the closest integer harmonic by computing an equivalent rms value.

By applying this technique, the equivalent integer current harmonics can be determined for SVM with the carrier frequency $f_c = 750$ Hz and for MPDPC. For the latter, we consider again the case with the switching horizon *eSESE* and the bounds $\delta_P = \delta_Q = 0.085$ pu. The resulting current harmonics are shown in Fig. 11.21. The limits are shown as light gray bars, harmonics that meet these limits are shown as dark gray bars, and harmonics violating their corresponding limit are indicated by black bars.

The first violation for SVM occurs at the 17th harmonic, which is an upper sideband of the carrier frequency. For MPDPC, the first violation occurs at the 20th harmonic. Despite SVM

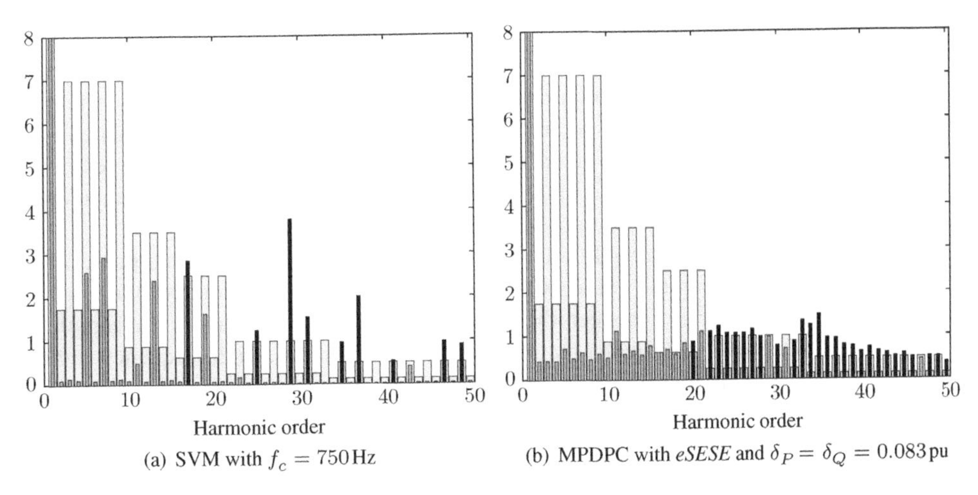

(a) SVM with $f_c = 750$ Hz

(b) MPDPC with *eSESE* and $\delta_P = \delta_Q = 0.083$ pu

Figure 11.21 Current harmonics (%) at the PCC, which correspond to the simulations in Fig. 11.16. The grid standard limits are shown as light gray bars, current harmonics that meet these limits are shown as dark gray bars, and harmonics violating their corresponding limit are indicated by black bars

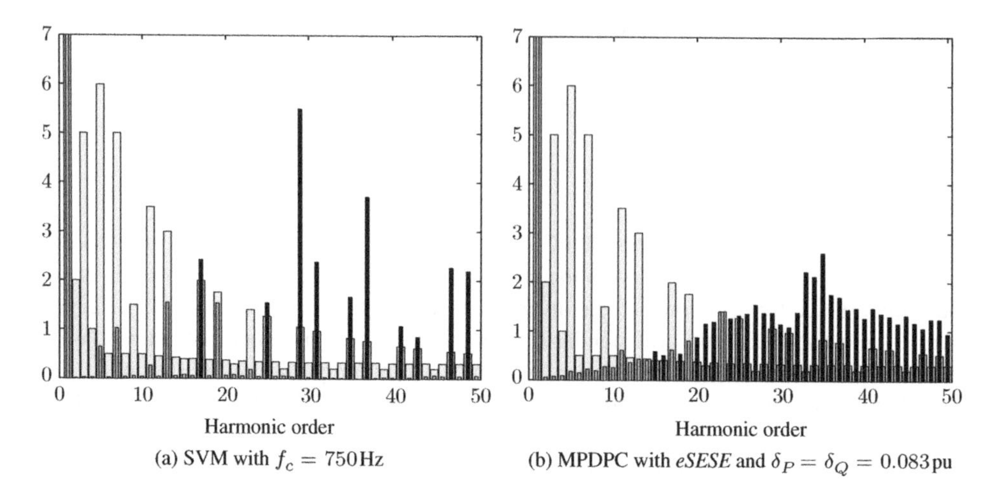

(a) SVM with $f_c = 750\,\text{Hz}$

(b) MPDPC with *eSESE* and $\delta_P = \delta_Q = 0.083\,\text{pu}$

Figure 11.22 Voltage harmonics (%) at the PCC, which correspond to the simulations in Fig. 11.16. The grid standard limits are shown as light gray bars, voltage harmonics that meet these limits are shown as dark gray bars, and harmonics violating their corresponding limit are indicated by black bars

violating the limits on the 29th and 37th harmonics by factors of 3.8 and 4, respectively, MPDPC causes even larger violations. The 22nd and the 34th current harmonics exceed their limits by factors of 4.4 and 10, respectively, as the limits on the even current harmonics are particularly tight. To meet the grid standards, both schemes require the addition of an *LC* filter that is placed between the converter and the transformer.

The voltage harmonics at the PCC are computed by separating the resistances and reactances within the lumped model shown in Fig. 11.15 into those on either side of the PCC according to Fig. 2.28. Limits on voltage harmonics are specified in the IEC 61000-2-4 standard. Assuming a Class 2 environment and repeating Fig. 3.2, these limits are shown in Fig. 11.22 as light gray bars. The limits on high-order voltage harmonics at triplen odd multiples of the fundamental frequency are particularly tight. Indeed, MPDPC violates the limit on the 33th harmonic 11-fold. SVM with $f_c = 750\,\text{Hz}$, in contrast, produces only non-triplen and odd harmonics. The most significant violation occurs at the 29th harmonic, with the limit being violated by a factor of 5.2. Despite the large 29th voltage harmonic of SVM, MPDPC features larger violations at lower frequencies. This implies that MPDPC would require an *LC* filter with a lower cut-off frequency and larger filter components than SVM.

We conclude that it is particularly difficult to meet grid standards on voltage harmonics. The amplitude of a voltage harmonic at the PCC is determined by the amplitude of the injected voltage harmonic and the ratio between the transformer and grid impedances. Indeed, these two impedances form a voltage divider that is frequency-independent. High-frequency *voltage* harmonics are thus not attenuated. High-frequency *current* harmonics are, on the other hand, greatly attenuated, because the impedance between the converter and grid voltage increases linearly with the frequency.

Despite their apparent differences, both control and modulation schemes poorly utilize the harmonic limits imposed by the grid standards. In the low-frequency region of up to the 20th harmonic, both schemes inject little harmonic power and rather concentrate it in the higher frequency range. Such a strategy is preferred when the aim is to minimize the overall current distortions, that is, the current TDD. To avoid violating the amplitude limits on the current harmonics, however, the harmonic power would have to be better distributed, with some content being shifted toward lower frequencies. This is particularly true for voltage grid standards, which penalize high-frequency harmonics in a disproportionately stringent manner.

11.2.5.3 Operation during Transients

Assume in the following that the grid-connected converter acts as an active rectifier unit in a VSD system. According to the sign of the converter currents in Fig. 11.15, positive real power implies power flowing from the dc-link stage to the grid, with the electrical machine operating in generator mode. In motoring operation, on the other hand, the power flow is reversed and the sign of the real power is negative. To investigate the dynamic performance of MPDPC, a step of amplitude 1 pu is applied to the reference of the instantaneous real power every 4 ms. With the electrical machine initially operating in motoring mode, the first power step is from -1 pu to zero. We choose the switching horizon *eSESE* and the penalty $\boldsymbol{q} = 1000 \cdot [1\,1\,0]^T$.

The resulting waveforms of the real and reactive power, three-phase converter current, and switch positions are shown in Fig. 11.23. The average settling time of the positive steps from -1 pu to zero is 2.2 ms. To halt the current flow into the converter, the converter voltage needs to match or exceed the grid voltage. As little voltage margin is available to accomplish this, the dynamic response of the converter is relatively sluggish. On the other hand, when stepping down the real power from zero to -1 pu, ample voltage margin is available. Indeed, by inverting the converter voltage, very fast transients can be achieved, and the average settling time of 0.5 ms is very short. The switching frequency remains low at 425 Hz, indicating that transients do not increase the switching frequency.

MPDPC achieves decoupled control of the real and reactive power and keeps the reactive power around its zero set point despite the transients in the real power (see Fig. 11.23(b)). This ensures that the converter operates under unity power factor regardless of the operating conditions. The neutral point potential, which is not shown here, remains well within its bounds.

We conclude that the machine-side inverter and the grid-side converter have opposite dynamic characteristics. The converter is capable of quickly ramping up the real power to the dc-link, whereas increasing the power (by increasing the torque) is a slow task for the inverter, particularly when operating at nominal speed and nominal voltage. In contrast, the inverter is able to quickly reduce the power (by lowering the torque), while this is a time-consuming task for the converter.

11.3 Summary and Discussion

11.3.1 Model Predictive Direct Current Control

MPDCC is conceptually slightly more involved than MPDTC because of its time-varying current bounds. A major advantage of MPDCC is, however, that it requires only one tuning parameter—the current bound width, unlike MPDTC, which is based on two tuning parameters—the bound widths on the torque and stator flux magnitude. Even more important, MPDCC has a small edge over MPDTC in terms of the harmonic current distortions.

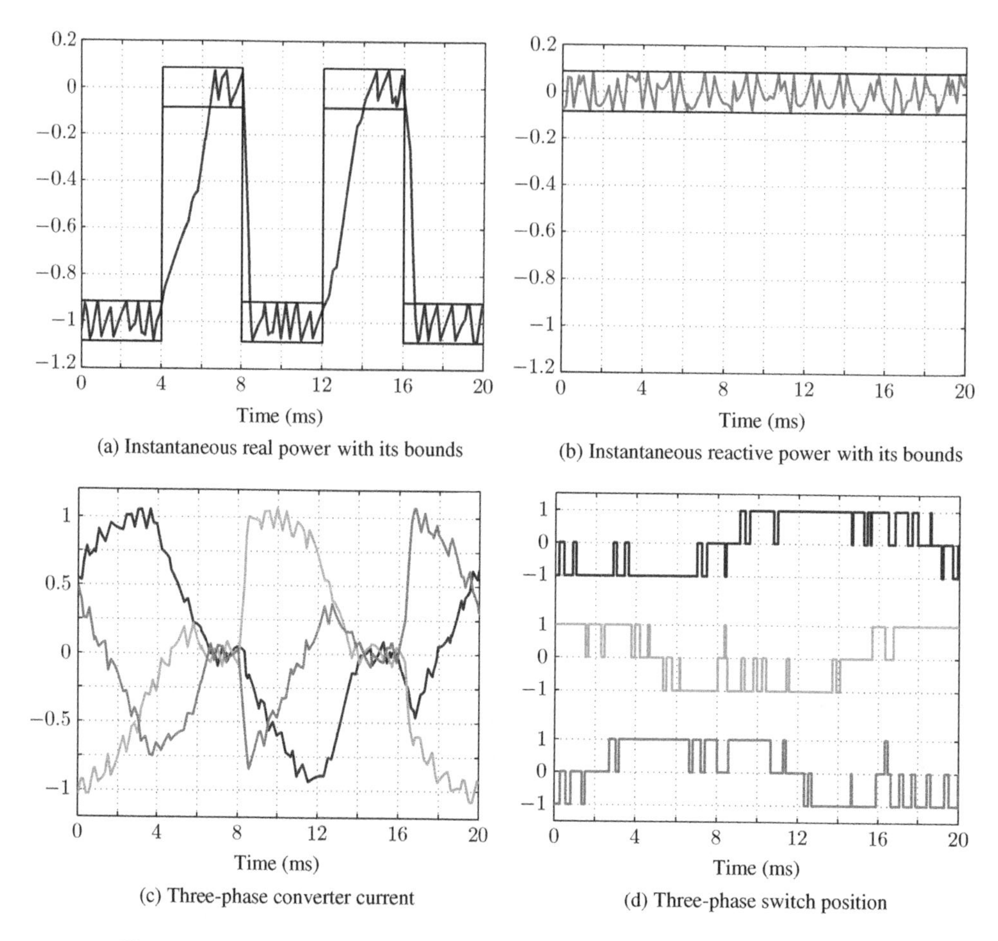

Figure 11.23 Dynamic response of MPDPC to reference steps in the real power

The hexagonal current bounds in MPDCC are ideally suited to achieving minimal current distortions for a given switching effort, as explained in Sect. 11.1.3.

When minimizing the switching losses and adopting long switching horizons, MPDCC is capable of outperforming OPPs in terms of the switching losses and the current TDD, at least when operating at very low switching frequencies. However, the switching frequency and the torque TDD of MPDCC tend to be higher. When minimizing the switching frequency in the cost function, MPDCC often achieves a performance close to that of OPPs, but it is unable to outperform them. This is shown in the extensive performance comparison [6].

A three-level NPC inverter with an induction machine was used in this chapter as an illustrative example for an MV drive system. Addressing other topologies and machines only requires a change of the internal controller model, and is thus a straightforward undertaking. Notably, MPDCC can be applied to both modular multilevel converters [30] and grid-connected converters [31]. In the presence of an *LC* filter, an active damping loop is required to suppress the filter resonance. To this end, a *virtual resistor* [32, 33] can be incorporated into the MPDCC algorithm [34]. Such a resistor emulates a physical damping resistor without incurring power losses. A virtual resistor is also effective in attenuating current harmonics that result from grid voltage harmonics, as indicated by the experimental results in [34].

An important aspect, which has not yet been discussed, is closed-loop stability. It can be shown for MPDCC that the algorithm guarantees closed-loop stability; the load currents are moved into the imposed bounds and are kept within them. This is formally proven in [35], although the switching constraint $||\Delta \boldsymbol{u}(\ell)||_\infty \leq 1$ in (11.24e) had to be neglected and the hexagonal bounds were approximated by a circle. It can also be shown that, by slightly modifying the MPDCC algorithm, robustness to bounded additive parameter uncertainties can be established. For more details on the stability and robustness of MPDCC, the reader is referred to [35].

11.3.2 Model Predictive Direct Power Control

MPDPC imposes upper and lower bounds on the instantaneous real and reactive power with the bound width parameters δ_P and δ_Q. As shown in [36], the grid current distortions are proportional to $\delta_P^2 + \delta_Q^2$, simplifying the tuning procedure. The lowest current distortions per switching frequency are achieved when the two bound widths are the same. Therefore, MPDPC has—like MPDCC—in effect only one tuning parameter. Furthermore, thanks to the fact that the bounds are constant within the prediction horizon, the control problem formulation is as simple as in MPDTC.

MPDPC with long switching horizons is capable of reducing the current distortions per switching losses compared to SVM. The harmonic spectrum of MPDPC is, however, ill suited to meet harmonic grid standards imposed at the PCC, because MPDPC produces harmonic content at even and triplen integer multiples of the fundamental frequency. Moreover, even if such a harmonic spectrum were acceptable, MPDCC might still be the preferred choice for grid-connected converters, as it can also be used for the grid side and—thanks to its hexagonal bounds—tends to produce slightly lower current distortions.

An *LC* filter is required in most cases to meet the relevant grid standards. To dampen the filter resonance, MPDPC can be augmented with a virtual resistor [32, 33] in a similar manner as for MPDCC. More specifically, as proposed in [37], virtual resistors can be placed in parallel to the filter inductor and capacitor. In the introduction to this section, we have seen that DPC can be formulated in terms of virtual converter and grid fluxes, giving rise to VF-DPC. Accordingly, MPDPC can be extended to VF-MPDPC, as shown in [38]. Experimental results for an NPC converter are provided in [36], using the MPDPC formulation explained in this section. Several implementation-related aspects of MPDPC are also discussed, which reduce the computational burden and simplify the implementation.

11.3.3 Target Sets

We have seen in this chapter that MPDCC and MPDPC are closely related to MPDTC. Indeed, the main differences between these three control methods are both their controller model and which output variables they control by imposing upper and lower bounds on them. These bounds form a set, within which the controlled variables are retained. In the following, we

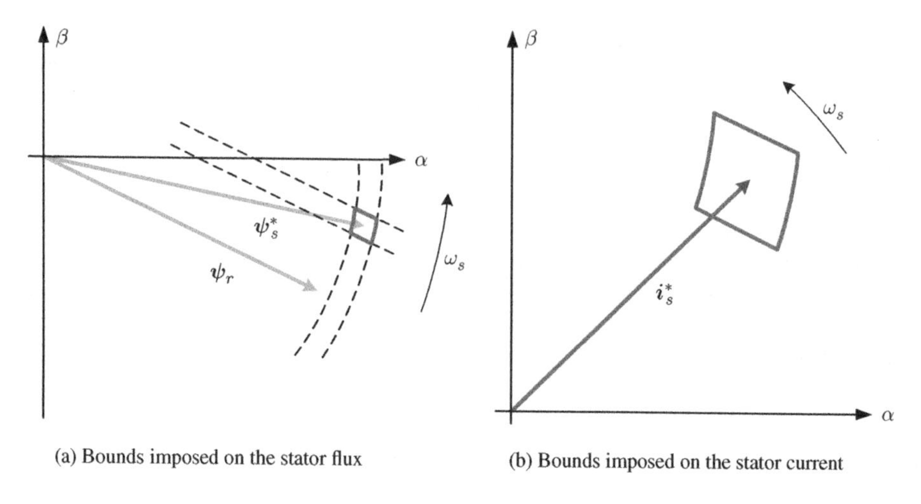

(a) Bounds imposed on the stator flux

(b) Bounds imposed on the stator current

Figure 11.24 MPDTC target set

will compare the characteristic target sets of MPDTC, MPDCC, and MPDPC by expressing them in terms of the (stator or converter) current.

MPDTC imposes upper and lower bounds on the torque and stator flux magnitude around their references T_e^* and Ψ_s^*. For a given rotor flux vector ψ_r, these references can be translated with the help of (9.1) into an equivalent reference ψ_s^* on the stator flux vector. The upper and lower torque and flux magnitude bounds can be translated accordingly into bounds on the stator flux vector. As shown in Fig. 11.24(a), the torque bounds form lines parallel to the rotor flux vector, whereas the stator flux magnitude bounds are concentric circles around the origin. Note that the example in this figure corresponds to an MV induction machine, which operates at nominal torque and whose rated values are provided in Table 2.9. The target set rotates with the angular stator frequency ω_s.

In a further step, the bounds on the stator *flux* vector can be translated into equivalent bounds on the stator *current* vector. To this end, consider the first row of (2.53), which is

$$\boldsymbol{i}_s = \frac{X_r}{D}\boldsymbol{\psi}_s - \frac{X_m}{D}\boldsymbol{\psi}_r = 3.92\boldsymbol{\psi}_s - 3.75\boldsymbol{\psi}_r. \tag{11.50}$$

The numerical values correspond to the 3.3 kV MV induction machine used throughout the book (see Table 2.10). The current vector is a linear transformation of the stator and rotor flux vectors. Specifically, the current vector is the scaled stator flux shifted by the scaled rotor flux. The resulting current bounds, which are shown in Fig. 11.24(b), are equivalent to those on the stator flux vector in Fig. 11.24(a). It can be seen that the target set is increased about fourfold and shifted upwards. The line segments of the torque bounds and the circular segments of the flux magnitude bounds are preserved thanks to the linear transformation in (11.50). In particular, the orientation of the bounds in the stationary orthogonal reference frame remains unchanged.

In MPDCC, upper and lower bounds are imposed on the three phase currents. This formulation of the bounds is, as discussed in Sect. 11.1.3, beneficial to achieve a favorable ratio

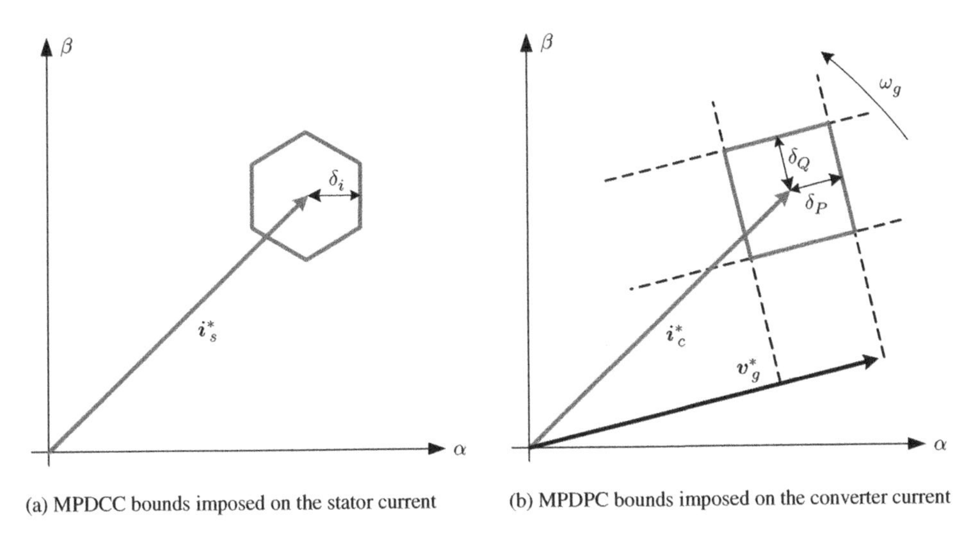

(a) MPDCC bounds imposed on the stator current

(b) MPDPC bounds imposed on the converter current

Figure 11.25 Target sets of MPDCC and MPDPC

between the current distortions and the switching losses. The bounds form a hexagonal target set in the stationary orthogonal reference frame. As depicted in Fig. 11.25(a), this target set is centered at the stator current reference vector $\boldsymbol{i}_s^*$. In contrast to MPDTC, the MPDCC target set does not rotate. In particular, the vertical edges, which correspond to the upper and lower bounds on the phase a current, remain parallel to the β-axis of the stationary orthogonal coordinate system.

MPDPC controls the instantaneous real and reactive power with the help of upper and lower bounds. Equation (11.B.11) implies that the bounds on the real power are orthogonal to the grid voltage, whereas the bounds on the reactive power are parallel to it. These bounds form a rectangular target set around the reference of the converter current $\boldsymbol{i}_c^*$ (see Fig. 11.25(b)). As shown in [36], the lowest current distortions per switching effort are achieved when the target set is square. The bounds rotate in the stationary orthogonal coordinate system with the angular grid frequency ω_g.

It is clear that MPDCC with hexagonal bounds achieves the lowest current distortions per switching effort, followed by the square bounds of MPDPC. The non-square bounds of MPDTC are less suitable to achieving very low current distortions. The difference between MPDCC and MPDTC in terms of the current distortions per switching *losses* is, however, small, as will be shown in Sect. 15.1 through extensive simulations. Similarly, only small differences arise when comparing the current distortions per switching *frequency*, as shown in [6].

Appendix 11.A: Controller Model used in MPDCC

The matrices (11.19) of the continuous-time MPDCC model (11.18) are derived in this appendix. Recall that the machine model (11.4) is described in the stationary $\alpha\beta$ reference frame. In order to guard against confusion, we will explicitly state in this appendix the coordinate systems in which the vectors are defined.

In a first step, we express the stator dynamic (11.4a) in terms of the three-phase stator current $\boldsymbol{i}_{s,abc}$. A left-multiplication of (11.4a) with the pseudo-inverse of the reduced Clarke transformation $\tilde{\boldsymbol{K}}^{-1}$ (see (2.13)) yields

$$\frac{\mathrm{d}}{\mathrm{d}t}\left(\tilde{\boldsymbol{K}}^{-1}\boldsymbol{i}_{s,\alpha\beta}\right) = -\frac{1}{\tau_s}\tilde{\boldsymbol{K}}^{-1}\boldsymbol{i}_{s,\alpha\beta} + \tilde{\boldsymbol{K}}^{-1}\left(\frac{1}{\tau_r}\boldsymbol{I}_2 - \omega_r\begin{bmatrix}0 & -1\\ 1 & 0\end{bmatrix}\right)\frac{X_m}{D}\boldsymbol{\psi}_{r,\alpha\beta} + \frac{X_r}{D}\tilde{\boldsymbol{K}}^{-1}\boldsymbol{v}_{s,\alpha\beta}. \tag{11.A.1}$$

In a next step, we insert the expression (11.1) for the stator voltage into (11.A.1) and write the differential equation in terms of the three-phase stator current. Doing the same for the rotor dynamic (11.4b) leads to the revised machine model

$$\frac{\mathrm{d}\boldsymbol{i}_{s,abc}}{\mathrm{d}t} = -\frac{1}{\tau_s}\boldsymbol{i}_{s,abc} + \tilde{\boldsymbol{K}}^{-1}\left(\frac{1}{\tau_r}\boldsymbol{I}_2 - \omega_r\begin{bmatrix}0 & -1\\ 1 & 0\end{bmatrix}\right)\frac{X_m}{D}\boldsymbol{\psi}_{r,\alpha\beta} + \frac{v_{\mathrm{dc}}}{2}\frac{X_r}{D}\tilde{\boldsymbol{K}}^{-1}\tilde{\boldsymbol{K}}\boldsymbol{u}_{abc} \tag{11.A.2a}$$

$$\frac{\mathrm{d}\boldsymbol{\psi}_{r,\alpha\beta}}{\mathrm{d}t} = \frac{X_m}{\tau_r}\tilde{\boldsymbol{K}}\boldsymbol{i}_{s,abc} - \frac{1}{\tau_r}\boldsymbol{\psi}_{r,\alpha\beta} + \omega_r\begin{bmatrix}0 & -1\\ 1 & 0\end{bmatrix}\boldsymbol{\psi}_{r,\alpha\beta}. \tag{11.A.2b}$$

The system and input matrices $\boldsymbol{F}_m$ and $\boldsymbol{G}_m$ of the continuous-time machine model can then be stated as

$$\boldsymbol{F}_m = \begin{bmatrix} -\frac{1}{\tau_s} & 0 & 0 & \frac{1}{\tau_r}\frac{X_m}{D} & \omega_r\frac{X_m}{D} \\ 0 & -\frac{1}{\tau_s} & 0 & -\frac{1}{2}(\frac{1}{\tau_r}+\sqrt{3}\omega_r)\frac{X_m}{D} & \frac{1}{2}(\frac{\sqrt{3}}{\tau_r}-\omega_r)\frac{X_m}{D} \\ 0 & 0 & -\frac{1}{\tau_s} & \frac{1}{2}(\sqrt{3}\omega_r-\frac{1}{\tau_r})\frac{X_m}{D} & -\frac{1}{2}(\frac{\sqrt{3}}{\tau_r}+\omega_r)\frac{X_m}{D} \\ \frac{2}{3}\frac{X_m}{\tau_r} & -\frac{1}{3}\frac{X_m}{\tau_r} & -\frac{1}{3}\frac{X_m}{\tau_r} & -\frac{1}{\tau_r} & -\omega_r \\ 0 & \frac{1}{\sqrt{3}}\frac{X_m}{\tau_r} & -\frac{1}{\sqrt{3}}\frac{X_m}{\tau_r} & \omega_r & -\frac{1}{\tau_r} \end{bmatrix}, \tag{11.A.3a}$$

$$\boldsymbol{G}_m = \frac{v_{\mathrm{dc}}}{6}\frac{X_r}{D}\begin{bmatrix} 2 & -1 & -1 \\ -1 & 2 & -1 \\ -1 & -1 & 2 \\ 0 & 0 & 0 \\ 0 & 0 & 0 \end{bmatrix}. \tag{11.A.3b}$$

Note that the three upper rows of $\boldsymbol{G}_m$ relate to the matrix product $\tilde{\boldsymbol{K}}^{-1}\tilde{\boldsymbol{K}}$, which can be written as

$$\tilde{\boldsymbol{K}}^{-1}\tilde{\boldsymbol{K}} = \frac{1}{3}\begin{bmatrix} 2 & -1 & -1 \\ -1 & 2 & -1 \\ -1 & -1 & 2 \end{bmatrix} = \begin{bmatrix} 1 & 0 & 0 \\ 0 & 1 & 0 \\ 0 & 0 & 1 \end{bmatrix} - \frac{1}{3}\begin{bmatrix} 1 & 1 & 1 \\ 1 & 1 & 1 \\ 1 & 1 & 1 \end{bmatrix}. \tag{11.A.4}$$

The last matrix in (11.A.4) removes the common-mode component of the three-phase switch position from each phase. This implies that—not surprisingly—only the differential-mode part of the three-phase switch position manipulates the stator current, while the common-mode part has no effect.

The dynamic of the neutral point potential is described by (11.3) in terms of the component-wise absolute value of the inverter switch position $|\boldsymbol{u}_{abc}| = [|u_a| \;\; |u_b| \;\; |u_c|]^T$ and the three-phase stator current. The input-dependent system vector of the inverter directly follows as

$$\boldsymbol{f}_i(\boldsymbol{u}_{abc}) = \frac{1}{2X_c}\left[|u_a| \;\; |u_b| \;\; |u_c| \;\; 0 \;\; 0\right]. \tag{11.A.5}$$

Appendix 11.B: Real and Reactive Power

The expressions for the instantaneous power of the three-phase system shown in Fig. 11.15 are derived in this appendix. We start with the instantaneous real power at the grid voltage sources. In phase a, the instantaneous real power is the product of the grid voltage v_{ga} and the converter current i_{ca}. By summing up the contributions from all three phases, the instantaneous three-phase real power in the SI system is given by

$$P = v_{ga}i_{ca} + v_{gb}i_{cb} + v_{gc}i_{cc} = \boldsymbol{v}_{g,abc}^T\boldsymbol{i}_{c,abc}. \tag{11.B.1}$$

Using the inverse Clarke transformation (2.10), the three-phase grid voltage can be expressed in stationary orthogonal coordinates as

$$\boldsymbol{v}_{g,abc} = \boldsymbol{K}^{-1}\boldsymbol{v}_{g,\alpha\beta0}, \tag{11.B.2}$$

where $\boldsymbol{v}_{g,\alpha\beta0} = [v_{g\alpha} \; v_{g\beta} \; v_{g0}]$. The same can be performed with the converter current, where $\boldsymbol{i}_{c,\alpha\beta0}$ is defined accordingly. Note that we included the 0-component here. We can then rewrite (11.B.1) as

$$P = \boldsymbol{v}_{g,\alpha\beta0}^T\boldsymbol{K}^{-T}\boldsymbol{K}^{-1}\boldsymbol{i}_{c,\alpha\beta0} = \boldsymbol{v}_{g,\alpha\beta0}^T\begin{bmatrix}1.5 & 0 & 0\\ 0 & 1.5 & 0\\ 0 & 0 & 3\end{bmatrix}\boldsymbol{i}_{c,\alpha\beta0}. \tag{11.B.3}$$

The zero-sequence current i_{c0} is zero in the absence of faults, and (11.B.3) reduces to

$$P = \frac{3}{2}(v_{g\alpha}i_{c\alpha} + v_{g\beta}i_{c\beta}). \tag{11.B.4}$$

Note that the factor 1.5 is a consequence of the Clarke transformation being amplitude-invariant rather than power-invariant.

The reactive power is commonly defined in single-phase systems to be at its maximum when the current leads the voltage by 90°. The reactive power can then be defined in phase a as the product of the current i_{ca} and the voltage $\breve{v}_{ga}$. The latter represents the phase a grid voltage v_{ga} that is phase-shifted by 90°. This definition can be extended to three-phase systems. According to [39], the instantaneous three-phase reactive power in the SI system is defined as

$$Q = \breve{v}_{ga}i_{ca} + \breve{v}_{gb}i_{cb} + \breve{v}_{gc}i_{cc} = \breve{\boldsymbol{v}}_{g,abc}^T\boldsymbol{i}_{c,abc}, \tag{11.B.5}$$

where the phase-shifted grid voltages in phases b and c, $\breve{v}_{gb}$ and $\breve{v}_{gc}$, are defined similar to $\breve{v}_{ga}$. The phase-shifted three-phase voltage is defined as $\breve{\boldsymbol{v}}_{g,abc} = [\breve{v}_{ga} \; \breve{v}_{gb} \; \breve{v}_{gc}]^T$.

We define

$$\breve{\boldsymbol{v}}_{g,abc} = \boldsymbol{K}^{-1}\breve{\boldsymbol{v}}_{g,\alpha\beta0} \tag{11.B.6}$$

in accordance with (11.B.2). It is easy in the orthogonal reference frame to relate the grid voltage to its phase-shifted counterpart. For the α- and β-components, this is achieved by a

90° rotation, whereas the 0-component remains the same.

$$\breve{\boldsymbol{v}}_{g,\alpha\beta 0} = \begin{bmatrix} 0 & -1 & 0 \\ 1 & 0 & 0 \\ 0 & 0 & 1 \end{bmatrix} \boldsymbol{v}_{g,\alpha\beta 0}. \tag{11.B.7}$$

Equation (11.B.5) can be rewritten with the help of (11.B.6) and (11.B.7) as

$$Q = \boldsymbol{v}_{g,\alpha\beta 0}^T \begin{bmatrix} 0 & -1 & 0 \\ 1 & 0 & 0 \\ 0 & 0 & 1 \end{bmatrix}^T \boldsymbol{K}^{-T}\boldsymbol{K}^{-1}\boldsymbol{i}_{c,\alpha\beta 0} = \boldsymbol{v}_{g,\alpha\beta 0}^T \begin{bmatrix} 0 & 1.5 & 0 \\ -1.5 & 0 & 0 \\ 0 & 0 & 3 \end{bmatrix} \boldsymbol{i}_{c,\alpha\beta 0}. \tag{11.B.8}$$

With i_{c0} being zero, (11.B.8) simplifies to

$$Q = \frac{3}{2}(v_{g\alpha} i_{c\beta} - v_{g\beta} i_{c\alpha}). \tag{11.B.9}$$

This definition is in line with [26]. The definitions of the instantaneous real and reactive power components are applicable to symmetric as well as asymmetric grid voltages, provided that the sum of the converter currents—the zero-sequence current—is zero.

Additional insight can be gained by adopting a rotating dq reference frame, which rotates with the angular grid frequency $\omega_g = 2\pi f_g$. In this reference frame, we define the grid voltage $\boldsymbol{v}_{g,dq} = [v_{gd}\ v_{gq}]^T$ and the converter current $\boldsymbol{i}_{c,dq} = [i_{cd}\ i_{cq}]^T$. With the help of the $\alpha\beta$ to dq transformation (2.24), which is based on the rotation matrix $\boldsymbol{R}$ (see (2.25)), the power components can be written as

$$P = \frac{3}{2}(v_{gd} i_{cd} + v_{gq} i_{cq}) \tag{11.B.10a}$$

$$Q = \frac{3}{2}(v_{gd} i_{cq} - v_{gq} i_{cd}). \tag{11.B.10b}$$

By aligning the d-axis with the grid voltage, the quadrature component of the grid voltage becomes zero, and the expressions for the real and reactive power components simplify to

$$P = \frac{3}{2} v_{gd} i_{cd} \tag{11.B.11a}$$

$$Q = \frac{3}{2} v_{gd} i_{cq}. \tag{11.B.11b}$$

This compact representation reveals that the d-component of the converter current produces real power while the quadrature component of the current produces reactive power. This fact is depicted in Fig. 11.B.1. We can also see that positive reactive power corresponds to the capacitive case, in which the current leads the voltage.

To express (11.B.4) and (11.B.9) in the pu system, we normalize the grid voltage and the converter current by the base voltage V_B and the base current I_B, respectively. The power is normalized by the base apparent power $S_B = 1.5 V_B I_B$ (see also Table 2.1 and Sect. 2.5.4). To this end, we introduce the pu quantities

$$\boldsymbol{v}_g' = \frac{\boldsymbol{v}_g}{V_B}, \boldsymbol{i}_c' = \frac{\boldsymbol{i}_c}{I_B}, P' = \frac{P}{S_B} \text{ and } Q' = \frac{Q}{S_B}. \tag{11.B.12}$$

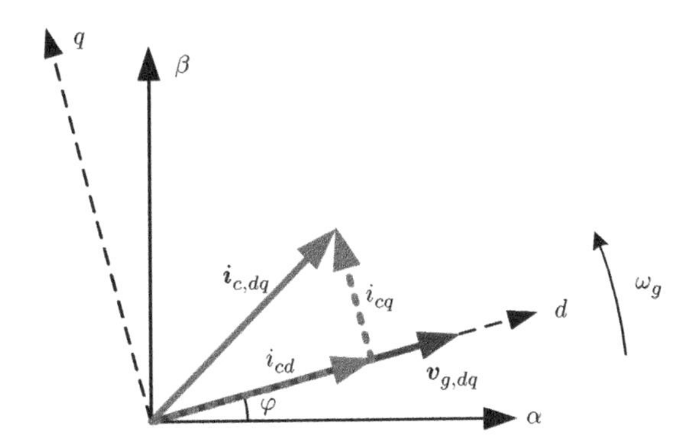

Figure 11.B.1 Converter current $\boldsymbol{i}_{c,dq}$ and grid voltage $\boldsymbol{v}_{g,dq}$ in an orthogonal reference frame that is aligned with the grid voltage and rotates with the angular grid frequency ω_g. The current component i_{cd} relates to the real power P, whereas the quadrature current component i_{cq} relates to the reactive power Q

Dividing (11.B.4) and (11.B.9) by V_B, I_B, and 1.5 leads to

$$P' = \frac{P}{1.5V_B I_B} = v'_{g\alpha} i'_{c\alpha} + v'_{g\beta} i'_{c\beta} \tag{11.B.13a}$$

$$Q' = \frac{Q}{1.5V_B I_B} = v'_{g\alpha} i'_{c\beta} - v'_{g\beta} i'_{c\alpha}. \tag{11.B.13b}$$

To simplify the notation, we will drop hereafter the superscript $'$. Per unit quantities are adopted throughout this book and all variables and parameters are normalized.

Appendix 11.C: Controller Model used in MPDPC

The continuous-time prediction model for MPDPC is derived in this appendix. The model is based on the converter current $\boldsymbol{i}_c$, the converter voltage $\boldsymbol{v}_c$, and the grid voltage $\boldsymbol{v}_g$ in the stationary orthogonal reference frame. The switch position $\boldsymbol{u}$ is a three-phase vector.

We start by inserting the converter voltage (11.30) into the differential equation (11.34) of the converter current. This leads to

$$\frac{\mathrm{d}\boldsymbol{i}_c}{\mathrm{d}t} = -\frac{R}{X}\boldsymbol{i}_c - \frac{1}{X}\boldsymbol{v}_g + \frac{v_{\mathrm{dc}}}{2X}\tilde{\boldsymbol{K}}\,\boldsymbol{u}. \tag{11.C.1}$$

By combining (11.C.1) with the evolution of the grid voltage (11.35), we obtain

$$\frac{\mathrm{d}}{\mathrm{d}t}\begin{bmatrix}\boldsymbol{i}_c\\ \boldsymbol{v}_g\end{bmatrix} = \boldsymbol{F}_g\begin{bmatrix}\boldsymbol{i}_c\\ \boldsymbol{v}_g\end{bmatrix} + \boldsymbol{G}_g\,\boldsymbol{u} \tag{11.C.2}$$

with the matrices

$$\boldsymbol{F}_g = \begin{bmatrix} -\frac{R}{X} & 0 & -\frac{1}{X} & 0\\ 0 & -\frac{R}{X} & 0 & -\frac{1}{X}\\ 0 & 0 & 0 & -\omega_g\\ 0 & 0 & \omega_g & 0\end{bmatrix} \text{ and } \boldsymbol{G}_g = \frac{v_{\mathrm{dc}}}{6X}\begin{bmatrix} 2 & -1 & -1\\ 0 & \sqrt{3} & -\sqrt{3}\\ 0 & 0 & 0\\ 0 & 0 & 0\end{bmatrix}. \tag{11.C.3}$$

The dynamic evolution of the neutral point potential depends on both the component-wise absolute value of the switch position and the three-phase converter current. The latter can be expressed in terms of the $\alpha\beta$ current, which allows us to rewrite (11.31) as

$$\frac{\mathrm{d}v_n}{\mathrm{d}t} = \frac{1}{2X_c}|\boldsymbol{u}|^T\,\tilde{\boldsymbol{K}}^{-1}\,\boldsymbol{i}_c \tag{11.C.4}$$

or equivalently as

$$\frac{\mathrm{d}v_n}{\mathrm{d}t} = \boldsymbol{f}_c(\boldsymbol{u})\begin{bmatrix}\boldsymbol{i}_c\\ \boldsymbol{v}_g\end{bmatrix}, \tag{11.C.5}$$

where we have introduced

$$\boldsymbol{f}_c(\boldsymbol{u}) = \frac{1}{4X_c}\left[2|u_a| - |u_b| - |u_c| \;\; \sqrt{3}(|u_b| - |u_c|) \;\; 0 \;\; 0\right]. \tag{11.C.6}$$

By recalling the definitions of the instantaneous real and reactive power (11.36), the output function directly follows, as

$$\boldsymbol{h}(\boldsymbol{x}) = \begin{bmatrix}x_1x_3 + x_2x_4\\ x_2x_3 - x_1x_4\\ x_5\end{bmatrix}. \tag{11.C.7}$$

References

[1] T. Geyer, G. Papafotiou, and M. Morari, "Model predictive direct torque control—Part I: Concept, algorithm and analysis," *IEEE Trans. Ind. Electron.*, vol. 56, pp. 1894–1905, Jun. 2009.

[2] J. C. R. Martinez, R. M. Kennel, and T. Geyer, "Model predictive direct current control," in *Proceedings of IEEE International Conference on Industrial Technology* (Viña del Mar, Chile), pp. 1808–1813, Mar. 2010.

[3] J. Holtz and S. Stadtfeld, "A predictive controller for the stator current vector of AC machines fed from a switched voltage source," in *Proceedings of IEEE International Power Electronics Conference* (Tokyo, Japan), pp. 1665–1675, Apr. 1983.

[4] J. Holtz and S. Stadtfeld, "Field-oriented control by forced motor currents in a voltage fed inverter drive," in *Proceedings of IFAC Symposium* (Lausanne, Switzerland), pp. 103–110, Sep. 1983.

[5] A. Khambadkone and J. Holtz, "Low switching frequency and high dynamic pulsewidth modulation based on field-orientation for high-power inverter drive," *IEEE Trans. Power Electron.*, vol. 7, pp. 627–632, Oct. 1992.

[6] J. Scoltock, T. Geyer, and U. K. Madawala, "A comparison of model predictive control schemes for MV induction motor drives," *IEEE Trans. Ind. Inf.*, vol. 9, pp. 909–919, May 2013.

[7] M. P. Kazmierkowski and L. Malesani, "Current control techniques for three-phase voltage-source PWM converters: A survey," *IEEE Trans. Ind. Electron.*, vol. 45, pp. 691–703, Oct. 1998.

[8] J. Rodríguez, S. Bernet, B. Wu, J. Pontt, and S. Kouro, "Multilevel voltage-source-converter topologies for industrial medium-voltage drives," *IEEE Trans. Ind. Electron.*, vol. 54, pp. 2930–2945, Dec. 2007.

[9] M. Hiller, R. Sommer, and M. Beuermann, "Medium-voltage drives," *IEEE Ind. Appl. Mag.*, vol. 16, pp. 22–30, Mar./Apr. 2010.

[10] J. M. Carrasco, L. G. Franquelo, J. T. Bialasiewicz, E. Galván, R. C. P. Guisado, A. M. Prats, J. I. León, and N. Moreno-Alfonso, "Power-electronic systems for the grid integration of renewable energy sources: A survey," *IEEE Trans. Ind. Electron.*, vol. 53, pp. 1002–1016, Aug. 2006.

[11] R. W. De Doncker, C. Meyer, R. U. Lenke, and F. Mura, "Power electronics for future utility applications," in *Proceedings of IEEE International Conference on Power Electronics and Drive Systems* (Bangkok, Thailand), Nov. 2007.

[12] M. Malinowski, M. P. Kazmierkowski, and A. M. Trzynadlowski, "A comparative study of control techniques for PWM rectifiers in AC adjustable speed drives," *IEEE Trans. Power Electron.*, vol. 18, pp. 1390–1396, Nov. 2003.

[13] H. Kohlmeier, O. Niermeyer, and D. F. Schröder, "Highly dynamic four-quadrant ac motor drive with improved power factor and on-line optimized pulse pattern with PROMC," *IEEE Trans. Ind. Appl.*, vol. IA-23, pp. 1001–1009, Nov./Dec. 1987.

[14] V. Blasko and V. Kaura, "A new mathematical model and control of a three-phase AC–DC voltage source converter," *IEEE Trans. Power Electron.*, vol. 12, pp. 116–123, Jan. 1997.

[15] T. Noguchi, H. Tomiki, S. Kondo, and I. Takahashi, "Direct power control of PWM converter without power-source voltage sensors," *IEEE Trans. Ind. Appl.*, vol. 34, pp. 473–479, May/Jun. 1998.

[16] S. Bhattacharya, A. Veltman, D. M. Divan, and R. D. Lorenz, "Flux-based active filter controller," *IEEE Trans. Ind. Appl.*, vol. 32, pp. 491–502, May/Jun. 1996.

[17] M. Malinowski, M. P. Kazmierkowski, S. Hansen, F. Blaabjerg, and G. D. Marques, "Virtual-flux-based direct power control of three-phase PWM rectifiers," *IEEE Trans. Ind. Appl.*, vol. 37, pp. 1019–1027, Jul./Aug. 2001.

[18] M. C. Chandorkar, D. M. Divan, and R. Adapa, "Control of parallel connected inverters in standalone AC supply systems," *IEEE Trans. Ind. Appl.*, vol. 29, pp. 136–143, Jan./Feb. 1993.

[19] L. A. Serpa and J. W. Kolar, "Virtual-flux direct power control for mains connected three-level NPC inverter systems," in *Proceedings of the Power Conversion Conference* (Nagoya, Japan), pp. 130–136, Apr. 2007.

[20] L. A. Serpa, P. M. Barbosa, P. K. Steimer, and J. W. Kolar, "Five-level virtual-flux direct power control for the active neutral-point clamped multilevel converter," in *Proceedings of IEEE Power Electronics Specialists Conference*, pp. 1668–1674, 2008.

[21] L. A. Serpa, S. Ponnaluri, P. M. Barbosa, and J. W. Kolar, "A modified direct power control strategy allowing the connection of three-phase inverters to the grid through LCL filters," *IEEE Trans. Ind. Appl.*, vol. 43, pp. 1388–1400, Sep./Oct. 2007.

[22] M. Malinowski, M. Jasinski, and M. P. Kazmierkowski, "Simple direct power control of three-phase PWM rectifier using space-vector modulation (DPC–SVM)," *IEEE Trans. Ind. Electron.*, vol. 51, pp. 447–454, Apr. 2004.

[23] A. Bouafia, J.-P. Gaubert, and F. Krim, "Predictive direct power control of three-phase pulsewidth modulation (PWM) rectifier using space-vector modulation (SVM)," *IEEE Trans. Power Electron.*, vol. 25, pp. 228–236, Jan. 2010.

[24] S. Aurtenechea, M. A. Rodríguez, E. Oyarbide, and J. R. Torrealday, "Predictive control strategy for DC/AC converters based on direct power control," *IEEE Trans. Ind. Electron.*, vol. 54, pp. 1261–1271, Jun. 2007.

[25] P. Cortés, J. Rodríguez, P. Antoniewicz, and M. Kazmierkowski, "Direct power control of an AFE using predictive control," *IEEE Trans. Power Electron.*, vol. 23, pp. 2516–2523, Sep. 2008.

[26] H. Akagi, Y. Kanazawa, and A. Nabae, "Instantaneous reactive power compensators comprising switching devices without energy storage components," *IEEE Trans. Ind. Appl.*, vol. IA-20, pp. 625–630, May/Jun. 1984.

[27] R. Teodorescu, M. Liserre, and P. Rodríguez, *Grid converters for photovoltaic and wind power systems*. Chichester, UK: John Wiley & Sons, Ltd, 2011.

[28] IEEE Std 519-1992, *"IEEE recommended practices and requirements for harmonic control in electrical power systems,"* Apr. 1993.

[29] IEC 61000-2-4, *"Electromagnetic compatibility (EMC)—part 2-4: Environment—compatibility levels in industrial plants for low-frequency conducted disturbances,"* Sep. 2002.

[30] B. S. Riar, T. Geyer, and U. K. Madawala, "Model predictive direct current control of modular multilevel converters: Modelling, analysis, and experimental evaluation," *IEEE Trans. Power Electron.*, vol. 30, pp. 431–439, Jan. 2015.

[31] J. Scoltock, T. Geyer, and U. K. Madawala, "Model predictive direct current control for a grid-connected converter: LCL-filter versus L-filter," in *Proceedings of IEEE International Conference on Industrial Technology* (Cape Town, South Africa), Feb. 2013.

[32] P. A. Dahono, Y. R. Bahar, Y. Sato, and T. Kataoka, "Damping of transient oscillations on the output LC filter of PWM inverters by using a virtual resistor," in *Proceedings of IEEE International Conference on Power Electronics and Drive Systems*, pp. 403–407, Oct. 2001.

[33] P. A. Dahono, "A control method to damp oscillation in the input LC filter of AC-DC PWM converters," in *Proceedings of IEEE Power Electronics Specialists Conference*, pp. 1630–1635, Jun. 2002.

[34] J. Scoltock, T. Geyer, and U. K. Madawala, "A model predictive direct current control strategy with predictive references for MV grid-connected converters with LCL-filters," *IEEE Trans. Power Electron.*, vol. 30, pp. 5926–5937, Oct. 2015.

[35] T. Geyer, R. P. Aguilera, and D. E. Quevedo, "On the stability and robustness of model predictive direct current control," in *Proceedings of IEEE International Conference on Industrial Technology* (Cape Town, South Africa), Feb. 2013.
[36] J. Scoltock, T. Geyer, and U. K. Madawala, "Model predictive direct power control for grid-connected neutral-point-clamped converters," *IEEE Trans. Ind. Electron.*, vol. 62, pp. 5319–5328, Sep. 2015.
[37] J. Scoltock, T. Geyer, and U. K. Madawala, "Model predictive direct power control for a grid-connected converter with an LCL-filter," in *Proceedings of IEEE International Conference on Industrial Technology* (Cape Town, South Africa), Feb. 2013.
[38] T. Geyer, J. Scoltock, and U. K. Madawala, "Model predictive direct power control for grid-connected converters," in *Proceedings of IEEE Industrial Electronics Society Annual Conference* (Melbourne, Australia), Nov. 2011.
[39] Y. Komatsu and T. Kawabata, "A control method of active power filter in unsymmetrical and distorted voltage system," in *Proceedings of the Power Conversion Conference* (Nagaoka, Japan), pp. 161–168, Aug. 1997.

Part Four

Model Predictive Control based on Pulse Width Modulation

12

Model Predictive Pulse Pattern Control

Offline-computed optimized pulse patterns (OPPs) allow the minimization of the current distortion for a given switching frequency. Conceptually, OPPs are a particularly attractive choice for medium-voltage (MV) drives. Traditionally, however, it has been possible to use OPPs only in a modulator driven by a very slow control loop. When the operating point is changed or when transitions between different pulse patterns occur, the absence of a fast controller leads to a poor dynamic performance and to large deviations of the stator currents from their references.

After recapitulating state-of-the-art control methods that utilize OPPs, this chapter describes a novel control and modulation strategy that combines the dynamic performance of a high-bandwidth controller such as direct torque control (DTC) with the superb harmonic performance of OPPs during steady-state operation. More specifically, the proposed pulse pattern controller achieves a nearly optimum ratio of harmonic current distortions per switching frequency at steady-state operation, and a fast rejection of disturbances. During transients, very fast current and torque response times are achieved that are similar to that of DTC, particularly when inserting additional pulses when required.

The underlying optimization problem constitutes a quadratic program (QP), which can be solved efficiently in real time. Alternatively, the pulse pattern controller can be simplified to a deadbeat (DB) controller. Simulation and experimental results of MV drive systems are provided in Chap. 13.

12.1 State-of-the-Art Control Methods

A common method to establish closed-loop control is to use field-oriented control (FOC) for machine-side inverters and voltage-oriented control (VOC) for grid-side converters. For a review of FOC and VOC, the reader is referred to Sect. 3.6.2 and to the introduction of Sect. 11.2, respectively. When using OPPs in the modulator, however, the performance of the overall control scheme is very limited—even in quasi-steady-state operation. Current excursions occur that may lead to overcurrent conditions [1]. Therefore, the application of FOC and VOC

Model Predictive Control of High Power Converters and Industrial Drives, First Edition. Tobias Geyer.

Companion Website: www.wiley.com/go/geyermodelpredictivecontrol

with OPPs has been typically limited to grid-connected setups, where the range of modulation indices is relatively small during nominal operation. When the goal is to use this method in applications with widely varying modulation indices, as is the case for variable-speed drives, the (inner) current control loop is tuned to be sufficiently slow such that its operation does not interfere with the optimal volt-second balance of the OPP. However, such a tuning significantly decreases the dynamic performance of the drive.

Furthermore, in this case, the offline optimization procedure of OPPs itself is compromised, because restrictions need to be added to the optimization algorithm that avoid discontinuities in the switching angles when changing the modulation index. Eliminating these discontinuities in the OPP improves operation at quasi steady state by eliminating *a priori* the possibility of current excursions when the operating point changes. However, the resulting currents are sub-optimal in terms of the total demand distortion (TDD) even at steady-state operation, because of the additional restrictions added during the optimization procedure.

As an improvement to FOC with OPPs, *current* trajectory tracking was proposed in [1–3]. This method derives the optimal steady-state stator current trajectory from the pulse pattern in use. The actual stator current space vector is forced to follow this target trajectory. A disadvantage is that the stator current trajectory depends on the parameters of the electrical machine, notably on the total leakage inductance [4]. Variations in the load conditions have also been found to influence the stator current trajectory.

A further improvement can be made by tracking the *stator flux* trajectory [5, 6], which is insensitive to parameter variations and is thus better suited for trajectory tracking control. To establish closed-loop control, the instantaneous fundamental components of the stator current and flux linkage vectors are required in real time. As the ripple current is nonzero at the sampling instants, these fundamental machine quantities cannot be directly sampled when using OPPs [4]. This makes the design of the closed-loop controller difficult, because these signals are required to achieve flux and torque control. For this reason, existing control schemes, such as [3, 7, 8], employ an observer to derive the instantaneous fundamental current and flux linkage values separately from their respective harmonic quantities.

12.2 Optimized Pulse Patterns

In this section, the notion and computation of OPPs is recapitulated from Sect. 3.4. The reference trajectory of the stator flux vector is computed, and the storing of OPPs in a look-up table is discussed.

12.2.1 Summary, Properties, and Computation

OPPs are computed in an offline procedure by calculating the optimal switching angles and switching transitions. Typically, the aim is to minimize the current TDD for a given switching frequency (or pulse number). Assuming a load with a predominantly inductive characteristic, the current TDD is proportional to the weighted sum of the squared differential-mode voltage harmonics

$$I_{\text{TDD}} \propto \sqrt{\sum_{n=5,7,11,\ldots} \left(\frac{\hat{v}_n}{n}\right)^2}, \tag{12.1}$$

where $\hat{v}_n$ is the magnitude of the *n*th voltage harmonic. In an induction machine, the quantity $\hat{v}_n/n$ is proportional to the magnitude of the *n*th current harmonic. It is therefore common practice to choose (12.1) as the cost function when computing OPPs, as explained in Sect. 3.4.2. Note that only the differential-mode harmonics are penalized in the cost function.

In the first step, a *single-phase* pulse pattern is computed that minimizes the cost function (12.1) for a given modulation index m. Owing to the imposed quarter-wave symmetry, the single-phase pulse pattern is fully characterized by the primary switching angles and the corresponding switch positions over the first quarter of the fundamental period. To this end, we define the vector of primary switching angles

$$A = [\alpha_1 \; \alpha_2 \; \dots \alpha_d]^T \tag{12.2}$$

and the vector of switch positions (the switching sequence)

$$U = [u_1 \; u_2 \dots u_d]^T. \tag{12.3}$$

Both vectors are of length d, where d denotes the *pulse number*. Note that the initial switch position u_0 is assumed to be zero. Switching is performed from u_{i-1} to u_i at the angular position α_i, where $i \in \{1, \dots, d\}$. This relationship is exemplified in Fig. 12.1(a).

It is clear that the pulse pattern is a piecewise constant signal. In the interval $0 \le \theta \le \frac{\pi}{2}$, the single-phase pulse pattern can be described by the statement

$$u(\theta) = \begin{cases} 0, & \text{if } 0 \le \theta < \alpha_1 \\ u_1, & \text{if } \alpha_1 \le \theta < \alpha_2 \\ u_2, & \text{if } \alpha_2 \le \theta < \alpha_3 \\ \vdots & \vdots \\ u_{d-1}, & \text{if } \alpha_{d-1} \le \theta < \alpha_d \\ u_d, & \text{if } \alpha_d \le \theta \le \frac{\pi}{2}. \end{cases} \tag{12.4}$$

The voltage harmonics $\hat{v}_n$ in (12.1) can be expressed as a function of the dc-link voltage, the primary switching angles, and the corresponding switching transitions (see (3.29) and

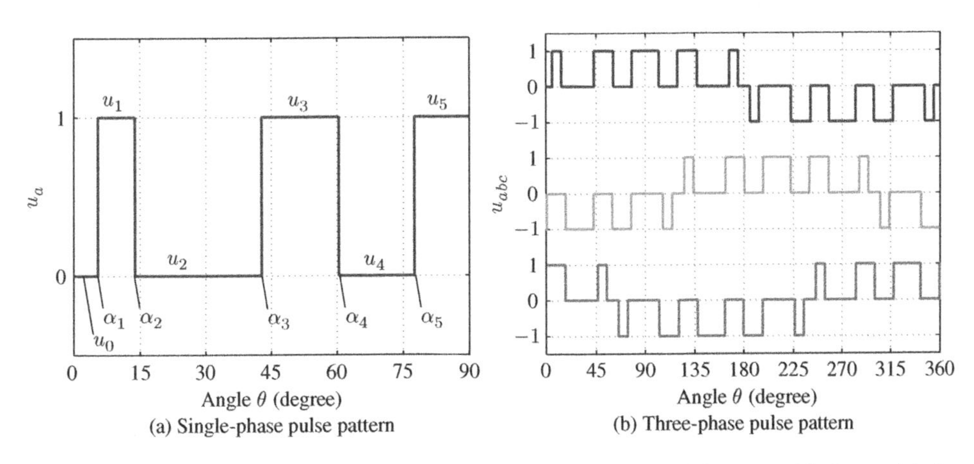

Figure 12.1 OPP with $d = 5$ primary switching angles for a three-level inverter with the modulation index $m = 0.6$

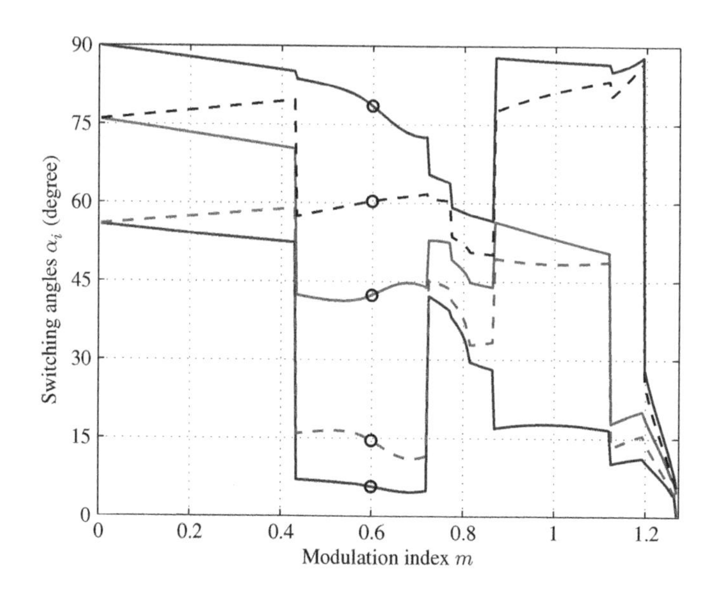

Figure 12.2 Primary switching angles as a function of the modulation index. The switching angles indicated by (black) circles correspond to the modulation index $m = 0.6$. The corresponding switching sequence is shown in Fig. 12.1

(3.24b)). Owing to the quarter-wave symmetry, the computation is restricted to one quarter of the fundamental period. The minimization of the cost function (12.1) for a given modulation index m results in the vector of primary switching angles A and the vector of switch positions U, whose corresponding pulse pattern minimizes the current TDD for an inductive load.

To derive the OPP for the whole range of modulation indices $[0, 4/\pi]$, this set is finely gridded (or discretized) resulting, for example, in the set $\{0, 0.005, 0.01, \dots 4/\pi\}$. For each discrete element in this set, the corresponding switching angles and switch positions are computed, resulting in a matrix of primary switching angles and switch positions.

The *three-phase* OPP can easily be constructed from the single-phase pulse pattern in two steps. First, by applying quarter-wave and half-wave symmetry, the remaining single-phase switching angles over $360°$ can be derived. Second, by shifting the single-phase pattern by $120°$ and $240°$, respectively, the three-phase OPP is constructed.

Example 12.1 *For a three-level converter, the switching sequence is always* $U = [1\ 0\ 1 \dots]^T$. *The primary switching angles are shown in Fig. 12.2, assuming the pulse number* $d = 5$. *The single-phase pulse pattern that corresponds to the modulation index* $m = 0.6$ *is shown in Fig. 12.1(a), while the three-phase OPP is shown in Fig. 12.1(b).*

12.2.2 Relationship between Flux Magnitude and Modulation Index

In this section, we establish the simple yet important relationship between the stator flux magnitude, nominal dc-link voltage V_{dc}, modulation index, and fundamental frequency. As this relationship is between fundamental and dc quantities, we neglect the phenomenon of switching and assume sinusoidal three-phase voltages with the angular fundamental frequency ω_1. According to the definition of the modulation index m in (3.12), the (ideal) three-phase voltage produced by the inverter is

$$\boldsymbol{v}_{abc}(t) = m\frac{V_{\mathrm{dc}}}{2}\begin{bmatrix}\sin(\omega_1 t)\\ \sin(\omega_1 t - \frac{2\pi}{3})\\ \sin(\omega_1 t - \frac{4\pi}{3})\end{bmatrix}. \tag{12.5}$$

Using the Clarke transformation (2.12), we can express the sinusoidal inverter voltages in orthogonal coordinates[1] as

$$\boldsymbol{v}(t) = \begin{bmatrix}v_\alpha(t)\\ v_\beta(t)\end{bmatrix} = \tilde{\boldsymbol{K}}\boldsymbol{v}_{abc}(t) = m\frac{V_{\mathrm{dc}}}{2}\begin{bmatrix}\sin(\omega_1 t)\\ -\cos(\omega_1 t)\end{bmatrix}. \tag{12.6}$$

When connecting a three-phase electrical machine to the inverter, the stator frequency equals the fundamental frequency, that is, $\omega_s = \omega_1$, and the stator voltage equals the inverter voltage, that is, $\boldsymbol{v}_s = \boldsymbol{v}$. Neglecting the stator resistance, the stator flux vector $\boldsymbol{\psi}_s = [\psi_{s\alpha}\ \psi_{s\beta}]^T$ at time t is given by

$$\boldsymbol{\psi}_s(t) = \boldsymbol{\psi}_s(0) + \int_0^t \boldsymbol{v}_s(\tau)\mathrm{d}\tau. \tag{12.7}$$

By inserting (12.6) into (12.7), the evolution of the ideal stator flux vector can be rewritten as

$$\boldsymbol{\psi}_s(t) = -\frac{m}{\omega_s}\frac{V_{\mathrm{dc}}}{2}\begin{bmatrix}\cos(\omega_s t)\\ \sin(\omega_s t)\end{bmatrix}. \tag{12.8}$$

It directly follows that the magnitude of the stator flux vector is

$$\Psi_s = \|\boldsymbol{\psi}_s\| = \frac{m}{\omega_s}\frac{V_{\mathrm{dc}}}{2}. \tag{12.9}$$

Recall that the influence of switching was neglected in the derivation of (12.9). Therefore, Ψ_s is a pure dc quantity without harmonic components.

We observe that, in order to maintain a desired stator flux magnitude for the machine in a variable speed drive, the modulation index should be adjusted proportionally to the stator frequency. The relationship (12.9) holds true both for SI quantities and in the per unit (pu) system.

12.2.3 *Relationship between Time and Angle*

Before proceeding, we relate the time t with the angle θ, which is used as an argument in the OPP. Assuming that the angle is given in radians and the time in seconds, the rule of proportions allows us to state

$$\frac{\theta}{2\pi} = \frac{t}{T_1}, \tag{12.10}$$

where $T_1 = 1/f_1$ is the period of the fundamental waveform. With the angular stator frequency ω_s being equal to $2\pi/T_1$, (12.10) can be rewritten in SI quantities as

$$\theta = \omega_s t. \tag{12.11}$$

[1] To simplify the notation, throughout this chapter we will drop the subindex $\alpha\beta$ from vectors in the stationary orthogonal coordinate system.

Recall that in Sect. 2.2.1 we had defined the stator frequency in the pu system as $\omega_s' = \omega_s/\omega_B$ and the time in pu as $t' = t\omega_B$, with ω_B denoting the angular base frequency. Therefore, (12.11) also holds true when adopting the pu system, assuming that the superscript $'$ is dropped as in Chap. 2.

12.2.4 Stator Flux Reference Trajectory

For the derivation of the reference trajectory of the stator flux vector, consider an electrical machine connected to the inverter terminals and neglect the machine's stator resistance. For a given OPP, the stator flux reference trajectory in stationary coordinates is obtained by integrating over time the switched voltage sequence of the OPP. For a three-level inverter with the switch positions $\mathcal{U} = \{-1, 0, 1\}$, the stator voltage in stationary coordinates is given by

$$\boldsymbol{v}_s^*(t) = \frac{V_{dc}}{2}\,\tilde{\boldsymbol{K}}\boldsymbol{u}_{abc}^*(\omega_s t) \tag{12.12}$$

according to (2.77), where we neglected the fluctuations of the neutral point potential and assumed the dc-link voltage to be at its nominal value. The argument of the three-phase switching waveform $\boldsymbol{u}_{abc}^*$ of the OPP is the angle θ. As per (12.11), the latter is equal to $\omega_s t$.

By inserting (12.12) into (12.7), the reference of the stator flux vector at time t can be stated as

$$\boldsymbol{\psi}_s^*(t) = \boldsymbol{\psi}_s^*(0) + \frac{V_{dc}}{2}\int_0^t \tilde{\boldsymbol{K}}\boldsymbol{u}_{abc}^*(\omega_s\tau)\mathrm{d}\tau. \tag{12.13}$$

Changing the integrand to $\vartheta = \omega_s\tau$, the reference can be stated, with the angle θ as an argument, as

$$\boldsymbol{\psi}_s^*\left(\frac{\theta}{\omega_s}\right) = \boldsymbol{\psi}_s^*(0) + \frac{V_{dc}}{2}\frac{1}{\omega_s}\int_0^{\frac{\theta}{\omega_s}} \tilde{\boldsymbol{K}}\boldsymbol{u}_{abc}^*(\vartheta)\mathrm{d}\vartheta. \tag{12.14}$$

With the help of (12.9), this statement can be further simplified to

$$\boldsymbol{\psi}_s^*(\theta) = \boldsymbol{\psi}_s^*(0) + \frac{\Psi_s^*}{m}\int_0^{\theta} \tilde{\boldsymbol{K}}\boldsymbol{u}_{abc}^*(\vartheta)\mathrm{d}\vartheta. \tag{12.15}$$

We have also redefined the stator flux reference as being a function of the angle θ instead of the time t. The benefit of the representation (12.15) is that the stator flux vector reference is independent of the drive parameters, most notably the dc-link voltage. It depends only on the desired flux magnitude and the modulation index.

Example 12.2 *Consider again the OPP with pulse number $d = 5$ and modulation index $m = 0.6$, as described in Example 12.1. By setting $\Psi_s^* = 1$, solving the integral (12.15), and choosing $\boldsymbol{\psi}_s^*(0)$ such that the trajectory is centered on the origin, the corresponding piecewise affine stator flux reference trajectory is obtained. The latter is shown in Fig. 12.3 over 90° in stationary coordinates. As the three-phase pulse pattern is piecewise constant, the stator flux reference trajectory is piecewise affine, that is, piecewise linear with offsets.*

Despite the average amplitude of the stator flux trajectory being 1, it is obvious from Fig. 12.3 that the instantaneous amplitude oscillates around 1, as shown in Fig. 12.4(a).

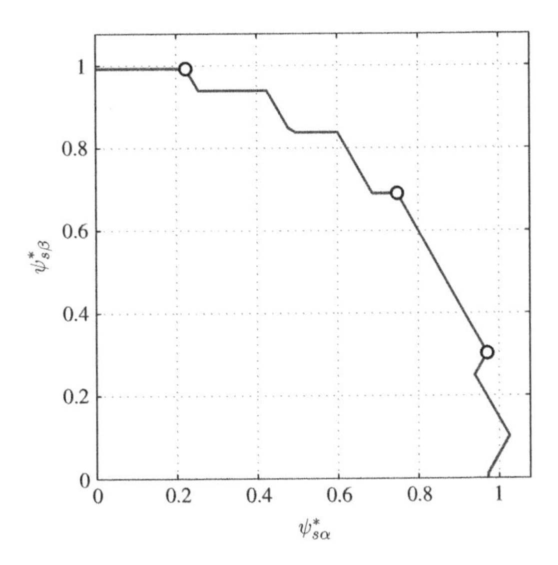

Figure 12.3 Stator flux reference trajectory in the $\alpha\beta$-plane for the OPP shown in Fig. 12.1

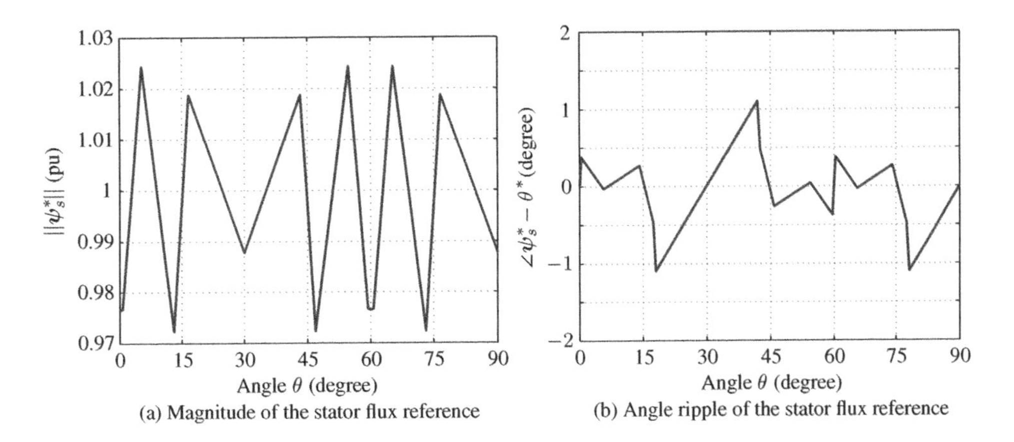

Figure 12.4 Magnitude and angle ripple of the stator flux reference for the OPP shown in Fig. 12.1

The instantaneous angle of the stator flux vector also oscillates around its nominal value. Figure 12.4(b) shows the angle ripple, that is, the difference between the angle of the stator flux vector and its argument θ. This ripple is the result of variations in the instantaneous angular speed of the stator flux vector, which necessarily arise when applying voltage vectors of different and discrete magnitudes. Zero vectors, for example, temporarily bring the stator flux vector to a halt. The zero vectors are indicated by small circles in Fig. 12.3.

To avoid confusion, we use $||\psi_s^*||$ to denote the instantaneous magnitude of the stator flux reference vector, which includes the ripple because of switching, while Ψ_s^* refers to the average (or dc) flux magnitude, which excludes the switching ripple. Similarly, for the angle of the stator flux reference, $\angle\psi_s^*$ denotes the instantaneous angle, while θ^* is the angle reference.

We also observe that the stator flux trajectory exhibits a high degree of symmetry. Specifically, it repeats itself every 60°. Within the 60° segments, the trajectory is symmetric at 30°, as can be seen in Fig. 12.3. The same observations apply to the ripples on the magnitude and angle of the stator flux vector. This ripple is optimal in the sense that it corresponds to the optimum ripple current and to minimum current distortions. The angular ripple leads to a torque ripple, while the amplitude ripple leads to a magnetization ripple.

12.2.5 Look-Up Table

OPPs can be conveniently stored in a look-up table. Because of the strong symmetry observed here, it suffices to store for each modulation index m and pulse number d the following information:

1. Primary switching angles α_i and switch positions u_i, with $i = 1, 2, \ldots, d$, of the single-phase pulse pattern over the first quarter of the fundamental period. Note that u_0 is always zero.
2. Corner points of the reference trajectory of the stator flux vector in stationary orthogonal coordinates with the stator flux amplitude 1 pu. It can be shown that $d + 1$ corner points over 30° suffice.

Based on this information, the three-phase OPP and the flux trajectory reference over the full fundamental period can easily be constructed.

12.3 Stator Flux Control

12.3.1 Control Objectives

The requirements for a control and modulation scheme for an MV drive are fourfold.

1. At steady-state operation, the overall current distortion in the stator windings should be minimized for a given switching frequency.
2. During transients and torque steps, fast dynamic control and a short torque response time should be achieved.
3. A simple implementation and a high degree of reliability should be ensured by avoiding the need for estimating the fundamental component of the stator flux or current in real time.
4. The controller should be insensitive to parameter variations and measurement noise.

12.3.2 Control Principle

By modifying the switching instants of the OPP, closed-loop control of an electrical machine based on OPPs can be achieved by controlling the stator flux vector along its reference trajectory. To illustrate this control principle, consider the stator flux in phase a. By inserting $v_{sa} = 0.5 v_{\text{dc}} u_a$ into the phase a component of (12.7), we obtain

$$\psi_{sa}(t) = \psi_{sa}(0) + \frac{v_{\text{dc}}}{2} \int_0^t u_a(\tau) \mathrm{d}\tau. \tag{12.16}$$

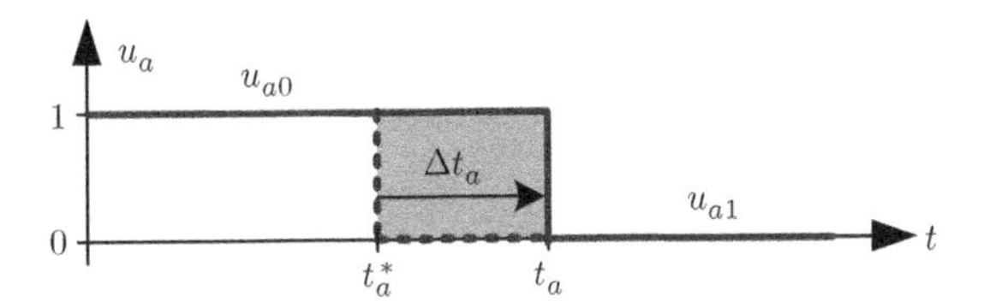

Figure 12.5 Delaying the negative switching transition $\Delta u_a = -1$ in phase a by Δt_a, with regard to the nominal switching time t_a^*, increases the stator flux component in this phase

Note that we use here the instantaneous dc-link voltage $v_{\rm dc}$ rather than the nominal one to account for dc-link voltage fluctuations.

Assume that one switching transition occurs in phase a in the time interval between zero and t. Let $\Delta u_a = u_{a1} - u_{a0}$ denote this switching transition, where Δu_a is a nonzero integer. The nominal switching time is t_a^*, while the actual or modified switching time is

$$t_a = t_a^* + \Delta t_a. \tag{12.17}$$

According to (12.16), the stator flux in phase a at time $t \geq t_a$ is equal to

$$\psi_{sa}(t) = \psi_{sa}(0) + \frac{v_{\rm dc}}{2}\left(\int_0^{t_a} u_{a0}{\rm d}\tau + \int_{t_a}^{t} u_{a1}{\rm d}\tau\right), \tag{12.18}$$

which can, by solving the integrals, be rewritten as

$$\psi_{sa}(t) = \psi_{sa}(0) + \frac{v_{\rm dc}}{2}((u_{a0} + \Delta u_a)t - \Delta u_a t_a^* - \Delta u_a \Delta t_a). \tag{12.19}$$

The stator flux at time t can be manipulated through the last term in (12.19) and the switching time modification Δt_a. We conclude that modifying the switching transition by Δt_a changes the phase a stator flux by

$$\Delta\psi_{sa}(\Delta t_a) = -\frac{v_{\rm dc}}{2}\Delta u_a \Delta t_a. \tag{12.20}$$

Example 12.3 *An example for this is shown in Fig. 12.5. Delaying the negative switching transition $\Delta u_a = -1$ by Δt_a increases the volt-seconds and thus the stator flux in this phase by $0.5 v_{\rm dc}\Delta t_a$. Advancing the switching event has the opposite effect, that is, it decreases the flux amplitude in the direction of phase a. The same holds true for phases b and c.*

12.3.3 Control Problem

The stator flux error is the difference between the reference of the stator flux vector and the actual stator flux of the machine. Even at steady-state operation, this flux error is generally nonzero because of non-idealities of the real-world drive system. These non-idealities include fluctuations in the dc-link voltage, the presence of the stator resistance, which was neglected in (12.15), and non-idealities of the power inverter, such as dead-time effects. For a summary of non-idealities in inverter drive systems, the reader is referred to [9].

During transient operation, the flux error accurately reflects the change in the operating point. A stepwise change in the torque setpoint, for example, results in a stepwise change in

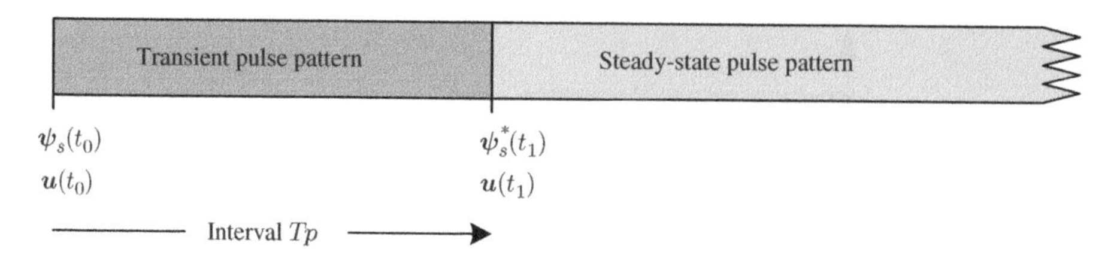

Figure 12.6 Boundary control problem formulated over the time interval T_p. The transient pulse pattern drives the stator flux vector from its current position $\psi_s(t_0)$ to the desired position on the reference trajectory $\psi_s^*(t_1)$ with the switch position $u(t_1)$

the flux error. By correcting this flux error, the torque is regulated to its new setpoint. The faster this correction is achieved, the shorter is the torque response time.

The stator flux control problem can be interpreted as a boundary control problem. For this, we introduce the three-phase switch position $u_{abc} = [u_a \; u_b \; u_c]^T$. To simplify the exposition, we will often drop the indices from u_{abc} and simply write u. The boundary control problem is illustrated in Fig. 12.6. Starting at time t_0 with the switch position $u(t_0)$ and the stator flux vector $\psi_s(t_0)$, a *transient* pulse pattern over the time interval T_p needs to be derived. This pulse pattern drives the stator flux vector to the terminal stator flux $\psi_s^*(t_1)$ and leads to the terminal switch position $u(t_1)$ at time $t_1 = t_0 + T_p$. In general, $\psi_s^*(t_1)$ is on the reference flux trajectory and $u(t_1)$ is the corresponding switch position. In this boundary control problem, $u(t_0)$ and $\psi_s(t_0)$ are the initial conditions while $u(t_1)$ and $\psi_s(t_1)$ are the terminal conditions.

12.3.4 Control Approach

The control problem can be formulated as a constrained optimal control problem with a receding horizon policy or, equivalently, as a model predictive control (MPC) problem. For a review of the concept of MPC, the reader is referred to Sect. 1.3 and the references therein.

The key idea of the proposed control approach is to associate the prediction horizon with the time interval $T_p = t_1 - t_0$, and to drive the stator flux vector over this horizon to its desired position, thus correcting the stator flux error. This is enforced by adding a terminal equality constraint on the state vector. From the end of the horizon onward, steady-state operation is assumed. In particular, the controller *assumes* that from t_1 onward, the nominal, that is, the *steady-state* pulse pattern, will be applied. Nevertheless, because of the receding horizon policy illustrated in Fig. 12.7, the steady-state OPP will *never* be applied. Instead, at every time step, the first part of the modified OPP, that is, the pulse pattern within the sampling interval $T_s < T_p$, will be applied to the drive system. More specifically, at time step k, the pulse pattern from kT_s to $(k+1)T_s$ is applied, and switching from $u_a = 0$ to 1 is performed in the example shown in Fig. 12.7(a). Similarly, at time step $k+1$, the pulse pattern from $(k+1)T_s$ to $(k+2)T_s$ is applied, which is constant at $u_a = 1$ in Fig. 12.7(b).

The stator flux error is small at steady-state operating conditions, typically amounting to 1% or 2% of the nominal flux magnitude. Only small corrections of the switching instants are thus required to remove the flux error. This allows us to use the steady-state OPP as a baseline pattern and to re-optimize around it locally to achieve closed-loop control. The resulting transient pulse pattern is not optimal in the strict sense, but its derivation is computationally much simpler than the computation of an entirely new transient pulse pattern.

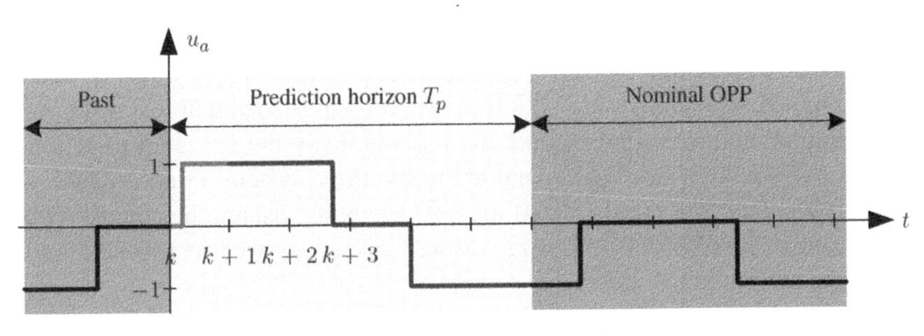

(a) Prediction horizon at time step k

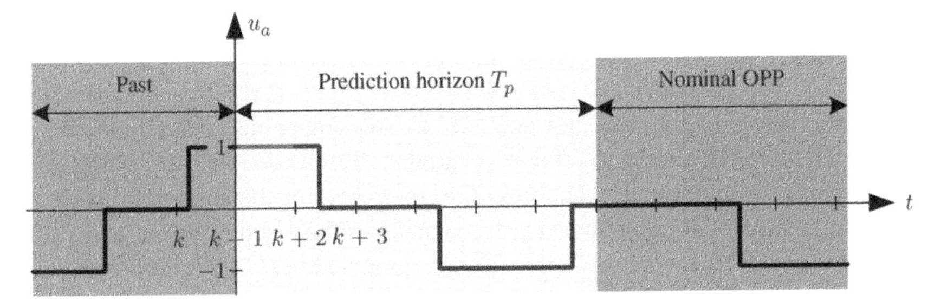

(b) Prediction horizon at time step $k + 1$

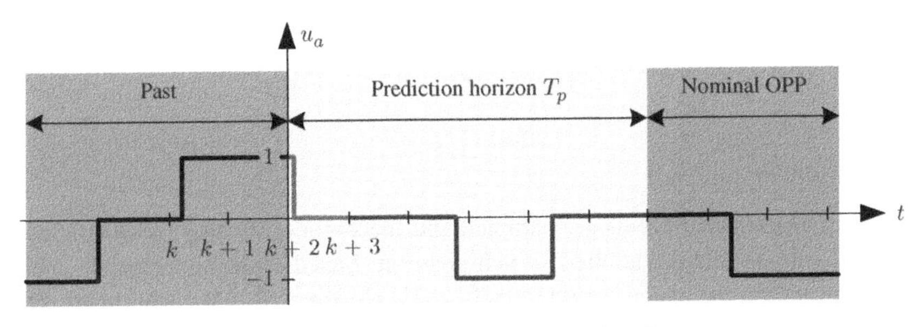

(c) Prediction horizon at time step $k + 2$

Figure 12.7 Illustration of the receding horizon policy for phase a. The modified pulse pattern is computed over the prediction horizon T_p. Out of this, only the first part of the pulse pattern over the sampling interval T_s is applied to the inverter

12.4 MP³C Algorithm

Following the introductory statements made in the previous section, the notion of stator flux trajectory tracking control is generalized in this section. A model predictive controller with the receding horizon policy is proposed, which we refer to as *model predictive pulse pattern control* (MP³C). This control scheme addresses in a unified manner the tasks of the inner control loop and modulator.

The internal model of this controller is based on two integrators of the form (12.16), one for each axis in the stationary orthogonal coordinate system. A prediction horizon of a finite length in time is used, and the switching instants of the pulse pattern are shifted such that the stator flux error is corrected within the horizon. From the end of the horizon onward, steady-state operation is assumed. The underlying optimization problem is solved in real time, yielding a sequence of optimal control actions within the horizon. Only the *first* control action of this sequence (the pulse pattern over T_s) is applied to the drive system, in accordance with the receding horizon policy (see Fig. 12.7). At the next sampling instant, the control sequence is recomputed over a shifted horizon, thus providing feedback as well as robustness to model inaccuracies and measurement noise.

To illustrate the MP³C concept, we will focus on an NPC inverter that produces at each phase the three voltage levels $\{-\frac{v_{dc}}{2}, 0, \frac{v_{dc}}{2}\}$. These voltages can be described by the integer variables $u_x \in \{-1, 0, 1\}$, with $x \in \{a, b, c\}$ denoting one of the three phases. The three-phase switch position has already been defined as $\boldsymbol{u} = \boldsymbol{u}_{abc} = [u_a\ u_b\ u_c]^T$.

The three-phase inverter is connected to an induction machine with the stator flux vector $\boldsymbol{\psi}_s = [\psi_{s\alpha}\ \psi_{s\beta}]^T$ and the rotor flux vector $\boldsymbol{\psi}_r = [\psi_{r\alpha}\ \psi_{r\beta}]^T$. Let $\angle\boldsymbol{\psi}$ denote the (instantaneous) angular position of a flux vector and $\|\boldsymbol{\psi}\|$ its (instantaneous) magnitude. We use the superscript * to denote the reference value of a variable. For the dc reference of the stator flux magnitude, we use the variable Ψ_s^*. The (average) reference for the stator flux angle is given by θ^*. The angular electrical stator and rotor frequencies of the machine are ω_s and ω_r, respectively. We use $t_0 = kT_s$ to denote the current time instant, where $k \in \mathbb{N}$ is the current time step and T_s is the sampling interval.

The proposed MP³C scheme is summarized by the block diagram in Fig. 12.8. The controller operates in the discrete-time domain and is activated at the equally spaced time instants kT_s. The control problem is formulated and solved in stationary orthogonal coordinates. The seven controller entities are described in the following.

12.4.1 Observer

In the first step, the stator currents are sampled and the stator voltage is reconstructed based on the dc-link voltage and the applied switch position. Based on these quantities, the stator and rotor flux vectors, $\boldsymbol{\psi}_s$ and $\boldsymbol{\psi}_r$, respectively, can be estimated in the stationary reference frame. According to (2.56), the torque estimate can be constructed based on these two flux vectors as

$$T_e = \frac{1}{\text{pf}} \frac{X_m}{D} \boldsymbol{\psi}_r \times \boldsymbol{\psi}_s. \tag{12.21}$$

Recall that X_m denotes the main (or magnetizing) reactance and D is the determinant as defined in (2.54). Preferably, a fast observer is used to avoid restricting the achievable bandwidth of MP³C.

Unlike in simulations, the implementation of a controller on hardware always entails a delay that is introduced by the time required to compute the control response. Specifically, one sampling interval typically elapses between the sampling of the currents and the availability of the new control output at the gate driver units of the inverter. To compensate for this delay, the estimated stator and rotor flux vectors can be rotated forward in time by $\omega_s T_s$. We set $\angle\boldsymbol{\psi}_s = \angle\boldsymbol{\psi}_s + \omega_s T_s$ for the stator flux and $\angle\boldsymbol{\psi}_r = \angle\boldsymbol{\psi}_r + \omega_s T_s$ for the rotor flux. Recall that during steady-state operation, both flux vectors rotate at the constant angular velocity ω_s.

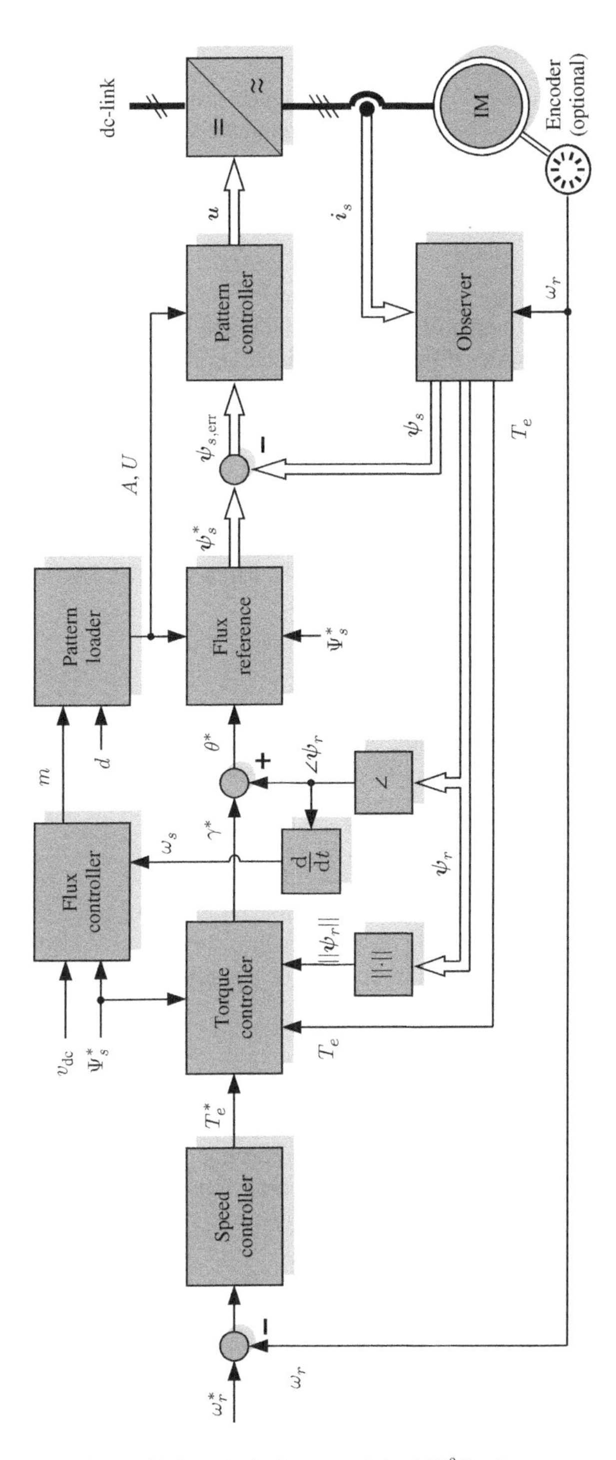

Figure 12.8 Block diagram of the MP³C scheme

12.4.2 Speed Controller

The speed controller regulates the (electrical) angular speed ω_r of the rotor along its reference ω_r^* by manipulating the setpoint of the electromagnetic torque. Recall the speed equation (2.51b) of the drive system. Using the number of pole pairs p, we replace the mechanical speed ω_m in (2.51b) by the electrical angular speed ω_r and obtain

$$\frac{M}{p}\frac{\mathrm{d}\omega_r}{\mathrm{d}t} = T_e - T_\ell. \tag{12.22}$$

M denotes the moment of inertia of the motor, shaft, and load, and T_ℓ is the load torque. As (12.22) constitutes a first-order differential equation, a proportional–integral (PI) controller usually suffices as speed controller. The latter is often augmented with an anti-windup scheme and a torque limiter. The speed is either measured by an encoder or estimated by the observer.

12.4.3 Torque Controller

The amplitudes of the flux vectors and their angular displacement γ determine the electromagnetic torque

$$T_e = \frac{1}{\mathrm{pf}}\frac{X_m}{D}\|\boldsymbol{\psi}_s\|\|\boldsymbol{\psi}_r\|\sin(\gamma) \tag{12.23}$$

that is produced at the rotor shaft of the machine by the interaction between the electromagnetic fields of the stator and the rotor windings through the air gap (see also (2.57)). Conversely, for a given torque reference T_e^*, the desired load angle between the stator and rotor flux vectors is calculated as

$$\gamma^* = \arcsin\left(\mathrm{pf}\frac{D}{X_m}\frac{T_e^*}{\Psi_s^*\|\boldsymbol{\psi}_r\|}\right). \tag{12.24}$$

When the machine is to be fully magnetized, the magnitude of the reference stator flux vector is effectively 1 pu, and the magnitude of the rotor flux vector is provided by the observer. In its simplest form, the torque controller consists of a feedforward term that maps the torque reference into a load angle reference.

12.4.4 Flux Controller

The stator flux magnitude is maintained at its reference Ψ_s^* by adjusting the amplitude of the fundamental voltage waveform applied to the machine. Rewriting (12.9), the modulation index is adjusted by the feedforward term

$$m = \frac{2}{v_{\mathrm{dc}}}\,\omega_s\Psi_s^* \tag{12.25}$$

in proportion to the inverse of the instantaneous dc-link voltage v_{dc} and the angular stator frequency ω_s. The instantaneous dc-link voltage is often low-pass-filtered to ensure that the modulation index is free of any significant dc-link voltage ripple. The feedforward term can be augmented by a conventional linear controller that regulates the error of the stator or rotor flux magnitude to zero by manipulating the modulation index.

12.4.5 Pulse Pattern Loader

The pattern loader provides the OPP required at the current operating point, which is determined by the modulation index m and the pulse number d. To determine the pulse number, recall that for a three-level inverter, for example, the device switching frequency is given by $f_{sw} = df_1$, where f_1 denotes the fundamental frequency. The switching frequency must not exceed its maximum, $f_{sw,\max}$. To ensure this, we select the pulse number d as the largest integer not greater than $f_{sw,\max}/f_1$, that is,

$$d = \text{floor}\left(\frac{f_{sw,\max}}{f_1}\right). \tag{12.26}$$

The pattern loader provides the primary switching angles A and the corresponding single-phase switching sequence U, which were defined in (12.2) and (12.3), respectively. Based on these two vectors, the three-phase OPP can easily be constructed. To reduce the number of mathematical operations that need to be performed in real time, the $\alpha\beta$-coordinates of the corner points of the stator flux reference trajectory can be read in from a look-up table instead of being calculated according to (12.15) (see also Sect. 12.2.5).

12.4.6 Flux Reference

In the next step, the stator flux reference vector $\boldsymbol{\psi}_s^*$ is computed. To achieve the desired torque, the (average) angular position of the stator flux reference must be equal to

$$\theta^* = \angle\boldsymbol{\psi}_r + \gamma^*. \tag{12.27}$$

To achieve the desired magnetization of the machine, the (average) magnitude of the stator flux reference vector should be equal to its reference Ψ_s^*. The stator flux reference $\boldsymbol{\psi}_s^*(\theta^*)$ corresponding to the angle θ^* and the magnitude Ψ_s^* can be derived by integrating up the OPP according to (12.15). The derivation of the reference flux vector is illustrated in the vector diagram in Fig. 12.9.

Alternatively, the stator flux reference can be computed from the corner points of the stator flux reference trajectory. Specifically, the primary switching angles are compared with θ^* to identify the two adjacent corner points of the stator flux reference. Between these two flux reference corner points, a linear interpolation is performed based on their corresponding primary switching angles and θ^*. This interpolation determines the flux reference vector $\boldsymbol{\psi}_s^*$. To account for flux magnitude references different from 1, the stator flux vectors of the corner points can be scaled with Ψ_s^*.

The resulting instantaneous stator flux reference $\boldsymbol{\psi}_s^*$ has, in general, a magnitude $\|\boldsymbol{\psi}_s^*\|$ and an angle $\angle\boldsymbol{\psi}_s^*$, which differ slightly from their desired (average) values Ψ_s^* and θ^*. These differences correspond to the optimal flux (and current) ripple of the OPP (see also the discussion at the end of Sect. 12.2.4).

12.4.7 Pulse Pattern Controller

The instantaneous stator flux error is defined as the difference between its reference and the estimated stator flux vector

$$\boldsymbol{\psi}_{s,\text{err}} = \boldsymbol{\psi}_s^* - \boldsymbol{\psi}_s. \tag{12.28}$$

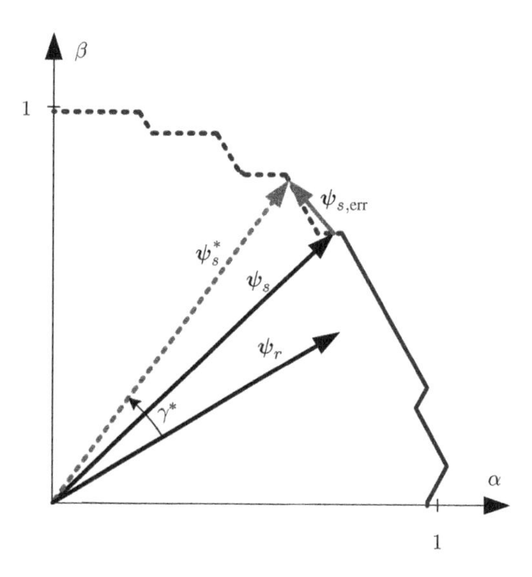

Figure 12.9 Rotor flux vector ψ_r, stator flux vector ψ_s, reference stator flux vector ψ_s^*, and stator flux error $\psi_{s,\mathrm{err}}$ in stationary coordinates

The MP³C control problem of removing the stator flux error within the prediction horizon by manipulating the switching instants of the OPP can be formulated as an optimization problem with a quadratic cost function and linear constraints, a so-called quadratic program (QP). The quadratic cost function

$$J(\Delta \boldsymbol{t}) = \|\boldsymbol{\psi}_{s,\mathrm{err}} - \Delta\boldsymbol{\psi}_s(\Delta \boldsymbol{t})\|_2^2 + \Delta \boldsymbol{t}^T \boldsymbol{Q} \Delta \boldsymbol{t} \tag{12.29}$$

penalizes two terms. The first term, the uncorrected flux error at the end of the prediction horizon, is the difference between the stator flux error $\boldsymbol{\psi}_{s,\mathrm{err}}$ in stationary $\alpha\beta$ coordinates at the current time step and the flux correction $\Delta\boldsymbol{\psi}_s(\Delta \boldsymbol{t})$ that will be achieved by the end of the prediction horizon. The flux correction term will be further examined in Sect. 12.5. The uncorrected flux error constitutes the controlled variable.

The second term penalizes the modifications $\Delta \boldsymbol{t}$ of the switching instants, using the diagonal penalty matrix $\boldsymbol{Q}$. This is the manipulated variable. Very small penalties are used in $\boldsymbol{Q}$. This ensures that the first term in the cost function is prioritized and that, as a result, the uncorrected flux error is close to zero.[2] The corrections of the switching instants are aggregated in the vector

$$\Delta \boldsymbol{t} = [\Delta t_{a1}\ \Delta t_{a2} \dots \Delta t_{an_a}\ \Delta t_{b1} \dots \Delta t_{bn_b}\ \Delta t_{c1} \dots \Delta t_{cn_c}]^T. \tag{12.30}$$

For phase a, for example, the correction of the ith transition time is given by

$$\Delta t_{ai} = t_{ai} - t_{ai}^*, \tag{12.31}$$

[2] In an alternative formulation, one could replace the first term in the cost function by the (terminal) equality constraint $\boldsymbol{\psi}_{s,\mathrm{err}} - \Delta\boldsymbol{\psi}_s(\Delta \boldsymbol{t}) = 0$. This, however, could lead to numerical difficulties, specifically to infeasibilities, in which no feasible solution exists for the optimization problem. It is thus preferable to relax the terminal equality constraint, by moving it to the cost function and by imposing a (comparably) large penalty on it.

where t_{ai} denotes the actual and t_{ai}^* the nominal switching instant of the ith switching transition Δu_{ai}. The latter is defined as

$$\Delta u_{ai} = u_a(t_{ai}^*) - u_a(t_{ai}^* - \mathrm{d}t) \tag{12.32}$$

with $\mathrm{d}t$ being an infinitesimally small time step. Moreover, in (12.30), n_a denotes the number of switching transitions in phase a that fall within the prediction horizon. The quantities for phases b and c are defined accordingly.

Several constraints need to be imposed on the switching instants. First, switching transitions may not be moved into the past. To avoid this, the constraint $kT_s \leq t_{a1}$ is imposed on the first switching instant in phase a. Second, the sequence of switching transitions must be maintained in each phase to avoid changes to the single-phase switching sequence and—in the worst case—the creation of switch positions that exceed the capability of the inverter. To this end, neighboring switching transitions in phase a are constrained by $t_{a1} \leq t_{a2} \leq \cdots \leq t_{an_a} \leq t_{a(n_a+1)}^*$, where $t_{a(n_a+1)}^*$ refers to the nominal switching time of the first switching transition beyond the horizon. Similar constraints are imposed on the switching instants in phases b and c.

The minimization of the cost function (12.29) subject to these constraints leads to the QP

$$\underset{\Delta t}{\text{minimize}} \quad J(\Delta t) \tag{12.33a}$$

$$\text{subject to} \quad kT_s \leq t_{a1} \leq t_{a2} \leq \cdots \leq t_{an_a} \leq t_{a(n_a+1)}^* \tag{12.33b}$$

$$kT_s \leq t_{b1} \leq t_{b2} \leq \cdots \leq t_{bn_b} \leq t_{b(n_b+1)}^* \tag{12.33c}$$

$$kT_s \leq t_{c1} \leq t_{c2} \leq \cdots \leq t_{cn_c} \leq t_{c(n_c+1)}^*. \tag{12.33d}$$

Note that the OPP uses the angle θ as an argument, while the MP^{3}C controller is formulated in the time domain. Assuming that the angular stator frequency remains constant within the prediction horizon, we can use (12.11) to translate switching angles into switching times.

Example 12.4 *Figure 12.10 provides an example to illustrate the constraints on the switching instants* (12.33b)–(12.33d). *The number of transitions that fall within the prediction horizon are $n_a = 2$, $n_b = 3$, and $n_c = 1$. The single switching transition in phase c within the prediction horizon, t_{c1}, is constrained by the current time instant kT_s and the nominal switching instant of the second transition in phase c, t_{c2}^*, which lies outside of the prediction horizon.*

The first switching transition in phase b, t_{b1}, is constrained to lie between kT_s and the (actual) switching instant of the second transition in phase b, t_{b2}. The latter is constrained by the first switching transition, t_{b1}, and the third one, t_{b3}, and so on. Note that the switching instants in a given phase can be modified independently from the transitions in the other phases.

The length of the prediction horizon T_p is a (time-invariant) design parameter. If, however, the horizon is overly short such that it does not comprise switching transitions in all three phases, it is temporarily increased until it includes switching transitions in all three phases. For an example, consider again Fig. 12.10. In the case where T_p is smaller than $t_{c1}^* - kT_s$, T_p is increased to this value. This adjustment is made to avoid numerical difficulties, as pointed out in the next section.

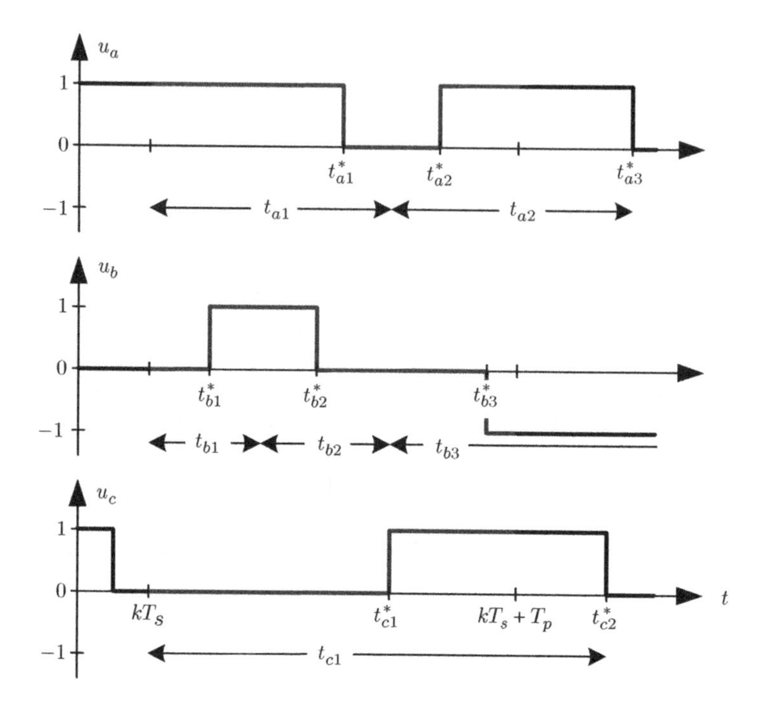

Figure 12.10 MP³C problem for a three-phase, three-level OPP. Six switching transitions fall within the horizon T_p, which is of a fixed length. The lower and upper bounds on the switching instants are depicted by arrows

In the last step, the actual switching instants within the sampling interval are identified. The switching sequence over the sampling interval, that is, the switching instants and their associated switch positions, is sent to the gate units of the semiconductor switches in the inverter.

In summary, by adopting the notion of stator flux trajectory tracking, an MPC scheme can be formulated that regulates the stator flux vector along its given reference trajectory by modifying the switching instants of the OPP within the prediction horizon. In doing so, the four control objectives stated in Sect. 12.3.1 can be met, as will be shown in the next chapter. Specifically, because the flux reference trajectory results from the integration of an offline-computed OPP voltage waveform that minimizes the current distortions, a nearly optimal ratio of harmonic current distortion per switching frequency is obtained during steady-state operation, provided that the flux reference trajectory is accurately tracked. Second, by directly manipulating the switching instants and inserting additional pulses if needed, which will be discussed in Sect. 12.6, a fast dynamic response during transients can be achieved. Third, thanks to the fact that the MPC formulation is based on the instantaneous flux including the ripple component, an estimator of the fundamental component is not required. This greatly simplifies the control scheme. Fourth, the use of a relatively long prediction horizon reduces the sensitivity of the controller to measurement and observer noise, by distributing the control effort over a long time interval. The influence of such noise on the closed-loop performance of MP³C will be investigated in Sect. 13.1.2.

12.5 Computational Variants of MP³C

The QP (12.33) formulated in the previous section constitutes the main computational stage of the MP³C algorithm. Before attempting to solve it, we reformulate it into a more convenient form.

Recall that by shifting the ith switching transition in phase a by the time Δt_{ai}, the volt-second correction

$$\Delta \psi_{sai}(\Delta t_{ai}) = -\frac{v_{\text{dc}}}{2} \Delta u_{ai} \Delta t_{ai} \tag{12.34}$$

is achieved, where we have generalized (12.20) to the ith switching transition. Assume that n_a switching transitions fall within the horizon in phase a. The total flux correction in phase a is obtained by summing up the individual volt-second modifications of the n_a switching transitions. Applying (12.34) n_a times leads to

$$\Delta \psi_{sa}(\Delta \boldsymbol{t}_a) = -\frac{v_{\text{dc}}}{2} \sum_{i=1}^{n_a} \Delta u_{ai} \Delta t_{ai} \tag{12.35}$$

with

$$\Delta \boldsymbol{t}_a = [\Delta t_{a1} \ \Delta t_{a2} \dots \Delta t_{an_a}]^T. \tag{12.36}$$

Similar expressions can be derived for phases b and c.

The stator flux correction in $\alpha\beta$ is obtained by aggregating the flux corrections in the phases a, b, and c and by using the Clarke transformation (2.13) to transform them to $\alpha\beta$.

$$\Delta \boldsymbol{\psi}_s(\Delta \boldsymbol{t}) = \tilde{\boldsymbol{K}} \begin{bmatrix} \Delta \psi_{sa}(\Delta \boldsymbol{t}_a) \\ \Delta \psi_{sb}(\Delta \boldsymbol{t}_b) \\ \Delta \psi_{sc}(\Delta \boldsymbol{t}_c) \end{bmatrix} = -\frac{v_{\text{dc}}}{2} \tilde{\boldsymbol{K}} \begin{bmatrix} \sum_i^{n_a} \Delta u_{ai} \Delta t_{ai} \\ \sum_i^{n_b} \Delta u_{bi} \Delta t_{bi} \\ \sum_i^{n_c} \Delta u_{ci} \Delta t_{ci} \end{bmatrix}. \tag{12.37}$$

Based on this, the QP can be rewritten in the standard form

$$\underset{\Delta \boldsymbol{t}}{\text{minimize}} \ \ \Delta \boldsymbol{t}^T \boldsymbol{H} \Delta \boldsymbol{t} + 2\boldsymbol{c}^T \Delta \boldsymbol{t} \tag{12.38a}$$

$$\text{subject to} \ \ \boldsymbol{G} \Delta \boldsymbol{t} \leq \boldsymbol{g}. \tag{12.38b}$$

The detailed derivation of the QP is provided in Appendix 12.A along with the matrices $\boldsymbol{H}$ and $\boldsymbol{G}$ and the vectors $\boldsymbol{c}$ and $\boldsymbol{g}$. Note that $\boldsymbol{H}$ is a function of the switching transitions and the penalty matrix $\boldsymbol{Q}$, $\boldsymbol{c}$ is a function of the switching transitions and the stator flux error $\boldsymbol{\psi}_{s,\text{err}}$, and $\boldsymbol{g}$ depends on the nominal switching instants. Mathematical programming in general and QPs in particular are reviewed in Sect. 3.8.

Now, we present two computational variations of MP³C, which differ in the penalty matrix $\boldsymbol{Q}$. These variations entail different approaches to solving the underlying QP.

12.5.1 MP³C based on Quadratic Program

The QP formulated in (12.33) and stated in vector representation in (12.38) can be greatly simplified when applying the same penalty to all switching transitions, that is, by setting

$$\boldsymbol{Q} = q\boldsymbol{I}. \tag{12.39}$$

The simplified QP can then be solved efficiently by adopting the so-called *active set method*. This is a standard approach to solve QPs of small to medium scale. The active set method is described in detail, for example, in [10, Sect. 16.4].

12.5.1.1 Unconstrained Solution

We start by computing the unconstrained solution, that is, we minimize (12.38a), while neglecting the timing constraints (12.38b). Assume for the time being that the step size of all switching transitions is ± 1, that is, $|\Delta u_{xi}| = 1$, with $x \in \{a, b, c\}$. In a given phase, each switching transition then provides the same volt-second correction provided that the modifications of the switching instants are the same. This fact can also be observed from (12.37). As all these modifications are penalized with the same weight q, it is clear that it is optimal in the unconstrained case to modify in each phase the switching instants by the same *absolute* value. Therefore, for phase a, for example, we have

$$|\Delta t_{a1}| = |\Delta t_{a2}| = \cdots = |\Delta t_{an_a}|. \tag{12.40}$$

Equivalently, the required per-phase volt-second correction is equally distributed among all switching transitions in a given phase. This implies that

$$\Delta \psi_{sa1} = \Delta \psi_{sa2} = \cdots = \Delta \psi_{san_a} \tag{12.41}$$

holds for phase a. Similar statements hold for phases b and c.

As a result, we can define *one* new variable that corresponds to the volt-second correction of *any* of the n_a switching transitions in phase a, that is,

$$\delta_a = \Delta \psi_{sai}, \quad i \in \{1, 2, \ldots, n_a\}. \tag{12.42}$$

With this, (12.35) can be simplified to

$$\Delta \psi_{sa}(\delta_a) = n_a \delta_a, \tag{12.43}$$

where we have also changed the argument of the volt-second correction. The same reasoning applies to phases b and c, with the variables δ_b and δ_c being defined accordingly.

The stator flux correction in $\alpha\beta$, (12.37), can then be rewritten as

$$\Delta \boldsymbol{\psi}_s(\boldsymbol{\delta}) = \tilde{\boldsymbol{K}} \begin{bmatrix} n_a \delta_a \\ n_b \delta_b \\ n_c \delta_c \end{bmatrix} = \tilde{\boldsymbol{K}}\ \boldsymbol{N}\ \boldsymbol{\delta}, \tag{12.44}$$

where we have aggregated the per-phase variables δ_a, δ_b, and δ_c to the three-phase vector $\boldsymbol{\delta} = [\delta_a\ \delta_b\ \delta_c]^T$. We have also introduced $\boldsymbol{N} = \text{diag}(n_a, n_b, n_c)$.

The second term in the cost function (12.29), which penalizes the modifications of the switching instants, can be rewritten with the help of (12.39) and (12.40) as

$$\Delta \boldsymbol{t}^T \boldsymbol{Q} \Delta \boldsymbol{t} = q(n_a(\Delta t_{a1})^2 + n_b(\Delta t_{b1})^2 + n_c(\Delta t_{c1})^2). \tag{12.45}$$

To relate the switching instant modifications with $\boldsymbol{\delta}$, we insert (12.34) into (12.42) and square it. As the step size of all switching transitions is 1, this leads to

$$\delta_a^2 = \left(\frac{v_{\text{dc}}}{2}\right)^2 (\Delta t_{ai})^2 \tag{12.46}$$

for phase a. Similar expressions can be obtained for phases b and c. With this, and by defining the scaled weight

$$q' = q\left(\frac{2}{v_{\text{dc}}}\right)^2, \tag{12.47}$$

(12.45) can be further simplified to

$$\Delta \boldsymbol{t}^T \boldsymbol{Q} \Delta \boldsymbol{t} = q'(n_a\delta_a^2 + n_b\delta_b^2 + n_c\delta_c^2) = q'\boldsymbol{\delta}^T \boldsymbol{N}\boldsymbol{\delta}. \tag{12.48}$$

The cost function (12.29) can then be restated as a function of $\boldsymbol{\delta}$ as

$$J(\boldsymbol{\delta}) = \|\boldsymbol{\psi}_{s,\text{err}} - \tilde{\boldsymbol{K}}\boldsymbol{N}\boldsymbol{\delta}\|_2^2 + q'\boldsymbol{\delta}^T \boldsymbol{N}\boldsymbol{\delta}, \tag{12.49}$$

where we have replaced the stator flux correction by (12.44) and the penalty on modifying the switching instants by (12.48).

As shown in Appendix 12.B, the unconstrained minimum of (12.49) is obtained for

$$\boldsymbol{\delta} = \boldsymbol{M}^{-1}\tilde{\boldsymbol{K}}^T \boldsymbol{\psi}_{s,\text{err}} \tag{12.50}$$

with

$$\boldsymbol{M} = \frac{2}{9}\begin{bmatrix} 2n_a + 4.5q' & -n_b & -n_c \\ -n_a & 2n_b + 4.5q' & -n_c \\ -n_a & -n_b & 2n_c + 4.5q' \end{bmatrix}. \tag{12.51}$$

The expression $\boldsymbol{M}^{-1}\tilde{\boldsymbol{K}}^T$ can be derived algebraically. In particular, the inverse of the matrix $\boldsymbol{M}$ does not need to be computed in real time.

12.5.1.2 Active Set Method

The active set method tailored to solving the QP involves a few iterations of the following three steps:

Step 1. The number of switching transitions that fall within the prediction horizon, n_a, n_b, and n_c, are determined for each phase. If required, the prediction horizon is extended until at least one switching transition per phase occurs.

Step 2. The timing constraints are neglected and the unconstrained volt-second correction $\boldsymbol{\delta}$ is computed according to (12.50). The volt-second correction is converted into unconstrained switching instants, taking the sign of the switching transition into account. For the ith transition in phase a, this implies $t_{ai} = t_{ai}^* + \Delta t_{ai}$ with

$$\Delta t_{ai} = -\frac{2}{v_{\text{dc}}}\frac{\delta_a}{\Delta u_{ai}}. \tag{12.52}$$

The latter expression is obtained from (12.34) and (12.42). The respective values for the unconstrained switching instants in phases b and c are obtained accordingly.

Step 3. The timing constraints (12.33b)–(12.33d) are imposed on the unconstrained switching instants. The switching instants that violate one or more of the constraints are the so-called *active* constraints, for which the following operations are performed:

1. The unconstrained switching instants are limited by imposing the constraints. This yields the final solution for these switching instants.
2. These switching instants and their associated switching transitions are removed from the optimization problem by reducing the relevant entries in $\boldsymbol{N}$ accordingly.
3. The flux correction that results from these modified switching instants is computed and the remaining (as-yet-uncorrected) flux error is updated accordingly.

Steps 2 and 3 are repeated until the solution remains unchanged. In general, two iterations suffice.

The active set method is computationally simple, and its complexity is independent of the number of switching transitions and thus of the length of the horizon. Specifically, the dimensions of the matrix $\boldsymbol{M}^{-1}\tilde{\boldsymbol{K}}^T$ is always 3×2 (see (12.B.7)). In the subsequent sections, we will refer to this MP^{3}C variation as *MP^{3}C based on QP*, or simply as *QP MP^{3}C*.

12.5.1.3 Optimality

At the beginning of this section, we had assumed that the step size of all switching transitions is ± 1, that is, $|\Delta u_{xi}| = 1$. Under this assumption, the active set method provides the optimal solution $\Delta \boldsymbol{t}$ to the QP (12.33) in the practical cases observed, even though a formal proof for this statement is not available. When the step size exceeds 1, a subtle difference in the solution arises, as shown in the following example.

Example 12.5 *Consider phase a, as illustrated in Fig. 12.11, with the second switching transition being $\Delta u_{a2} = -2$. The switching transitions in phase b and c along with their impact on the flux error and thus on phase a are ignored to simplify the exposition. In the original QP formulation* (12.33), *the modifications to the switching instants are penalized, regardless of the step size of the switching transition. As a result, as shown in Fig. 12.11(a), the switching instants are modified by the same absolute value provided that the timing constraints are inactive. In the figure, we have $\Delta t_{a1} = -\Delta t_{a2}$.*

When formulating and solving the active set method, however, the volt-second modifications are penalized. In any given phase, the switching transitions are modified by the same volt-second modification, as exemplified in Fig. 12.11(b). When the absolute value of the switching transition exceeds 1, the time modification is reduced according to (12.52). *In this example, we have $\Delta t_{a1} = -2\Delta t_{a2}$.*

This difference is subtle and arises only for step sizes exceeding ± 1. In effect, the active set method solves a different QP, in which the modifications to the switching instants are also weighted by the absolute value of their step size. Apart from this, the volt-second modification per phase is the same in this example for both approaches. When considering all three phases, however, additional small differences arise.

We conclude that the proposed active set method provides the optimal solution to the QP (12.33), provided that all switching transitions are limited to $\Delta u_{xi} = \pm 1$.

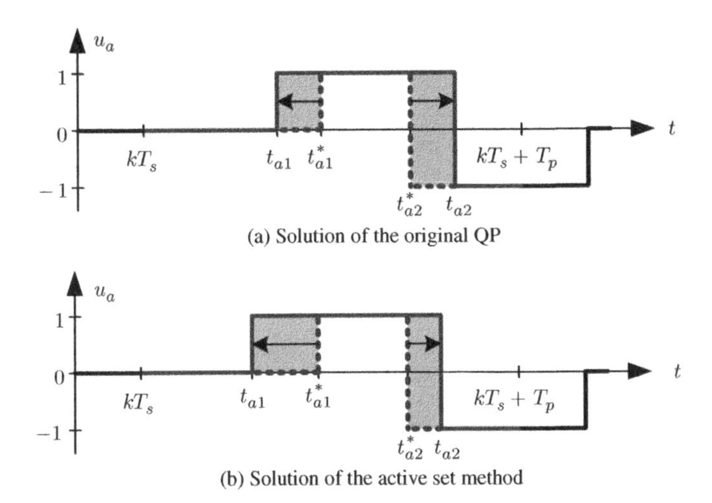

(a) Solution of the original QP

(b) Solution of the active set method

Figure 12.11 Comparison of the phase a switching instants that result from the original QP formulation and the active set method, when ignoring the phases b and c

12.5.2 MP³C based on Deadbeat Control

Another alternative is to set the penalty matrix $\boldsymbol{Q}$ in (12.29) to zero. As a result, the amount by which the switching instants are modified is not penalized. The unconstrained solution to (12.33) is then to fully remove the flux error. This is achieved by the flux correction $\Delta\boldsymbol{\psi}_s(\Delta\boldsymbol{t}) = \boldsymbol{\psi}_{s,\text{err}}$.

The horizon is kept as short as possible. Specifically, the horizon is redefined as the minimum time interval starting at the current time instant such that at least two phases exhibit switching transitions. This leads to a pulse pattern controller with DB characteristic and a time-varying prediction horizon. The control algorithm is computationally and conceptually simple, as summarized in the following.

Step 1. The two phases that have the next scheduled switching transitions are determined. We refer to those as the *active* phases. There are three pairs of active phases, namely ab, bc, or ac. All switching transitions within the horizon are determined. In Fig. 12.10, for example, phases a and b have the next switching transitions and are thus the active phases. Their nominal switching instants are t_{b1}^*, t_{b2}^*, and t_{a1}^*. The horizon thus spans the time interval from kT_s to t_{a1}^*.

Step 2. The required stator flux correction is translated from $\alpha\beta$ to abc by mapping the flux correction into the two active phases. The flux correction of the third phase is set to zero. In example shown in Fig. 12.5, with the active phases a and b, the mapping is given by

$$\Delta\boldsymbol{\psi}_{s,abc} = \tilde{\boldsymbol{K}}_{ab}^{-1}\Delta\boldsymbol{\psi}_s \tag{12.53}$$

with

$$\tilde{\boldsymbol{K}}_{ab}^{-1} = \frac{1}{2}\begin{bmatrix} 3 & \sqrt{3} \\ 0 & 2\sqrt{3} \\ 0 & 0 \end{bmatrix}. \tag{12.54}$$

The derivation of this matrix, along with the matrices $\tilde{\boldsymbol{K}}_{bc}^{-1}$ and $\tilde{\boldsymbol{K}}_{ac}^{-1}$ of the other two pairs of active phases, is provided in Appendix 12.C.

Step 3. The flux correction is scaled by the inverse of the instantaneous dc-link voltage to make it independent thereof. To this end, we introduce $\Delta\boldsymbol{\psi}'_{s,abc} = [\Delta\psi'_{sa}\ \Delta\psi'_{sb}\ \Delta\psi'_{sc}]^T$ and define

$$\Delta\boldsymbol{\psi}'_{s,abc} = \frac{2}{v_{\mathrm{dc}}}\Delta\boldsymbol{\psi}_{s,abc}. \tag{12.55}$$

For phase a, for example, this implies

$$\Delta\psi'_{sa} = -\sum_{i=1}^{n_a} \Delta u_{ai}\Delta t_{ai}, \tag{12.56}$$

which extends (12.35). Similar statements hold for phases b and c.

Step 4. The first active phase x, with $x \in \{a, b, c\}$, is identified. For its first switching transition, set $i = 1$. The DB controller aims at translating all of the required volt-second correction $\Delta\psi'_{sx}$ into a modification of the *first* switching instant in phase x. Specifically, for the ith switching transition in this phase with the nominal switching instant t^*_{xi} and the switching transition Δu_{xi}, the following operations are performed:

1. The desired modification $\Delta t_{xi} = -\Delta\psi'_{sx}/\Delta u_{xi}$ is computed.
2. The switching instant is modified to $t_{xi} = t^*_{xi} + \Delta t_{xi}$.
3. The switching instant t_{xi} is constrained by imposing the respective timing constraints on t_{xi}, according to (12.33b)–(12.33d).
4. The phase x component of the desired volt-second correction is updated by replacing $\Delta\psi'_{sx}$ with $\Delta\psi'_{sx} + \Delta u_{xi}(t_{xi} - t^*_{xi})$.

While the desired volt-second correction $\Delta\psi'_{sx}$ in this phase is nonzero, this procedure is repeated for the next switching transition and $i = i + 1$ is set.

Step 5. The second active phase is identified, and the procedure in Step 4 is repeated for this phase.

Note that $t_{xi} - t^*_{xi}$ equals the desired modification Δt_{xi} only when the associated constraints are not active. For the DB controller, no assumption has been made regarding the step size of the switching transitions. In particular, the DB version of MP³C is applicable to arbitrary (nonzero integer) switching transitions $\Delta u_{xi} \in \mathbb{Z}\setminus 0$.

As the DB controller aims to remove the stator flux error as quickly as possible, and because corrections in the switching instants are not penalized, the DB controller tends to be very fast and aggressive. There is, however, no guarantee that the flux error is fully removed within the horizon, because the horizon is as short as possible and the constraints on the switching instants have to be met.

12.6 Pulse Insertion

In an OPP, the switching transitions are not evenly distributed in time. Particularly at very low switching frequencies of a few hundred hertz, long time intervals might arise between two switching transitions. When a reference torque step is applied at the beginning of such an interval, a significant amount of time might elapse before the torque starts to change, resulting in a long initial time delay and often also in a prolonged settling time.

Once the controlled variable has started to change, the torque response might be sluggish and significantly slower than when using DB control such as DTC (see Section 3.6.3). The sluggish response is typically due to the absence of a suitable voltage vector that moves the controlled flux vector in the direction with the velocity that ensures the fastest possible compensation of the flux error. In order to ensure a very fast torque response during transients, at least one phase needs to be switched to the upper or lower dc-link rail. In a low-voltage ride-through setting, for example, this might imply reversing the voltage in at least one phase during the majority of the transient.

Directly related to the issue of sluggish torque responses is the risk of current excursions. Such excursions might occur when the switching transitions, which are to be shifted in time so as to remove the flux error, are spread over a long time interval. This increases the risk that the flux vector is not moved along the shortest path from its current position to its new and desired one. Instead, the flux vector might temporarily deviate from this path, exceeding its nominal magnitude. This is equivalent to a large current, which might result in an overcurrent trip.

This section proposes a solution to improve the performance of MP^{3}C during transients and (quasi) steady-state operation. When the flux error exceeds a given threshold, additional switching transitions are added to the OPP. As will be shown in Sect. 13.1.3, with the insertion of additional switching transitions, the merits of OPPs and DB control can be combined.

The concept of introducing additional switching transitions has been previously mentioned in the literature in the context of trajectory tracking control, albeit only very briefly in the form of short remarks in [2, 5], and [11]. This section formally introduces the notion of inserting switching transitions and generalizes this concept by performing closed-loop rather than (open-loop) feedforward transition insertion.

12.6.1 Definitions

As previously, we refer to $\Delta u_x(t) = u_x(t) - u_x(t - \mathrm{d}t)$ as a single-phase switching *transition* in phase x, with $x \in \{a, b, c\}$. Any nonzero integer step size is allowed. Three-phase switching transitions are defined accordingly as $\Delta \boldsymbol{u}(t) = \boldsymbol{u}(t) - \boldsymbol{u}(t - \mathrm{d}t)$. A *pulse* consists of two consecutive switching transitions in the same phase. The two switching transitions have opposite signs but do not necessarily have the same magnitude, as will be shown in the next section.

To illustrate the notion of pulse insertion, we will focus on a five-level inverter that produces at each phase the five voltage levels $\{0, \pm\frac{v_{\mathrm{dc}}}{4}, \pm\frac{v_{\mathrm{dc}}}{2}\}$. These voltages can be described by the integer variables $u_x \in \{0, \pm 1, \pm 2\}$, with $x \in \{a, b, c\}$ denoting one of the three phases.

12.6.2 Algorithm

The standard MP^{3}C algorithm proposed in Sect. 12.4 is augmented by an additional unit that inserts additional switching transitions when required. This unit is added as a preprocessing stage to the pulse pattern controller. It consists of four computational steps.

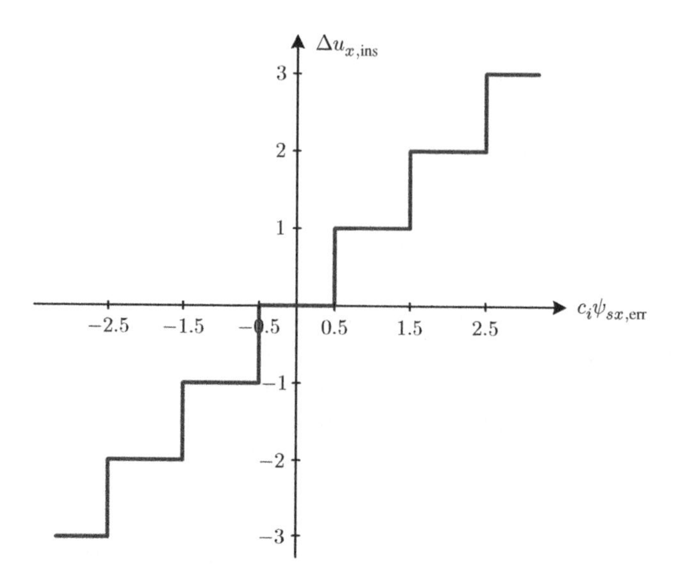

Figure 12.12 Definition of the per-phase error bands on the stator flux error in phase x, with $x \in \{a, b, c\}$. The switching transition to be inserted is denoted by $\Delta u_{x,\text{ins}}$

Step 1. After the computation of the stator flux error $\boldsymbol{\psi}_{s,\text{err}}$ in (12.28), the flux error is mapped from the orthogonal $\alpha\beta$ coordinate system into the three-phase abc system

$$\boldsymbol{\psi}_{s,abc,\text{err}} = \tilde{\boldsymbol{K}}^{-1}\,\boldsymbol{\psi}_{s,\text{err}}, \tag{12.57}$$

where $\tilde{\boldsymbol{K}}^{-1}$ denotes the 3×2 matrix of the reduced inverse Clarke transformation (2.13).

Step 2. In each phase, error bands on the stator flux error are introduced, as shown in Fig. 12.12. Based on these bands, it is determined whether an incremental switching vector, the three-phase switching transition $\Delta\boldsymbol{u}_{\text{ins}}$, is to be inserted. If this is the case, the magnitude and sign of the switching transition is determined for each phase. These two statements can be expressed in a compact way as

$$\Delta\boldsymbol{u}_{\text{ins}} = \text{round}(c_i\boldsymbol{\psi}_{s,abc,\text{err}}), \tag{12.58}$$

where the gain c_i is a user-defined scalar parameter. Note that the gain and rounding operation implicitly define the error bands. As the stator flux is the integral of the inverter switch positions weighted with half the dc-link voltage (see (12.13)) the term $0.5v_{\text{dc}}$ is implicitly included in the gain c_i.

As shown in Fig. 12.12, the magnitude and sign of the flux error in abc determine the magnitude and sign of the additional switching transition $\Delta\boldsymbol{u}_{\text{ins}}$. This is performed for each phase separately. If the switching transition is zero in all three phases, that is, when $|c_i\psi_{sx,\text{err}}| < 0.5$, with $x \in \{a, b, c\}$, then no additional switching transition is inserted.

If the flux error is positive, which is caused by too small a stator flux, additional volt-second needs to be added, which is equivalent to adding a positive switching transition and hence a positive pulse. Specifically, an additional switching transition of magnitude 1 is required in phase x, $\Delta u_{x,\text{ins}} = 1$, if $0.5 \leq c_i \psi_{sx,\text{err}} < 1.5$. Correspondingly, a transition of magnitude $\Delta u_{x,\text{ins}} = 2$ is added in the case where $1.5 \leq c_i \psi_{sx,\text{err}} < 2.5$, and so on. Negative switching transitions are added in the presence of negative flux errors.

Step 3. The repeated insertion of short pulses might produce a chattering phenomenon and unnecessarily increase the switching frequency. This issue can be avoided by imposing the restriction that, when switching transitions are inserted, the magnitude of the inserted transitions decreases in each phase while their sign is maintained. Specifically, for each phase, the required additional switching transition is modified when required, according to the following three rules:

1. If $\|\Delta \boldsymbol{u}_{\text{ins}}(k-1)\| > 0$ and $\Delta u_{x,\text{ins}}(k-1) = 0$, then $\Delta u_{x,\text{ins}}(k) = 0$.
2. If $\Delta u_{x,\text{ins}}(k-1) > 0$, then $\Delta u_{x,\text{ins}}(k) = \min(\max(\Delta u_{x,\text{ins}}(k), 0), \Delta u_{x,\text{ins}}(k-1))$.
3. If $\Delta u_{x,\text{ins}}(k-1) < 0$, then $\Delta u_{x,\text{ins}}(k) = \max(\min(\Delta u_{x,\text{ins}}(k), 0), \Delta u_{x,\text{ins}}(k-1))$.

The first rule ensures that, when a pulse insertion campaign has ended in phase x but is still going on in another phase, it is not to be restarted in phase x, before it has ended in all three phases. The second and third rules impose that the absolute values of the magnitudes of the inserted switching transitions decrease monotonically until they reach zero.

Step 4. The additional switching transition $\Delta \boldsymbol{u}_{\text{ins}}$ is added to the nominal pulse pattern (with the nominal switch positions and the nominal switching instants). This process is shown in Fig. 12.13(a) and entails the following three steps:

1. The nominal OPP is read out from the look-up table, and the nominal switching sequence starting at time instant t_0 is built sufficiently far into the future. In Fig. 12.13(a), the nominal switching sequence in phase x is shown as the dotted line.
2. The value of the switch position at time t_0 is determined, which is given by $\boldsymbol{u}(t_0) = \boldsymbol{u}(t_0 - \text{d}t) + \Delta \boldsymbol{u}_{\text{ins}}(t_0)$. Here, the switch position currently applied to the inverter is denoted by $\boldsymbol{u}(t_0 - \text{d}t)$. In the case where $\boldsymbol{u}(t_0)$ exceeds the set of available switch positions of the inverter, $\boldsymbol{u}(t_0)$ is saturated at the maximum or minimum attainable switch position.

   This implies that it might not be possible to implement the inserted switching transition to the full extent requested. As an example for this, assume that the currently applied switch position in phase x is $u_x(t_0 - \text{d}t) = 1$ and that the additional switching transition $\Delta u_x(t_0) = 3$ has been requested. For a five-level inverter, for example, it is only possible to implement the switch position $u_x(t_0) = 2$, which corresponds to an inserted transition of $\Delta u_x(t_0) = 1$.
3. A pulse of the infinitesimally small width $\text{d}t$ is inserted, by adding a switching transition at time t_0 from $\boldsymbol{u}(t_0 - \text{d}t)$ to $\boldsymbol{u}(t_0)$ and another switching transition with the opposite sign at time $t_0 + \text{d}t$ from $\boldsymbol{u}(t_0)$ to $\boldsymbol{u}(t_0 + \text{d}t)$.

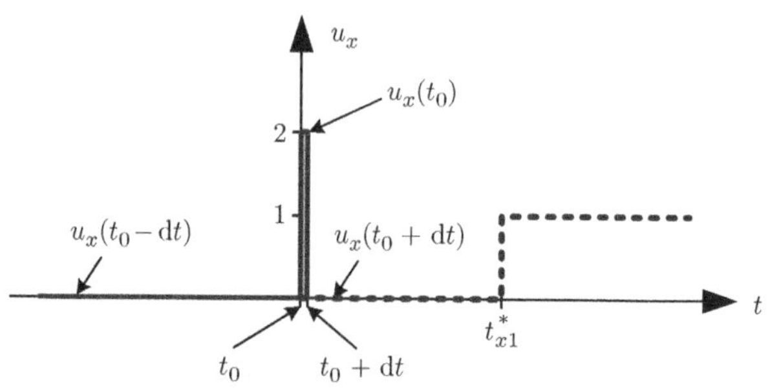

(a) Insertion of an additional pulse to the nominal OPP. The width of the pulse is zero

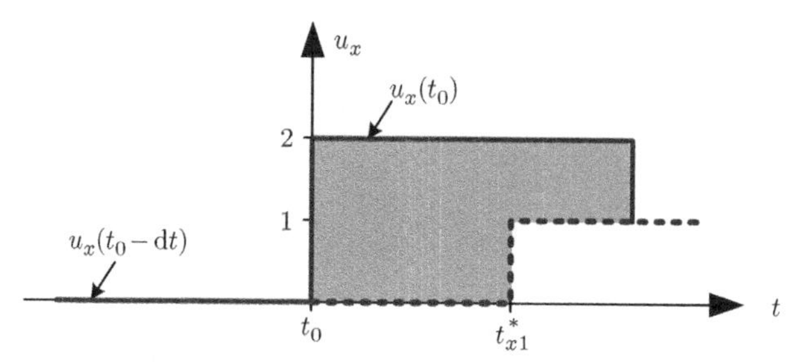

(b) Modification of the additional pulse by MP^{3}C to generate the required volt-seconds

Figure 12.13 Insertion of a pulse of amplitude 2 and width $\mathrm{d}t$ at the current time instant t_0 and modification by MP^{3}C to achieve fast closed-loop control

Special care must be taken to ensure that the magnitude of the second switching transitions (at time $t_0 + \mathrm{d}t$) is correct, because the first and second switching transitions do not necessarily sum up to zero. This case arises, for example, when a nominal switching transition is scheduled at t_0. The switch position $\boldsymbol{u}(t_0 + \mathrm{d}t)$ must match the nominal switch position at time $t_0 + \mathrm{d}t$.

The resulting switching sequence consists of the nominal switching transitions of the OPP and an additional pulse of width $\mathrm{d}t$ at time t_0. The volt-second of the inserted pulse is zero. The added pulse is shown as a solid line in Fig. 12.13(a).

Lastly, the pulse pattern controller is executed by formulating and solving the QP as shown in Sect. 12.4. Alternatively, the DB version of MP^{3}C can be used with pulse insertion. Starting at time t_0, the pattern controller modifies the switching instants of the three-phase switching sequence (including inserted pulses) such that the required volt-second correction is generated that removes the flux error. This process is illustrated in Fig. 12.13(b). By inserting pulses, the stator flux vector can be driven back to its reference trajectory as quickly as possible. This characteristic will be shown in Sect. 13.1.3 when evaluating and discussing the closed-loop performance of pulse insertion.

Appendix 12.A: Quadratic Program

In this appendix, the matrices $\boldsymbol{H}$ and $\boldsymbol{G}$ and the vectors $\boldsymbol{c}$ and $\boldsymbol{g}$ of the QP (12.38) are derived. The stator flux correction (12.37) in the orthogonal coordinate system can be simplified to

$$\Delta\boldsymbol{\psi}_s(\Delta\boldsymbol{t}) = -\boldsymbol{V}\ \Delta\boldsymbol{t} \tag{12.A.1}$$

with the voltage matrix

$$\boldsymbol{V} = \frac{v_{\mathrm{dc}}}{6}\begin{bmatrix} 2\Delta u_{a1} & 0 \\ \vdots & \vdots \\ 2\Delta u_{an_a} & 0 \\ -\Delta u_{b1} & \sqrt{3}\Delta u_{b1} \\ \vdots & \vdots \\ -\Delta u_{bn_b} & \sqrt{3}\Delta u_{bn_b} \\ -\Delta u_{c1} & -\sqrt{3}\Delta u_{c1} \\ \vdots & \vdots \\ -\Delta u_{cn_c} & -\sqrt{3}\Delta u_{cn_c} \end{bmatrix}^T . \tag{12.A.2}$$

Using (12.A.1), the cost function (12.29) can be rewritten as

$$J(\Delta\boldsymbol{t}) = (\boldsymbol{\psi}_{s,\mathrm{err}} + \boldsymbol{V}\Delta\boldsymbol{t})^T(\boldsymbol{\psi}_{s,\mathrm{err}} + \boldsymbol{V}\Delta\boldsymbol{t}) + \Delta\boldsymbol{t}^T\boldsymbol{Q}\Delta\boldsymbol{t}, \tag{12.A.3}$$

which can be further simplified to

$$J(\Delta\boldsymbol{t}) = \Delta\boldsymbol{t}^T(\boldsymbol{V}^T\boldsymbol{V} + \boldsymbol{Q})\Delta\boldsymbol{t} + 2\boldsymbol{\psi}_{s,\mathrm{err}}^T\boldsymbol{V}\Delta\boldsymbol{t} + \boldsymbol{\psi}_{s,\mathrm{err}}^T\boldsymbol{\psi}_{s,\mathrm{err}}. \tag{12.A.4}$$

Comparing (12.A.4) with (12.38a), the terms

$$\boldsymbol{H} = \boldsymbol{V}^T\boldsymbol{V} + \boldsymbol{Q} \text{ and } \boldsymbol{c} = \boldsymbol{V}^T\boldsymbol{\psi}_{s,\mathrm{err}} \tag{12.A.5}$$

directly follow. Note that the third term in (12.A.4) constitutes a constant offset and thus can be neglected in (12.38a).

With the definition (12.31), it is straightforward to rewrite the constraints (12.33b) on the switching instants for phase a in matrix form. This results in

$$\boldsymbol{G}_a\Delta\boldsymbol{t}_a \le \boldsymbol{g}_a \tag{12.A.6}$$

with

$$\boldsymbol{G}_a = \begin{bmatrix} -1 & 0 & \dots & & & & \\ 1 & -1 & 0 & \dots & & & \\ 0 & 1 & -1 & 0 & \dots & & \\ & \ddots & \ddots & \ddots & \ddots & & \\ & \dots & 0 & 1 & -1 & 0 & \\ & & & \dots & 0 & 1 & -1 \\ & & & & \dots & 0 & 1 \end{bmatrix}, \quad \boldsymbol{g}_a = \begin{bmatrix} t_{a1}^* - kT_s \\ t_{a2}^* - t_{a1}^* \\ t_{a3}^* - t_{a2}^* \\ \vdots \\ t_{a(n_a-1)}^* - t_{a(n_a-2)}^* \\ t_{an_a}^* - t_{a(n_a-1)}^* \\ t_{a(n_a+1)}^* - t_{an_a}^* \end{bmatrix}. \tag{12.A.7}$$

$\boldsymbol{G}_a$ is of dimension $(n_a + 1) \times n_a$, while $\boldsymbol{g}_a$ is a row vector of length n_a. The vector of switching time modifications in phase a, $\Delta \boldsymbol{t}_a$, was defined in (12.36).

Similarly, the constraints associated with phases b and c, (12.33c) and (12.33d), can be expressed by

$$\boldsymbol{G}_b \Delta \boldsymbol{t}_b \leq \boldsymbol{g}_b \tag{12.A.8a}$$

$$\boldsymbol{G}_c \Delta \boldsymbol{t}_c \leq \boldsymbol{g}_c. \tag{12.A.8b}$$

The matrices $\boldsymbol{G}_b$, $\boldsymbol{G}_c$ and vectors $\boldsymbol{g}_b$, $\boldsymbol{g}_c$ are defined according to (12.A.7).

The single-phase constraints (12.A.6) and (12.A.9) can be aggregated to (12.38b) with

$$\boldsymbol{G} = \begin{bmatrix} \boldsymbol{G}_a & \boldsymbol{0} & \boldsymbol{0} \\ \boldsymbol{0} & \boldsymbol{G}_b & \boldsymbol{0} \\ \boldsymbol{0} & \boldsymbol{0} & \boldsymbol{G}_c \end{bmatrix} \text{ and } \boldsymbol{g} = \begin{bmatrix} \boldsymbol{g}_a \\ \boldsymbol{g}_b \\ \boldsymbol{g}_c \end{bmatrix}, \tag{12.A.9}$$

where $\boldsymbol{0}$ denotes zero matrices of appropriate dimensions.

Appendix 12.B: Unconstrained Solution

The unconstrained solution to the minimization of the cost function (12.49) is derived in this appendix. The cost function can be rewritten as

$$J(\boldsymbol{\delta}) = (\boldsymbol{\psi}_{s,\text{err}} - \tilde{\boldsymbol{K}}\boldsymbol{N}\boldsymbol{\delta})^T(\boldsymbol{\psi}_{s,\text{err}} - \tilde{\boldsymbol{K}}\boldsymbol{N}\boldsymbol{\delta}) + q'\boldsymbol{\delta}^T\boldsymbol{N}\boldsymbol{\delta} \tag{12.B.1a}$$

$$= \boldsymbol{\delta}^T\boldsymbol{H}\boldsymbol{\delta} + 2\boldsymbol{c}^T\boldsymbol{\delta} + \boldsymbol{\psi}_{s,\text{err}}^T\boldsymbol{\psi}_{s,\text{err}}, \tag{12.B.1b}$$

where we have introduced

$$\boldsymbol{H} = \boldsymbol{N}^T\tilde{\boldsymbol{K}}^T\tilde{\boldsymbol{K}}\boldsymbol{N} + q'\boldsymbol{N} \tag{12.B.2a}$$

$$\boldsymbol{c} = -\boldsymbol{N}^T\tilde{\boldsymbol{K}}^T\boldsymbol{\psi}_{s,\text{err}}. \tag{12.B.2b}$$

As the prediction horizon is required to be long enough to cover switching transitions in all three phases, the diagonal entries of $\boldsymbol{N} = \text{diag}(n_a, n_b, n_c)$ are nonzero. This implies that $\boldsymbol{N}$ is invertible.

In Sect. 3.8, we had recalled the definition of positive definite matrices. $\boldsymbol{H}$ is positive definite if $\boldsymbol{\xi}^T\boldsymbol{H}\boldsymbol{\xi} > 0$ holds for all nonzero $\boldsymbol{\xi} \in \mathbb{R}^3$. We can rewrite this term as

$$\boldsymbol{\xi}^T\boldsymbol{H}\boldsymbol{\xi} = \|\tilde{\boldsymbol{K}}\boldsymbol{N}\boldsymbol{\xi}\|_2^2 + q'(n_a\xi_1^2 + n_b\xi_2^2 + n_c\xi_3^2). \tag{12.B.3}$$

The first term in (12.B.3) is positive definite, while the second term is positive semidefinite because $q' \geq 0$. We conclude that $\boldsymbol{H}$ is positive definite.

As shown in Sect. 3.8, the unconstrained minimum of (12.B.1) is obtained for

$$\boldsymbol{H}\boldsymbol{\delta} = -\boldsymbol{c}. \tag{12.B.4}$$

As $\boldsymbol{N}$ is symmetric and invertible, (12.B.4) is equivalent to

$$(\tilde{\boldsymbol{K}}^T\tilde{\boldsymbol{K}}\boldsymbol{N} + q'\boldsymbol{I}_3)\boldsymbol{\delta} = \tilde{\boldsymbol{K}}^T\boldsymbol{\psi}_{s,\text{err}}, \tag{12.B.5}$$

which leads to the unconstrained solution

$$\boldsymbol{\delta} = \boldsymbol{M}^{-1} \tilde{\boldsymbol{K}}^T \, \boldsymbol{\psi}_{s,\text{err}} \tag{12.B.6}$$

with

$$\boldsymbol{M} = \tilde{\boldsymbol{K}}^T \tilde{\boldsymbol{K}} \boldsymbol{N} + q' \boldsymbol{I}_3 = \frac{2}{9} \begin{bmatrix} 2 & -1 & -1 \\ -1 & 2 & -1 \\ -1 & -1 & 2 \end{bmatrix} \boldsymbol{N} + q' \boldsymbol{I}_3$$

$$= \frac{2}{9} \begin{bmatrix} 2n_a + 4.5q' & -n_b & -n_c \\ -n_a & 2n_b + 4.5q' & -n_c \\ -n_a & -n_b & 2n_c + 4.5q' \end{bmatrix}.$$

The expression $\boldsymbol{M}^{-1} \tilde{\boldsymbol{K}}^T$ can be computed algebraically as

$$\boldsymbol{M}^{-1} \tilde{\boldsymbol{K}}^T = \frac{2\sqrt{3}}{\det} \begin{bmatrix} \sqrt{3}(n_b + n_c + 3q') & n_b - n_c \\ -\sqrt{3}(n_c + 1.5q') & 2n_a + n_c + 4.5q' \\ -\sqrt{3}(n_b + 1.5q') & -2n_a - n_b - 4.5q' \end{bmatrix} \tag{12.B.7}$$

with the determinant

$$\det = 4n_a(n_b + n_c + 3q') + 4n_b(n_c + 3q') + 12n_c q' + 27(q')^2. \tag{12.B.8}$$

Therefore, to compute the unconstrained solution (12.B.6), the matrix inversion of $\boldsymbol{M}$ can be avoided. Only one division along with a few multiplications, additions, and shift operations needs to be performed in a real-time implementation.

Appendix 12.C: Transformations for Deadbeat MP^{3}C

In the DB MP^{3}C algorithm, the required flux correction is mapped from the orthogonal $\alpha\beta$ coordinate system to the two active phases. The three pairs of active phases are *ab*, *bc*, and *ac*.

Consider the first case, in which the quantity ξ_α, ξ_β is mapped into ξ_a and ξ_b, with $\xi_c = 0$. According to the reduced Clarke transformation (2.12), we can write

$$\begin{bmatrix} \xi_\alpha \\ \xi_\beta \end{bmatrix} = \frac{2}{3} \begin{bmatrix} 1 & -\frac{1}{2} & -\frac{1}{2} \\ 0 & \frac{\sqrt{3}}{2} & -\frac{\sqrt{3}}{2} \end{bmatrix} \begin{bmatrix} \xi_a \\ \xi_b \\ 0 \end{bmatrix} = \frac{1}{3} \begin{bmatrix} 2 & -1 \\ 0 & \sqrt{3} \end{bmatrix} \begin{bmatrix} \xi_a \\ \xi_b \end{bmatrix}. \tag{12.C.1}$$

Its inverse is

$$\begin{bmatrix} \xi_a \\ \xi_b \end{bmatrix} = \frac{1}{2} \begin{bmatrix} 3 & \sqrt{3} \\ 0 & 2\sqrt{3} \end{bmatrix} \begin{bmatrix} \xi_\alpha \\ \xi_\beta \end{bmatrix}, \tag{12.C.2}$$

which directly leads to

$$\tilde{\boldsymbol{K}}_{ab}^{-1} = \frac{1}{2} \begin{bmatrix} 3 & \sqrt{3} \\ 0 & 2\sqrt{3} \\ 0 & 0 \end{bmatrix}. \tag{12.C.3}$$

The transformations from $\alpha\beta$ to bc and ac can be derived in a similar way. They are given by

$$\tilde{\boldsymbol{K}}_{bc}^{-1} = \frac{1}{2}\begin{bmatrix} 0 & 0 \\ -3 & \sqrt{3} \\ -3 & -\sqrt{3} \end{bmatrix} \quad \text{and} \quad \tilde{\boldsymbol{K}}_{ac}^{-1} = \frac{1}{2}\begin{bmatrix} 3 & -\sqrt{3} \\ 0 & 0 \\ 0 & -2\sqrt{3} \end{bmatrix}. \tag{12.C.4}$$

References

[1] J. Holtz and B. Beyer, "Fast current trajectory tracking control based on synchronous optimal pulsewidth modulation," *IEEE Trans. Ind. Appl.*, vol. 31, pp. 1110–1120, Sep./Oct. 1995.

[2] J. Holtz and B. Beyer, "Off-line optimized synchronous pulsewidth modulation with on-line control during transients," *EPE Journal*, vol. 1, pp. 193–200, Dec. 1991.

[3] J. Holtz and B. Beyer, "The trajectory tracking approach—A new method for minimum distortion PWM in dynamic high-power drives," *IEEE Trans. Ind. Appl.*, vol. 30, pp. 1048–1057, Jul./Aug. 1994.

[4] B. Beyer, *Schnelle Stromregelung für Hochleistungsantriebe mit Vorgabe der Stromtrajektorie durch off-line optimierte Pulsmuster*. PhD thesis, Wuppertal University, 1998.

[5] J. Holtz and N. Oikonomou, "Synchronous optimal pulsewidth modulation and stator flux trajectory control for medium-voltage drives," *IEEE Trans. Ind. Appl.*, vol. 43, pp. 600–608, Mar./Apr. 2007.

[6] J. Holtz and N. Oikonomou, "Fast dynamic control of medium voltage drives operating at very low switching frequency—An overview," *IEEE Trans. Ind. Electron.*, vol. 55, pp. 1005–1013, Mar. 2008.

[7] N. Oikonomou and J. Holtz, "Stator flux trajectory tracking control for high-performance drives," in *Proceedings of IEEE Industry Applications Society Annual Meeting* (Tampa, FL, USA), pp. 1268–1275, Oct. 2006.

[8] N. Oikonomou and J. Holtz, "Estimation of the fundamental current in low-switching-frequency high dynamic medium-voltage drives," *IEEE Trans. Ind. Appl.*, vol. 44, pp. 1597–1605, Sep./Oct. 2008.

[9] J. Holtz and J. Quan, "Drift- and parameter-compensated flux estimator for persistent zero-stator-frequency operation of sensorless-controlled induction motors," *IEEE Trans. Ind. Appl.*, vol. 39, pp. 1052–1060, Jul./Aug. 2003.

[10] J. Nocedal and S. J. Wright, *Numerical optimization*. New York: Springer, 1999.

[11] N. Oikonomou, *Control of medium-voltage drives at very low switching frequency*. PhD thesis, University of Wuppertal, 2008.

13

Performance Evaluation of Model Predictive Pulse Pattern Control

The performance of the model predictive pulse pattern controller (MP^{3}C) is evaluated in this chapter through simulations and experiments on medium-voltage (MV) drives. Key performance criteria, such as the switching frequency and the current distortions, are compared with those of state-of-the-art control and modulation schemes, including carrier-based pulse width modulation (CB-PWM), space vector modulation (SVM), and direct torque control (DTC).

More specifically, simulation results are provided for a neutral-point-clamped (NPC) inverter drive system during steady-state operation and transients. When compared to SVM operating at the same switching frequency, MP^{3}C reduces the current distortions by up to 50%. The benefit of inserting pulses during transients is illustrated. Experimental results for a five-level active NPC inverter drive system are shown in the second part of the chapter, with the MV induction machine operating at up to 1 MVA. A summary and discussion of the main benefits and characteristics of MP^{3}C is provided at the end of this chapter.

13.1 Performance Evaluation for the NPC Inverter Drive System

The steady-state performance of MP^{3}C is evaluated through simulations of an NPC inverter drive system. The impact of flux observer noise and machine parameter variations on the drive's performance is also investigated. The performance of deadbeat (DB) and quadratic programming (QP) MP^{3}C are compared with each other. The closed-loop response of MP^{3}C during transients is evaluated, and the benefit of inserting pulses is shown.

13.1.1 Simulation Setup

As a case study, consider a three-level NPC voltage source inverter driving an induction machine with a constant mechanical load, as shown in Fig. 13.1. A 3.3 kV, 50 Hz squirrel-cage induction machine rated at 2 MVA with a total leakage reactance of 0.25 per unit (pu) is used

Model Predictive Control of High Power Converters and Industrial Drives, First Edition. Tobias Geyer.

Companion Website: www.wiley.com/go/geyermodelpredictivecontrol

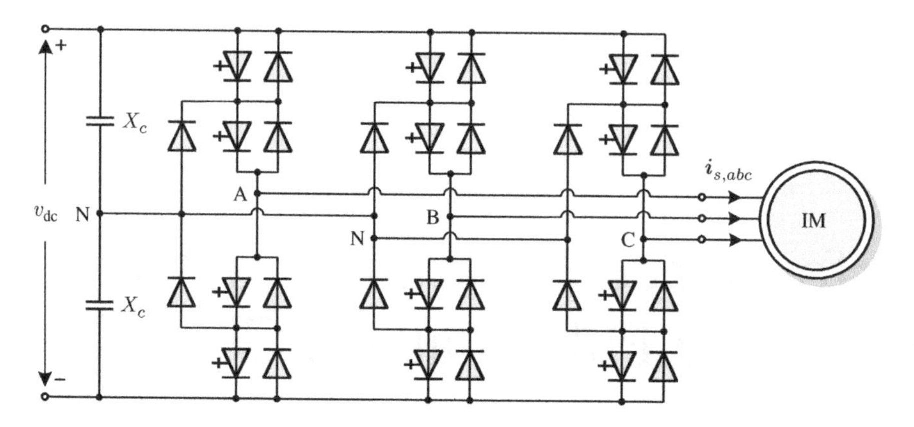

Figure 13.1 NPC inverter driving an induction machine (IM)

as an example of a typical MV induction machine. The nominal dc-link voltage is $V_{dc} = 5.2$ kV. The drive system case study is summarized and the pu system is defined in Sect. 2.5.1. The detailed parameters of the machine and the inverter are provided in Table 2.10. The floating neutral point potential is not actively controlled. The sampling interval $T_s = 25$ μs is used.

13.1.2 Steady-State Operation

13.1.2.1 Operation under Nominal Conditions

At nominal speed and rated torque, closed-loop simulations were run to evaluate the performance of MP^{3}C under steady-state operating conditions. The key performance criteria are the harmonic distortions of the current and torque for a given switching frequency. Nominal conditions are assumed, that is, the stator flux observations are free of noise and the machine parameters are precisely known. DB MP^{3}C is considered, which yields closed-loop results at steady-state operation and nominal conditions, which are effectively identical to that of the QP version of MP^{3}C. Optimized pulse patterns (OPPs) were calculated offline, following the procedure described in Sect. 3.4. The modulation index is equal to $m = 1.04$.

The performance of MP^{3}C is compared with that of CB-PWM and SVM. Specifically, a three-level, asymmetric, regularly sampled CB-PWM was implemented with two triangular carriers that are in phase, the so-called *phase disposition*. It is generally accepted that for multilevel inverters CB-PWM with phase disposition results in the lowest harmonic distortion. According to (3.14), a third harmonic component is added to the modulating reference signals to boost the differential-mode voltage. The SVM is obtained by adopting the approach proposed in [1]: A common-mode voltage, which is of the min/max type plus a modulus operation, is added to the reference voltage (see (3.16)). For a review of CB-PWM, third harmonic injection, and the equivalence between CB-PWM and SVM, the reader is referred to Sect. 3.3.

We start by comparing the steady-state performance of MP^{3}C with that of SVM. Operation is at a device switching frequency of 250 Hz, which is typically used in MV drives. This

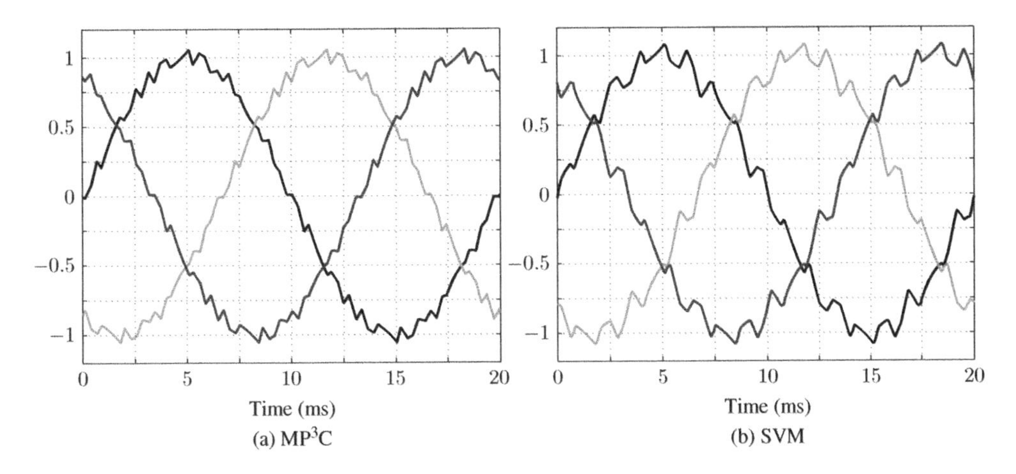

Figure 13.2 Stator currents for MP^{3}C and SVM with the switching frequency 250 Hz

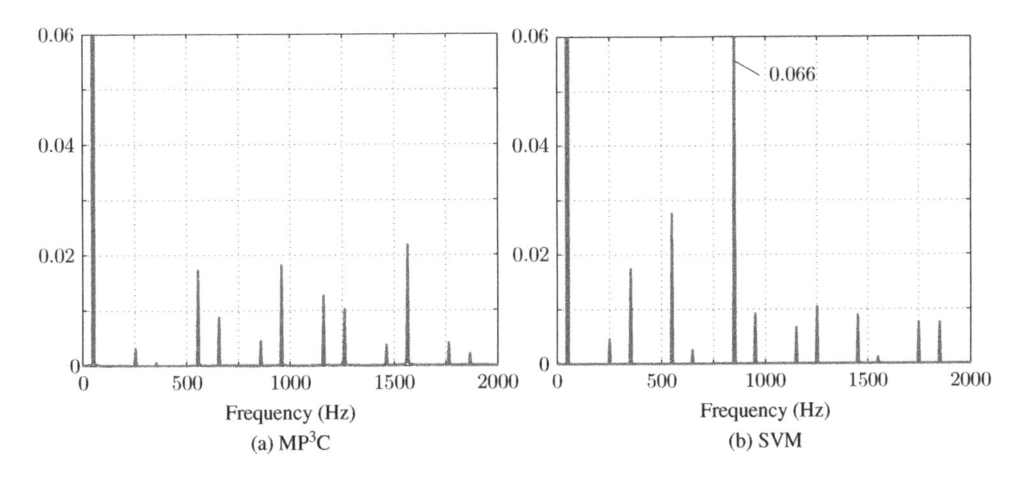

Figure 13.3 Stator current spectrum for MP^{3}C and SVM with the switching frequency 250 Hz

switching frequency is achieved by choosing the pulse number $d = 5$ for MP^{3}C and the carrier frequency $f_c = 450$ Hz for SVM. The resulting stator current waveforms and spectra of MP^{3}C and SVM along with the phase leg switch positions are shown in Figs. 13.2–13.4.

It is apparent from the current waveforms that MP^{3}C produces a much lower current ripple. Correspondingly, the harmonic components of the MP^{3}C current spectrum are much reduced, particularly the harmonics around f_c and the 17th harmonic. This improvement is also reflected in the total demand distortion (TDD) of the current, which is reduced from 7.71% for SVM to 4.13% for MP^{3}C—a reduction of 46%.

Table 13.1 compares the current distortions that result from CB-PWM, SVM, and MP^{3}C for three different switching frequencies. The data shows that for low switching frequencies of a few hundred hertz, MP^{3}C effectively halves the current distortions for the same switching

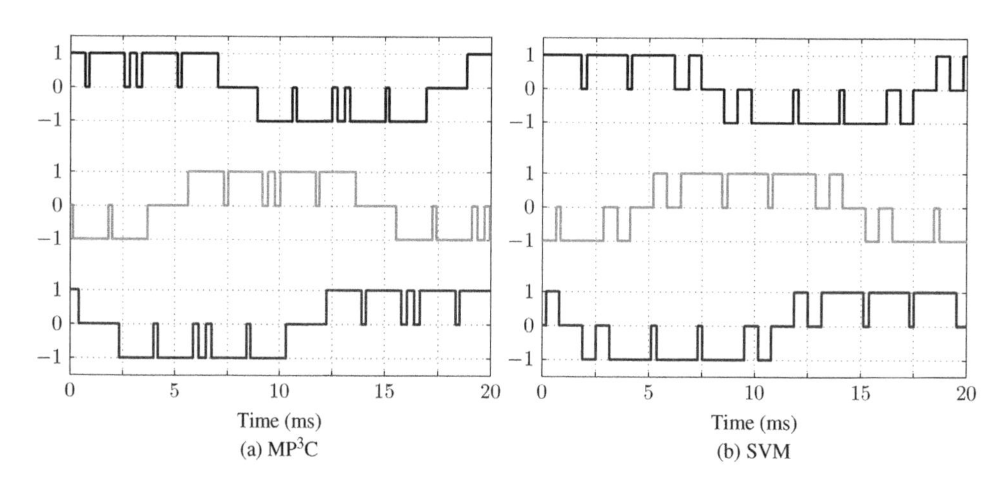

Figure 13.4 Switch positions for MP³C and SVM with the switching frequency 250 Hz

Table 13.1 Comparison of DB MP³C with CB-PWM and SVM in terms of the switching frequency f_{sw}, the stator current TDD I_{TDD}, and the torque TDD T_{TDD}

Scheme	Setting	f_{sw} (Hz)	I_{TDD} (%)	T_{TDD} (%)	I_{TDD} (%)	T_{TDD} (%)
CB-PWM	$f_c = 250$ Hz	150	16.1	11.0	100	100
SVM	$f_c = 250$ Hz	150	15.5	9.83	96.8	89.6
MP³C	$d = 3$	150	7.29	6.54	45.4	59.6
CB-PWM	$f_c = 450$ Hz	250	7.94	5.79	100	100
SVM	$f_c = 450$ Hz	250	7.71	5.35	97.1	92.4
MP³C	$d = 5$	250	4.13	3.41	52.0	58.9
CB-PWM	$f_c = 750$ Hz	400	4.68	3.41	100	100
SVM	$f_c = 750$ Hz	400	4.52	3.06	96.6	89.7
MP³C	$d = 8$	400	2.94	2.75	62.8	80.9

The center section shows absolute values, while the values in the right section are relative to the CB-PWM baseline. The pulse number is given by d and the carrier frequency by f_c. The operating point is at nominal speed and rated torque. Nominal conditions (noise-free flux estimates and accurate machine parameters) are assumed.

frequency, when compared to CB-PWM or SVM. It can also be seen that the harmonic performance of CB-PWM is similar to that of SVM. When increasing the switching frequency, the performance benefit of MP³C abates. At a switching frequency of 400 Hz, for example, the reduction of the current TDD that MP³C achieves with respect to SVM amounts to 35%.

To further investigate the switching-frequency-dependent performance improvement of MP³C, simulations were run for DB MP³C during steady-state operation at nominal speed and rated torque for different OPPs with pulse numbers ranging from $d = 2$ to $d = 10$. The individual simulations are indicated by diamonds in Fig. 13.5. Similarly, several simulations were run for SVM with the carrier frequencies $f_c = 250, 300, \ldots, 950$ Hz, which are integer

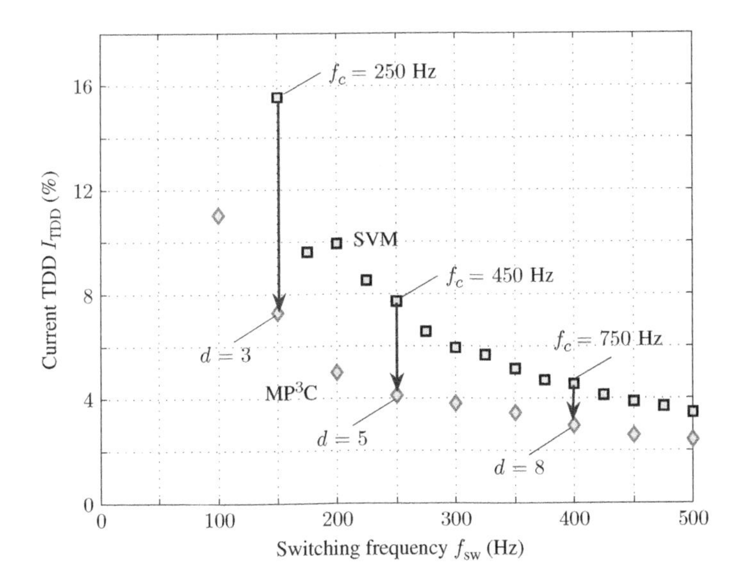

Figure 13.5 Stator current TDD for SVM and MP^{3}C as a function of the switching frequency, when operating the NPC inverter drive system at nominal speed and rated torque

multiples of the fundamental frequency. These simulation results are shown by squares. The three sets of comparisons made in Table 13.1 at $d = 3$, 5, and 8 are indicated in the figure by vertical arrows.

The performance benefit of MP^{3}C is at its maximum when the device switching frequency compared to the fundamental frequency is low, that is, for low pulse numbers. As the pulse number and thus the switching frequency increases, the performance benefit is reduced. Nevertheless, even at the (hypothetical) switching frequency of 1 kHz, MP^{3}C reduces the current distortions by 26% when compared to SVM. More specifically, SVM with the carrier frequency $f_c = 1950$ Hz and MP^{3}C with an OPP of pulse number $d = 20$ yield current TDDs of 1.69% and 1.25%, respectively. Yet, computing OPPs for such high pulse numbers is numerically challenging. This is reflected in the fact that earlier results shown in [2], which were based on a preliminary set of OPPs, indicated that OPPs and SVM have effectively the same harmonic performance for pulse numbers exceeding 15.

13.1.2.2 Operation with Flux Observer Noise

MP^{3}C requires an accurate estimate of the stator flux vector, which is provided by a flux observer, as shown in Fig. 12.8. The flux estimate is typically affected by noise, which is the difference between the actual flux and its estimate. This section investigates the impact observer noise has on the closed-loop performance of MP^{3}C, particularly with regard to the current TDD.

To this end, DB MP^{3}C was run at nominal speed and torque under steady-state operating conditions on a 1 MVA MV drive in the laboratory. The evolution of the stator flux vector was measured along with that of the stator flux reference vector. The difference between the two

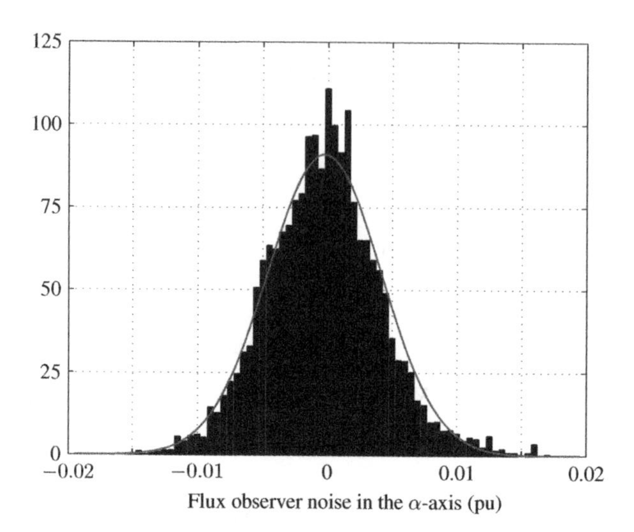

Figure 13.6 Probability density function of the flux observer noise as measured in the MV laboratory

stator flux vectors was defined in (12.28) as the stator flux error $\psi_{s,\text{err}}$. At steady-state operation, MP^{3}C removes the flux error almost completely—the residual error typically amounts to less than 1% of the nominal flux magnitude. This residual error is dominated by noise from the flux observer. In the following analysis, we therefore refer to $\psi_{s,\text{err}}$ as the flux observer noise, which includes various noise sources in the path of the stator flux observer. These include the noise in the current and voltage measurements, the discretization noise introduced by the analog-to-digital conversion, drifts in the current measurement probes, and a ripple in the angular velocity signal. Uncompensated for non-idealities of the power inverter also contribute to the residual noise.

Figure 13.6 shows the probability density function of the flux observer noise in the α-axis, with the noise in the β-axis being very similar. Note that the integral of the probability density function is 1. The noise can be well approximated as Gaussian noise with a zero mean value and a standard deviation of $\sigma = 0.0044$ pu, as shown by the solid line in Fig. 13.6. However, the noise exhibits a certain degree of auto-correlation, implying that the noise amplitude at time step k somewhat depends on the noise amplitude at the previous time step $k-1$. This auto-correlation is not described by the Gaussian noise. Therefore, we distinguish between *Gaussian* noise and *measured* noise. The Gaussian noise is characterized by a given standard deviation and exhibits zero auto-correlation. The measured noise is the measured flux error $\psi_{s,\text{err}}$, whose probability density function is effectively Gaussian, but features a nonzero auto-correlation.

Before proceeding, we define the (angular) prediction horizon

$$\theta_p = \frac{180}{\pi}\omega_s T_p, \tag{13.1}$$

which is given in degree and refers to the stator flux angle over which MP^{3}C looks into the future. In contrast, the (time) prediction horizon T_p is given in the pu system. Equation (13.1) follows directly from (12.11).

Using the same setting as in the previous section (nominal speed, rated torque, pulse number $d = 5$), the impact of Gaussian observer noise on the current TDD was evaluated through simulations. The result is shown in Fig. 13.7(a). Without noise, the current TDD is 4.13%, which is also stated in the sixth row in Table 13.1. When increasing the standard deviation of the observer noise, the harmonic performance quickly deteriorates for DB MP^{3}C. For $\sigma = 0.0044$ pu, for example, the current TDD increases by 10% to 4.57%. When using QP MP^{3}C, however, the deterioration because of flux observer noise can be significantly reduced, particularly when adopting long prediction horizons. For the horizon of $\theta_p = 60°$ and $\sigma = 0.0044$ pu, for example, the noise deterioration can effectively be avoided altogether—the current TDD is 4.19%, which is equivalent to a deterioration of only 1.5% with respect to the noise-free baseline of 4.13%. Note that at nominal speed, $\theta_p = 60°$ is equivalent to $T_p = 1.047$ pu or 3.33 ms.

The impact of the measured noise on the current TDD is shown in Fig. 13.7(b). The noise measured in the MV laboratory is scaled with the so-called *noise scaling factor*. This allows us to study the effect of different noise intensities. A scaling factor of 1 implies that the original MV laboratory noise is applied to the flux observer. In this case, DB MP^{3}C results in a current TDD of 4.65%, which is 13% worse than the nominal case. For QP MP^{3}C with $\theta_p = 60°$ and the same noise scaling factor, the current TDD can be brought down to 4.24%, which corresponds to a mere 2.7% deterioration.

It can be seen in Fig. 13.7 that the resilience to flux observer noise changes significantly for QP MP^{3}C when the (angular) prediction horizon is increased from $\theta_p = 20°$ to 30°. The reason for this is that for $\theta_p = 20°$, in 14% of the cases the prediction horizon needs to be extended to capture switching transitions in all three phases. This implies that in several cases only one switching transition per phase is available. As a result, the switching instants need to be modified significantly to achieve the required flux correction.

However, when multiple switching transitions per phase are considered, the flux error compensation mechanism is less vulnerable to noise, because the required flux correction is achieved by manipulating many switching transitions by small amounts. The intuitive hypothesis that a longer prediction horizon makes the control scheme more robust to noise is thus confirmed.

Summing up, on one hand, the DB version is affected by flux observer noise, which is a common characteristic of aggressive control schemes. The QP approach, on the other hand, is less susceptible to noise, particularly for long horizons, because the controller carefully weighs in the objective of removing the flux error within the horizon versus the penalty on modifying the switching transitions. This is a fundamental characteristic of the so-called *optimal control* schemes, such as QP MP^{3}C, which are based on the trade-off between good tracking performance and low control effort. In this case, this trade-off is determined by the length of the horizon. Note that the penalty on modifying the switching instants implicitly determines the penalty on the terminal soft constraint, while the penalty has only a mild impact on this trade-off.

In the investigations done previously, we assumed $\sigma = 0.0044$ pu and the noise scaling factor of 1 to be representative for flux observer noise in a real-world MV drive setting. This assumption might be pessimistic, because the recorded noise also includes uncompensated stator flux errors. The real observer noise is thus probably smaller. When assuming Gaussian noise with $\sigma = 0.003$ pu, the corresponding deterioration of the current TDD is reduced to 5%

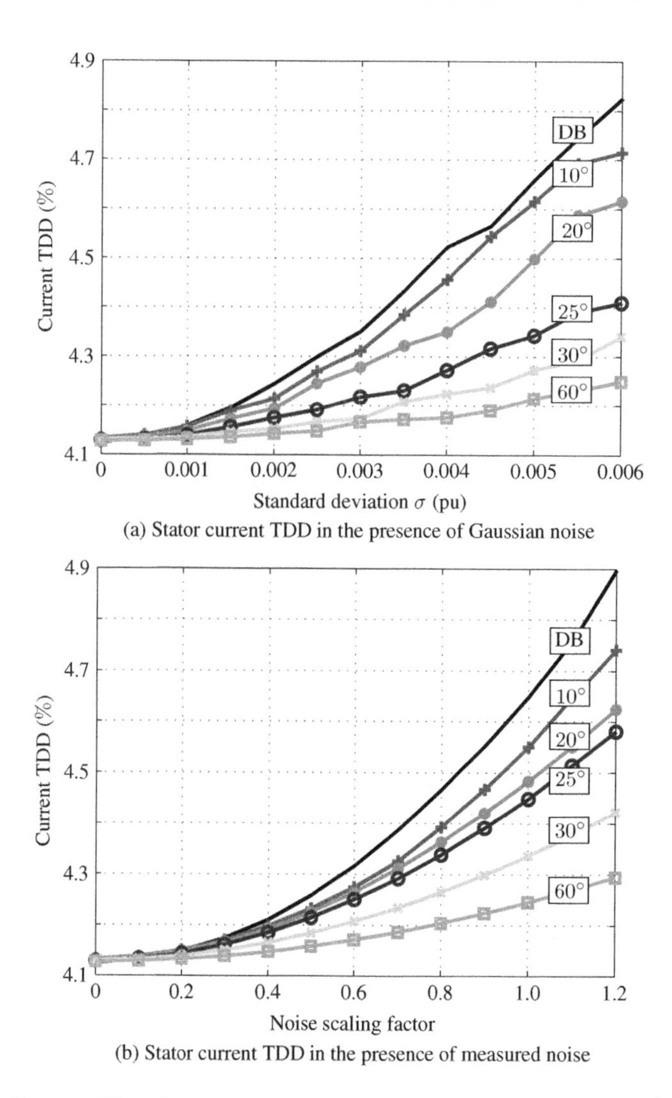

(a) Stator current TDD in the presence of Gaussian noise

(b) Stator current TDD in the presence of measured noise

Figure 13.7 Influence of flux observer noise on the closed-loop performance of MP^{3}C. For Gaussian noise, the stator current TDD is shown as a function of the standard deviation, while for the measured noise the current TDD is shown as a function of the noise scaling factor. The lines refer to DB MP^{3}C and to QP MP^{3}C with the angular prediction horizons $\theta_p = 10°, 20°, 25°, 30°,$ and $60°$

for DB MP^{3}C and to 1% for the QP controller with a long horizon. It can be concluded that MP^{3}C is robust to flux observer noise.

13.1.2.3 Operation under Machine Parameter Variations

Another potential source of control performance degradation is variations in the machine parameters that are unaccounted for by the controller. In the following, we investigate the

Table 13.2 Robustness of MP^{3}C to machine parameter variations under steady-state operating conditions, using DB and QP MP^{3}C with $\theta_p = 30°$

Control scheme	Variation of R_s (%)	Variation of R_r (%)	Deviation of T_e (%)	Deviation of $\|\psi_s\|$ (%)	Deviation of $\|\psi_r\|$ (%)
DB	75	100	0.19	0.03	0.02
DB	125	100	−0.18	−0.04	−0.02
QP	75	100	0.24	0.12	0.11
QP	125	100	−0.24	−0.12	−0.11
DB	100	75	0.13	0.00	−0.02
DB	100	125	−0.08	0.00	0.02
QP	100	75	0.42	0.06	0.03
QP	100	125	−0.36	−0.05	−0.03

The deviations of the electromagnetic torque T_e, stator flux magnitude $\|\psi_s\|$, and rotor flux magnitude $\|\psi_r\|$ from their nominal values are shown in percent, when altering the stator and rotor resistance by $\pm 25\%$, respectively.

impact that changes in the stator and rotor resistances, R_s and R_r, have on the steady-state tracking accuracy of MP^{3}C. As previously, operation at nominal speed and rated torque with an OPP of pulse number $d = 5$ is assumed. The resistances are altered by $\pm 25\%$. The performance of DB MP^{3}C is compared with that of the QP version with the angular horizon $\theta_p = 30°$. The steady-state deviations of the electromagnetic torque, stator flux magnitude, and rotor flux magnitude from their nominal values are used as performance indices. To highlight the performance of MP^{3}C, the outer flux and torque control loops in Fig. 12.8 are disabled.

As shown in Table 13.2, the steady-state errors are barely measurable. For DB MP^{3}C, the errors are below 0.2%, while for QP MP^{3}C they are below 0.5%. In general, DB MP^{3}C performs better in the presence of machine parameter variations than QP MP^{3}C. The steady-state tracking errors because of parameter variations affect only the fundamental components, not the ripple components. As a result, these variations neither have an impact on the harmonic distortions nor do they influence the device switching frequency.

The model used in MP^{3}C consists of two integrators—one integrator for the component of the stator flux vector in the α-axis and another one for the β-axis. The stator resistance is neglected. As can be seen in Table 13.2, variations in the stator resistance are indeed of minor importance, because the resulting voltage drop is in any case small for MV applications. Variations in the rotor resistance also have only a minor impact, because they merely alter the time-constant of the coupling between the stator and rotor. By forcing the stator flux vector along its desired trajectory, both errors can be mostly compensated for. To compensate for the residual small errors, the outer control loops and their integral terms are used.

13.1.3 Operation during Transients

The dynamic performance of MP^{3}C during torque reference steps is investigated now. At 50% speed, steps of magnitude 1 pu are imposed on the torque reference. An OPP with pulse number $d = 10$ is used, which results in a device switching frequency of $f_{sw} = 250$ Hz. In a first step, the performance of DB MP³C is compared with that of QP MP³C for different prediction horizons. In a second step, the benefit of inserting pulses during transients is investigated.

13.1.3.1 Operation without Pulse Insertion

Figure 13.8 shows the performance of DB MP³C during torque steps, with the torque reference steps being applied at time instants $t = 0$ and 20 ms. As can be seen in Fig. 13.8(a), the settling time is less than 2 ms and thus similar to that of hysteresis control schemes, such as DTC (see Sect. 3.6.3). Over- and undershoots in the torque response are avoided. The torque and the stator flux magnitude are perfectly decoupled, with the stator flux magnitude in Fig. 13.8(b) remaining unaffected by the torque steps. To achieve this fast torque response, the stator currents are driven quickly to their new values. This can be seen in Fig. 13.8(c) for the

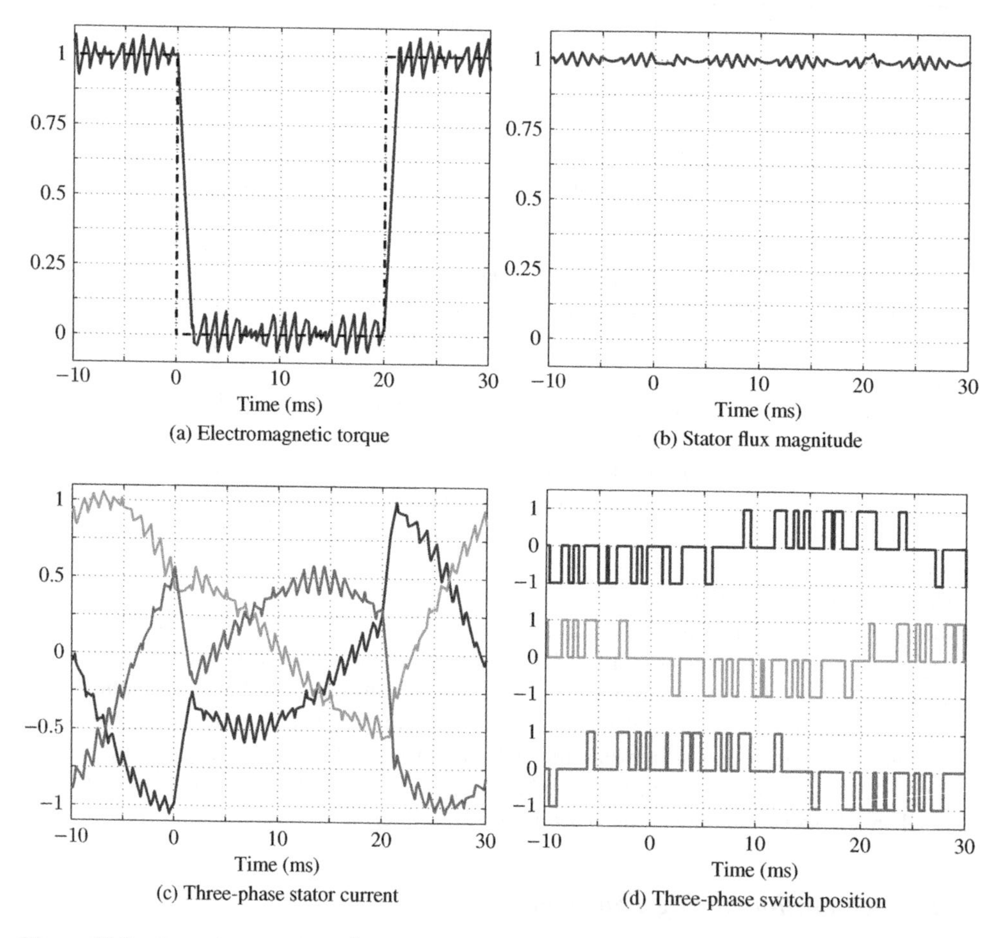

Figure 13.8 Operation of DB MP³C during torque reference steps. The torque, stator flux magnitude, stator currents, and switch positions are shown versus the time axis in milliseconds. Operation is at 50% speed with an OPP of pulse number $d = 10$, which yields a device switching frequency of 250 Hz

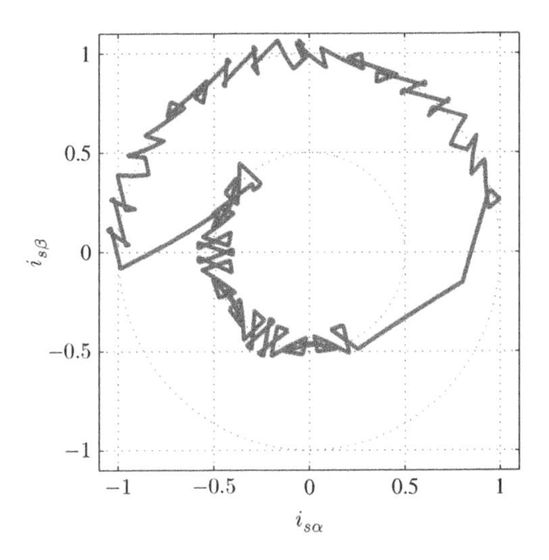

Figure 13.9 Stator currents in stationary orthogonal coordinates during the torque steps shown in Fig. 13.8

three-phase currents and in Fig. 13.9 for the currents in orthogonal coordinates plotted against each other.

When applying the negative torque reference step at $t = 0$ ms, the reference angle of the stator flux vector must be reduced by 13.7°, as per (12.24). At 50% speed, this reduction is equivalent to delaying the nominal OPP by 1.52 ms. As can be seen in Fig. 13.9, the stator current's α-component must be increased by almost 0.7 pu, whereas the β-component needs to be increased by about 0.45 pu. To achieve this, additional volt-second contributions are required—a large positive contribution from phase a and a significant negative contribution from phase c. Phase b is almost orthogonal to the required volt-second modification and thus contributes little. The same can be concluded by inspecting the three-phase currents in Fig. 13.8(c).

Figure 13.10(a) shows the switch positions around the negative torque reference step. The open-loop switch positions of the nominal OPP are shown as dash-dotted lines, while the closed-loop and modified switch positions are shown as solid lines. As can be seen, DB MP^{3}C achieves the required volt-second modifications in a DB fashion by removing the first pulse in phase a and by significantly shortening the second pulse. As the original pulses would reduce the volt-seconds in phase a, their removal adds positive volt-seconds. Similarly, in phase c, the first pulse is removed and the second one is shortened, thus reducing the volt-second contribution in phase c. As no switching transition is available in phase b, only the phases a and c are modified.

In the next step, the performance of DB MP^{3}C is compared with that of QP MP^{3}C for different (angular) prediction horizons. Figure 13.11 shows the respective torque responses for the negative torque reference step. The corresponding switch positions are shown in Fig. 13.10. The time axis is shown between −4 and 12 ms.

As expected, QP MP^{3}C leads to slower torque responses than DB MP^{3}C. For DB MP^{3}C, the torque becomes zero 1.6 ms after the torque reference step, while for QP MP^{3}C with $\theta_p = 10°$

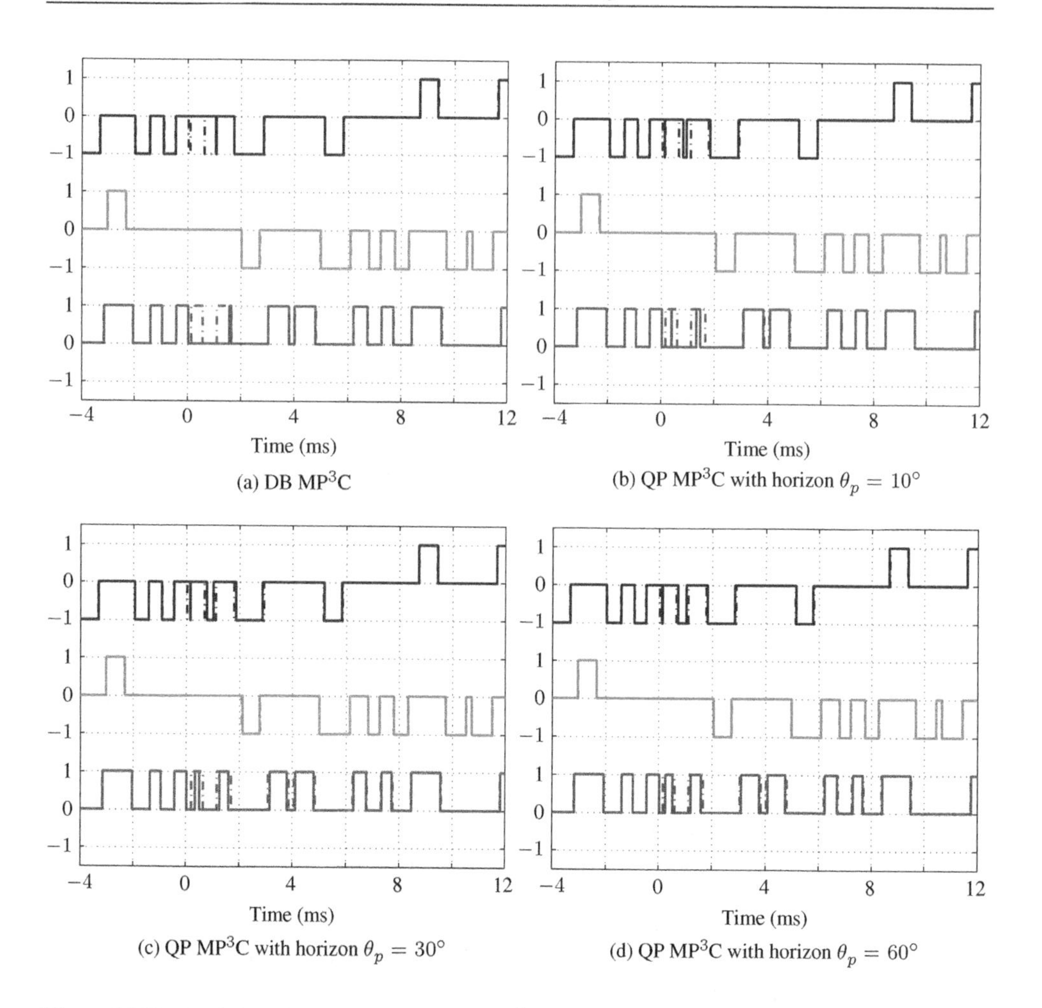

Figure 13.10 Switch positions for DB and QP MP^{3}C, corresponding to the torque step response in Fig. 13.11. The dash-dotted lines refer to the switching sequence of the unmodified original OPP, whereas the solid lines correspond to the closed-loop switching sequence modified by MP^{3}C

it takes 3.1 ms for the torque to achieve this. Nevertheless, in practice, the difference between DB and QP MP^{3}C with short horizons is small. For longer angular horizons, however, the settling time is significantly longer, amounting to 8.4 ms for $\theta_p = 30°$ and 10.6 ms for $\theta_p = 60°$. As the prediction horizon is increased, the required volt-second correction is distributed evenly over multiple switching transitions—this becomes evident when comparing the closed-loop switch positions in Figs. 13.10(b), 13.10(c), and 13.10(d).

13.1.3.2 Operation with Pulse Insertion

The merits of pulse insertion during torque reference steps are investigated now. We compare the torque response of MP^{3}C with pulse insertion with that of standard MP^{3}C. The closed-loop

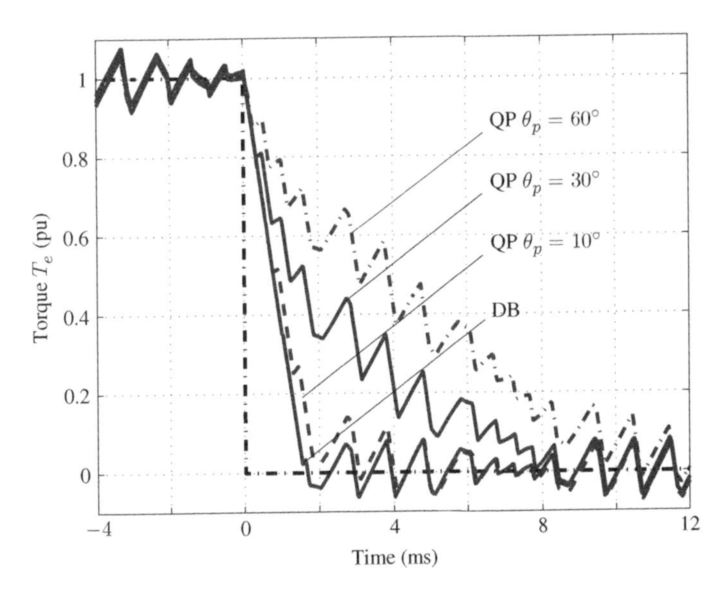

Figure 13.11 Torque responses to a torque reference step at time $t = 0$ ms for DB and QP MP³C with the angular prediction horizons $\theta_p = 10°, 30°$, and $60°$. The drive operates at 50% speed

response of the latter is shown in Fig. 13.8. To facilitate a direct comparison with the former, we adopt the previously used simulation settings: operation at 50% speed, torque steps of magnitude ± 1 pu and pulse number $d = 10$. Pulse insertion is turned on with the insertion gain $c_i = 25$. For DB MP³C, the closed-loop results are shown in Fig. 13.12.

For the negative torque step at $t = 0$ ms, pulse insertion reduces the torque settling time from 1.6 to 0.6 ms. This can be seen in Fig. 13.13, which zooms in on the time axis between −2 and 4 ms. Figure 13.14 compares the closed-loop switch positions of DB MP³C with pulse insertion to the switch positions without it. The nominal OPP at $t = 0$ ms corresponds to the switch position $\boldsymbol{u} = [-1\ 0\ 0]^T$. As shown in Fig. 13.14(b), without pulse insertion, DB MP³C applies the zero vector $\boldsymbol{u} = [0\ 0\ 0]^T$ instead, which momentarily halts the rotating stator flux vector.

With pulse insertion, however, as shown in Fig. 13.14(a), a positive pulse is inserted in phase a at $t = 0$ ms. Similarly, but with the opposite sign, a negative pulse is inserted in phase c. A short pulse is also added to phase b. This results in the switch position $\boldsymbol{u} = [1\ 1\ -1]^T$, which corresponds to a voltage vector of full magnitude in a direction that is almost opposite to that of the nominal voltage vector with $\boldsymbol{u} = [-1\ 0\ 0]^T$. In doing so, pulse insertion temporarily inverts the voltage applied to the machine during the negative torque step, thus fully utilizing the available dc-link voltage and achieving the fastest possible torque response.

The drastic speed-up of the torque response is at the expense of a temporary increase in the switching effort. Over the time window of 40 ms, the switching frequency is temporarily increased from 250 to 275 Hz. Such a short and relatively modest increase in the switching frequency is usually tolerable, because pulse insertion is required only during torque steps of significant magnitude and thus is rarely exercised.

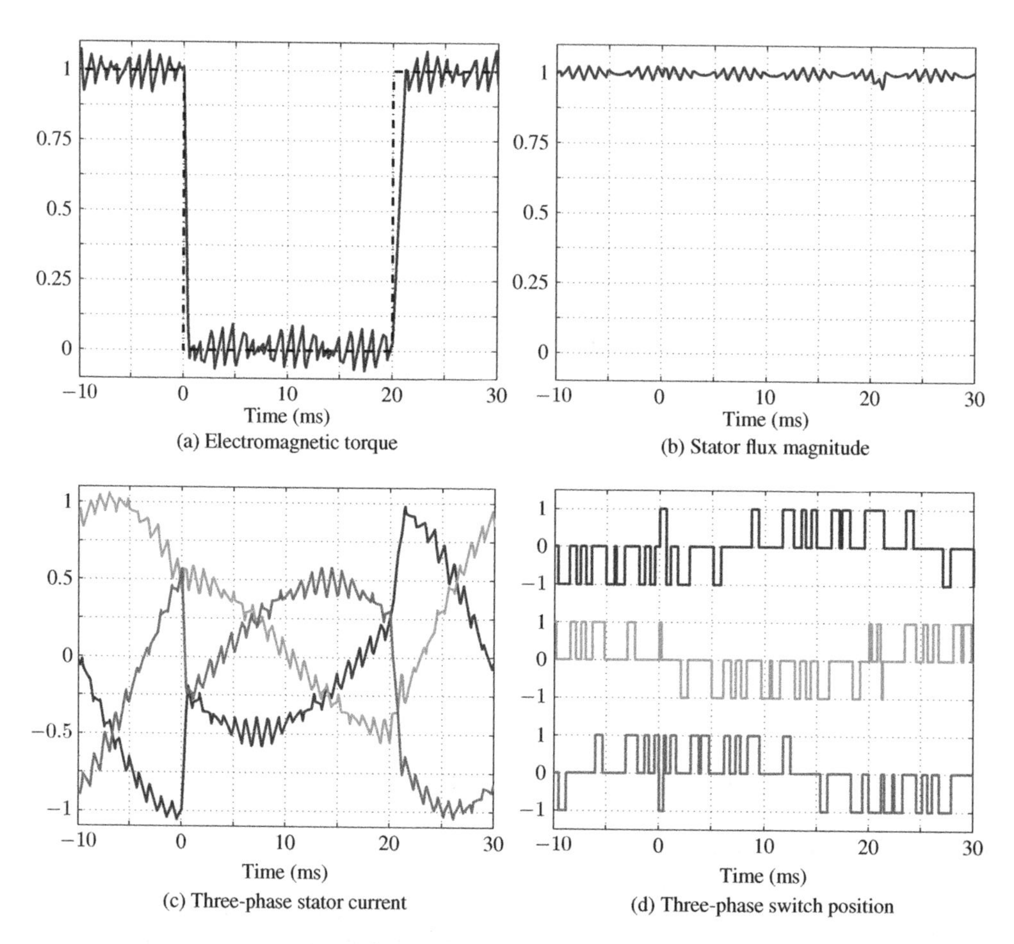

Figure 13.12 Operation of DB MP^{3}C with pulse insertion during torque reference steps. The figures use the same scaling and simulation setup as in Fig. 13.8 to allow a direct comparison with DB MP^{3}C without pulse insertion

During the positive torque reference step at $t = 20$ ms, the stator flux vector needs to be rotated forward. This is achieved by maintaining the angle of the voltage vector while increasing its magnitude. As the voltage margin is small, the insertion of additional pulses has only a minor effect. As a result, the torque settling time of 1.2 ms remains almost unchanged when compared to DB MP^{3}C without pulse insertion (see Fig. 13.8(a)).

Pulse insertion can lead to large voltage steps, for example, from $v_{dc}/2$ to $-v_{dc}/2$. In a practical converter setting, such large steps are neither desirable nor feasible. Restrictions are usually imposed in order to limit the rate of change of the voltage, dv/dt. This is due to the fact that high dv/dt values are detrimental to the lifetime of the machine windings. Apart from that, per-phase switching by more than one step up or down is prohibited for the NPC converter.

To impose these switching restrictions, the switching commands issued by MP^{3}C are post-processed before being applied to the inverter. For example, the large steps of magnitude 2 that MP^{3}C requests for phase c in Fig. 13.14(a) are reduced to steps of magnitude 1 in the

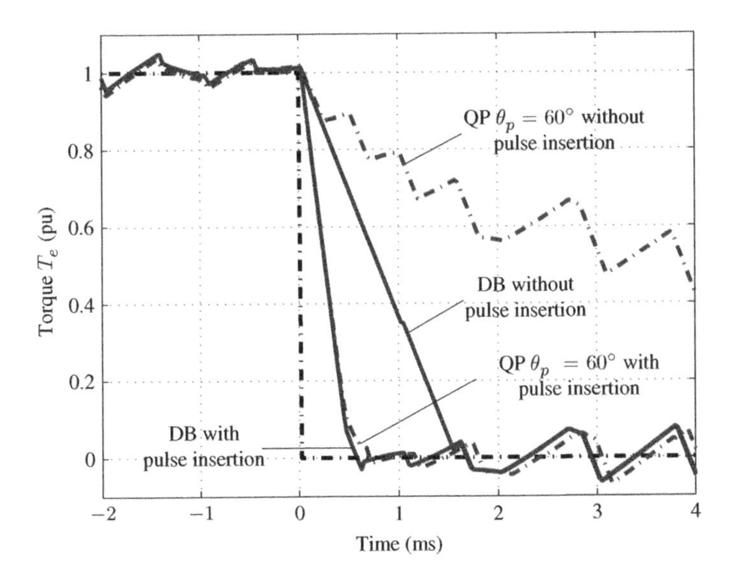

Figure 13.13 Torque responses to a torque reference step at time $t = 0$ ms with and without pulse insertion for DB and QP MP^{3}C with the angular prediction horizon $\theta_p = 60°$. The drive operates at 50% speed

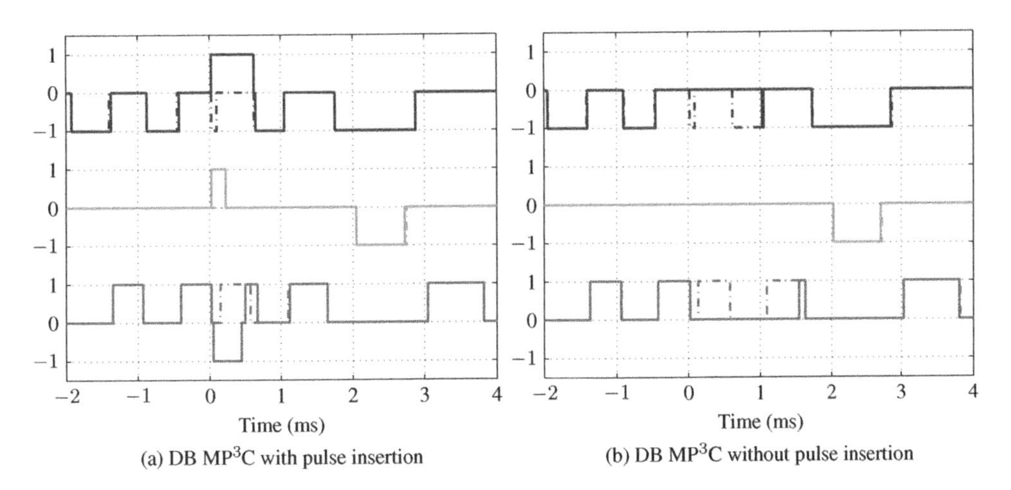

Figure 13.14 Three-phase switch positions for DB MP^{3}C with and without pulse insertion, corresponding to the torque step responses in Fig. 13.13. The dash-dotted lines refer to the unmodified and nominal OPP, whereas the solid lines correspond to the OPP modified by MP^{3}C

post-processing stage. Specifically, instead of switching from $u_c = 1$ to -1 at time $t = 0$ ms as requested by MP^{3}C, the switch position $u_c = 0$ is applied. At the next sampling instant, at $t = 25$ μs, MP^{3}C requests again to switch to $u_c = -1$, which is then permitted by the post-processing stage. As the width of the pulse amounts to several hundred microseconds and thus exceeds the sampling interval by an order of magnitude, the impact of these modifications on MP^{3}C and the torque response are minor. Importantly, the feedback characteristic inherent to MP^{3}C is well suited to address these modifications in the switching commands.

The performance improvement that pulse insertion carries is even more pronounced for QP MP^{3}C. As can be seen in Fig. 13.13 for the angular prediction horizon of 60°, pulse insertion reduces the torque settling time during the negative torque reference step from 10.6 to 0.7 ms. This is a 15-fold reduction. As a result, even for long horizons, QP MP^{3}C with pulse insertion achieves closed-loop performances during transients similar to that of DB MP^{3}C. As with DB MP^{3}C, the switching frequency is temporarily increased from 250 to 275 Hz over the time window of 40 ms.

13.2 Experimental Results for the ANPC Inverter Drive System

DB MP^{3}C has been successfully implemented and tested on a five-level MV drive system, which is shown in Fig. 13.15. Experimental results that were obtained with the MV drive being operated at power levels of up to 1 MVA are provided in this section.

13.2.1 Experimental Setup

Consider the five-level active neutral-point-clamped (ANPC) inverter drive system shown in Fig. 13.16. The dc-link is comprised of two identical dc-link capacitors with the pu equivalence

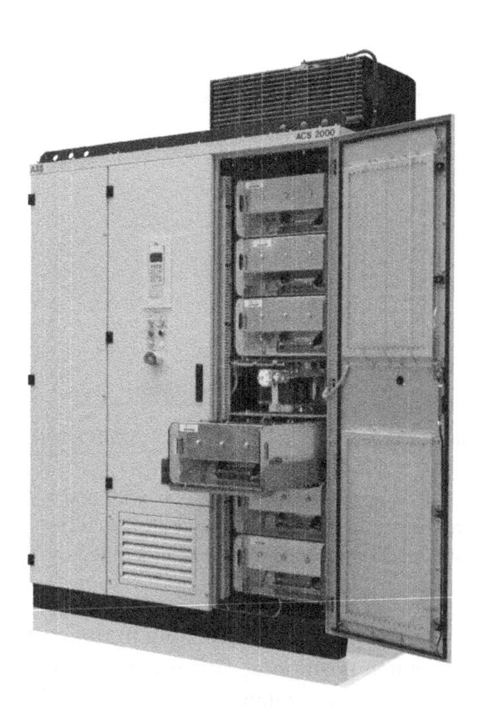

Figure 13.15 Five-level ANPC inverter with a rated voltage of up to 6.6 kV and a rated current of 98 A. The three phase models of the inverter and of the active front end, respectively, are shown on the right-hand side. Source: ABB Image Bank. Reproduced with permission of ABB Ltd

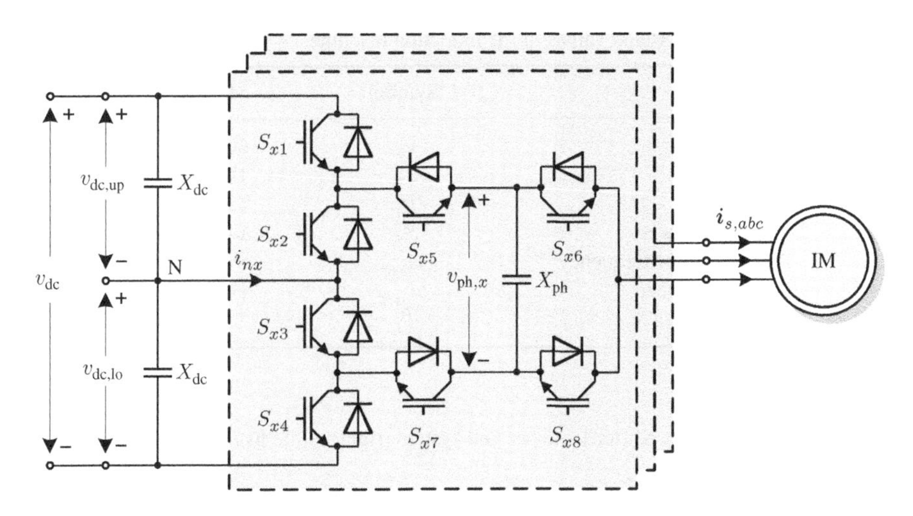

Figure 13.16 Five-level ANPC inverter drive system with an MV induction machine (IM)

X_{dc}. These dc-link capacitors form the neutral point N in between them. The total instantaneous dc-link voltage is $v_{dc} = v_{dc,up} + v_{dc,lo}$, where $v_{dc,up}$ and $v_{dc,lo}$ denote the voltages over the upper and lower dc-link capacitors, respectively. The ANPC inverter comprises the neutral point potential $v_n = \frac{1}{2}(v_{dc,lo} - v_{dc,up})$ and three flying phase capacitors X_{ph} with the voltages $v_{ph,x}$, where $x \in \{a, b, c\}$ denotes the three phases. The neutral point potential and the phase capacitor voltages are floating and must be actively controlled.

The inverter produces the five voltage levels $\pm v_{dc}/2$, $\pm v_{dc}/4$, and 0 at its phase terminals. Owing to the floating phase capacitor voltages and the generally nonzero neutral point potential, the phase voltages exhibit significant voltage variations around their nominal voltage levels. We refer to the nominal voltage levels by their phase levels $u_x \in \{-2, -1, 0, 1, 2\}$. The three-phase vector of phase levels is defined as $\boldsymbol{u} = \boldsymbol{u}_{abc} = [u_a \; u_b \; u_c]^T$.

Owing to the single-phase redundancy in the inverter, eight (rather than five) feasible switching combinations exist per phase. The single-phase switch positions are denoted by $s_x \in \{0, 1, \ldots, 7\}$, and the corresponding three-phase switch position is given by $\boldsymbol{s} = \boldsymbol{s}_{abc} = [s_a \; s_b \; s_c]^T$. For an in-depth review of the ANPC topology, the redundancy it offers, and the switching constraints it entails, the reader is referred to Sect. 2.4.2.

The drive encompasses a 6 kV, 50 Hz squirrel-cage induction machine rated at 1.424 MVA. The rated values of the machine are summarized in Table 13.3. The pu system is established using the base quantities $V_B = \sqrt{2/3} V_R = 4899\,\text{V}$, $I_B = \sqrt{2} I_R = 193.7\,\text{A}$, and $\omega_B = \omega_{sR} = 2\pi 50\,\text{rad/s}$. The machine and inverter parameters are summarized in Table 13.4 as SI quantities and pu values, along with their respective symbols. Note that the induction machine used for the experimental results is overrated and differs from the machine of the five-level ANPC case study in Sect. 2.5.3.

13.2.2 Hierarchical Control Architecture

The control problem of the ANPC inverter drive system is complicated by the requirement for balancing the neutral point potential and the three phase capacitor voltages around their

Table 13.3 Rated values of the induction machine

Parameter	Symbol	SI value
Voltage	V_R	6000 V
Current	I_R	137 A
Real power	P_R	1.2 MW
Apparent power	S_R	1.424 MVA
Angular stator frequency	ω_{sR}	$2\pi 50$ rad/s
Rotational speed	ω_{mR}	1488 rpm
Air-gap torque	T_R	7.976 kNm

Table 13.4 Parameters in the SI (left) and per unit system (right) of the five-level ANPC inverter drive system

Parameter	SI symbol	SI value	pu symbol	pu value
Stator resistance	R_s	203 mΩ	R_s	0.008
Rotor resistance	R_r	203 mΩ	R_r	0.008
Stator leakage inductance	L_{ls}	8.579 mH	X_{ls}	0.107
Rotor leakage inductance	L_{lr}	8.579 mH	X_{lr}	0.107
Main inductance	L_m	246.8 mH	X_m	3.066
Number of pole pairs	p	2		
dc-link voltage	V_{dc}	9.8 kV	V_{dc}	2.000
dc-link capacitor	C_{dc}	200 μF	X_{dc}	1.589
Phase capacitor	C_{ph}	140 μF	X_{ph}	1.112

nominal values. To achieve this, MP^{3}C is augmented by a subsequent balancing controller. Specifically, a hierarchical control architecture is adopted, which is shown in Fig. 13.17. The MP^{3}C block refers to the controller structure shown in the block diagram in Fig. 12.8 excluding the speed controller and the flux observer.

In the top layer, MP^{3}C regulates the stator flux vector along its optimal trajectory. In doing so, the electromagnetic torque T_e and the stator flux magnitude Ψ_s are controlled. MP^{3}C issues a switching sequence that starts at the current time instant kT_s and covers the sampling interval T_s. This switching sequence can be described by the sequence of phase levels $\boldsymbol{U} = [\boldsymbol{u}_1^T \; \boldsymbol{u}_2^T \ldots]^T$ and the corresponding vector of the switching instants $\boldsymbol{t}_u = [t_{u1} \; t_{u2} \ldots]^T$. In its simplest form, one phase level vector is issued at the current time instant, that is, $\boldsymbol{U} = \boldsymbol{u}_1$ and $\boldsymbol{t}_u = kT_s$.

The balancing controller of the bottom layer maintains the neutral point potential v_n and the three phase capacitor voltages $\boldsymbol{v}_{\text{ph}} = [v_{\text{ph},a} \; v_{\text{ph},b} \; v_{\text{ph},c}]^T$ within upper and lower bounds around their reference values. The reference of the neutral point potential is zero, while the reference for the phase capacitor voltages is, in general, a quarter of the nominal dc-link voltage, that is, $V_{\text{dc}}/4$.

To achieve its objective, the balancing controller exploits the single-phase and three-phase redundancies of the ANPC inverter. Specifically, the single-phase redundancy is used to control

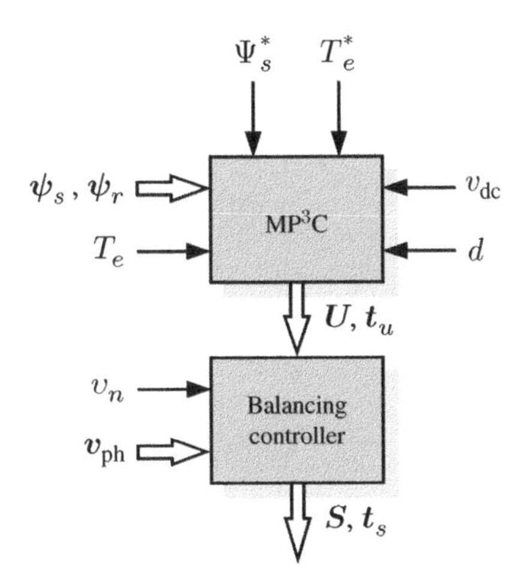

Figure 13.17 Hierarchical control architecture used for the five-level ANPC inverter drive system

the phase capacitor voltages and the neutral point potential, while the three-phase redundancy can be exploited to control the neutral point potential. The balancing controller also imposes switching constraints that are induced by the topology. In doing so, the reference phase level sequence $\{\boldsymbol{U}, \boldsymbol{t}_u\}$ is translated into the switching sequence $\{\boldsymbol{S}, \boldsymbol{t}_s\}$, where $\boldsymbol{S} = [\boldsymbol{s}_1^T \; \boldsymbol{s}_2^T \ldots]^T$ denotes the sequence of three-phase switch positions and $\boldsymbol{t}_s = [t_{s1} \; t_{s2} \ldots]^T$ is the vector of the switching instants.

In the majority of cases, a switching sequence $\{\boldsymbol{S}, \boldsymbol{t}_s\}$ is available that achieves the requested differential-mode voltage, which corresponds to $\{\boldsymbol{U}, \boldsymbol{t}_u\}$. If required, the common-mode voltage is adjusted to control the neutral point potential. Matching the desired differential-mode voltage is, however, not always possible, for example, when switching constraints are active. In this case, the differential-mode voltage requested by MP³C can be synthesized only in an approximate manner. These mismatches constitute output disturbances in the MP³C stator flux control loop, which are captured by the flux observer and fed back to MP³C. Through the use of feedback, the differential-mode voltage disturbances are corrected for by MP³C.

For more details on the redundancy offered by the ANPC topology, the reader is referred to Sect. 2.4.2. Some details on the balancing controller can be found in [3]. Alternatively, a derivative of the model predictive direct torque control (MPDTC) methodology—model predictive direct balancing control (MPDBC)—can be used to address the balancing control problem, as proposed in [4].

13.2.3 Steady-State Operation

The steady-state performance of MP³C is compared with that of DTC using experimental results on the five-level ANPC inverter MV drive system. For DB MP³C, an OPP with pulse number $d = 9$ and the balancing controller shown in Fig. 13.17 is used. For DTC, the same control hierarchy is adopted, but the MP³C block in Fig. 13.17 is replaced by DTC.

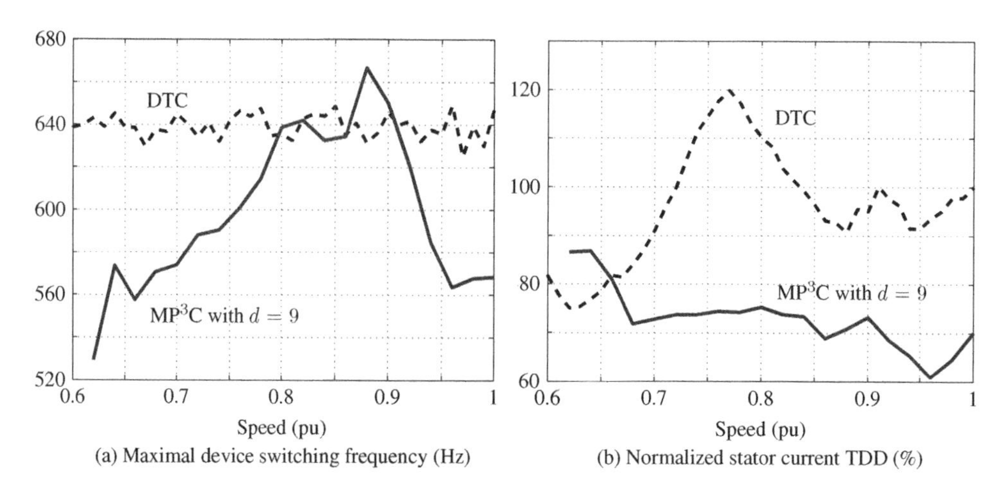

(a) Maximal device switching frequency (Hz)

(b) Normalized stator current TDD (%)

Figure 13.18 Performance comparison between DTC and DB MP³C in terms of the switching frequency and the current distortions, when operating at steady-state with a quadratic torque profile

Figure 13.18(a) shows the performance comparison of DTC and DB MP³C for fundamental frequencies between 0.6 and 1 pu, or equivalently between 30 and 50 Hz. A quadratic torque profile was used that reaches its maximum torque at 0.8 pu fundamental frequency. As the rated current corresponds to the 137 A machine current while the inverter phase current is limited to 98 A, the maximum torque achievable with this drive system is 0.6 pu.

Figure 13.18(a) shows the maximum device switching frequency $f_{\mathrm{sw,max}}$, which is obtained by monitoring the respective device switching frequencies of the semiconductor devices and recording the switching frequency with the highest value. Recall that the insulated-gate bipolar transistors (IGBTs) of the five-level ANPC topology (c.f. Fig. 13.16) can be divided into two groups. The switches S_{x1}–S_{x4} form the ANPC part, while the IGBTs S_{x5}–S_{x8} constitute the flying capacitor (FC) part in phase x, with $x \in \{a, b, c\}$. Switching in the FC part is not only required to synthesize the desired phase voltages but is also instrumental in maintaining the phase capacitor voltages close to their nominal values. The IGBTs S_{x5}–S_{x8} carry the majority of the switching burden and thus constitute the limiting factor in terms of the maximal switching frequency. Their switching frequency $f_{\mathrm{sw,FC}}$ is defined as the average switching frequency of the IGBTs in the FC part. Assuming that the switching burden is well balanced,

$$f_{\mathrm{sw,FC}} \approx f_{\mathrm{sw,max}} \tag{13.2}$$

holds.

Each transition in a phase level u_x entails exactly one *on* transition in the FC part, as can be seen in Fig. 2.24. Given that there are four IGBTs in the FC part, the switching frequency contribution of MP³C is expected to be $f_{\mathrm{sw,FC}} = df_1$, where f_1 denotes the fundamental frequency in hertz. As per (13.2), one would expect $f_{\mathrm{sw,max}} \approx df_1$.

As can be seen in Fig. 13.18(a), the maximum device switching frequency is significantly higher than df_1. Indeed, for the range of fundamental frequencies shown in this figure, the switching frequency is increased by a factor of 1.25 at nominal speed, while it is a factor of almost 2 at 60% speed. These additional switching transitions are required to balance the phase capacitor voltages and the neutral point potential, particularly at low fundamental frequencies and at high phase currents. To gain more insight into the closed-loop characteristics of the phase capacitor voltages, see the simulation results of MPDTC discussed in Sect. 8.2.

Figure 13.18(b) compares the stator current distortions that result from applying the two control methods. Normalized current distortions are shown, which are obtained by dividing the measured current TDD by DTC's current TDD at nominal speed. A particularly difficult operating point is around 80% speed, at which full current is provided, albeit at a fundamental frequency of 40 Hz. In this region, MP^{3}C provides the largest performance benefit when compared to DTC. At 77% speed, for example, the current distortions are reduced by 38%. At nominal speed, the reduction is by 30%. Note that by using an OPP with the single pulse number $d = 9$, MP^{3}C does not fully utilize the available device switching frequency. By selecting OPPs with higher pulse numbers, the current distortions could be further reduced.

A more detailed comparison between MP^{3}C and DTC is provided at 80% speed and 60% torque, which corresponds to the maximum inverter current of 98 A rms. The maximum device switching frequency is in both cases 640 Hz. The stator currents are shown in Fig. 13.19 over one fundamental period, and their harmonic spectra are shown in Fig. 13.20. The (peak) magnitude of the fundamental component is in both cases 0.715 pu, which corresponds to the full 98 A (rms) inverter current. Compared to DTC, MP^{3}C reduces the current distortions by 31% and exhibits significantly smaller fifth and seventh harmonics.

Despite the characteristic current harmonics of the OPP at odd and non-triplen integer multiples of the fundamental frequency, the harmonic spectrum exhibits a noise floor, which worsens the current TDD. This noise floor is due to non-idealities of the inverter system, such as dead-time effects [5], delays in the feedback loop, and disturbances that result from the balancing controller when enforcing switching constraints. Another root cause for the noise floor is the fluctuating voltages at the inverter's phase terminals, which are shown in Fig. 13.21 with respect to the dc-link midpoint. As discussed, these voltage fluctuations are due to deviations of the phase capacitor voltages and neutral point potential from their reference values. These non-idealities and disturbances manifest themselves as deviations of the stator flux vector from

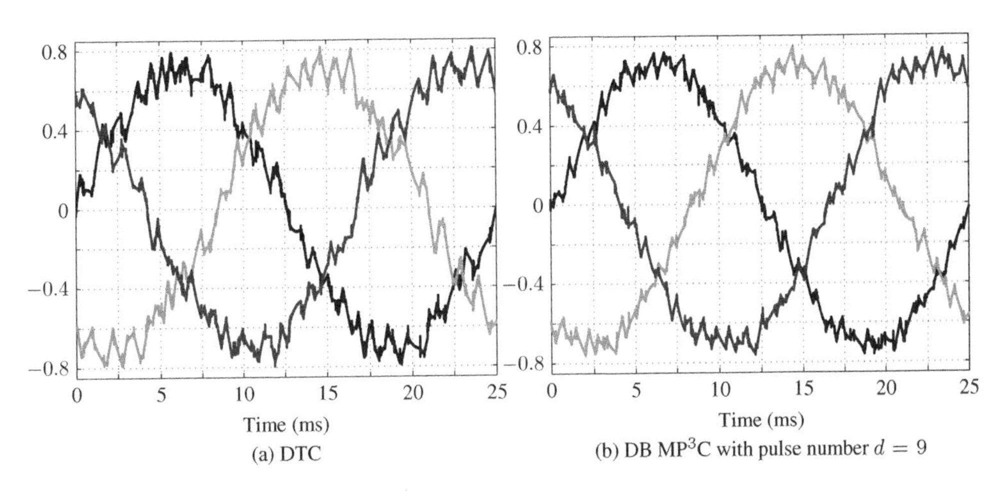

Figure 13.19 Stator currents of DTC and MP^{3}C, when operating at 80% speed and 60% torque in an MV laboratory

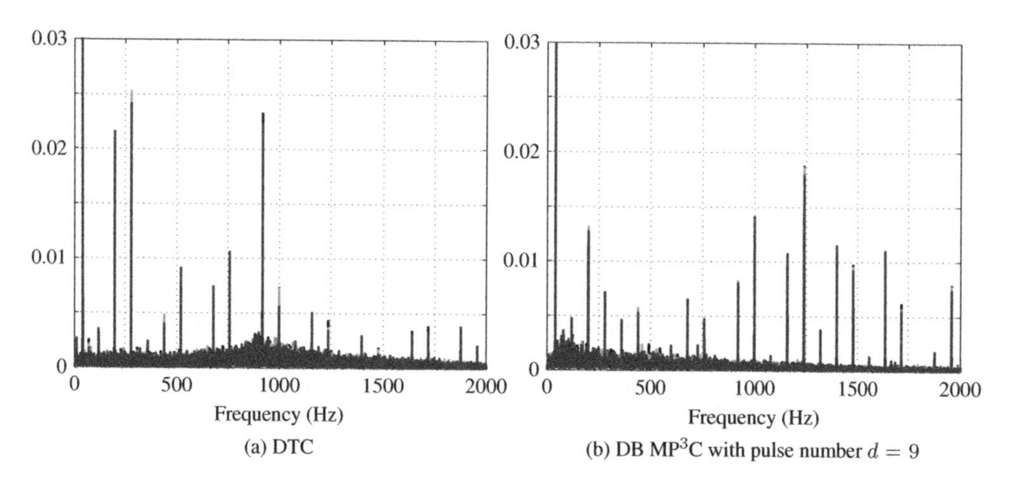

Figure 13.20 Spectra of the stator currents of DTC and MP^{3}C shown in Fig. 13.19

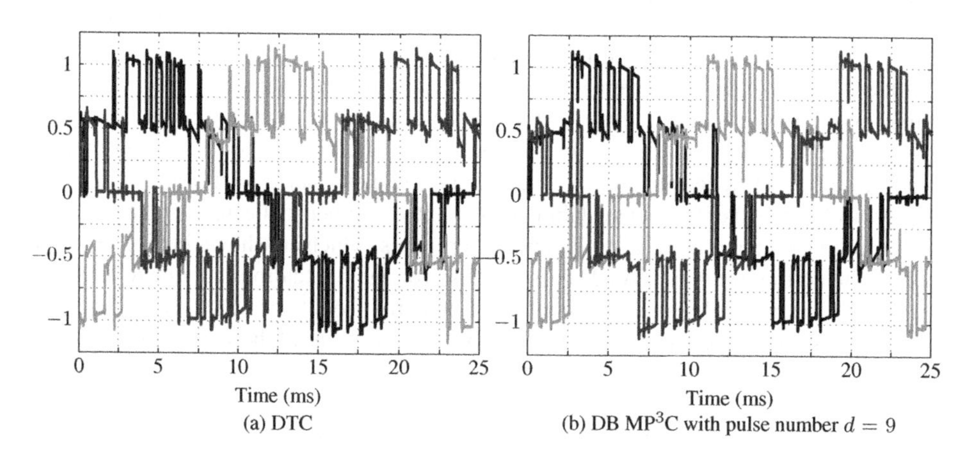

Figure 13.21 Phase voltages of DTC and MP^{3}C, corresponding to the experimental results in Fig. 13.19

its reference trajectory. This results in an increased harmonic content in the stator currents of the machine despite the utilization of OPPs and the fast volt-second corrections of MP^{3}C.

13.3 Summary and Discussion

Chapter 12 proposed a new model predictive control (MPC) method based on OPPs, which resolves the classic contradiction in power electronics control—very fast control during transients on one hand, and optimal performance at steady-state operation on the other; that is, minimum current distortions for a given switching frequency. The former is typically achieved by DB control schemes and DTC, while the latter is in the realm of OPPs.

The proposed controller, MP^{3}C, achieves both objectives by adopting the principles of constrained optimal and receding horizon control. The latter provides feedback and a high degree of robustness to disturbances and inverter non-idealities, ensuring that the optimal volt-second balance of the OPP is maintained at (quasi) steady-state operation and during transients. As a result, MP^{3}C achieves both very low harmonic distortion levels per switching frequency at steady-state operating conditions and fast current and torque responses during transients.

To achieve the latter, the notion of pulse insertion is instrumental, as it provides the controller with an additional degree of freedom when required to remove the flux error as quickly as possible. Switching transitions can be inserted and switching patterns synthesized, which correspond to voltage vectors with magnitudes and angles that differ greatly from the voltage vectors inherent to the precalculated OPP. Specifically, the phase voltage applied to the stator windings of the machine can be temporarily increased to its maximum value and its sign can be inverted, thus fully exploiting the available dc-link voltage of the inverter.

General-purpose drive applications, such as fans and pumps, increasingly impose stringent harmonic performance requirements. Owing to the use of OPPs, this goal is easily achieved with MP^{3}C. Special-purpose drive applications, such as rolling steel mills, require a high dynamic performance and very fast torque responses. To achieve this, pulse insertion is beneficial. This makes MP^{3}C an ideal choice for both general-purpose and special-purpose drives.

Moreover, MP^{3}C can be extended to inverters with *LC* filters, which require an active damping loop to attenuate the filter resonance. To this end, a linear quadratic regulator can be added, which injects a damping signal to the stator flux error [6]. The principle of MP^{3}C is also applicable to grid-side converters [7] and modular multilevel converters [8].

13.3.1 Differences to the State of the Art

The proposed MP^{3}C controller differs from state-of-the-art trajectory controllers in the following important aspects.

13.3.1.1 Fundamental Flux Component

State-of-the-art techniques such as [9] control the *fundamental* component of the stator flux separately from its *ripple* component. To achieve this, the fundamental flux component must be estimated, which greatly complicates the task of the flux observer [10].

The separation between fundamental and ripple components is avoided in MP^{3}C. Instead, the *instantaneous* value of the stator flux vector, which encompasses both its fundamental and ripple components, is regulated along its optimal trajectory. The stator flux error (12.28) is then simply the difference between the reference and the estimated (instantaneous) stator flux vector. As a result, a standard flux observer can be used. For the experimental results shown in Sect. 13.2, for example, the same flux observer was employed that is commonly used for DTC.

In summary, MP^{3}C treats the flux error (12.28) as a single quantity that encompasses both the harmonic flux error and the fundamental flux error. The *harmonic* flux error relates to ripple current errors at quasi-steady-state operation. These errors are caused, for example, by dc-link voltage fluctuations, discontinuities in the OPP, and changes in the pulse number. *Fundamental* flux errors result from changes in the operating point, such as changes in the load torque or the angular velocity of the machine.

13.3.1.2 Penalty on the Manipulated Variable

State-of-the-art trajectory control methods [9, 11] only penalize the controlled variable, that is, the stator flux error, according to the principle of DB control. QP MP^{3}C, however, also penalizes *changes* in the manipulated variable, that is, modifications to the nominal time instants of the switching transitions. In doing so, the required volt-second correction is distributed over the prediction horizon, thus attempting to preserve the volt-second balance of the precalculated OPP. This also reduces the sensitivity of QP MP^{3}C to flux observer noise.

To ensure that the predicted stator flux error at the end of the prediction horizon is effectively zero, the penalty on modifying the switching instants is set to a small value, thereby prioritizing the correction of the stator flux error. Penalizing both the controlled and the manipulated variable is performed in accordance with the principle that changes to the manipulated variable also incur a cost. Indeed, in optimal control, a trade-off between good tracking performance and little use of the manipulated variable arises. This trade-off is adjusted by the penalty matrix $\boldsymbol{Q}$.

13.3.1.3 Receding Horizon Policy

Even though a sequence of switch positions is planned over a long prediction horizon, only the switching sequence over the sampling interval is applied to the inverter. The predictions are recomputed at the next sampling instant using new measurements; a shifted—and if necessary revised—sequence of switch positions is derived. This is referred to as the *receding horizon policy* (see Fig. 12.7). The receding horizon policy provides feedback and makes MP^{3}C robust to flux observer noise and modeling errors. Longer horizons reduce the controller sensitivity to flux observer errors, as shown in Sect. 13.1.2.

As a result, when operating on an experimental drive setup with measurement and observer noise, the steady-state current distortions are expected to be lower, when compared with an overly aggressive controller. Examples for the latter include controllers that operate with a very short prediction horizon or a DB controller that does not penalize the corrective actions at all. Under idealized simulation conditions without noise and without a significant voltage ripple on the neutral point potential (and the phase capacitor voltages), all MP^{3}C variations provide almost identical current distortions.

13.3.1.4 Deadbeat Controller

The DB version of MP^{3}C might appear to bear some similarities with state-of-the-art methods [9, 10]. These, however, typically construct the flux error every 500 μs, map the $\alpha\beta$ flux error into all three phases (using the inverse Clarke transformation (2.11)), modify the time instants of the switching transitions within these 500 μs, and send the modified switching sequence over the entire subsequent 500 μs to the inverter. Feedback is thus only applied every 500 μs.

In contrast to this, the proposed DB MP^{3}C method adopts the receding horizon policy. The modified switching sequence is applied to the inverter over a very short time interval, typically 25 μs. Therefore, feedback is provided every 25 μs. Even for DB MP^{3}C, the prediction horizon is typically quite long and in the range of 500 μs to 1 ms. Moreover, the $\alpha\beta$ flux error is mapped only into two phases, namely the two phases that feature the first switching transitions. This type of mapping tends to reduce the time required to correct the stator flux error.

13.3.1.5 Pulse Insertion

When inserting a pulse in MP^{3}C, a *virtual* pulse is added with zero volt-second contribution. While applying the inserted pulse, the volt-second generated by the pulse is modified in a *closed-loop* fashion by adjusting the time instant of the second switching transition. According to the receding horizon policy, at subsequent sampling instants, the pulse width is readjusted to account for flux observer noise, disturbances affecting the stator flux, further changes in the torque reference, and restrictions on the allowed dv/dt.

As a consequence, the step size of the second switching transition leading back to the nominal OPP is not determined at the time the pulse is inserted. This can be seen, for example, in Fig. 12.13(b). When shifting the second transition of the pulse beyond the nominal switching instant of the next transition t^*_{x1}, the step size is reduced from 2 to 1.

The closed-loop control paradigm of the proposed pulse insertion method is in stark contrast to the method previously mentioned in the literature [12–14]. The latter appears to rely on *open-loop* pulse insertion, using a feedforward approach, in which both switching transitions are determined at the time they are inserted. In addition to this, depending on the flux error and the error bounds, in the proposed method switching transitions are inserted in one, two, or three phases, rather than always in two phases, as described in the literature.

13.3.2 Discussion

In the following section, further aspects and features of the proposed MP^{3}C scheme are clarified and discussed.

13.3.2.1 Optimality

It is important to point out that optimality, that is, minimum current TDD, is achieved when the reference stator flux trajectory is accurately tracked. Optimality is thus defined in terms of the flux trajectory rather than in terms of the voltage waveform of the OPP. These two quantities match only at steady-state operation under ideal conditions. Optimality can also be achieved at quasi-steady-state conditions, by ensuring that the reference flux trajectory is closely tracked. By manipulating the width of the pulses, the optimal volt-second balance can be maintained despite voltage fluctuations in the dc-link, neutral point, and phase capacitors.

During transients, however, defining optimality in terms of minimum current TDD is not meaningful, because the notion of harmonic distortion is based on frequency analysis, requiring (quasi) steady-state operation. Large transients typically arise when large torque steps are imposed, switching between different OPPs is performed, or when the operating point is moved across discontinuities in the switching angles. The stator flux error tends to be large in all three cases, necessitating significant corrections of the switching instants. When such transients occur, the controller aims to achieve a very fast dynamic response by rapidly regulating the stator flux vector to its new reference. Strictly speaking, however, re-optimizing around the existing OPP or inserting pulses might be suboptimal, in the sense that a more suitable switching sequence might exist that achieves the same torque response but at a lower switching burden. Owing to the additional computational and implementation-related complexity such an approach might entail, it has not been further pursued.

13.3.2.2 Operation at Low Modulation Indices

The MP^{3}C method is conceptually applicable to the whole speed range. At the upper end of the modulation index, OPPs can reach six-step operation. At the lower end, however, practical considerations limit the applicability of OPPs. Specifically, low modulation indices and low fundamental frequencies lead to very high pulse numbers, making the computational procedure of deriving such OPPs challenging. Moreover, the advantage of OPPs over CB-PWM becomes minor in terms of harmonic distortions. Therefore, the standard practice is to switch to CB-PWM with field-oriented control at low modulation indices, as explained in detail in [10].

13.3.2.3 Computational Requirements

In general, the computational burden is often high for MPC methods, requiring a powerful control platform to solve the underlying optimization problem within the sampling interval. By precomputing OPPs, however, the majority of the computations is moved offline for MP^{3}C at the expense of increased memory requirements to store these patterns. During runtime, the OPP is modified by the controller so as to compensate for non-idealities and to achieve fast control during transients.

DB MP^{3}C is well suited to implementation on a field-programmable gate array (FPGA). Additions, shift operations and if-then-else statements are computationally cheap, because they require little space on the FPGA and just one clock cycle. Divisions and multiplications, however, are computationally expensive on an FPGA, requiring multiple clock cycles and dedicated multiplier units, which are often scarce. Of those expensive computations, DB MP^{3}C requires only one division and a few multiplications to compute the controller output. Therefore, the computational effort required by DB MP^{3}C to modify the precomputed OPPs is roughly the same as the effort required to establish control by field orientation.

The QP version of MP^{3}C is computationally more demanding than the DB version. To facilitate implementation, an active set method can be adopted that is tailored to the optimization problem at hand, as explained in Sect. 12.5.1. The computational burden of the active set method is only slightly higher than that of DB MP^{3}C. Alternatively, a fast gradient solver can be implemented on an FPGA to solve the QP, as proposed in [15].

References

[1] B. P. McGrath, D. G. Holmes, and T. Lipo, "Optimized space vector switching sequences for multilevel inverters," *IEEE Trans. Power Electron.*, vol. 18, pp. 1293–1301, Nov. 2003.

[2] T. Geyer, N. Oikonomou, G. Papafotiou, and F. Kieferndorf, "Model predictive pulse pattern control," *IEEE Trans. Ind. Appl.*, vol. 48, pp. 663–676, Mar./Apr. 2012.

[3] F. Kieferndorf, M. Basler, L. Serpa, J.-H. Fabian, A. Coccia, and G. Scheuer, "A new medium voltage drive system based on ANPC-5L technology," in *Proceedings of IEEE International Conference on Industrial Technology* (Viña del Mar, Chile), pp. 605–611, Mar. 2010.

[4] F. Kieferndorf, P. Karamanakos, P. Bader, N. Oikonomou, and T. Geyer, "Model predictive control of the internal voltages of a five-level active neutral point clamped converter," in *Proceedings of IEEE Energy Conversion Congress and Exposition* (Raleigh, NC, USA), Sep. 2012.

[5] J. Holtz and J. Quan, "Drift- and parameter-compensated flux estimator for persistent zero-stator-frequency operation of sensorless-controlled induction motors," *IEEE Trans. Ind. Appl.*, vol. 39, pp. 1052–1060, Jul./Aug. 2003.

[6] P. Al Hokayem, T. Geyer, and N. Oikonomou, "Active damping for model predictive pulse pattern control," in *Proceedings of IEEE Energy Conversion Congress and Exposition* (Pittsburgh, PA, USA), pp. 1220–1227, Sep. 2014.

[7] E. Rohr, T. Geyer, and P. Al Hokayem, *"Control of electrical converter based on optimized pulse patterns,"* EP patent application 15 176 085 A1, 2015.

[8] V. Spudic, T. Geyer, and N. Oikonomou, *"Optimized pulse patterns for MMC control,"* EP patent application 15 184 836 A1, 2015.

[9] N. Oikonomou and J. Holtz, "Stator flux trajectory tracking control for high-performance drives," in *Proceedings of the IEEE Industry Applications Society Annual Meeting* (Tampa, FL, USA), pp. 1268–1275, Oct. 2006.

[10] N. Oikonomou and J. Holtz, "Estimation of the fundamental current in low-switching-frequency high dynamic medium-voltage drives," *IEEE Trans. Ind. Appl.*, vol. 44, pp. 1597–1605, Sep./Oct. 2008.

[11] B. Beyer, *Schnelle Stromregelung für Hochleistungsantriebe mit Vorgabe der Stromtrajektorie durch off-line optimierte Pulsmuster*. PhD thesis, Wuppertal University, 1998.

[12] J. Holtz and B. Beyer, "Off-line optimized synchronous pulsewidth modulation with on-line control during transients," *EPE Journal*, vol. 1, pp. 193–200, Dec. 1991.

[13] J. Holtz and N. Oikonomou, "Synchronous optimal pulsewidth modulation and stator flux trajectory control for medium-voltage drives," *IEEE Trans. Ind. Appl.*, vol. 43, pp. 600–608, Mar./Apr. 2007.

[14] N. Oikonomou, *Control of medium-voltage drives at very low switching frequency*. PhD thesis, University of Wuppertal, 2008.

[15] H. Peyrl, J. Liu, and T. Geyer, "An FPGA implementation of the fast gradient method for solving the model predictive pulse pattern control problem," in *Workshop on Predictive Control of Electrical Drives and Power Electronics* (Munich, Germany), Oct. 2013.

14

Model Predictive Control of a Modular Multilevel Converter

14.1 Introduction

The modular multilevel converter (MMC) topology [1] has recently received significant interest in the literature [2, 3]. This converter has many attractive properties, chiefly the series connection of modules. Series connection allows one to scale the converter output voltage and to achieve high output voltages, despite the relatively low voltages typically used for the modules. Moreover, increasing the number of series-connected modules directly increases the number of voltage levels available at the converter's output terminals, which significantly reduces the harmonic distortions in the output voltages and currents as compared to standard two- or three-level voltage source converters [4]. As a result, the size of the harmonic filter can be reduced and the switching frequency can be lowered in order to reduce the total losses in the converter. These properties make the MMC topology well suited to high-power and high-voltage applications, such as high-voltage direct current (HVDC) transmission systems [5, 6]. In the medium-voltage (MV) area, promising applications include static VAR compensators (STATCOMs) [7, 8] and railway interties [9], which do not require a transformer.

The control and modulation problem of the MMC is to regulate the load currents along their time-varying references, to balance the capacitor voltages around their nominal values, to minimize the converter and switching losses, and to meet harmonic requirements of the load. The MMC must be operated within its safe operating limits, particularly with regard to the branch currents and capacitor voltages. Owing to the multiple-input multiple-output (MIMO) structure of the converter and its various internal dynamics, this control problem is intrinsically difficult to solve. State-of-the-art hierarchical controllers can be grouped into current, average, and energy-based schemes.

In the upper layer of a hierarchical current controller, as developed and implemented in [10], the voltage references of the branches are computed in view of the desired time-varying output load currents and the elimination of circulating currents. A subsequent carrier-based pulse width modulator (CB-PWM) or space vector modulator (SVM) in conjunction with a capacitor

Model Predictive Control of High Power Converters and Industrial Drives, First Edition. Tobias Geyer.

Companion Website: www.wiley.com/go/geyermodelpredictivecontrol

voltage balancing controller translates the references signals of the branch currents into gating commands for the converter. In [11], a hierarchical scheme that combines averaging and balancing control is presented. The controller splits the effort between the number of modules per branch. A subsequent pulse width modulator (PWM) is used to generate the switching sequences. The hierarchical energy-regulating controller drives the total branch energies along time-varying reference values, thus guaranteeing the equalization of capacitor voltages within and among the branches. The reference signals are derived as functions of the desired time-varying load currents and the measured dc-link voltage. The energy balance is achieved by exploiting the circulating currents and the common-mode voltage. Examples of hierarchical energy-regulating controllers can be found in [12] and [13].

Hierarchical schemes with multiple proportional–integral (PI) control loops tend to perform poorly when fast control actions are required during transient operation or when the switching frequency is very low. This shortcoming motivates the investigation of modern control methods formulated in the time domain, most notably model predictive control (MPC) [14]. The literature on MPC schemes for the MMC topology is scarce and restricted to direct MPC methods that do not use a modulator. Direct MPC methods with a prediction horizon of one step were proposed in [15] for the single-phase ac–ac MMC topology and in [5] for a back-to-back HVDC system. Both approaches follow the enumeration-based direct MPC paradigm, which is summarized in Chap. 4. Longer prediction horizons were achieved in [16] for a three-phase dc–ac MMC.

This chapter proposes an *indirect* model predictive current controller for the MMC, which is based on a PWM. A prediction horizon of 5–10 steps is used. The underlying optimization problem is a quadratic program (QP), which can be solved efficiently using off-the-shelf solvers. The control problem is addressed in a hierarchical manner. The MPC scheme constitutes the upper layer that provides voltage references to the subsequent PWM stage. The capacitor voltages are balanced within the branches by balancing controllers that operate on the lower layer.

This MPC scheme provides optimal control actions both at steady-state operation as well as during transients, such as power up, load steps, and faults. Owing to the ability of MPC to address constraints, the proposed controller achieves fast transient responses while respecting constraints on the branch currents and capacitor voltages, thus ensuring that the converter operates under safe conditions even during transient operation. This stands in stark contrast to the traditionally used hierarchical controllers with multiple PI loops (see, e.g., [11] and [17]) whose dynamic response tends to be slow in order to avoid violations of the safe operating area.

14.2 Preliminaries

14.2.1 Topology

The three-phase dc–ac MMC topology under investigation is shown in Fig. 14.1. Each phase leg $x \in \{a, b, c\}$ of the converter is divided into an upper and a lower branch. Each of the six branches $r \in \{1, \dots, 6\}$ consists of n series-connected unipolar modules M_{rj}, with $j \in \{1, \dots, n\}$. A (small) branch inductor L_{br} is added to each branch. Its ohmic resistance and the conduction losses in the branches are modeled by the resistor R_{br}.

As shown in Fig. 14.2, each module M_{rj} consists of the capacitor C_m with the voltage v_{rj} and two insulated-gate bipolar transistors (IGBTs). The IGBTs form a half-bridge with two

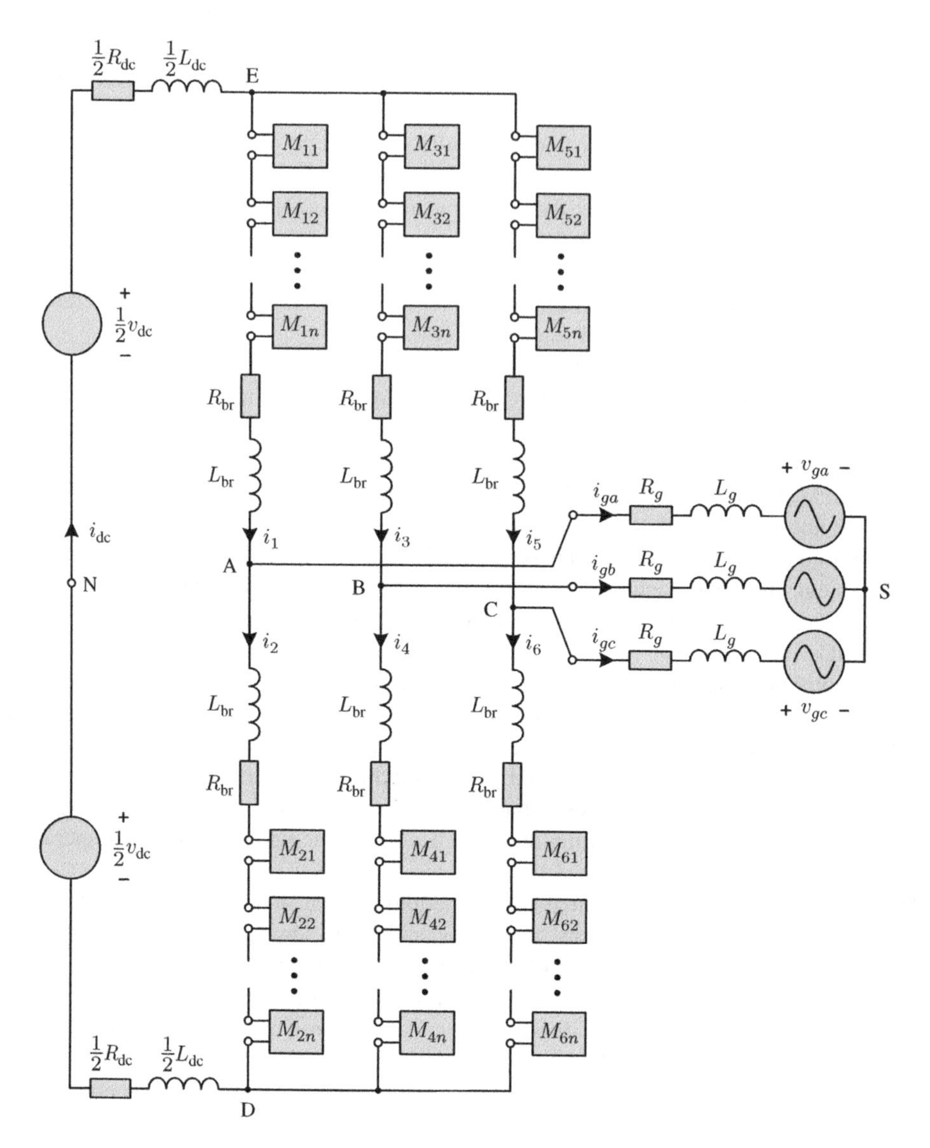

Figure 14.1 Three-phase dc–ac MMC topology with n series-connected unidirectional modules per branch

switching states. During the *on* state of the module, when the upper IGBT is turned on and the lower one is turned off, the capacitor C_m is connected to the branch, and the terminal voltage of the module is equal to the capacitor voltage v_{rj}. During the *off* state, when the lower IGBT is turned on and the upper one is off, the module is bypassed and the terminal voltage of the module is zero. Note that in order to avoid a potential shoot-through, a short blanking time is introduced in which both IGBTs are off.

The MMC is fed by a dc source with the supply voltage v_{dc}. Parasitic inductances and resistances in the dc supply are modeled by the inductor L_{dc} and the resistor R_{dc}, respectively,

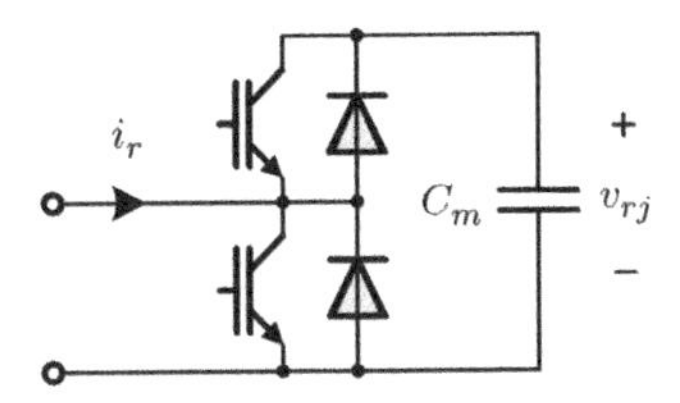

Figure 14.2 Unipolar MMC module M_{rj} consisting of a half-bridge and the capacitor C_m

which are connected in series to the dc supply. The three-phase output terminals of the MMC are connected to the grid, which is represented by the grid inductor L_g in series with the grid resistor R_g and the three-phase grid voltage $\boldsymbol{v}_g = [v_{ga}\ v_{gb}\ v_{gc}]^T$. Alternatively, the grid parameters can be interpreted to represent an active resistive–inductive load, such as an electrical machine.

14.2.2 Nonlinear Converter Model

For each branch r, we define the branch current i_r. The grid currents are linear combinations of the branch currents. The phase a grid current, for example, is given by

$$i_{ga} = i_1 - i_2. \tag{14.1}$$

The grid currents for the phases b and c are defined accordingly. The circulating current $i_{\text{circ},x}$ is defined as the current circulating through the upper and lower branches of the phase leg x and the dc-link. For phase leg a, for example, the circulating current is defined as

$$i_{\text{circ},a} = \frac{i_1 + i_2}{2} - \frac{i_{\text{dc}}}{3}, \tag{14.2}$$

where i_{dc} denotes the dc-link current. The circulating currents in phase legs b and c are defined accordingly.

We also define the (discrete) insertion index $\nu_r \in \{0, \frac{1}{n}, \frac{2}{n}, \ldots, 1\}$, which specifies the proportion of modules inserted into the rth branch [10]. Specifically, $\nu_r = 0$ implies that none of the modules in the rth branch is inserted (all are bypassed), whereas $\nu_r = 1$ means that all n modules in the branch are inserted.

We assume that all modules have the same capacitance C_m and that all voltages across the capacitors are the same; that is, the capacitor voltages are balanced. Following [18], this assumption allows us to describe the series connection of the modules inserted into branch r by the (time-varying) branch capacitance

$$C_r = \frac{1}{\nu_r}\frac{C_m}{n} \tag{14.3}$$

with the voltage

$$v_r = \nu_r v_r^{\Sigma}. \tag{14.4}$$

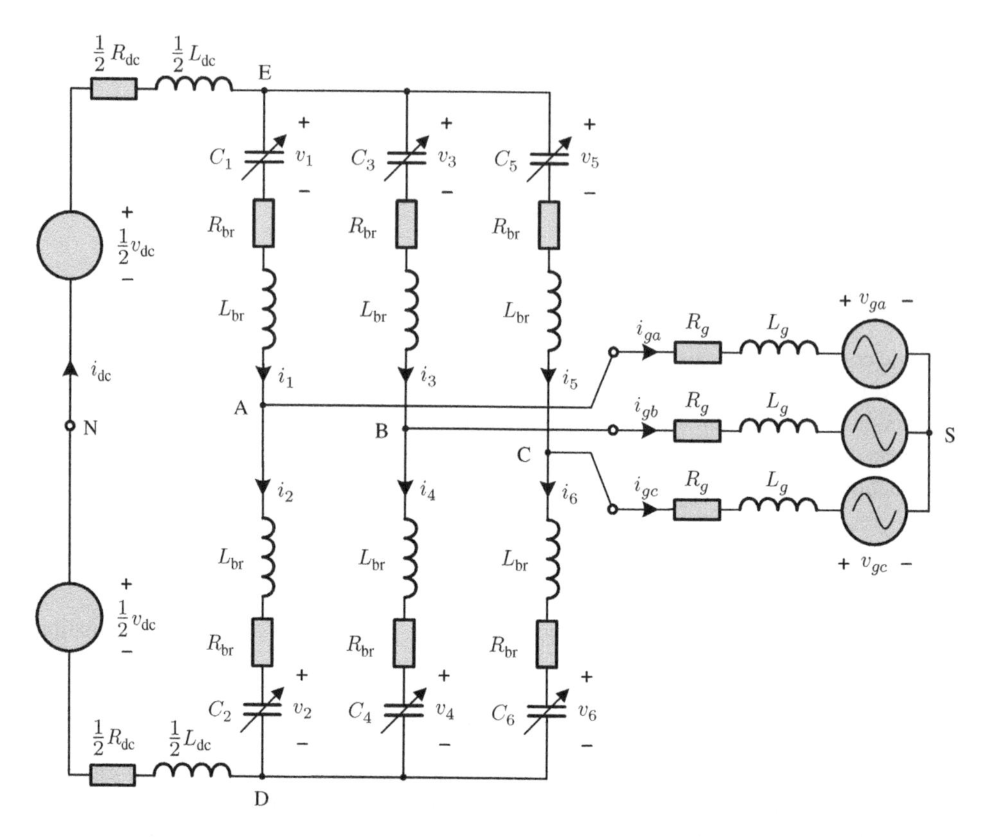

Figure 14.3 Equivalent representation of the three-phase dc–ac MMC topology, in which the series-connected branch modules were replaced by the equivalent controlled capacitors C_r with the voltages v_r. The index $r \in \{1, \ldots, 6\}$ denotes the branch number

With this, the MMC topology shown in Fig. 14.1 can be simplified to the equivalent representation shown in Fig. 14.3, in which the inserted modules in the rth branch are replaced with a controlled capacitor with the capacitance C_r and the voltage v_r.

The last term in (14.4) is the sum over all capacitor voltages in branch r, which is defined as

$$v_r^{\Sigma} = \sum_{j=1}^{n} v_{rj}. \tag{14.5}$$

Note that v_r^{Σ} is independent of the number of modules that are actually inserted into the branch. The evolution of v_r^{Σ} is a function of the branch current i_r and the inserted branch capacitance C_r, that is,

$$\frac{\mathrm{d}v_r^{\Sigma}}{\mathrm{d}t} = \frac{i_r}{C_r} = \frac{n}{C_m}\nu_r i_r, \tag{14.6}$$

where we have used (14.3).

For a sufficiently large number of modules and/or a high switching frequency, the insertion index ν_r can be approximated by the real-valued and bounded variable $\nu_r \in [0, 1]$ (see also [19]). This allows the derivation of a nonlinear, continuous-time dynamic model of the MMC with real-valued variables.

In the sequel of this chapter, we use the definition $\nu_r \in [0, 1]$ for the insertion indices of the six branches. These insertion indices constitute the input vector to the nonlinear system model. As there are five linearly independent currents, we choose as state variables the upper and lower branch currents of phase legs a and b (i.e., i_1, i_2, i_3, and i_4), the dc-link current i_{dc}, the sums of the capacitor voltages v_r^Σ of the six branches, and the grid voltages $v_{g\alpha}$, $v_{g\beta}$ in the orthogonal $\alpha\beta$ coordinate system. The outputs of the model are the grid currents $i_{g\alpha}$, $i_{g\beta}$ in the orthogonal coordinate system, along with the six sums of the capacitor voltages per branch v_r^Σ. To this end, we define the state and output vectors as

$$\boldsymbol{x} = [i_1 \dots i_4 \; i_{\text{dc}} \; v_1^\Sigma \dots v_6^\Sigma \; v_{g\alpha} \; v_{g\beta}]^T \tag{14.7a}$$

$$\boldsymbol{y} = [i_{g\alpha} \; i_{g\beta} \; v_1^\Sigma \dots v_6^\Sigma]^T. \tag{14.7b}$$

The grid currents are defined as

$$\begin{bmatrix} i_{g\alpha} \\ i_{g\beta} \end{bmatrix} = \tilde{\boldsymbol{K}} \begin{bmatrix} i_{ga} \\ i_{gb} \\ i_{gc} \end{bmatrix} = \tilde{\boldsymbol{K}} \begin{bmatrix} i_1 - i_2 \\ i_3 - i_4 \\ i_5 - i_6 \end{bmatrix}, \tag{14.8}$$

where $\tilde{\boldsymbol{K}}$ denotes the reduced Clarke transformation as defined in (2.13).

Inspired by the modeling procedure described in [16], the differential equations of the five independent currents can easily be derived by applying Kirchhoff's voltage law to five meshes in the circuit. We choose the meshes EADNE, EBDNE, ECDNE, DASBD, and DASCD, which are defined by their corresponding nodes as indicated in Fig. 14.3. The five differential equations are provided in Appendix 14.A.

The six differential equations of the sums of the capacitor voltages per branch v_r^Σ are stated in (14.6). The evolution of the grid voltages in the $\alpha\beta$ frame is given by

$$\frac{\mathrm{d}}{\mathrm{d}t} \begin{bmatrix} v_{g\alpha} \\ v_{g\beta} \end{bmatrix} = \omega_g \begin{bmatrix} 0 & -1 \\ 1 & 0 \end{bmatrix} \begin{bmatrix} v_{g\alpha} \\ v_{g\beta} \end{bmatrix}, \tag{14.9}$$

where $\omega_g = 2\pi f_g$ is the angular electrical frequency of the grid.

We conclude that, by representing the number of modules inserted into a branch by the insertion index ν_r, each branch can be described by its time-varying branch capacitance C_r and its time-varying voltage v_r. This greatly simplifies the modeling and subsequent controller design.

14.3 Model Predictive Control

14.3.1 Control Problem

A controller is to be developed for the MMC topology that regulates the grid currents along their sinusoidal references and maintains the capacitor voltages close to their nominal values. At steady-state operation, a low total demand distortion (TDD) of the grid current and a low device switching frequency are required, with the latter being in the range of a few hundred hertz. During transients, a very fast current response is to be achieved. Furthermore, the branch currents, dc-link current, and the capacitor voltages must be kept within given bounds, which are due to the physical limitations of the switching devices and passive components.

14.3.2 Controller Structure

To address the control problem of the MMC, a hierarchical control scheme with three layers is proposed, as depicted in Fig. 14.4. The MPC scheme on the upper layer regulates the grid currents and the sums of the capacitor voltages in the six branches. The MPC scheme is based on the principle of constrained optimal control. Given the state vector $\boldsymbol{x}$ and the output reference vector $\boldsymbol{y}^*$, a quadratic cost function is minimized subject to constraints and the evolution of a linearized state-space model of the MMC. The resulting optimization problem is a QP, which is formulated and solved in real time. The result of this optimization stage is a sequence of manipulated variables over the prediction horizon. In accordance with the receding horizon policy (see Sect. 1.3.2), only the first element of this sequence is applied to the MMC, namely the real-valued insertion indices ν_r for the six branches.

On the middle level, six multilevel CB-PWM units translate the insertion indices into the six integer variables n_r. The latter denote the number of modules to be inserted per branch. On the lower layer, six independently operating balancing controllers balance the capacitor voltages within each branch, by equally distributing the energy per branch. These controllers exploit the redundancy present within the branches and choose the gating commands for the individual MMC modules.

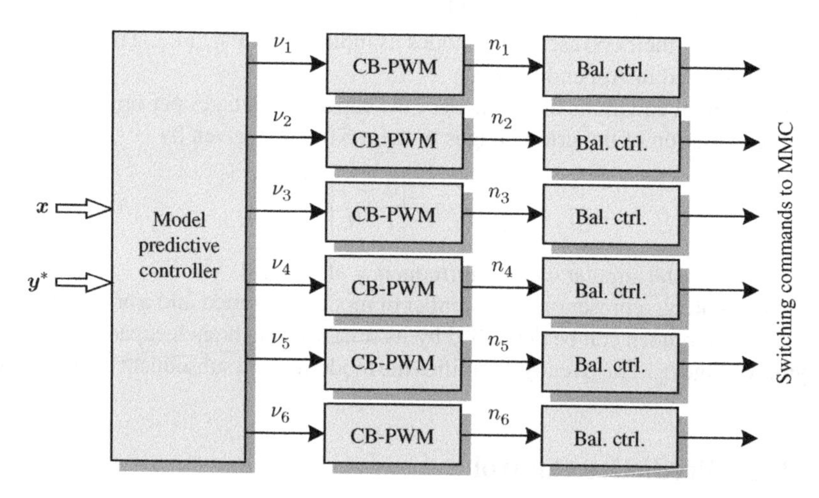

Figure 14.4 Structure of the proposed control scheme. Based on the state vector $\boldsymbol{x}$ and the output reference vector $\boldsymbol{y}^*$, the model predictive controller regulates the grid currents and the sums of the capacitor voltages in the six branches by manipulating the six real-valued insertion indices $\nu_r \in [0, 1]$. Carrier-based PWM translates these indices into the discrete number of modules $n_r \in \{0, 1, \ldots, n\}$ to be inserted into each branch. Six balancing controllers maintain the capacitor voltages within each branch at their nominal levels by deciding which module to turn on or off

14.3.3 Linearized Prediction Model

The differential equations describing the dynamic behavior of the MMC contain nonlinear terms, namely the multiplications $\nu_r v_r^\Sigma$ in the current equations (14.A.1) (see Appendix 14.A) and $\nu_r i_r$ in the capacitor voltage equations (14.6). At time $t = t_0$, a linearization of these nonlinear terms can be done around the current operating point of the system, which is given by $\nu_r(t_0)$, $v_r^\Sigma(t_0)$, and $i_r(t_0)$. Specifically, the first-order Taylor series expansions

$$\begin{aligned}\nu_r(t)v_r^\Sigma(t) &= \nu_r(t_0)v_r^\Sigma(t_0) + \Delta\nu_r(t)v_r^\Sigma(t_0) + \nu_r(t_0)(v_r^\Sigma(t) - v_r^\Sigma(t_0)) \\ &= \nu_r(t_0)v_r^\Sigma(t) + \Delta\nu_r(t)v_r^\Sigma(t_0) \qquad (14.10a)\\ \nu_r(t)i_r(t) &= \nu_r(t_0)i_r(t) + \Delta\nu_r(t)i_r(t_0) \qquad (14.10b)\end{aligned}$$

are performed, where $\Delta\nu_r(t) = \nu_r(t) - \nu_r(t_0)$ is the variation in the insertion index.

The resulting linearized, continuous-time prediction model is of the form

$$\frac{\mathrm{d}\boldsymbol{x}(t)}{\mathrm{d}t} = \boldsymbol{F}(t_0)\boldsymbol{x}(t) + \boldsymbol{G}(t_0)\boldsymbol{u}(t) + \boldsymbol{g}(t_0) \qquad (14.11a)$$

$$\boldsymbol{y}(t) = \boldsymbol{C}\boldsymbol{x}(t) \qquad (14.11b)$$

with the state, input, and output vectors

$$\boldsymbol{x} = [i_1 \dots i_4 \; i_{\mathrm{dc}} \; v_1^\Sigma \dots v_6^\Sigma \; v_{g\alpha} \; v_{g\beta}]^T \qquad (14.12a)$$

$$\boldsymbol{u} = [\Delta\nu_1 \dots \Delta\nu_6]^T \qquad (14.12b)$$

$$\boldsymbol{y} = [i_{g\alpha} \; i_{g\beta} \; v_1^\Sigma \dots v_6^\Sigma]^T . \qquad (14.12c)$$

Note that the state and output vectors remain the same as for the nonlinear model (see (14.7)). The input vector $\boldsymbol{u}$, however, comprises the *variations* in the insertion indices, rather than the insertion indices. As $\nu_r \in [0, 1]$, the input vector is constrained to $\boldsymbol{u} \in [-\nu_1(t_0), 1 - \nu_1(t_0)] \times \dots \times [-\nu_6(t_0), 1 - \nu_6(t_0)]$. Note that $\boldsymbol{x} \in \mathbb{R}^{13}$, $\boldsymbol{u} \in \mathbb{R}^6$ and $\boldsymbol{y} \in \mathbb{R}^8$. The time-varying matrices $\boldsymbol{F}(t_0)$, $\boldsymbol{G}(t_0)$, and $\boldsymbol{C}$, and the vector $\boldsymbol{g}(t_0)$ are provided in Appendix 14.B.

Using Euler's exact discretization method (see, e.g., (5.5)) with the sampling interval T_s, the discrete-time representation of the linearized model

$$\boldsymbol{x}(k+1) = \boldsymbol{A}(t_0)\boldsymbol{x}(k) + \boldsymbol{B}(t_0)\boldsymbol{u}(k) + \boldsymbol{b}(t_0) \qquad (14.13a)$$

$$\boldsymbol{y}(k) = \boldsymbol{C}\boldsymbol{x}(k) \qquad (14.13b)$$

can be derived.

14.3.4 Cost Function

The cost function maps the control objectives into a scalar cost. The proposed cost function consists of four terms. The first three terms are defined in this section, while the fourth term is introduced in the following section.

The first cost function term penalizes the predicted evolution of the tracking error, which is the difference between the time-varying reference of the output vector $\boldsymbol{y}^*$ and the predicted output vector $\boldsymbol{y}$. The output reference is defined as

$$\boldsymbol{y}^*(\ell) = [i^*_{g\alpha}(\ell)\ i^*_{g\beta}(\ell)\ V_{\rm dc}\ V_{\rm dc}\ V_{\rm dc}\ V_{\rm dc}\ V_{\rm dc}\ V_{\rm dc}]^T, \tag{14.14}$$

where $i^*_{g\alpha}$ and $i^*_{g\beta}$ are the desired grid currents expressed in the $\alpha\beta$ coordinate system. The nominal dc-link voltage $V_{\rm dc}$ is the reference for the sums of the capacitor voltages per branch.

The predicted tracking errors $\boldsymbol{y}^* - \boldsymbol{y}$ are squared and weighted with the penalty matrix $\boldsymbol{Q}_y$. Summing up these quadratic terms from time step k until the end of the prediction horizon N_p yields the cost function term

$$J_1(\boldsymbol{x}(k), \boldsymbol{U}) = \sum_{\ell=k}^{k+N_p-1} (\boldsymbol{y}^*(\ell) - \boldsymbol{y}(\ell))^T \boldsymbol{Q}_y\,(\boldsymbol{y}^*(\ell) - \boldsymbol{y}(\ell)). \tag{14.15}$$

We require $\boldsymbol{Q}_y$ to be positive semidefinite and symmetric. The future output vectors $\boldsymbol{y}(\ell)$ are a function of the state vector $\boldsymbol{x}(k)$ and the sequence of manipulated variables

$$\boldsymbol{U} = [\boldsymbol{u}^T(k)\ \boldsymbol{u}^T(k+1)\ \ldots\ \boldsymbol{u}^T(k+N_p-1)]^T \tag{14.16}$$

over the prediction horizon N_p. This dependency is explained in Sect. 5.6 (see (5.38)).

The second term of the cost function

$$J_2(\boldsymbol{x}(k), \boldsymbol{U}) = \sum_{\ell=k}^{k+N_p-1} (\boldsymbol{i}(\ell))^T \boldsymbol{Q}_i(k)\boldsymbol{i}(\ell) \tag{14.17}$$

adds an operating-point-dependent penalty on the branch current vector $\boldsymbol{i} = [i_1\ i_2\ \ldots\ i_6]^T$. The penalty matrix

$$\boldsymbol{Q}_i(k) = (1 - \|\boldsymbol{i}^*_{g,\alpha\beta}(k)\|_2)\,\boldsymbol{Q}'_i \tag{14.18}$$

inversely depends on the amplitude of the grid current reference $\boldsymbol{i}^*_{g,\alpha\beta} = [i^*_{g\alpha}\ i^*_{g\beta}]^T$, where $\boldsymbol{Q}'_i$ is a constant penalty matrix. Note that we adopt the per unit (pu) system as defined in Sect. 14.4.1. Therefore, the rated grid current is of the magnitude 1.

A reference-dependent penalty is beneficial for the following reason. As the grid currents are equal to the differences between the respective upper and lower branch currents, the branch currents constitute an uncontrolled quantity when controlling only the grid currents. When operating at low-load or no-load conditions, the desired small grid currents can be achieved for a wide range of branch currents, including some large ones. By imposing the penalty $\boldsymbol{Q}'_i$ on the branch currents, these currents are minimized, and the switching and conduction losses are reduced. When the converter operates at full load, however, the converter must provide high grid currents, which require high branch currents. In this case, trying to minimize the branch currents would conflict with the cost function term J_1, resulting in a grid current tracking error.

The third term of the cost function penalizes changes in the manipulated variable within the prediction horizon N_p:

$$J_3(\boldsymbol{u}(k-1), \boldsymbol{U}) = \sum_{\ell=k}^{k+N_p-1} (\Delta\boldsymbol{u}(\ell))^T \boldsymbol{R}\,\Delta\boldsymbol{u}(\ell). \tag{14.19}$$

Specifically, changes $\Delta\boldsymbol{u}(\ell) = \boldsymbol{u}(\ell) - \boldsymbol{u}(\ell-1)$ in the insertion index are penalized with the matrix $\boldsymbol{R}$. Penalizing changes in the manipulated variable rather than the manipulated variable itself is preferred, because time-varying references need to be tracked that require nonzero manipulated variables during steady-state operation. Because we penalize changes in the manipulated variable, J_3 is a function of the previously applied manipulated variable $\boldsymbol{u}(k-1)$. Note that the trade-off between tracking accuracy and control effort is determined by the ratio between $\boldsymbol{Q}_y$ and $\boldsymbol{R}$.

14.3.5 Hard and Soft Constraints

A considerable advantage of the proposed control framework is its ability to address hard or soft constraints during the controller synthesis. Hard constraints relate to strict physical limitations of the converter, such as limits on the modulation or bounds on the safe operating area. The latter directly relate to trip levels, such as overvoltage or overcurrent trip levels. Hard constraints are added as inequality constraints to the optimization problem and limit the admissible state-input space. We impose the hard constraints

$$0 \le \nu_r(\ell) \le 1 \tag{14.20}$$

on the six insertion indices ν_r.

As hard constraints on state and output variables might lead to feasibility issues, we impose soft rather than hard constraints on these variables to restrict the operation of the MMC to the limits of the safe operating area. A soft constraint is an (in)equality constraint that can be relaxed using the so-called *slack variable*. The degree of the constraint violation corresponds to the value of the nonnegative slack variable. As such, the slack variable maps the constraint violation into a nonnegative real number. To minimize the constraint violation, the slack variable is penalized heavily in the cost function.

As shown in Fig. 14.5(a), we impose soft constraints on the branch currents i_r. Specifically, we introduce upper and lower constraints at $i_{\max}$ and $-i_{\max}$ using the slack variable ξ_r and the three inequality constraints

$$\xi_r(\ell) \ge i_r(\ell) - i_{\max} \tag{14.21a}$$

$$\xi_r(\ell) \ge -\left(i_r(\ell) + i_{\max}\right) \tag{14.21b}$$

$$\xi_r(\ell) \ge 0. \tag{14.21c}$$

The three constraints are indicated by the three thick lines in Fig. 14.5(a), while the feasible space of the slack variable is indicated by the shaded area. Although the slope of the

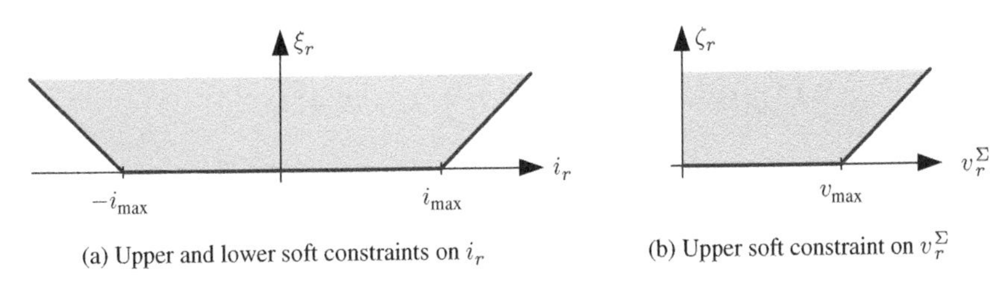

(a) Upper and lower soft constraints on i_r

(b) Upper soft constraint on v_r^{Σ}

Figure 14.5 Soft constraints on the branch current i_r and on the sum v_r^{Σ} of the capacitor voltages of the rth branch, using the slack variable ξ_r and ζ_r, respectively

soft constraints is only $45°$, we will weigh the slack variables with a large penalty, thus achieving—in effect—very steep slopes. As a result of this, the soft constraints keep the branch currents almost as strictly within their bounds as hard constraints do, while avoiding potential numerical and feasibility issues.

Similar to (14.21), upper and lower soft constraints are imposed on the dc-link current i_{dc}, using the slack variable ξ_{dc}

$$\xi_{dc}(\ell) \geq i_{dc}(\ell) - i_{max} \tag{14.22a}$$

$$\xi_{dc}(\ell) \geq -\left(i_{dc}(\ell) + i_{max}\right) \tag{14.22b}$$

$$\xi_{dc}(\ell) \geq 0. \tag{14.22c}$$

Moreover, an upper soft constraint at v_{max} is imposed on the sum of the capacitor voltages per branch v_r^{Σ}, by introducing the slack variable ζ_r and the two inequality constraints

$$\zeta_r(\ell) \geq v_r^{\Sigma}(\ell) - v_{max} \tag{14.23a}$$

$$\zeta_r(\ell) \geq 0. \tag{14.23b}$$

This soft constraint is depicted in Fig. 14.5(b).

By aggregating the slack variables in the two vectors

$$\boldsymbol{\xi} = [\xi_1 \dots \xi_6 \ \xi_{dc}]^T \in \mathbb{R}_+^7 \tag{14.24a}$$

$$\boldsymbol{\zeta} = [\zeta_1 \dots \zeta_6]^T \in \mathbb{R}_+^6, \tag{14.24b}$$

the fourth term of the cost function can be written as

$$J_4(\boldsymbol{x}(k), \boldsymbol{U}, \Xi) = \sum_{\ell=k}^{k+N_p-1} \lambda_\xi \|\boldsymbol{\xi}(\ell)\|_1 + \lambda_\zeta \|\boldsymbol{\zeta}(\ell)\|_1, \tag{14.25}$$

in which the slack variables are penalized using the 1-norm and the scalar penalties λ_ξ and λ_ζ. Large positive values are chosen for these penalties. The sequence of slack variables over the prediction horizon is defined as

$$\Xi = [\boldsymbol{\xi}^T(k) \ \boldsymbol{\zeta}^T(k) \ \boldsymbol{\xi}^T(k+1) \ \boldsymbol{\zeta}^T(k+1) \ \dots \ \boldsymbol{\xi}^T(k+N_p-1) \ \boldsymbol{\zeta}^T(k+N_p-1)]^T \tag{14.26}$$

similar to $\boldsymbol{U}$.

14.3.6 Optimization Problem

Minimizing the sum of the four cost function terms leads to the optimization problem

$$\boldsymbol{U}_{opt}(k) = \arg \underset{\boldsymbol{U}(k), \Xi(k)}{\text{minimize}} \ J_1 + J_2 + J_3 + J_4 \tag{14.27a}$$

$$\text{subject to (14.13), (14.20)–(14.23)} \tag{14.27b}$$

$$\forall \ell = k, \dots, k + N_p - 1. \tag{14.27c}$$

As the cost function is quadratic and minimized subject to the evolution of a linear state-space model with linear inequality constraints, the resulting optimization problem (14.27a) is a QP.

To write the QP in the standard form (3.94), the procedure summarized in Sect. 5.2.1 is followed. Specifically, the cost function is expressed as a function of the sequence of manipulated variables $\boldsymbol{U}(k)$ and the sequence of slack variables $\Xi(k)$. These two vectors constitute the optimization vector. As there are 6 manipulated variables and 13 slack variables at each time step, the optimization vector is of the dimension $19N_p$. The cost function also depends on the initial state vector $\boldsymbol{x}(k)$ and the previously chosen manipulated variable $\boldsymbol{u}(k-1)$.

The QP can be solved efficiently, for example, by using an active set method or an interior point solver. For more details on QPs and optimization techniques to solve them, the reader is referred to Sect. 3.8 and the references therein. The result of the optimization stage is the sequence of optimal manipulated variables $\boldsymbol{U}$ at time step k.

The first element of this sequence is implemented at time step k and sent to the PWM. At the next time step $k+1$, new measurements are obtained and the optimization problem is solved again over a shifted prediction horizon. This so-called receding horizon policy provides feedback and ensures that the controller is robust—both to parameter uncertainties and to the linearization errors because of the Taylor series approximation.

14.3.7 Multilevel Carrier-Based Pulse Width Modulation

The middle level of the hierarchical control scheme shown in Fig. 14.4 performs the CB-PWM. Specifically, regularly sampled multilevel CB-PWM with phase disposition is used with the carrier frequency f_c, as summarized in Sect. 3.3. The different triangular carrier waveforms are not interleaved within each branch, but a phase shift of 180° is applied between the carrier waveforms of the upper and lower branches.

The MPC controller is executed at the (upper and lower) peaks of the triangular carrier waveform, that is, at the time instants $t = kT_s$, with $T_s = \frac{1}{2f_c}$. New insertion indices are provided to the modulator at these time instants. The insertion index ν_r can be interpreted as the modulation index of a multilevel PWM scheme. A scaling by the sum of all capacitor voltages per branch, v_r^Σ, is not required, because variations in v_r^Σ are captured in the prediction model thanks to the choice of v_r^Σ as a state variable (see (14.12a)). Based on ν_r and v_r^Σ, the branch voltages v_r in the current equations (14.A.1) are manipulated.

Each insertion index is compared with n triangular carrier signals, which are of the magnitude $1/n$. In doing so, the PWM translates the real-valued insertion index $\nu_r \in [0, 1]$ into the integer $n_r \in \{0, 1, \ldots, n\}$, which specifies the number of modules to be inserted into the rth branch. Using six independent PWM stages, each branch is modulated independently of the others, which results in six decision variables for the control scheme.

Note that many state-of-the-art controllers, such as the ones reported in [12] and [20], add a dependency between the control signals of the upper and lower branches of each phase leg, by imposing that the number of modules inserted *per phase leg* is always equal to n. For the phase leg a, for example, this constraint can be written as $n_1 + n_2 = n$. As a result, the number of manipulated variables is reduced to 3.

In the case of MPC, however, imposing such a per-phase constraint is deemed to be an unnecessary and conservative choice. Using six instead of three manipulated variables enables MPC to control the energies in the branch capacitors independently of the grid currents. Soft constraints on the branch currents and on the sums of the capacitor voltages ensure that the MMC is operated within its safe operating limits. The benefit of adopting six independently operating PWM stages becomes evident particularly during transients, as will be shown and discussed in Sect. 14.4.

14.3.8 Balancing Control

The lower control layer exploits the per-branch redundancy of the modules to balance the capacitor voltages within the branches while minimizing the switching events. Each branch uses its own balancing controller, which receives as input from the modulation stage the number of modules n_r to be inserted into the rth branch. The balancing algorithm is executed at the switching events of the modulator.

The adopted balancing method is based on the sorting algorithm presented in [1]. For each branch, two ordered sets are maintained that contain the branch modules that are currently on (i.e., inserted into the branch) and off (i.e., bypassed), respectively. We denote these two sets by $\mathcal{L}_{\text{on}}$ and $\mathcal{L}_{\text{off}}$. The intersection of the sets is empty. Let n_r^{on} and n_r^{off} denote the number of modules contained in $\mathcal{L}_{\text{on}}$ and $\mathcal{L}_{\text{off}}$, respectively. It holds that $n_r^{\text{on}} + n_r^{\text{off}} = n$ for each branch r. The modules contained in each set are sorted in an ascending order of their capacitor voltages.

The switching of modules is performed following the requested n_r and the polarity of the branch current. If $n_r > n_r^{\text{on}}$, then $n_r - n_r^{\text{on}}$ modules in $\mathcal{L}_{\text{off}}$ are switched on. If the branch current is positive (negative), then the $n_r - n_r^{\text{on}}$ modules with the smallest (highest) capacitor voltages are chosen from $\mathcal{L}_{\text{off}}$ and moved to $\mathcal{L}_{\text{on}}$.

Conversely, if $n_r < n_r^{\text{on}}$, then $n_r^{\text{on}} - n_r$ modules in $\mathcal{L}_{\text{on}}$ are switched off. If the branch current is positive (negative), then the $n_r^{\text{on}} - n_r$ modules with the highest (smallest) capacitor voltages are chosen from $\mathcal{L}_{\text{on}}$ and moved to $\mathcal{L}_{\text{off}}$.

14.4 Performance Evaluation

The performance of the proposed control scheme is evaluated at steady-state operating conditions and during transients. At steady state, the grid current TDD and the device switching frequency of the MMC modules serve as performance metrics. During transient operation, the dynamic response of the converter is used as a metric. To this end, quantities such as the overshoot, rise time, and settling time of the response to step changes are examined.

14.4.1 System and Control Parameters

Consider a three-phase MV MMC with $n = 8$ modules per branch that is connected to the grid. The MMC operates in dc–ac inverter mode, and has a rated apparent power of 4.28 MVA. The input of the MMC is connected to a dc supply with the constant and nominal voltage $V_{\text{dc}} = 6.8\,\text{kV}$. The rated values of the MMC are provided in Table 14.1. The pu system is established using the base voltage $V_B = \sqrt{2/3}V_R = 3.10\,\text{kV}$, the base current $I_B = \sqrt{2}I_R = 919\,\text{A}$, and the angular base frequency $\omega_B = \omega_{gR} = 2\pi 50\,\text{rad/s}$. The MMC parameters are summarized in Table 14.2. The MMC, grid, MPC scheme, CB-PWM, and balancing controller were implemented in MATLAB/SIMULINK and PLECS. The full nonlinear model of the MMC shown in Fig. 14.1 was simulated in PLECS. This model contains 48 modules and captures their switching characteristic.

Table 14.1 Rated values of the MMC

Parameter	Symbol	SI value
Voltage	V_R	3800 V
Current	I_R	650 A
Apparent power	S_R	4.278 MVA
Angular grid frequency	ω_{gR}	$2\pi 50$ rad/s

Table 14.2 Parameters in the SI (left) and per unit system (right) of the grid-connected MMC system

Parameter	SI symbol	SI value	pu value
dc-link voltage	V_{dc}	6800 V	2.192
dc-link resistance	R_{dc}	0.1 mΩ	$2.963 \cdot 10^{-5}$
dc-link inductance	L_{dc}	50 μH	$4.654 \cdot 10^{-3}$
Branch resistance	R_{br}	0.25 mΩ	$7.407 \cdot 10^{-5}$
Branch inductance	L_{br}	1 mH	0.093
Module capacitance	C_m	8.2 mF	8.695
Number of modules per branch	n	8	
Grid voltage	V_g	3800 V	1
Grid resistance	R_g	67.51 mΩ	0.02
Grid inductance	L_g	1.61 mH	0.15

The carrier frequency $f_c = 2.5$ kHz is used, unless otherwise specified. The MPC scheme is executed at the (upper and lower) peaks of the triangular carrier, implying a sampling interval of $T_s = 200$ μs for MPC. The state vector $\boldsymbol{x}$ is assumed to be available to the controller along with the time-varying reference signal $\boldsymbol{y}^*$. Measurement and computational delays are assumed to be fully compensated for. The computed control actions are kept constant between the time steps k and $k+1$ and are sent to the PWM stage.

When choosing the weights in the cost function, the *ratio* between these weights rather than their absolute value determines the control actions. For the penalty matrix $\boldsymbol{R}$ on the manipulated variable, without loss of generality, we choose the identity matrix. To achieve accurate tracking of the grid currents, the latter are prioritized over the balancing of the capacitor voltages by imposing large penalties on the current error tracking terms in $\boldsymbol{Q}_y$. Moreover, a relatively small weight is associated with the penalty on the branch currents $\boldsymbol{Q}'_i$ to avoid a deterioration in the tracking performance of the grid currents. This choice leads to the penalty matrices

$$\boldsymbol{Q}_y = \begin{bmatrix} 10\boldsymbol{I}_2 & \boldsymbol{0}_{2\times 6} \\ \boldsymbol{0}_{6\times 2} & \boldsymbol{I}_6 \end{bmatrix}, \quad \boldsymbol{Q}'_i = 0.1\boldsymbol{I}_6 \text{ and } \boldsymbol{R} = \boldsymbol{I}_6. \tag{14.28}$$

To ensure that the soft constraints are met, the very high penalties $\lambda_\xi = \lambda_\zeta = 10^5$ are used. The soft constraints are activated at $i_{max} = 1.1$ pu and $v_{max} = 1.1V_{dc}$.

The prediction horizon is selected as $N_p = 6$. As discussed in Sect. 14.5.2, shorter horizons might impact on the system stability, while very long horizons achieve diminishing returns in terms of performance gains. Long horizons also increase the computational burden when solving the QP. For the prediction horizon $N_p = 6$, the optimizer of the QP is of the dimension 114, facilitating a relatively fast solution process. The Multi-Parametric Toolbox 3.0 [21] and the Gurobi Optimizer [22] were used to formulate and solve the QP problem.

14.4.2 *Steady-State Operation*

The performance of the MPC scheme at steady-state operating conditions is evaluated in this section. All quantities are provided in the pu system. At full active power and zero reactive power, Fig. 14.6(a) shows the three-phase grid currents over two fundamental periods. The grid currents are effectively sinusoidal waveforms. The discrete Fourier transform (DFT) of the three-phase grid currents is computed over a time window of 100 ms. The resulting grid current spectrum is shown in Fig. 14.6(b). Note that the magnitude of the harmonics is given in percent of 1 pu, that is, their amplitudes are below 0.1% of the amplitude of the rated current and thus negligible. The sidebands around the carrier frequency $f_c = 2.5$ kHz can be identified. Additional harmonics of low magnitude are present, which are a result of the fluctuations in the capacitor voltages. The TDD of the grid current is 0.40%, and the average device switching frequency of the IGBTs is 351 Hz. To compute the latter, the number of *on* transitions is counted per switching device, and it is divided by the length of the simulated time interval. This yields the device switching frequency of each IGBT. By averaging over these switching frequencies, the average device switching frequency is obtained.

Figures 14.6(c) and 14.6(d) show the upper and lower branch currents of phase leg a. The upper and lower soft constraints at ± 1.1 pu are indicated by straight lines. At steady-state operation, these constraints are typically inactive. The branch currents of the phase legs b and c are the same as that of phase leg a, albeit phase-shifted by 120° and 240°, respectively. To facilitate a direct comparison, the time-domain current waveforms shown in Fig. 14.6 are all scaled to the interval between -1.25 and 1.25 pu.

The circulating current of phase leg a is shown in Fig. 14.6(e), with the circulating currents of phase legs b and c matching that shown in the figure, except for a phase shift. The circulating currents are not directly controlled by the MPC scheme—they result from the branch currents, which are controlled by MPC to synthesize the demanded grid currents and to achieve the required energy balancing between the capacitors of the upper and the lower branches.

The dc-link current is shown in Fig. 14.6(f) with the straight lines indicating the upper and lower bounds at ± 1.1 pu. Unlike [18], for example, it is notable that the dc-link current is not constant but exhibits a ripple with a significant sixth harmonic component. This sixth harmonic tends to ease the provision of a constant grid power, thus slightly reducing the grid current TDD, while maintaining the energy stored in the module capacitors at a level as constant as possible. If the dc-link ripple current proved to be undesirable, the term

$$J_5(\boldsymbol{x}(k), \boldsymbol{x}(k-1), \boldsymbol{U}) = \sum_{\ell=k}^{k+N_p-1} (\Delta i_{\mathrm{dc}}(\ell))^T \boldsymbol{Q}_{\mathrm{dc}} \, \Delta i_{\mathrm{dc}}(\ell) \tag{14.29}$$

could be added to the cost function. Penalizing the changes $\Delta i_{\mathrm{dc}}(\ell) = i_{\mathrm{dc}}(\ell) - i_{\mathrm{dc}}(\ell - 1)$ with the positive semidefinite and symmetric penalty matrix $\boldsymbol{Q}_{\mathrm{dc}}$ reduces the ripple on the dc-link current.

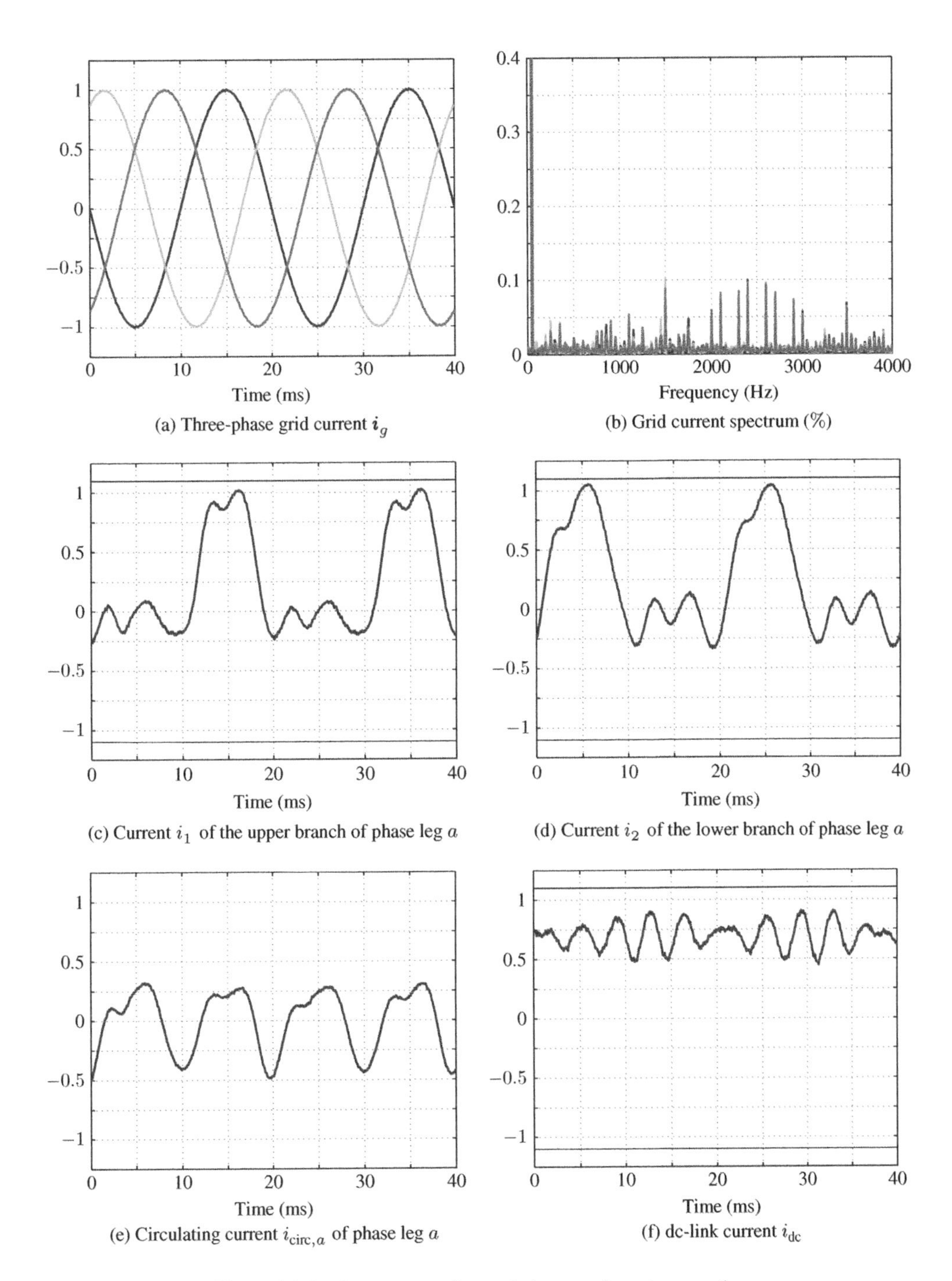

Figure 14.6 Current waveforms during steady-state operation

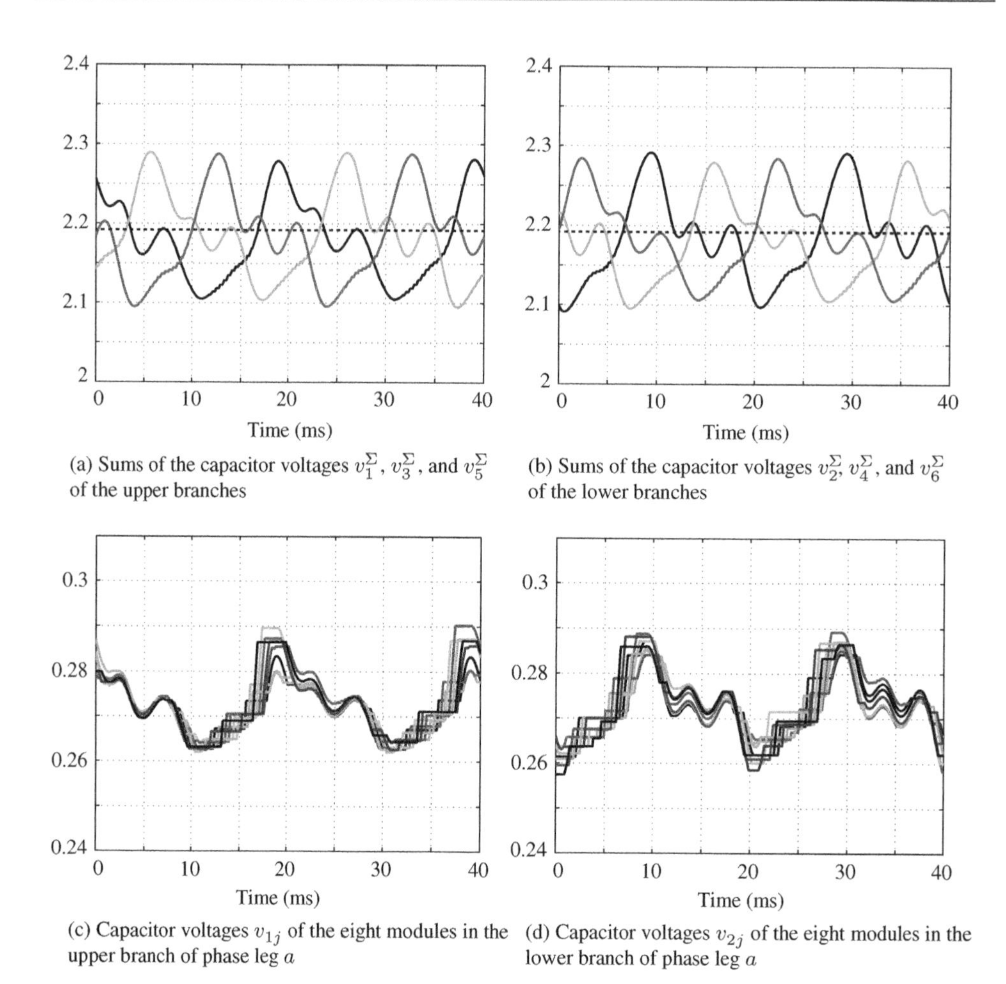

(a) Sums of the capacitor voltages v_1^Σ, v_3^Σ, and v_5^Σ of the upper branches

(b) Sums of the capacitor voltages v_2^Σ, v_4^Σ, and v_6^Σ of the lower branches

(c) Capacitor voltages v_{1j} of the eight modules in the upper branch of phase leg a

(d) Capacitor voltages v_{2j} of the eight modules in the lower branch of phase leg a

Figure 14.7 Capacitor voltage waveforms during steady-state operation

Figures 14.7(a) and 14.7(b) depict the sums of the capacitor voltages of the three upper and the three lower branches, respectively. In each branch, the MPC scheme maintains the sum of the capacitor voltages v_r^Σ around the dash-dotted reference V_{dc}. The ripple amounts to less than $\pm 5\%$. The upper soft constraint at $1.1 V_{\text{dc}} = 2.412$ pu is inactive during steady-state operation.

The effective operation of the lower layer balancing algorithm can be observed in Fig. 14.7(c), which shows the capacitor voltages v_{1j} of the eight modules of the upper branch of phase leg a. The capacitor voltages of these modules closely match each other. The small differences between the capacitor voltages are due to the relatively low switching frequency and the fact that the balancing controller is activated only when a new switching command is issued by the PWM stage. These differences are pronounced around 15 and 35 ms, when the corresponding branch current is high (see Fig. 14.6(c)). This also leads to steep capacitor voltage slopes when inserting modules into the branch. Nevertheless, the capacitor voltage ripple is about $\pm 5\%$ and thus relatively small. The same observations hold for the capacitor voltages of the other five branches, with the capacitor voltages of the lower branch of phase leg a shown in Fig. 14.7(d).

The insertion indices of the three upper branches are shown in Fig. 14.8(a), while the ones of the lower branches are displayed in Fig. 14.8(b). The upper and lower hard constraints at 1 and 0, respectively, are shown as straight lines. The insertion indices are translated by the PWM stage into the numbers of modules to be inserted into the branches. These integers are shown in Figs. 14.8(c) and 14.8(d) for the phase legs of the upper and lower branches, respectively. To better visualize the switching signals, all waveforms in Fig. 14.8 are shown over the first 20 ms of the 40 ms interval used in the previous two figures.

As discussed in Sect. 14.3.7, the number of modules inserted per phase leg is not restricted to the number of branch modules n. Lifting this restriction provides the controller with an additional degree of freedom, which can be used, for example, to control the circulating current. When considering phase leg a and comparing the sum $n_1 + n_2$ shown in Fig. 14.8(e) with that of the corresponding circulating current in Fig. 14.6(e), it can be seen that the MPC scheme uses this degree of freedom to control the circulating current. Nevertheless, as shown in Figs. 14.8(e) and 14.8(f) for the phase legs a and b, the number of modules inserted per phase leg is approximately n when operating at steady state.

14.4.3 Operation during Transients

To investigate the dynamic behavior of the closed-loop system, grid current reference steps of magnitude 1 pu are applied. At 40 ms, the grid current reference is changed from 1 to 0 pu and back to 1 pu at $t = 100$ ms. These transients are equivalent to steps in the real power from 1 to 0 pu and back to 1 pu. The reactive power reference is kept at zero.

The resulting dynamic response of the three-phase grid currents is shown in Fig. 14.9(a). The MPC scheme achieves very fast current responses without overshoots. For the negative reference step, when sufficient voltage margin is available, the settling time of the current transient is 0.75 ms. To step up the current from zero to the rated current takes 2.1 ms, which is still an impressive result.

The upper branch current of the phase leg a is shown in Fig. 14.9(b). The upper soft constraint at 1.1 pu is activated during the positive current step shortly after $t = 100$ ms. During no-load operation, the branch current is very small, thanks to the cost function term J_2. As a result, the circulating current shown in Fig. 14.9(c) is effectively zero during no-load operation.

The dc-link current shown in Fig. 14.9(d) exhibits significant overshoots. During the negative current step, the branch inductors in the lower branch of phase leg b and in the upper branch of phase leg c must be demagnetized, necessitating a negative dc-link current during the transient. Similarly, during the positive current step, the branch inductors in the upper branch of phase leg a and in the two branches of phase leg b need to be magnetized, requiring a significant current contribution from the dc-link. As the latter is limited to 1.1 pu, the controller compensates for the missing current contribution by drawing some reactive current from the grid.

This can be seen in Fig. 14.10, which shows in detail all relevant currents during the positive current step around $t = 100$ ms. Note that when considering a dq reference frame rotating in synchronism with the grid voltage and when aligning the d-axis with the grid voltage phasor, the real and reactive power components shown in Fig. 14.10(b) directly correspond to the d- and q-components of the grid current.

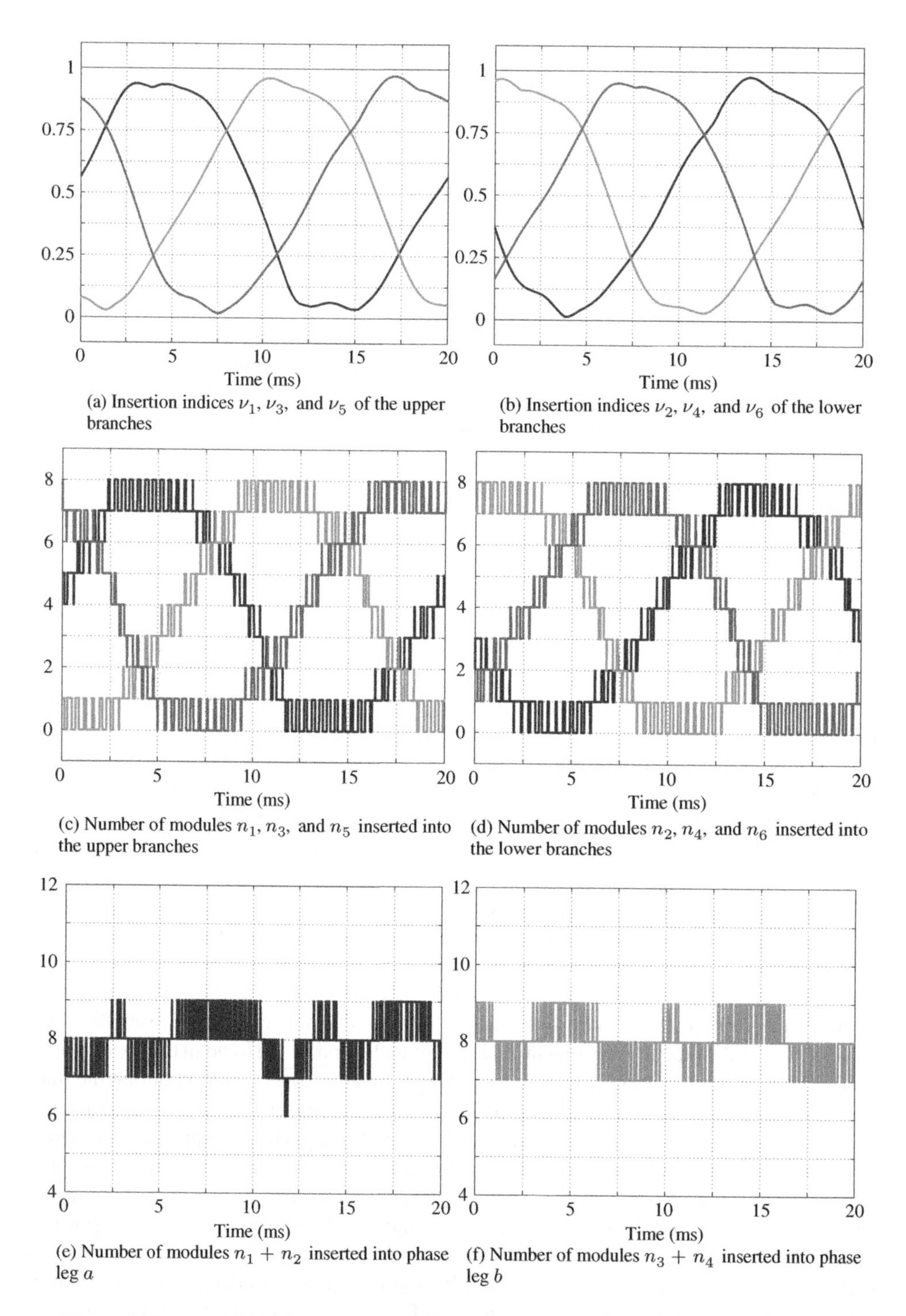

(a) Insertion indices ν_1, ν_3, and ν_5 of the upper branches

(b) Insertion indices ν_2, ν_4, and ν_6 of the lower branches

(c) Number of modules n_1, n_3, and n_5 inserted into the upper branches

(d) Number of modules n_2, n_4, and n_6 inserted into the lower branches

(e) Number of modules $n_1 + n_2$ inserted into phase leg a

(f) Number of modules $n_3 + n_4$ inserted into phase leg b

Figure 14.8 Insertion indices and number of modules inserted during steady-state operation

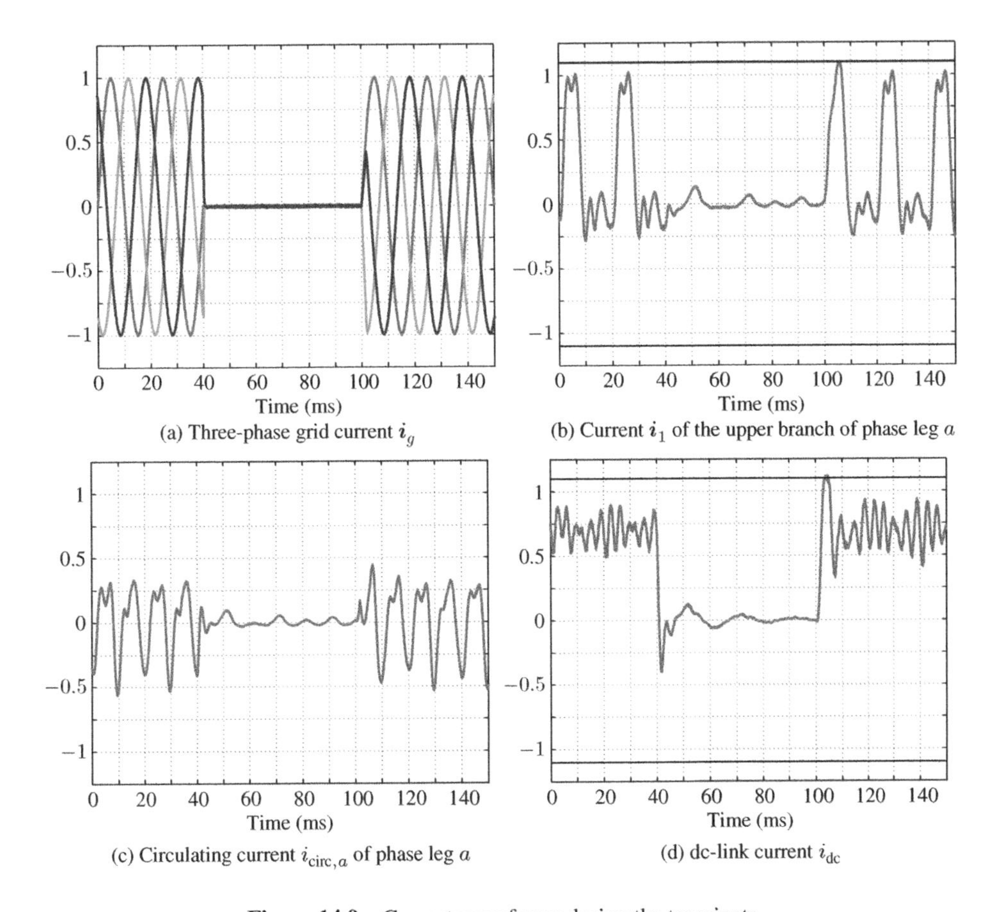

(a) Three-phase grid current i_g

(b) Current i_1 of the upper branch of phase leg a

(c) Circulating current $i_{\text{circ},a}$ of phase leg a

(d) dc-link current i_{dc}

Figure 14.9 Current waveforms during the transients

The ability of the MPC scheme to respect operational constraints on the branch currents and the dc-link current can be appreciated in Fig. 14.10(c), (d), and (f). In particular, the dc-link current is in effect clamped by the controller to its upper limit at 1.1 pu for about 2 ms during the transient.

Figure 14.11(a) and (b) shows the sums of the capacitor voltages of the upper and the lower branches, respectively. In general, these voltages are kept close to their dash-dotted reference V_{dc}. The negative current step occurs, however, when the ripple on the capacitor voltages in the phase leg a is high. This leads to a visible offset in v_1^{Σ} and v_2^{Σ}, which is corrected within about two fundamental periods. This shift of energy from the capacitors in the upper branch to those in the lower branch of phase leg a can also be observed in Fig. 14.11(c) and (d). During the positive current step, a short transient in v_4^{Σ} is visible in Fig. 14.11(a). Its peak remains below the soft constraint at $1.1V_{\text{dc}} = 2.412$ pu, which remains inactive during this transient.

The insertion indices, which are manipulated by the MPC scheme, are shown in Fig. 14.11(e) for the upper branches and in Fig. 14.11(f) for the lower branches. The hard constraints at 0 and 1 are activated during the current steps, particularly during the second one.

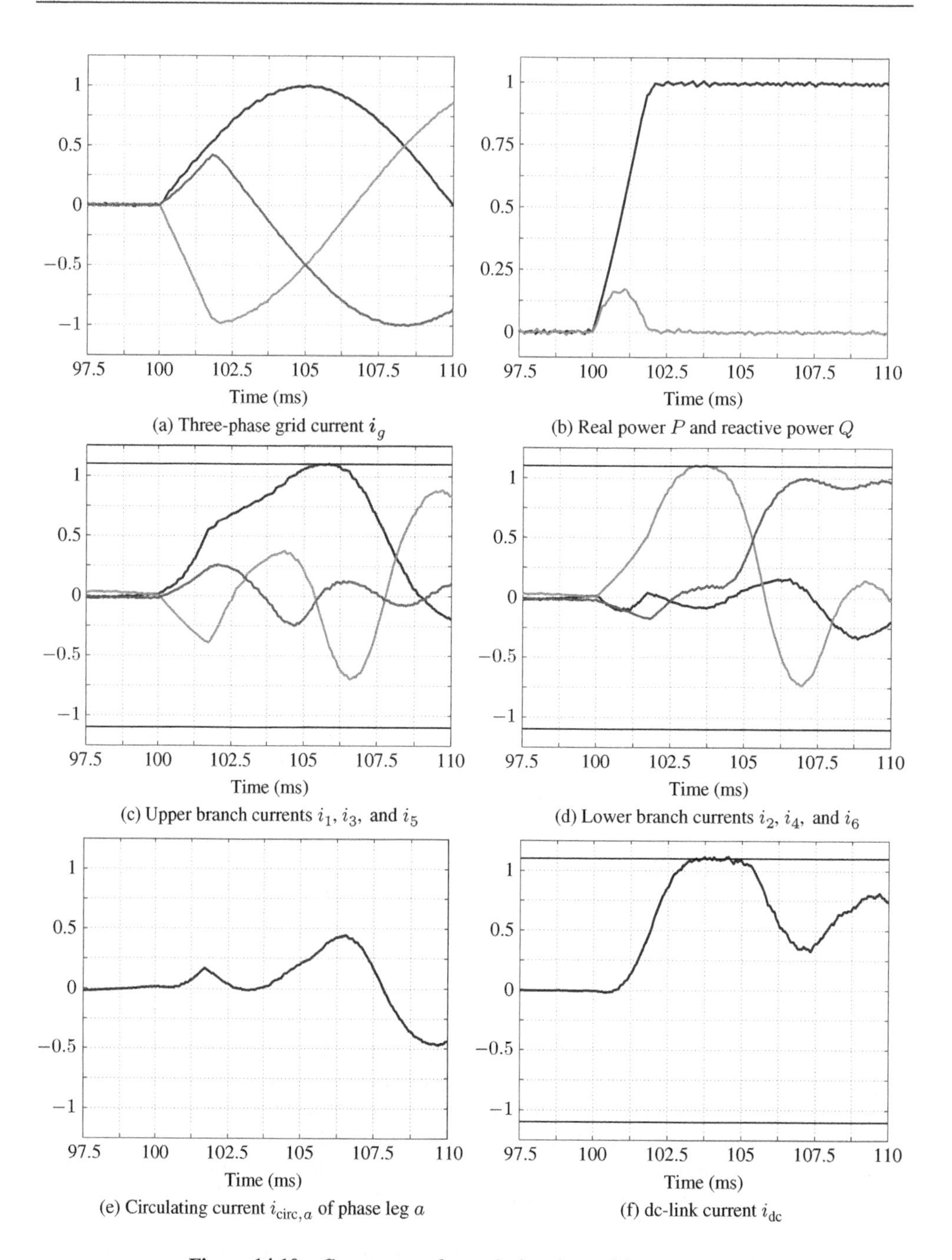

(a) Three-phase grid current $\boldsymbol{i}_g$

(b) Real power P and reactive power Q

(c) Upper branch currents i_1, i_3, and i_5

(d) Lower branch currents i_2, i_4, and i_6

(e) Circulating current $i_{\mathrm{circ},a}$ of phase leg a

(f) dc-link current i_{dc}

Figure 14.10 Current waveforms during the positive power step

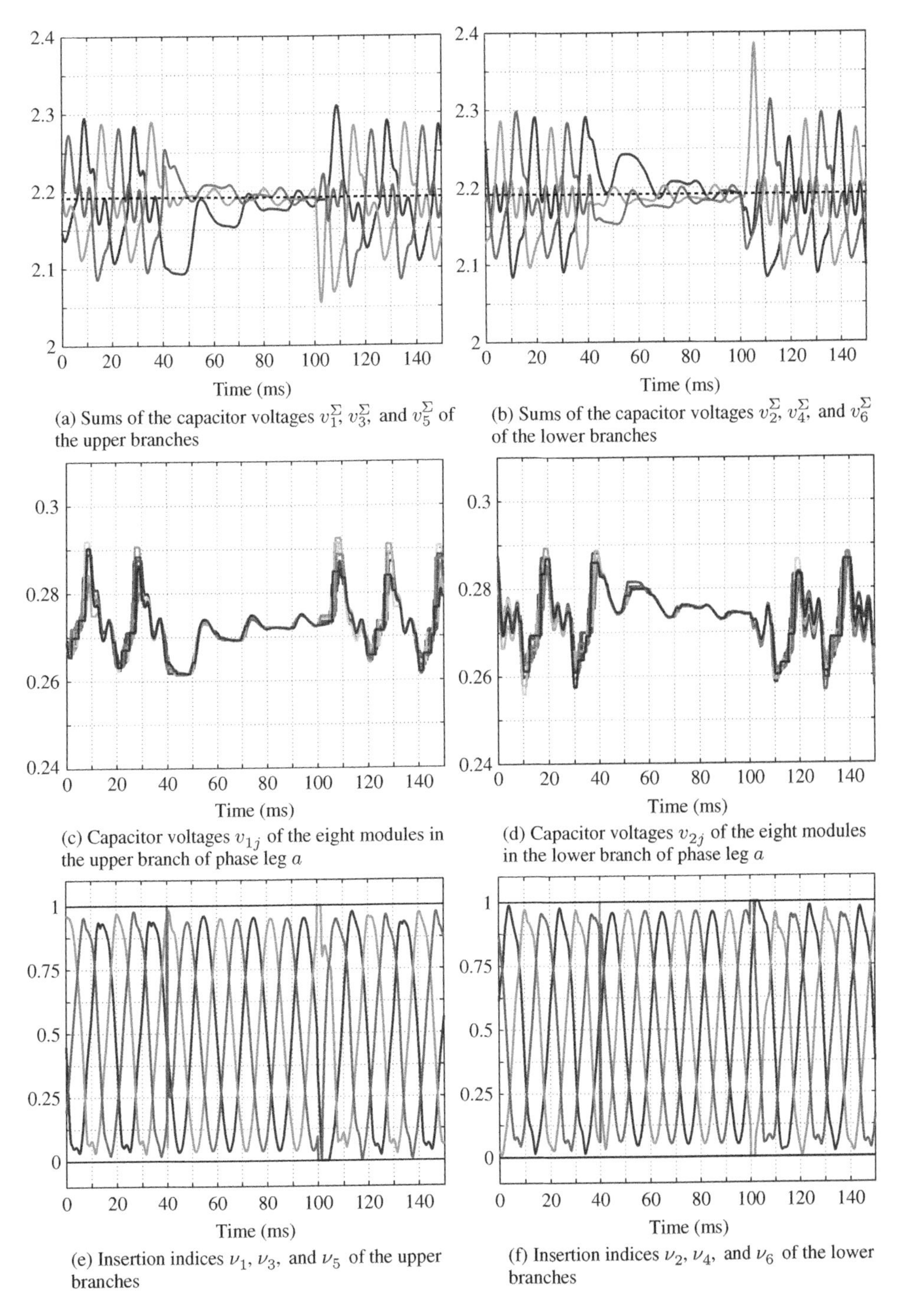

(a) Sums of the capacitor voltages v_1^Σ, v_3^Σ, and v_5^Σ of the upper branches

(b) Sums of the capacitor voltages v_2^Σ, v_4^Σ, and v_6^Σ of the lower branches

(c) Capacitor voltages v_{1j} of the eight modules in the upper branch of phase leg a

(d) Capacitor voltages v_{2j} of the eight modules in the lower branch of phase leg a

(e) Insertion indices ν_1, ν_3, and ν_5 of the upper branches

(f) Insertion indices ν_2, ν_4, and ν_6 of the lower branches

Figure 14.11 Capacitor voltage waveforms and insertion indices during the transients

To achieve the very fast responses shown in Figs. 14.9 and 14.10, the MPC scheme modifies the insertion indices in a stepwise manner. This behavior is demonstrated in Fig. 14.12(a) and (b), which zoom in on the insertion indices during the positive current step. When the reference step is applied at $t = 100$ ms, all six insertion indices are modified by 0.1 or more within one sampling interval, which is of the length $T_s = 200$ μs. The effect this modification has on the number of modules inserted into the branches is shown in Fig. 14.12(c) and (d).

We had seen in Fig. 14.11(a) and (b) that the positive current step results in a short imbalance in the capacitor voltages between the upper and the lower branch of phase leg b. To account for this, the total number of modules inserted into phase leg b differs significantly during the transient, as can be seen in Fig. 14.12(f). This degree of freedom is also used by the controller to drive a circulating current that compensates for the capacitor voltage imbalance. In phase leg a, the capacitor voltages remain balanced, not requiring any such control action (see Fig. 14.12(e)).

14.5 Design Parameters

The impact of the prediction horizon on the closed-loop performance will be investigated and discussed in this section. As will be shown, because of the linearization of the prediction model, long prediction horizons lead to significant open-loop prediction errors, with the actual state variables deviating considerably from the predicted ones. As a result, very long horizons achieve diminishing returns in terms of steady-state performance gains. Nevertheless, because of some slow MMC dynamics, such as the capacitor voltages, a relatively long prediction horizon is required to ensure a good steady-state and transient performance. In particular, very short horizons tend to lead to system instabilities.

14.5.1 Open-Loop Prediction Errors

To investigate the accuracy of the linearized prediction model, open-loop simulations based on the prediction model were compared with closed-loop simulations using the nonlinear MMC model. While operating at rated power, more than 1000 insertion indices of the six branches were randomly generated at different time steps within the fundamental period. We define the open-loop prediction error of a state variable as the difference between the predicted value and the actual one.

The prediction errors were computed for the branch currents, the dc-link current, and the sums of the capacitor voltages. Figure 14.13 shows the errors for the branch current and the sum of the capacitor voltages in the upper branch of phase leg a. More specifically, the median and the 50% confidence intervals of the errors at the end of the prediction horizon are shown for a set of prediction horizons ranging from 1 to 9. The range between the whiskers captures all errors except for a few outliers, that is, 99.3% of the data points. As the errors have an approximately Gaussian distribution, the upper and lower whiskers correspond to approximately ± 2.7 standard deviations from the mean value.

The errors in the six branch currents are similar to that shown in Fig. 14.13(a), while the error in the dc-link current is about twice as large. Accordingly, the prediction errors in the six sums of the capacitor voltages are similar to that shown in Fig. 14.13(b).

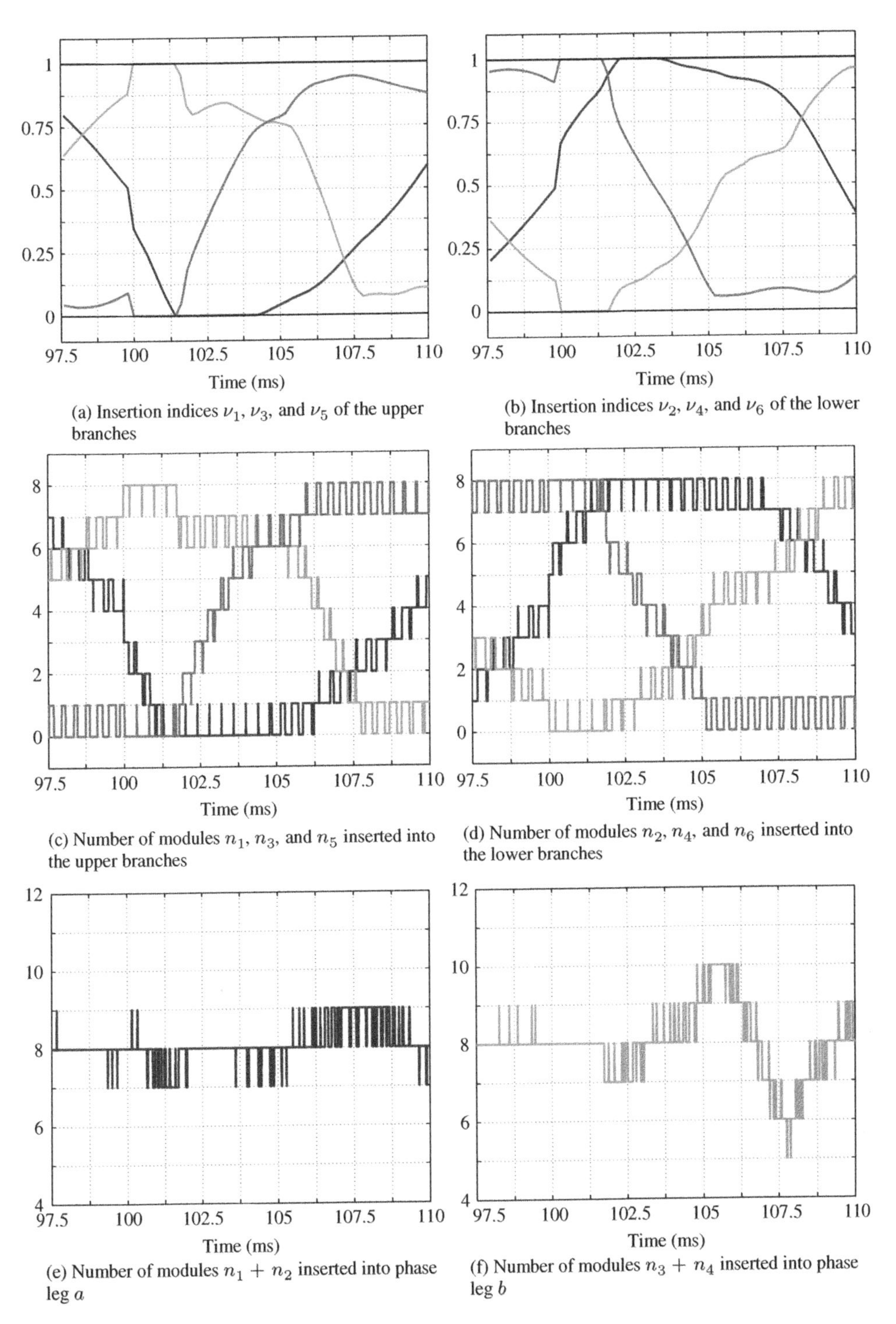

(a) Insertion indices ν_1, ν_3, and ν_5 of the upper branches

(b) Insertion indices ν_2, ν_4, and ν_6 of the lower branches

(c) Number of modules n_1, n_3, and n_5 inserted into the upper branches

(d) Number of modules n_2, n_4, and n_6 inserted into the lower branches

(e) Number of modules $n_1 + n_2$ inserted into phase leg a

(f) Number of modules $n_3 + n_4$ inserted into phase leg b

Figure 14.12 Insertion indices and number of modules inserted during the positive power step at $t = 100$ ms

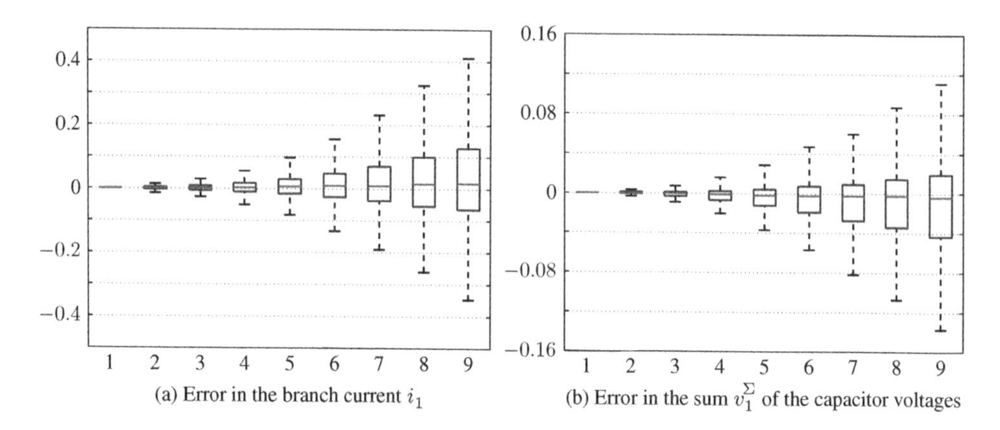

(a) Error in the branch current i_1

(b) Error in the sum v_1^{Σ} of the capacitor voltages

Figure 14.13 Open-loop prediction errors (difference between the predicted and the actual value) of the branch current i_1 and the sum v_1^{Σ} of the capacitor voltages as a function of the length of the prediction horizon N_p. The central line in the box is the median, the lower (upper) edge of the box corresponds to the 25th (75th) percentile, and the range between the whiskers captures all errors except for outliers

Obviously, the predictions obtained from the linearized prediction model begin deviating considerably from the actual state variables when extending the prediction horizon. These errors are pronounced for the dc-link and the branch currents, while they are small for the capacitor voltages. Figure 14.13 indicates that a prediction horizon exceeding 10 steps should be avoided. For a prediction horizon of $N_p = 6$, however, the errors in the capacitor voltages are almost negligible. The relatively large current errors at the time steps $k + 5$ and $k + 6$ can be tolerated by the MPC scheme because of its use of the receding horizon policy, and because the branch current errors at the time steps $k + 1$, $k + 2$, and $k + 3$ are very small.

14.5.2 Closed-Loop Performance

The sensitivity of the closed-loop performance to variations in the carrier frequency and the length of the prediction horizon will be investigated in this section. To this end, simulation results obtained at rated power and steady-state operating conditions will be analyzed and discussed.

First, for the prediction horizon $N_p = 6$, the carrier frequency of the multilevel PWM is varied between 500 and 2500 Hz with steps of 500 Hz. The average device switching frequency f_{sw} and the grid current TDD are used as performance metrics, which are shown in Table 14.3. This data is illustrated in Fig. 14.14 along with a linear function that approximates the switching frequency and a cubic polynomial curve that approximates the current TDD.

On one hand, as expected, the current TDD increases significantly when lowering the carrier frequency. On the other hand, the use of a regularly sampled PWM scheme implies that the switching frequency depends linearly on the carrier frequency. It is notable that the device switching frequency can be lowered down to 167 Hz with the current TDD remaining as low as 1.05%. This is a promising result, because it shows that the proposed MPC scheme can be used in applications that require very low device switching frequencies.

Table 14.3 Influence of the PWM carrier frequency f_c on the device switching frequency f_{sw} and the grid current TDD I_{TDD} for the prediction horizon $N_p = 6$

f_c (Hz)	f_{sw} (Hz)	I_{TDD} (%)
500	106	2.55
1000	167	1.05
1500	231	0.66
2000	290	0.49
2500	350	0.40

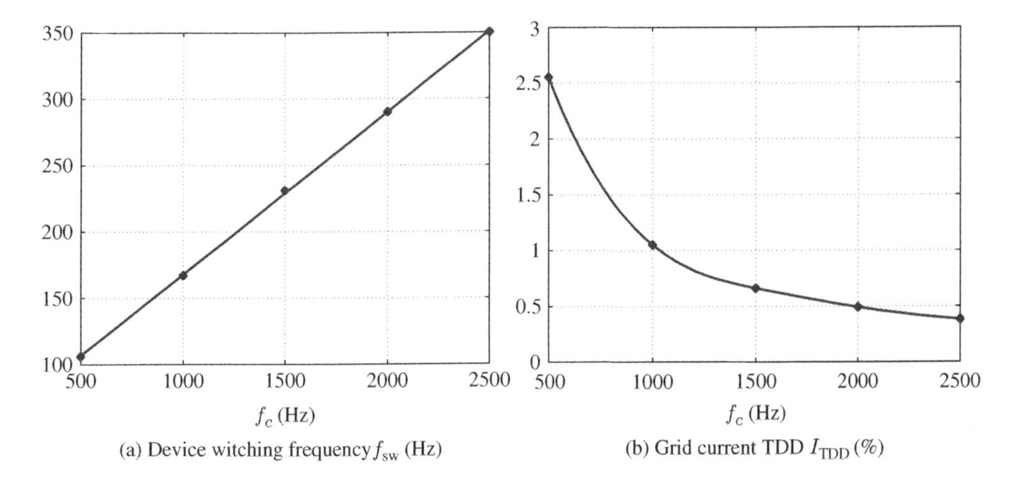

(a) Device witching frequency f_{sw} (Hz)

(b) Grid current TDD I_{TDD} (%)

Figure 14.14 Device switching frequency and grid current TDD as a function of the carrier frequency when using the prediction horizon $N_p = 6$

Second, the carrier frequency is set to 2.5 kHz and the prediction horizon is varied between 3 and 10. As previously, the average device switching frequency and the current TDD are used as performance metrics. Table 14.4 confirms that the length of the prediction horizon has no influence on the switching frequency and that it has only a minor influence on the current TDD. Note that the MPC scheme tends to become unstable for short horizons, such as four or less. This is due to the presence of the slow capacitor voltage dynamics, which require a relatively long prediction horizon.

14.6 Summary and Discussion

An MPC scheme with a subsequent PWM stage was proposed in this chapter for the MMC topology. This versatile control approach is applicable to any MMC topology regardless of its

Table 14.4 Influence of the prediction horizon N_p on the device switching frequency f_{sw} and the grid current TDD I_{TDD} for the carrier frequency $f_c = 2500$ Hz

N_p (time steps)	f_{sw} (Hz)	I_{TDD} (%)
3	Unstable	Unstable
4	Unstable	Unstable
5	350	0.42
6	350	0.40
7	350	0.40
8	350	0.40
9	350	0.38
10	350	0.37

circuit parameters, phase configuration, or number of modules. The controller is conceptually simple with a linearized converter prediction model based on first principles, constraints on the main physical quantities, and an easy-to-devise cost function. The underlying optimization problem is a QP, which can be solved efficiently using off-the-shelf solvers.

The proposed MPC scheme is formulated in such a way that it is independent of the number of modules per branch. The use of a modulator ensures an effectively constant device switching frequency and a deterministic harmonic spectrum of the grid currents. Along with the low device switching frequency of a few hundred hertz, which the MPC scheme can operate at, these features make the developed framework suitable for high-power MMC applications.

The performance benefits of the MPC scheme over existing approaches, which are mostly based on multiple PI control loops, can be attributed to the following three key features. First, MPC is a MIMO control method that is ideally suited to control multiple variables simultaneously, even when they relate to conflicting control objectives. Second, MPC is capable of handling soft and hard constraints on state, input, and output variables. Third, by adopting the receding horizon policy, a significant degree of robustness to modeling errors is achieved.

These features entail that the MPC scheme is capable of simultaneously regulating the grid currents along their references, balancing the module capacitor voltages and respecting the safety constraints on the branch currents, dc-link current, and capacitor voltages—even during large transients. This is achieved by independently manipulating the number of modules that are inserted into each branch and by adopting a scalar cost function, which captures the (conflicting) control objectives. By minimizing this cost function, an optimal consensus is achieved among the various control objectives.

As a result, the MPC scheme tends to outperform most of the existing control approaches for the MMC, particularly during transients such as power up, load steps, and faults [17, 20, 23]. Very fast responses close to the physical limits of the MMC are achieved, with settling times in the range of 2 ms and below. Overshoots in the branch currents are avoided, and the operation of the converter within safe operating limits is ensured under all circumstances. At steady-state operation, a very low grid current TDD of about 0.40% is achieved while operating the semiconductor devices at a switching frequency of about 350 Hz.

Appendix 14.A: Dynamic Current Equations

The differential equations of the five independent currents of the nonlinear MMC system model are provided here. These five equations relate to the meshes EADNE, EBDNE, ECDNE, DASBD, and DASCD of the equivalent representation of the MMC shown in Fig. 14.3.

$$L_{\text{br}}\left(\frac{di_1(t)}{dt}+\frac{di_2(t)}{dt}\right)+L_{\text{dc}}\frac{di_{\text{dc}}(t)}{dt}$$
$$=-R_{\text{br}}\left(i_1(t)+i_2(t)\right)-R_{\text{dc}}i_{\text{dc}}(t)-\nu_1(t)v_1^{\Sigma}(t)-\nu_2(t)v_2^{\Sigma}(t)+v_{\text{dc}} \quad (14.\text{A}.1\text{a})$$

$$L_{\text{br}}\left(\frac{di_3(t)}{dt}+\frac{di_4(t)}{dt}\right)+L_{\text{dc}}\frac{di_{\text{dc}}(t)}{dt}$$
$$=-R_{\text{br}}\left(i_3(t)+i_4(t)\right)-R_{\text{dc}}i_{\text{dc}}(t)-\nu_3(t)v_3^{\Sigma}(t)-\nu_4(t)v_4^{\Sigma}(t)+v_{\text{dc}} \quad (14.\text{A}.1\text{b})$$

$$-L_{\text{br}}\left(\frac{di_1(t)}{dt}+\frac{di_2(t)}{dt}+\frac{di_3(t)}{dt}+\frac{di_4(t)}{dt}\right)+(L_{\text{dc}}+2L_{\text{br}})\frac{di_{\text{dc}}(t)}{dt}$$
$$=R_{\text{br}}\left(i_1(t)+i_2(t)+i_3(t)+i_4(t)\right)-(R_{\text{dc}}+2R_{\text{br}})i_{\text{dc}}(t)$$
$$-\nu_5(t)v_5^{\Sigma}(t)-\nu_6(t)v_6^{\Sigma}(t)+v_{\text{dc}} \quad (14.\text{A}.1\text{c})$$

$$L_g\frac{di_1(t)}{dt}-(L_{\text{br}}+L_g)\frac{di_2(t)}{dt}-L_g\frac{di_3(t)}{dt}+(L_{\text{br}}+L_g)\frac{di_4(t)}{dt}$$
$$=-R_gi_1(t)+(R_{\text{br}}+R_g)i_2(t)+R_gi_3(t)-(R_{\text{br}}+R_g)i_4(t)$$
$$+\nu_2(t)v_2^{\Sigma}(t)-\nu_4(t)v_4^{\Sigma}(t)-v_{ga}(t)+v_{gb}(t) \quad (14.\text{A}.1\text{d})$$

$$2L_g\frac{di_1(t)}{dt}-2(L_{\text{br}}+L_g)\frac{di_2(t)}{dt}+L_g\frac{di_3(t)}{dt}-(L_{\text{br}}+L_g)\frac{di_4(t)}{dt}+L_{\text{br}}\frac{di_{\text{dc}}(t)}{dt}$$
$$=-2R_gi_1(t)+2(R_{\text{br}}+R_g)i_2(t)-R_gi_3(t)+(R_{\text{br}}+R_g)i_4(t)$$
$$-R_{\text{br}}i_{\text{dc}}(t)+\nu_2(t)v_2^{\Sigma}(t)-\nu_6(t)v_6^{\Sigma}(t)-v_{ga}(t)+v_{gc}(t) \quad (14.\text{A}.1\text{e})$$

Appendix 14.B: Controller Model of the Converter System

The derivation of the matrices $\boldsymbol{F}(t_0)$, $\boldsymbol{G}(t_0)$, and $\boldsymbol{C}$, and the vector $\boldsymbol{g}(t_0)$ of the linearized continuous-time state-space model (14.11) is briefly outlined here. The five differential equations (14.A.1) of the currents are linearized using the Taylor series (14.10a). Similarly, the six differential equations (14.6) of the sums v_r^{Σ} of the capacitor voltages are linearized using (14.10b).

The dynamic evolution of the grid voltages is modeled in (14.9) in stationary and orthogonal $\alpha\beta$ coordinates, while the three-phase representation of the grid voltages is used in the differential equations of the currents (14.A.1d) and (14.A.1e). We use the reduced Clarke transformation $\tilde{\boldsymbol{K}}$ and its pseudo-inverse $\tilde{\boldsymbol{K}}^{-1}$ to transform the $\alpha\beta$ grid voltages into their three-phase representation v_{ga}, v_{gb}, and v_{gc}, and vice versa, according to (2.13).

We define the auxiliary time-invariant matrices as

$$\boldsymbol{L} = \begin{bmatrix} L_{\text{br}} & L_{\text{br}} & 0 & 0 & L_{\text{dc}} \\ 0 & 0 & L_{\text{br}} & L_{\text{br}} & L_{\text{dc}} \\ -L_{\text{br}} & -L_{\text{br}} & -L_{\text{br}} & -L_{\text{br}} & L_{\text{dc}} + 2L_{\text{br}} \\ L_g & -(L_{\text{br}} + L_g) & -L_g & L_{\text{br}} + L_g & 0 \\ 2L_g & -2(L_{\text{br}} + L_g) & L_g & -(L_{\text{br}} + L_g) & L_{\text{br}} \end{bmatrix} \tag{14.B.1a}$$

$$\boldsymbol{R} = \begin{bmatrix} -R_{\text{br}} & -R_{\text{br}} & 0 & 0 & -R_{\text{dc}} \\ 0 & 0 & -R_{\text{br}} & -R_{\text{br}} & -R_{\text{dc}} \\ R_{\text{br}} & R_{\text{br}} & R_{\text{br}} & R_{\text{br}} & -(R_{\text{dc}} + 2R_{\text{br}}) \\ -R_g & R_{\text{br}} + R_g & R_g & -(R_{\text{br}} + R_g) & 0 \\ -2R_g & 2(R_{\text{br}} + R_g) & -R_g & R_{\text{br}} + R_g & -R_{\text{br}} \end{bmatrix} \tag{14.B.1b}$$

$$\boldsymbol{E}_1 = \begin{bmatrix} 0 & 0 & 0 \\ 0 & 0 & 0 \\ 0 & 0 & 0 \\ -1 & 1 & 0 \\ -1 & 0 & 1 \end{bmatrix} \quad \boldsymbol{E}_2 = \begin{bmatrix} 1 & -1 & 0 & 0 & 0 \\ 0 & 0 & 1 & -1 & 0 \\ -1 & 1 & -1 & 1 & 0 \end{bmatrix} \tag{14.B.1c}$$

and the auxiliary time-varying matrices as

$$\boldsymbol{N}_1(t_0) = \begin{bmatrix} \nu_1(t_0) & 0 & 0 & 0 & 0 \\ 0 & \nu_2(t_0) & 0 & 0 & 0 \\ 0 & 0 & \nu_3(t_0) & 0 & 0 \\ 0 & 0 & 0 & \nu_4(t_0) & 0 \\ -\nu_5(t_0) & 0 & -\nu_5(t_0) & 0 & \nu_5(t_0) \\ 0 & -\nu_6(t_0) & 0 & -\nu_6(t_0) & \nu_6(t_0) \end{bmatrix} \tag{14.B.2a}$$

$$\boldsymbol{N}_2(t_0) = \begin{bmatrix} -\nu_1(t_0) & -\nu_2(t_0) & 0 & 0 & 0 & 0 \\ 0 & 0 & -\nu_3(t_0) & -\nu_4(t_0) & 0 & 0 \\ 0 & 0 & 0 & 0 & -\nu_5(t_0) & -\nu_6(t_0) \\ 0 & \nu_2(t_0) & 0 & -\nu_4(t_0) & 0 & 0 \\ 0 & \nu_2(t_0) & 0 & 0 & 0 & -\nu_6(t_0) \end{bmatrix} \tag{14.B.2b}$$

$$\boldsymbol{V}_1(t_0) = \begin{bmatrix} v_{\text{dc}}(t_0) & v_{\text{dc}}(t_0) & v_{\text{dc}}(t_0) & 0 & 0 \end{bmatrix}^T \tag{14.B.2c}$$

$$\boldsymbol{V}_2(t_0) = \begin{bmatrix} -v_1^{\Sigma}(t_0) & -v_2^{\Sigma}(t_0) & 0 & 0 & 0 & 0 \\ 0 & 0 & -v_3^{\Sigma}(t_0) & -v_4^{\Sigma}(t_0) & 0 & 0 \\ 0 & 0 & 0 & 0 & -v_5^{\Sigma}(t_0) & -v_6^{\Sigma}(t_0) \\ 0 & v_2^{\Sigma}(t_0) & 0 & -v_4^{\Sigma}(t_0) & 0 & 0 \\ 0 & v_2^{\Sigma}(t_0) & 0 & 0 & 0 & -v_6^{\Sigma}(t_0) \end{bmatrix} \tag{14.B.2d}$$

$$\boldsymbol{I}_i(t_0) = \text{diag}(i_1(t_0), i_2(t_0), i_3(t_0), i_4(t_0), i_5(t_0), i_6(t_0)) \tag{14.B.2e}$$

with

$$i_5(t_0) = i_{\text{dc}}(t_0) - i_1(t_0) - i_3(t_0) \tag{14.B.3a}$$

$$i_6(t_0) = i_{\text{dc}}(t_0) - i_2(t_0) - i_4(t_0). \tag{14.B.3b}$$

We can then write the system matrix, input matrix, and offset vector of the state-space model as

$$\boldsymbol{F}(t_0) = \begin{bmatrix} \boldsymbol{L}^{-1}\boldsymbol{R} & \boldsymbol{L}^{-1}\boldsymbol{N}_2(t_0) & \boldsymbol{L}^{-1}\boldsymbol{E}_1\tilde{\boldsymbol{K}}^{-1} \\ \frac{n}{C_m}\boldsymbol{N}_1(t_0) & \boldsymbol{0}_{6\times 6} & \boldsymbol{0}_{6\times 2} \\ \boldsymbol{0}_{2\times 5} & \boldsymbol{0}_{2\times 6} & \omega_g \begin{bmatrix} 0 & -1 \\ 1 & 0 \end{bmatrix} \end{bmatrix} \tag{14.B.4a}$$

$$\boldsymbol{G}(t_0) = \begin{bmatrix} \boldsymbol{L}^{-1}\boldsymbol{V}_2(t_0) \\ \frac{n}{C_m}\boldsymbol{I}_i(t_0) \\ \boldsymbol{0}_{2\times 6} \end{bmatrix} \quad \boldsymbol{g}(t_0) = \begin{bmatrix} \boldsymbol{L}^{-1}\boldsymbol{V}_1(t_0) \\ \boldsymbol{0}_{6\times 1} \\ \boldsymbol{0}_{2\times 1} \end{bmatrix}. \tag{14.B.4b}$$

The output matrix is given by

$$\boldsymbol{C} = \begin{bmatrix} \tilde{\boldsymbol{K}}\boldsymbol{E}_2 & \boldsymbol{0}_{2\times 6} & \boldsymbol{0}_{2\times 2} \\ \boldsymbol{0}_{6\times 5} & \boldsymbol{I}_6 & \boldsymbol{0}_{6\times 2} \end{bmatrix}. \tag{14.B.5}$$

Recall that $\boldsymbol{0}_{n\times m}$ denotes the zero matrix of the dimensions $n \times m$, and $\boldsymbol{I}_n$ is the n-dimensional identity matrix.

This controller model needs to be translated into the pu system, following the steps explained in Sects. 2.2.3 and 2.4.1. The pu system is established using the base quantities stated in Sect. 14.4.1.

References

[1] A. Lesnicar and R. Marquardt, "An innovative modular multilevel converter topology suitable for a wide power range," in *Proceedings of the IEEE Power Tech Conference* (Bologna, Italy), Jun. 2003.

[2] M. A. Pérez, S. Bernet, J. Rodríguez, S. Kouro, and R. Lizana, "Circuit topologies, modeling, control schemes, and applications of modular multilevel converters," *IEEE Trans. Power Electron.*, vol. 30, pp. 4–17, Jan. 2015.

[3] S. Debnath, J. Qin, B. Bahrani, M. Saeedifard, and P. Barbosa, "Operation, control, and applications of the modular multilevel converter: A review," *IEEE Trans. Power Electron.*, vol. 30, pp. 37–53, Jan. 2015.

[4] J. Rodríguez, J.-S. Lai, and F. Peng, "Multilevel inverters: A survey of topologies, controls, and applications," *IEEE Trans. Ind. Electron.*, vol. 49, pp. 727–738, Aug. 2002.

[5] J. Qin and M. Saeedifard, "Predictive control of a modular multilevel converter for a back-to-back HVDC system," *IEEE Trans. Power Delivery*, vol. 27, pp. 1538–1547, Jul. 2012.

[6] A. Nami, J. Liang, F. Dijkhuizen, and G. D. Demetriadis, "Modular multilevel converters for HVDC applications: Review on converter cells and functionalities," *IEEE Trans. Power Electron.*, vol. 30, pp. 18–36, Jan. 2015.

[7] H. M. Pirouz and M. T. Bina, "A transformerless medium-voltage STATCOM topology based on extended modular multilevel converters," *IEEE Trans. Power Electron.*, vol. 26, pp. 1534–1545, May 2011.

[8] J. I. Y. Ota, Y. Shibano, N. Niimura, and H. Akagi, "A phase-shifted-PWM D-STATCOM using a modular multilevel cascade converter (SSBC)—Part I: Modeling, analysis, and design of current control," *IEEE Trans. Ind. Appl.*, vol. 51, pp. 279–288, Jan./Feb. 2015.

[9] M. Winkelnkemper, A. Korn, and P. Steimer, "A modular direct converter for transformerless rail interties," in *Proceedings of the IEEE International Symposium on Industrial Electronics* (Bari, Italy), pp. 562–567, Jul. 2010.

[10] A. Antonopoulos, L. Ängquist, and H.-P. Nee, "On dynamics and voltage control of the modular multilevel converter," in *Proceedings European Power Electronics Conference* (Barcelona, Spain), Sep. 2009.

[11] M. Hagiwara and H. Akagi, "Control and experiment of pulsewidth-modulated modular multilevel converters," *IEEE Trans. Power Electron.*, vol. 24, pp. 1737–1746, Jul. 2009.

[12] L. Ängquist, A. Antonopoulos, D. Siemaszko, K. Ilves, M. Vasiladiotis, and H.-P. Nee, "Open-loop control of modular multilevel converters using estimation of stored energy," *IEEE Trans. Ind. Appl.*, vol. 47, pp. 2516–2524, Nov./Dec. 2011.

[13] A. J. Korn, M. Winkelnkemper, P. Steimer, and J. W. Kolar, "Capacitor voltage balancing in modular multilevel converters," in *Proceedings of International Conference on Power Electronics, Machine and Drives*, Mar. 2012.

[14] J. B. Rawlings and D. Q. Mayne, *Model predictive control: Theory and design*. Madison, WI: Nob Hill Pub., 2009.

[15] M. A. Pérez, J. Rodríguez, E. J. Fuentes, and F. Kammerer, "Predictive control of AC–AC modular multilevel converters," *IEEE Trans. Ind. Electron.*, vol. 59, pp. 2832–2839, Jul. 2012.

[16] B. S. Riar, T. Geyer, and U. K. Madawala, "Model predictive direct current control of modular multilevel converters: Modelling, analysis, and experimental evaluation," *IEEE Trans. Power Electron.*, vol. 30, pp. 431–439, Jan. 2015.

[17] J. Kolb, F. Kammerer, F. Gommeringer, and M. Braun, "Cascaded control system of the modular multilevel converter for feeding variable-speed drives," *IEEE Trans. Power Electron.*, vol. 30, pp. 349–357, Jan. 2015.

[18] K. Ilves, A. Antonopoulos, S. Norrga, and H.-P. Nee, "Steady-state analysis of interaction between harmonic components of arm and line quantities of modular multilevel converters," *IEEE Trans. Power Electron.*, vol. 27, pp. 57–68, Jan. 2012.

[19] L. Harnefors, A. Antonopoulos, S. Norrga, L. Ängquist, and H.-P. Nee, "Dynamic analysis of modular multilevel converters," *IEEE Trans. Ind. Electron.*, vol. 60, pp. 2526–2537, Jul. 2013.

[20] M. Saeedifard and R. Iravani, "Dynamic performance of a modular multilevel back-to-back HVDC system," *IEEE Trans. Power Delivery*, vol. 25, pp. 2903–2912, Oct. 2010.

[21] M. Herceg, M. Kvasnica, C. Jones, and M. Morari, "Multi-parametric toolbox 3.0," in *Proceedings of European Control Conference* (Zurich, Switzerland), pp. 502–510, Jul. 2013. http://control.ee.ethz.ch/ mpt.

[22] Gurobi Optimization, Inc., *"Gurobi optimizer reference manual,"* 2014. www.gurobi.com.

[23] D. Siemaszko, A. Antonopoulos, K. Ilves, M. Vasiladiotis, L. Ängquist, and H.-P. Nee, "Evaluation of control and modulation methods for modular multilevel converters," in *Proceedings of IEEE International Power Electronics Conference* (Sapporo, Japan), pp. 746–753, Jun. 2010.

Part Five
Summary

15

Summary and Conclusion

In this last chapter, we review the steady-state performance of the proposed model predictive control (MPC) schemes and benchmark them with space vector modulation (SVM). To facilitate the selection of a control and modulation method for a given power electronics problem, we provide a critical assessment of the strengths and weaknesses of each of the proposed methods. To provide a comprehensive view, we extend this assessment to direct torque control (DTC) and field-oriented control (FOC) with SVM. To conclude the book, we outline the commercial benefits MPC is expected to bring to power electronics products. We also discuss requirements for, and obstacles to, the successful commercialization of MPC. In lieu of an outlook, we suggest a few research directions which we consider to be particularly relevant.

15.1 Performance Comparison of Direct Model Predictive Control Schemes

In this section, the direct MPC schemes proposed in this book are compared with each other when applied to a machine-side inverter. These MPC schemes include one-step predictive current control with reference tracking, model predictive direct torque control (MPDTC), model predictive direct current control (MPDCC), and model predictive pulse pattern control (MP^3C). For MPDTC and MPDCC, both short and long switching horizons will be considered. MP^3C is based on optimized pulse patterns (OPPs) that minimize the current distortions for a given switching frequency. The comparison is extended to FOC with SVM, which is arguably the most commonly used control and modulation scheme. The comparison is performed during steady-state operation. The key performance criteria are the switching losses in the inverter, and the harmonic current and torque distortions in the machine. We chose the total demand distortion (TDD), see the definitions (3.1) and (3.2), to quantify the harmonic distortions.

The trade-off between switching losses and harmonic distortions is well understood. Indeed, as was shown in Sect. 3.5 for carrier-based pulse width modulation (CB-PWM) and SVM, the product of these two quantities is equal to a constant. This gives rise to

Model Predictive Control of High Power Converters and Industrial Drives, First Edition. Tobias Geyer.

Companion Website: www.wiley.com/go/geyermodelpredictivecontrol

a hyperbolic trade-off curve that characterizes the steady-state performance of a given modulation scheme. For modulation techniques other than CB-PWM and SVM, the notion of the trade-off curve is equally applicable, but the shape of the curve matches a hyperbola only in an approximate manner.

Other possible performance criteria include the controller's sensitivity to parameter variations and flux observer noise. These aspects were examined in Sect. 3.1.2 for MP^3C, and a related analysis was performed in [1] for MPDCC. The dynamic control performance, which includes the torque settling time during torque steps, was investigated for each of the control schemes in the relevant chapters. Specifically, we have seen that all direct MPC schemes achieve torque and current transients that are similar to those of DTC and deadbeat control. Most notably, these MPC schemes all fully exploit the available dc-link voltage. During negative torque steps, for example, the controllers temporarily invert the voltage applied to the stator windings of the machine to achieve a torque transient that is as short as possible.

In the following, we will first determine the characteristic trade-off curve for each modulation method. This will be done through simulations at steady-state operation. In a second step, by comparing the trade-off curves of the different control and modulation schemes with each other, we will provide insight into the steady-state performance of these methods. It will be shown that long predictions horizons significantly enhance the control performance by shifting the performance trade-off curves towards the origin, thus lowering both the switching losses and the harmonic distortions. Conversely, overly short horizons often lead to a worse performance than SVM.

15.1.1 Case Study

The case study is intended to be as general as possible to ensure that the conclusions drawn are both meaningful and sufficiently generalizable to be of value. The comparison focuses on the core performance characteristic of the different control and modulation methods, in order to establish their theoretical baseline performance. To achieve this, we will neglect non-idealities and second-order effects that typically arise in a real-world drive setting, such as dc-link voltage ripples, neutral point potential fluctuations, interlocking times, minimum on and off times, saturation of the machine's magnetic material, controller delays, measurement errors, flux observer noise, and load torque variations. In an industrial controller implementation, techniques are available to largely compensate for these adverse effects. This applies both to traditional as well as predictive control and modulation methods.

To this end, we select as a case study a three-level neutral-point-clamped (NPC) voltage source inverter that drives a medium-voltage (MV) induction machine. This drive setup is shown in Fig. 15.1. In the arena of MV drives, this drive configuration is the one that is most commonly used. The total dc-link voltage is $V_{\rm dc} = 5.2\,{\rm kV}$. Switching between the upper and the lower rail is prohibited, but all other switching transitions are allowed. A summary of the machine and inverter parameters is provided in Sect. 2.5.1.

15.1.2 Performance Trade-Off Curves

With the help of simulations run at steady-state operating conditions, the trade-off curves are derived for the investigated control and modulation schemes. All simulations were run at 60%

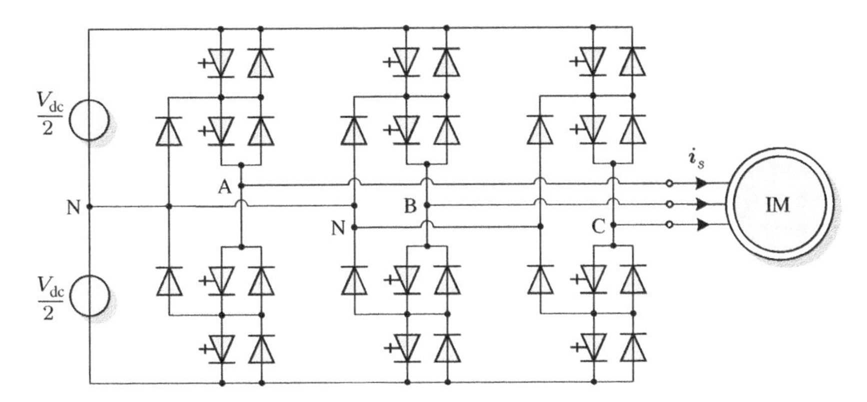

Figure 15.1 Three-level NPC voltage source inverter driving an induction machine with a fixed neutral point potential

speed with a 100% torque setpoint. As we have seen previously, the performance differences between the different control and modulation methods tend to be pronounced at this operating point. The sampling interval of the controller is set to $T_s = 25$ μs, unless otherwise noted.

15.1.2.1 Rotor FOC with SVM

As explained in Sect. 3.6.2, rotor FOC is formulated in an orthogonal reference frame that rotates synchronously with the rotor flux vector. Two (orthogonal) control loops are used—one for the flux and another one for the torque-producing current. A subsequent modulator translates the reference of the stator voltage into gating commands for the inverter. A three-level, asymmetric, regularly sampled CB-PWM with two triangular carrier signals is used. The two carrier waveforms are in phase, and a common-mode voltage is added to the reference voltage. By deriving this common-mode voltage through a min/max plus modulo operation, the same gating signals as with SVM are generated. For more details on CB-PWM and the relationship with SVM, the reader is referred to Sect. 3.3.

The switching frequency was varied between 100 and 500 Hz in this analysis. Synchronous modulation was used, with the carrier frequency being an integer multiple of the fundamental frequency. After reaching steady-state operating conditions, the stator currents, stator voltages, electromagnetic torque, and the switch positions were recorded. The inverter switching losses P_{sw} were computed based on these measurements and in accordance with Sect. 2.5.1. The current and torque TDDs were derived using Fourier transformations over integer multiples of the fundamental period. The switching losses were normalized using the rated apparent power $S_R = 2.035$ MVA. Normalized switching losses of 0.1%, for example, correspond to the switching losses 2.035 kW.

Figure 15.2 shows the resulting harmonic distortions of the stator currents and of the torque as a function of the normalized switching losses. The individual simulation results are denoted by circles, which can be approximated by hyperbolic functions of the form

$$I_{\text{TDD}} \cdot \frac{P_{\text{sw}}}{S_R} = 1.3 \text{ and } T_{\text{TDD}} \cdot \frac{P_{\text{sw}}}{S_R} = 0.55. \tag{15.1}$$

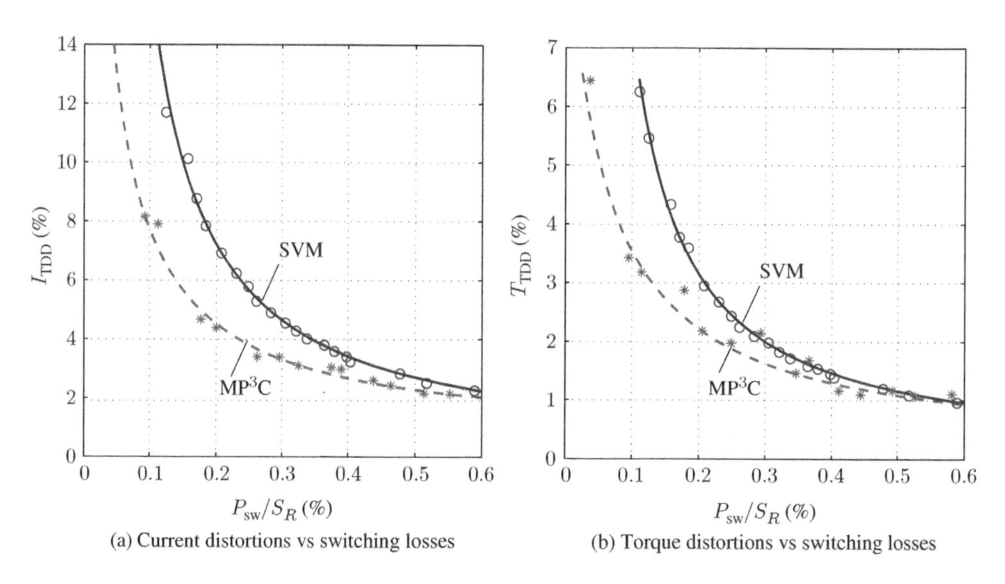

Figure 15.2 Performance trade-off for FOC with SVM (○ data points) and MP^{3}C using OPPs (⋆ data points)

This implies that when reducing the switching frequency with the objective to reduce the switching losses, for example, by 50%, the current and torque TDDs are increased by 50%, and vice versa. A small offset in the normalized switching losses is neglected in (15.1). This offset accounts for the fact that the switching losses cannot be reduced to zero. In order to synthesize the desired amplitude and phase of the voltage waveform, at least one switching transition per phase and quarter of the fundamental period is required. This modulation regime is commonly referred to as a *six-step operation*. Its switching frequency is equal to the fundamental frequency.

Alternatively, the FOC loops can be replaced by an MPC scheme, which is sometimes referred to as *indirect* MPC, because it manipulates the switch positions indirectly via an intermediate modulator. Examples for such MPC schemes include [2] and [3], which replace the inner current control loop by MPC. During steady-state operation, when neglecting second-order effects, the harmonic performance of indirect MPC schemes is determined solely by the modulator. When using SVM, the harmonic performance of indirect MPC is thus the same as stated in (15.1) and shown in Fig. 15.2.

15.1.2.2 MP^{3}C

Alternatively, OPPs can be calculated in an offline procedure for a range of modulation indices and pulse numbers. Recall that the pulse number of a single-phase switching pattern is defined as the number of switching transitions within a quarter of the fundamental period. The optimization criterion is the minimization of the weighted voltage distortion, which is proportional to the current distortion of an inductive load. For a given pulse number and modulation index, the minimization procedure leads to the optimal switching angles (see Sect. 3.4). These sets of switching angles can be stored in look-up tables.

OPPs are typically used in slow control loops such as V/f control or non-aggressively tuned FOC loops. By combining the notion of trajectory control with the receding horizon policy, an OPP-based controller with a very high dynamic performance can be designed. The proposed control method, MP^{3}C, is explained in detail in Chap. 12. Figure 15.2 shows the steady-state performance of MP^{3}C in terms of the harmonic distortions versus the normalized switching losses. The individual simulations are denoted by stars, and the resulting trade-off curves are indicated by dashed lines.

15.1.2.3 MPDTC

Similar to DTC, MPDTC directly controls the electromagnetic torque and the stator flux magnitude by imposing upper and lower bounds on them. Using a dynamic model and an optimization stage, MPDTC predicts candidate switching sequences that maintain the torque and the stator flux magnitude within their respective bounds. When considering NPC inverters with floating neutral point potentials, this principle can be extended to the neutral point potential, in order to balance it around zero.

The principle of MPDTC is to freeze the switch positions and extend the output trajectories until a bound violation is predicted to occur. In doing so, very long prediction horizons can be achieved at a modest computational burden. By minimizing a cost function, which captures the predicted switching frequency or the switching losses, the optimal switching sequence is selected. The MPDTC algorithm is described in detail in Chap. 7. Branch-and-bound techniques can be used to further reduce the computation time (see Chap. 10).

For the performance analysis, we penalize the switching losses in the cost function. Additional terms in the cost function, such as penalties on bound violations or weights on terminal states, are not used and are set to zero. The bounds on the torque and stator flux magnitude were randomly selected within large intervals. Hundreds of simulations were run for the switching horizons *eSE* and *eSESESE*. Each data point in Fig. 15.3 corresponds to one such simulation with a certain combination of torque and flux bounds.

The envelopes of the data points can again be described by hyperbolic functions, despite them being shifted further along the horizontal axis. Owing to the random selection of the bound widths, many points lie far away from their envelope and are thus suboptimal. It is clear that the existence of two tuning parameters—the width of the torque bounds and the width of the stator flux magnitude bounds—complicates the tuning process, particularly if minimum current distortions are to be achieved. When choosing very wide bounds, MPDTC locks into fundamental frequency switching and provides a switching pattern akin to the six-step operation. The corresponding data points correspond to normalized switching losses of about 0.04%.

Sets of bounds that achieve low torque distortions do not necessarily achieve low current distortions. To highlight this, we define the stator ripple current

$$\boldsymbol{i}_{\text{rip}} = \boldsymbol{i}_s^* - \boldsymbol{i}_s \tag{15.2}$$

in stationary orthogonal coordinates as the difference between the stator current reference $\boldsymbol{i}_s^*$ and the actual stator current $\boldsymbol{i}_s$. Recall from (11.6) that the torque can be expressed in terms of the rotor flux vector and the stator current vector as

$$T_e = \frac{1}{\text{pf}}\frac{X_m}{X_r}\boldsymbol{\psi}_r \times \boldsymbol{i}_s = T_e^* - \frac{1}{\text{pf}}\frac{X_m}{X_r}\boldsymbol{\psi}_r \times \boldsymbol{i}_{\text{rip}}, \tag{15.3}$$

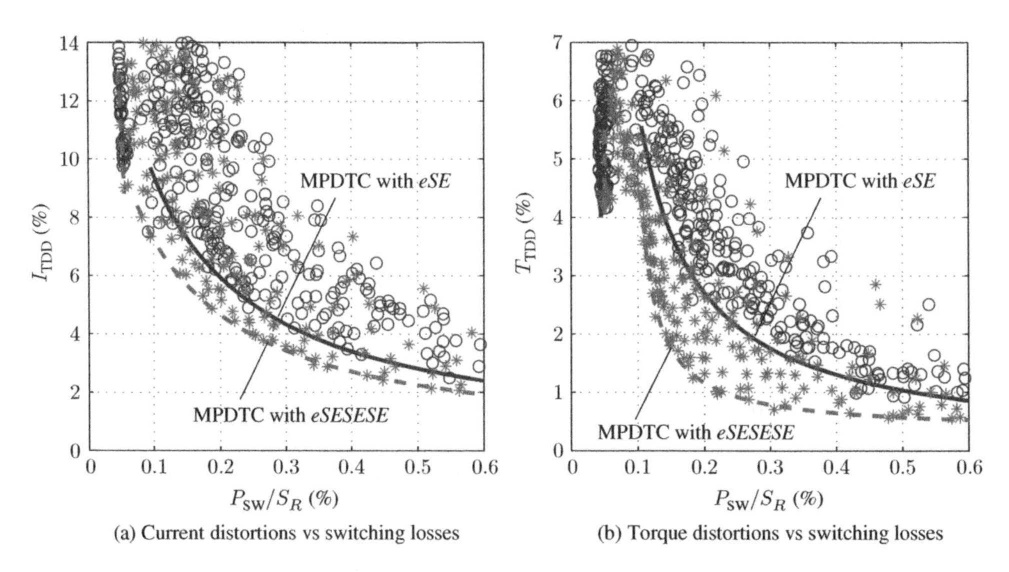

(a) Current distortions vs switching losses

(b) Torque distortions vs switching losses

Figure 15.3 Performance trade-off for MPDTC with the switching horizons *eSE* (∘ data points) and *eSESESE* (⋆ data points)

where we have used (15.2) and introduced the torque reference $T_e^* = \frac{1}{\text{pf}}\frac{X_m}{X_r}\boldsymbol{\psi}_r \times \boldsymbol{i}_s^*$. The torque ripple is then given by

$$T_{\text{rip}} = T_e^* - T_e = \frac{1}{\text{pf}}\frac{X_m}{X_r}\boldsymbol{\psi}_r \times \boldsymbol{i}_{\text{rip}} = \frac{1}{\text{pf}}\frac{X_m}{X_r}(\psi_{r\alpha} i_{\text{rip},\beta} - \psi_{r\beta} i_{\text{rip},\alpha}). \tag{15.4}$$

Zero torque ripple is (hypothetically) achieved when the right-hand side of (15.4) is zero. Neglecting the special case when any of the β-components is zero allows us to write

$$\frac{i_{\text{rip},\alpha}}{i_{\text{rip},\beta}} = \frac{\psi_{r\alpha}}{\psi_{r\beta}}. \tag{15.5}$$

The torque ripple is zero when the α- and β-components of the ripple current have the same ratio as the α- and β-components of the rotor flux vector. When the aim is to minimize the torque ripple, the stator ripple current vector must rotate synchronously with the rotor flux vector. On the other hand, in order to achieve minimum current distortions, the ripple current components should be of a similar magnitude. As these are conflicting requirements, it is clear that the torque ripple cannot be minimized without impacting on the current distortions.

This analysis suggests that very low torque distortions can be achieved in MPDTC, albeit at the expense of pronounced current distortions. As a result, for a specific switching loss, the data points in Fig. 15.3(b) that minimize the torque TDD do not, in general, also minimize the current TDD in Fig. 15.3(a). On the other hand, minimum current TDDs usually also imply low torque TDDs.

15.1.2.4 MPDCC

A derivative of MPDTC, MPDCC imposes upper and lower bounds on the three-phase stator currents. These bounds are symmetrical around the three-phase current references. The width of the bounds directly determines the current ripple, which in turn is proportional to the current TDD (see Sect. 11.1.8). The control objectives are to keep the instantaneous currents within their bounds and to minimize the switching losses. The bound width is the (only) tuning parameter. It determines the point on the trade-off curve between the current distortions and the switching losses. The existence of one (instead of two) bound widths greatly simplifies the tuning process.

Figure 15.4 depicts the simulation results for the switching horizons *eSE* and *eSESESE*, respectively. The widths of the current bounds were varied between 0.035 and 0.22 pu. These numbers indicate the difference between the upper (or lower) bound and the reference—they thus correspond to half the current ripple. The data points can be described by hyperbolic trade-off functions, particularly when operating at high switching losses. Six-step operation is represented by a second set of trade-off curves, which are given by the almost vertical lines in Fig. 15.4.

MPDCC tends to lock into certain fixed values of switching losses, despite significant variations in the bound width. This phenomenon is particularly apparent with the short switching horizon *eSE* at 0.135%, 0.18%, and 0.225% of the normalized switching losses. The corresponding switching frequencies are 90, 120, and 150 Hz, that is, 3, 4, and 5 times the fundamental frequency. A similar phenomenon has previously also been identified for predictive current control with reference tracking—both for the single-phase case (see Fig. 4.7(b)) and for the three-phase case in Sect. 6.1.4.

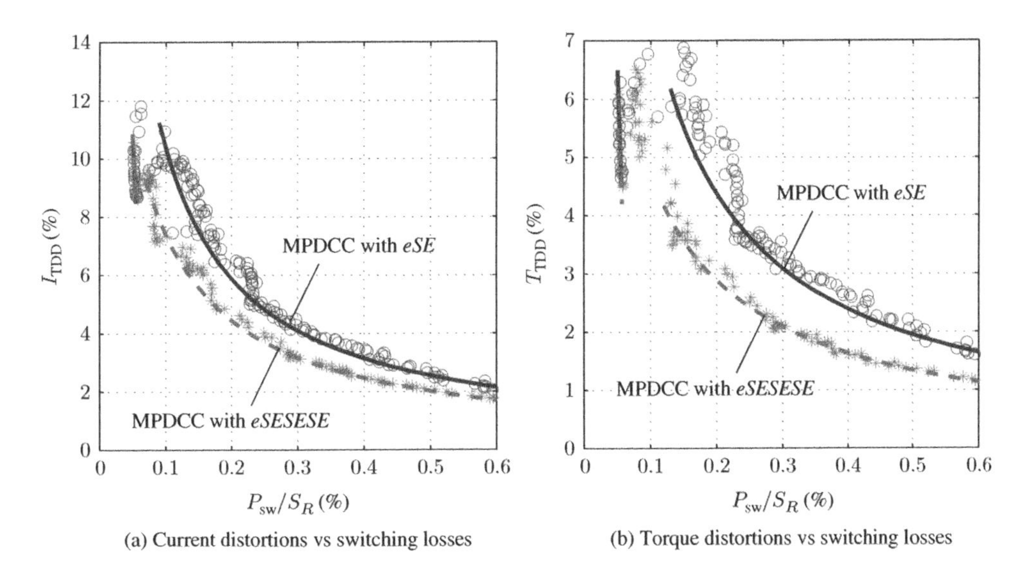

(a) Current distortions vs switching losses

(b) Torque distortions vs switching losses

Figure 15.4 Performance trade-off for MPDCC with the switching horizons *eSE* (○ data points) and *eSESESE* (⋆ data points)

15.1.2.5 Predictive Current Control with Reference Tracking

The optimization problem underlying MPDCC can be greatly simplified by setting the bound width to zero, using a prediction horizon of length 1 and minimizing the number of switching transitions. This scheme operates in the stationary $\alpha\beta$ reference frame and regulates the α- and β-current components along their references. The tuning parameter λ_u is used to adjust the trade-off between tracking accuracy and the number of switching transitions. This control scheme is described in detail in Sect. 4.2.

Note that, as stated in (4.19), the current error is penalized using the squared 2-norm, instead of the 1-norm as initially proposed in [4] and [5]. The 2-norm avoids stability issues and reduces the current and torque distortions significantly. At 0.3% of the normalized switching losses, for example, the 2-norm reduces the current TDD from about 5% to 4.5%.

Hundreds of simulations were performed, while varying the tuning parameter λ_n between 0.001 and 0.03, and the controller sampling interval between 25 and 100 μs. Figure 15.5 depicts the resulting current and torque distortions versus the normalized switching losses for each simulation. As previously, the envelope of these data points can be accurately approximated by hyperbolic trade-off curves. The phenomenon of the current controller locking into fixed values of the switching losses is again visible, particularly at 0.135% of the normalized switching losses. This operating point corresponds to a switching frequency of about 90 Hz, which is equal to 3 times the fundamental frequency.

Alternatively, long prediction horizons can be adopted for the predictive current controller. A branch-and-bound method based on sphere decoding was proposed in Chap. 5 to solve the underlying integer optimization problem. Long horizons provide a significantly better steady-state performance by lowering the current distortions per switching losses. The resulting performance is between that of SVM and OPPs, as indicated in Sect. 6.1.4.

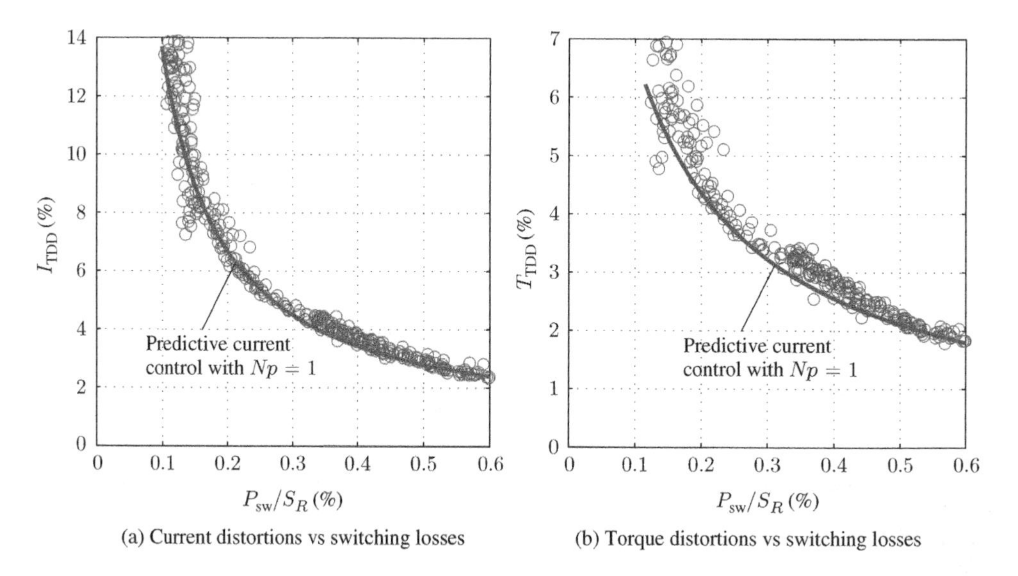

(a) Current distortions vs switching losses (b) Torque distortions vs switching losses

Figure 15.5 Performance trade-off for predictive current control with reference tracking and the prediction horizon $N_p = 1$

15.1.3 Summary and Discussion

The characteristic trade-off curves of five control and modulation schemes were derived in the previous section. These schemes include FOC with SVM, MP^{3}C based on OPPs, MPDTC, MPDCC, and one-step predictive current control. Short and long switching horizons were investigated for MPDTC and MPDCC. Figure 15.6(a) summarizes the obtained trade-off curves in terms of the (absolute) current TDD and the normalized switching losses.

To better illustrate the differences between the trade-off curves, we recall from (6.2) the definition of the *relative* current TDD

$$I_{\text{TDD}}^{\text{rel}} = \frac{I_{\text{TDD}} - I_{\text{TDD,OPP}}}{I_{\text{TDD,OPP}}}. \tag{15.6}$$

This measure represents the current TDD degradation with respect to OPPs. Under idealized simulation conditions, the closed-loop harmonic performance of MP^{3}C and the nominal (open-loop) performance of OPPs are the same. Figure 15.7(a) depicts the trade-off curves in terms of the relative current TDD (in percent) and the normalized switching losses.

Between 0.1% and 0.6% of the normalized switching losses, SVM uses switching frequencies between 90 and 480 Hz. With the fundamental frequency being 30 Hz, these switching frequencies correspond to the pulse numbers 3–15. In this range, MP^{3}C significantly reduces the current distortions thanks to the use of OPPs. Notably, the current distortions are halved at pulse number 3 with respect to SVM. For high pulse numbers of 15 and beyond, the differences in the harmonic performance between SVM and MP^{3}C are less significant, but they do persist. However, the computational effort required to solve offline the nonconvex optimization problems of OPPs with pulse numbers in excess of 15 increases significantly, making it difficult to derive the optimal switching angles of such OPPs.

At high switching losses, MPDCC with the long switching horizon *eSESESE* achieves current distortions that are similar to those of MP^{3}C. MPDCC also slightly outperforms MPDTC in this operating regime. When approaching six-step operation, MPDCC and MPDTC both outperform MP^{3}C. This was demonstrated for MPDCC in Sect. 11.1.7. This somewhat surprising result is due to the difference in the cost functions. OPPs minimize the current distortions for a given switching frequency, whereas MPDCC minimizes the switching losses for a given current ripple, which implies a certain current distortion. Provided that the torque and flux bounds in MPDTC are set such that they approximate the bounds in MPDCC, this statement also holds true for MPDTC.

Short switching horizons are less effective at achieving low current distortions, as can be seen in Fig. 15.7(a). Nevertheless, MPDCC with the switching horizon *eSE* consistently outperforms SVM. Predictive current control with reference tracking, the squared 2-norm, and a horizon of one step exhibits a harmonic performance that is not dissimilar to that of SVM. For switching frequencies below 280 Hz, this predictive controller tends to have a small advantage over SVM, while the opposite is true above that threshold.

Figures 15.6(b) and 15.7(b) depict the torque trade-off curves in terms of the absolute and the relative torque TDDs, respectively, versus the normalized switching losses. The relative torque TDD is defined similar to (15.6). MP^{3}C halves the torque distortions of SVM when operating at very low switching losses. At high switching losses, however, MP^{3}C provides no harmonic benefit in terms of the torque distortions when compared to SVM. This is in contrast to the current distortions, in which MP^{3}C always outperforms SVM.

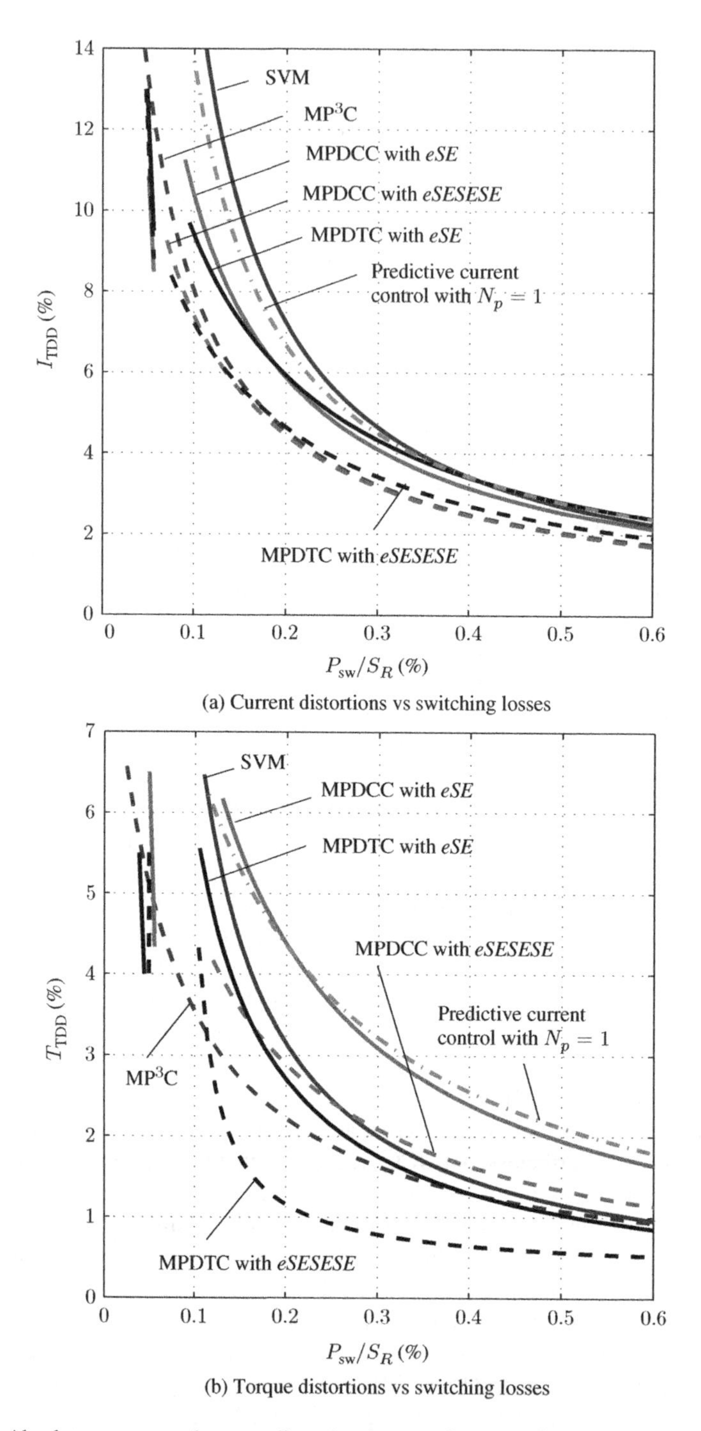

Figure 15.6 Absolute current and torque distortions versus the normalized switching losses, summarizing the trade-off curves of the investigated control and modulation schemes

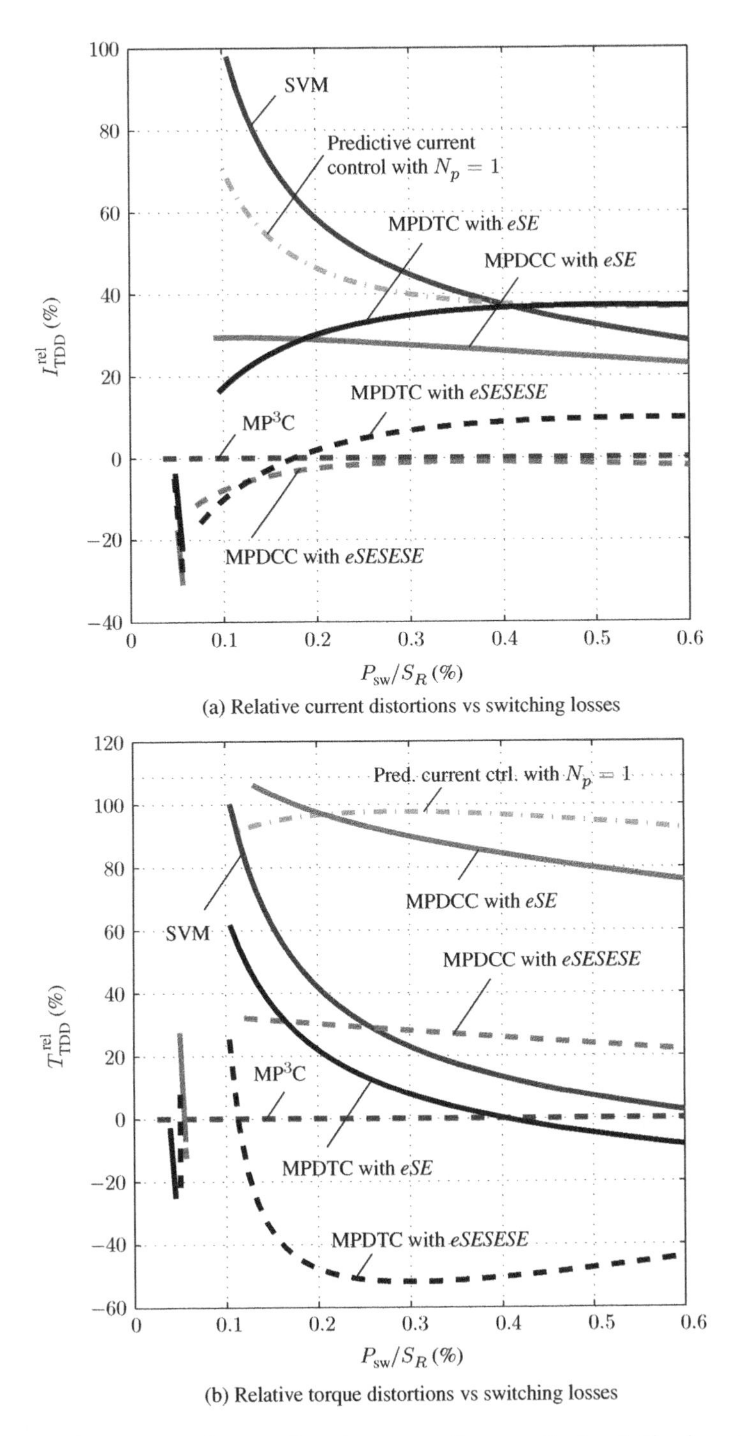

(a) Relative current distortions vs switching losses

(b) Relative torque distortions vs switching losses

Figure 15.7 Relative current and torque distortions versus the normalized switching losses, summarizing the trade-off curves of the investigated control and modulation schemes

MPDTC with long switching horizons achieves a large reduction in the torque TDD of up to 50% compared to MP^{3}C. Even MPDTC with the short switching horizon *eSE* slightly outperforms MP^{3}C, provided that operation at high switching losses is considered. Yet, these improvements in the torque distortions come at the expense of inferior current TDDs, as discussed in the previous section. Close to six-step operation, MPDTC is capable of performing as well as MP^{3}C.

The torque distortions of MPDCC are, in general, much larger than those of MPDTC. In Fig. 15.7(b), the torque trade-off curves of MPDCC are shifted by about 50 percentage points upward compared to those of MPDTC. As a result, MPDCC with the short switching horizon *eSE* entails almost twice the torque distortions of MP^{3}C. Nevertheless, long switching horizons significantly reduce the torque TDD, with MPDCC with *eSESESE* achieving torque distortions that are not dissimilar to those of SVM. The predictive current controller with reference tracking and the prediction horizon $N_p = 1$ causes even higher torque distortions than MPDCC with the short switching horizon. As a result, the control and modulation scheme with the worst torque distortions in this analysis is one-step predictive current control.

In summary, and unsurprisingly, predictive current control schemes tend to excel at reducing the current distortions. To a lesser extent, they also reduce the torque distortions, because low current distortions imply low torque distortions. The reverse statement, however, does not hold true. Very low torque distortions can be achieved, albeit at the price of pronounced current distortions. The differences between the control approaches are emphasized at low switching frequencies and low switching losses. At high switching losses, the differences in the current distortions become smaller in both absolute and relative terms. For the torque distortions, however, significant differences also persist at higher switching losses. When the aim is to minimize both the current and the torque distortions, MP^{3}C outperforms all the other control and modulation schemes that were assessed. SVM provides a good overall balance between acceptable current and torque distortions.

One might argue that the main benefit of MPC schemes is the performance improvement they bring when compared to traditional schemes such as FOC with SVM or DTC. To achieve this, long prediction horizons are mandatory to enable the optimization-based controller to make well-informed decisions when choosing the next switching state. In contrast, short prediction horizons appear to be often less effective than established methods. The one-step predictive control family is nevertheless attractive in light of its conceptual and computational simplicity.

The derived trade-off curves are effectively independent of the machine and inverter parameters used, because only the *relative* performance of the control and modulation schemes matters in this comparison. For a machine with a smaller total leakage inductance, for example, the *absolute* TDD values would be higher, thus stretching the trade-off curves on the vertical axis in Fig. 15.6. Yet, the percentage-wise (relative) differences between the curves in Fig. 15.7 would remain the same.

In an alternative performance analysis, the current and torque distortions are depicted versus the switching *frequency* rather than the switching losses. Such an analysis was performed in [6], using the same drive system with the same parameters and running the drive at the same operating point (60% speed and rated torque). The cost function of MPDTC and MPDCC was modified accordingly, in that the switching frequency was minimized instead of the switching losses. The overall result is similar to the trade-off curves provided here, but the advantage of MPDTC and MPDCC over SVM and MP^{3}C is reduced. In particular,

MPDTC and MPDCC are unable—as one would expected—to outperform MP^3C in terms of harmonic distortions per switching frequency.

15.2 Assessment of the Control and Modulation Methods

In the previous section, we compared the nominal steady-state performance of the main direct MPC schemes with each other and with the classic methods. This was done in an idealized setup. We will next assess the advantages and disadvantages of the various methods in an industrial setup. Despite the somewhat subjective nature of this classification, it should nevertheless serve as a rough guide to narrow down the set of candidate control and modulation methods that are most suitable for a specific problem at hand. To this end, the main determining characteristics are the converter topology, the type of load connected to it, whether a machine-side inverter or a grid-connected converter is considered, and the pulse number. Recall that the latter is defined as the number of switching transitions in the single-phase switching pattern within a quarter of the fundamental period.

15.2.1 FOC and VOC with SVM

The most commonly used control and modulation method is SVM with a current controller that is formulated in a rotating and orthogonal dq reference frame. On the machine side, the reference frame is aligned with a flux vector, giving rise to FOC, while on the grid side the reference frame is typically aligned with the voltage vector at the point of common coupling (PCC). The latter concept is commonly referred to as voltage-oriented control (VOC). These control methods are well understood and provide a good dynamic performance during current transients, provided that the control loops in the d- and q-axes are fully decoupled not only during steady-state operation but also during transients [7].

In general, SVM provides acceptable current and torque distortions and lower harmonic distortions than CB-PWM with third-harmonic injection. Numerous variants of CB-PWM and SVM have been proposed, many of which aim to reduce the switching losses. However, because of the fixed-length modulation cycle, the scope of these improvements is limited. At low pulse numbers, say below 15, the harmonic performance of SVM is inferior to that of OPPs.

SVM provides a harmonic spectrum with exclusively discrete frequencies, making it a suitable modulation technique for grid-connected inverters. In the presence of LC filters, the current control loop can be augmented with an active damping loop to dampen the filter resonance. For the damping loop to be effective, the switching frequency must be significantly higher than the resonance frequency of the filter.

15.2.2 DTC and DPC

DTC presents for machine-side converters an alternative to FOC with SVM. DTC provides a superior robustness to machine parameter variations, dc-link voltage ripples, measurement noise, and flux observer noise. DTC also achieves an unmatched dynamic performance during torque transients and faults. The torque distortions are typically relatively low.

However, DTC also carries a few notable disadvantages. Operation at very high modulation indices is challenging, the switching frequency is not directly controlled, and the current distortions are often pronounced. Except for a few non-triplen odd harmonics that are of a low order, such as the fifth and seventh harmonics, the harmonic spectrum of DTC is flat and exhibits even as well as non-integer harmonics. The grid-side equivalent of DTC, direct power control (DPC) [8], is therefore ill suited to meet grid standards on harmonics, such as the ones summarized in Sect. 3.1.2. DPC is thus rarely used in practice.

Nevertheless, the DTC concept represents an attractive alternative to FOC and remains an active research topic. Several extensions have been proposed to alleviate some of the shortcomings of DTC. In particular, by imposing a fixed-length switching interval, constant switching frequency operation is achieved. For a brief summary of the main extensions of DTC, the reader is referred to Sect. 3.6.3.

15.2.3 Direct MPC with Reference Tracking

Direct MPC with reference tracking is a control and modulation method that has become popular in academia. When adopting a prediction horizon of 1, exhaustive enumeration can be used to solve the underlying optimization problem. This facilitates the use of a nonlinear prediction model that may be adapted online when required—for example, during faults and changing grid conditions. Various and diverse control objectives can be represented in the cost function. As a result, this MPC scheme is easy to devise, implement, and use. Its dynamic performance during transients is superb, resembling that of deadbeat control and DTC.

To achieve acceptable current distortions for a given switching frequency, a penalty on the switching effort is mandatory. As switching is restricted to the regularly spaced discrete time instants at which the controller is executed, experience indicates that the sampling frequency must exceed the switching frequency by at least two orders of magnitude to avoid a deterioration of the harmonic performance. This requirement becomes a limiting factor when operating at high switching frequencies. One way to overcome this issue is to adopt the technique of variable switching points, as proposed in [9].

The current distortions of one-step predictive control are similar to those of SVM, whereas its torque distortions are, in general, up to twice as high (see Fig. 15.7). The harmonic spectrum is flat and exhibits even and non-integer harmonics. Hence, this concept is generally not suitable for grid-connected converters. To reduce the harmonic distortions, long prediction horizons are required. Indeed, when operating at low pulse numbers, long-horizon direct MPC with reference tracking outperforms SVM, as shown in Sect. 6.1. Solving the underlying optimization problem is, however, a nontrivial task, even when using branch-and-bound techniques such as sphere decoding, as proposed in Sect. 5.3.

Direct MPC with reference tracking is a suitable choice for complicated systems such as back-to-back converter systems [10] and inverters that are connected via an intermediate *LC* filter to a machine or the grid. It allows one to address such systems through *one* control loop, which can be relatively easily designed. Specifically, cascaded control loops or additional active damping loops can be avoided. To ensure stability and low harmonic distortions, however, long prediction horizons are typically required (see Sect. 6.3).

The tuning of direct MPC schemes with reference tracking is difficult, particularly when the cost function includes conflicting terms and multiple weighting factors. The weighting factors in such multi-criterion optimization problems can be tuned by exploring the trade-off surface (see [11, Sect. 4.7]). It is often overlooked that the choice of the sampling interval has a profound impact on the closed-loop performance, particularly when the ratio between the sampling frequency and the switching frequency is below 100.

15.2.4 Direct MPC with Bounds

Instead of regulating the controlled variables along time-varying references, this requirement can be relaxed by imposing upper and lower bounds on the controlled variables. The bound widths determine the ripple on the controlled quantities. In the case of MPDCC, this allows one to set the current ripple and thus the current TDD with one parameter. The second control objective, that is, the minimization of the switching effort, is captured by the cost function, which minimizes either the switching frequency or the switching losses. As the sampling interval is not a tuning parameter, a sampling interval as short as is permitted by the control hardware is usually selected. As it involves only one parameter, the tuning procedure is straightforward. Other examples of the direct MPC family with bounds include MPDTC and model predictive direct power control (MPDPC).

The family of direct MPC with bounds shares a number of additional advantages with direct MPC with reference tracking. These advantages include the possibility of using nonlinear and time-varying prediction models, as well as the excellent dynamic performance during load transients, faults, and reference steps, as demonstrated in Sects. 8.1.3, 8.2.5, 11.1.7, and 11.2.5. Furthermore, thanks to the use of bounds, these direct MPC schemes are robust to model parameter mismatches [1], measurement noise, and flux observer noise. This high degree of robustness is inherited from DTC and DPC.

Direct MPC with bounds achieves long prediction horizons thanks to the concept of extending the predicted output trajectories between the switching transitions. As a result, the harmonic distortions are generally low. For short horizons, it often outperforms SVM, while for long horizons, current distortions per switching losses are achieved that are akin to those of OPPs.

As with the other direct MPC schemes, however, the harmonic spectrum contains even and non-integer harmonics. Another disadvantage is that direct MPC with bounds is conceptually and computationally more involved than predictive control with reference tracking. In particular, solving the underlying optimization problem in real time for long switching horizons is computationally demanding. Branch-and-bound techniques lend themselves to reduce this computational burden (see Chap. 10), but they require additional tuning parameters.

Imposing bounds on the controlled variables in conjunction with integer manipulated variables leads to one conceptual disadvantage—the emergence of deadlocks. Deadlocks refer to situations in which the control problem is infeasible. To nevertheless facilitate the derivation of a suitable manipulated variable in the case of a deadlock, the control algorithm is augmented by a deadlock resolution mechanism. As explained in Sect. 9.4, in the case of a deadlock, the bounds on the controlled variables are relaxed and treated as soft constraints. Techniques are also available to reduce the likelihood of deadlocks (see Sect. 9.5).

15.2.5 MP^{3}C based on OPPs

The two direct MPC schemes discussed here derive their switching patterns exclusively by means of *online* computations. In an alternative approach, optimal switching patterns can be precomputed *offline* in the form of OPPs for all relevant pulse numbers and modulation indices.

To achieve fast closed-loop control, the switching pattern is modified online using MP^{3}C. In particular, MP^{3}C compensates for phase voltage fluctuations and preserves the optimal volt-second balance of the OPP during steady-state operation. During transients and load steps, by temporarily modifying the volt-second balance, very fast transient response times akin to DTC can be achieved.

Basing the controller on precomputed OPPs carries three major advantages. First, an excellent harmonic performance is achieved during steady-state operation, provided that the power electronic system fulfills the following requirements.

- The voltage steps at the inverter phase terminals should be of equal magnitude. A ripple on the phase voltages can be tolerated, because it only mildly affects the harmonic performance, but a bias or a persistent offset leads to a significant harmonic deterioration.
- The noise levels in the current measurements, voltage measurements, and flux observer should be low. For higher noise levels, the bandwidth of the controller has to be reduced, for example, by resorting to MP^{3}C based on a quadratic program (QP) with a long horizon.
- The three-phase load should be balanced, symmetric, and free of significant harmonics. Three-phase machines comply with this requirement unless they are faulty. The PCC voltages on the grid-side, however, are often imbalanced and typically include low-order harmonics.
- The delays in the closed-loop system should be either small or fully compensated for. Relevant delays include the measurement, computation, and actuation delays.

Second, thanks to the use of OPPs, the harmonic spectrum of MP^{3}C features only harmonics at odd and non-triplen integer multiples of the fundamental frequency. This makes MP^{3}C well suited for grid-connected inverters. Furthermore, the harmonic spectrum is predetermined and known in advance. In particular, OPPs not only allow one to minimize the current TDD but also facilitate the shaping of the harmonic spectrum. For grid-connected inverters, for example, OPPs can be computed with the aim of meeting specific grid standards. This can be accomplished by imposing upper bounds on specific harmonics while minimizing the voltage or current TDD of the remaining harmonics. Similarly, in the presence of an LC filter, harmonics close to the filter resonance can be minimized.

Note that OPPs provide a more versatile modulation framework than the commonly used concept of selective harmonic elimination (SHE), in which the low-frequency harmonics are set to zero (see also Sect. 3.4). For grid-connected inverters with an LC filter, an OPP allows one to optimally distribute the harmonic energy over the frequency range. The grid standards can be met *and* the size and weight of the filter can be minimized by placing a high proportion of the harmonic content above the filter cut-off frequency.

Third, the online modification of the OPPs requires only a few computations. A streamlined deadbeat version of MP^{3}C can be based on a few multipliers, adders, and logic operations. As a result, the execution time of MP^{3}C is in the range of a few microseconds.

On the other hand, the use of OPPs entails several disadvantages. First, and foremost, basing the controller on an offline-computed OPP limits the controller's flexibility and adaptability, particularly in the presence of nonuniformly spaced phase voltages, imbalances in the PCC voltages, and unforeseen events such as faults. In such cases, the performance of long-horizon direct MPC schemes might match, if not surpass, the performance of MP^{3}C. Basing the control decisions exclusively on online optimization facilitates a superior adaptation to changing operating conditions, parameter changes, disturbances, or even faults. Furthermore, direct

MPC schemes also provide the optimal switching pattern during transients, while OPPs fail to do so, as they were computed assuming steady-state operating conditions. Nevertheless, the notion of pulse insertion largely mitigates this issue.

Second, the computation of nontrivial OPPs constitutes a challenging task. To this end, a dedicated and versatile toolbox is required, which should be based on state-of-the-art nonlinear optimization tools. Computing OPPs for pulse numbers exceeding 20 or for inverters with a large number of voltage levels remains a time-consuming undertaking even when employing a powerful computer.

Third, as with synchronous PWM, the switching frequency of OPPs is restricted to integer multiples of the fundamental frequency. This prevents one from fully exploiting the available switching frequency, particularly when operating at very low pulse numbers.

15.2.6 Indirect MPC

In the two direct MPC schemes and $\mathrm{MP^3C}$, the inner (current, flux, or torque) control loop and the modulation of the switching signal are addressed in one computational stage. Alternatively, the modulator can be kept as a separate entity, and an MPC scheme can be devised that manipulates the input to the modulator. This input is the voltage reference, which is a real-valued variable.

The modulation stage conceals the switching nature of the inverter from the controller, allowing one to design a control loop based on averaging. This greatly simplifies the controller design, because the control problem includes only real-valued variables. Even though the system dynamics are often linear, constraints on states and manipulated variables are usually present. MPC is ideally suited to address such constrained linear systems, and it is relatively straightforward to formulate a corresponding MPC scheme. Solving the underlying optimization problem in real time is, however, a challenging task, particularly in the presence of short sampling intervals. Arguably, this is the reason why indirect MPC for power electronic systems remains largely unexplored.

A few notable exceptions for three-phase systems include [12], in which the optimization problem is solved online using a fast gradient solver, and [2, 13–15], which solves the MPC problem offline by computing the corresponding piecewise affine state-feedback control law, that is, the explicit solution of MPC. In the case of a linear system without constraints, MPC can be replaced by a linear quadratic regulator (LQR), as proposed in [16].

All these methods address either the machine-side or the grid-side converter. Alternatively, the back-to-back converter with its load can be treated as one large system, for which one controller is formulated. To this end, a nonlinear MPC scheme was proposed in [17] for a back-to-back load-commutated inverter with a synchronous machine. The problem was formulated and solved using the ACADO toolbox [18].

The harmonic performance of indirect MPC schemes is determined by their modulator. Indirect MPC schemes are thus best suited to either relatively high pulse numbers or when the system to be controlled is complex, requires a superior performance during transients, and includes constraints. These conditions are met for the modular multilevel converter (MMC) in Chap. 14, making indirect MPC a promising control technique for MMCs with a high module count. Similarly, indirect MPC can be applied to MMC-based static VAR compensators (see [19]).

15.3 Conclusion

This book has focused predominantly on high-performance MPC schemes for high-power converter systems. These systems are characterized by low pulse numbers, multilevel converters, and demanding control problems. Examples for the latter include multilevel converters with internal voltages to be balanced (five-level active NPC inverter), multilevel converters with internal voltages and currents to be controlled (MMCs), and inverter systems with additional passive components such as *LC* filters.

The use of standard modulation techniques, such as CB-PWM or SVM, the separation of the control and modulation problem into two distinct tasks, and the use of linear single-input single-output (SISO) controllers limits the performance that can be achieved for such systems. During steady-state operation, a suboptimal ratio between harmonic distortions and switching losses results. During transients and faults, the dynamic response is either slow, or it is poorly decoupled, exhibiting overshoots and often violating safety constraints.

The proposed high-performance MPC schemes are ideally suited to address such systems. Their commercial benefits can be summarized as follows.

- Minimal harmonic distortions per switching losses, or vice versa.
- Superior performance during transients, load steps, and faults. Operation within the safe operating area can be ensured thanks to the imposition of constraints.
- Minimization of the size, weight, and cost of passive components, such as *LC* filters and dc-link capacitors.
- Model-based design that reduces the controller design effort.

These advantages all translate into cost savings. The MPC methods have, however, one disadvantage in common—they are conceptually and often also computationally demanding. This slows down the adoption rate by industry mainly for the following reasons.

- Knowledge, skills, and an in-depth experience related to the design, analysis, and implementation of MPC schemes must be built up through training, education, and hiring. This is a lengthy process that requires a significant long-term commitment. A successful team combines expertise in optimal control, MPC, power electronics, electrical machines, mathematical programming, numerical optimization, and embedded systems. The latter includes proficiency in Assembler, C, VHDL, communication protocols, and hardware architectures.
- Any new MPC-based control and modulation method must be available for the whole product range. For a drive product, for example, this includes machine-side and grid-side converters, multiple converter units operating in parallel, different types of machines such as induction machines and (permanent magnet) synchronous machines, and *LC* filters, which often require active damping.

The most significant obstacle, however, is the fact that decision makers and industrial R&D personnel are intrinsically risk-averse and favor established and well-proven control and modulation methods over emerging and unproven ones. This is compounded by the observation that researchers often overestimate the benefit of their favorite technique. Any new method must therefore promise to provide massive commercial benefits and a significant positive return on investment. Without such a performance promise, the investment required and the risk involved in the development and productization of the new control and modulation method cannot be justified.

This observation also indicates that computational and conceptual simplicity is desirable but not sufficient. This statement is underlined in [20], in which the authors assess the technology readiness levels of different emerging MPC methods for high-power applications. Their analysis highlights that MPC methods that promise significant performance benefits have achieved higher technology readiness levels—and are thus closer to commercialization—than the simpler methods.

15.4 Outlook

MPC emerged from the process industry in the 1970s. It has since matured into a well-established control paradigm and has become the method of choice to address constrained linear and nonlinear systems. Today, model predictive controllers are used in thousands of industrial applications [21]. It is difficult to envision MPC not also playing a significant role in power electronics in the future. As such, the open question is not so much *if* MPC will be adopted by the power electronics industry, but which *variety* of MPC will prove to be particularly successful.

Despite a recent surge in research activities and publications, the field of MPC for power electronics remains largely unexplored. Challenges that could form the basis for future research activities abound. A few challenges are summarized in the following, which we consider to be particularly meaningful and important.

- Fast solvers for the optimization problems underlying MPC, including quadratic, nonlinear, and mixed-integer programs. These solvers must provide real-time guarantees and must run on small, inexpensive embedded systems.
- Direct MPC methods with discrete harmonic spectra, which avoid even and non-integer harmonics.
- Control methods with performance and stability guarantees for power electronic systems with high-dimensional state vectors and multiple constraints.
- Estimation schemes with performance and stability guarantees for state variables and time-varying parameters.
- OPPs that address a wide range of control objectives and can be adapted to varying operating conditions.

This book is intended to serve as a starting point for this quest and aims to encourage the reader to advance the exciting field of high-performance MPC for power electronics.

References

[1] T. Geyer, R. P. Aguilera, and D. E. Quevedo, "On the stability and robustness of model predictive direct current control," in *Proceedings of IEEE International Conference on Industrial Technology* (Cape Town, South Africa), Feb. 2013.

[2] A. Linder and R. Kennel, "Model predictive control for electrical drives," in *Proceedings of IEEE Power Electronics Specialists Conference* (Recife, Brasil), pp. 1793–1799, 2005.

[3] S. Mariéthoz, A. Domahidi, and M. Morari, "Sensorless explicit model predictive control of permanent synchronous motors," in *Proceedings of IEEE International Electric Machines & Drives Conference* (Miami, Florida, USA), pp. 1492–1499, May 2009.

[4] R. Vargas, P. Cortés, U. Ammann, J. Rodríguez, and J. Pontt, "Predictive control of a three-phase neutral-point-clamped inverter," *IEEE Trans. Ind. Electron.*, vol. 54, pp. 2697–2705, Oct. 2007.
[5] T. Geyer, "A comparison of control and modulation schemes for medium-voltage drives: emerging predictive control concepts versus PWM-based schemes," *IEEE Trans. Ind. Appl.*, vol. 47, pp. 1380–1389, May/Jun. 2011.
[6] J. Scoltock, T. Geyer, and U. K. Madawala, "A comparison of model predictive control schemes for MV induction motor drives," *IEEE Trans. Ind. Inf.*, vol. 9, pp. 909–919, May 2013.
[7] J. Holtz, J. Quan, J. Pontt, J. Rodríguez, P. Newman, and H. Miranda, "Design of fast and robust current regulators for high-power drives based on complex state variables," *IEEE Trans. Ind. Appl.*, vol. 40, pp. 1388–1397, Sep./Oct. 2004.
[8] T. Noguchi, H. Tomiki, S. Kondo, and I. Takahashi, "Direct power control of PWM converter without power-source voltage sensors," *IEEE Trans. Ind. Appl.*, vol. 34, pp. 473–479, May/Jun. 1998.
[9] P. Karamanakos, P. Stolze, R. M. Kennel, S. Manias, and H. du Toit Mouton, "Variable switching point predictive torque control of induction machines," *J. Emerging Sel. Top. Power Electron.*, vol. 2, pp. 285–295, Jun. 2014.
[10] Z. Zhang, F. Wang, T. Sun, J. Rodríguez, and R. Kennel, "FPGA-based experimental investigation of a quasi-centralized model predictive control for back-to-back converters," *IEEE Trans. Power Electron.*, vol. 31, pp. 662–674, Jan. 2016.
[11] S. Boyd and L. Vandenberghe, *Convex optimization*. Cambridge, UK: Cambridge Univ. Press, 2004.
[12] S. Richter, S. Mariéthoz, and M. Morari, "High-speed online MPC based on fast gradient method applied to power converter control," in *Proceedings of the American Control Conference* (Baltimore, MD, USA), 2010.
[13] M. Cychowski, K. Szabat, and T. Orlowska-Kowalska, "Constrained model predictive control of the drive system with mechanical elasticity," *IEEE Trans. Ind. Electron.*, vol. 56, pp. 1963–1973, Jun. 2009.
[14] S. Bolognani, S. Bolognani, L. Peretti, and M. Zigliotto, "Design and implementation of model predictive control for electrical motor drives," *IEEE Trans. Ind. Electron.*, vol. 56, pp. 1925–1936, Jun. 2009.
[15] S. Mariéthoz, A. Domahidi, and M. Morari, "High-bandwidth explicit model predictive control of electrical drives," *IEEE Trans. Ind. Appl.*, vol. 48, pp. 1980–1992, Nov./Dec. 2012.
[16] T. Murata, T. Tsuchiya, and I. Takeda, "Vector control for induction machine on the application of optimal control theory," *IEEE Trans. Ind. Electron.*, vol. 37, pp. 283–290, Aug. 1990.
[17] T. J. Besselmann, S. van de Moortel, S. Almér, P. Jörg, and H. J. Ferreau, "Model predictive control in the multi-megawatt range," *IEEE Trans. Ind. Electron.*, vol. 63, pp. 4641–4648, Jul. 2016.
[18] B. Houska, H. J. Ferreau, and M. Diehl, "An auto-generated real-time iteration algorithm for nonlinear MPC in the microsecond range," *Automatica*, vol. 47, pp. 2279–2285, Oct. 2011.
[19] T. Geyer, G. Darivianakis, and W. van der Merwe, "Model predictive control of a STATCOM based on a modular multilevel converter in delta configuration," in *Proceedings of European on Power Electronics Conference* (Geneva, Switzerland), Sep. 2015.
[20] G. Papafotiou, G. Demetriades, and V. Agelidis, "Integration of model-predictive control in medium and high-voltage power electronics products: An industrial perspective on gaps and progress required," in *Proceedings of IEEE Industrial Electronics Society Annual Conference* (Yokohama, Japan), Nov. 2015.
[21] S. J. Qin and T. A. Badgwell, "A survey of industrial model predictive control technology," *Control Eng. Pract.*, vol. 11, pp. 733–764, Jul. 2003.

Index

Model Predictive Control of High Power Converters and Industrial Drives, First Edition. Tobias Geyer.

Companion Website: www.wiley.com/go/geyermodelpredictivecontrol

Printed and bound by CPI Group (UK) Ltd, Croydon, CR0 4YY
16/04/2025
14658384-0004